VW Jetta & Golf Automotive Repair Manual

by Jeff Killingsworth

Models covered:
Jetta - 2011 (A6 chassis only) through 2018
Golf - 2015 through 2019
Includes GLI, GTI, Jetta Sportwagen,
Golf Sportwagen, Golf R 4Motion and Golf Alltrack
Does not include information specific to diesel or hybrid models

Haynes Group Limited (96021-5AA1)
Sparkford Nr Yeovil   ABCDE
Somerset BA22 7JJ England   FGHIJ
                                              KLMNO
                                              PQ

Haynes North America, Inc.
2801 Townsgate Road, Suite 340
Thousand Oaks, CA 91361 USA

www.haynes.com

Disclaimer

There are risks associated with automotive repairs. The ability to make repairs depends on the individual's skill, experience and proper tools. Individuals should act with due care and acknowledge and assume the risk of performing automotive repairs.

The purpose of this manual is to provide comprehensive, useful and accessible automotive repair information, to help you get the best value from your vehicle. However, this manual is not a substitute for a professional certified technician or mechanic.

This repair manual is produced by a third party and is not associated with an individual vehicle manufacturer. If there is any doubt or discrepancy between this manual and the owner's manual or the factory service manual, please refer to the factory service manual or seek assistance from a professional certified technician or mechanic.

Even though we have prepared this manual with extreme care and every attempt is made to ensure that the information in this manual is correct, neither the publisher nor the author can accept responsibility for loss, damage or injury caused by any errors in, or omissions from, the information given.

Acknowledgements

Wiring diagrams provided exclusively for Haynes North America, Inc. by HaynesPro BV. Technical writers who contributed to this project include Demian Hurst and Scott "Gonzo" Weaver. Mechanical work and photography was provided by Mark Henderson.

© **Haynes North America, Inc. 2020**

With permission from Haynes Group Limited

A book in the Haynes Automotive Repair Manual Series

Printed in India

All rights reserved. No part of this book may be reproduced or transmitted in any form or by any means, electronic or mechanical, including photocopying, recording or by any information storage or retrieval system, without permission in writing from the copyright holder.

ISBN-10: 1-62092-366-1
ISBN-13: 978-1-62092-366-5

Library of Congress Control Number: 2020942413

Disclaimer

There are risks associated with automotive repairs. The ability to make repairs depends on individual skill, experience and proper tools. Individuals should act with due care and acknowledge and assume the risk of making automotive repairs. While every attempt is made to ensure that the information in this manual is correct, no liability can be accepted by the authors or publishers for loss, damage or injury caused by any errors in, or omissions from, the information given.

Contents

Introductory pages

About this manual	0-5
Introduction	0-5
Vehicle identification numbers	0-6
Recall information	0-7
Buying parts	0-12
Maintenance techniques, tools and working facilities	0-12
Jacking and towing	0-20
Booster battery (jump) starting	0-21
Conversion factors	0-22
Fraction/decimal/millimeter equivalents	0-23
Automotive chemicals and lubricants	0-24
Safety first!	0-25
Troubleshooting	0-26

Chapter 1
Tune-up and routine maintenance — 1-1

Chapter 2 Part A
4-cylinder engines — 2A-1

Chapter 2 Part B
5-cylinder engines — 2B-1

Chapter 2 Part C
General engine overhaul procedures — 2C-1

Chapter 3
Cooling, heating and air conditioning systems — 3-1

Chapter 4
Fuel and exhaust systems — 4-1

Chapter 5
Engine electrical systems — 5-1

Chapter 6
Emissions and engine control systems — 6-1

Chapter 7 Part A
Manual transaxle — 7A-1

Chapter 7 Part B
Automatic transaxle — 7B-1

Chapter 8
Clutch and driveline — 8-1

Chapter 9
Brakes — 9-1

Chapter 10
Suspension and steering systems — 10-1

Chapter 11
Body — 11-1

Chapter 12
Chassis electrical system — 12-1

Wiring diagrams — 12-22

Index — IND-1

Haynes mechanic and photographer with a 2017 VW Jetta SE

About this manual

Its purpose

The purpose of this manual is to provide comprehensive, useful and accessible automotive repair information, to help you get the best value from your vehicle. It can do so in several ways. It can help you decide what work must be done, even if you choose to have it done by a dealer service department or a repair shop; it provides information and procedures for routine maintenance and servicing; and it offers diagnostic and repair procedures to follow when trouble occurs.

We hope you use the manual to tackle the work yourself. For many simpler jobs, doing it yourself may be quicker than arranging an appointment to get the vehicle into a shop and making the trips to leave it and pick it up. More importantly, a lot of money can be saved by avoiding the expense the shop must pass on to you to cover its labor and overhead costs. An added benefit is the sense of satisfaction and accomplishment that you feel after doing the job yourself. However, this manual is not a substitute for a professional certified technician or mechanic. There are risks associated with automotive repairs. The ability to make repairs on a vehicle depends on individual skill, experience, and proper tools. Individuals should act with due care and acknowledge and assume the risk of performing automotive repairs.

Using the manual

The manual is divided into Chapters. Each Chapter is divided into numbered Sections, which are headed in bold type between horizontal lines. Each Section consists of consecutively numbered paragraphs.

The reference numbers used in illustration captions pinpoint the pertinent Section and the Step within that Section. That is, illustration 3.2 means the illustration refers to Section 3 and Step (or paragraph) 2 within that Section.

Procedures, once described in the text, are not normally repeated. When it's necessary to refer to another Chapter, the reference will be given as Chapter and Section number. Cross references given without use of the word "Chapter" apply to Sections and/or paragraphs in the same Chapter. For example, "see Section 8" means in the same Chapter. References to the left or right side of the vehicle assume you are sitting in the driver's seat, facing forward.

This repair manual is produced by a third party and is not associated with an individual car manufacturer. If there is any doubt or discrepancy between this manual and the owner's manual or the factory service manual, please refer to factory service manual or seek assistance from a professional certified technician or mechanic. Even though we have prepared this manual with extreme care, neither the publisher nor the author can accept responsibility for any errors in, or omissions from, the information given.

NOTE

A **Note** provides information necessary to properly complete a procedure or information which will make the procedure easier to understand.

CAUTION

A **Caution** provides a special procedure or special steps which must be taken while completing the procedure where the Caution is found. Not heeding a Caution can result in damage to the assembly being worked on.

WARNING

A **Warning** provides a special procedure or special steps which must be taken while completing the procedure where the Warning is found. Not heeding a Warning can result in personal injury.

Introduction

Volkswagen Jetta models are available in either a four-door sedan or 5-door sports wagon body style. The Golf is available in a two-door, four-door hatchback model or 5-door sportswagen body style, the GTI and Golf R 4Motion, Golf Alltrack are only available in a four-door hatchback model.

There are five different engines available depending on year and model: a 1.4L DOHC, 16-valve turbocharged four-cylinder, a 1.8L DOHC, 16-valve turbocharged four-cylinder, a 2.0L SOHC, 8-valve non-turbocharged four-cylinder, a 2.0L DOHC, 16-valve turbocharged four-cylinder engine, or a 2.5L DOHC, 20-valve five-cylinder engine.

Automatic transmission models come equipped with either a Tiptronic six-speed automatic transmission or a 6-speed Direct Shift Gearbox (DSG) (which is a cross between an automatic transmission and a manual transmission). Manual transmission models are equipped with either a five-speed or six-speed manual transmission. The Golf Sportwagen, R 4Motion and Alltrack are All Wheel Drive (AWD) models.

The front suspension on all models is independent, using McPherson struts and a stabilizer bar. The rear suspension on all models except the 2011 Jetta is a four-link design which consists of an upper transverse link (control arm), a lower control arm and tie-rod, a trailing arm, individual coil springs, shock absorbers and a stabilizer bar. 2011 Jetta models use a torsion beam axle supported by coil springs and shock aborbers.

The steering system consists of a rack-and-pinion steering gear and two adjustable tie-rods. Power steering is provided by an electro-mechanical motor and control module mounted directly to the steering gear.

The brakes are disc at the front and rear, with power assist standard. An Anti-lock Brake System (ABS) is standard equipment.

Vehicle identification numbers

Modifications are a continuing and unpublicized process in vehicle manufacturing. Since spare parts lists and manuals are compiled on a numerical basis, the individual vehicle numbers are necessary to correctly identify the component required.

Vehicle Identification Number (VIN)

This very important identification number is stamped on a plate attached to the dashboard inside the windshield on the driver's side of the vehicle (see illustration). The VIN also appears on the Vehicle Certificate of Title and Registration. It contains information such as where and when the vehicle was manufactured, the model year and the body style.

VIN engine and model year codes

Two particularly important pieces of information found in the VIN are the engine code and the model year code. Counting from the left, the engine code letter designation is the 5th digit and the model year code letter designation is the 10th digit.

On the vehicles covered by this manual the model year codes are:

B 2011
C 2012
D 2013
E 2014
F 2015
G 2016
H 2017
J 2018
K 2019

On the vehicles covered by this manual the VIN engine codes are:

Jetta VIN Z, L5 2.5L (CBUA) - 2011
Jetta VIN X, L5 2.5L (CBTA) - 2011 through 2014
Jetta VIN K, 2.0L (CBPA) - 2011 through 2015
Jetta VIN P, L5 2.5L (CBUA) - 2012 through 2014
Jetta VIN 8, 2.0L T (CCTA) - 2012
Jetta VIN A, 2.0L T (CBFA) - 2012 and 2013
Jetta VIN 6, 2.0L T (CCTA) - 2012 and 2013
Jetta VIN S, 2.0L T (CPLA) - 2013 through 2017
Jetta VIN T, 2.0L T (CPPA) - 2013 through 2018
Jetta VIN 0, 1.8L T (CPKA) - 2014 through 2016
Jetta VIN 1, 1.8L T (CPRA) - 2014 through 2018
Jetta VIN 6, 1.4L T (CZTA) - 2016 through 2018
Jetta VIN B, 1.4L T (CZTA) - 2017 and 2018
Golf VIN 1, 1.8L T (CNSA, CNSB) - 2015
Golf VIN 1, 1.8L T (CXBA, CXBB) - 2015 through 2019
GTI VIN T, 2.0L T (CXCA) - 2015 through 2017
GTI VIN 4, 2.0L T (CXCB) - 2015 through 2018
Golf VIN T, 1.4L T (DGXA) - 2019
GTI VIN T, 2.0L T (DKFA) - 2019
Golf R 4Motion VIN F, 2.0L (CYFB) - 2015 through 2017
Golf R 4Motion VIN F, 2.0L (DJJA) - 2018 and 2019
Golf R 4Motion VIN A, 2.0L (DLRA) - 2019

On the vehicles covered by this manual the transmission codes are:

0A4   5-speed manual transmission
0AF   5-speed manual transmission
02Q, A, H, K, S, 0BB
          6-speed manual transmission
09G/09GL 6-speed automatic transmission
09S   8-speed automatic transmission
02E/09D   6-speed direct shift gearbox (DSG) transmission
0AM/0CW   7-speed direct shift gearbox (DSG) transmission

Vehicle Certification Label

The Vehicle Certification Label is attached to the driver's side door post (see illustration). Information on this label includes the name of the manufacturer, the month and year of production and the Vehicle Identification Number.

Engine identification number

Locations of the engine identification numbers:

Four-cylinder engines: The engine code and serial number are stamped onto a pad on the left end of the engine block (see illustration) where the transaxle and engine meet, and the engine code letters are stamped onto the front left corner of the cylinder head.

L5 five-cylinder engines: The engine code letters and serial number are stamped onto the back side of the engine block just above the oil pan, and the engine code letters are on a sticker attached to the top of the valve cover.

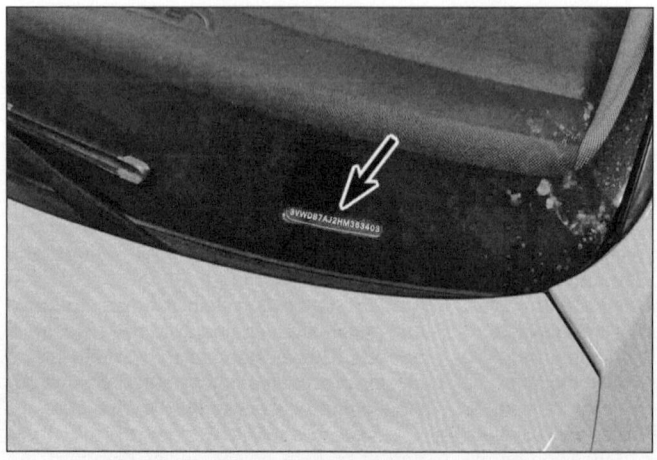

The VIN number is visible through the windshield on the driver's side

All transmission codes are stamped onto the top, or a sticker is affixed to the side of the transmission.

The Vehicle Certification Label is affixed to the driver's side door post

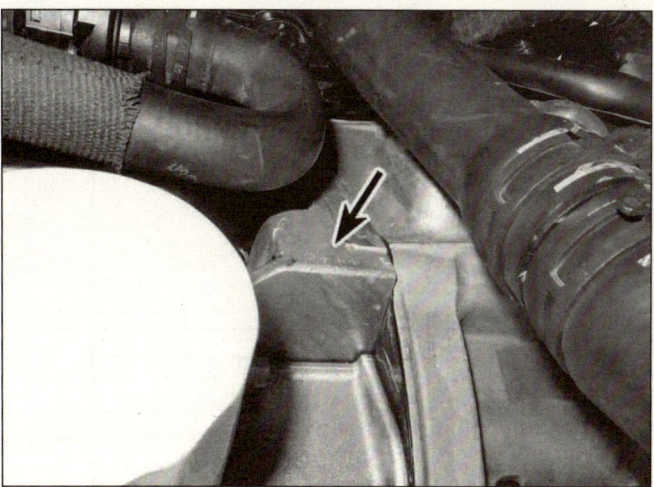

On four-cylinder engine, the engine identification number is stamped onto a pad on rear corner of the engine block - 1.4L (CZTA) engine shown

Recall information

Vehicle recalls are carried out by the manufacturer in the rare event of a possible safety-related defect. The vehicle's registered owner is contacted at the address on file at the Department of Motor Vehicles and given the details of the recall. Remedial work is carried out free of charge at a dealer service department.

If you are the new owner of a used vehicle which was subject to a recall and you want to be sure that the work has been carried out, it's best to contact a dealer service department and ask about your individual vehicle - you'll need to furnish them your Vehicle Identification Number (VIN).

The table below is based on information provided by the National Highway Traffic Safety Administration (NHTSA), the body which oversees vehicle recalls in the United States. The recall database is updated constantly. For the latest information on vehicle recalls, check the NHTSA website at www.nhtsa.gov, www.safercar.gov, or call the NHTSA hotline at 1-888-327-4236.

Recall date	Recall campaign number	Model(s) affected	Concern
March 29, 2011	11V196000	2011 Jetta	Some models may have an electrical wiring and fuse layout where the converter box is protected by the same fuse used by the signal horn and the anti-theft alarm system. Should that fuse be blown, the converter box will be disconnected from the power supply which, in turn, will shut off applications such as the engine management system, lighting system, and wipers. Should this happen while the vehicle is being driven, the engine could stall, or the headlights or wipers could turn off unexpectedly, potentially leading to a crash without warning.
September 7, 2011	11E036000	2011, 2012 Jetta	Volkswagen is recalling certain stainless steel exhaust tips, part number 1K0 071 910 U, sold as accessory equipment for use on model year 2011-2012 Jetta sedan vehicles manufactured from March 18, 2010, through August 22, 2011. These exhaust tips may extend beyond the original length of the factory-installed exhaust pipes. It is possible for inadvertent contact to occur with a person's leg. If the tailpipe extension is hot during inadvertent contact, a burn could occur.

Recall information

Recall date	Recall campaign number	Model(s) affected	Concern
September 8, 2011	11V466000	2011, 2012 Jetta	Some models had a stainless steel exhaust tip installed at the port during importation. These exhaust tips may extend beyond the original length of the factory-installed exhaust pipes. It is possible for inadvertent contact to occur with a person's leg. If the tailpipe extension is hot during inadvertent contact, a burn could occur.
March 7, 2014	14E007000	2011, 2012, 2013 Jetta	Volkswagen is recalling certain aftermarket water pumps. In the affected water pumps, the pulley or sprocket that turns the timing belt may develop microfractures causing the timing belt to fail. A failure of the timing belt may cause the engine to shut down, potentially increasing the risk of a vehicle crash.
April 15, 2014	14V182000	2014 Jetta	Volkswagen is recalling certain models equipped with a 1.8T engine and torque converter automatic transmission. In the affected vehicles, the O-ring seals between the oil cooler and the transmission may leak fluid. The leaking transmission fluid could contact a hot surface and result in a vehicle fire.
July 7, 2014	14V412000	2015 Golf, GTI	Volkswagen is recalling certain models where the stabilizer link fasteners may come loose and possibly interfere with the steering of the vehicle. A loose stabilizer link may interfere with the vehicle's steering, requiring additional effort to control the vehicle, increasing the risk of a crash.
October 20, 2014	14V656000	2011, 2012, 2013 Jetta	Volkswagen is recalling certain models do to the durability of the rear trailing arms may be reduced in vehicles whose rear trailing arms have been previously deformed, such as a result of a rear or side-rear impact crash. The reduced durabiliy of the trailing arm may result in its sudden fracture, possibly causing loss of vehicle control and increasing the risk of a crash.
October 24, 2014	14V670000	2015 Jetta	Volkswagen is recalling certain models equipped with manual front seatback recliners. In the affected vehicles, the seatback recliner retaining bracket may not engage correctly, resulting in unexpected movement of the seatback. Unexpected movement of the driver's seatback may distract the driver, increasing the risk of a crash.
December 15, 2014	14V790000	2015 Jetta	Volkswagen is recalling certain models due to incorrect software within the headlight control module; the low beam headlights may turn off when high beam lights are turned on. As a result, the light output from the headlights may be insufficient. Insufficient headlight output may increase the risk of a crash.

Recall information

Recall date	Recall campaign number	Model(s) affected	Concern
January 23, 2015	15V028000	2014, 2015 Jetta, Golf and GTI	Volkswagen is recalling certain models where a sealing cap at the fuel rail may fail, allowing fuel to leak into the engine compartment. Leaking fuel in the presence of an ignition source, may lead to a fire.
April 17, 2015	15V229000	2015 Golf, GTI	Volkswagen is recalling certain models that have improper nickel plating of components within the fuel pump may result in the fuel pump failing. If the fuel pump fails, the vehicle will not start, or if the engine is running, it will stop and the vehicle will stall, increasing the risk of a crash.
August 4, 2015	15V483000	2011, 2012, 2013, 2014 Golf, Jetta and GTI	In the affected vehicles, debris may contaminate the airbag clock spring, a spiral wound, flat cable that keeps the airbag powered while the steering wheel is being turned. This contamination may tear the cable and result in a loss of electrical connection to the driver's frontal airbag. A loss of electrical connection to the driver's frontal airbag will prevent the airbag from deploying in the event of a vehicle crash, increasing the risk of injury.
October 7, 2015	15V627000	2015 Jetta, Golf, GTI	Volkswagen is recalling certain models equipped a Passenger Occupant Detection System (PODS) that may have been manufactured improperly. As a result, the front passenger seat occupant may be improperly classified or may not be detected. In the event of a crash, if the front passenger seat occupant is incorrectly classified or non-detected, the passenger frontal air bag may deploy improperly or not at all, increasing the risk of occupant injury.
October 26, 2015	15V705000	2015, 2016 Jetta, Golf, GTI	Volkswagen is recalling certain models, where the camshaft lobe that drives the brake vacuum pump may shear off, resulting in a loss of brake assist. If the camshaft lobe shears off there would be a loss of brake assist, lengthening the distance needed to stop the vehicle and increasing the risk of a crash.
February 10, 2016	16V078000	2010, 2011, 2012, 2013, 2014 Jetta, Golf	Volkswagen is recalling certain models, where the deployment of the driver's frontal airbag, excessive internal pressure may cause the inflator to rupture. In the event of a crash necessitating deployment of the driver's frontal airbag, the inflator could rupture with metal fragments striking the vehicle occupants potentially resulting in serious injury or death.
September 7, 2016	16V647000	2015, 2016 Golf, GTI	Volkswagen is recalling certain models due to a problem with the suction pump inside the fuel tank; fuel may flow into the evaporative emissions (EVAP) system. As fuel accumulates in the EVAP system, it may leak out through the charcoal canister filter element. A fuel leak in the presence of an ignition source increases the risk of a fire.

Recall information

Recall date	Recall campaign number	Model(s) affected	Concern
December 29, 2016	16V955000	2017 Golf	Volkswagen is recalling certain models where the driver frontal airbags, passenger frontal airbags or head airbags may not deploy properly. In the event of a crash, if the airbags do not inflate or function properly, the vehicle occupants have an increased risk of injury.
February 3, 2017	17V070000	2017 Jetta	Volkswagen is recalling certain models equipped with a 1.4L engine. These vehicles have an engine that may seize due to an improper casting of the block. If the engine seizes, the wheels may suddenly lock up, causing a loss of vehicle control and increasing the risk of a crash.
March 2, 2017	17V136000	2017 Jetta	Volkswagen is recalling certain models that have incorrect information on the tire information label, possibly causing the operator to overload the vehicle. Overloading the vehicle may affect vehicle handling or result in tire damage, increasing the risk of a crash.
March 2, 2017	17V137000	2017 Jetta	Volkswagen is recalling certain models that have incorrect information on the tire information label, possibly causing the operator to overload the vehicle. Overloading the vehicle may affect vehicle handling or result in tire damage, increasing the risk of a crash.
May 31, 2017	17V352000	2017 Jetta	Volkswagen is recalling certain models that have Vehicle Identification Number (VIN) markings on the body that do not match the VIN plate near the windshield. The mismatched VINs make the vehicle non-compliant with regulatory requirements.
March 6, 2018	18V148000	2010, 2011, 2012, 2013, 2014 Golf	Volkswagen is recalling certain model that upon deployment of the driver's frontal airbag, excessive internal pressure may cause the inflator to explode. In the event of a crash necessitating deployment of the driver's frontal airbag, the inflator could explode with metal fragments striking the vehicle occupants potentially resulting in serious injury or death.
May 16, 2018	18V329000	2011, 2012, 2013, 2014, 2015, 2016 Golf	Volkswagen is recalling certain models where modifications made while the vehicles were in an internal evaluation period may cause the affected vehicles to not comply with all of the applicable regulatory requirements. If the vehicles do not meet all regulatory requirements, there could be an increased risk of a crash, fire or injury.
June 1, 2018	18V369000	2018 Golf, SportWagen, GTI	Volkswagen is recalling certain models where the brake caliper pistons may have insufficient coating, potentially reducing the brake performance. A reduction of braking performance can increase the risk of a crash.

Recall information

Recall date	Recall campaign number	Model(s) affected	Concern
July 11, 2018	18V464000	2015, 2016, 2017, 2018 Golf	Volkswagen is recalling certain models where the build up of silicate on the shift lever micro switch contacts may enable the key to be removed from the ignition while the vehicle shift lever is not in "Park." Removing the key while the shift lever is in a position other than "PARK" increases the risk of an unintended vehicle rollaway that may result in personal injury or a crash.
September 26, 2018	18V671000	2019 Jetta	Volkswagen is recalling certain models equipped with LED headlights. The passenger side headlight may be incorrectly positioned, reducing the driver's visibility when it is dark. A reduction in the driver's visibility can increase the risk of a crash.
November 14, 2018	18V803000	2018 Golf	Volkswagen is recalling certain models where the rear seat frame head restraint guide sleeves may be incorrectly welded to the seat frame. If the guide sleeves are incorrectly welded, in the event of a crash, the rear seat head restraints may have reduced stability, increasing the risk of injury.
November 26, 2018	18V824000	2018, 2019 Jetta	Volkswagen is recalling certain models that do not have keyless entry. The instrument cluster may not provide an audible warning to let the driver know that the key is still in the ignition when the door is open. If the driver is not notified by an audible noise that the key is left in the ignition, it can increase the risk of vehicle theft or crash.
December 19, 2018	18V904000	2015 Golf SportWagen, 2018, 2019 Jetta	Volkswagen is recalling certain models where the rear coil springs may prematurely fracture. If a coil spring fractures while driving, it may damage a rear tire causing a loss of vehicle control, increasing the risk of a crash.
February 20, 2019	19V110000	2019 Jetta	Volkswagen is recalling certain models that may have an incorrect driver frontal airbag may have been installed during a service/repair visit. In the event of a crash necessitating air bag deployment, the wrong air bag may not work correctly, increasing the risk of injury.
March 6, 2019	19V188000	2017, 2018, 2019 Jetta, Golf	Volkswagen is recalling certain models where the rear coil springs may prematurely fracture. If a coil spring fractures while driving, it may damage a rear tire causing a loss of vehicle control, increasing the risk of a crash.

Buying parts

Replacement parts are available from many sources. Our advice concerning them is as follows:

Retail auto parts stores: Good auto parts stores will stock frequently needed components which wear out relatively fast, such as clutch components, exhaust systems, brake parts, tune-up parts, etc. These stores often supply new or reconditioned parts on an exchange basis, which can save a considerable amount of money. Discount auto parts stores are often very good places to buy materials and parts needed for general vehicle maintenance such as oil, grease, filters, spark plugs, belts, touch-up paint, bulbs, etc. They also usually sell tools and general accessories, have convenient hours, charge lower prices and can give you knowledgeable answers to your questions. To be sure of obtaining the correct parts, have engine and chassis numbers available and, if possible, take the old parts along for positive identification.

Authorized dealer parts department: This is the best source for parts which are unique to the vehicle and not generally available elsewhere. Prices for most parts tend to be higher than at retail auto parts stores.

Auto recyclers or salvage yards: Auto recyclers and salvage yards are good sources for components that are specific to the vehicle and not subject to wear, such as fenders, bumpers, trim pieces, etc. You can expect substantial savings by going this route, and self-service salvage yards offer still more savings if you're willing to bring your own tools and pull the part(s) yourself.

Warranty information: If the vehicle is still covered under warranty, be sure that any replacement parts purchased - regardless of the source - do not invalidate the warranty! In most cases, replacement parts, even from aftermarket suppliers, are designed to meet manufacturer specifications. If in doubt, check with the parts supplier.

Because of a Federally mandated extended warranty that covers the emissions control system components, check with your dealer about warranty coverage before working on any emissions-related systems.

Discount auto parts stores provide a wide variety of parts, and the knowledgeable counter-people found there are an excellent resource

You'll find the details of your vehicle's warranty coverage in the warranty book provided by the manufacturer

Maintenance techniques, tools and working facilities

Maintenance techniques

There are a number of techniques involved in maintenance and repair that will be referred to throughout this manual. Application of these techniques will enable the home mechanic to be more efficient, better organized and capable of performing the various tasks properly, which will ensure that the repair job is thorough and complete.

Fasteners

Fasteners are nuts, bolts, studs and screws used to hold two or more parts together. There are a few things to keep in mind when working with fasteners. Almost all of them use a locking device of some type, either a lockwasher, locknut, locking tab or thread adhesive. All threaded fasteners should be clean and straight, with undamaged threads and undamaged corners on the hex head where the wrench fits. Develop the habit of replacing all damaged nuts and bolts with new ones. Special locknuts with nylon or fiber inserts can only be used once. If they are removed, they lose their locking ability and must be replaced with new ones.

Rusted nuts and bolts should be treated with a penetrating fluid to ease removal and prevent breakage. Some mechanics use turpentine in a spout-type oil can, which works quite well. After applying the rust penetrant, let it work for a few minutes before trying to loosen the nut or bolt. Badly rusted fasteners may have to be chiseled or sawed off or removed with a special nut breaker, available at tool stores.

If a bolt or stud breaks off in an assembly, it can be drilled and removed with a special tool commonly available for this purpose. Most automotive machine shops can perform this task, as well as other repair procedures, such as the repair of threaded holes that have been stripped out.

Flat washers and lockwashers, when removed from an assembly, should always be replaced exactly as removed. Replace any damaged washers with new ones. Never use a lockwasher on any soft metal surface (such as aluminum), thin sheet metal or plastic.

Maintenance techniques, tools and working facilities

Fastener sizes

For a number of reasons, automobile manufacturers are making wider and wider use of metric fasteners. Therefore, it is important to be able to tell the difference between standard (sometimes called U.S. or SAE) and metric hardware, since they cannot be interchanged.

All bolts, whether standard or metric, are sized according to diameter, thread pitch and length. For example, a standard 1/2 - 13 x 1 bolt is 1/2 inch in diameter, has 13 threads per inch and is 1 inch long. An M12 - 1.75 x 25 metric bolt is 12 mm in diameter, has a thread pitch of 1.75 mm (the distance between threads) and is 25 mm long. The two bolts are nearly identical, and easily confused, but they are not interchangeable.

In addition to the differences in diameter, thread pitch and length, metric and standard bolts can also be distinguished by examining the bolt heads. To begin with, the distance across the flats on a standard bolt head is measured in inches, while the same dimension on a metric bolt is sized in millimeters (the same is true for nuts). As a result, a standard wrench should not be used on a metric bolt and a metric wrench should not be used on a standard bolt. Also, most standard bolts have slashes radiating out from the center of the head to denote the grade or strength of the bolt, which is an indication of the amount of torque that can be applied to it. The greater the number of slashes, the greater the strength of the bolt. Grades 0 through 5 are commonly used on automobiles. Metric bolts have a property class (grade) number, rather than a slash, molded into their heads to indicate bolt strength. In this case, the higher the number, the stronger the bolt. Property class numbers 8.8, 9.8 and 10.9 are commonly used on automobiles.

Strength markings can also be used to distinguish standard hex nuts from metric hex nuts. Many standard nuts have dots stamped into one side, while metric nuts are marked with a number. The greater the number of dots, or the higher the number, the greater the strength of the nut.

Metric studs are also marked on their ends according to property class (grade). Larger studs are numbered (the same as metric bolts), while smaller studs carry a geometric code to denote grade.

It should be noted that many fasteners, especially Grades 0 through 2, have no distinguishing marks on them. When such is the case, the only way to determine whether it is standard or metric is to measure the thread pitch or compare it to a known fastener of the same size.

Standard fasteners are often referred to as SAE, as opposed to metric. However, it should be noted that SAE technically refers to a non-metric fine thread fastener only. Coarse thread non-metric fasteners are referred to as USS sizes.

Since fasteners of the same size (both standard and metric) may have different strength ratings, be sure to reinstall any bolts, studs or nuts removed from your vehicle in their original locations. Also, when replacing a fastener with a new one, make sure that the new one has a strength rating equal to or greater than the original.

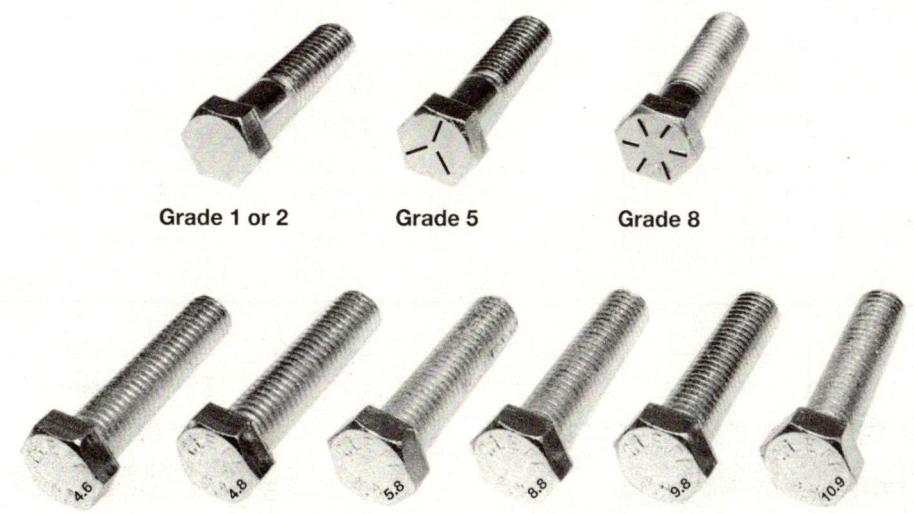

Bolt strength marking (standard/SAE/USS; bottom - metric)

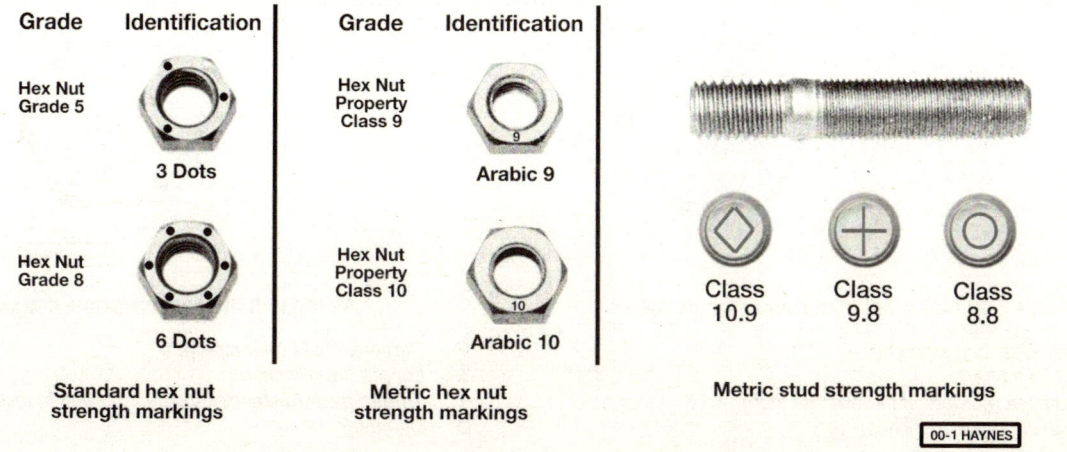

Standard hex nut strength markings

Metric hex nut strength markings

Metric stud strength markings

Tightening sequences and procedures

Most threaded fasteners should be tightened to a specific torque value (torque is the twisting force applied to a threaded component such as a nut or bolt). Overtightening the fastener can weaken it and cause it to break, while undertightening can cause it to eventually come loose. Bolts, screws and studs, depending on the material they are made of and their thread diameters, have specific torque values, many of which are noted in the Specifications at the beginning of each Chapter. Be sure to follow the torque recommendations closely. For fasteners not assigned a specific torque, a general torque value chart is presented here as a guide. These torque values are for dry (unlubricated) fasteners threaded into steel or cast iron (not aluminum). As was previously mentioned, the size and grade of a fastener determine the amount of torque that can safely be applied to it. The figures listed here are approximate for Grade 2 and Grade 3 fasteners. Higher grades can tolerate higher torque values.

Fasteners laid out in a pattern, such as cylinder head bolts, oil pan bolts, differential cover bolts, etc., must be loosened or tightened in sequence to avoid warping the component. This sequence will normally be shown in the appropriate Chapter. If a specific pattern is not given, the following procedures can be used to prevent warping.

Initially, the bolts or nuts should be assembled finger-tight only. Next, they should be tightened one full turn each, in a criss-cross or diagonal pattern. After each one has been tightened one full turn, return to the first one and tighten them all one-half turn, following the same pattern. Finally, tighten each of them one-quarter turn at a time until each fastener has been tightened to the proper torque. To loosen and remove the fasteners, the procedure would be reversed.

Metric thread sizes	Ft-lbs	Nm
M-6	6 to 9	9 to 12
M-8	14 to 21	19 to 28
M-10	28 to 40	38 to 54
M-12	50 to 71	68 to 96
M-14	80 to 140	109 to 154

Pipe thread sizes		
1/8	5 to 8	7 to 10
1/4	12 to 18	17 to 24
3/8	22 to 33	30 to 44
1/2	25 to 35	34 to 47

U.S. thread sizes		
1/4 - 20	6 to 9	9 to 12
5/16 - 18	12 to 18	17 to 24
5/16 - 24	14 to 20	19 to 27
3/8 - 16	22 to 32	30 to 43
3/8 - 24	27 to 38	37 to 51
7/16 - 14	40 to 55	55 to 74
7/16 - 20	40 to 60	55 to 81
1/2 - 13	55 to 80	75 to 108

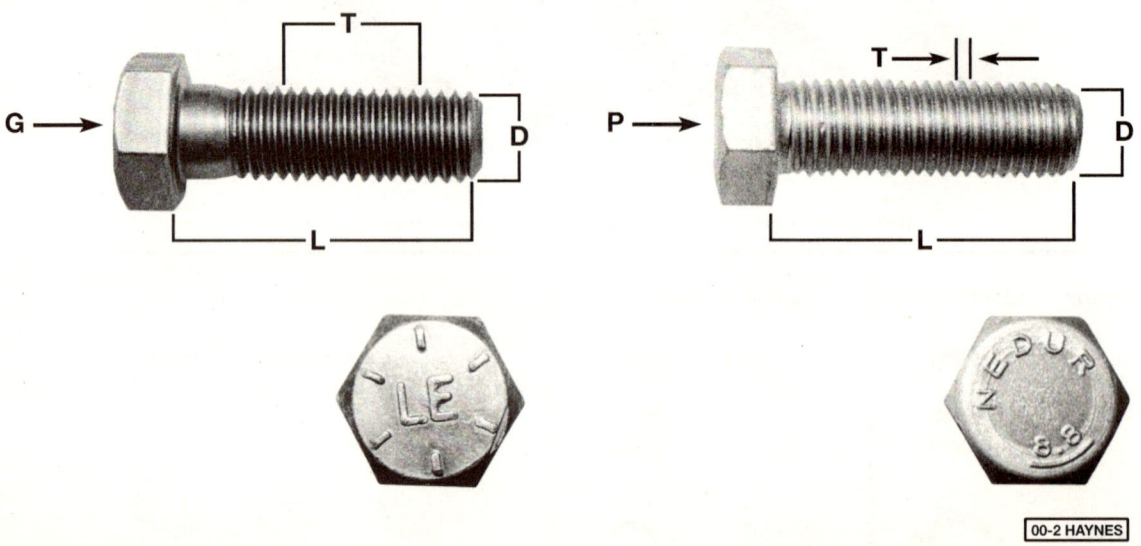

Standard (SAE and USS) bolt dimensions/grade marks

- G   Grade marks (bolt strength)
- L   Length (in inches)
- T   Thread pitch (number of threads per inch)
- D   Nominal diameter (in inches)

Metric bolt dimensions/grade marks

- P   Property class (bolt strength)
- L   Length (in millimeters)
- T   Thread pitch (distance between threads in millimeters)
- D   Diameter

Maintenance techniques, tools and working facilities

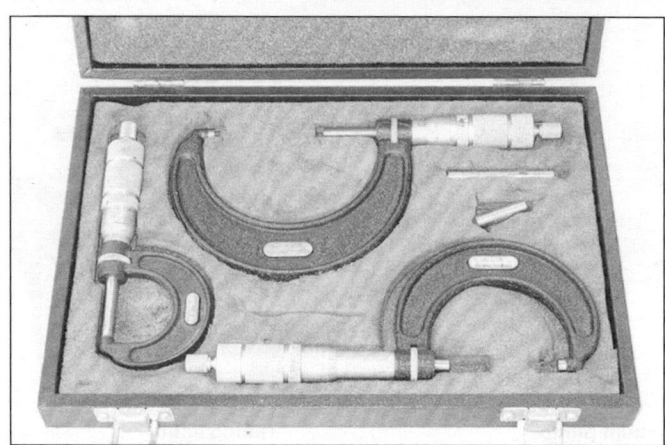

Micrometer set

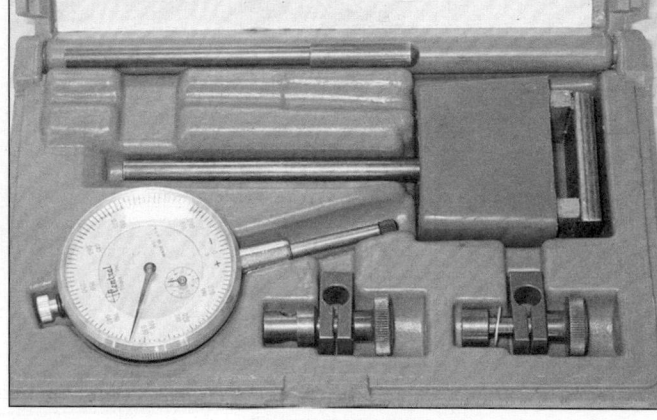

Dial indicator set

Component disassembly

Component disassembly should be done with care and purpose to help ensure that the parts go back together properly. Always keep track of the sequence in which parts are removed. Make note of special characteristics or marks on parts that can be installed more than one way, such as a grooved thrust washer on a shaft. It is a good idea to lay the disassembled parts out on a clean surface in the order that they were removed. It may also be helpful to make sketches or take instant photos of components before removal.

When removing fasteners from a component, keep track of their locations. Sometimes threading a bolt back in a part, or putting the washers and nut back on a stud, can prevent mix-ups later. If nuts and bolts cannot be returned to their original locations, they should be kept in a compartmented box or a series of small boxes. A cupcake or muffin tin is ideal for this purpose, since each cavity can hold the bolts and nuts from a particular area (i.e. oil pan bolts, valve cover bolts, engine mount bolts, etc.). A pan of this type is especially helpful when working on assemblies with very small parts, such as the carburetor, alternator, valve train or interior dash and trim pieces. The cavities can be marked with paint or tape to identify the contents.

Whenever wiring looms, harnesses or connectors are separated, it is a good idea to identify the two halves with numbered pieces of masking tape so they can be easily reconnected.

Gasket sealing surfaces

Throughout any vehicle, gaskets are used to seal the mating surfaces between two parts and keep lubricants, fluids, vacuum or pressure contained in an assembly.

Many times these gaskets are coated with a liquid or paste-type gasket sealing compound before assembly. Age, heat and pressure can sometimes cause the two parts to stick together so tightly that they are very difficult to separate. Often, the assembly can be loosened by striking it with a soft-face hammer near the mating surfaces. A regular hammer can be used if a block of wood is placed between the hammer and the part. Do not hammer on cast parts or parts that could be easily damaged. With any particularly stubborn part, always recheck to make sure that every fastener has been removed.

Avoid using a screwdriver or bar to pry apart an assembly, as they can easily mar the gasket sealing surfaces of the parts, which must remain smooth. If prying is absolutely necessary, use an old broom handle, but keep in mind that extra clean up will be necessary if the wood splinters.

After the parts are separated, the old gasket must be carefully scraped off and the gasket surfaces cleaned. Stubborn gasket material can be soaked with rust penetrant or treated with a special chemical to soften it so it can be easily scraped off. **Caution:** *Never use gasket removal solutions or caustic chemicals on plastic or other composite components.* A scraper can be fashioned from a piece of copper tubing by flattening and sharpening one end. Copper is recommended because it is usually softer than the surfaces to be scraped, which reduces the chance of gouging the part. Some gaskets can be removed with a wire brush, but regardless of the method used, the mating surfaces must be left clean and smooth. If for some reason the gasket surface is gouged, then a gasket sealer thick enough to fill scratches will have to be used during reassembly of the components. For most applications, a non-drying (or semi-drying) gasket sealer should be used.

Hose removal tips

Warning: *If the vehicle is equipped with air conditioning, do not disconnect any of the A/C hoses without first having the system depressurized by a dealer service department or a service station.*

Hose removal precautions closely parallel gasket removal precautions. Avoid scratching or gouging the surface that the hose mates against or the connection may leak. This is especially true for radiator hoses. Because of various chemical reactions, the rubber in hoses can bond itself to the metal spigot that the hose fits over. To remove a hose, first loosen the hose clamps that secure it to the spigot. Then, with slip-joint pliers, grab the hose at the clamp and rotate it around the spigot. Work it back and forth until it is completely free, then pull it off. Silicone or other lubricants will ease removal if they can be applied between the hose and the outside of the spigot. Apply the same lubricant to the inside of the hose and the outside of the spigot to simplify installation.

As a last resort (and if the hose is to be replaced with a new one anyway), the rubber can be slit with a knife and the hose peeled from the spigot. If this must be done, be careful that the metal connection is not damaged.

If a hose clamp is broken or damaged, do not reuse it. Wire-type clamps usually weaken with age, so it is a good idea to replace them with screw-type clamps whenever a hose is removed.

Tools

A selection of good tools is a basic requirement for anyone who plans to maintain and repair his or her own vehicle. For the owner who has few tools, the initial investment might seem high, but when compared to the spiraling costs of professional auto maintenance and repair, it is a wise one.

To help the owner decide which tools are needed to perform the tasks detailed in this manual, the following tool lists are offered: *Maintenance and minor repair, Repair/overhaul* and *Special*.

The newcomer to practical mechanics should start off with the *maintenance and minor repair* tool kit, which is adequate for the simpler jobs performed on a vehicle. Then, as confidence and experience grow, the owner can tackle more difficult tasks, buying additional tools as they are needed. Eventually the basic kit will be expanded into the *repair and overhaul* tool set. Over a period of time, the experienced do-it-yourselfer will assemble a tool set complete enough for most repair and overhaul procedures and will add tools from the special category when it is felt that the expense is justified by the frequency of use.

0-16 Maintenance techniques, tools and working facilities

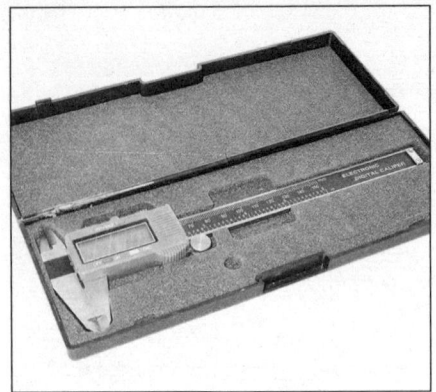
Dial caliper

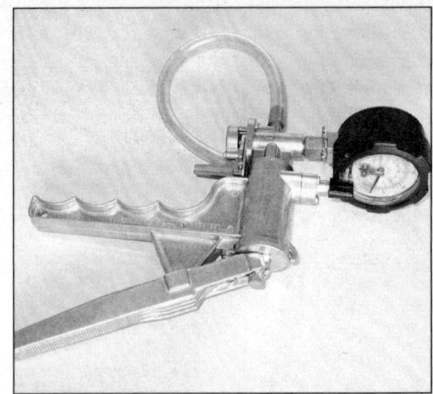
Hand-operated vacuum pump

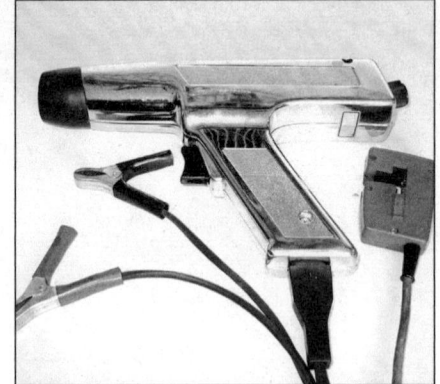
Timing light

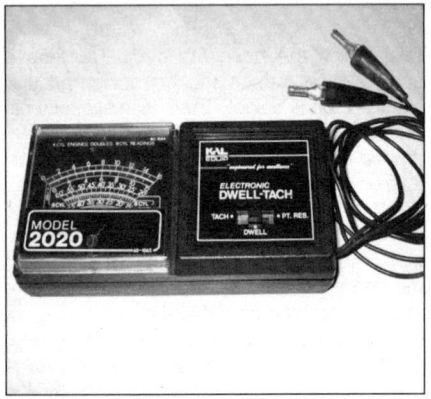
Tachometer and dwellmeter

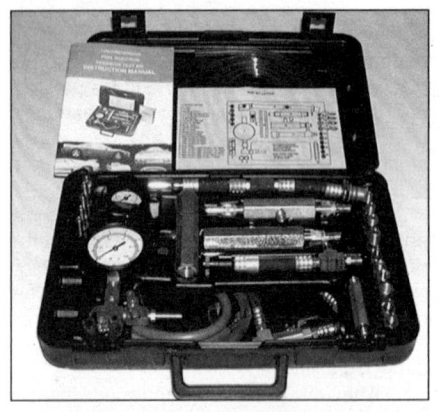
Fuel pressure guage set

Trouble code reader

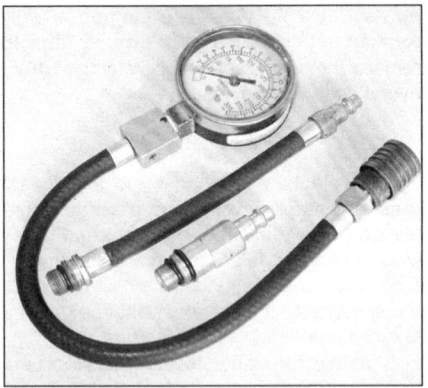
Compression gauge

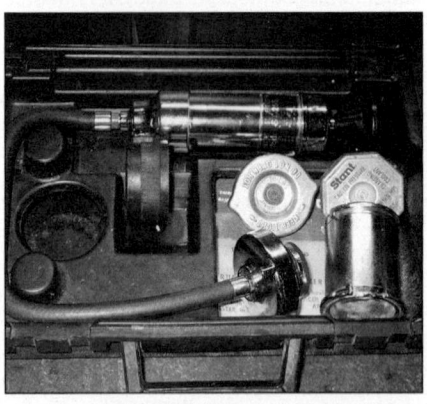
Cooling system pressure tester

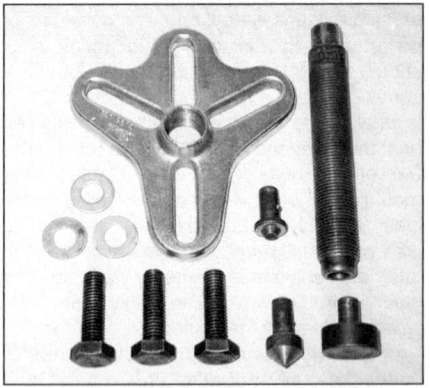
Damper and steering wheel puller

Maintenance and minor repair tool kit

The tools in this list should be considered the minimum required for performance of routine maintenance, servicing and minor repair work. We recommend the purchase of combination wrenches (box-end and open-end combined in one wrench). While more expensive than open end wrenches, they offer the advantages of both types of wrench.

*Combination wrench set
  (1/4-inch to 1 inch or 6 mm to 19 mm)
Adjustable wrench, 8 inch*

*Spark plug wrench with rubber insert
Spark plug gap adjusting tool
Feeler gauge set
Brake bleeder wrench
Standard screwdriver
  (5/16-inch x 6 inch)
Phillips screwdriver (No. 2 x 6 inch)
Combination pliers - 6 inch
Hacksaw and assortment of blades
Tire pressure gauge
Grease gun
Oil can
Fine emery cloth
Wire brush*

Battery post and cable cleaning tool
Oil filter wrench
Funnel (medium size)
Safety goggles
Jackstands (2)
Drain pan

Note: *If basic tune-ups are going to be part of routine maintenance, it will be necessary to purchase a good quality stroboscopic timing light and combination tachometer/dwell meter. Although they are included in the list of special tools, it is mentioned here because they are absolutely necessary for tuning most vehicles properly.*

Maintenance techniques, tools and working facilities   0-17

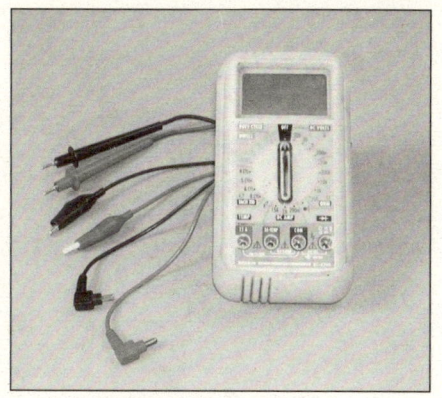
Electrical multimeter

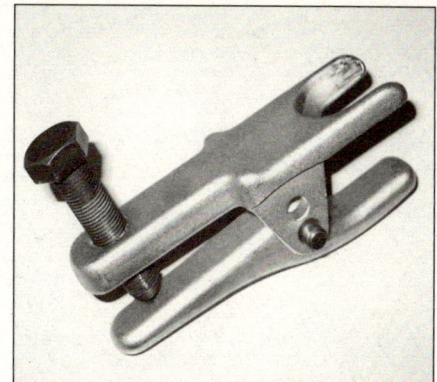
Balljoint separator

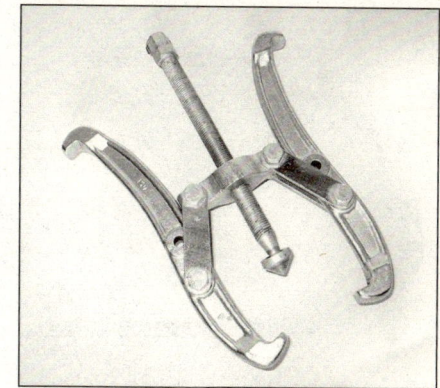
General purpose puller

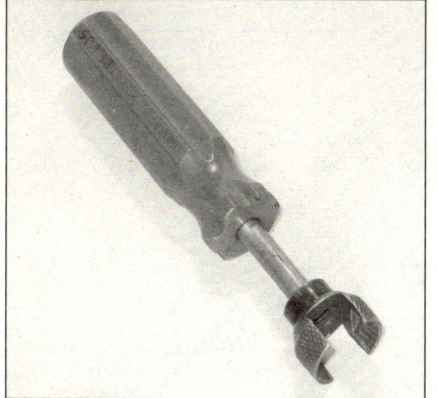
Brake hold-down spring tool

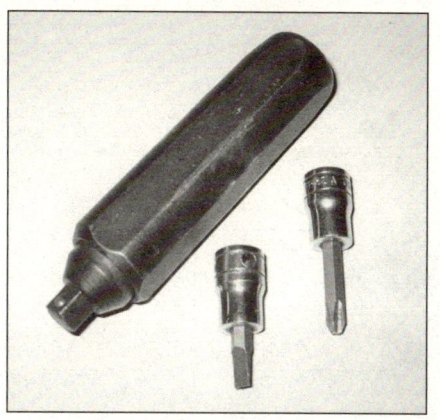
Impact screwdriver

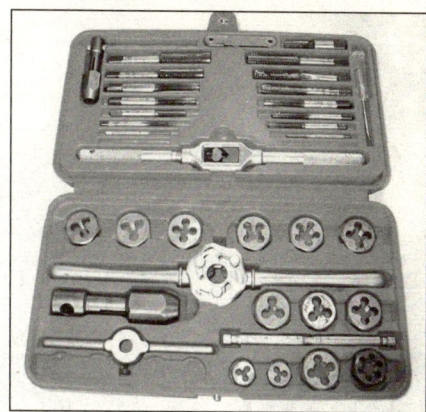
Tap and die set

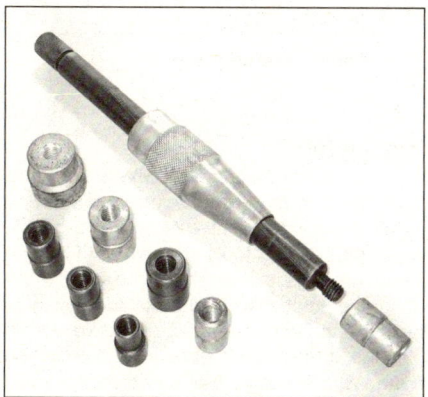
Clutch plate alignment tool

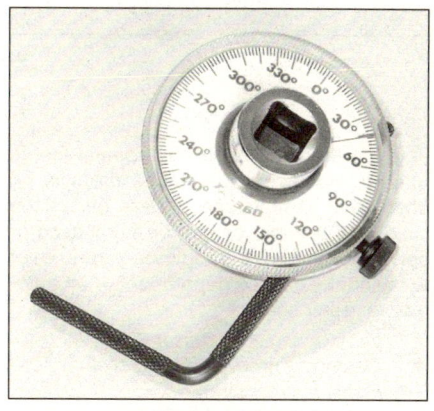
Torque angle gauge

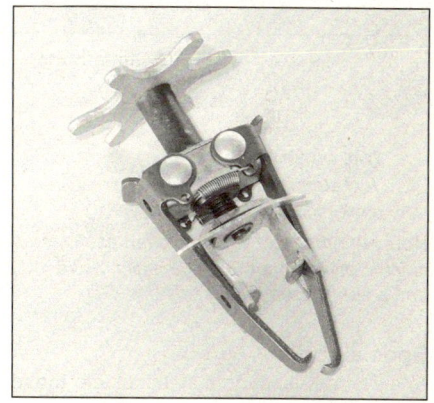
Valve spring compressor

Repair and overhaul tool set

These tools are essential for anyone who plans to perform major repairs and are in addition to those in the maintenance and minor repair tool kit. Included is a comprehensive set of sockets which, though expensive, are invaluable because of their versatility, especially when various extensions and drives are available. We recommend the 1/2-inch drive over the 3/8-inch drive. Although the larger drive is bulky and more expensive, it has the capacity of accepting a very wide range of large sockets. Ideally, however, the mechanic should have a 3/8-inch drive set and a 1/2-inch drive set.

Socket set(s)
Reversible ratchet
Extension - 10 inch
Universal joint
Torque wrench
 (same size drive as sockets)
Ball peen hammer - 8 ounce
Soft-face hammer (plastic/rubber)
Standard screwdriver (1/4-inch x 6 inch)
Standard screwdriver (stubby - 5/16-inch)
Phillips screwdriver (No. 3 x 8 inch)
Phillips screwdriver (stubby - No. 2)
Pliers - vise grip

Pliers - lineman's
Pliers - needle nose
Pliers - snap-ring (internal and external)
Cold chisel (1/2-inch)
Scribe
Scraper (made from flattened
 copper tubing)
Centerpunch
Pin punches (1/16, 1/8, 3/16-inch)
Steel rule/straightedge (12 inch)
Allen wrench set (1/8 to 3/8-inch or
 4 mm to 10 mm)
A selection of files

0-18　Maintenance techniques, tools and working facilities

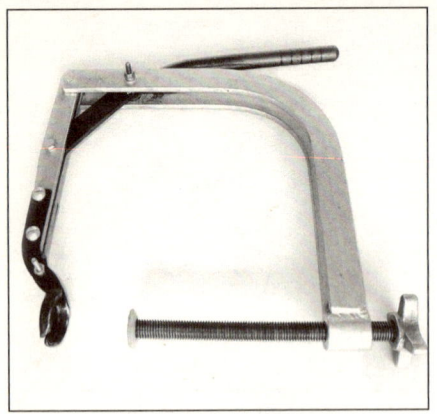
Valve spring compressor

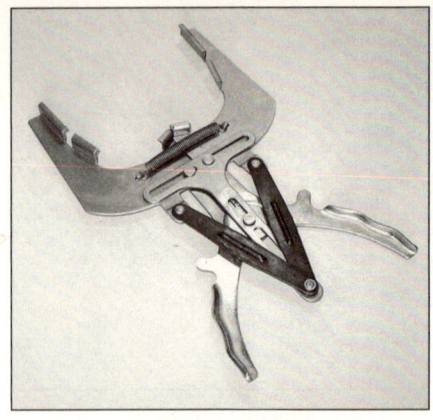
Piston ring removal and installation tool

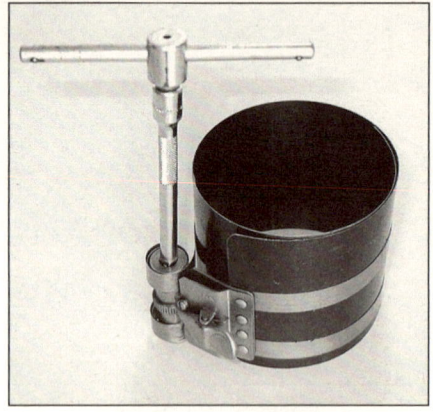
Piston ring compressor

Cylinder hone

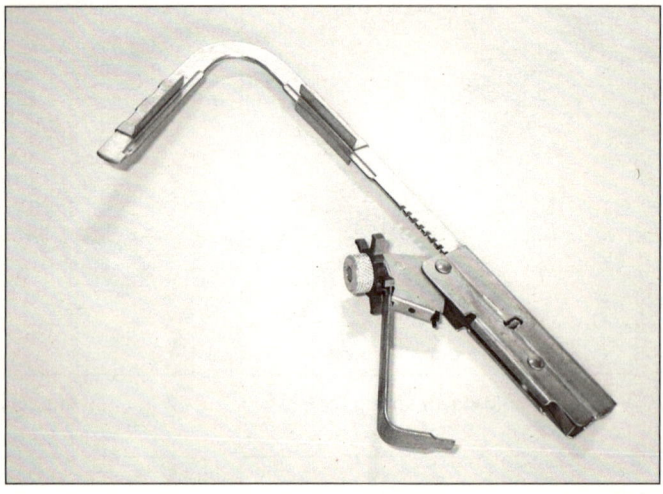
Piston ring groove cleaning tool

Wire brush (large)
Jackstands (second set)
Jack (scissor or hydraulic type)

Note: *Another tool which is often useful is an electric drill with a chuck capacity of 3/8-inch and a set of good quality drill bits.*

Special tools

The tools in this list include those which are not used regularly, are expensive to buy, or which need to be used in accordance with their manufacturer's instructions. Unless these tools will be used frequently, it is not very economical to purchase many of them. A consideration would be to split the cost and use between yourself and a friend or friends. In addition, most of these tools can be obtained from a tool rental shop on a temporary basis.

This list primarily contains only those tools and instruments widely available to the public, and not those special tools produced by the vehicle manufacturer for distribution to dealer service departments. Occasionally, references to the manufacturer's special tools are included in the text of this manual. Generally, an alternative method of doing the job without the special tool is offered. However, sometimes there is no alternative to their use. Where this is the case, and the tool cannot be purchased or borrowed, the work should be turned over to the dealer service department or an automotive repair shop.

Valve spring compressor
Piston ring groove cleaning tool
Piston ring compressor
Piston ring installation tool
Cylinder compression gauge
Cylinder ridge reamer
Cylinder surfacing hone
Cylinder bore gauge
Micrometers and/or dial calipers
Hydraulic lifter removal tool
Balljoint separator
Universal-type puller
Impact screwdriver
Dial indicator set
Stroboscopic timing light (inductive pickup)
Hand operated vacuum/pressure pump
Tachometer/dwell meter
Universal electrical multimeter
Cable hoist
Brake spring removal and installation tools
Floor jack

Buying tools

For the do-it-yourselfer who is just starting to get involved in vehicle maintenance and repair, there are a number of options available when purchasing tools. If maintenance and minor repair is the extent of the work to be done, the purchase of individual tools is satisfactory. If, on the other hand, extensive work is planned, it would be a good idea to purchase a modest tool set from one of the large retail chain stores. A set can usually be bought at a substantial savings over the individual tool prices, and they often come with a tool box. As additional tools are needed, add-on sets, individual tools and a larger tool box can be purchased to expand the tool selection. Building a tool set gradually allows the cost of the tools to be spread over a longer period of time and gives the mechanic the freedom to choose only those tools that will actually be used.

Tool stores will often be the only source of some of the special tools that are needed, but regardless of where tools are bought, try to avoid cheap ones, especially when buying screwdrivers and sockets, because they won't last very long. The expense involved in replacing cheap tools will eventually be greater than the initial cost of quality tools.

Maintenance techniques, tools and working facilities

0-19

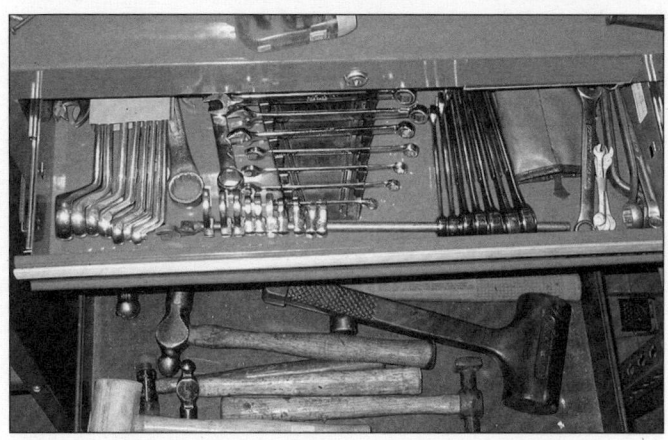

Keep your tools clean and organized – you'll spend less time under the hood, and the time spent will be more pleasant

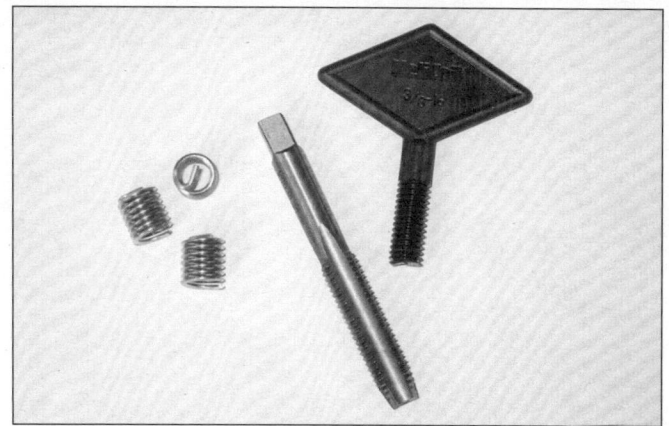

Thread repair kits like this one can be purchased at your local auto parts store

Care and maintenance of tools

Good tools are expensive, so it makes sense to treat them with respect. Keep them clean and in usable condition and store them properly when not in use. Always wipe off any dirt, grease or metal chips before putting them away. Never leave tools lying around in the work area. Upon completion of a job, always check closely under the hood for tools that may have been left there so they won't get lost during a test drive.

Some tools, such as screwdrivers, pliers, wrenches and sockets, can be hung on a panel mounted on the garage or workshop wall, while others should be kept in a tool box or tray. Measuring instruments, gauges, meters, etc. must be carefully stored where they cannot be damaged by weather or impact from other tools.

When tools are used with care and stored properly, they will last a very long time. Even with the best of care, though, tools will wear out if used frequently. When a tool is damaged or worn out, replace it. Subsequent jobs will be safer and more enjoyable if you do.

How to repair damaged threads

Sometimes, the internal threads of a nut or bolt hole can become stripped, usually from overtightening. Stripping threads is an all-too-common occurrence, especially when working with aluminum parts, because aluminum is so soft that it easily strips out.

Usually, external or internal threads are only partially stripped. After they've been cleaned up with a tap or die, they'll still work. Sometimes, however, threads are badly damaged. When this happens, you've got three choices:

1) Drill and tap the hole to the next suitable oversize and install a larger diameter bolt, screw or stud.
2) Drill and tap the hole to accept a threaded plug, then drill and tap the plug to the original screw size. You can also buy a plug already threaded to the original size. Then you simply drill a hole to the specified size, then run the threaded plug into the hole with a bolt and jam nut. Once the plug is fully seated, remove the jam nut and bolt.
3) *The third method uses a patented thread repair kit like Heli-Coil or Slimsert. These easy-to-use kits are designed to repair damaged threads in straight-through holes and blind holes. Both are available as kits which can handle a variety of sizes and thread patterns. Drill the hole, then tap it with the special included tap. Install the Heli-Coil and the hole is back to its original diameter and thread pitch.*

Regardless of which method you use, be sure to proceed calmly and carefully. A little impatience or carelessness during one of these relatively simple procedures can ruin your whole day's work and cost you a bundle if you wreck an expensive part.

Working facilities

Not to be overlooked when discussing tools is the workshop. If anything more than routine maintenance is to be carried out, some sort of suitable work area is essential.

It is understood, and appreciated, that many home mechanics do not have a good workshop or garage available, and end up removing an engine or doing major repairs outside. It is recommended, however, that the overhaul or repair be completed under the cover of a roof.

A clean, flat workbench or table of comfortable working height is an absolute necessity. The workbench should be equipped with a vise that has a jaw opening of at least four inches.

As mentioned previously, some clean, dry storage space is also required for tools, as well as the lubricants, fluids, cleaning solvents, etc. which soon become necessary.

Sometimes waste oil and fluids, drained from the engine or cooling system during normal maintenance or repairs, present a disposal problem. To avoid pouring them on the ground or into a sewage system, pour the used fluids into large containers, seal them with caps and take them to an authorized disposal site or recycling center. Plastic jugs, such as old antifreeze containers, are ideal for this purpose.

Always keep a supply of old newspapers and clean rags available. Old towels are excellent for mopping up spills. Many mechanics use rolls of paper towels for most work because they are readily available and disposable. To help keep the area under the vehicle clean, a large cardboard box can be cut open and flattened to protect the garage or shop floor.

Whenever working over a painted surface, such as when leaning over a fender to service something under the hood, always cover it with an old blanket or bedspread to protect the finish. Vinyl covered pads, made especially for this purpose, are available at auto parts stores.

A sturdy workbench is an essential part of any workshop

Jacking and towing

Jacking

Warning: *The jack supplied with the vehicle should only be used for changing a tire or placing jackstands under the frame. Never work under the vehicle or start the engine while this jack is being used as the only means of support.*

The vehicle should be on level ground. Place the shift lever in Park, and block the wheel diagonally opposite the wheel being changed. Set the parking brake.

Remove the spare tire and jack from stowage. Remove the wheel cover and trim ring (if so equipped) with the tapered end of the wheel bolt wrench by inserting and twisting the handle and then prying against the back of the wheel cover. Loosen the wheel bolts about 1/4-to-1/2 turn each. On models equipped with alloy wheels, using the hook from the tool bag, insert the hook into the wheel bolt cover and pull the covers off of the wheel bolts **(see illustration)**.

Place the jack under the side of the vehicle and adjust the jack height until it engages the vertical rocker panel flange nearest the wheel to be changed. There is a front and rear jacking point on each side of the vehicle **(see illustration)**.

Turn the jack handle clockwise until the tire clears the ground. Remove the wheel bolts and pull the wheel off, then install the spare.

Install the wheel bolts and tighten them snugly. Don't attempt to tighten them completely until the vehicle is lowered or it could slip off the jack. Turn the jack handle counterclockwise to lower the vehicle. Remove the jack and tighten the wheel bolts in a diagonal pattern.

Install the cover (and trim ring, if used) and be sure it's snapped into place all the way around.

Stow the tire, jack and wrench. Unblock the wheels.

Towing

The manufacturer states that the only safe way to tow these vehicles is with a flatbed-type car carrier or, on front-wheel drive models, a wheel-lift type tow truck with the front wheels raised. Other methods could cause damage to the drivetrain.

On alloy wheel models, use the hook from the tool bag to remove the wheel bolt covers

Place the jack so it engages the reinforced area of the rocker panel nearest the wheel to be raised

Booster battery (jump) starting

Observe these precautions when using a booster battery to start a vehicle:

a) Before connecting the booster battery, make sure the ignition switch is in the Off position.
b) Turn off the lights, heater and other electrical loads.
c) Your eyes should be shielded. Safety goggles are a good idea.
d) Make sure the booster battery is the same voltage as the dead one in the vehicle.
e) The two vehicles MUST NOT TOUCH each other!
f) Make sure the transaxle is in Neutral (manual) or Park (automatic).
g) If the booster battery is not a maintenance-free type, remove the vent caps and lay a cloth over the vent holes.

Connect the red jumper cable to the positive (+) terminals of each vehicle (see illustrations).
Note: *On 2016 and later models, the battery is located in the trunk or rear hatch area. A remote positive terminal for jump starting is located under the hood, at the rear of the fuse/relay box near the coolant expansion tank. A grounding lug for the negative jumper cable is located on the left (driver's side) strut tower.*

Connect one end of the black jumper cable to the negative (-) terminal of the booster battery. The other end of this cable should be connected to a good ground on the vehicle to be started, such as a bolt or bracket on the body.

Start the engine using the booster battery, then, with the engine running at idle speed, disconnect the jumper cables in the reverse order of connection.

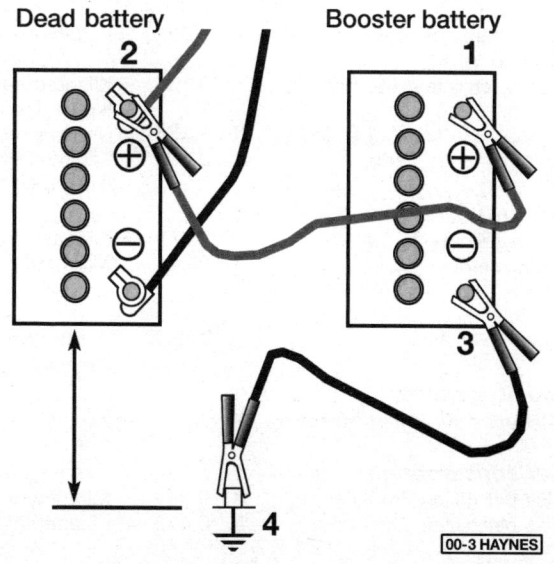

Make the booster battery cable connections in the numerical order shown (note that the negative cable of the booster battery is NOT attached to the negative terminal of the dead battery)

Conversion factors

Length (distance)
Inches (in)	X	25.4 = Millimeters (mm)	X	0.0394	= Inches (in)
Feet (ft)	X	0.305 = Meters (m)	X	3.281	= Feet (ft)
Miles	X	1.609 = Kilometers (km)	X	0.621	= Miles

Volume (capacity)
Cubic inches (cu in; in^3)	X	16.387 = Cubic centimeters (cc; cm^3)	X	0.061	= Cubic inches (cu in; in^3)
Imperial pints (Imp pt)	X	0.568 = Liters (l)	X	1.76	= Imperial pints (Imp pt)
Imperial quarts (Imp qt)	X	1.137 = Liters (l)	X	0.88	= Imperial quarts (Imp qt)
Imperial quarts (Imp qt)	X	1.201 = US quarts (US qt)	X	0.833	= Imperial quarts (Imp qt)
US quarts (US qt)	X	0.946 = Liters (l)	X	1.057	= US quarts (US qt)
Imperial gallons (Imp gal)	X	4.546 = Liters (l)	X	0.22	= Imperial gallons (Imp gal)
Imperial gallons (Imp gal)	X	1.201 = US gallons (US gal)	X	0.833	= Imperial gallons (Imp gal)
US gallons (US gal)	X	3.785 = Liters (l)	X	0.264	= US gallons (US gal)

Mass (weight)
Ounces (oz)	X	28.35 = Grams (g)	X	0.035	= Ounces (oz)
Pounds (lb)	X	0.454 = Kilograms (kg)	X	2.205	= Pounds (lb)

Force
Ounces-force (ozf; oz)	X	0.278 = Newtons (N)	X	3.6	= Ounces-force (ozf; oz)
Pounds-force (lbf; lb)	X	4.448 = Newtons (N)	X	0.225	= Pounds-force (lbf; lb)
Newtons (N)	X	0.1 = Kilograms-force (kgf; kg)	X	9.81	= Newtons (N)

Pressure
Pounds-force per square inch (psi; lbf/in^2; lb/in^2)	X	0.070 = Kilograms-force per square centimeter (kgf/cm^2; kg/cm^2)	X	14.223	= Pounds-force per square inch (psi; lbf/in^2; lb/in^2)
Pounds-force per square inch (psi; lbf/in^2; lb/in^2)	X	0.068 = Atmospheres (atm)	X	14.696	= Pounds-force per square inch (psi; lbf/in^2; lb/in^2)
Pounds-force per square inch (psi; lbf/in^2; lb/in^2)	X	0.069 = Bars	X	14.5	= Pounds-force per square inch (psi; lbf/in^2; lb/in^2)
Pounds-force per square inch (psi; lbf/in^2; lb/in^2)	X	6.895 = Kilopascals (kPa)	X	0.145	= Pounds-force per square inch (psi; lbf/in^2; lb/in^2)
Kilopascals (kPa)	X	0.01 = Kilograms-force per square centimeter (kgf/cm^2; kg/cm^2)	X	98.1	= Kilopascals (kPa)

Torque (moment of force)
Pounds-force inches (lbf in; lb in)	X	1.152 = Kilograms-force centimeter (kgf cm; kg cm)	X	0.868	= Pounds-force inches (lbf in; lb in)
Pounds-force inches (lbf in; lb in)	X	0.113 = Newton meters (Nm)	X	8.85	= Pounds-force inches (lbf in; lb in)
Pounds-force inches (lbf in; lb in)	X	0.083 = Pounds-force feet (lbf ft; lb ft)	X	12	= Pounds-force inches (lbf in; lb in)
Pounds-force feet (lbf ft; lb ft)	X	0.138 = Kilograms-force meters (kgf m; kg m)	X	7.233	= Pounds-force feet (lbf ft; lb ft)
Pounds-force feet (lbf ft; lb ft)	X	1.356 = Newton meters (Nm)	X	0.738	= Pounds-force feet (lbf ft; lb ft)
Newton meters (Nm)	X	0.102 = Kilograms-force meters (kgf m; kg m)	X	9.804	= Newton meters (Nm)

Vacuum
Inches mercury (in. Hg)	X	3.377 = Kilopascals (kPa)	X	0.2961	= Inches mercury
Inches mercury (in. Hg)	X	25.4 = Millimeters mercury (mm Hg)	X	0.0394	= Inches mercury

Power
Horsepower (hp)	X	745.7 = Watts (W)	X	0.0013	= Horsepower (hp)

Velocity (speed)
Miles per hour (miles/hr; mph)	X	1.609 = Kilometers per hour (km/hr; kph)	X	0.621	= Miles per hour (miles/hr; mph)

Fuel consumption*
Miles per gallon, Imperial (mpg)	X	0.354 = Kilometers per liter (km/l)	X	2.825	= Miles per gallon, Imperial (mpg)
Miles per gallon, US (mpg)	X	0.425 = Kilometers per liter (km/l)	X	2.352	= Miles per gallon, US (mpg)

Temperature
Degrees Fahrenheit = (°C x 1.8) + 32 Degrees Celsius (Degrees Centigrade; °C) = (°F - 32) x 0.56

*It is common practice to convert from miles per gallon (mpg) to liters/100 kilometers (l/100km), where mpg (Imperial) x l/100 km = 282 and mpg (US) x l/100 km = 235

O-23

DECIMALS to MILLIMETERS

Decimal	mm	Decimal	mm
0.001	0.0254	0.500	12.7000
0.002	0.0508	0.510	12.9540
0.003	0.0762	0.520	13.2080
0.004	0.1016	0.530	13.4620
0.005	0.1270	0.540	13.7160
0.006	0.1524	0.550	13.9700
0.007	0.1778	0.560	14.2240
0.008	0.2032	0.570	14.4780
0.009	0.2286	0.580	14.7320
0.010	0.2540	0.590	14.9860
0.020	0.5080		
0.030	0.7620		
0.040	1.0160	0.600	15.2400
0.050	1.2700	0.610	15.4940
0.060	1.5240	0.620	15.7480
0.070	1.7780	0.630	16.0020
0.080	2.0320	0.640	16.2560
0.090	2.2860	0.650	16.5100
		0.660	16.7640
0.100	2.5400	0.670	17.0180
0.110	2.7940	0.680	17.2720
0.120	3.0480	0.690	17.5260
0.130	3.3020		
0.140	3.5560		
0.150	3.8100		
0.160	4.0640	0.700	17.7800
0.170	4.3180	0.710	18.0340
0.180	4.5720	0.720	18.2880
0.190	4.8260	0.730	18.5420
		0.740	18.7960
0.200	5.0800	0.750	19.0500
0.210	5.3340	0.760	19.3040
0.220	5.5880	0.770	19.5580
0.230	5.8420	0.780	19.8120
0.240	6.0960	0.790	20.0660
0.250	6.3500		
0.260	6.6040		
0.270	6.8580	0.800	20.3200
0.280	7.1120	0.810	20.5740
0.290	7.3660	0.820	21.8280
		0.830	21.0820
0.300	7.6200	0.840	21.3360
0.310	7.8740	0.850	21.5900
0.320	8.1280	0.860	21.8440
0.330	8.3820	0.870	22.0980
0.340	8.6360	0.880	22.3520
0.350	8.8900	0.890	22.6060
0.360	9.1440		
0.370	9.3980		
0.380	9.6520		
0.390	9.9060	0.900	22.8600
0.400	10.1600	0.910	23.1140
0.410	10.4140	0.920	23.3680
0.420	10.6680	0.930	23.6220
0.430	10.9220	0.940	23.8760
0.440	11.1760	0.950	24.1300
0.450	11.4300	0.960	24.3840
0.460	11.6840	0.970	24.6380
0.470	11.9380	0.980	24.8920
0.480	12.1920	0.990	25.1460
0.490	12.4460	1.000	25.4000

FRACTIONS to DECIMALS to MILLIMETERS

Fraction	Decimal	mm	Fraction	Decimal	mm
1/64	0.0156	0.3969	33/64	0.5156	13.0969
1/32	0.0312	0.7938	17/32	0.5312	13.4938
3/64	0.0469	1.1906	35/64	0.5469	13.8906
1/16	0.0625	1.5875	9/16	0.5625	14.2875
5/64	0.0781	1.9844	37/64	0.5781	14.6844
3/32	0.0938	2.3812	19/32	0.5938	15.0812
7/64	0.1094	2.7781	39/64	0.6094	15.4781
1/8	0.1250	3.1750	5/8	0.6250	15.8750
9/64	0.1406	3.5719	41/64	0.6406	16.2719
5/32	0.1562	3.9688	21/32	0.6562	16.6688
11/64	0.1719	4.3656	43/64	0.6719	17.0656
3/16	0.1875	4.7625	11/16	0.6875	17.4625
13/64	0.2031	5.1594	45/64	0.7031	17.8594
7/32	0.2188	5.5562	23/32	0.7188	18.2562
15/64	0.2344	5.9531	47/64	0.7344	18.6531
1/4	0.2500	6.3500	3/4	0.7500	19.0500
17/64	0.2656	6.7469	49/64	0.7656	19.4469
9/32	0.2812	7.1438	25/32	0.7812	19.8438
19/64	0.2969	7.5406	51/64	0.7969	20.2406
5/16	0.3125	7.9375	13/16	0.8125	20.6375
21/64	0.3281	8.3344	53/64	0.8281	21.0344
11/32	0.3438	8.7312	27/32	0.8438	21.4312
23/64	0.3594	9.1281	55/64	0.8594	21.8281
3/8	0.3750	9.5250	7/8	0.8750	22.2250
25/64	0.3906	9.9219	57/64	0.8906	22.6219
13/32	0.4062	10.3188	29/32	0.9062	23.0188
27/64	0.4219	10.7156	59/64	0.9219	23.4156
7/16	0.4375	11.1125	15/16	0.9375	23.8125
29/64	0.4531	11.5094	61/64	0.9531	24.2094
15/32	0.4688	11.9062	31/32	0.9688	24.6062
31/64	0.4844	12.3031	63/64	0.9844	25.0031
1/2	0.5000	12.7000	1	1.0000	25.4000

Automotive chemicals and lubricants

A number of automotive chemicals and lubricants are available for use during vehicle maintenance and repair. They include a wide variety of products ranging from cleaning solvents and degreasers to lubricants and protective sprays for rubber, plastic and vinyl.

Cleaners

Carburetor cleaner and choke cleaner is a strong solvent for gum, varnish and carbon. Most carburetor cleaners leave a dry-type lubricant film which will not harden or gum up. Because of this film it is not recommended for use on electrical components.

Brake system cleaner is used to remove brake dust, grease and brake fluid from the brake system, where clean surfaces are absolutely necessary. It leaves no residue and often eliminates brake squeal caused by contaminants.

Electrical cleaner removes oxidation, corrosion and carbon deposits from electrical contacts, restoring full current flow. It can also be used to clean spark plugs, carburetor jets, voltage regulators and other parts where an oil-free surface is desired.

Demoisturants remove water and moisture from electrical components such as alternators, voltage regulators, electrical connectors and fuse blocks. They are non-conductive and non-corrosive.

Degreasers are heavy-duty solvents used to remove grease from the outside of the engine and from chassis components. They can be sprayed or brushed on and, depending on the type, are rinsed off either with water or solvent.

Lubricants

Motor oil is the lubricant formulated for use in engines. It normally contains a wide variety of additives to prevent corrosion and reduce foaming and wear. Motor oil comes in various weights (viscosity ratings) from 0 to 50. The recommended weight of the oil depends on the season, temperature and the demands on the engine. Light oil is used in cold climates and under light load conditions. Heavy oil is used in hot climates and where high loads are encountered. Multi-viscosity oils are designed to have characteristics of both light and heavy oils and are available in a number of weights from 0W-20 to 20W-50.

Gear oil is designed to be used in differentials, manual transmissions and other areas where high-temperature lubrication is required.

Chassis and wheel bearing grease is a heavy grease used where increased loads and friction are encountered, such as for wheel bearings, balljoints, tie-rod ends and universal joints.

High-temperature wheel bearing grease is designed to withstand the extreme temperatures encountered by wheel bearings in disc brake equipped vehicles. It usually contains molybdenum disulfide (moly), which is a dry-type lubricant.

White grease is a heavy grease for metal-to-metal applications where water is a problem. White grease stays soft under both low and high temperatures (usually from -100 to +190-degrees F), and will not wash off or dilute in the presence of water.

Assembly lube is a special extreme pressure lubricant, usually containing moly, used to lubricate high-load parts (such as main and rod bearings and cam lobes) for initial start-up of a new engine. The assembly lube lubricates the parts without being squeezed out or washed away until the engine oiling system begins to function.

Silicone lubricants are used to protect rubber, plastic, vinyl and nylon parts.

Graphite lubricants are used where oils cannot be used due to contamination problems, such as in locks. The dry graphite will lubricate metal parts while remaining uncontaminated by dirt, water, oil or acids. It is electrically conductive and will not foul electrical contacts in locks such as the ignition switch.

Moly penetrants loosen and lubricate frozen, rusted and corroded fasteners and prevent future rusting or freezing.

Heat-sink grease is a special electrically non-conductive grease that is used for mounting electronic ignition modules where it is essential that heat is transferred away from the module.

Sealants

RTV sealant is one of the most widely used gasket compounds. Made from silicone, RTV is air curing, it seals, bonds, waterproofs, fills surface irregularities, remains flexible, doesn't shrink, is relatively easy to remove, and is used as a supplementary sealer with almost all low and medium temperature gaskets.

Anaerobic sealant is much like RTV in that it can be used either to seal gaskets or to form gaskets by itself. It remains flexible, is solvent resistant and fills surface imperfections. The difference between an anaerobic sealant and an RTV-type sealant is in the curing. RTV cures when exposed to air, while an anaerobic sealant cures only in the absence of air. This means that an anaerobic sealant cures only after the assembly of parts, sealing them together.

Thread and pipe sealant is used for sealing hydraulic and pneumatic fittings and vacuum lines. It is usually made from a Teflon compound, and comes in a spray, a paint-on liquid and as a wrap-around tape.

Chemicals

Anti-seize compound prevents seizing, galling, cold welding, rust and corrosion in fasteners. High-temperature anti-seize, usually made with copper and graphite lubricants, is used for exhaust system and exhaust manifold bolts.

Anaerobic locking compounds are used to keep fasteners from vibrating or working loose and cure only after installation, in the absence of air. Medium strength locking compound is used for small nuts, bolts and screws that may be removed later. High-strength locking compound is for large nuts, bolts and studs which aren't removed on a regular basis.

Oil additives range from viscosity index improvers to chemical treatments that claim to reduce internal engine friction. It should be noted that most oil manufacturers caution against using additives with their oils.

Gas additives perform several functions, depending on their chemical makeup. They usually contain solvents that help dissolve gum and varnish that build up on carburetor, fuel injection and intake parts. They also serve to break down carbon deposits that form on the inside surfaces of the combustion chambers. Some additives contain upper cylinder lubricants for valves and piston rings, and others contain chemicals to remove condensation from the gas tank.

Miscellaneous

Brake fluid is specially formulated hydraulic fluid that can withstand the heat and pressure encountered in brake systems. Care must be taken so this fluid does not come in contact with painted surfaces or plastics. An opened container should always be resealed to prevent contamination by water or dirt.

Weatherstrip adhesive is used to bond weatherstripping around doors, windows and trunk lids. It is sometimes used to attach trim pieces.

Undercoating is a petroleum-based, tar-like substance that is designed to protect metal surfaces on the underside of the vehicle from corrosion. It also acts as a sound-deadening agent by insulating the bottom of the vehicle.

Waxes and polishes are used to help protect painted and plated surfaces from the weather. Different types of paint may require the use of different types of wax and polish. Some polishes utilize a chemical or abrasive cleaner to help remove the top layer of oxidized (dull) paint on older vehicles. In recent years many non-wax polishes that contain a wide variety of chemicals such as polymers and silicones have been introduced. These non-wax polishes are usually easier to apply and last longer than conventional waxes and polishes.

Safety first!

Regardless of how enthusiastic you may be about getting on with the job at hand, take the time to ensure that your safety is not jeopardized. A moment's lack of attention can result in an accident, as can failure to observe certain simple safety precautions. The possibility of an accident will always exist, and the following points should not be considered a comprehensive list of all dangers. Rather, they are intended to make you aware of the risks and to encourage a safety conscious approach to all work you carry out on your vehicle.

Essential DOs and DON'Ts

DON'T rely on a jack when working under the vehicle. Always use approved jackstands to support the weight of the vehicle and place them under the recommended lift or support points.
DON'T attempt to loosen extremely tight fasteners (i.e. wheel lug nuts) while the vehicle is on a jack - it may fall.
DON'T start the engine without first making sure that the transmission is in Neutral (or Park where applicable) and the parking brake is set.
DON'T remove the radiator cap from a hot cooling system - let it cool or cover it with a cloth and release the pressure gradually.
DON'T attempt to drain the engine oil until you are sure it has cooled to the point that it will not burn you.
DON'T touch any part of the engine or exhaust system until it has cooled sufficiently to avoid burns.
DON'T siphon toxic liquids such as gasoline, antifreeze and brake fluid by mouth, or allow them to remain on your skin.
DON'T inhale brake lining dust - it is potentially hazardous (see *Asbestos* below).
DON'T allow spilled oil or grease to remain on the floor - wipe it up before someone slips on it.
DON'T use loose fitting wrenches or other tools which may slip and cause injury.
DON'T push on wrenches when loosening or tightening nuts or bolts. Always try to pull the wrench toward you. If the situation calls for pushing the wrench away, push with an open hand to avoid scraped knuckles if the wrench should slip.
DON'T attempt to lift a heavy component alone - get someone to help you.
DON'T rush or take unsafe shortcuts to finish a job.
DON'T allow children or animals in or around the vehicle while you are working on it.
DO wear eye protection when using power tools such as a drill, sander, bench grinder, etc. and when working under a vehicle.
DO keep loose clothing and long hair well out of the way of moving parts.
DO make sure that any hoist used has a safe working load rating adequate for the job.
DO get someone to check on you periodically when working alone on a vehicle.
DO carry out work in a logical sequence and make sure that everything is correctly assembled and tightened.
DO keep chemicals and fluids tightly capped and out of the reach of children and pets.
DO remember that your vehicle's safety affects that of yourself and others. If in doubt on any point, get professional advice.

Steering, suspension and brakes

These systems are essential to driving safety, so make sure you have a qualified shop or individual check your work. Also, compressed suspension springs can cause injury if released suddenly - be sure to use a spring compressor.

Airbags

Airbags are explosive devices that can **CAUSE** injury if they deploy while you're working on the vehicle. Follow the manufacturer's instructions to disable the airbag whenever you're working in the vicinity of airbag components.

Asbestos

Certain friction, insulating, sealing, and other products - such as brake linings, brake bands, clutch linings, torque converters, gaskets, etc. - may contain asbestos or other hazardous friction material. Extreme care must be taken to avoid inhalation of dust from such products, since it is hazardous to health. If in doubt, assume that they do contain asbestos.

Fire

Remember at all times that gasoline is highly flammable. Never smoke or have any kind of open flame around when working on a vehicle. But the risk does not end there. A spark caused by an electrical short circuit, by two metal surfaces contacting each other, or even by static electricity built up in your body under certain conditions, can ignite gasoline vapors, which in a confined space are highly explosive. Do not, under any circumstances, use gasoline for cleaning parts. Use an approved safety solvent.

Always disconnect the battery ground (-) cable at the battery before working on any part of the fuel system or electrical system. Never risk spilling fuel on a hot engine or exhaust component. It is strongly recommended that a fire extinguisher suitable for use on fuel and electrical fires be kept handy in the garage or workshop at all times. Never try to extinguish a fuel or electrical fire with water.

Fumes

Certain fumes are highly toxic and can quickly cause unconsciousness and even death if inhaled to any extent. Gasoline vapor falls into this category, as do the vapors from some cleaning solvents. Any draining or pouring of such volatile fluids should be done in a well ventilated area.

When using cleaning fluids and solvents, read the instructions on the container carefully. Never use materials from unmarked containers.

Never run the engine in an enclosed space, such as a garage. Exhaust fumes contain carbon monoxide, which is extremely poisonous. If you need to run the engine, always do so in the open air, or at least have the rear of the vehicle outside the work area.

The battery

Never create a spark or allow a bare light bulb near a battery. They normally give off a certain amount of hydrogen gas, which is highly explosive.

Always disconnect the battery ground (-) cable at the battery before working on the fuel or electrical systems.

If possible, loosen the filler caps or cover when charging the battery from an external source (this does not apply to sealed or maintenance-free batteries). Do not charge at an excessive rate or the battery may burst.

Take care when adding water to a non maintenance-free battery and when carrying a battery. The electrolyte, even when diluted, is very corrosive and should not be allowed to contact clothing or skin.

Always wear eye protection when cleaning the battery to prevent the caustic deposits from entering your eyes.

Household current

When using an electric power tool, inspection light, etc., which operates on household current, always make sure that the tool is correctly connected to its plug and that, where necessary, it is properly grounded. Do not use such items in damp conditions and, again, do not create a spark or apply excessive heat in the vicinity of fuel or fuel vapor.

Secondary ignition system voltage

A severe electric shock can result from touching certain parts of the ignition system (such as the spark plug wires) when the engine is running or being cranked, particularly if components are damp or the insulation is defective. In the case of an electronic ignition system, the secondary system voltage is much higher and could prove fatal.

Hydrofluoric acid

This extremely corrosive acid is formed when certain types of synthetic rubber, found in some O-rings, oil seals, fuel hoses, etc. are exposed to temperatures above 750-degrees F (400-degrees C). The rubber changes into a charred or sticky substance containing the acid. *Once formed, the acid remains dangerous for years. If it gets onto the skin, it may be necessary to amputate the limb concerned.*

When dealing with a vehicle which has suffered a fire, or with components salvaged from such a vehicle, wear protective gloves and discard them after use.

Troubleshooting

Contents

Symptom	Section
Engine	
Engine backfires	15
Engine continues to run after switching off	18
Engine hard to start when cold	3
Engine hard to start when hot	4
Engine lacks power	14
Engine lopes while idling or idles erratically	8
Engine misses at idle speed	9
Engine misses throughout driving speed range	10
Engine rotates but will not start	2
Engine runs with oil pressure light on	17
Engine will not rotate when attempting to start	1
Engine stalls	13
Engine starts but stops immediately	6
Engine stumbles on acceleration	11
Engine surges while holding accelerator steady	12
Oil puddle under engine	7
Pinging or knocking engine sounds during acceleration or uphill	16
Starter motor noisy or excessively rough in engagement	5
Engine electrical systems	
Alternator light fails to come on when key is turned on	21
Alternator light fails to go out	20
Battery will not hold a charge	19
Fuel system	
Excessive fuel consumption	22
Fuel leakage and/or fuel odor	23
Cooling system	
Coolant loss	28
External coolant leakage	26
Internal coolant leakage	27
Overcooling	25
Overheating	24
Poor coolant circulation	29
Clutch	
Clutch pedal stays on floor	39
Clutch slips (engine speed increases with no increase in vehicle speed)	35
Fluid in area of master cylinder dust cover and on pedal	31
Fluid on release cylinder	32
Grabbing (chattering) as clutch is engaged	36
High pedal effort	40
Noise in clutch area	38
Pedal travels to floor - no pressure or very little resistance	30
Pedal feels spongy when depressed	33
Transaxle rattling (clicking)	37
Unable to select gears	34

Symptom	Section
Manual transaxle	
Clicking noise in turns	44
Clunk on acceleration or deceleration	43
Knocking noise at low speeds	41
Leaks lubricant	50
Locked in gear	51
Noise most pronounced when turning	42
Noisy in all gears	48
Noisy in neutral with engine running	46
Noisy in one particular gear	47
Slips out of gear	49
Vibration	45
Automatic transaxle	
Fluid leakage	52
General shift mechanism problems	54
Transaxle slips, shifts roughly, is noisy or has no drive in forward or reverse gears	55
Transmission fluid brown or has a burned smell	53
Driveaxles	
Clicking noise in turns	56
Shudder or vibration during acceleration	57
Vibration at highway speeds	58
Brakes	
Brake pedal feels spongy when depressed	66
Brake pedal travels to the floor with little resistance	67
Brake roughness or chatter (pedal pulsates)	61
Dragging brakes	64
Excessive brake pedal effort required to stop vehicle	62
Excessive brake pedal travel	63
Grabbing or uneven braking action	65
Noise (high-pitched squeal or grinding when the brakes are applied)	60
Parking brake does not hold	68
Vehicle pulls to one side during braking	59
Suspension and steering systems	
Abnormal or excessive tire wear	70
Abnormal noise at the front end	75
Cupped tires	80
Erratic steering when braking	77
Excessive pitching and/or rolling around corners or during braking	78
Excessive play or looseness in steering system	84
Excessive tire wear on inside edge	82
Excessive tire wear on outside edge	81
Hard steering	73
Poor returnability of steering to center	74
Rattling or clicking noise in steering gear	85
Shimmy, shake or vibration	72
Suspension bottoms	79
Tire tread worn in one place	83
Vehicle pulls to one side	69
Wander or poor steering stability	76
Wheel makes a thumping noise	71

Troubleshooting   0-27

This section provides an easy reference guide to the more common problems which may occur during the operation of your vehicle. These problems and their possible causes are grouped under headings denoting various components or systems, such as Engine, Cooling system, etc. They also refer you to the chapter and/or section which deals with the problem.

Remember that successful troubleshooting is not a mysterious black art practiced only by professional mechanics. It is simply the result of the right knowledge combined with an intelligent, systematic approach to the problem. Always work by a process of elimination, starting with the simplest solution and working through to the most complex - and never overlook the obvious. Anyone can run the gas tank dry or leave the lights on overnight, so don't assume that you are exempt from such oversights.

Finally, always establish a clear idea of why a problem has occurred and take steps to ensure that it doesn't happen again. If the electrical system fails because of a poor connection, check the other connections in the system to make sure that they don't fail as well. If a particular fuse continues to blow, find out why - don't just replace one fuse after another. Remember, failure of a small component can often be indicative of potential failure or incorrect functioning of a more important component or system.

Engine

1 Engine will not rotate when attempting to start

1   Battery terminal connections loose or corroded (Chapter 1).
2   Battery discharged or faulty (Chapters 1 and 5).
3   Automatic transaxle not completely engaged in Park (Chapter 7A) or clutch pedal not completely depressed (Chapter 6).
4   Broken, loose or disconnected wiring in the starting circuit (Chapters 5 and 12).
5   Starter motor pinion jammed in flywheel ring gear (Chapter 5).
6   Starter solenoid faulty (Chapter 5).
7   Starter motor faulty (Chapter 5).
8   Ignition switch faulty (Chapter 12).
9   Starter pinion or flywheel teeth worn or broken (Chapter 5).

2 Engine rotates but will not start

1   Fuel tank empty.
2   Battery discharged (engine rotates slowly) (Chapter 5).
3   Battery terminal connections loose or corroded (Chapter 1).
4   Leaking fuel injector(s), faulty fuel pump, pressure regulator, etc. (Chapter 4).
5   Broken timing chain (Chapter 2A).
6   Ignition components damp or damaged (Chapter 5).
7   Worn, faulty or incorrectly-gapped spark plugs (Chapter 1).
8   Broken, loose or disconnected wiring in the starting circuit (Chapter 5).
9   Broken, loose or disconnected wires at the ignition coil or faulty coil (Chapter 5).
10   Defective crankshaft or camshaft sensor (Chapter 6).

3 Engine hard to start when cold

1   Battery discharged or low (Chapter 1).
2   Malfunctioning fuel system (Chapter 4).
3   Faulty coolant temperature sensor or intake air temperature sensor (Chapter 6).
4   Faulty ignition system (Chapter 5).

4 Engine hard to start when hot

1   Air filter clogged (Chapter 1).
2   Fuel not reaching the fuel injection rail (Chapter 4).
3   Corroded battery connections (Chapter 1).
4   Faulty coolant temperature sensor or intake air temperature sensor (Chapter 6).

5 Starter motor noisy or excessively rough in engagement

1   Pinion or flywheel gear teeth worn or broken (Chapter 5).
2   Starter motor mounting bolts loose or missing (Chapter 5).

6 Engine starts but stops immediately

1   Insufficient fuel reaching the fuel injector(s) (Chapters 1 and 4).
2   Vacuum leak at the gasket between the intake manifold/plenum and throttle body (Chapter 4).

7 Oil puddle under engine

1   Oil pan gasket and/or oil pan drain bolt washer leaking (Chapter 2A).
2   Oil pressure sending unit leaking (Chapter 2A).
3   Valve cover leaking (Chapter 2A).
4   Engine oil seals leaking (Chapter 2A).
5   Timing chain cover leaking (Chapter 2A).

8 Engine lopes while idling or idles erratically

1   Vacuum leakage (Chapters 2A and 4).
2   Leaking EGR valve (Chapter 6).
3   Air filter clogged (Chapter 1).
4   Malfunction in the fuel injection or engine control system (Chapters 4 and 6).
5   Leaking head gasket (Chapter 2A).
6   Timing chain and/or sprockets worn (Chapter 2A).
7   Camshaft lobes worn (Chapter 2A).

9 Engine misses at idle speed

1   Spark plugs worn or not gapped properly (Chapter 1).
2   Faulty coil(s) (Chapter 1).
3   Vacuum leaks (Chapter 1).
4   Uneven or low compression (Chapter 2A).
5   Problem with the fuel injection system (Chapter 4).

10 Engine misses throughout driving speed range

1   Fuel filter clogged and/or impurities in the fuel system (Chapters 1 and 4).
2   Low fuel pressure (Chapter 4).
3   Faulty or incorrectly gapped spark plugs (Chapter 1).
4   Faulty engine management system components (Chapter 6).
5   Low or uneven cylinder compression pressures (Chapter 2A).
6   Weak or faulty ignition system (Chapter 5).
7   Vacuum leak in fuel injection system, intake manifold, air control valve or vacuum hoses (Chapters 4 and 6).

11 Engine stumbles on acceleration

1   Spark plugs fouled (Chapter 1).
2   Problem with fuel injection or engine control system (Chapters 4 and 6).
3   Fuel filter clogged (Chapters 1 and 4).
4   Intake manifold air leak (Chapters 2A and 4).
5   Problem with the emissions control system (Chapter 6).

12 Engine surges while holding accelerator steady

1   Intake air leak (Chapter 4).
2   Fuel pump or fuel pressure regulator faulty (Chapter 4).
3   Problem with the fuel injection system (Chapter 4).
4   Problem with the emissions control system (Chapter 6).

13 Engine stalls

1 Idle speed incorrect (Chapter 1).
2 Fuel filter clogged and/or water and impurities in the fuel system (Chapters 1 and 4).
3 Faulty emissions system components (Chapter 6).
4 Faulty or incorrectly gapped spark plugs (Chapter 1).
5 Vacuum leak in the fuel injection system, intake manifold or vacuum hoses (Chapters 2A and 4).

14 Engine lacks power

1 Obstructed exhaust system (Chapter 4).
2 Faulty or incorrectly gapped spark plugs (Chapter 1).
3 Problem with the fuel injection system (Chapter 4).
4 Dirty air filter (Chapter 1).
5 Brakes binding (Chapter 9).
6 Automatic transaxle fluid level incorrect (Chapter 1).
7 Clutch slipping (Chapter 8).
8 Fuel filter clogged and/or impurities in the fuel system (Chapters 1 and 4).
9 Emission control system not functioning properly (Chapter 6).
10 Low or uneven cylinder compression pressures (Chapter 2A).

15 Engine backfires

1 Emission control system not functioning properly (Chapter 6).
2 Problem with the fuel injection system (Chapter 4).
3 Vacuum leak at fuel injector(s), intake manifold or vacuum hoses (Chapters 2A and 4).
4 Valve sticking (Chapter 2A).

16 Pinging or knocking engine sounds during acceleration or uphill

1 Incorrect grade of fuel.
2 Fuel injection system faulty (Chapter 4).
3 Improper or damaged spark plugs or wires (Chapter 1).
4 Knock sensor defective (Chapter 6).
5 EGR valve not functioning (Chapter 6).
6 Vacuum leak (Chapters 2A and 4).

17 Engine runs with oil pressure light on

1 Low oil level (Chapter 1).
2 Idle rpm below specification (Chapter 1).
3 Short in wiring circuit (Chapter 12).
4 Faulty oil pressure sender (Chapter 2A).
5 Worn engine bearings and/or oil pump (Chapter 2A).

18 Engine continues to run after switching off

1 Defective ignition switch (Chapter 12).
2 Faulty Powertrain Control Module (Chapter 6).
3 Faulty Body Control Module.
4 Leaking fuel injector (Chapter 4).

Engine electrical systems

19 Battery will not hold a charge

1 Drivebelt or tensioner defective (Chapter 1).
2 Battery electrolyte level low (Chapter 1).
3 Battery terminals loose or corroded (Chapter 1).
4 Alternator not charging properly (Chapter 5).
5 Loose, broken or faulty wiring in the charging circuit (Chapter 5).
6 Short in vehicle wiring (Chapter 12).
7 Internally defective battery (Chapters 1 and 5).

20 Alternator light fails to go out

1 Faulty alternator or charging circuit (Chapter 5).
2 Drivebelt or tensioner defective (Chapter 1).

21 Alternator light fails to come on when key is turned on

1 Instrument cluster defective (Chapter 12).
2 Fault in the wiring harness (Chapter 12).

Fuel system

22 Excessive fuel consumption

1 Dirty air filter element (Chapter 1).
2 Emissions system not functioning properly (Chapter 6).
3 Fuel injection system not functioning properly (Chapter 4).
4 Low tire pressure or incorrect tire size (Chapter 1).

23 Fuel leakage and/or fuel odor

1 Leaking fuel line (Chapters 1 and 4).
2 Tank overfilled.
3 Evaporative emissions control system problem (Chapters 1 and 6).
4 Problem with the fuel injection system (Chapter 4).

Cooling system

24 Overheating

1 Insufficient coolant in system (Chapter 1).
2 Water pump drivebelt defective or out of adjustment (Chapter 1).
3 Radiator core blocked or grille restricted (Chapter 3).
4 Thermostat faulty (Chapter 3).
5 Electric cooling fan inoperative or blades broken (Chapter 3).
6 Expansion tank cap not maintaining proper pressure (Chapter 3).

25 Overcooling

1 Faulty thermostat (Chapter 3).
2 Inaccurate temperature gauge sending unit (Chapter 3).

26 External coolant leakage

1 Deteriorated/damaged hoses; loose clamps (Chapters 1 and 3).
2 Water pump defective (Chapter 3).
3 Leakage from radiator core or coolant expansion tank (Chapter 3).
4 Engine drain or water jacket core plugs leaking (Chapter 2A).

27 Internal coolant leakage

1 Leaking cylinder head gasket (Chapter 2A).
2 Cracked cylinder bore or cylinder head (Chapter 2A).

28 Coolant loss

1 Too much coolant in reservoir (Chapter 1).
2 Coolant boiling away because of overheating (Chapter 3).
3 Internal or external leakage (Chapter 3).
4 Faulty radiator cap (Chapter 3).

29 Poor coolant circulation

1 Inoperative water pump (Chapter 3).
2 Restriction in cooling system (Chapters 1 and 3).
3 Drivebelt or tensioner defective (Chapter 1).
4 Thermostat sticking (Chapter 3).

Clutch

30 Pedal travels to floor - no pressure or very little resistance

1. Master or release cylinder faulty (Chapter 8).
2. Hose/pipe burst or leaking (Chapter 8).
3. Connections leaking (Chapter 8).
4. No fluid in reservoir (Chapter 8).
5. If fluid level in reservoir rises as pedal is depressed, master cylinder center valve seal is faulty (Chapter 8).
6. Broken release bearing or fork (Chapter 8).
7. Faulty pressure plate diaphragm spring (Chapter 8).

31 Fluid in area of master cylinder dust cover and on pedal

Piston primary seal failure in master cylinder (Chapter 8).

32 Fluid on release cylinder

Release cylinder plunger seal faulty (Chapter 8).

33 Pedal feels spongy when depressed

Air in system (Chapter 8).

34 Unable to select gears

1. Faulty transaxle (Chapter 7A).
2. Faulty clutch disc or pressure plate (Chapter 8).
3. Faulty release lever or release bearing (Chapter 8).
4. Faulty shift lever assembly or control cables (Chapter 8).

35 Clutch slips (engine speed increases with no increase in vehicle speed)

1. Clutch plate worn (Chapter 8).
2. Clutch plate is oil soaked by leaking rear main seal (Chapters 2A and 8).
3. Clutch plate not seated (Chapter 8).
4. Warped pressure plate or flywheel (Chapter 8).
5. Weak diaphragm springs (Chapter 8).
6. Clutch plate overheated. Allow to cool.

36 Grabbing (chattering) as clutch is engaged

1. Oil on clutch plate lining, burned or glazed facings (Chapter 8).
2. Worn or loose engine or transaxle mounts (Chapter 2A).
3. Worn splines on clutch plate hub (Chapter 8).
4. Warped pressure plate or flywheel (Chapter 8).
5. Burned or smeared resin on flywheel or pressure plate (Chapter 8).

37 Transaxle rattling (clicking)

1. Release lever loose (Chapter 8).
2. Clutch plate damper spring failure (Chapter 8).

38 Noise in clutch area

1. Fork shaft improperly installed (Chapter 8).
2. Faulty bearing (Chapter 8).

39 Clutch pedal stays on floor

1. Clutch master cylinder piston binding in bore (Chapter 8).
2. Broken release bearing or fork (Chapter 8).

40 High pedal effort

1. Piston binding in bore (Chapter 8).
2. Pressure plate faulty (Chapter 8).
3. Incorrect size master or release cylinder (Chapter 8).

Manual transaxle

41 Knocking noise at low speeds

Worn input shaft bearing (Chapter 7A).*

42 Noise most pronounced when turning

Rear differential gear noise (Chapter 10).*

43 Clunk on acceleration or deceleration

1. Loose engine or transaxle mounts (Chapter 2A).
2. Worn differential pinion shaft in case.*
3. Worn side gear shaft counterbore in rear differential case (Chapter 10).*

44 Clicking noise in turns

Worn or damaged outboard CV joint (Chapter 8).

45 Vibration

1. Rough wheel bearing (Chapter 10).
2. Damaged driveshaft (Chapter 8).
3. Out-of-round tires (Chapter 1).
4. Tire out of balance (Chapters 1 and 10).
5. Worn driveshaft joints (Chapter 8).

46 Noisy in neutral with engine running

1. Damaged input gear bearing (Chapter 7A).*
2. Damaged clutch release bearing (Chapter 8).

47 Noisy in one particular gear

1. Damaged or worn constant mesh gears (Chapter 7A).*
2. Damaged or worn synchronizers (Chapter 7A).*
3. Bent reverse fork (Chapter 7A).*
4. Damaged fourth/fifth speed gear or output gear (Chapter 7A).*
5. Worn or damaged reverse idler gear or idler bushing (Chapter 7A).*

48 Noisy in all gears

1. Insufficient lubricant (Chapter 7A).
2. Damaged or worn bearings (Chapter 7A).*
3. Worn or damaged input gear shaft and/or output gear shaft (Chapter 7A).*

49 Slips out of gear

1. Worn or improperly adjusted linkage (Chapter 7A).
2. Shift linkage does not work freely, binds (Chapter 7A).
3. Input gear bearing retainer broken or loose (Chapter 7A).*
4. Worn or bent shift fork (Chapter 7A).*

50 Leaks lubricant

1 Side gear shaft seals worn (Chapter 7A).
2 Excessive amount of lubricant in transaxle (Chapters 1 and 7A).
3 Loose or broken input gear shaft bearing retainer (Chapter 7A).*
4 Input gear bearing retainer O-ring and/or lip seal damaged (Chapter 7A).*

51 Locked in gear

1 Lock pin or interlock pin missing (Chapter 7A).*
*Although the corrective action necessary to remedy the symptoms described is beyond the scope of this manual, the above information should be helpful in isolating the cause of the condition so that the owner can communicate clearly with a professional mechanic.

Automatic transaxle

52 Fluid leakage

1 Automatic transmission fluid is a deep red color. Fluid leaks should not be confused with engine oil, which can easily be blown onto the transaxle by air flow.
2 To pinpoint a leak, first remove all built-up dirt and grime from the transaxle housing with degreasing agents and/or steam cleaning. Then drive the vehicle at low speeds so air flow will not blow the leak far from its source. Raise the vehicle and determine where the leak is coming from. Common areas of leakage are:
  Transaxle oil lines (Chapter 7A).
  Speed sensor (Chapter 6).
  Driveaxle oil seal (Chapter 7A).

53 Transmission fluid brown or has a burned smell

Transmission fluid overheated (Chapter 1).

54 General shift mechanism problems

1 Chapter 7A, Part B, deals with checking and adjusting the shift linkage on automatic transaxles. Common problems which may be attributed to poorly adjusted linkage are:
  Engine starting in gears other than Park or Neutral.
  Indicator on shifter pointing to a gear other than the one actually being used.
  Vehicle moves when in Park.
2 Refer to Chapter 7B for the shift linkage adjustment procedure.

55 Transaxle slips, shifts roughly, is noisy or has no drive in forward or reverse gears

There are many probable causes for the above problems, but the home mechanic should be concerned with only one possibility - fluid level. Before taking the vehicle to a repair shop, check the level and condition of the fluid as described in Chapter 1. Correct the fluid level as necessary or change the fluid and filter if needed. If the problem persists, have a professional diagnose the cause.

Driveaxles

56 Clicking noise in turns

Worn or damaged outboard CV joint (Chapter 8).

57 Shudder or vibration during acceleration

1 Excessive toe-in (Chapter 10).
2 Worn or damaged inboard or outboard CV joints (Chapter 8).
3 Sticking inboard CV joint assembly (Chapter 8).

58 Vibration at highway speeds

1 Out-of-balance front wheels and/or tires (Chapters 1 and 10).
2 Out-of-round front tires (Chapters 1 and 10).
3 Worn CV joint(s) (Chapter 8).

Brakes

59 Vehicle pulls to one side during braking

1 Incorrect tire pressures (Chapter 1).
2 Front end out of alignment (have the front end aligned).
3 Front, or rear, tire sizes not matched to one another.
4 Restricted brake lines or hoses (Chapter 9).
5 Malfunctioning caliper assembly (Chapter 9).
6 Loose suspension parts (Chapter 10).
7 Excessive wear of pad material or disc on one side (Chapter 9).
8 Contamination (grease or brake fluid) of brake pad material or disc on one side (Chapter 9).

60 Noise (high-pitched squeal or grinding when the brakes are applied)

Brake pads or shoes worn out. Replace pads or shoes with new ones immediately. Also inspect the discs/drums (Chapter 9).

61 Brake roughness or chatter (pedal pulsates)

1 Excessive lateral runout (Chapter 9).
2 Uneven pad wear (Chapter 9).
3 Defective disc (Chapter 9).

62 Excessive brake pedal effort required to stop vehicle

1 Malfunctioning power brake booster (Chapter 9).
2 Malfunctioning vacuum pump (Chapter 9).
3 Partial system failure (Chapter 9).
4 Excessively worn pads (Chapter 9).
5 Piston in caliper stuck or sluggish (Chapter 9).
6 Brake pads contaminated with oil or grease (Chapter 9).
7 Brake disc grooved and/or glazed (Chapter 9).

63 Excessive brake pedal travel

1 Partial brake system failure (Chapter 9).
2 Insufficient fluid in master cylinder (Chapters 1 and 9).
3 Air trapped in system (Chapter 9).

64 Dragging brakes

1 Incorrect adjustment of brake light switch (Chapter 9).
2 Master cylinder pistons not returning correctly (Chapter 9).
3 Caliper piston stuck (Chapter 9).
4 Restricted brakes lines or hoses (Chapter 9).
5 Incorrect parking brake adjustment (Chapter 9).

65 Grabbing or uneven braking action

1 Malfunction of proportioning valve (Chapter 9).
2 Binding brake pedal mechanism (Chapter 9).
3 Contaminated brake linings (Chapter 9).

Troubleshooting 0-31

66 Brake pedal feels spongy when depressed

1 Air in hydraulic lines (Chapter 9).
2 Master cylinder mounting bolts loose (Chapter 9).
3 Master cylinder defective (Chapter 9).

67 Brake pedal travels to the floor with little resistance

1 Little or no fluid in the master cylinder reservoir caused by leaking caliper piston(s) or wheel cylinder(s) (Chapter 9).
2 Loose, damaged or disconnected brake lines (Chapter 9).

68 Parking brake does not hold

Parking brake improperly adjusted (Chapter 9).

Suspension and steering systems

69 Vehicle pulls to one side

1 Mismatched or uneven tires (Chapter 10).
2 Broken or sagging springs (Chapter 10).
3 Wheel alignment incorrect. Have the wheels professionally aligned.
4 Front brake dragging (Chapter 9).

70 Abnormal or excessive tire wear

1 Wheel alignment out-of-specification. Have the wheels aligned.
2 Sagging or broken springs (Chapter 10).
3 Tire out-of-balance (Chapter 10).
4 Worn strut damper or shock absorber (Chapter 10).
5 Overloaded vehicle.
6 Tires not rotated regularly.

71 Wheel makes a thumping noise

1 Blister or bump on tire (Chapter 10).
2 Improper strut damper action (Chapter 10).

72 Shimmy, shake or vibration

1 Tire or wheel out-of-balance or out-of-round (Chapter 10).
2 Loose or worn wheel bearings (Chapter 10).
3 Worn tie-rod ends (Chapter 10).
4 Worn balljoints (Chapters 1 and 10).
5 Excessive wheel runout (Chapter 10).
6 Blister or bump on tire (Chapter 10).

73 Hard steering

1 Worn balljoints and/or tie-rod ends (Chapter 10).
2 Wheel alignment out-of-specifications. Have the wheels professionally aligned.
3 Low tire pressure(s) (Chapter 1).
4 Worn steering gear (Chapter 10).

74 Poor returnability of steering to center

1 Worn balljoints or tie-rod ends (Chapter 10).
2 Worn steering gear assembly (Chapter 10).
3 Wheel alignment out-of-specifications. Have the wheels professionally aligned.

75 Abnormal noise at the front end

1 Worn balljoints or tie-rod ends (Chapter 10).
2 Damaged shock absorber mounting (Chapter 10).
3 Worn control arm bushings or tie-rod ends (Chapter 10).
4 Loose stabilizer bar (Chapter 10).
5 Loose wheel nuts (Chapter 1).
6 Loose suspension bolts (Chapter 10).

76 Wander or poor steering stability

1 Mismatched or uneven tires (Chapter 10).
2 Worn balljoints or tie-rod ends (Chapters 1 and 10).
3 Worn struts or shock absorbers (Chapter 10).
4 Broken or sagging springs (Chapter 10).
5 Wheels out of alignment. Have the wheels professionally aligned.

77 Erratic steering when braking

1 Wheel bearings worn (Chapter 10).
2 Broken or sagging springs (Chapter 10).
3 Leaking wheel cylinder or caliper (Chapter 10).
4 Excessive brake disc runout (Chapter 9).

78 Excessive pitching and/or rolling around corners or during braking

1 Loose stabilizer bar or worn stabilizer bar bushings (Chapter 10).
2 Worn strut dampers or mountings (Chapter 10).
3 Broken or sagging springs (Chapter 10).
4 Overloaded vehicle.

79 Suspension bottoms

1 Overloaded vehicle.
2 Sagging springs (Chapter 10).

80 Cupped tires

1 Wheel alignment out-of-specifications. Have the wheels professionally aligned.
2 Worn shock absorbers (Chapter 10).
3 Wheel bearings worn (Chapter 10).
4 Excessive tire or wheel runout (Chapter 10).
5 Worn balljoints (Chapter 10).

81 Excessive tire wear on outside edge

1 Inflation pressures incorrect (Chapter 1).
2 Excessive speed in turns.
3 Wheel alignment incorrect (excessive toe-in). Have professionally aligned.
4 Suspension arm bent or twisted (Chapter 10).

82 Excessive tire wear on inside edge

1 Inflation pressures incorrect (Chapter 1).
2 Wheel alignment incorrect (toe-out). Have professionally aligned.
3 Loose or damaged steering components (Chapter 10).

83 Tire tread worn in one place

1 Tires out-of-balance.
2 Damaged or buckled wheel. Inspect and replace if necessary.
3 Defective tire (Chapter 1).

84 Excessive play or looseness in steering system

1 Wheel bearing(s) worn (Chapter 10).
2 Tie-rod end loose (Chapter 10).
3 Steering gear loose (Chapter 10).
4 Worn or loose steering intermediate shaft U-joint (Chapter 10).

85 Rattling or clicking noise in steering gear

1 Steering gear loose (Chapter 10).
2 Steering gear defective.

Notes

Chapter 1
Tune-up and routine maintenance

Contents

	Section		Section
Air filter check and replacement	7	Interior ventilation filter and fresh air intake filter replacement	15
Automatic transaxle fluid change	22	Introduction	2
Battery check, maintenance and charging	9	Maintenance schedule	1
Brake fluid change	20	Manual transaxle lubricant change	24
Brake system check	16	Seat belt check	11
Cooling system check	13	Spark plug check and replacement	21
Cooling system servicing	26	Suspension, steering and driveaxle boot check	17
Direct Shift Gearbox (DSG) transaxle fluid and filter change	23	Tire and tire pressure checks	5
Drivebelt check and replacement	19	Tire rotation	10
Engine oil and filter change	6	Transfer case (bevel box), AWD coupler (Haldex) and rear differential (AWD models) - check and lubricant replacement	27
Exhaust system check	14	Tune-up general information	3
Fluid level checks	4	Underhood hose check and replacement	12
Fuel filter replacement	25	Windshield wiper blade inspection and replacement	8
Fuel system check	18		

Specifications

Recommended lubricants and fluids

Note: *Listed here are manufacturer recommendations at the time this manual was written. Manufacturers occasionally upgrade their fluid and lubricant specifications, so check with your local auto parts store for current recommendations.*

Engine oil type and viscosity	Castrol EDGE® Professional 5W-40 fully synthetic, or equivalent oil meeting VW 502 00 specification
Fuel	
Four-cylinder engines	Unleaded gasoline, 91 octane minimum
Five-cylinder engines	Regular unleaded gasoline
Automatic transaxle fluid	VW part No. G 055 540 A2
Manual transaxle lubricant	
5-speed (0A4)	VW part No. G 052 512 A2
6-speed (02Q)	VW part No. G 052 171 A2
6-speed (DSG)	VW part No. G 052 182 A2
Transfer case (bevel box) lubricant (AWD models)	VW part no. G 052 145 S2
Rear differential lubricant (AWD models)	VW part no. G 052 145 S2
Rear AWD coupling (Haldex) lubricant	VW part no. G 060 175 A2
Brake and clutch fluid	DOT 4 Super or DOT 4+ brake fluid
Engine coolant	Phosphate-free coolant 50/50 mixture of G13 (Violet - VW part No. G13 A8J 1G) antifreeze and distilled water
Hood and trunk hinge lubricant	Lubriplate aerosol spray lubricant
Door hinge and check spring grease	NLGI no. 2 multi-purpose grease
Key lock cylinder lubricant	Graphite spray
Hood latch assembly lubricant	NLGI no. 2 multi-purpose grease
Door latch lubricant	NLGI no. 2 multi-purpose grease or equivalent

Capacities*

Engine oil (including filter)
  Four-cylinder engines
    1.4L (CZTA, DGXA) models... 4.2 quarts (4.0 liters)
    1.8L (CPKA, CPRA, CXBA, CXBB).................................. 5.8 quarts (5.5 liters)
    2.0L (CBPA)... 4.9 quarts (4.6 liters)
    2.0L (CBFA, CCTA, CPLA, CPPA, CNTA, CXCA, CXCB)........... 5.8 quarts (5.5 liters)
  Five-cylinder engines (CBTA, CBUA)................................... 6.10 quarts (5.8 liters)
Manual transaxle
  5-speed (0A4, 0AF)... 2.0 quarts (1.9 liters)
  6-speed (02Q)... 2.4 quarts (2.3 liters)
Automatic transaxle
  6-speed (09G)
    Drain and refill.. Not specified. The best way to determine the amount to add is to measure the amount drained
    From dry .... 7.4 quarts (7.0 liters)
  6-speed (DSG)
    Drain and refill.. 5.5 quarts (5.2 liters)
    From dry .... 7.6 quarts (7.2 liters)
  8-speed (09S)
    Drain and refill
    From dry
Transfer case (bevel box) lubricant .... 0.9 quart (0.85 liter)
Rear differential lubricant (AWD models) .... 1 quart (0.95 liter)
Rear AWD coupling (Haldex) lubricant .... 0.8 quart (0.75 liter)
Cooling system
  Four-cylinder engines.. 8.5 quarts (8.0 liters)
  Five-cylinder engines .... 10 quarts (9.5 liters)

Note: *All capacities approximate. Add as necessary to bring to appropriate level.*

Brakes

Disc brake pad wear limit (lining only)..................................... 0.079 inch (2.0 mm)
Drum brake shoe wear limit (lining only) .... 0.098 inch (2.5 mm)

Ignition system

Spark plug type
  Four-cylinder engines
    1.4L (CZTA, DGXA) engines.. Bosch 8180
    1.8L (CPRA, CPKA, CNSA, CNSB, CXBA, CXBB) engines .... Bosch 8160
    2.0L (CBPA) timing belt engine...................................... Bosch FR7HE02
    2.0L (CBFA, CCTA, CPLA, CPPA, CXCA, CXCB) engines......... Bosch 8160
  Five-cylinder engines (CBUA and CBTA)............................. Bosch FR7HE02
Spark plug gap
  Four-cylinder engines
    1.4L (CZTA, DGXA) engines.. 0.025 to 0.028 inch (0.65 to 0.75 mm)
    2.0L (CBPA) timing belt engine...................................... 0.035 to 0.043 inch (0.9 to 1.1 mm)
    1.8L (CPRA, CPKA, CNSA, CNSB, CXBA, CXBB) engines .... 0.028 to 0.031 inch (0.7 to 0.8 mm)
    2.0L (CBFA, CCTA, CPLA, CPPA, CXCA, CXCB) engines......... 0.028 to 0.031 inch (0.7 to 0.8 mm)
  Five-cylinder engine (CBTA and CBUA) .... 0.039 to 0.043 inch (1.0 to 1.1 mm)
Firing order
  Four-cylinder engines.. 1-3-4-2
  Five-cylinder engines 1-2-4-5-3

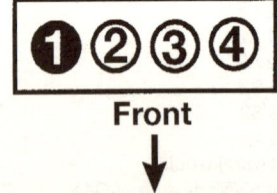

Cylinder locations - four-cylinder engine

Cylinder locations - five-cylinder engine

Chapter 1 Tune-up and routine maintenance

Torque specifications

Ft-lbs (unless otherwise indicated)   Nm

Note: *One foot-pound (ft-lb) of torque is equivalent to 12 inch-pounds (in-lbs) of torque. Torque values below approximately 15 ft-lbs are expressed in inch-pounds, because most foot-pound torque wrenches are not accurate at these smaller values.*

	Ft-lbs	Nm
Engine oil drain plug*	22	30
Oil filter housing-to-engine	18	25
Automatic transaxle check/fill plug	135 in-lbs	15
Automatic transaxle overflow tube-to-transmission	27 in-lbs	3
DSG transaxle check/drain plug	33	45
DSG transaxle overflow tube-to-transmission	27 in-lbs	3
DSG transaxle filter housing-to-transmission	15	20
Manual transaxle		
6-speed models		
Check/fill plug		
Six point (hex) plug	22	30
Twelve point (triple square) plug	33	45
Drain plug		
Six point (hex) plug	22	30
Twelve point (triple square) plug	33	45
Spark plugs		
1.4L engines	17	22
1.8L and 2.0L (CBPA) timing belt engines	18	25
2.0L timing chain engines	22	30
2.5L engines	18	25
Drivebelt tensioner mounting bolt(s)		
1.4L Four-cylinder timing belt engine		
Step 1	15	20
Step 2	Tighten an additional 90-degrees	
2.0L Four-cylinder timing belt engine	17	23
1.8L and 2.0L Four-cylinder timing chain engine		
Step 1	72 in-lbs	8
Step 2	Tighten an additional 90-degrees	
2.5L Five-cylinder engine	26	35
Transfer case (bevel box) check/fill plug*	135 in-lbs	15
Transfer case (bevel box) drain plug*	135 in-lbs	15
Rear differential (AWD models) check/fill plug	168 in-lbs	19
Rear differential (AWD models) drain plug	168 in-lbs	19
Rear AWD coupling (Haldex) check/fill plug	135 in-lbs	15
Rear AWD coupling (Haldex) drain plug	24	32
Wheel bolts	90	120

*Replace the drain plug after each use.

1 Maintenance schedule

1 The maintenance intervals in this manual are provided with the assumption that you, not the dealer, will be doing the work. These are the minimum maintenance intervals recommended by the factory for vehicles that are driven daily. If you wish to keep your vehicle in peak condition at all times, you may wish to perform some of these procedures even more often. Because frequent maintenance enhances the efficiency, performance and resale value of your car, we encourage you to do so. If you drive in dusty areas, tow a trailer, idle or drive at low speeds for extended periods or drive for short distances (less than four miles) in below freezing temperatures, shorter intervals are also recommended.

2 When your vehicle is new, it should be serviced by a factory authorized dealer service department to protect the factory warranty. In many cases, the initial maintenance check is done at no cost to the owner.

Every 250 miles (400 km) or weekly, whichever comes first

Check the engine oil level (see Section 4)
Check the engine coolant level (see Section 4)
Check the brake and clutch fluid level (see Section 4)
Check the windshield washer fluid level (see Section 4)
Check the tires and tire pressures (see Section 5)
Check the operation of all lights
Check the horn operation

Every 5,000 miles (8000 km) or 5 months, whichever comes first

Note: *All items listed above, plus:*
Change the engine oil and filter (see Section 6)
Check and replace, if necessary, the air filter element (see Section 7)

Every 6,000 miles (9600 km) or 6 months, whichever comes first

Note: *All items listed above, plus:*
Check the wiper blade condition (see Section 8)
Check and clean the battery and terminals (see Section 9)
Rotate the tires (see Section 10)
Check the seat belts (see Section 11)
Inspect underhood hoses (see Section 12)
Check the cooling system hoses and connections for leaks and damage (see Section 13)
Check the exhaust system (see Section 14)

Every 15,000 miles (24,000 km) or 12 months, whichever comes first

Note: *All items listed above, plus:*
Replace the interior ventilation filter (see Section 15)
Check the brake system (see Section 16)
Check the suspension/steering components and driveaxle boots (see Section 17)
Check the fuel system hoses and connections for leaks and damage (see Section 18)
Check the drivebelts and replace if necessary (see Section 19)

Every 30,000 miles (48,000 km) or 24 months, whichever comes first

Note: *All items listed above, plus:*
Replace the air filter element (see Section 7)
Change the brake fluid (see Section 20)
Replace the spark plugs (see Section 21)
Check the ignition coil(s) (see Chapter 5)

Every 40,000 miles (64,000 km) or 36 months, whichever comes first

Note: *All items listed above, plus:*
Change the Direct Shift Gearbox (DSG) transaxle fluid and filter (see Section 23)

Every 60,000 miles (96,000 km) or 48 months, which ever comes first

Replace the fuel filter (see Section 25)

Every 75,000 miles (120,000 km)

Replace the timing belt (four-cylinder CBPA, and CZTA and DGXA engines; see Chapter 2A)

Chapter 1 Tune-up and routine maintenance

2 Introduction

1 This Chapter is designed to help the home mechanic maintain their Jetta, GTI or Golf with the goals of maximum performance, economy, safety and reliability in mind.

2 Included is a master maintenance schedule, followed by procedures dealing specifically with each item on the schedule. Visual checks, adjustments, component replacement and other helpful items are included. Refer to the **accompanying illustrations** of the engine compartment and the underside of the vehicle for the locations of various components.

3 Servicing your vehicle in accordance with the mileage/time maintenance schedule and the step-by-step procedures will result in a planned maintenance program that should produce a long and reliable service life. Keep in mind that it's a comprehensive plan, so maintaining some items but not others at the specified intervals will not produce the same results.

2.2a Engine compartment layout (1.4L four-cylinder models)

1 Brake fluid reservoir
2 Battery
3 Fuse and relay block
4 Air filter housing
5 Windshield washer fluid reservoir
6 Coolant expansion tank
7 Engine oil dipstick
8 Engine oil filler cap
9 Ignition coil and spark plug

2.2b Typical front underside components (1.4L four-cylinder, others similar)

1 Lower radiator hose
2 Oil filter
3 Drivebelt
4 Engine oil drain plug
5 Front disc brake caliper
6 Automatic transaxle fluid level check/fill plug
7 Inner driveaxle boot
8 Tie-rod end
9 Exhaust pipe

Chapter 1 Tune-up and routine maintenance   1-7

2.2c Typical rear underside components

1 Fuel filter
2 Center muffler
3 Fuel tank
4 Rear brake caliper
5 Stabilizer bar link
6 Rear brake disc
7 Rear muffler

1-8   Chapter 1   Tune-up and routine maintenance

4   As you service your vehicle, you will discover that many of the procedures can - and should - be grouped together because of the nature of the particular procedure you're performing or because of the close proximity of two otherwise unrelated components to one another.

5   For example, if the vehicle is raised for chassis lubrication, you should inspect the exhaust, suspension, steering and fuel systems while you're under the vehicle. When you're rotating the tires, it makes good sense to check the brakes since the wheels are already removed. Finally, let's suppose you have to borrow or rent a torque wrench. Even if you only need it to tighten the spark plugs, you might as well check the torque of as many critical fasteners as time allows.

6   The first step in this maintenance program is to prepare yourself before the actual work begins. Read through all the procedures you're planning to do, then gather up all the parts and tools needed. If it looks like you might run into problems during a particular job, seek advice from a mechanic or an experienced do-it-yourselfer.

Owner's manual and VECI label information

7   Your vehicle owner's manual was written for your year and model and contains very specific information on component locations, specifications, fuse ratings, part numbers, etc. The owner's manual is an important resource for the do-it-yourselfer to have; if one was not supplied with your vehicle, it can generally be ordered from a dealer parts department.

8   Among other important information, the Vehicle Emissions Control Information (VECI) label contains specifications and procedures for applicable tune-up adjustments and, in some instances, spark plugs. The information on this label is the exact maintenance data recommended by the manufacturer. This data often varies by intended operating altitude, local emissions regulations, month of manufacture, etc.

9   This Chapter contains procedural details, safety information and more ambitious maintenance intervals than you might find in manufacturer's literature. However, you may also find procedures or specifications in your owner's manual or VECI label that differ with what's printed here. In these cases, the owner's manual or VECI label can be considered correct, since it is specific to your particular vehicle.

3   Tune-up general information

1   The term tune-up is used in this manual to represent a combination of individual operations rather than one specific procedure.

2   If, from the time the vehicle is new, the routine maintenance schedule is followed closely and frequent checks are made of fluid levels and high wear items, as sug-

4.2 Engine oil dipstick - 1.4L (CZTA) engine shown, others similar

gested throughout this manual, the engine will be kept in relatively good running condition and the need for additional work will be minimized.

3   More likely than not, however, there will be times when the engine is running poorly due to lack of regular maintenance. This is even more likely if a used vehicle, which has not received regular and frequent maintenance checks, is purchased. In such cases, an engine tune-up will be needed outside of the regular routine maintenance intervals.

4   The first step in any tune-up or diagnostic procedure to help correct a poor running engine is a cylinder compression check. A compression check (see Chapter 2C) will help determine the condition of internal engine components and should be used as a guide for tune-up and repair procedures. If, for instance, a compression check indicates serious internal engine wear, a conventional tune-up will not improve the performance of the engine and would be a waste of time and money. Because of its importance, the compression check should be done by someone with the right equipment and the knowledge to use it properly.

Minor tune-up

5   Check all engine-related fluids (Section 4)
6   Check the air filter (Section 7)
7   Clean, inspect and test the battery (Section 9)
8   Check all underhood hoses (Section 12)
9   Check the cooling system (Section 13)
10   Check the drivebelt (Section 19)
11   Check the charging system (Chapter 5)

Major tune-up

Note: *All items listed under Minor tune-up, plus...*
12   Replace the air filter (Section 7)
13   Replace the spark plugs (Section 21)
14   Replace the fuel filter (Section 25)

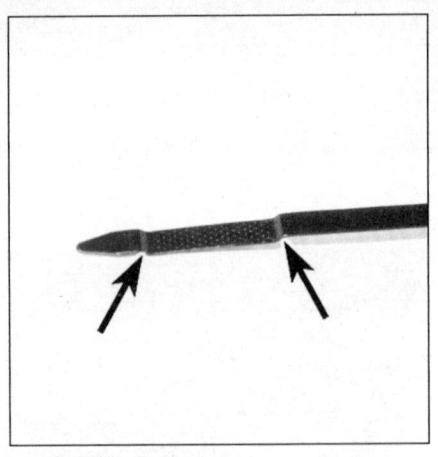

4.4 The oil level must be maintained between the MIN and MAX marks on the dipstick

4   Fluid level checks (every 250 miles [400 km] or weekly)

Note: *The vehicle must be on level ground when fluid levels are checked.*

1   Fluids are an essential part of the lubrication, cooling, brake and windshield washer systems. Because the fluids gradually become depleted and/or contaminated during normal operation of the vehicle, they must be periodically replenished. See *Recommended lubricants and fluids* in this Chapter's Specifications before adding fluid to any of the following components.

Engine oil

2   The oil level is checked with a dipstick, which is located on the side of the engine **(see illustration)**.

3   The oil level should be checked before the vehicle has been driven, or about five minutes after the engine has been shut off. If the oil is checked immediately after driving the vehicle, some of the oil will remain in the upper part of the engine, resulting in an inaccurate reading on the dipstick.

4   Pull the dipstick out of the tube and wipe all the oil from the end with a clean rag or paper towel. Insert the clean dipstick all the way back into the engine and pull it out again. Note the oil at the end of the dipstick; the level should be between the MIN and MAX marks **(see illustration)**.

5   Do not allow the level to drop below the minimum mark, or oil starvation may cause engine damage. Conversely, overfilling the engine (adding oil above the MAX mark) may cause oil fouled spark plugs, oil leaks or oil seal failures. The oil could also be whipped by the crankshaft, causing it to foam, which could cause accelerated wear of the friction surfaces in the engine due to lack of proper lubrication.

Caution: *Damage to the catalytic converter may result if the oil level is overfilled.*

Chapter 1 Tune-up and routine maintenance

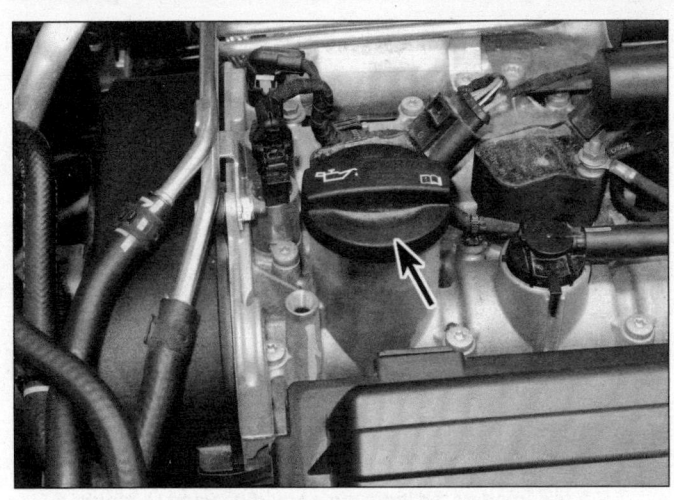

4.6 Engine oil filler cap - 1.4L (CZTA) engine shown, others similar

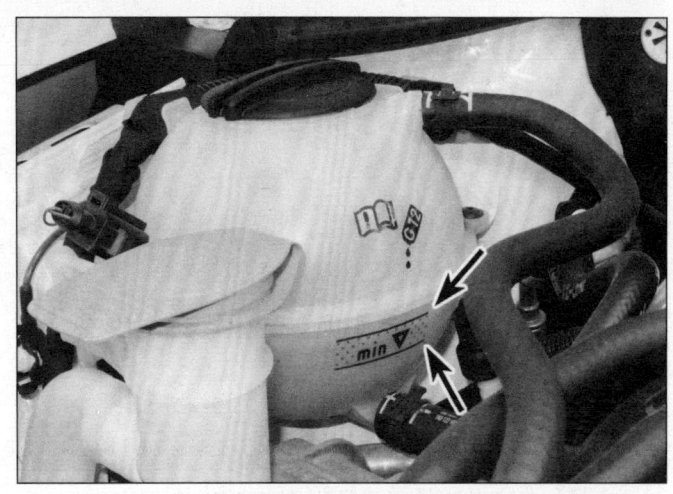

4.9 When the engine is cold, the coolant level should be between the MIN and MAX marks

6   To add oil, remove the filler cap from the valve cover **(see illustration)**. After adding oil, wait a few minutes to allow the level to stabilize, then pull out the dipstick and check the level again. Add more oil if required. Install the filler cap and tighten it by hand only.

7   Checking the oil level is an important preventive maintenance step. A consistently low oil level indicates oil leakage through damaged seals, defective gaskets or past worn rings or valve guides. If the oil looks milky in color or has water droplets in it, the cylinder head gasket may be blown or the head or block may be cracked. The engine should be checked immediately. The condition of the oil should also be checked. Whenever you check the oil level, slide your thumb and index finger up the dipstick before wiping off the oil. If you see small dirt or metal particles clinging to the dipstick, the oil should be changed (see Section 6).

Engine coolant

Warning: *Do not allow antifreeze to come in contact with your skin or painted surfaces of the vehicle. Flush contaminated areas immediately with plenty of water. Don't store new coolant or leave old coolant lying around where it's accessible to children or pets - they're attracted by its sweet smell. Ingestion of even a small amount of coolant can be fatal! Wipe up garage floor and drip pan spills immediately. Keep antifreeze containers covered and repair cooling system leaks as soon as they're noticed.*

8   All vehicles covered by this manual are equipped with a coolant expansion tank, located in the right side of the engine compartment, and connected by hoses to the cooling system.

9   The coolant level in the tank should be checked regularly. The level in the tank varies with the temperature of the engine. When the engine is cold, the coolant level should be between the MIN and MAX marks on the tank **(see illustration)**. If it isn't, remove the cap from the tank and add coolant to the tank.

Warning: *Never remove the pressure cap when the engine is warm!*

Warning: *Remove the cap slowly. If you hear a hissing sound when unscrewing the cap, wait until it stops, then proceed.*

10   Drive the vehicle and recheck the coolant level. If only a small amount of coolant is required to bring the system up to the proper level, water can be used. However, repeated additions of water will dilute the antifreeze and water solution. In order to maintain the proper ratio of antifreeze and water, always top up the coolant level with the correct mixture. Don't use rust inhibitors or additives. An empty plastic milk jug or bleach bottle makes an excellent container for mixing coolant.

11   If the coolant level drops consistently, there may be a leak in the system. Inspect the radiator, hoses, filler cap, drain plugs and water pump (see Section 13). If no leaks are noted, have the expansion tank cap pressure tested by a service station.

12   If you have to remove the pressure cap, wait until the engine has cooled completely, then wrap a thick cloth around the cap and turn it to the first stop. If coolant or steam escapes, or if you hear a hissing noise, let the engine cool down longer, then remove the cap.

13   Check the condition of the coolant as well. It should be relatively clear. If it's brown or rust colored, the system should be drained, flushed and refilled. Even if the coolant appears to be normal, the corrosion inhibitors wear out, so it must be replaced at the specified intervals.

Brake and clutch fluid

14   The brake master cylinder is located in the driver's side of the engine compartment, under the cowl cover near the firewall. On vehicles with manual transaxles, the clutch master cylinder is connected by a hose to the brake master cylinder reservoir.

4.15 Never let the brake fluid level drop below the MIN mark

15   To check the fluid level, remove the access cover from the left cowl cover, then look at the MAX and MIN marks on the reservoir **(see illustration)**. The level should be within the specified distance from the maximum fill line.

16   If the level is low, wipe the top of the reservoir cover with a clean rag to prevent contamination of the brake or clutch system before lifting the cover.

17   Add only the specified brake fluid to the brake reservoir (refer to *Recommended lubricants and fluids* in this Chapter's Specifications, or to your owner's manual). Mixing different types of brake fluid can damage the system. Fill the brake master cylinder reservoir only to the MAX line.

Warning: *Use caution when filling either reservoir - brake fluid can harm your eyes and damage painted surfaces. Do not use brake fluid that is more than one year old or has been left open. Brake fluid absorbs moisture from the air. Excess moisture can cause a dangerous loss of braking.*

1-10 Chapter 1 Tune-up and routine maintenance

4.22 The windshield washer fluid reservoir is located in the passenger's side of the engine compartment

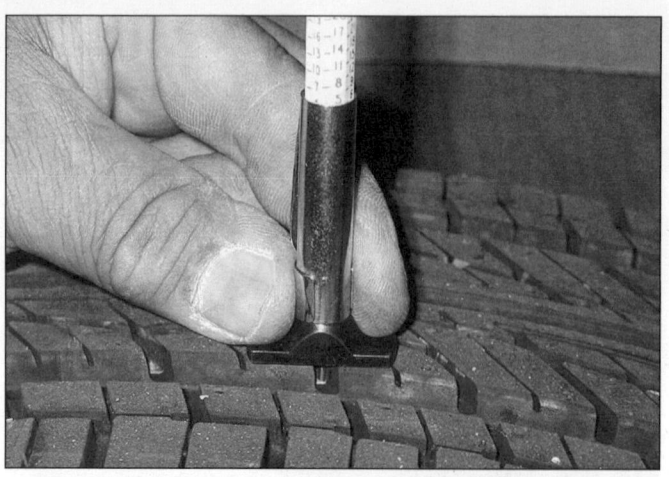

5.2 A tire tread depth indicator should be used to monitor tire wear - they are available at auto parts stores and service stations and cost very little

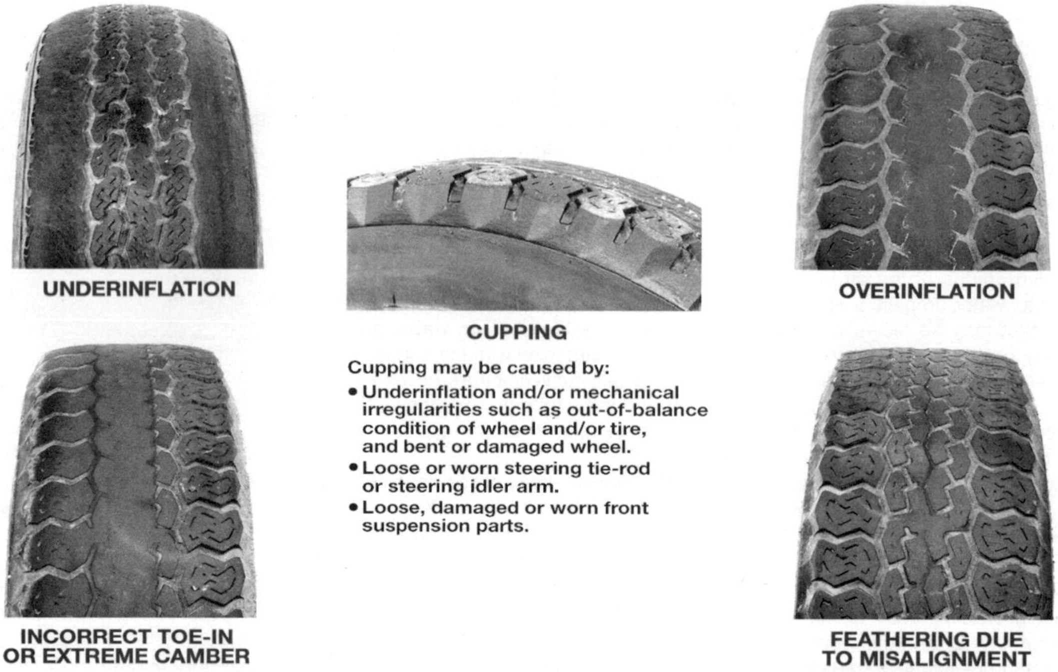

UNDERINFLATION

OVERINFLATION

CUPPING

Cupping may be caused by:
- Underinflation and/or mechanical irregularities such as out-of-balance condition of wheel and/or tire, and bent or damaged wheel.
- Loose or worn steering tie-rod or steering idler arm.
- Loose, damaged or worn front suspension parts.

INCORRECT TOE-IN OR EXTREME CAMBER

FEATHERING DUE TO MISALIGNMENT

5.3 This chart will help you determine the condition of your tires, the probable cause(s) of abnormal wear and the corrective action necessary

18 While the reservoir cap is removed, inspect the master cylinder reservoir for contamination. If deposits, dirt particles or water droplets are present, the fluid should be changed (see Section 8 for the brake fluid replacement procedure, or Chapter 8 for the clutch hydraulic system bleeding procedure).

19 After filling the reservoir to the proper level, make sure the lid is properly seated to prevent fluid leakage and/or system pressure loss.

20 The fluid in the brake master cylinder will drop slightly as the brake pads at each wheel wear down during normal operation. If the master cylinder requires repeated replenishing to keep it at the proper level, this is an indication of leakage in the brake system, which should be corrected immediately. If the brake system shows an indication of leakage, check all brake lines and connections, along with the calipers and master cylinder (see Section 16 for more information).

21 If, upon checking the brake master cylinder fluid level, you discover the reservoir empty or nearly empty, the brake and clutch systems should be thoroughly inspected (see Chapters 9 and 8).

Windshield washer fluid

22 Fluid for the windshield washer system is stored in a plastic reservoir located at the right rear of the engine compartment **(see illustration)**.

23 In milder climates, plain water can be used in the reservoir, but it should be kept no more than 2/3 full to allow for expansion if the water freezes. In colder climates, use windshield washer system antifreeze, available at any auto parts store, to lower the freezing point of the fluid. Mix the antifreeze with water in accordance with the manufacturer's directions on the container.

Caution: *Do not use cooling system antifreeze - it will damage the vehicle's paint.*

Chapter 1 Tune-up and routine maintenance   1-11

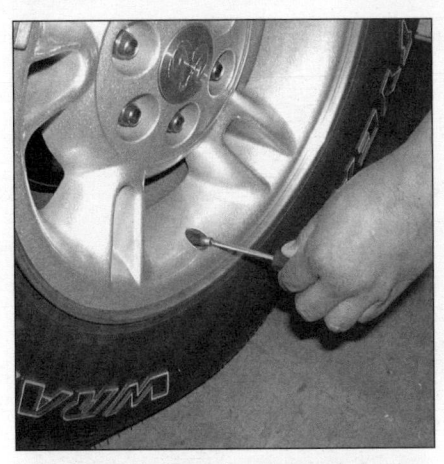

5.4a If a tire loses air on a steady basis, check the valve core first to make sure it's snug (special inexpensive wrenches are commonly available at auto parts stores)

5.4b If the valve core is tight, raise the corner of the vehicle with the low tire and spray a soapy water solution onto the tread as the tire is turned slowly - slow leaks will cause small bubbles to appear

5.8 To extend the life of your tires, check the air pressure at least once a week with an accurate gauge (don't forget the spare!)

5 Tire and tire pressure checks (every 250 miles [400 km] or weekly)

1 Periodic inspection of the tires may spare you the inconvenience of being stranded with a flat tire. It can also provide you with vital information regarding possible problems in the steering and suspension systems before major damage occurs.

2 The original tires on this vehicle are equipped with 1/2-inch wide bands that will appear when tread depth reaches 1/16-inch, at which point they can be considered worn out. Tread wear can be monitored with a simple, inexpensive device known as a tread depth indicator **(see illustration)**.

3 Note any abnormal tread wear **(see illustration)**. Tread pattern irregularities such as cupping, flat spots and more wear on one side than the other are indications of front end alignment and/or balance problems. If any of these conditions are noted, take the vehicle to a tire shop or service station to correct the problem.

4 Look closely for cuts, punctures and embedded nails or tacks. Sometimes a tire will hold air pressure for a short time or leak down very slowly after a nail has embedded itself in the tread. If a slow leak persists, check the valve stem core to make sure it is tight **(see illustration)**. Examine the tread for an object that may have embedded itself in the tire or for a plug that may have begun to leak (radial tire punctures are repaired with a plug that is installed in a puncture). If a puncture is suspected, it can be easily verified by spraying a solution of soapy water onto the puncture area **(see illustration)**. The soapy solution will bubble if there is a leak. Unless the puncture is unusually large, a tire shop or service station can usually repair the tire.

5 Carefully inspect the inner sidewall of each tire for evidence of brake fluid leakage. If you see any, inspect the brakes immediately.

6 Correct air pressure adds miles to the life span of the tires, improves mileage and enhances overall ride quality. Tire pressure cannot be accurately estimated by looking at a tire, especially if it's a radial. A tire pressure gauge is essential. Keep an accurate gauge in the glove compartment. The pressure gauges attached to the nozzles of air hoses at gas stations are often inaccurate.

7 Always check tire pressure when the tires are cold. Cold, in this case, means the vehicle has not been driven over a mile in the three hours preceding a tire pressure check. A pressure rise of four to eight pounds is not uncommon once the tires are warm.

8 Unscrew the valve cap protruding from the wheel or hubcap and push the gauge firmly onto the valve stem **(see illustration)**. **Note** *the reading on the gauge and compare the figure to the recommended tire pressure shown on the tire placard on the driver's side door. Be sure to reinstall the valve cap to keep dirt and moisture out of the valve stem mechanism. Check all four tires and, if necessary, add enough air to bring them up to the recommended pressure.*

9 Don't forget to keep the spare tire inflated to the specified pressure (refer to the pressure molded into the tire sidewall).

6 Engine oil and filter change (every 5000 miles [8000 km] or 5 months)

1 Frequent oil changes are the best preventive maintenance the home mechanic can give the engine, because aging oil becomes diluted and contaminated, which leads to premature engine wear.

2 Make sure you have all the necessary tools before you begin this procedure **(see illustration)**.

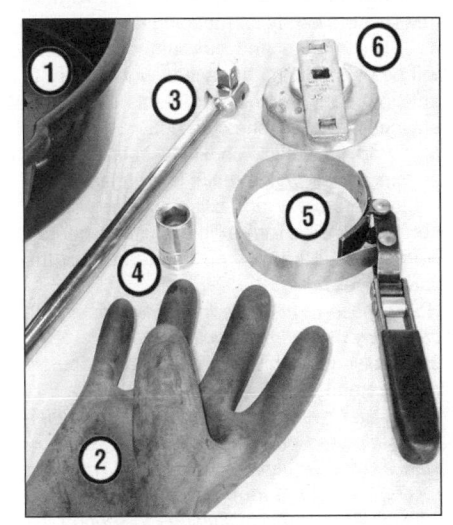

6.2 These tools are required when changing the engine oil and filter

1 **Drain pan** - It should be fairly shallow in depth, but wide to prevent spills
2 **Rubber gloves** - When removing the drain plug and filter, you will get oil on your hands (the gloves will prevent burns)
3 **Breaker bar** - Sometimes the oil drain plug is tight, and a long breaker bar is needed to loosen it
4 **Socket** – To be used with the breaker bar or a ratchet (must be the correct size to fit the drain plug - six-point preferred)
5 **Filter wrench** - This is a metal band-type wrench, which requires clearance around the filter to be effective
6 **Filter wrench** - This type fits on the bottom of the filter and can be turned with a ratchet or breaker bar (different-size wrenches are available for different types of filters)

1-12   Chapter 1   Tune-up and routine maintenance

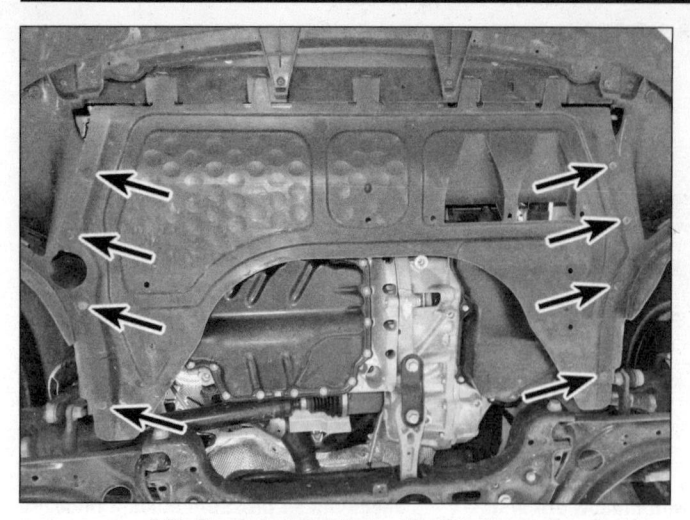

6.7a Remove the splash shield fasteners

6.7b Slide the shield back towards the rear, then lower the shield and remove it

3 You should also have plenty of rags or newspapers handy for mopping up any spills.

4 The engine and exhaust components will be warm during the actual work, so try to anticipate any potential problems before the engine and accessories are hot.

5 Park the vehicle on a level spot. Start the engine and allow it to reach its normal operating temperature. Warm oil and sludge will flow out more easily. Turn off the engine when it's warmed up. Remove the filler cap from the valve cover.

Note: *Place rags around the filter housing to catch any oil that might spill out.*

6 Raise the vehicle and support it securely on jackstands.

Warning: *Never get beneath the vehicle when it is supported only by a jack. The jack provided with your vehicle is designed solely for raising the vehicle to remove and replace the wheels. Always use jackstands to support the vehicle when it becomes necessary to place your body underneath the vehicle.*

7 Remove the under-vehicle splash shield fasteners, then slide the shield back and out from under the vehicle **(see illustrations)**.

8 Being careful not to touch the hot exhaust components, place the drain pan under the drain plug in the bottom of the pan and remove the plug **(see illustration)**. You may want to wear gloves while unscrewing the plug the final few turns if the engine is hot. **Caution:** *The manufacturer states the drain plug must be replaced after it has been removed.*

9 Allow the old oil to drain into the pan. It may be necessary to move the pan farther under the engine as the oil flow slows to a trickle.

10 Inspect the old oil for the presence of metal shavings and chips.

11 After all the oil has drained, wipe off the drain plug with a clean rag. Even minute metal particles clinging to the plug would immediately contaminate the new oil.

12 Clean the area around the drain plug opening, install a new drain plug and tighten it to the torque listed in this Chapter's Specifications.

13 Locate the oil filter/housing and clean the area where the housing meets the engine.

All models except 1.4L CZTA, DGXA and 2.0L CBPA engines

Note: *On 1.8L and 2.0L engines the oil filter housing is located at the top of the engine. On 2.5L engines the housing is located at the bottom of the engine.*

14 On 2.5L engines, remove the valve dust cap from the bottom of the filter housing **(see illustration)**.

15 On 2.5L engines, use a screwdriver to depress the valve at the bottom of the housing, allowing the oil in the filter housing to drain **(see illustration)**.

16 Unscrew the oil filter housing **(see illustration)**. The element is withdrawn with the oil filter housing, and can then be separated and discarded.

6.8 Use a proper size box-end wrench or socket to remove the oil drain plug and avoid rounding it off

6.14 On 2.5L engines unscrew the valve cap from the bottom of the filter housing and locate the drain valve in the filter housing

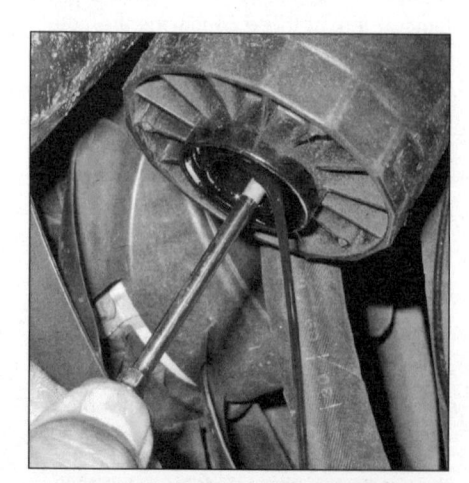

6.15 On 2.5L engines, use a screwdriver to depress the valve and drain the oil from the housing

Chapter 1 Tune-up and routine maintenance

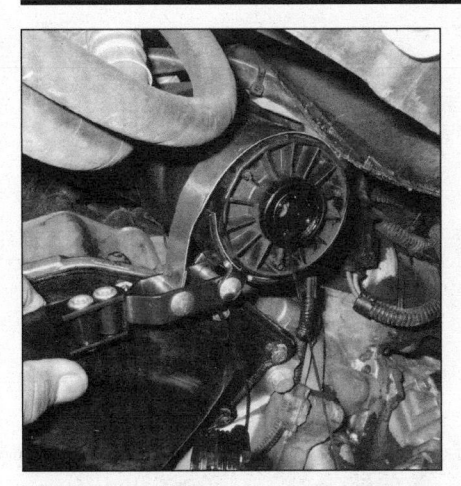

1-13

6.16 Use an oil filter wrench to remove the filter housing - 2.5L engine shown, other models similar

6.17 Pull the used element off of the filter housing stem and install a new O-ring in the housing - 2.5L engine shown, other models similar

6.18 Install the new filter element - 2.5L engine shown, other models similar

17 Wipe out the oil filter housing and cap using a clean rag, then install a new O-ring on the housing **(see illustration)**.
18 Lubricate the O-ring(s) with clean engine oil, then install the filter element into the housing **(see illustration)**.
19 Clean the surface where the filter housing meets the engine, then screw the filter housing onto the engine and tighten the housing securely.
20 Screw the valve dust cap into the bottom of the housing and tighten the cap securely, if equipped.

1.4L (CZTA, DGXA) and 2.0L (CBPA) engines

21 Loosen the oil filter by turning it counterclockwise with an oil filter wrench **(see illustration)**. Once the filter is loose, use your hands to unscrew it from the block. Keep the open end pointing up to prevent the oil inside the filter from spilling out.

22 With a clean rag, wipe off the mounting surface on the block. If a residue of old oil is allowed to remain, it will smoke when the block is heated up. Also make sure that none of the old gasket remains stuck to the mounting surface. It can be removed with a scraper if necessary.
23 Compare the old filter with the new one to make sure they are the same type. Smear some clean engine oil on the rubber gasket of the new filter **(see illustration)**.
24 Attach the new filter to the engine, following the tightening directions printed on the filter canister or packing box. Most filter manufacturers recommend against using a filter wrench due to the possibility of overtightening and damaging the seal.

All models

25 Remove all tools, rags, etc., from under the vehicle, being careful not to spill the oil in the drain pan, then lower the vehicle.

26 Add new oil to the engine through the oil filler cap in the valve cover. Use a funnel, if necessary, to prevent oil from spilling onto the top of the engine. Pour the specified type and amount of oil into the engine (refer to *Recommended lubricants and fluids* and *Capacities* in this Chapter's Specifications). Wait a few minutes to allow the oil to drain into the pan, then check the level (see Section 4). If the oil level is correct, install the filler cap hand-tight, start the engine and allow the new oil to circulate.
Caution: *Do not rev the engine.*
27 Allow the engine to run for about a minute, then turn it off.
28 Wait a few minutes to allow the oil to flow back into the pan, then recheck the level on the dipstick and, if necessary, add enough oil to bring it to the correct level.
29 During the first few trips after an oil change, make it a point to check frequently for leaks and proper oil level.

6.21 Use an oil filter wrench to remove the filter

6.23 Lubricate the oil filter gasket with clean engine oil before installing the filter on the engine

1-14   Chapter 1   Tune-up and routine maintenance

7.1 Engine cover details - four-cylinder (CZTA) model shown, others similar

1. Vent hose
2. Inlet air resonator pipe spring clamp
3. Inlet air duct spring clamp

7.4 Lift the engine cover up and off of the ballstuds

30   The old oil drained from the engine cannot be reused in its present state and should be disposed of. Check with your local auto parts store, disposal facility or environmental agency to see if they will accept the oil for recycling. After the oil has cooled, it can be drained into a container (capped plastic jugs, topped bottles, milk cartons, etc.) for transport to one of these disposal sites. Don't dispose of the oil by pouring it on the ground or down a drain!

Service indicator resetting

Note: After changing the engine oil, it's important to reset the service indicator so it can keep an accurate record of engine operating time/vehicle mileage.

Using the windshield wiper switch or steering wheel multi-function switch

31   Using the rocker switch on the windshield wiper lever switch or (if equipped) the steering wheel multi-function switch can also be used to reset the service indicator.
32   Turn the ignition key to the Off position.
33   Using the windshield wiper switch or steering wheel, select the "settings" menu and look for the "service menu."
34   Once in the "service menu" select "reset" and press the OK button on the steering wheel or on the bottom of the windshield wiper switch.
35   Press the OK button one more time for confirmation.

Using the buttons on the instrument cluster

36   Turn the ignition key to the Off position.
37   Press and hold the trip-odometer reset button, then turn the ignition key to the On position, but do not start the engine. Release the button.

38   Press and release the clock minute set button one time, then turn the ignition key to the Off position.

7   Air filter check and replacement (every 5000 miles [8000 km] or 5 months)

Note: On four-cylinder BPY and five-cylinder engines, the air filter is incorporated in the engine cover.

1.4L four-cylinder (CZTA and DGXA) engines

1   Release the inlet hose clamps at the front and rear of the engine cover **(see illustration)**.
2   Disconnect the inlet air resonator pipe

7.5 Rotate the water drain pipe clockwise, then pull the pipe straight up and off

from the rear of the engine cover then disconnect the vent hose from the housing.
3   Disconnect the inlet air duct from the front of the engine cover.
4   Lift the engine cover up from the right side of the ballstuds **(see illustration)**, then the left side, and remove the cover.
5   Place the cover upside down and remove the water drain pipe by rotating the pipe clockwise **(see illustration)**, then lift the pipe off to prevent accidental damaged to the drain pipe.
6   Remove the filter housing cover screws **(see illustration)**.
7   Depress the locking tabs **(see illustration)** and unhook it from the engine cover, then remove the filter from the housing.
8   Lift the filer cover up and off of the engine cover, then remove the filter **(see illustrations)**.

7.6 Remove the air filter housing cover screws

Chapter 1 Tune-up and routine maintenance

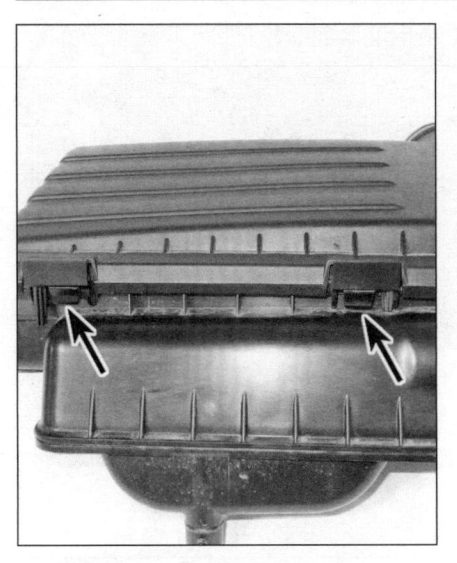

7.7 Depress the locking tabs and separate the housing halves

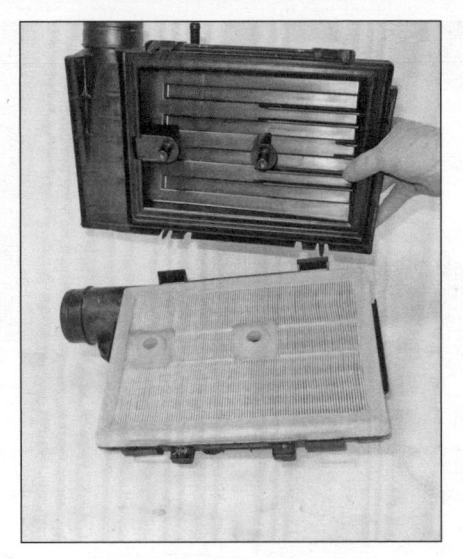

7.8a Lift the cover up and off of the engine cover. . .

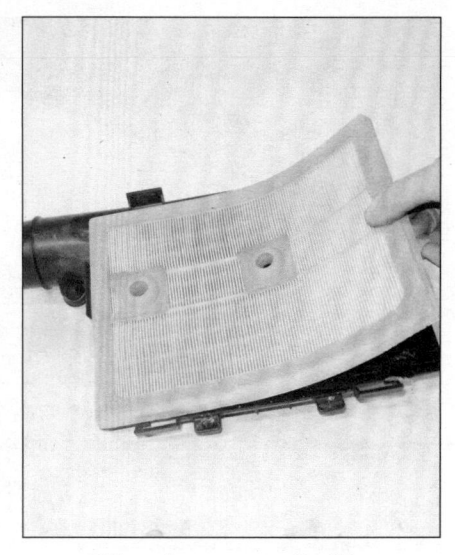

7.8b. . . then remove the filter

9 Clean the inside of the air filter housing with a damp cloth.
10 Installation is the reverse of removal.

1.8L and 2.0L four-cylinder engines

11 Release the air guide top cover clips and remove the top cover (all engines except 2.0L CBPA engines).
12 Release the air guide lower clips and remove the guide.
13 Remove the fasteners that secure the cover of the filter housing. On models so equipped, also detach the smaller hose from the front of the filter housing.
14 On all except 2.0L CBPA engines, lift the cover up and remove the air filter element.
15 On 2.0L CBPA engines, lift the cover up, then remove the filter clip fastener, the filter clip and remove the air filter element.
16 Clean the inside of the air filter housing with a damp cloth.
17 Installation is the reverse of removal.

2.5L five-cylinder engines (CBTA and CBUA)

18 Release the intake hose clamps at the rear of the engine cover and disconnect the intake hose from the Mass Air Flow (MAF) sensor **(see illustration)**.
19 Disconnect the Mass Air Flow (MAF) sensor electrical connector. On models so equipped, also detach the smaller hose from the housing.
20 Remove the fresh air intake air duct mounting screws at the front of the engine cover.
21 Lift the cover up from the right side off of the ballstuds, then the left side, and remove the cover.
22 Place the cover upside down and remove the filter housing mounting fasteners **(see illustration)**.

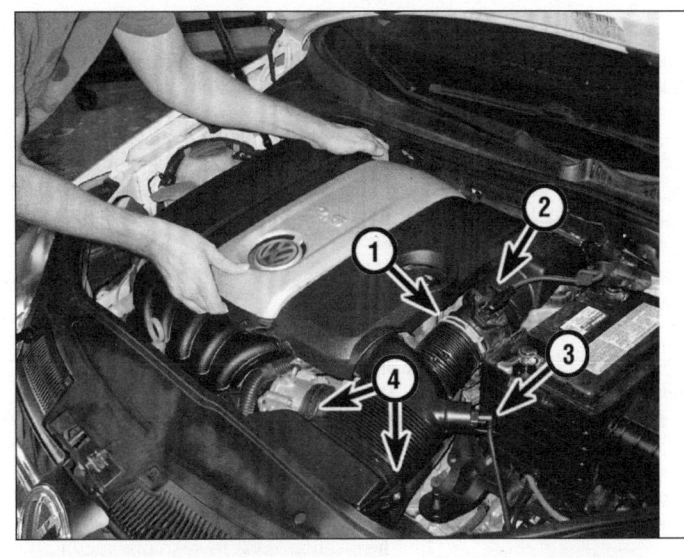

7.18 Engine cover details - five-cylinder engines

1 Intake hose clamp
2 Mass Air Flow (MAF) sensor electrical connector
3 Smaller hose
4 Intake air duct screws

7.22 Airfilter housing fastener locations - five-cylinder engine

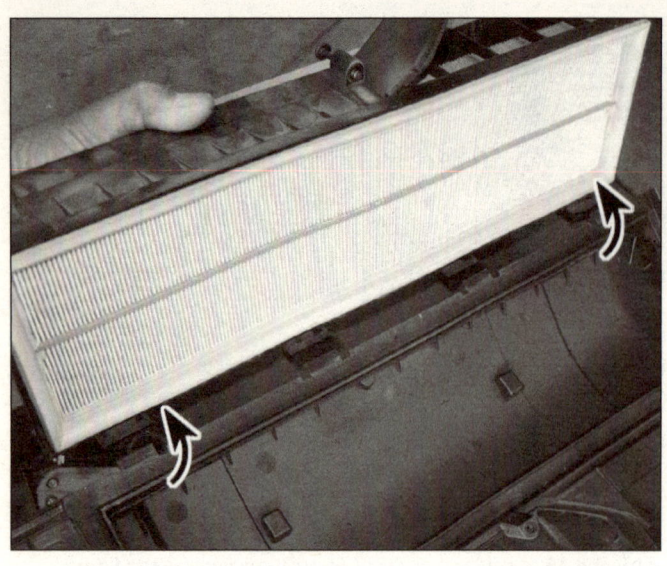

1-16   Chapter 1   Tune-up and routine maintenance

7.23 Rotate the housing up, then remove the filter element

8.5 Place a rag between the hood and the wiper arm to protect the paint

23 Rotate the housing up and unhook it from the engine cover, then remove the filter from the housing **(see illustration)**.
24 Clean the inside of the air filter housing with a damp cloth.
25 Installation is the reverse of removal.

8 Windshield wiper blade inspection and replacement (every 6000 miles [9600 km] or 6 months)

Note: *The rear wiper arm on station wagon models is removed the same way the front wiper blades are moved.*

1 The windshield wiper and blade assembly should be inspected periodically for damage, loose components and cracked or worn blade elements.

2 Road film can build up on the wiper blades and affect their efficiency, so they should be washed regularly with a mild detergent solution.
3 The action of the wiping mechanism can loosen bolts, nuts and fasteners, so they should be checked and tightened, as necessary, at the same time the wiper blades are checked.
4 If the wiper blade elements are cracked, worn or warped, or no longer clean adequately, they should be replaced with new ones.
5 Place a rag between the hood and where the wiper arm could contact the hood, then lift the arm assembly away from the glass **(see illustration)**.
6 Depress the locking tab and slide it upwards **(see illustrations)**, pivot the wiper blade until it reaches its stop, then slide the blade off the arm.
7 Installation is the reverse of removal.

9 Battery check, maintenance and charging (every 6000 miles [9600 km] or 6 months)

Warning: *Certain precautions must be followed when checking and servicing the battery. Hydrogen gas, which is highly flammable, is always present in the battery cells, so keep lighted tobacco and all other open flames and sparks away from the battery. The electrolyte inside the battery is actually diluted sulfuric acid, which will cause injury if splashed on your skin or in your eyes. It will also ruin clothes and painted surfaces. When removing the battery cables, always detach the negative cable first and hook it up last!*
Caution: *If the battery is disconnected, several systems must be re-learned before they will work properly (see Chapter 5, Section 3). Also, if the audio system is equipped with an anti-theft system, make sure you have the correct activation code before disconnecting the battery.*

1 A routine preventive maintenance program for the battery in your vehicle is the only way to ensure quick and reliable starts. But before performing any battery maintenance, make sure that you have the proper equipment necessary to work safely around the battery **(see illustration)**.
2 There are also several precautions that should be taken whenever battery maintenance is performed. Before servicing the battery, always turn the engine and all accessories off and disconnect the cable from the negative terminal of the battery (see Chapter 5).
3 The battery produces hydrogen gas, which is both flammable and explosive. Never create a spark, smoke or light a match around the battery. Always charge the battery in a ventilated area.
4 Electrolyte contains poisonous and corrosive sulfuric acid. Do not allow it to get in your eyes, on your skin or on your clothes.

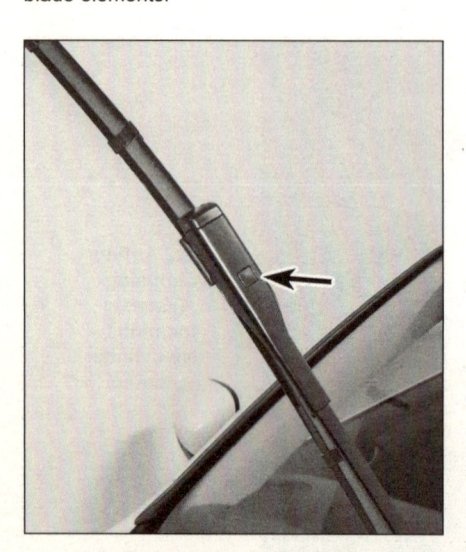

8.6a Depress the locking tab...

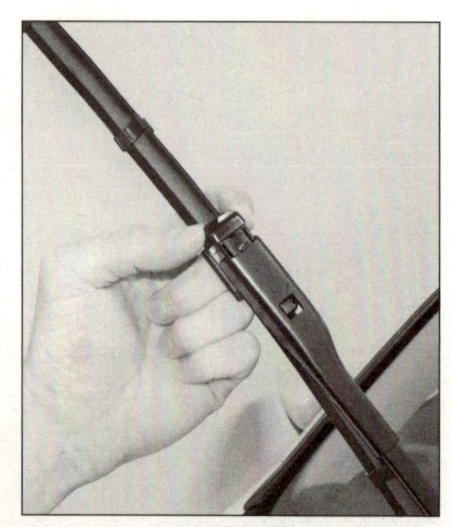

8.6b... then slide the locking tabs upwards

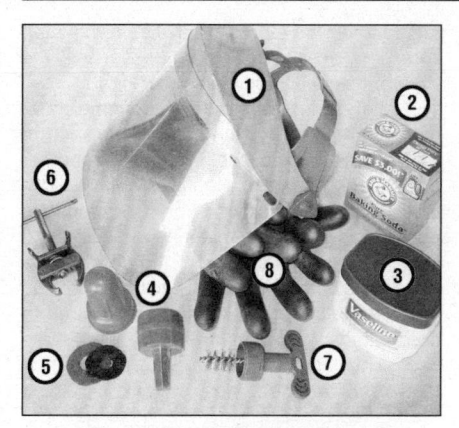

Chapter 1 Tune-up and routine maintenance

1-17

9.1 Tools and materials required for battery maintenance

1. **Face shield/safety goggles** - When removing corrosion with a brush, the acidic particles can easily fly up into your eyes
2. **Baking soda** - A solution of baking soda and water can be used to neutralize corrosion
3. **Petroleum jelly** - A layer of this on the battery posts will help prevent corrosion
4. **Battery post/cable cleaner** - This wire brush cleaning tool will remove all traces of corrosion from the battery posts and cable clamps
5. **Treated felt washers** - Placing one of these on each post, directly under the cable clamps, will help prevent corrosion
6. **Puller** - Sometimes the cable clamps are very difficult to pull off the posts, even after the nut/bolt has been completely loosened. This tool pulls the clamp straight up and off the post without damage
7. **Battery post/cable cleaner** - Here is another cleaning tool which is a slightly different version of Number 4 above, but it does the same thing
8. **Rubber gloves** - Another safety item to consider when servicing the battery; remember that's acid inside the battery!

9.6a Battery terminal corrosion usually appears as light, fluffy powder

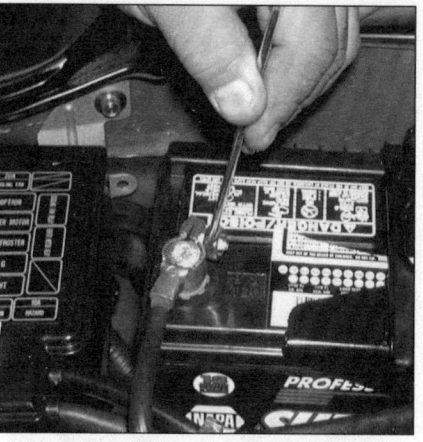

9.7a When cleaning the cable clamps, all corrosion must be removed

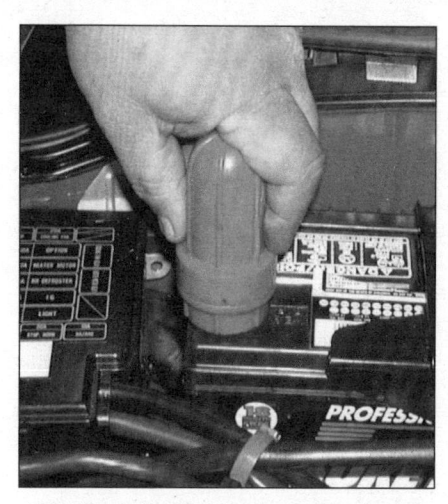

9.6b Removing a cable from the battery post with a wrench - sometimes a pair of special battery pliers are required for this procedure if corrosion has caused deterioration of the nut hex (always remove the negative cable first and hook it up last!)

9.7b Regardless of the type of tool used to clean the battery posts, a clean, shiny surface should be the result

9.8 Make sure the battery hold-down fastener is tight

Never ingest it. Wear protective safety glasses when working near the battery. Keep children away from the battery.

5 Note the external condition of the battery. If the positive terminal and cable clamp on your vehicle's battery is equipped with a rubber protector, make sure that it's not torn or damaged. It should completely cover the terminal. Look for any corroded or loose connections, cracks in the case or cover or loose hold-down clamps. Also check the entire length of each cable for cracks and frayed conductors.

6 If corrosion, which looks like white, fluffy deposits **(see illustration)** is evident, particularly around the terminals, the battery should be removed for cleaning. Loosen the cable clamp bolts with a wrench, being careful to remove the ground cable first, and slide them off the terminals **(see illustration)**. Then disconnect the hold-down clamp bolt and nut, remove the clamp and lift the battery from the engine compartment.

7 Clean the cable clamps thoroughly with a battery brush or a terminal cleaner and a solution of warm water and baking soda **(see illustration)**. Wash the terminals and the top of the battery case with the same solution but make sure that the solution doesn't get into the battery. When cleaning the cables, terminals and battery top, wear safety goggles and rubber gloves to prevent any solution from coming in contact with your eyes or hands. Wear old clothes too - even diluted, sulfuric acid splashed onto clothes will burn holes in them. If the terminals have been extensively corroded, clean them up with a terminal cleaner **(see illustration)**. Thoroughly wash all cleaned areas with plain water.

8 Make sure that the battery tray is in good condition and the hold-down clamp fasteners are tight **(see illustration)**. If the battery is removed from the tray, make sure no parts remain in the bottom of the tray when the battery is reinstalled. When reinstalling the hold-down clamp bolts, do not overtighten them.

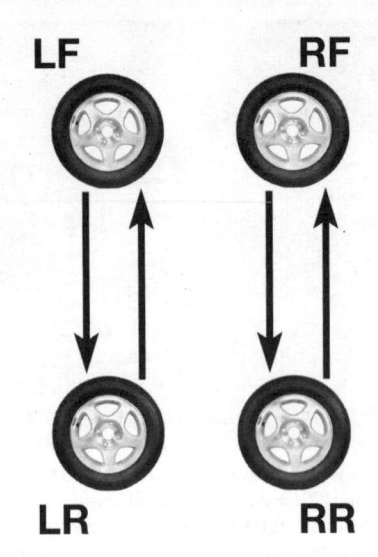

10.2 The recommended four-tire rotation pattern for these vehicles

9 Information on removing and installing the battery can be found in Chapter 5. If you disconnected the cable(s) from the negative and/or positive battery terminals, see Chapter 5. Information on jump starting can be found at the front of this manual. For more detailed battery checking procedures, refer to the *Haynes Automotive Electrical Manual*.

Cleaning

10 Corrosion on the hold-down components, battery case and surrounding areas can be removed with a solution of water and baking soda. Thoroughly rinse all cleaned areas with plain water.
11 Any metal parts of the vehicle damaged by corrosion should be covered with a zinc-based primer, then painted.

Charging

Warning: *When batteries are being charged, hydrogen gas, which is very explosive and flammable, is produced. Do not smoke or allow open flames near a charging or a recently charged battery. Wear eye protection when near the battery during charging. Also, make sure the charger is unplugged before connecting or disconnecting the battery from the charger.*

12 Slow-rate charging is the best way to restore a battery that's discharged to the point where it will not start the engine. It's also a good way to maintain the battery charge in a vehicle that's only driven a few miles between starts. Maintaining the battery charge is particularly important in the winter when the battery must work harder to start the engine and electrical accessories that drain the battery are in greater use.
13 It's best to use a one or two-amp battery charger (sometimes called a "trickle" charger). They are the safest and put the least strain on the battery. They are also the least expensive. For a faster charge, you can use a higher amperage charger, but don't use one rated more than 1/10th the amp/hour rating of the battery. Rapid boost charges that claim to restore the power of the battery in one to two hours are hardest on the battery and can damage batteries not in good condition. This type of charging should only be used in emergency situations.
14 The average time necessary to charge a battery should be listed in the instructions that come with the charger. As a general rule, a trickle charger will charge a battery in 12 to 16 hours.

10 Tire rotation (every 6000 miles [9600 km] or 6 months)

1 The tires should be rotated at the specified intervals and whenever uneven wear is noticed.
2 Refer to the accompanying **illustration** for the preferred tire rotation pattern.
3 Loosen the wheel bolts. Refer to the information in *Jacking and towing* at the front of this manual for the proper procedures to follow when raising the vehicle and changing a tire. If the brakes are to be checked, don't apply the parking brake as stated. Make sure the tires are blocked to prevent the vehicle from rolling as it's raised.
4 Preferably, the entire vehicle should be raised at the same time. This can be done on a hoist or by jacking up each corner, then lowering the vehicle onto jackstands placed under the frame rails. Always use four jackstands and make sure the vehicle is safely supported.
5 After rotation, check and adjust the tire pressures as necessary. Tighten the wheel bolts to the torque listed in this Chapter's Specifications.

11 Seat belt check (every 6000 miles [9600 km] or 6 months)

1 Check seat belts, buckles, latch plates and guide loops for obvious damage and signs of wear.
2 Where the seat belt receptacle bolts to the floor of the vehicle, check that the bolts are secure.
3 See if the seat belt reminder light comes on when the key is turned to the Run or Start position.

12 Underhood hose check and replacement (every 6000 miles [9600 km] or 6 months)

General

Caution: *Replacement of air conditioning hoses must be left to a dealer service department or air conditioning shop that has the equipment to depressurize the system safely and recover the refrigerant. Never remove air conditioning components or hoses until the system has been depressurized.*

1 High temperatures in the engine compartment can cause the deterioration of the rubber and plastic hoses used for engine, accessory and emission systems operation. Periodic inspection should be made for cracks, loose clamps, material hardening and leaks. Information specific to the cooling system hoses can be found in Section 13.
2 Some, but not all, hoses are secured to their fittings with clamps. Where clamps are used, check to be sure they haven't lost their tension, allowing the hose to leak. If clamps aren't used, make sure the hose has not expanded and/or hardened where it slips over the fitting, allowing it to leak.

Vacuum hoses

3 It's quite common for vacuum hoses, especially those in the emissions system, to be color-coded or identified by colored stripes molded into them. Various systems require hoses with different wall thickness, collapse resistance and temperature resistance. When replacing hoses, be sure the new ones are made of the same material.
4 Often the only effective way to check a hose is to remove it completely from the vehicle. If more than one hose is removed, be sure to label the hoses and fittings to ensure correct installation.
5 When checking vacuum hoses, be sure to include any plastic T-fittings in the check. Inspect the fittings for cracks and the hose where it fits over the fitting for distortion, which could cause leakage.
6 A small piece of vacuum hose (1/4-inch inside diameter) can be used as a stethoscope to detect vacuum leaks. Hold one end of the hose to your ear and probe around vacuum hoses and fittings, listening for the hissing sound characteristic of a vacuum leak.
Warning: *When probing with the vacuum hose stethoscope, be very careful not to come into contact with moving engine components such as the drivebelt, cooling fan, etc.*

Fuel hose

Warning: *There are certain precautions that must be taken when inspecting or servicing fuel system components. Work in a well-ventilated area and do not allow open flames (cigarettes, appliances, etc.) or bare light bulbs near the work area. Mop up any spills immediately and do not store fuel soaked rags where they could ignite. The fuel system is under high pressure, so if any fuel lines are to be disconnected, the pressure in the system must be relieved first (see Chapter 4 for more information).*

7 Check all rubber fuel lines for deterioration and chafing. Check especially for cracks in areas where the hose bends and just before fittings, such as where a hose attaches to the fuel filter.
8 High quality fuel line, made specifically for high-pressure fuel injection systems, must

Chapter 1 Tune-up and routine maintenance

Check for a chafed area that could fail prematurely.

Check for a soft area indicating the hose has deteriorated inside.

Overtightening the clamp on a hardened hose will damage the hose and cause a leak.

Check each hose for swelling and oil-soaked ends. Cracks and breaks can be located by squeezing the hose.

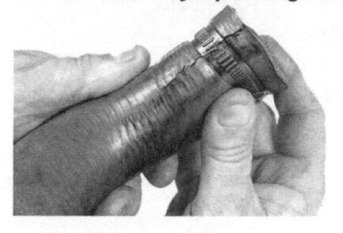

13.4 Hoses, like drivebelts, have a habit of failing at the worst possible time - to prevent the inconvenience of a blown radiator or heater hose, inspect them carefully as shown here

14.2 Inspect the muffler (A) and all hangers (B) for signs of deterioration

12 Check the metal brake lines where they enter the master cylinder and brake proportioning unit for cracks in the lines or loose fittings. Any sign of brake fluid leakage calls for an immediate and thorough inspection of the brake system.

13 Cooling system check (every 6000 miles [9600 km] or 6 months)

Warning: *Wait until the engine is completely cool before performing this procedure.*

1 Many major engine failures can be attributed to a faulty cooling system. If the vehicle is equipped with an automatic transmission, the cooling system also cools the transmission fluid and thus plays an important role in prolonging transmission life.
2 The cooling system should be checked with the engine cold. Do this before the vehicle is driven for the day or after it has been shut off for at least three hours.
3 Remove the cooling system pressure cap and thoroughly clean the cap, inside and out, with clean water. Also clean the opening in the expansion tank. All traces of corrosion should be removed. The coolant inside the expansion tank should be relatively transparent. If it is rust-colored, the system should be drained, flushed and refilled (see Section 26). If the coolant level is not up to the MIN mark, add additional antifreeze/coolant mixture (see Section 4).
4 Carefully check the large upper and lower radiator hoses along with the smaller diameter heater hoses that run from the engine to the firewall. Inspect each hose along its entire length, replacing any hose that is cracked, swollen or shows signs of deterioration. Cracks may become more apparent if the hose is squeezed **(see illustration)**.
5 Make sure all hose connections are tight. A leak in the cooling system will usually show up as white or rust-colored deposits on the areas adjoining the leak. If spring-type clamps are used at the ends of the hoses, it may be a good idea to replace them with more secure screw-type clamps.
6 Use compressed air or a soft brush to remove bugs, leaves, etc., from the front of the radiator or air conditioning condenser. Be careful not to damage the delicate cooling fins or cut yourself on them.
7 Every other inspection, or at the first indication of cooling system problems, have the cap and system pressure tested. If you don't have a pressure tester, most repair shops will do this for a minimal charge.

14 Exhaust system check (every 6000 miles [9600 km] or 6 months)

1 With the engine cold (at least three hours after the vehicle has been driven), check the complete exhaust system from the manifold to the end of the tailpipe. Be careful around the catalytic converter, which may be hot even after three hours. The inspection should be done with the vehicle on a hoist to permit unrestricted access. If a hoist isn't available, raise the vehicle and support it securely on jackstands.
2 Check the exhaust pipes and connections for signs of leakage and/or corrosion indicating a potential failure. Make sure that all brackets and hangers are in good condition and tight **(see illustration)**.
3 Inspect the underside of the body for holes, corrosion, open seams, etc. which may allow exhaust gasses to enter the passenger compartment. Seal all body openings with silicone sealant or body putty.
4 Rattles and other noises can often be traced to the exhaust system, especially the hangers, mounts and heat shields. Try to move the pipes, mufflers and catalytic converter. If the components can come in contact with the body or suspension parts, secure the exhaust system with new brackets and hangers.

be used for fuel line replacement. Never, under any circumstances, use unreinforced vacuum line, clear plastic tubing or water hose for fuel lines.
9 Spring-type clamps are commonly used on fuel lines. These clamps often lose their tension over a period of time, and can be sprung during removal. Replace all spring-type clamps with screw clamps whenever a hose is replaced.

Metal lines

10 Sections of metal line are routed along the frame, between the fuel tank and the engine. Check carefully to be sure the line has not been bent or crimped and that cracks have not started in the line.
11 If a section of metal fuel line must be replaced, only seamless steel tubing should be used, since copper and aluminum tubing don't have the strength necessary to withstand normal engine vibration.

1-20 Chapter 1 Tune-up and routine maintenance

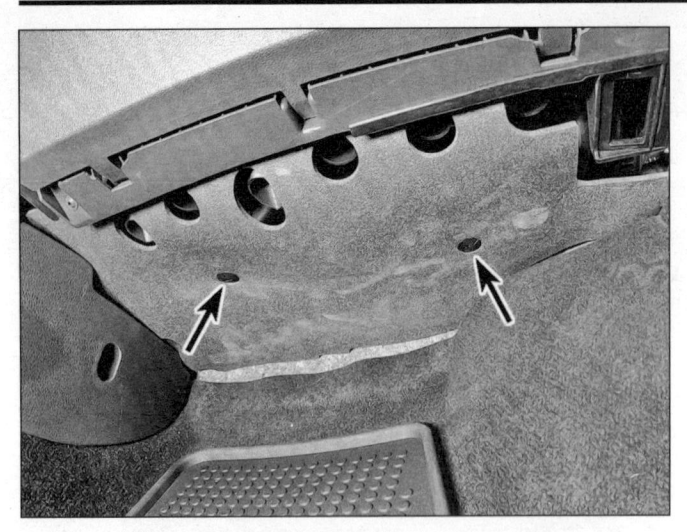

15.2 Remove the trim panel fasteners and panel

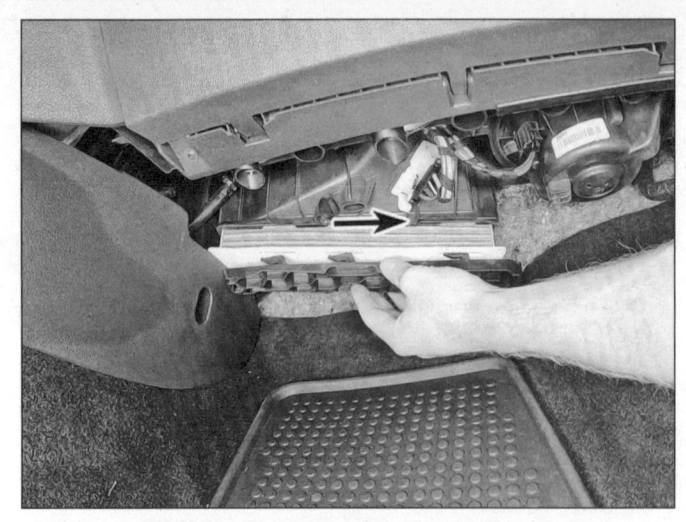

15.3 Slide the cover to the right until the locking tabs are free and remove the cover

15 Interior ventilation filter and fresh air intake filter replacement (every 15,000 miles [24,000 km] or 12 months)

Interior ventilation filter

Note: *Some models may be equipped with a dust and pollen filter and an activated charcoal insert.*

1 The interior ventilation filter is located inside the heater and evaporator core housing under the right (passenger's) side of the instrument panel.
2 Remove the trim panel fasteners and trim panel from under the glove box **(see illustration)**.
3 Remove the ventilation filter cover fastener and slide the cover to the right until the locking tabs are free, then pull the cover down and out of the housing **(see illustration)**.
4 Slide the filter out of the housing.
5 Install the filter with the slots in the filter on the correct side; four slots to the left and two slots to the right **(see illustration)**. If the filter is installed incorrectly the cover can't be installed.
6 Install the cover then push the cover towards the driver's side until it locks into place.
7 Installation is the reverse of the removal procedure.

Fresh air intake filter

8 Working in the engine compartment, remove the cowl panel (see Chapter 11).
9 Unclip the top of the fresh air intake cover and pull it outwards from the top to remove it.
10 Disconnect the fresh air sensor electrical connector to disengage the sensor retainer and remove the sensor, if equipped.
11 Remove the fresh air filter screen retaining nuts and remove the screen.

15.5 The filter has six slots that the cover tabs slide into. They must be oriented properly before sliding the filter into the housing

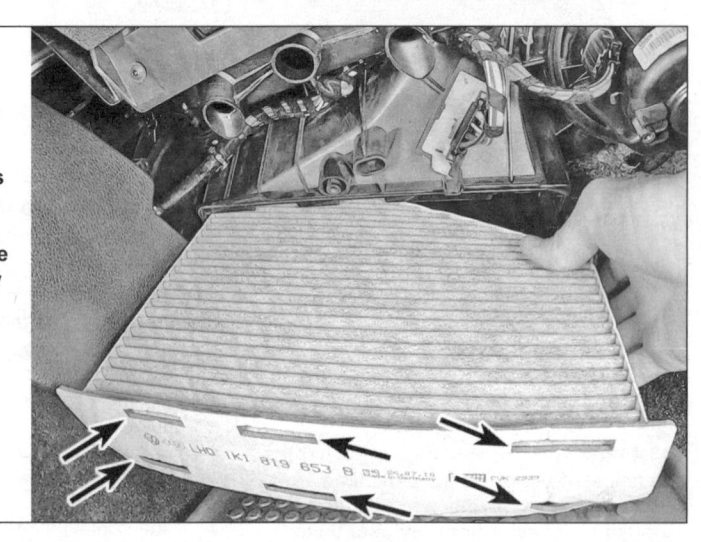

12 Disengage the fresh air intake filter clips. Starting from the bottom of the filter, rotate the filter up and out of the intake opening.
Note: *Be sure to install the filter in the same orientation it was in before it was removed.*
13 If the filter is dirty it can be vacuumed out, but if too much debris remains, the filter will have to be replaced.
14 Install the filter into the opening making sure the clips on the outer edges of the filter lock the filter into place.
15 Installation is the reverse of the removal procedure.

16 Brake system check (every 15,000 miles [24,000 km] or 12 months)

Warning: *The dust created by the brake system is harmful to your health. Never blow it out with compressed air and don't inhale any of it. An approved filtering mask should be worn when working on the brakes. Do not, under any circumstances, use petroleum-based solvents to clean brake parts. Use brake system cleaner only!*
Note: *For detailed photographs of the brake system, refer to Chapter 9.*

1 In addition to the specified intervals, the brakes should be inspected every time the wheels are removed or whenever a defect is suspected.
2 Any of the following symptoms could indicate a potential brake system defect: The vehicle pulls to one side when the brake pedal is depressed; the brakes make squealing or dragging noises when applied; brake pedal travel is excessive; the pedal pulsates; or brake fluid leaks, usually onto the inside of the tire or wheel. A brake pedal that sinks slowly to the floor, with no apparent external fluid leakage, indicates a faulty master cylinder.
Note: *A faulty master cylinder can leak fluid into the power brake booster.*
3 Loosen the wheel bolts.
4 Raise the vehicle and place it securely on jackstands.
5 Remove the wheels.

Chapter 1 Tune-up and routine maintenance   1-21

16.7 With the wheel off, check the thickness of the inner brake pad through the inspection hole (front brake shown, rear brake similar)

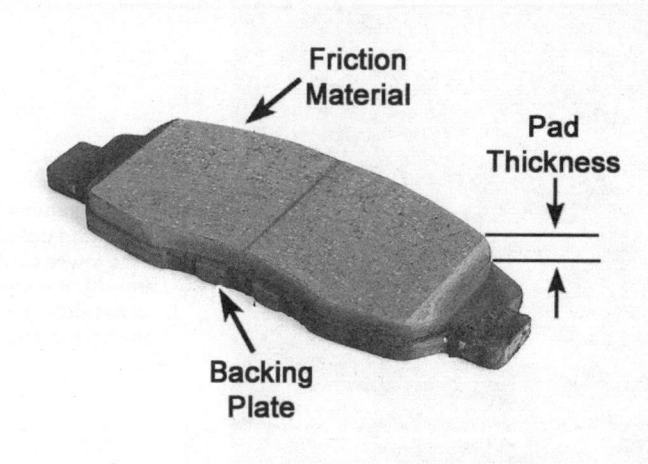

16.9 If a more precise measurement of pad thickness is necessary, remove the pads and measure the remaining friction material

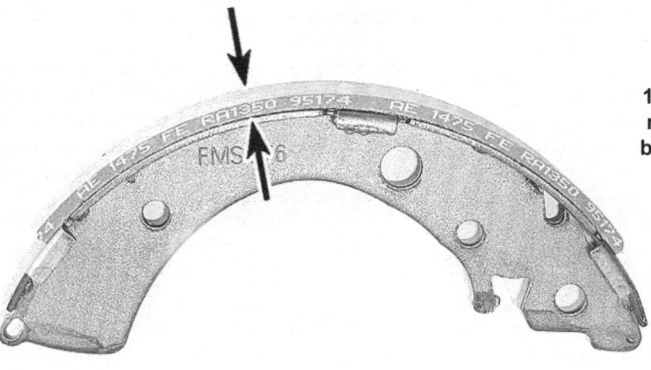

16.17 A more precise measurement of the brake shoe thickness can be made by measuring the shoes directly with the brake drum removed

Disc brakes (front and rear)

6 There are two pads (an outer and an inner) in each caliper. The pads are visible with the wheels removed.

7 Check the pad thickness by looking at each end of the caliper through the inspection window in the caliper body **(see illustration)**. If the lining material is less than the thickness listed in this Chapter's Specifications, replace the pads.

Note: *Keep in mind that the lining material is riveted or bonded to a metal backing plate and the metal portion is not included in this measurement.*

8 If it is difficult to determine the exact thickness of the remaining pad material by the above method, or if you are at all concerned about the condition of the pads, remove the caliper(s), then remove the pads from the calipers for further inspection (see Chapter 9).

9 Once the pads are removed from the calipers, clean them with brake cleaner and re-measure them with a ruler or a vernier caliper **(see illustration)**.

10 Measure the disc thickness with a micrometer to make sure that it still has service life remaining. If any disc is thinner than the specified minimum thickness, replace it (see Chapter 9). Even if the disc has service life remaining, check its condition. Look for scoring, gouging and burned spots. If these conditions exist, remove the disc and have it resurfaced (see Chapter 9).

11 Before installing the wheels, check all brake lines and hoses for damage, wear, deformation, cracks, corrosion, leakage, bends and twists, particularly in the vicinity of the rubber hoses at the calipers. Check the clamps for tightness and the connections for leakage. Make sure that all hoses and lines are clear of sharp edges, moving parts and the exhaust system. If any of the above conditions are noted, repair, reroute or replace the lines and/or fittings as necessary (see Chapter 9).

Drum brakes

12 Some models are equipped with rear drum brakes. Loosen the wheel bolts, raise the vehicle and support it securely on jackstands. Block the front tires to prevent the vehicle from rolling; however, don't apply the parking brake or it will lock the drums in place.

13 Remove the wheels, referring to *Jacking and towing* at the front of this manual if necessary.

14 Mark the drum and hub so it can be reinstalled in the same position. Use a scribe, chalk, etc., on the drum and hub.

15 Remove the brake drums. If the drum still won't come off, refer to Chapter 9.

16 With the drums removed, carefully clean the brake assembly with brake system cleaner.

Warning: *Don't blow the dust out with compressed air and don't inhale any of it (it is harmful to your health).*

17 Note the thickness of the lining material on both front and rear brake shoes **(see illustration)**. Compare the measurement with the limit given in this Chapter's Specifications; if any lining thickness is less than specified, then all of the brake shoes must be replaced (see Chapter 9). The shoes should also be replaced if they're cracked, glazed (shiny areas), or covered with brake fluid.

18 Make sure all the brake assembly springs are connected and in good condition.

19 Check the brake components for signs of fluid leakage. With your finger or a small screwdriver, carefully pry back the rubber cups on the wheel cylinder located at the top of the brake shoes. Any leakage here is an indication that the wheel cylinders should be replaced immediately (see Chapter 9). Also, check all hoses and connections for signs of leakage.

20 Wipe the inside of the drum with a clean rag and denatured alcohol or brake cleaner. Again, be careful not to breathe the dangerous brake dust.

21 Check the inside of the drum for cracks, score marks, deep scratches and hard spots which will appear as small discolored areas. If imperfections cannot be removed with fine emery cloth, the drum must be taken to an automotive machine shop for resurfacing.

22 Repeat the procedure for the remaining wheel. If the inspection reveals that all parts are in good condition, reinstall the brake drums, install the wheels and lower the vehicle to the ground.

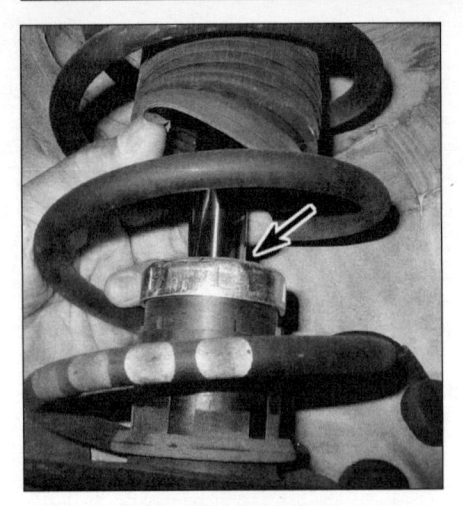

1-22 Chapter 1 Tune-up and routine maintenance

17.6 Check the shocks/struts for leakage at the indicated area

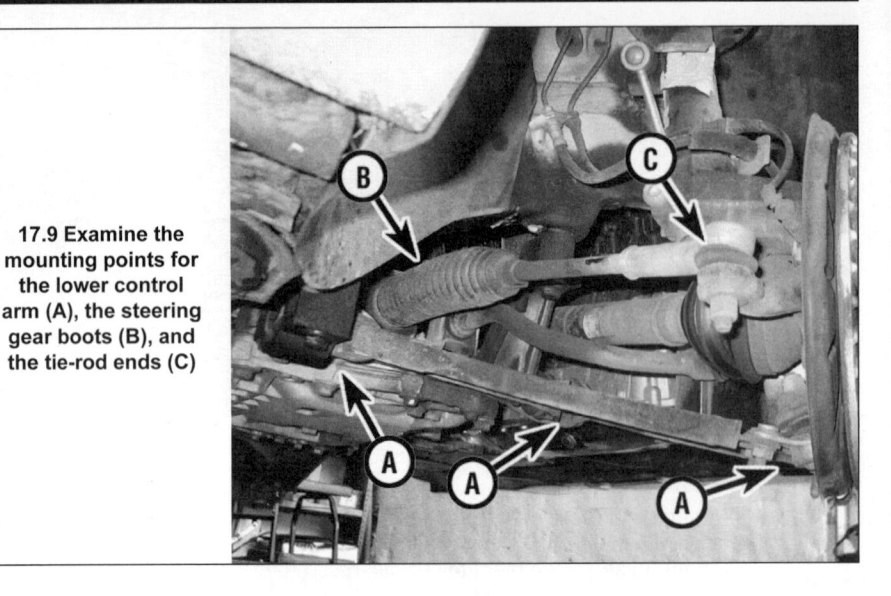

17.9 Examine the mounting points for the lower control arm (A), the steering gear boots (B), and the tie-rod ends (C)

Brake booster check

23 Sit in the driver's seat and perform the following sequence of tests.
24 With the brake fully depressed, start the engine - the pedal should move down a little when the engine starts.
25 With the engine running, depress the brake pedal several times - the travel distance should not change.
26 Depress the brake, stop the engine and hold the pedal in for about 30 seconds - the pedal should neither sink nor rise.
27 Restart the engine, run it for about a minute and turn it off. Then firmly depress the brake several times - the pedal travel should decrease with each application.
28 If your brakes do not operate as described, the brake booster has failed. Refer to Chapter 9 for the replacement procedure.

Parking brake

Note: *The rear brakes are self-adjusting, and do not require normal maintenance adjustments. Adjustment is only required after replacing brake discs, brake pads, brake calipers or parking brake cables.*
29 One method of checking the parking brake is to park the vehicle on a steep hill with the parking brake set and the transmission in Neutral (be sure to stay in the vehicle for this check!). If the parking brake cannot prevent the vehicle from rolling, it's in need of adjustment (see Chapter 9).

17 Suspension, steering and driveaxle boot check (every 15,000 miles [24,000 km] or 12 months)

Note: *The steering linkage and suspension components should be checked periodically. Worn or damaged suspension and steering linkage components can result in excessive and abnormal tire wear, poor ride quality and vehicle handling and reduced fuel economy. For more information on the steering and suspension components, refer to Chapter 10.*

Shock absorber check

1 Park the vehicle on level ground, turn the engine off and set the parking brake. Check the tire pressures.
2 Push down at one corner of the vehicle, then release it while noting the movement of the body. It should stop moving and come to rest in a level position within one or two bounces.
3 If the vehicle continues to move up-and-down or if it fails to return to its original position, a worn or weak shock absorber is probably the reason.
4 Repeat the above check at each of the three remaining corners of the vehicle.
5 Raise the vehicle and support it securely on jackstands.
6 Check the shock absorbers for evidence of fluid leakage (see illustration). A light film of fluid is no cause for concern. Make sure that any fluid noted is from the shocks and not from some other source. If leakage is noted, replace the shocks as a set.
7 Check the shocks to be sure that they are securely mounted and undamaged. Check the upper mounts for damage and wear. If damage or wear is noted, replace the shocks as a set (front or rear).
8 If the shocks must be replaced, refer to Chapter 10 for the procedure.

Steering and suspension check

9 Visually inspect the steering and suspension components (front and rear) for damage and distortion. Look for damaged seals, boots and bushings and leaks of any kind. Examine the bushings where the control arms meet the chassis (see illustration).
10 Clean the lower end of the steering knuckle. Have an assistant grasp the lower edge of the tire and move the wheel in-and-out while you look for movement at the steering knuckle-to-control arm balljoint. If there is any movement the suspension balljoint(s) must be replaced.
11 Grasp each front tire at the front and rear edges, push in at the front, pull out at the rear and feel for play in the steering system components (see illustration). If any freeplay is noted, check the steering gear mounts and the tie-rod ends for looseness.
12 Additional steering and suspension system information and illustrations can be found in Chapter 10.

Driveaxle boot check

13 The driveaxle boots are very important because they prevent dirt, water and foreign material from entering and damaging the Constant Velocity (CV) joints. Because it constantly pivots back and forth following the steering action of the front hub, the outer CV boot wears out sooner and should be inspected regularly.

17.11 With the steering wheel in the locked position and the vehicle raised, grasp the front tire as shown and try to move it back-and-forth - if any play is noted, check the steering gear mounts and tie-rod ends for looseness

Chapter 1 Tune-up and routine maintenance   1-23

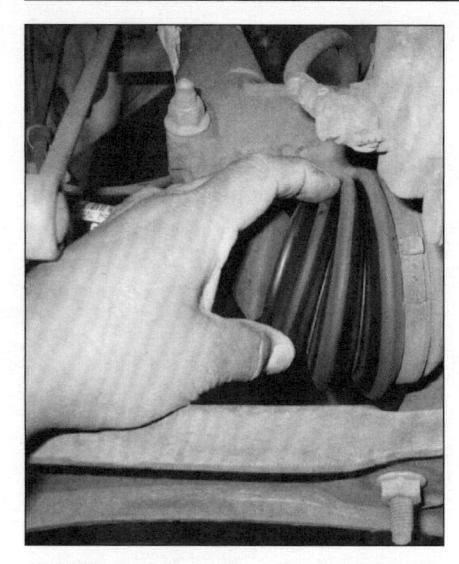

17.14 Inspect the inner and outer driveaxle boots for loose clamps, cracks or signs of leaking lubricant

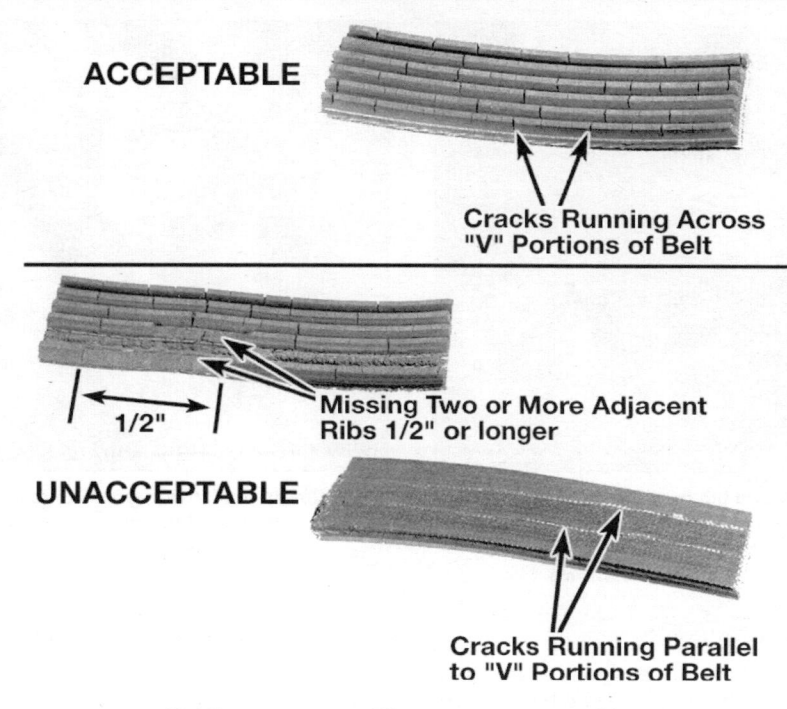

19.4 Here are some of the more common problems associated with drivebelts (check the belts very carefully to prevent an untimely breakdown)

14 Inspect the boots for tears and cracks as well as loose clamps **(see illustration)**. If there is any evidence of cracks or leaking lubricant, they must be replaced as described in Chapter 8.

18 Fuel system check (every 15,000 miles [24,000 km] or 12 months)

Warning: *Gasoline is flammable, so take extra precautions when you work on any part of the fuel system. Don't smoke or allow open flames or bare light bulbs near the work area, and don't work in a garage where a gas-type appliance (such as a water heater or clothes dryer) is present. Since fuel is carcinogenic, wear fuel-resistant gloves when there's a possibility of being exposed to fuel, and, if you spill any fuel on your skin, rinse it off immediately with soap and water. Mop up any spills immediately and do not store fuel-soaked rags where they could ignite. When you perform any kind of work on the fuel system, wear safety glasses and have a Class B type fire extinguisher on hand. The fuel system is under constant pressure, so, before any lines are disconnected, the fuel system pressure must be relieved (see Chapter 4).*

1 If you smell fuel while driving or after the vehicle has been sitting in the sun, inspect the fuel system immediately.

2 Remove the fuel filler cap and inspect it for damage and corrosion. The gasket should have an unbroken sealing imprint. If the gasket is damaged, install a new cap.

3 Inspect the fuel feed line for cracks. Make sure that the connections in the fuel line are free of leaks.

Warning: *Your vehicle is fuel injected, so you must relieve the fuel system pressure before servicing fuel system components. The fuel system pressure relief procedure is outlined in Chapter 4.*

4 Since some components of the fuel system - the fuel tank and feed lines, for example - are underneath the vehicle, they can be inspected more easily with the vehicle raised on a hoist. If that's not possible, raise the vehicle and support it on jackstands.

5 With the vehicle raised and safely supported, inspect the fuel tank and filler neck for punctures, cracks and other damage. The connection between the filler neck and the tank is particularly critical. Sometimes a rubber filler neck will leak because of loose clamps or deteriorated rubber. Inspect all fuel tank mounting brackets and straps to be sure that the tank is securely attached to the vehicle.

6 Carefully check all rubber hoses and metal lines leading away from the fuel tank. Check for loose connections, deteriorated hoses, crimped lines and other damage. Replace damaged sections as necessary (see Chapter 4).

19 Drivebelt check and replacement (every 15,000 miles [24,000 km] or 12 months)

1 The drivebelt(s) is/are located at the front of the engine and play(s) an important role in the overall operation of the vehicle and its components. Due to their function and material make-up, drivebelts are prone to failure after a period of time and should be inspected and adjusted periodically to prevent major engine damage.

2 Four-cylinder engines are equipped with a single self-adjusting serpentine drivebelt, which is used to drive all of the accessory components such as the alternator, power steering pump, water pump and air conditioning compressor. Five-cylinder engines use two serpentine drivebelts; the outer belt is used to drive the air conditioning compressor, and the inner belt (driven by the air conditioning compressor) is used to drive the alternator and water pump.

Inspection

3 With the engine off, open the hood and locate the drivebelt at the front of the engine. Using your fingers (and a flashlight, if necessary), move along the belt checking for cracks and separation of the belt plies. Also check for fraying and glazing, which gives the belt a shiny appearance. Both sides of the belt should be inspected, which means you will have to twist the belt to check the underside.

4 Check the ribs on the underside of the belt. They should all be the same depth, with none of the surface uneven **(see illustration)**.

5 The tension of the belt is automatically adjusted by the belt tensioner and does not require any adjustments.

19.7 Rotate the tensioner arm, then insert an Allen wrench or drill bit into the cast hole to hold it in place - 1.4L engine shown

19.8 Rotate the tensioner arm to relieve belt tension - typical

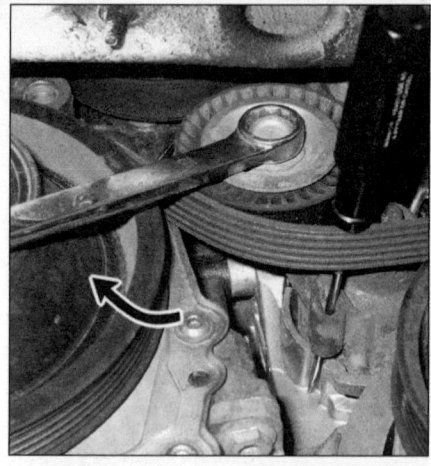

19.10 Rotate the tensioner clockwise and lock the tensioner into place

Replacement

6 Remove the under-vehicle splash shield fasteners then slide the shield back and out from under the vehicle **(see illustration 6.7a and 6.7b)**.

Four-cylinder engines

Note: For water pump belt replacement on 1.4L four-cylinder engines, see Chapter 3, Section 7.

7 On 1.8L and 2.0L timing chain-equipped engines and 1.4L timing belt engines, use a box-end wrench to rotate the tensioner clockwise, then insert a drill bit into the cast hole to hold the tensioner in place **(see illustration)**.

8 On 2.0L timing belt-equipped engines, rotate the tensioner clockwise using the cast square on the top of the tensioner to relieve the tension on the belt **(see illustration)**.

9 Route the new belt over the various pulleys - again rotating the tensioner to allow the belt to be installed on timing belt engines - then release the belt tensioner. On 1.8L and 2.0L timing chain models and 1.4L timing belt engines, rotate the tensioner enough to allow the drill bit to be removed, then slowly release the belt tensioner. Make sure the belt fits properly into the pulley grooves; it must be completely engaged.

Five-cylinder engines

Air conditioning compressor (outer) belt

10 Use a box-end wrench and rotate the tensioner clockwise, then insert a drill bit or screwdriver into the cast hole to hold the tensioner in place **(see illustration)**.

11 If the belt is to be re-used, mark the direction of rotation and remove the belt.

12 Route the new belt over the various pulleys, rotate the tensioner enough to allow the drill bit or screwdriver to be removed, then slowly release the belt tensioner. Make sure the belt fits properly into the pulley grooves - it must be completely engaged.

Alternator and water pump (inner) belt

13 Remove the air conditioning compressor outer belt (see Steps 10 and 11). Rotate the tensioner clockwise enough to remove the drill bit or screwdriver and slowly release the tensioner.

14 Using a box-end wrench or swivel wrench **(see illustration)**, rotate the tensioner clockwise and insert a drill bit or screwdriver into the cast hole to hold the tensioner in place.

15 If the belt is to be re-used, mark the direction of rotation and remove the belt.

16 Route the new belt over the various pulleys, rotate the tensioner enough to allow the drill bit or screwdriver to be removed, then slowly release the belt tensioner. Make sure the belt fits properly into the pulley grooves - it must be completely engaged.

17 The remainder of installation is the reverse of removal.

Tensioner replacement

18 Remove the drivebelt (four-cylinder engines, see Steps 7 and 8; five cylinder engines, see Steps 9 through 15).

2.0L Four-cylinder timing belt engines

19 Remove the fasteners that secure the tensioner to the engine, then remove the tensioner.

20 Installation is the reverse of removal. Tighten the mounting bolt(s) to the torque listed in this Chapter's Specifications.

1.8L and 2.0L Four-cylinder timing chain engines and 1.4L timing belt engines

21 Apply the parking brake and block the rear wheels. Loosen the right front wheel lug bolts, then raise and support the vehicle on jackstands. Remove the under-vehicle splash shield **(see illustration 6.7a and 6.7b)**.

22 Remove the right front wheel and front half of the inner fender liner (see Chapter 11).

23 Remove the tensioner mounting bolts and remove the tensioner from the front of the accessory bracket.

24 Installation is the reverse of removal. Tighten the mounting bolt(s) to the torque listed in this Chapter's Specifications.

Five-cylinder 2.5L engines

Note: On 2.5L models, two tensioners are used, one for the air conditioning belt and one for the alternator, power steering and water pump drive belt.

25 Rotate the tensioner pulley mounting bolt and remove the 12-point tensioner mounting bolt and tensioner.

Note: The air conditioning belt tensioner must be removed first to gain access to the alternator and water pump tensioner.

26 Installation is the reverse of removal. Tighten the mounting bolt(s) to the torque listed in this Chapter's Specifications.

19.14 Rotate the tensioner clockwise, then insert a drill bit to lock the tensioner body

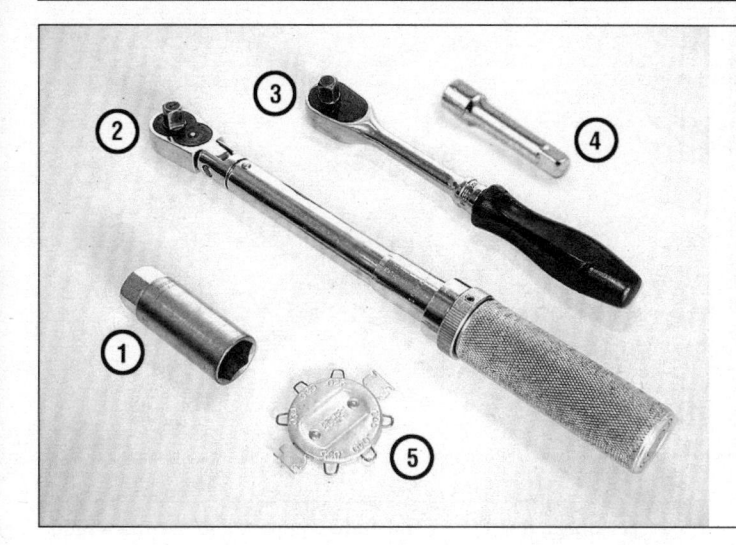

Chapter 1 Tune-up and routine maintenance   1-25

21.2 Tools required for changing spark plugs
1 **Spark plug socket** - This will have special padding inside to protect the spark plug's porcelain insulator
2 **Torque wrench** - Although not mandatory, using this tool is the best way to ensure the plugs are tightened properly
3 **Ratchet** - Standard hand tool to fit the spark plug socket
4 **Extension** - Depending on model and accessories, you may need special extensions and universal joints to reach one or more of the plugs
5 **Spark plug gap gauge** - This gauge for checking the gap comes in a variety of styles. Make sure the gap for your engine is included

20 Brake fluid change (every 30,000 miles [48,000 km] or 24 months)

Warning: *Brake fluid can harm your eyes and damage painted surfaces, so use extreme caution when handling or pouring it. Do not use brake fluid that has been standing open or is more than one year old. Brake fluid absorbs moisture from the air. Excess moisture can cause a dangerous loss of braking effectiveness.*

Note: *Used brake fluid is considered a hazardous waste and it must be disposed of in accordance with federal, state and local laws. DO NOT pour it down the sink, into septic tanks or storm drains, or on the ground.*

Note: *On models equipped with a manual transmission, the clutch master cylinder gets its fluid from the brake master cylinder reservoir. The clutch system fluid should be changed at the same time (see Chapter 8 for clutch bleeding).*

1   At the specified intervals, the brake fluid should be replaced. Since the brake fluid may drip or splash when pouring it, place plenty of rags around the master cylinder to protect any surrounding painted surfaces.
2   Before beginning work, purchase the specified brake fluid (see *Recommended lubricants and fluids* in this Chapter's Specifications).
3   Disconnect the fluid level sensor connector from the top of the master cylinder reservoir cap and unscrew the cap.
4   Using a hand-held suction pump or similar device, withdraw the fluid from the master cylinder reservoir.
5   Add new fluid to the master cylinder until the level of the fluid rises to the base of the filler neck. Install the cap.
6   Bleed the brake system as described in Chapter 9 at all four brakes until new and uncontaminated fluid is expelled from the bleeder screw. Be sure to maintain the fluid level in the master cylinder as you perform the bleeding process. If you allow the master cylinder to run dry, air will enter the system.

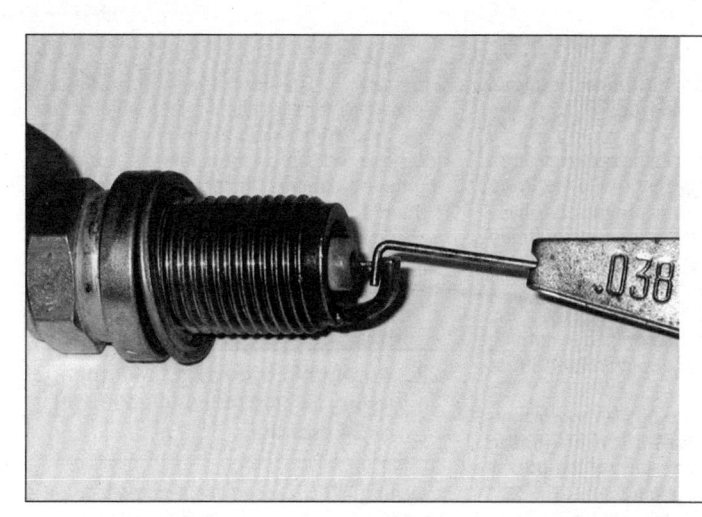

7   Refill the master cylinder with fluid, reconnect the fluid level sensor and check the operation of the brakes. The pedal should feel solid when depressed, with no sponginess.
Warning: *Do not operate the vehicle if you are in doubt about the effectiveness of the brake system.*

21 Spark plug check and replacement (every 30,000 miles [48,000 km] or 24 months)

1   The spark plugs are located in the cylinder head.
2   In most cases, the tools necessary for spark plug replacement include a spark plug socket which fits onto a ratchet (this special socket is padded inside to protect the porcelain insulators on the new plugs and hold them in place), various extensions and a feeler gauge to check and adjust the spark plug gap **(see illustration)**. Since these engines are equipped with an aluminum cylinder head, a torque wrench should be used when tightening the spark plugs.

21.5 When checking the spark plug gap, the wire should slide between the electrodes with a slight drag

3   The best approach when replacing the spark plugs is to purchase the new spark plugs beforehand, adjust them to the proper gap, then replace each plug one at a time. When buying the new spark plugs, be sure to obtain the correct plug for your specific engine. This information can be found in this Chapter's Specifications.
4   Allow the engine to cool completely before attempting to remove any of the plugs. During this cooling off time, each of the new spark plugs can be inspected for defects and the gaps can be checked.
5   The gap is checked by inserting the proper thickness gauge between the electrodes at the tip of the plug **(see illustration)**. The gap between the electrodes should be as listed in this Chapter's Specifications or in your owner's manual. Also check for cracks in the spark plug body (if any are found, the plug must not be used).
6   Cover the fender to prevent damage to the paint. Fender covers are available from auto parts stores but an old blanket will work just fine.
7   There is one centrally mounted spark plug per cylinder. The ignition coils are

1-26 Chapter 1 Tune-up and routine maintenance

21.9 Use a socket and extension to unscrew the spark plugs

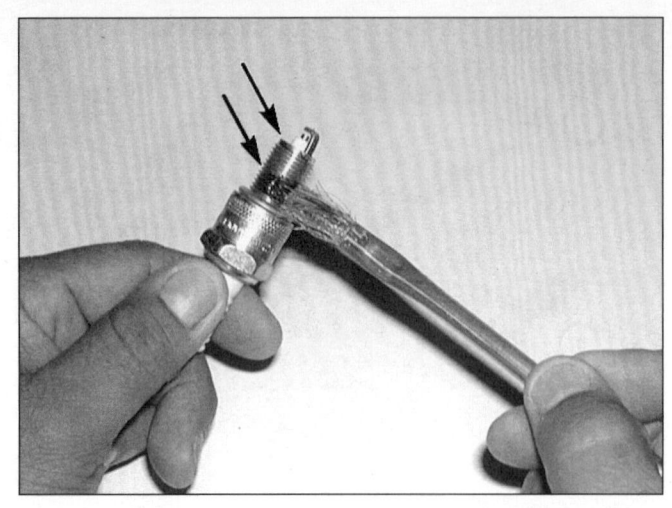

21.11 Apply a thin film of anti-seize compound to the spark plug threads to prevent damage to the cylinder head

mounted directly over the plugs. Remove each ignition coil from the spark plug (see Chapter 5).

8 If compressed air is available, use it to blow any dirt or foreign material away from the spark plug area to eliminate the possibility of material falling into the cylinder through the spark plug hole as the spark plug is removed. **Warning:** *Wear eye protection!*

9 Place the spark plug socket over the plug and remove it from the engine by turning it in a counterclockwise direction **(see illustration)**.

10 Compare the spark plug with the accompanying chart **(see illustration on inside rear cover)** to get an indication of the overall running condition of the engine.

11 It's a good idea to lightly coat the threads of the spark plugs with an anti-seize compound **(see illustration)** to insure that the spark plugs do not seize in the aluminum cylinder head.

12 It's often difficult to insert spark plugs into their holes without cross-threading them. To avoid this possibility, fit a piece of rubber hose over the end of the spark plug **(see illustration)**. The flexible hose acts as a universal joint to help align the plug with the plug hole. Should the plug begin to cross-thread, the hose will slip on the spark plug, preventing thread damage. Install the spark plug and tighten it to the torque listed in this Chapter's Specifications.

13 Repeat the procedure for the remaining spark plugs.

22 Automatic transaxle fluid change (every 30,000 miles [48,000 km] or 24 months)

Warning: *This procedure is potentially dangerous and is best left to a professional shop. The vehicle must be kept level while being safely raised high enough for access to the plugs on the transmission.*

Note: *When the vehicle is serviced by a dealer or other qualified repair shop, a scan tool is used to read the transaxle fluid temperature. To perform the job at home, you will have to estimate the temperature of the fluid.*

Note: *The engine must be running when checking the fluid level and when adding fluid.*

1 At the specified intervals, the transmission fluid should be drained and replaced. Since the fluid will remain hot long after driving, perform this procedure only after everything has cooled down completely.

2 Before beginning work, purchase the specified transmission fluid (see *Recommended lubricants and fluids* in this Chapter's Specifications).

3 Other tools necessary for this job include jackstands to support the vehicle in a raised position, a drain pan capable of holding several quarts, newspapers and clean rags.

4 Raise and support the vehicle on jackstands. Remove the under-vehicle splash shield **(see illustration 6.7a)**.

5 Place the drain pan underneath the transmission. Remove the check/fill plug **(see illustration)**, then unscrew the overflow tube through the check/fill plug opening and allow the fluid to drain.

6 Reinsert the overflow tube and tighten it to the torque listed in this Chapter's Specifications. Measure the amount of fluid drained

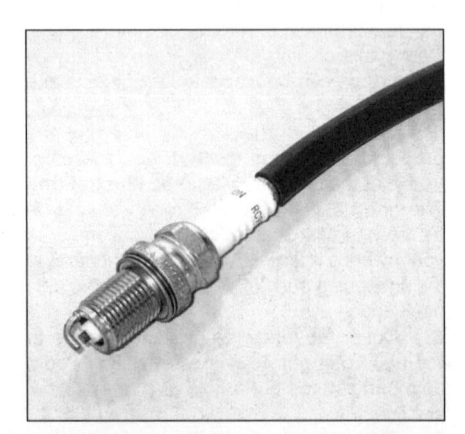

21.12 A length of snug-fitting rubber hose will save time and prevent damaged threads when installing the spark plugs

22.5 Automatic transaxle fluid level check/fill plug location

Chapter 1 Tune-up and routine maintenance   1-27

22.7 The nozzle of the tool being used to add fluid to the transaxle must pass through the window in the overflow tube

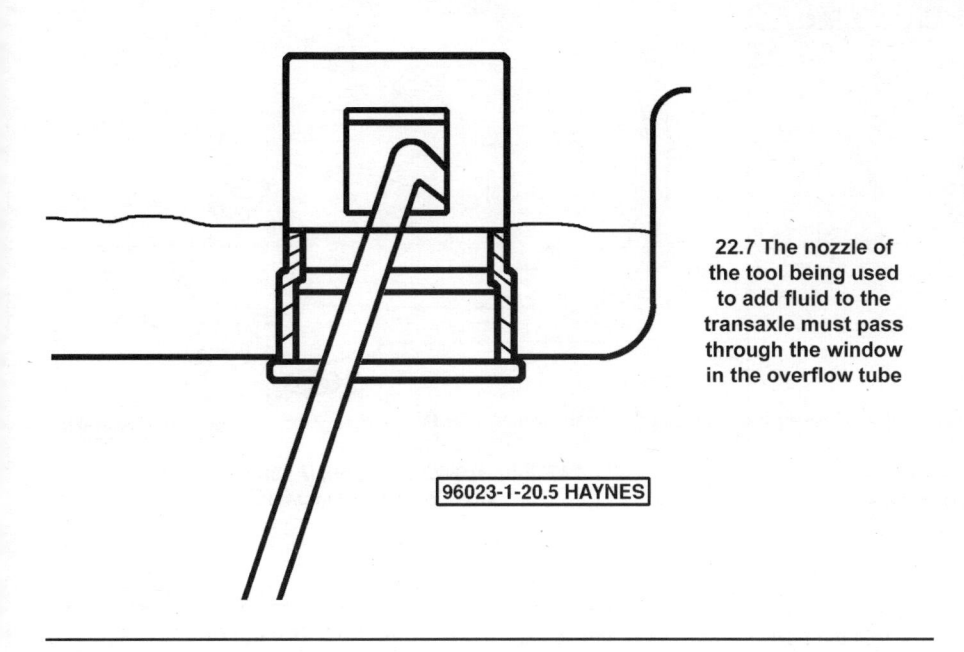

(the same amount will be added to the transmission later).

7   On these transaxles, the fluid is added through the check/fill plug hole. A special tool is available for this purpose, but a suction gun equipped with an angled nozzle can be used to pump the fluid up into the transaxle. The nozzle must pass through the opening in the overflow tube. Remove the check/fill plug **(see illustration 22.5)** and slowly add fluid through the check/fill hole until it runs out **(see illustration)**. Temporarily reinstall the check/fill plug.

8   With the shifter in Park and the parking brake set, run the engine at a fast idle.

9   Move the gear selector through each range, pausing for about two seconds in each range, then back to Park. Let the engine idle for a few minutes, then remove the check/fill plug. If fluid runs out of the hole, allow it to run out until it just drips. If no fluid runs out, add fluid through the hole until it does run out, allowing it to flow out until it just drips.

Note: *The ideal temperature of the fluid when adjusting the level should be 95 to 113-degrees F.*

10   Install a new seal on the check/fill plug, then reinsert the plug and tighten it to the torque listed in this Chapter's Specifications.

11   Reinstall the under-vehicle splash shield.

23   Direct Shift Gearbox (DSG) transaxle fluid and filter change (every 40,000 miles [64,000 km] or 36 months)

Warning: *This procedure is potentially dangerous and is best left to a professional shop. The vehicle must be kept level while being safely raised high enough for access to the plugs on the transmission.*

Caution: *If the battery is disconnected, several systems must be re-learned before they will work properly (see Chapter 5, Section 3).*

Note: *When the vehicle is serviced by a dealer or other qualified repair shop, a scan tool is used to read the transaxle fluid temperature. To perform the job at home, you will have to estimate the temperature of the fluid.*

Note: *A minimum of 5.5 qts of DSG fluid must be added prior to starting the engine, and the engine must be running when checking the fluid level or when adding any more fluid.*

1   At the specified intervals, the transmission fluid should be drained and replaced. Since the fluid will remain hot long after driving, perform this procedure only after everything has cooled down completely.

2   Before beginning work, purchase the specified transmission fluid (see *Recommended lubricants and fluids* in this Chapter's Specifications).

3   Other tools necessary for this job include jackstands to support the vehicle in a raised position, a drain pan capable of holding several quarts, newspapers and clean rags.

4   Raise the front and rear of the vehicle and support it securely on jackstands.

Note: *The vehicle must be level when checking or filling the transaxle, so the rear of the vehicle must be raised and supported as well.*

5   Remove the splash shield under the engine **(see illustration 6.7a)**.

6   Remove the engine cover/air filter housing (see Section 7).

7   Disconnect and remove the battery and battery tray (see Chapter 5).

8   Move a drain pan and rags under the transmission.

9   Using an open end wrench, loosen the filter housing from the top of the transaxle but do not remove the filter housing. Once the filter housing is loosened allow it to sit for a few minutes; this will allow the oil in the housing to drain back into the transaxle and prevent excess fluid from spilling.

10   Remove the filter housing, pull the filter element out from the housing and take off the O-ring.

11   Clean out the filter housing and install a new O-ring to the housing.

12   Coat the end of the new filter element with DSG oil and install the filter, with the locating arrow pointing down into the transaxle.

13   Coat the O-ring on the filter housing with DSG oil, install the filter housing over the filter element and on to the transaxle. Tighten the filter housing to the torque listed in this Chapter's Specifications.

14   Working under the vehicle, remove the transaxle shield mounting bolts and shield, if equipped.

15   Move the drain pan under the transaxle check/drain plug and remove the plug. Once the plug is removed, use an Allen wrench to unscrew and remove the overflow tube, allowing the fluid to drain out of the transaxle.

16   After the lubricant has drained completely, reinstall the overflow tube and tighten the tube to the torque listed in this Chapter's Specifications. Measure the amount of fluid drained (the same amount will be added later).

17   On these transaxles, the fluid is added through the check/fill plug hole. A special tool is available for this purpose, but a suction gun equipped with an angled nozzle can be used to pump the fluid up into the transaxle. The nozzle must pass through the opening in the overflow tube. Slowly add fluid through the check/fill hole until it runs out **(see illustration 22.7)**. Temporarily reinstall the check/drain plug.

Note: *Add at least 5.5 qts of DSG oil before starting the engine.*

18   With the transmission in Park and the parking brake set, start the engine.

19   Move the gear selector through each range, pausing for about three seconds in each range, then back to Park. Let the engine idle for a few minutes, then remove the check/drain plug. If fluid runs out of the hole, allow it to run out until it just drips. If no fluid runs out, add fluid through the hole until it does run out, allowing it to flow out until it just drips.

Note: *Every few seconds a small amount of oil will surge out due to the clutch operations - this does not mean there is too much oil in the transmission. Also, the ideal temperature of the fluid when adjusting the level should be 95 to 113-degrees F.*

20   Install a new seal on the check/drain plug, then reinsert the plug and tighten it to the torque listed in this Chapter's Specifications.

21   Reinstall the under-vehicle splash shield.

22   Lower the vehicle. The remainder of installation is the reverse of removal.

23   Drive the vehicle for a short distance, then check the check/drain plug and filter housing for leakage.

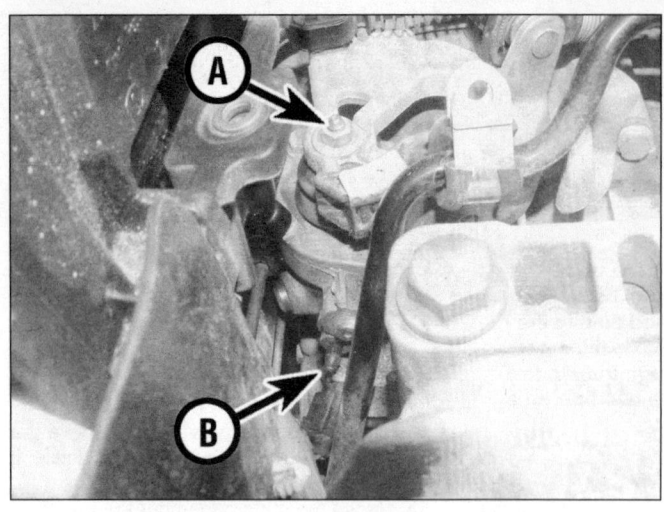

1-28   **Chapter 1 Tune-up and routine maintenance**

24.5 Five-speed selector shaft lock details

A   Selector shaft   B   Locking pin

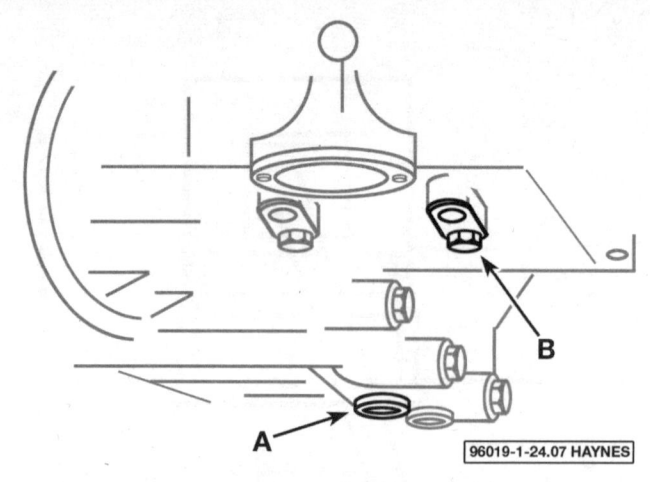

24.7 Five-speed drain plug (A) and selector shaft journal (B) locations

24 Manual transaxle lubricant change (every 60,000 miles [96,000 km] or 48 months)

5-speed models

Caution: *If the battery is disconnected, several systems must be re-learned before they will work properly (see Chapter 5, Section 3).*
Note: *The fill plug located on the side of the transaxle is used to fill the transaxle only while it is out of the vehicle. Due to the angle of the engine, once the transaxle is installed, the fluid level is above the fill plug.*

1   Raise the vehicle and support it securely on jackstands.
2   Remove the splash shield under the engine **(see illustration 6.7a)**.
3   Remove the engine cover/air filter housing (see Section 7).
4   Disconnect and remove the battery and battery tray (see Chapter 5).
5   Push down on the selector shaft. While holding the shaft down, rotate the locking lever clockwise. The selector shaft is now locked and should not be able to be moved **(see illustration)**.
6   Move a drain pan, rags, newspapers and wrenches under the transaxle.
7   Remove the drain plug, selector shaft journal fastener and journal from the bottom of the transaxle case **(see illustration)**, and allow the lubricant to drain into the pan.
8   After the lubricant has drained completely, reinstall the drain plug and selector shaft journal and tighten them securely.
Note: *Replace the O-ring on the selector shaft journal with a new one.*
9   Rotate the locking lever counterclockwise to unlock the selector shaft **(see illustration 24.5)**.
10   Disconnect and remove the back-up light switch at the top of the transaxle. Using a hand pump, syringe or squeeze bottle, fill the transaxle with the specified amount of lubri-

cant listed in this Chapter's Specifications. Reinstall the back-up light switch and tighten it securely.
11   Lower the vehicle.
12   Drive the vehicle for a short distance, then check the drain plug and selector shaft journal for leakage.

6-speed models

Warning: *This procedure is potentially dangerous and is best left to a professional shop. The vehicle must be kept level while being safely raised high enough for access to the plugs on the transaxle.*

13   This procedure should be performed after the vehicle has been driven so the lubricant will be warm and therefore will flow out of the transaxle more easily.
14   Raise the vehicle and support it securely on jackstands.
Note: *The vehicle must be level when checking or filling the transaxle, so the rear of the vehicle must be raised and supported as well.*
15   Remove the splash shield under the engine **(see illustration 6.7)**.
16   Move a drain pan, rags, newspapers and wrenches under the transaxle.
17   Remove the fill plug from the side of the transaxle case, then remove the transaxle drain plug at the bottom of the transaxle and allow the lubricant to drain into the pan.
Note: *Two types of drain and fill plugs are used, a six point (hex) plug or a twelve point (triple square) plug. Both have different tightening specifications.*
18   After the lubricant has drained completely, reinstall the drain plug and tighten it securely.
19   Using a hand pump, syringe or squeeze bottle, fill the transaxle with the specified lubricant until it just reaches the bottom edge of the hole. Reinstall the fill plug and tighten it securely.
20   Lower the vehicle.
21   Drive the vehicle for a short distance, then check the drain and fill plugs for leakage.

25 Fuel filter replacement (every 60,000 miles [96,000 km] or 48 months)

Warning: *Gasoline is extremely flammable, so take extra precautions when you work on any part of the fuel system. Don't smoke or allow open flames or bare light bulbs near the work area, and don't work in a garage where a gas-type appliance (such as a water heater or clothes dryer) is present. Since gasoline is carcinogenic, wear fuel-resistant gloves when there's a possibility of being exposed to fuel, and, if you spill any fuel on your skin, rinse it off immediately with soap and water. Mop up any spills immediately and do not store fuel-soaked rags where they could ignite. When you perform any kind of work on the fuel system, wear safety glasses and have a Class B type fire extinguisher on hand. The fuel system is under pressure, so if any lines must be disconnected, the pressure in the system must be relieved first (see Chapter 4 for more information).*
Caution: *If the battery is disconnected, several systems must be re-learned before they will work properly (see Chapter 5, Section 3).*
Caution: *On five-cylinder engines, the fuel system must be bled once the lines have been opened (see Chapter 4). Damage to the catalytic converter may result if the system is not bled.*

1   The fuel filter is located under the rear of the vehicle, ahead of the right-rear wheel, adjacent to the fuel tank **(see illustration)**.
2   The manufacturer does not give a replacement interval, but our suggested interval is based on experience with many other vehicles; replacing it is like inexpensive insurance and may prevent an untimely breakdown.
3   Depressurize the fuel system (see Chapter 4), then disconnect the cable from the negative terminal of the battery. Raise the vehicle and support it securely on jackstands.

Chapter 1 Tune-up and routine maintenance

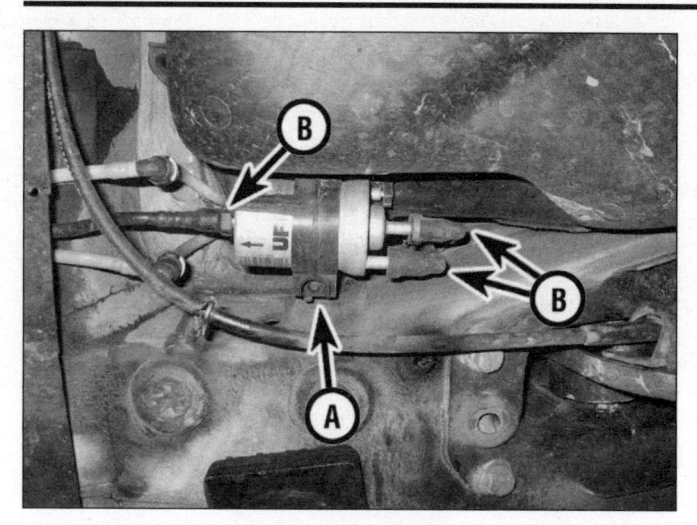

25.1 Fuel filter details

A Clamp screw
B Fuel line fittings

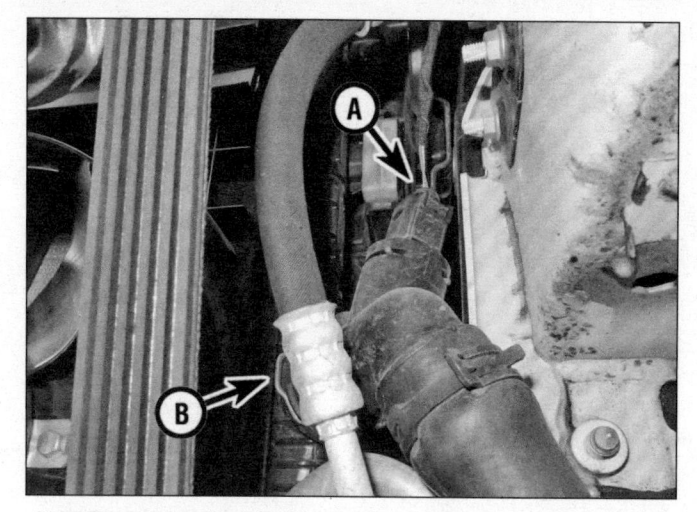

26.6a Disconnect the electrical connector from the engine temperature sensor (A) then remove the radiator hose retaining clip (B)

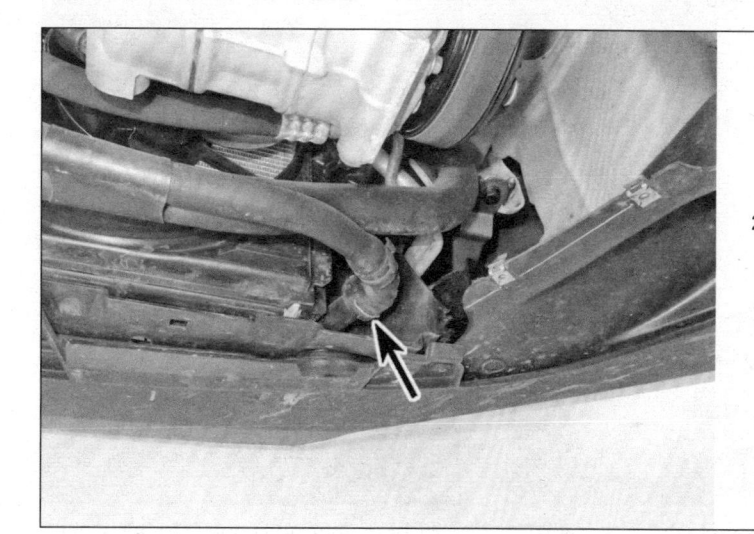

26.6b Remove the clip for the coupler at the charge air cooler

4   Use compressed air or brake system cleaner to clean any dirt surrounding the fuel inlet and outlet line fittings.
Note: *On some models, it is necessary to remove a plastic splash shield for access.*
5   Disconnect the fuel lines from the filter; refer to Chapter 4 for information on how to disconnect quick-connect fittings.
Note: *Have rags ready to catch or wipe up gasoline that will spill from the filter.*
6   Remove the clamp screw and detach the filter.
7   Remove the retaining clip and remove the pressure regulator from the filter body.
8   Replace the gasket and O-ring to the pressure regulator, install the regulator to the filter body and insert the retaining clip.
9   Installation is the reverse of removal. Make sure the arrow on the side of the new filter is pointing toward the engine side of the fuel system, and check for leaks after running the vehicle.

26 Cooling system servicing (draining, flushing and refilling)

Warning: *Do not allow antifreeze to come in contact with your skin or painted surfaces of the vehicle. Rinse off spills immediately with plenty of water. Antifreeze is highly toxic if ingested. Never leave antifreeze lying around in an open container or in puddles on the floor; children and pets are attracted by its sweet smell and may drink it. Check with local authorities about disposing of used antifreeze. Many communities have collection centers which will see that antifreeze is disposed of safely.*
Warning: *Wait until the engine is completely cool before beginning this procedure.*
1   Periodically, the cooling system should be drained, flushed and refilled to replenish the antifreeze mixture and prevent formation of rust and corrosion, which can impair the performance of the cooling system and cause engine damage. When the cooling system is serviced, all hoses and the expansion tank cap should be checked and replaced if necessary.

Draining

2   Apply the parking brake and block the wheels. If the vehicle has just been driven, wait several hours to allow the engine to cool down before beginning this procedure.
3   Raise and support the vehicle on jackstands. Remove the under-vehicle splash shield **(see illustration 6.7a and 6.7b)**.
4   Once the engine is completely cool, remove the expansion tank cap.
5   Move a large container under the radiator hose to catch the coolant.
6   **Four-cylinder engines:** Disconnect the electrical connector from the engine temperature sensor, then remove the radiator hose retaining clip and hose **(see illustration)**. Allow the coolant to drain. From under the vehicle remove the clip for the coupler at the charge air cooler **(see illustration)**, then remove the hose, allowing the remaining coolant to drain.
7   **Five-cylinder engines:** Loosen and remove the hose clamp from the lower radiator hose and allow the coolant to drain.
8   While the coolant is draining, check the condition of the radiator hoses, heater hoses and clamps (refer to Section 9 if necessary). Replace any damaged clamps or hoses.

Flushing

9   Reconnect the lower radiator hoses. Fill the cooling system with clean water, following the *Refilling* procedure (see Steps 15 through 26).
10 Start the engine and allow it to reach normal operating temperature, then rev up the engine a few times.
11 Turn the engine off and allow it to cool completely, then drain the system as described earlier.
12 Repeat Steps 9 through 11 until the water being drained is free of contaminants.
13 In severe cases of contamination or clogging of the radiator, remove the radiator (see Chapter 3) and have a radiator repair facility clean and repair it if necessary.

1-30 Chapter 1 Tune-up and routine maintenance

26.15 Add coolant until the level reaches the MAX mark on the expansion tank

26.16a Pry up the clip and disconnect the upper radiator hose. . .

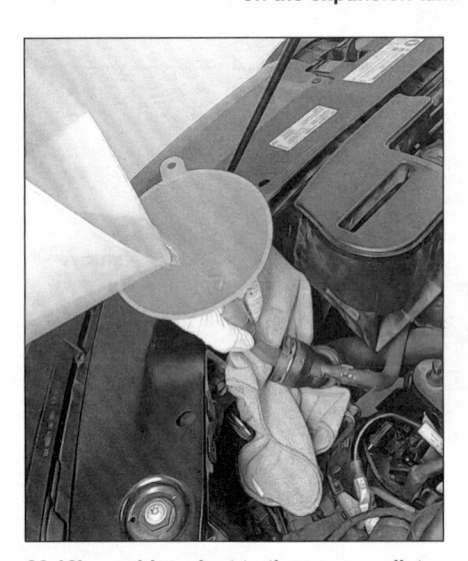

26.16b. . .add coolant to the upper radiator hose until coolant emerges. . .

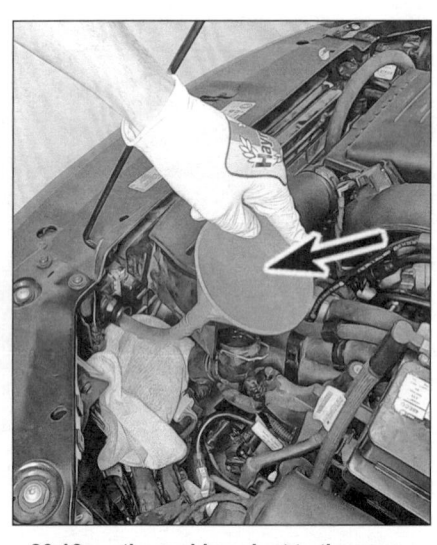

26.16c . .then add coolant to the upper radiator port until coolant emerges.

26.17 Squeeze the radiator hoses to purge the cooling system of air

14 Many deposits can be removed by the chemical action of a cleaner available at auto parts stores. Follow the procedure outlined in the manufacturer's instructions.
Note: *When the coolant is regularly drained and the system refilled with the correct antifreeze/water mixture, there should be no need to use chemical cleaners or descalers.*

Refilling

Note: *Always refill the cooling system with a 50/50 percent coolant and water mixture.*
15 Add coolant to the expansion tank until the level reaches the MAX mark. Keep adding coolant until the level stabilizes **(see illustration)**. Loosely install the cap.
16 Disconnect the upper radiator hose from the radiator. Using a funnel with a hose attached, add coolant to the radiator hose until coolant emerges. Then add coolant to the radiator port until coolant emerges **(see illustrations)**. Use shop rags to catch the coolant spillage. Reconnect the upper radiator hose.
17 Squeeze the radiator hoses several times to initially purge the system of air that has been collected **(see illustration)**.
18 Start the engine and let it idle for a minute.
19 Place the heater temperature control in the maximum heat position, with the fan speed on high and the A/C turned off.
20 Run the engine in a well-ventilated area at 2000 rpm for three minutes.
21 Turn off the engine and let it cool down a bit. Check the coolant level in the expansion tank, filling it as necessary between the MIN and MAX marks. Install the expansion tank cap securely.
22 Start the engine and allow the engine to idle for several minutes until both radiator hoses are warm (indicating that the thermostat has opened) **(see illustration 26.17)**. Keep an eye on the coolant level during this process - if at any time it drops too far below the MIN mark, turn off the engine, allow it to cool, then add more coolant as necessary.
23 Raise the engine speed to 3000 rpm for one minute.
24 Turn the engine off and let it cool. Check the coolant level; add more coolant mixture, if necessary, to bring it between the MIN and MAX marks on the expansion tank. The level should be between these marks on the tank when the engine is cold. Install the expansion tank cap.
25 Repeat this process of adding coolant, running the engine, turning it off and letting it cool, then checking the level until it remains stable in the expansion tank.
26 Start the engine, allow it to reach normal operating temperature and check for leaks. Also, set the heater and blower controls to the maximum setting (with the A/C button off) and check to see that the heater output from the air ducts is warm. This is a good indication that all air has been purged from the cooling system.

Chapter 1 Tune-up and routine maintenance

27 Transfer case (bevel box), AWD coupler (Haldex) and rear differential (AWD models) - check and lubricant replacement

1 Raise the front and rear of the vehicle, making sure the vehicle is level and support it securely on jackstands.

Transfer case (bevel box)
Fluid check
2 With the fluid warm - 68 to 104-degrees F (20 to 40-degrees C) - remove the check/fill plug towards the top side of the transfer case (bevel box).
3 The fluid level should be just at the bottom of the check/fill plug hole. If the level is not correct, add the specified clutch fluid (see this Chapter's Specifications) until the fluid is even with the bottom of the plug opening.
4 Install a new check/fill plug and tighten it to the torque listed in this Chapter's Specifications.

Fluid replacement
5 Place a drain pan under the transfer case (bevel box) drain plug.
6 Place a piece of cardboard, or plastic wrap over the subframe to prevent the clutch oil from draining onto the subframe and directing the oil to the drain pan.
7 Remove the transfer case (bevel box) drain plug and allow the clutch oil to drain into the drain pan.
8 Once the fluid has completely drained, install the new drain plug and tighten it to the torque listed in this Chapter's Specifications.
9 Clean the area around the transfer case (bevel box) check/fill plug, then remove the plug from the upper side of the transfer case (bevel box).
10 Using a hand pump and hose, insert the hose into the check/fill plug opening and fill the transfer case (bevel box) with the appropriate clutch fluid listed in this Chapter's Specifications until the fluid starts to drip out of the hole. Reinstall the check/fill plug and tighten it securely.
11 Check the fluid level (see Steps 2 through 4).

AWD coupler (Haldex)
Fluid check
12 Place a drain pan under the AWD coupler.
13 With the fluid warm - 68 to 104-degrees F (20 to 40-degrees C) - remove the check/fill plug (5 mm Allen head) from the AWD coupler (Haldex) housing.
14 The fluid level should be even with the bottom edge of the check fill/plug hole or slightly dripping out. If the level is not correct, add the specified fluid (see this Chapter's Specifications) until the fluid is even with the bottom of the plug opening.
15 Install the check/fill plug and tighten it to the torque listed in this Chapter's Specifications.
16 Lower the vehicle then drive the vehicle until the fluid is warm. Raise the front and rear of the vehicle, making sure the vehicle is level, and support it securely on jackstands. With the fluid warm, remove the check fill plug (5 mm Allen head) from the AWD coupler (Haldex) housing and let any excess fluid drain out.
17 Install the check/fill plug and tighten it to the torque listed in this Chapter's Specifications.

Fluid replacement
18 With the fluid warm, remove the check/fill plug (5 mm Allen head) from the AWD coupler (Haldex) housing.
19 Place a drain pan under the AWD coupler drain plug (8 mm allen head) and remove the drain plug.
Note: *The AWD coupler drain plug is located towards the front lower corner of the AWD coupler to the right of the Haldex control unit.*
20 Allow the fluid to drain, then reinstall the drain plug and tighten it to the torque listed in this Chapter's Specifications.
21 Using a fluid pump, insert the pump hose into the fill plug opening and fill the AWD coupler with the specified fluid listed this Chapter's Specifications, until the fluid starts to pour out of the check/fill plug opening. Reinstall the check/fill plug and tighten it to the torque listed in this Chapter's Specifications.
22 Check the fluid level (see Steps 12 through 17).

Rear differential (AWD models)
23 Place a drain pan under the rear differential.

Fluid check
24 Locate the oil filler/check on plug the driver's side of the rear differential, at approximately the 3 o'clock position.
25 Remove the oil filler/check plug from the rear differential.
26 The fluid level should be even with the bottom edge of the oil filler/check plug hole. If the level is not correct add the specified fluid (see this Chapter's Specifications) until the fluid is even with the bottom of the plug opening.
27 Install the oil filler/check plug and tighten it to the torque listed in this Chapter's Specifications.

Fluid replacement
28 Remove the oil filler/check plug.
29 Remove the drain plug and drain the lubricant.
Note: *The drain plug is located on the driver's side of the rear differential, at approximately the 6 o'clock position.*
30 Reinstall the drain plug and tighten it to the torque listed in this Chapter's Specifications.
31 Add new lubricant until it is up to the proper level (see Step 26). See this Chapter's Specifications for the specified lubricant type.
32 Reinstall the oil filler/check plug and tighten it to the torque listed in this Chapter's Specifications.

Notes

Chapter 2 Part A
Four-cylinder engines

Contents

	Section		Section
Balance shafts - removal and installation	16	Oil pan(s) - removal and installation	12
Camshaft housing, camshafts, sprockets, roller rocker arms and lash adjusters - removal, inspection and installation	8	Oil pump - removal and installation	13
		Oil separator and crankcase ventilator - removal and installation	17
Crankshaft front oil seal and housing - replacement	11	Rear main oil seal - replacement	15
Crankshaft pulley - removal and installation	10	Repair operations possible with the engine in the vehicle	2
Cylinder head - removal and installation	9	Timing belt and sprockets - removal, inspection and installation	5
Engine mounts - check and replacement	18		
Engine oil cooler - removal and installation	19	Timing chain cover, timing chains and tensioners - removal and installation	6
Flywheel/driveplate - removal and installation	14		
General Information	1	Top Dead Center (TDC) for number one piston - locating	3
Intake manifold - removal and installation	7	Valve cover - removal and installation	4

Specifications

General

Timing belt engine designations
- 1.4L models.. CZTA, DGXA
- 2.0L models.. CBPA

Timing chain engine designations
- 1.8L models.. CPKA, CPRA, CXBA, CXBB, CNSA, CNSB
- 2.0L models.. CCTA, CBFA, CXCA, CXCB, CPLA, CPPA, CNTA, CYFB, CKFA, DJJA, DRLA

Firing order 1-3-4-2

Driveplate

Driveplate installed height
- Without intermediate plate .... 0.76 to 0.83 inch (19.5 to 21.1 mm)
- With intermediate plate .... 0.74 to 0.80 inch (18.8 to 20.4 mm)

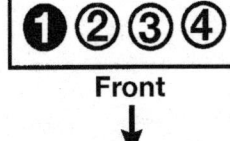

Camshafts

Endplay (maximum)
- Timing belt models
  - 2.0L (CBPA) engine .... 0.008 inch (0.20 mm)
  - 1.4L (CZTA, DGXA) engine .... 0.009 inch (0.25 mm)
- Timing chain models .... Not available

Cylinder locations

① ② ③ ④ Front ↓

Chapter 2 Part A Four-cylinder engines

Torque specifications — Ft-lbs (unless otherwise specified) — Nm

Note: One foot-pound (ft-lb) of torque is equivalent to 12 inch-pounds (in-lbs) of torque. Torque values below approximately 15 ft-lbs are expressed in inch-pounds, because most foot-pound torque wrenches are not accurate at these smaller values.

Timing belt engines

Fastener	Ft-lbs	Nm
Camshaft adjuster sprocket bolt 1.4L (CZTA, DGXA)* models		
Step 1	15	20
Step 2	37	50
Step 3	Tighten an additional 135-degrees	
Intake camshaft adjuster plug 1.4L (CZTA, DGXA) models	15	20
Exhaust camshaft adjuster sprocket bolt cover fasteners 1.4L (CZTA, DGXA)* models		
Step 1	71 in-lbs	8
Step 2	Tighten an additional 45-degrees	
Camshaft housing bolts - 1.4L (CZTA, DGXA) models (in sequence - see illustration 8.93)		
Step 1	88 in-lbs	10
Step 2	Tighten an additional 180-degrees	
Camshaft cap nuts 2.0L (CBPA) models (in sequence - see illustration 8.101)	15	20
Camshaft sprocket bolt 2.0L (CBPA) models	74	100
Crankshaft pulley bolts (CBPA) models*	35	25
Crankshaft sprocket bolt*		
1.4L (CZTA, DGXA) models		
Step 1	110.5	150
Step 2	Tighten an additional 180-degrees	
Step 1	66	90
Step 2	Tighten an additional 90-degrees	
Crankshaft seal retainer flange bolts		
Front	132 in-lbs	15
Rear		
All except 1.4L (CZTA, DGXA) models	132 in-lbs	15
1.4L (CZTA, DGXA) models, in sequence (see illustration 15.29)		
Step 1	Hand-tight	
Step 2	89 in-lbs	10
Cylinder head bolts*		
1.4L (CZTA, DGXA) engines (in sequence - see illustration 9.41a or 9.41b)		
Step 1	29.5	40
Step 2	Tighten an additional 90-degrees	
Step 3	Tighten an additional 90-degrees	
Step 4	Tighten an additional 90-degrees	
2.0L (CBPA) engines (in sequence - see illustration 9.41c)		
Step 1	29.5	40
Step 2	Tighten an additional 90-degrees	
Step 3	Tighten an additional 90-degrees	
Exhaust manifold/turbocharger nuts	15	20
Intake manifold		
1.4L (CZTA, DGXA) models	71 in-lbs	8
2.0L (CBPA) models	17	23
Flywheel/driveplate bolts		
Step 1	44	60
Step 2	Tighten an additional 90 degrees	
Engine mount (passenger's side)*		
Bracket-to-mount and body bolts		
Step 1	15	20
Step 2	Tighten an additional 90-degrees	
Engine mount-to-body bolts		
Step 1	29.5	40
Step 2	Tighten an additional 90-degrees	
Engine mount-to-engine bracket bolts		
Step 1	44	60
Step 2	Tighten an additional 90-degrees	
Transaxle mount		
Mount-to-transaxle bracket bolts* (driver's side)		
Step 1	29.5	40
Step 2	Tighten an additional 90-degrees	
Mount-to-body bracket bolts* (driver's side)		
Step 1	44	60
Step 2	Tighten an additional 90-degrees	

Chapter 2 Part A Four-cylinder engines

Torque specifications (continued)

	Ft-lbs (unless otherwise specified)	Nm
Pendulum mount		
Mount-to-transaxle bolts*		
Step 1		
CBPA models	29.5	40
CZTA, DGXA models	37	50
Step 2	Tighten an additional 90-degrees	
Mount-to-subframe bolt*		
Step 1		
CBPA models	74	100
CZTA, DGXA	96	130
Step 2	Tighten an additional 90-degrees	
Oil cooler, 1.4L (CZTA, DGXA) engines		
Step 1	71 in-lbs	8
Step 2	Tighten an additional 90 degrees	
Lower oil pan bolts*		
1.4L (CZTA, DGXA) models, in sequence **(see illustration 12.21a)**		
Step 1	Hand-tight	
Step 2	108 in-lbs	12
2.0L (CBPA) models	16.5	22
Upper oil pan bolts		
2.0L (CBPA) models		
Step 1, pan-to-engine block bolts	Hand-tight	
Step 2, transaxle-to-pan bolts	Hand-tight	
Step 3, pan-to-engine block bolts	Hand-tighten again	
Step 4, transaxle-to-pan bolts	33	45
Step 5, pan-to-engine block bolts	132 in-lbs	15
1.4L (CZTA, DGXA) models, in sequence **(see illustration 12.51a)**.		
Step 1, bolts 1 through 19	Hand-tight	
Step 2, transaxle-to-upper pan bolts	Hand-tight	
Step 3, pan-to-engine block bolts	71 in-lbs	8
Step 4, pan-to-engine block bolts	Tighten an additional 90-degrees	
Step 5, transaxle-to-upper pan bolts	29.5	40
Engine oil cooler fasteners		
1.4L (CZTA, DGXA) models		
Step 1	71 in-lbs	8
Step 2	Tighten an additional 90-degrees	
TDC bolt, 1.4L (CZTA, DGXA) engines	22	30
Timing belt cover bolts (upper and lower)		
1.4L (CZTA, DGXA) models	71 in-lbs	8
2.0L (CBPA) models	88 in-lbs	10
Timing belt tensioner bolt (CZTA, DGXA) models	18.5	25
Timing belt tensioner nut (CBPA) models	15	20
Idler roller bolt (CZTA, DGXA) models	18.5	45
Valve cover bolts	88 in-lbs	10
Oil pump pick-up tube fasteners		
1.4L (CZTA, DGXA)		
Step 1	47 in-lbs	5
Step 2	Tighten an additional 90-degrees	
2.0L (CBPA)	89 in-lbs	10
Oil pump sprocket bolt		
2.0L (CBPA)		
Step 1	15	20
Step 2	Tighten an additional 90-degrees	
Oil pump chain tensioner bolt - 2.0L (CBPA) models	132 in-lbs	15
Upper oil pan mounting fasteners		
1.4L (CZTA, DGXA) - (in sequence, **see illustration 12.51a**)		
Step 1, bolts 1 through 19	Tighten hand-tight	
Step 2, transaxle-to-oil pan bolts	Tighten hand-tight	
Step 3, bolts 1 through 19	72 in-lbs	8
Step 4, bolts 1 through 19	Tighten an additional 90-degrees	
Step 5, transaxle-to-oil pan bolts	See Chapter 7A or Chapter 7B	
2.0L (CBPA)	132 in-lbs	15
1.4L (CZTA, DGXA)	106 in-lbs	12
2.0L (CBPA)	16.5	22

* Replace with new bolt(s)

Torque specifications (continued)

Note: *One foot-pound (ft-lb) of torque is equivalent to 12 inch-pounds (in-lbs) of torque. Torque values below approximately 15 ft-lbs are expressed in inch-pounds, because most foot-pound torque wrenches are not accurate at these smaller values.*

Timing chain engines

	Ft-lbs (unless otherwise specified)	Nm
Oil cooler sub assembly mounting bracket bolts (see illustration 19.22)		
Step 1	15	20
Step 2	Tighten an additional 90-degrees	
Balance shaft mounting bolt	80 in-lbs	9
Balance shaft chain tensioner	63	85
Bearing bracket-to-cylinder head bolts, in sequence (see illustrations 6.77a or 6.77b)		
With steel bolts		
Step 1	26 in-lbs	3
Step 2	80 in-lbs	9
With aluminum bolts		
Step 1	36 in-lbs	4
Step 2	Tighten an additional 180-degrees	
Bearing bracket-to-exhaust camshaft bolt		
M6 size bolt		
Step 1	71 in-lbs	8
Step 2	Tighten an additional 90-degrees	
M8 size bolt		
Step 1	15	20
Step 2	Tighten an additional 90-degrees	
Camshaft adjustment valve bolt	80 in-lbs	9
Camshaft chain tensioner bolts	80 in-lbs	9
Camshaft sprocket control valve/regulator valve	26	35
Crankshaft damper/pulley bolt*		
Step 1	110	150
Step 2	Tighten an additional 90-degrees	
Crankshaft real seal housing flange bolts		
2.0L (CBPA) models	132 in-lbs	15
1.8L and 2.0L (all except CBPA) models, in sequence (see illustration 15.11)		
Step 1	Hand-tight	
Step 2	80 in-lbs	9
Cylinder head bolts* (in sequence - see illustration 9.90a or 9.90b)		
Bolt groups 1 through 5		
Step 1	29.5	40
Step 2	Tighten an additional 90-degrees	
Step 3	Tighten an additional 90-degrees	
Bolts "A" (see illustrations 9.90a or 9.90b)		
Step 1		
CBFA, CCTA models	71 in-lbs	8
All models except CBFA, CCTA models	36 in-lbs	4
Step 2	Tighten an additional 90-degrees	
Cylinder head cover bolts* (in sequence - see illustration 8.117)		
Step 1	Hand-tight	
Step 2	71 in-lbs	8
Step 3	Tighten an additional 90-degrees	
Drivebelt tensioner nut	88 in-lbs	10
Dual mass flywheel bolts		
Step 1	44	60
Step 2	Tighten an additional 90-degrees	
Driveplate bolts		
Step 1	44	60
Step 2	Tighten an additional 90-degrees	
Engine mount		
Engine bracket bolts*		
Step 1	65 in-lbs	7
Step 2	29.5	40
Step 3		
All models except CCTA, CBFA	Tighten an additional 90-degrees	
CCTA, CBFA models	Tighten an additional 180-degrees	
Mount-to-engine bracket bolts* (passenger's side)		
Step 1	44	60
Step 2	Tighten an additional 90-degrees	
Mount-to-body bolts (passenger's side)		
Step 1		
Horizontal bolt	15	20
Vertical bolts	29.5	40
Step 2	Tighten an additional 90-degrees	
Support bracket-to-body and mount bolts		
Step 1	15	20
Step 2	Tighten an additional 90-degrees	

Chapter 2 Part A Four-cylinder engines

Torque specifications (continued)	Ft-lbs (unless otherwise specified)	Nm
Transmission mount*		
Mount-to-transmission bracket bolts (driver's side)		
Step 1	37	50
Step 2	Tighten an additional 90-degrees	
Mount-to-body bracket bolts (driver's side)		
Step 1	44	60
Step 2	Tighten an additional 90-degrees	
Pendulum mount*		
Mount-to-transmission bolts		
Step 1	37	50
Step 2	Tighten an additional 90-degrees	
Mount-to-subframe bolt		
Step 1		
CNTA, CXCB, CNSA, CXBA, CXBB, CNSB, CXCA, CYFB, DJJA, DRLA	96	130
CCTA, CBFA, CPLA, CPPA, CPKA, CPRA models	74	100
Step 2	Tighten an additional 90-degrees	
Intake manifold		
Step 1	Hand-tight	
Step 2	80 in-lbs	9
Intake camshaft control valve bolt (left-hand thread)	26	35
Lower oil pan bolts* (in sequence - **see illustration 12.21b or 12.21c**)		
Plastic oil pans		
Step 1	71 in-lbs	8
Step 2	Tighten an additional 90-degrees	
Metal oil pans		
Step 1	Hand-tight	
Step 2	71 in-lbs	8
Step 3	Tighten an additional 45-degrees	
Upper oil pan bolts* (in sequence - **see illustration 12.51b, 12.51c or 15.51d**)		
2.0L (CBFA, CCTA) models		
Step 1	132 in-lbs	15
Step 2	Tighten an additional 90-degrees	
All models except CBFA, CCTA with 14 bolt upper oil pans		
Step 1, bolts 1 through 14	71 in-lbs	8
Step 2, bolts 1 and 2	Tighten an additional 180-degrees	
Step 3, bolts 3 through 9	Tighten an additional 45-degrees	
Step 4, bolt 10	Tighten an additional 180-degrees	
Step 5, bolts 11 through 14	Tighten an additional 90-degrees	
All models except CBFA, CCTA with 18 bolt upper oil pans		
Step 1, bolts 1 through 18	71 in-lbs	8
Step 2, bolts 1 and 2	Tighten an additional 180-degrees	
Step 3, bolts 3 through 13	Tighten an additional 45-degrees	
Step 4, bolts 14 through 18	Tighten an additional 90-degrees	
Oil baffle bolts	80 in-lbs	9
Oil cooler bolts		
Step 1	71 in-lbs	8
Step 2	Tighten an additional 90-degrees	
Oil dipstick bolts	80 in-lbs	9
Oil filter element	16.5	22
Oil pump-to-upper oil pan bolts		
CCTA, CBFA models		
M6 size bolt	80 in-lbs	9
M8 size bolt	15	20
All except CCTA, CBFA models		
Step 1	36 in-lbs	4
Step 2	Tighten an additional 90-degrees	
Oil pump chain tensioner guide pin-to-cylinder block bolt	80 in-lbs	9
Oil separator fasteners	80 in-lbs	9
Timing chain cover bolts		
Upper cover	80 in-lbs	9
Lower cover		
Step 1		
Aluminum bolts	34 in-lbs	4
Steel bolts	71 in-lbs	8
Step 2	Tighten an additional 45-degrees	
Step 3 (on all models except CBFA, CCTA models, after the crankshaft pulley is installed)	Tighten an additional 360-degrees	

Replace with new bolt(s)

Torque specifications (continued)

Note: *One foot-pound (ft-lb) of torque is equivalent to 12 inch-pounds (in-lbs) of torque. Torque values below approximately 15 ft-lbs are expressed in inch-pounds, because most foot-pound torque wrenches are not accurate at these smaller values.*

	Ft-lbs (unless otherwise specified)	Nm
Timing chain guide pins	15	20
Timing chain tensioner bolts		
2.0L CBFA, CCTA models	80 in-lbs	9
All except 2.0L CBFA, CCTA models		
Step 1	34	4
Step 2	Tighten an additional 90-degrees	
Intermediate shaft sprocket bolts		
Step 1	89 in-lbs	10
Step 2	Rotate the sprocket. **Note:** *No play must be felt when rotating*	
Step 3	22	30
Step 4	Tighten an additional 90-degrees	
Water pump toothed belt drive gear bolt-to-intake balance shaft		
Step 1	89 in-lbs	10
Step 2	Tighten an additional 90-degrees	
Vacuum pump mounting bolts	80 in-lbs	9
Oil pump fasteners		
1.8L (CPKA, CPRA, CXBA, CXBB, CXCA, CXCB)		
Step 1	72 in-lbs	8
Step 2	Tighten an additional 90-degrees	
2.0L (CBFA, CCTA, CPLA, CPPA)		
M6 bolt	80 in-lbs	9
M8 bolt	15	20
Oil pump pick-up tube fasteners		
1.8L (CPKA, CPRA, CXBA, CXBB, CXCA, CXCB)		
Step 1	36 in-lbs	4
Step 2	Tighten an additional 90-degrees	
2.0L (CBFA, CCTA, CPLA, CPPA)	80 in-lbs	9
Oil pump tensioner fastener	89 in-lbs	10
Upper oil pan mounting fasteners		
1.8L (CPKA, CPRA, CXBA, CXBB, CXCA, CXCB)		
Models with 14 mounting bolts		
Step 1, bolts 1 through 14	72 in-lbs	8
Step 2, bolts 1 and 2	Tighten an additional 180-degrees	
Step 3, bolts 3 through 9	Tighten an additional 45-degrees	
Step 4, bolt 10	Tighten an additional 180-degrees	
Step 5, bolts 11 through 14	Tighten an additional 90-degrees	
Models with 18 mounting bolts		
Step 1, bolts 1 through 14	72 in-lbs	8
Step 2, bolts 1 and 2	Tighten an additional 180-degrees	
Step 3, bolts 3 through 13	Tighten an additional 45-degrees	
Step 4, bolts 14 through 18	Tighten an additional 90-degrees	
2.0L (CPLA, CPPA)		
Step 1, bolts 1 through 14	72 in-lbs	8
Step 2, bolts 1 and 2	Tighten an additional 180-degrees	
Step 3, bolts 3 through 9	Tighten an additional 45-degrees	
Step 4, bolt 10	Tighten an additional 180-degrees	
Step 5, bolts 11 through 14	Tighten an additional 90-degrees	
2.0L (CBFA, CCTA)		
Step 1, bolts 1 through 14	Hand-tighten	
Step 2, bolts 1 through 14	132 in-lbs	15
Step 3, bolts 1 through 14	Tighten an additional 90-degrees	

1 General Information

1 This Part of Chapter 2A is devoted to in-vehicle repair procedures for the four-cylinder gasoline engines. There are several engines offered: two timing belt models, a 1.4L DOHC (CZTA, DGXA) engine or a 2.0L SOHC (CBPA) engine, which uses a timing belt and an oil pump/balance shaft drive that is externally mounted to the crankshaft and driven by a chain. The 1.4L engine has a turbocharger that is incorporated with the exhaust manifold and utilizes dual overhead camshafts. The 1.4L engine utilizes an aluminum engine block, and both engines use an aluminum cylinder head. The 1.8L (CPKA, CPRA, CXBA, CXBB, CNSA, CNSB) or 2.0L (CBFA, CCTA, CPLA, CPPA, CNTA, CXCA, CXCB, CYFB, DJJA, DLRA) engines use a timing chain with two balance shafts mounted in the engine block. These engines utilize an aluminum engine block with an aluminum cylinder head. Both engines have a turbocharger that is incorporated with the exhaust manifold and utilize dual overhead camshafts. The two camshafts are driven by a timing belt or chain, each operating eight valves via roller rocker arms. Valve clearance is maintained automatically at zero lash by hydraulic lash adjusters. The aluminum cylinder head is equipped with pressed-in valve guides and hardened valve seats. The oil pump is mounted below the front of the

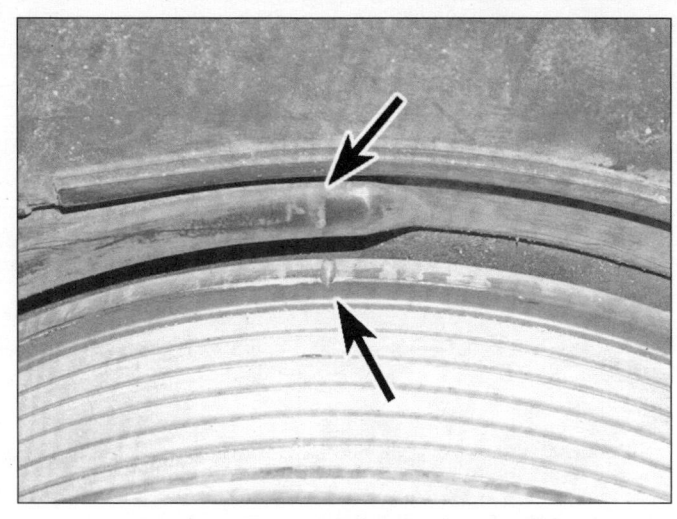

3.6a Align the notch on the crankshaft drivebelt pulley with the arrow on the timing belt cover - 2.0L timing belt (CBPA) engines

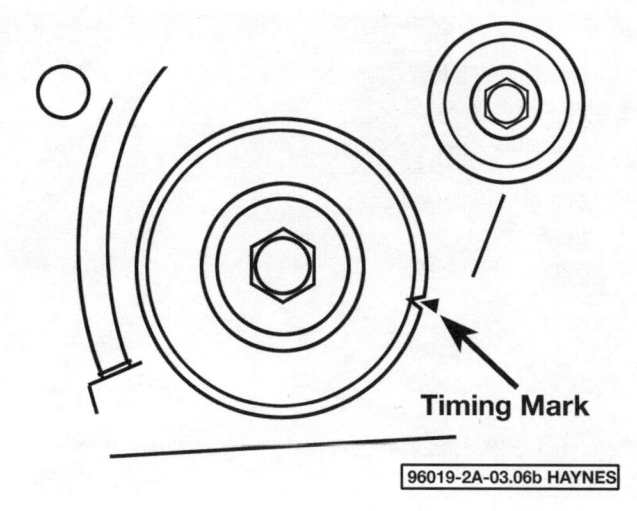

3.6b Align the notch on the crankshaft pulley with the mark on the timing belt cover - timing chain engines

engine and is chain driven from the crankshaft. The engine identification is stamped on a machined flat spot at the top of the cylinder block where the engine and transmission meet. On some models, an additional engine identification sticker is attached to the upper half of the timing cover.

2   Information concerning engine removal and installation and engine overhaul can be found in Chapter 2C.

3   The following repair procedures are based on the assumption that the engine is installed in the vehicle. If the engine has been removed from the vehicle and mounted on a stand, many of the steps outlined in this Chapter will not apply.

2   Repair operations possible with the engine in the vehicle

1   Many major repair operations can be accomplished without removing the engine from the vehicle.

2   Clean the engine compartment and the exterior of the engine with some type of degreaser before any work is done. It will make the job easier and help keep dirt out of the internal areas of the engine.

3   Depending on the components involved, it may be helpful to remove the hood to improve access to the engine as repairs are performed (refer to Chapter 11 if necessary). Cover the fenders to prevent damage to the paint. Special pads are available, but an old bedspread or blanket will also work.

4   If vacuum, exhaust, oil or coolant leaks develop, indicating a need for gasket or seal replacement, the repairs can generally be made with the engine in the vehicle. The intake and exhaust manifold gaskets, oil pan gasket, crankshaft oil seals and cylinder head gasket are all accessible with the engine in place.

5   Exterior engine components, such as the intake and exhaust manifolds, the oil pan, the oil pump, the water pump, the starter motor, the alternator and the fuel system components can be removed for repair with the engine in place.

6   Since the cylinder head can be removed without pulling the engine, camshaft and valve component servicing can also be accomplished with the engine in the vehicle. Replacement of the timing belt or chain and sprockets is also possible with the engine in the vehicle.

7   In extreme cases caused by a lack of necessary equipment, repair or replacement of piston rings, pistons, connecting rods and rod bearings is possible with the engine in the vehicle. However, this practice is not recommended because of the cleaning and preparation work that must be done to the components involved.

3   Top Dead Center (TDC) for number one piston - locating

All models

1   Top Dead Center (TDC) is the highest point in the cylinder that each piston reaches as it travels up the cylinder bore. Each piston reaches TDC on the compression stroke and again on the exhaust stroke, but TDC generally refers to piston position on the compression stroke.

2   Positioning the piston(s) at TDC is an essential part of many procedures such as valve adjustment, camshaft, timing belt or timing chain/sprocket removal.

3   Before beginning this procedure, be sure to place the transmission in Neutral and apply the parking brake or block the rear wheels.

4   Remove the spark plugs (see Chapter 1) and install a compression gauge in the number one spark plug hole (see Chapter 2C). It should be a gauge with a screw-in fitting and a hose at least six inches long.

5   Rotate the crankshaft using a socket and ratchet attached to the bolt threaded into the front of the crankshaft. Turn the bolt in a clockwise direction only. The moment the gauge shows pressure indicates that the number one cylinder has begun the compression stroke.

6   Once the compression stroke has begun, TDC for the compression stroke is reached by bringing the piston to the top of the cylinder. On all models except 1.4L (CZTA, DGXA) models, continue turning the crankshaft until the notch in the crankshaft damper is aligned with the "TDC" or the "0" mark on the timing chain cover **(see illustrations)**. At this point, the number one cylinder is at TDC on the compression stroke. If the marks are aligned but there was no compression, the piston was on the exhaust stroke; continue rotating the crankshaft 360-degrees (1-turn).

Note: *If a compression gauge is not available, you can simply place a blunt object over the spark plug hole and listen for compression as the engine is rotated. Once compression at the No.1 spark plug hole is noted the remainder of the Step is the same.*

7   On 2.0L (CBPA) timing belt models, verify the timing mark is in the view opening on the transaxle; on manual transaxles the TDC mark on the flywheel and the V-notch on the transaxle case should be aligned or, on automatic transaxles, the TDC 0-mark on the torque converter should just be showing at the bottom of the view opening.

8   On timing chain engines, remove the upper timing chain cover (see Section 6). With the crankshaft pulley notch aligned with the mark on the timing chain cover **(see illustration 3.6b)**, the camshaft sprocket timing marks should be up at approximately the 11 o'clock and 1 o'clock positions (see Section 6).

2A-8   **Chapter 2 Part A   Four-cylinder engines**

3.12 Remove the turbocharger ventilation hose fasteners, then disconnect the hose from the turbocharger and the camshaft housing

3.15 Remove the camshaft housing cap fasteners, then remove the cap and replace the O-ring

3.19 TDC plug location - 1.4L CZTA engine shown, DGXA models identical

3.22a The grooves on the exhaust camshaft are viewed through the holes in the sprocket for the water pump belt

3.22b The grooves on the intake camshaft are above the centerline of the camshaft

9   After the number one piston has been positioned at TDC on the compression stroke, TDC for any of the remaining cylinders can be located by turning the crankshaft 180-degrees clockwise and following the firing order (refer to the Specifications). Rotating the engine 180 degrees past TDC #1 will put the engine at TDC compression for cylinder #3.

1.4L (CZTA, DGXA) models

10   Follow steps 1 through 5.
11   Remove the air filter housing, the resonator and the air inlet duct from the throttle body and the turbocharger (see Chapter 4).
12   Disconnect the EVAP hose from the turbocharger crankcase ventilation hose, then remove the vent hose fasteners and vent hose **(see illustration)**. Once the hose is removed, cover the openings.
13   Remove the spark plugs (see Chapter 1), if not already removed.
14   Release the harness clips and remove the water pump belt cover (see Chapter 3, Section 7).
15   Remove the cap fasteners and cap from the camshaft housing **(see illustration)**.
Caution: *Cover the water pump with a rag to prevent any oil that may leak out once the cap is removed. If oil comes in contact with the water pump seal it will damage the seal.*
16   Using a screwdriver that has a shaft that is least a 10 inches (250 mm) long, insert the screwdriver into the number one spark plug hole. Rotate the engine clockwise until cylinder one is at "BDC" (Bottom Dead Center) - the screwdriver should move down as the engine is rotated.
17   With the number one cylinder one at "BDC" (Bottom Dead Center), mark the screwdriver and rotate the engine clockwise until the screwdriver has come up about 1.18 inches (30 mm).
18   Raise the vehicle and support it securely on jackstands then remove the engine splash shield fasteners and shield.
19   Working under the vehicle, locate the TDC" plug on the lower right rear side of the engine **(see illustration)**.
20   Remove the "TDC" plug and insert locking pin (VW special tool #T10340) all the way into the hole and tighten the it to 22 ft-lbs (30 Nm). With the locking pin in place, slowly rotate the engine clockwise, by hand, until the crankshaft counterweight touches the pin.
21   If the locking pin can't be installed all the way, the crankshaft is not in the correct position. If this happens, remove the locking pin, rotate the crankshaft 90-degrees clockwise and try to reinstall the pin, then repeat Step 20.
22   The asymmetrical grooves on both camshaft ends must now be facing up **(see illustrations)**.
23   If the camshafts are not positioned as shown, remove the locking pin and rotate the crankshaft one more turn and repeat Step 20.

Chapter 2 Part A Four-cylinder engines

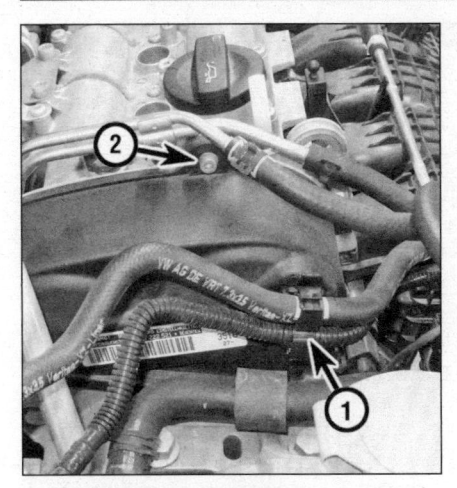

5.10 Disengage the retaining clip at the front of the cover, then remove the upper cover mounting bolt

1 Hose retaining clip
2 Upper cover mounting bolt

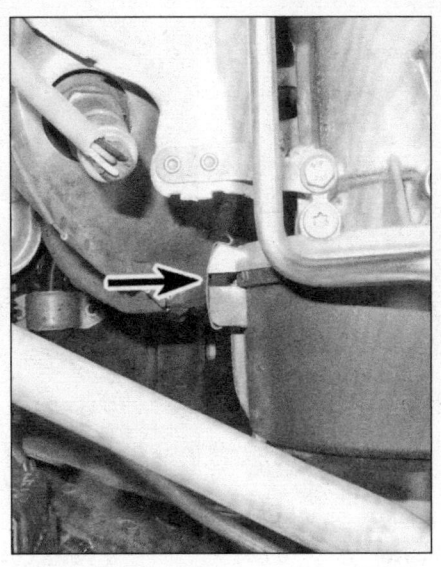

5.11a Using a small screwdriver, release the clip from the rear. . .

5.11b . . . and the front side of the cover

4 Valve cover - removal and installation

Note: *This procedure applies to the 2.0L timing belt (CBPA) engine only.*

Removal

1 Remove the engine cover (see Chapter 1, Section 7).
2 Remove the timing belt upper cover retaining fasteners (see Section 5).
3 Remove the ignition coils (see Chapter 5).
4 Remove the PCV valve fasteners and pull the PCV housing away from the cover.
5 Remove the crankcase hose heat shield retaining fasteners from the turbocharger, and remove the shield.
6 Disconnect the electrical connectors and harness clipped to the valve cover.
7 Loosen the fasteners a few turns at a time starting from the center and work in a circular pattern towards the ends until all the bolts are loose, then remove the retaining fasteners and detach the valve cover from the cylinder head.
8 If the cover is stuck to the head, bump the end with a block of wood and a hammer to jar it loose. If that doesn't work, try to slip a flexible putty knife between the head and cover to break the seal.
Caution: *Don't pry at the cover-to-head joint or damage to the sealing surfaces may occur, leading to oil leaks after the cover is re-installed.*

Installation

9 The mating surfaces of the cylinder head and cover must be clean when the cover is installed. Remove all traces of sealant and old gasket material including the spark plug tube seal gaskets, then clean the mating surfaces with brake system cleaner. If there's residue or oil on the mating surfaces when the cover is installed, oil leaks may develop. Also inspect the rubber end plug at the rear of the cylinder head for cracks and damage. Now would be a good time to replace it, if damage has occurred.
10 The rubber gasket can be reused if not damaged.
11 Install the valve cover and any brackets removed, then tighten the fasteners in a circular sequence to the torque listed in this Chapter's Specifications.
12 Reinstall the remaining parts, run the engine and check for oil leaks.

5 Timing belt and sprockets - removal, inspection and installation

Warning: *Wait until the engine is completely cool before beginning this procedure.*
Caution: *Do not rotate the crankshaft or the camshaft separately during this procedure with the timing belt removed, as damage to valves will occur. Only rotate the camshaft a few degrees as necessary to align the camshaft sprocket marks with the marks on the rear timing belt cover.*
Caution: *If the battery is disconnected, several systems must be re-learned before they will work properly (see Chapter 5, Section 3).*
Note: *This procedure applies to the 2.0L (CBPA) engine and 1.4L (CZTA, DGXA) engines.*

Removal

Caution: *The timing system is complex. Severe engine damage will occur if you make any mistakes. Do not attempt this procedure unless you are highly experienced with this type of repair. If you are at all unsure of your abilities, consult an expert. Double-check all your work and be sure everything is correct before you attempt to start the engine.*
Caution: *The contact points for the timing belt on the camshaft sprockets, crankshaft sprocket, tensioner and idler pulley must be kept free from oil to prevent damage to the belt and possible premature failure.*

1 Disconnect the cable from the negative terminal of the battery (see Chapter 5).
2 Loosen the right-front wheel bolts, raise the front of the vehicle and support it securely on jackstands. Block the wheels at the opposite end and remove the right-front wheel.
3 Working under the vehicle, remove the splash shield below the engine (see Chapter 1, Section 6).
4 Remove the driver's side inner fender liner (see Chapter 11).
5 Mark the rotation direction of the drive belt, then remove the drive belt (see Chapter 1) and spark plugs.
Note: *After the belt is removed, remove the locking pin from the tensioner and slowly release the tensioner.*
6 Set the engine to TDC (see Section 3).
7 Remove the coolant expansion tank (see Chapter 3).

1.4L (CZTA, DGXA) engines

8 Remove the air filter housing and the turbocharger charge air cooler pipes (see Chapter 4).
9 Disconnect the quick-connect fittings from the fuel hose and from the hose to the EVAP canister (see Chapter 6).
10 Disengage the retaining clip attaching the hoses to the front of the cover then remove the upper cover mounting bolt **(see illustration)**.
11 Remove the upper timing belt cover clips **(see illustrations)**.

2A-10   Chapter 2 Part A   Four-cylinder engines

5.12 Remove the upper timing belt cover

5.14 Lower timing cover fastener locations

5.15 If you intend to re-use the timing belt, apply directional marks on the belt

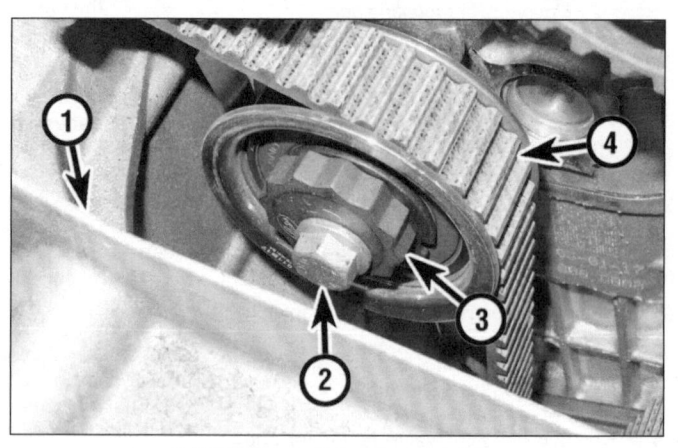

5.16 Timing belt tensioner details

1	Engine mount bracket	3	Timing belt tensioner eccentric pulley
2	Tensioner adjusting bolt	4	Timing belt

12   Remove the upper timing belt cover **(see illustration)**.
13   Remove the crankshaft pulley (see Section 10).
14   Remove the timing belt lower cover fasteners **(see illustration)**.
15   If you plan to re-use the timing belt, apply an arrow indicating direction of travel on the belt **(see illustration)**.
16   Working in the engine mount bracket opening, hold the timing belt tensioner eccentric pulley with VW special tool #T10499A or equivalent, and loosen the tensioner mounting bolt using VW special tool #T10500 or equivalent **(see illustration)** to release the timing belt tension.
17   Remove the timing belt from the camshafts, then slide the belt down through the engine mount bracket and out from below.
Caution: *The timing belt is made from a fiberglass cord webbing material that can be damaged by bending it too sharply. Never bend the toothed belt with a radius less than 1 inch (25 mm).*
Caution: *With the timing belt removed, do not rotate the engine out of TDC or severe engine damage may occur.*

2.0L (CBPA) engine

18   Remove the timing belt upper cover fasteners and cover.
19   Rotate the engine in the normal direction of rotation (clockwise) until the No. 1 cylinder is located at TDC (see Section 3). Verify that the camshaft sprocket mark is aligned with the "OT" mark on the timing belt rear cover at approximately the 11 o'clock position.
20   Support the engine from below **(see illustration 18.3)**, then remove the passenger's side engine mount (see Section 18).
Note: *The engine will have to be raised and lowered to remove the front engine mount fasteners.*
21   Remove the passenger's side engine mount bracket bolts and remove the bracket from the front of the engine.
22   Use an offset box-end wrench on the crankshaft center bolt to keep the crankshaft from rotating while removing the pulley retaining bolts. Loosen the crankshaft drive sprocket retaining bolt (only if the sprocket is to be removed) and the crankshaft pulley bolts, then remove the pulley (see Section 10). After the bolts are loosened, verify that the crankshaft has not moved from TDC.
Note: *Loosening the drive sprocket bolt is only required if the crankshaft drive sprocket is expected to be removed. It is not typically necessary to remove the drive sprocket when you're simply replacing a timing belt, but it will have to be removed if you are replacing the crankshaft front oil seal or housing. If you do remove the drive sprocket, obtain a new bolt (the manufacturer doesn't recommend re-using it).*

Chapter 2 Part A Four-cylinder engines

5.28a With the crankshaft drive sprocket retaining bolt removed...

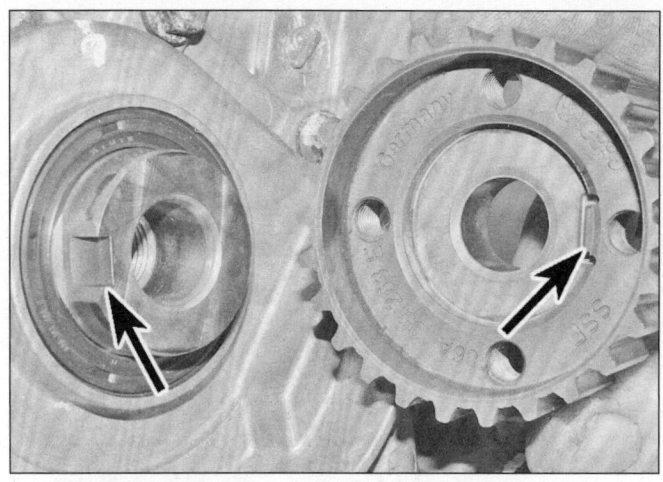

5.28b... the crankshaft sprocket is easily detached from the crankshaft. Note how the lug on the sprocket engages with the notch in the crankshaft

5.29 If necessary, the camshaft sprocket bolt can be loosened while holding the sprocket in place with a pin spanner wrench - typical camshaft sprocket and spanner tool shown

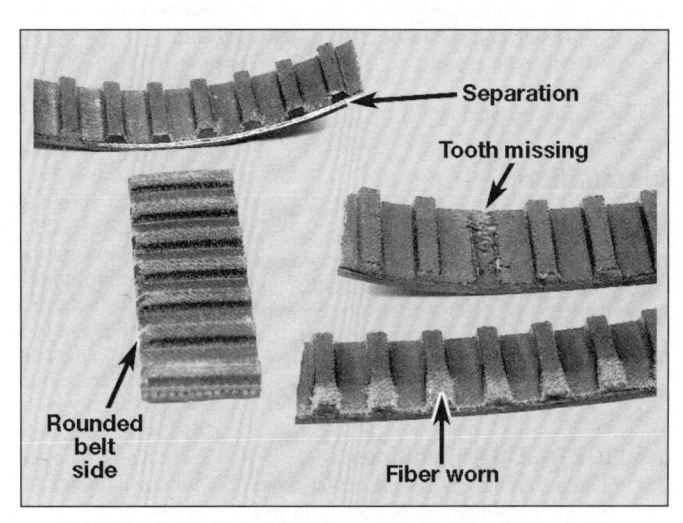

5.31 Check the timing belt for cracked and missing teeth - wear on one side of the belt indicates sprocket misalignment problems

23 Remove the timing belt center cover fasteners and cover.

24 If you plan to re-use the timing belt, apply match marks on the sprocket and belt, and make an arrow indicating direction of travel on the belt **(see illustration 5.15)**

25 Remove the timing belt lower cover retaining fasteners and remove the cover.

26 Loosen the tensioner pulley nut, then insert special tool #T10020 into the slot in the tensioner pulley and turn the tensioner counterclockwise until the timing belt can be removed from the sprockets.

27 Remove the timing belt from the engine, taking care to avoid twisting or kinking it excessively.

28 If the crankshaft sprocket is worn or damaged, or if you have to replace the crankshaft front oil seal, remove the drive sprocket retaining bolt which was loosened in Step 22 and detach the crankshaft sprocket from the crankshaft **(see illustrations)**.

29 If the camshaft sprocket is damaged or needs to be removed for other procedures such as oil seal replacement, use a pin spanner wrench or similar tool to hold the sprocket in place as the sprocket retaining bolt is loosened (see illustration), then, use a two-jaw puller to remove the camshaft sprocket.

Inspection

Caution: *Do not bend, twist or turn the timing belt inside out. Do not allow it to come in contact with oil, coolant or fuel. Do not turn the crankshaft or camshaft more than a few degrees (if necessary for tooth alignment) while the timing belt is removed.*

30 Spin the timing belt tensioner pulley (which is the large pulley bolted to the engine block) and the idler wheel (which is the small roller mounted on the tensioner body) and check the bearings for smooth operation and excessive play. Also inspect the remaining timing belt sprockets for any obvious damage. Replace all worn parts as necessary.

31 Examine the belt for evidence of contamination by coolant or oil. If this is the case, find the source of the contamination before progressing any further. Check the belt for signs of wear or damage, particularly around the leading edges of the belt teeth **(see illustration)**.

Caution: *If the belt appears to be in good condition and can be re-used, it is essential that it is reinstalled the same way around, otherwise accelerated wear will result, leading to premature failure.*

32 Replace the belt if its condition is in doubt; the cost of belt replacement is negligible compared with potential cost of the engine repairs, should the belt fail in service. Similarly, if the belt is known to have covered more than 60,000 miles, it is a good idea to replace it regardless of condition, as a precautionary measure.

2A-12   **Chapter 2 Part A   Four-cylinder engines**

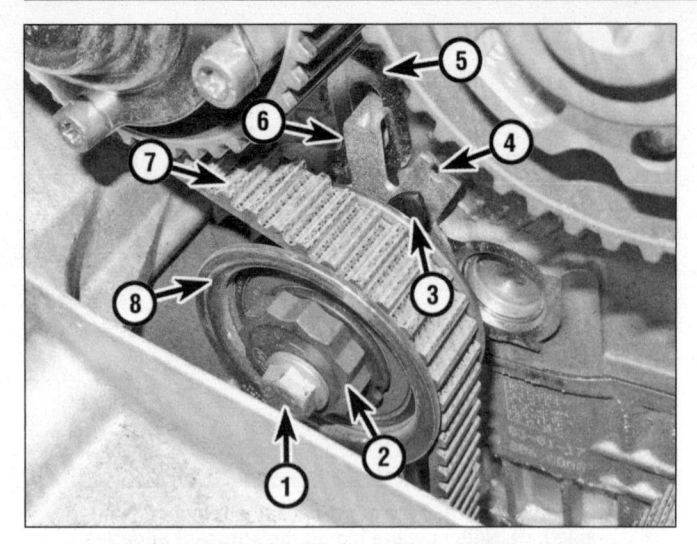

5.37 Tensioner pulley details - 1.4L (CZTA, DGXA) models shown

1. Tensioner adjuster bolt
2. Timing belt tensioner eccentric pulley
3. Pointer
4. Setting window
5. Cylinder head cast hole
6. Metal tab
7. Timing belt
8. Tensioner pulley

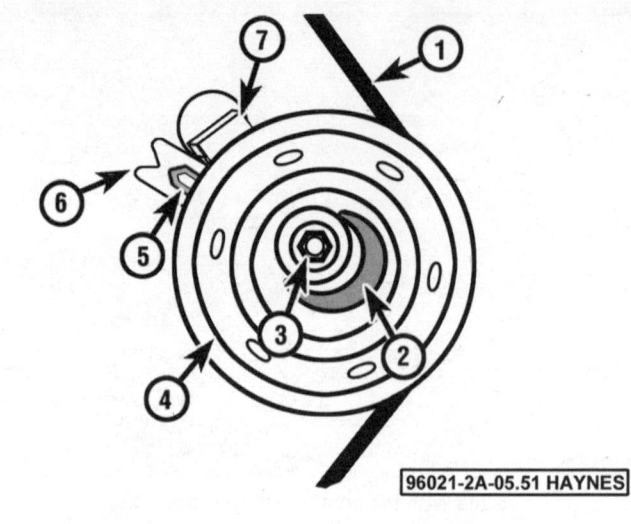

5.51 Timing belt tensioner details - 2.0L (CBPA) models

1. Timing belt
2. Half moon adjusting tool slot
3. Tenioner lock nut
4. Tensioner pulley
5. Tenioner pointer
6. Alignment notch
7. Metal locating tab

Installation

Caution: *Before starting the engine, carefully rotate the crankshaft by hand through at least two full revolutions (use a socket and breaker bar on the crankshaft pulley center bolt). If you feel any resistance, STOP! There is something wrong - most likely, valves are contacting the pistons. You must find the problem before proceeding. Check your work and see if any updated repair information is available.*

1.4L (CZTA, DGXA) engines

33 Ensure that the crankshaft is still set to TDC on No. 1 cylinder, as described in this Chapter's Specifications. If any of the camshaft sprockets were removed for inspection or needed replacement, install them back onto the camshafts now (see Section 8). If the timing belt idler was removed, reinstall the idler and tighten the idler bolt to the torque listed in this Chapter's Specifications. If the tensioner was removed, reinstall it now, making sure the metal tab of the setting window is seated into the cast hole of the cylinder head **(see illustration)**, then tighten the adjustment bolt loosely.

34 Working at the rear of the camshaft housing, in place of the camshaft cap **(see illustration 3.16)** attach VW special tool #T10494 camshaft positioning tool to the intake camshaft with the opposite end of the tool seating on the water pump mounting bolt on the exhaust camshaft.

35 If the camshaft positioning tool is easy to install (lining up with the asymmetrical grooves on intake camshaft) the valve timing is okay - proceed to he next Step. If the tool can't be installed, the valve timing is incorrect. If the valve timing is incorrect, the camshaft sprockets must be removed and reinstalled in the correct timed position (see Section 8).

36 Loop the timing belt loosely under the crankshaft sprocket and through the engine mount bracket - the outside of the timing belt must be around the tensioner and the idler pulley - then place the belt up and on both camshaft sprockets.

Caution: *Observe the direction of rotation markings on the belt.*

37 Turn the timing belt tensioner eccentric pulley clockwise using VW special tool #T10499 or equivalent until the pointer on the back side of the tensioner pulley is approximately 0.39 inch (10 mm) to the right of the metal tab of the setting window **(see illustration)**, then turn back the eccentric until the pointer is exactly inside the setting window and tighten the adjuster bolt to the torque listed in this Chapter's Specifications.

38 Remove the locking pin (VW special tool #T10340) from the TDC hole on the lower corner of the engine, then remove the camshaft positioing tool (VW special tool #T10494) from the ends of the camshaft.

39 Rotate the engine two complete revolutions, by hand in the direction of normal engine rotation, then insert locking pin (VW special tool #T10340) all the way into the hole and tighten it to 22 ft-lbs (30 Nm).

40 Slowly rotate the crankshaft until it contacts the locking pin and install VW special tool #T10494 camshaft positioning tool onto the end of the camshaft housing.

41 If the camshaft positioning tool can be easily inserted onto the intake camshaft, have an assistant use VW special tool #T10487, or equivalent, to push the timing belt down slightly between the sprockets while you insert the camshaft positioning tool. If the tool still cannot be installed, the timing is incorrect and the camshaft sprockets must be removed (see Section 8).

42 If the tool can be properly installed in Step 40, the timing is correct.

43 Reinstall the lower timing belt cover and tighten the fasteners to the torque listed in this Chapter's Specifications.

44 Install the crankshaft pulley (see Section 10).

45 Remainder of installation is the reverse of removal.

2.0L (CBPA) engines

46 Ensure that the crankshaft is still set to TDC on No. 1 cylinder, as described in this Chapter's Specifications. If any of the timing sprockets or the tensioner pulley were removed for inspection or needed replacement, install them back onto the engine now. If the timing belt tensioner was removed, reinstall it now and tighten the bolts to the torque listed in this Chapter's Specifications, then install the tensioner pulley adjustment bolt loosely.

47 Make sure the camshaft sprocket mark is still in alignment with the mark on the rear timing belt cover.

48 Loop the timing belt loosely under the crankshaft sprocket.

Caution: *Observe the direction of rotation markings on the belt.*

49 Route the belt clockwise in this order: around the water pump sprocket, over the camshaft sprocket, then over the tensioner pulley. Ensure the belt teeth seat completely on the sprockets.

Note: *Slight adjustments to the position of the camshaft sprocket may be necessary to achieve this.*

50 Verify the timing marks at the camshaft and the transaxle are all at TDC.

51 Insert VW special tool #T10020 or equivalent into the half-moon shaped slot on the tensioner, then rotate the tensioner counter-

Chapter 2 Part A Four-cylinder engines   2A-13

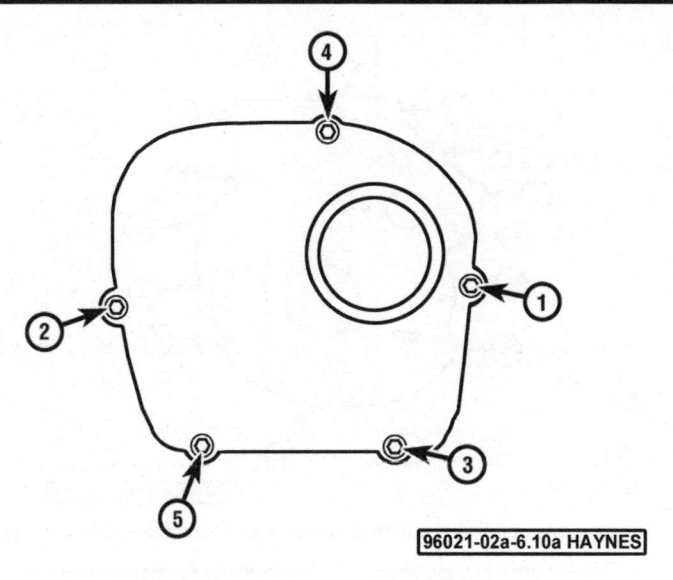

6.10a Upper timing chain cover tightening sequence - 2.0L (CBFA, CCTA) engines

6.10b Upper timing chain cover tightening sequence - 1.8L (CXBA, CXBB, CNSA, CNSB, CPPA, CPKA) and 2.0L (CPLA, CPPA) models

6.10c Upper timing chain cover tightening sequence - 2.0L (CXCA, CNTA, CXCB, DJJA, CYFB) models

clockwise until the pointer and the notch are aligned **(see illustration)**. This applies tension to the belt. Loosen the tensioner again, rotate the tensioner until the pointer and the notch are aligned again, then tighten the nut to the torque listed in this Chapter's Specifications.

52   Install the lower timing belt cover and the crankshaft pulley (see Section 10).

53   Rotate the crankshaft two complete revolutions, bringing it back to TDC on the compression stroke. Check the alignment marks again. Also re-check that the tensioner pointer is positioned by the middle of the notch **(see illustration 5.51)**.

54   The remainder of installation is the reverse of removal.

6 Timing chain cover, timing chains and tensioners - removal and installation

Warning: *Wait until the engine is completely cool before beginning this procedure.*
Caution: *Do not rotate the crankshaft or the camshafts separately during this procedure with the timing chain removed, as damage to valves may occur.*
Caution: *If the battery is disconnected, several systems must be re-learned before they will work properly (see Chapter 5, Section 3).*

Timing chain covers
Upper cover
Removal

1   Pull the engine cover up and off of the ballstuds, then remove the cover.

2   On models equipped with a noise generator, open the locking ring on the charge air pipe, pull the fuel lines from the retaining clip, then remove the EVAP canister retaining bolts and canister (see Chapter 6). Move the charge air pipe to the side.

3   Remove the oil dipstick tube retaining fasteners, unclip the top of the dipstick tube and move the tube out of the way.
Note: *If the dipstick tube is removed, be sure to replace the O-ring at the base of the tube.*

4   Push the coolant hoses away from the cover and tie them out of the way.

5   On CXCA, CNTA, CXCB, DJJA, CYFB models, disconnect the electrical connector from the intake camshaft adjustment valve 1 and the exhaust camshaft adjustment valve 1, then remove the adjustment valve(s) retaining fasteners, valve(s) and seal(s).
Note: *Always replace the adjustment valve seal and O-ring.*

6   Remove the upper timing chain cover retaining fasteners in a star pattern and remove the cover.

7   Discard the gasket; it must be replaced whenever it is removed. Check that the sealing faces are undamaged.

Installation

8   Clean the cover and the cylinder head faces carefully, then install a new gasket onto the upper cover.

9   Install the cover to the cylinder head, ensuring that the gasket remains seated as the cover is tightened.

10   Working in sequence **(see illustrations)**, first tighten the cover bolts by hand only. Once all the bolts are hand-tight, go around once more in sequence, tightening the retaining fasteners to the torque listed in this Chapter's Specifications.

11   The remainder of installation is the reverse of removal. Run the engine and check for leaks.

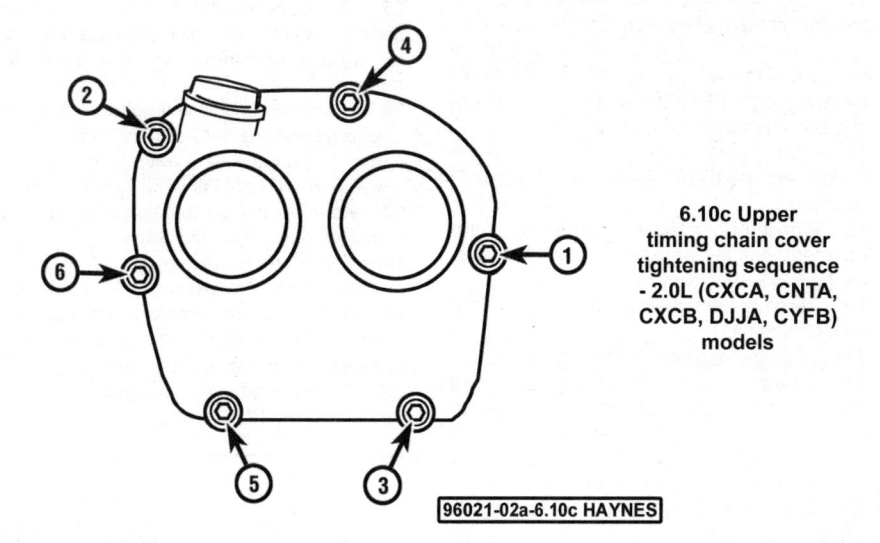

2A-14   Chapter 2 Part A Four-cylinder engines

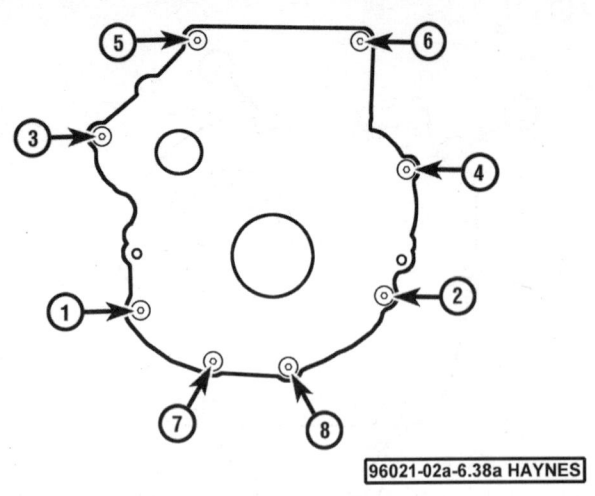

6.38a Lower timing chain cover tightening sequence - 8-bolt cover

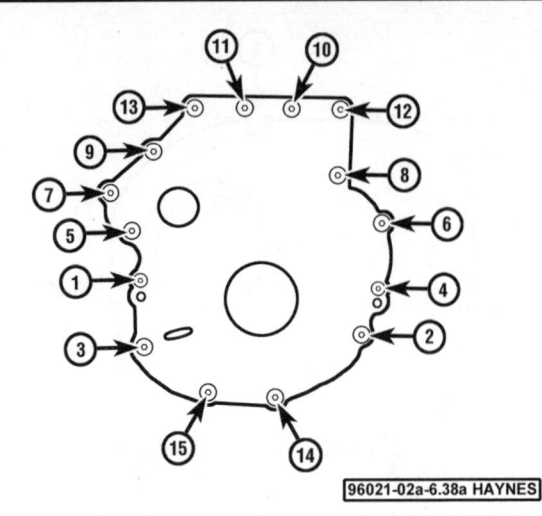

6.38b Lower timing chain cover tightening sequence - 15-bolt cover

Lower cover

Removal

12 Disconnect the cable from the negative terminal of the battery (see Chapter 5).

13 Loosen the right-front wheel bolts, raise the front of the vehicle and support it securely on jackstands. Block the wheels at the opposite end and remove the right-front wheel.

14 Working under the vehicle, remove the under-vehicle splash shield (see Chapter 1, Section 6).

15 Working from above in the engine compartment, pull the engine cover up and off of the ballstuds then remove the cover.

16 Remove the right-front inner fender liner (see Chapter 11).

17 Drain the engine oil (see Chapter 1).

18 Remove the air charge pipe retaining fasteners, the clamps at each end and the air charge pipe.

19 Loosen both hose clamps on the lower air charge hose and remove the hose.

20 Remove the drivebelt and tensioner (see Chapter 1).

21 Use an offset box-end wrench on the crankshaft center bolt and rotate the engine in the normal direction of rotation (clockwise) until the No. 1 cylinder is located at TDC (see Section 3). Verify that the notch on the crankshaft pulley is aligned with the mark on the lower timing chain cover.

22 Remove the crankshaft pulley (see Section 10) and reinstall the crankshaft pulley bolt.

Caution: *To prevent damaging the splines on the end of the crankshaft, install VW special tool #T10355 spacer or equivalent.*

23 On models equipped with a noise generator, open the locking ring on the charge air pipe, then move the charge air pipe to the side.

24 Support the engine from above with an engine support fixture (see Chapter 2C) or position a floor jack under the engine oil pan and place a large block of wood between the jack head and the oil pan, then carefully raise the engine just enough to take the weight off the mounts (see Section 18).

Caution: *DO NOT place any part of your body under the engine when it's supported only by a jack!*

25 Remove the front engine mount-to-mount bracket bolts, raise the engine enough to remove the mount-to-body bolts and remove the mount.

26 Lower the engine enough to remove the engine mount bracket bolts and bracket.

Note: *The engine may have to be raised or lowered to gain access or remove the bolts.*

27 Remove the oil dipstick tube retaining fasteners and the dipstick tube.

Note: *Be sure to replace the O-ring at the base of the dipstick tube.*

28 Disconnect the electrical connector from the oil pressure regulator valve, then remove the valve mounting bolt and remove the valve from the upper oil pan, if equipped.

29 On 2.0L (CBFA, CCTA) engines, disconnect the electrical connector and vacuum hoses from the turbocharger wastegate bypass regulator valve. Remove the regulator valve retaining fasteners and valve from the turbocharger.

30 On 2.0L (CBFA, CCTA) engines, remove the turbocharger support bracket retaining fasteners and bracket.

31 Remove the harness retainer bracket bolt and remove the bracket from the front side of the lower oil pan, if equipped.

32 Remove the lower timing chain cover retaining fasteners in the reverse of the tightening sequence **(see illustration 6.38a or 6.38b)** and carefully pry the cover from the engine block.

Note: *There can be two different timing chain covers used depending on the year and model: a cover with 8 retaining bolts or cover with 15 retaining bolts.*

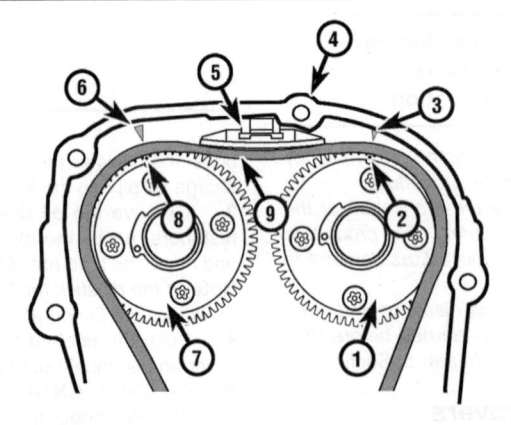

6.41 Camshaft timing mark identifications

1 Intake camshaft sprocket
2 Intake camshaft sprocket timing mark
3 Cylinder head cover timing mark (intake)
4 Cylinder head cover
5 Timing chain upper guide rail
6 Cylinder head cover timing mark (exhaust)
7 Exhaust camshaft sprocket
8 Exhaust camshaft sprocket timing mark
9 Timing chain

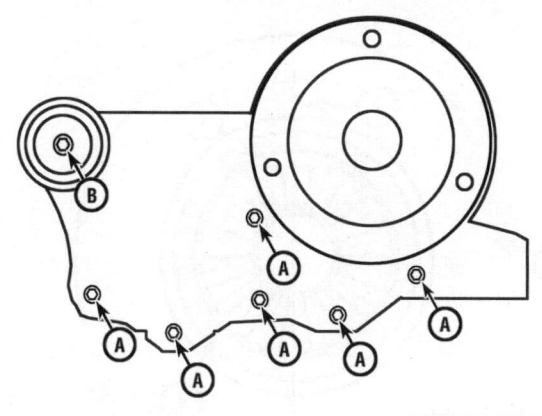

Chapter 2 Part A Four-cylinder engines　　2A-15

6.44a Remove the bearing retaining fasteners (A), then the bracket-to-exhaust camshaft bolt (B) - 1.8L (CXBA, CXBB, CNSA, CNSB, CPRA, CPKA) and 2.0L (CBFA, CCTA, CPLA, CPPA) models

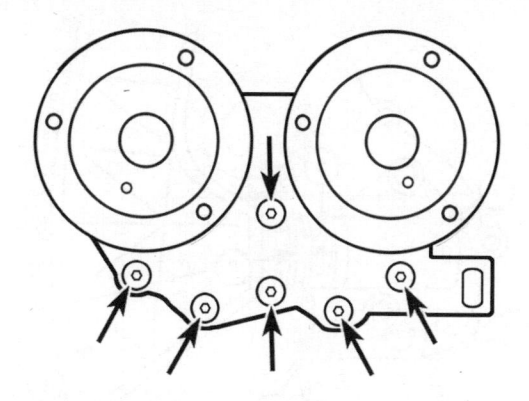

6.44b Remove the bearing bracket retaining fasteners - 2.0L (CXCA, CXCB, CNTA, DJJA, DLRA, CYFB) models

Installation

33 Use a scraper to remove all traces of old sealant from the block and timing chain cover. Clean the mating surfaces with brake system cleaner.
34 Make sure the threaded bolt holes in the block are clean.
35 Check the timing chain cover flange for distortion, particularly around the bolt holes. Remove any nicks or burrs as necessary.
36 Apply a 3/16-inch wide bead of RTV sealant to the mating surface of the timing cover.
Note: *Be sure to follow the sealant manufacturers recommendations for assembly and sealant curing times.*
37 Carefully position the timing chain cover on the engine block and install the timing cover retaining fasteners loosely.

38 Working in sequence **(see illustrations)**, first tighten the cover bolts by hand only. Once all the bolts are hand-tight, tighten the retaining fasteners in sequence to the torque listed in this Chapter's Specifications.
39 The remainder of installation is the reverse of removal. Run the engine and check for leaks.

Timing chain and tensioners

Caution: *The timing system is complex, and severe engine damage will occur if you make any mistakes. Do not attempt this procedure unless you are highly experienced with this type of repair. If you are at all unsure of your abilities, be sure to consult an expert. Double-check all your work and be sure everything is correct before you attempt to start the engine.*

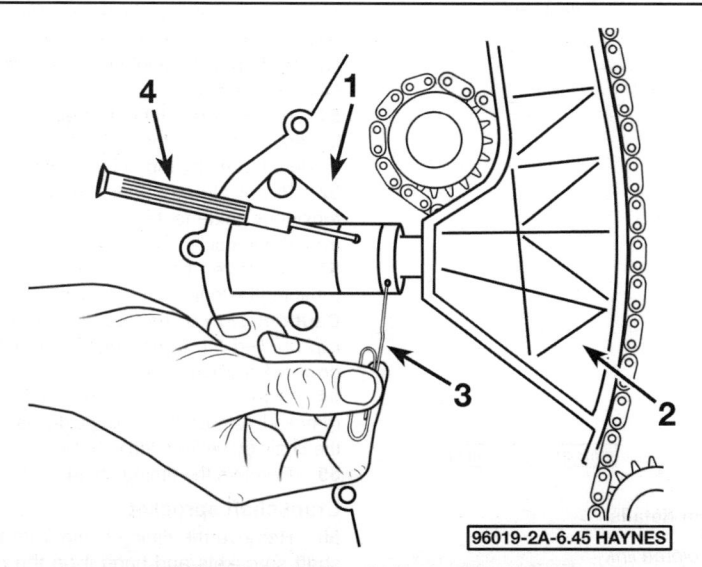

Removal

40 Remove the upper timing chain cover (see Steps 1 through 6).
41 Use an offset box-end wrench on the crankshaft center bolt and rotate the engine in the normal direction of rotation (clockwise) until the No. 1 cylinder is located at TDC (see Section 3). Verify that the notch on the crankshaft pulley is aligned with the mark on the lower timing chain cover **(see illustration 3.6b)** and the marks on the camshafts are pointing to the marks on the cylinder head cover **(see illustration)**.
42 Remove the lower timing chain cover (see Steps 12 through 32).
43 Insert VW special tool #10352 or equivalent onto the control valve(s), then rotate the valve clockwise and pull the valve out from the end of the camshaft. 2.0L (CXCA, CXCB, CNTA, DJJA, CYFB) models are equipped with an intake and exhaust camshaft control valve; both valve are removed the same way.
Note: *The control valve has left hand threads.*
44 Remove the bearing bracket retaining fasteners, bearing bracket-to-exhaust camshaft bolt (on in reverse order of the tightening sequence then the bearing bracket **(see illustrations)**.

Oil pump timing chain

45 Press the oil pump chain tensioner bracket away from the chain and place a pin (a drill bit or paper clip will work) into the hole of the tensioner bracket to hold it in the compressed position.
46 Remove the chain tensioner bracket bolt, the tensioner and chain guide, then remove the chain. **Caution:** *Do not remove the pin holding the tensioner spring in the compressed position once it's removed.*

Camshaft timing chain and tensioner

47 On Version 1 tensioners, use a small screwdriver or pick to hold the chain tensioner ratchet arm away (UP) from the ratchet stem, then slowly compress the timing chain tensioner piston and place a pin (a drill bit or paper clip will work) into the hole to hold it in the compressed position **(see illustration)**.

6.47 Version 1 timing chain tensioner details

1 Timing chain tensioner
2 Timing chain guide
3 Drill bit or paper clip
4 Small screwdriver

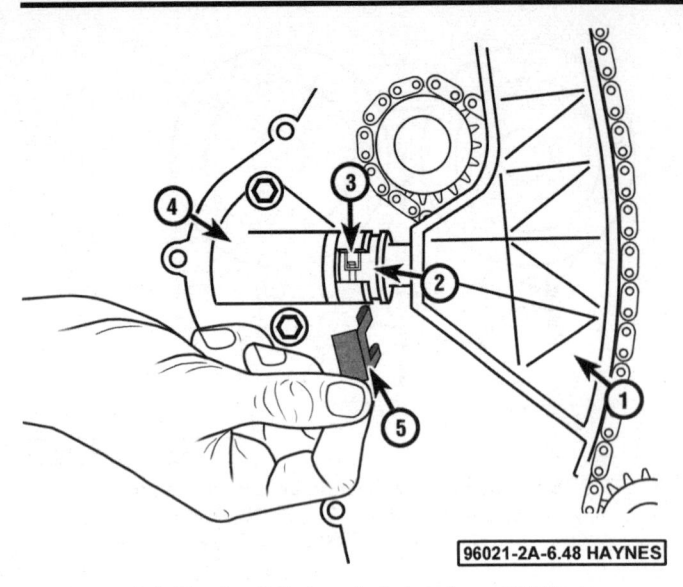

6.48 Version 2 timing chain tensioner details

1. Timing chain tensioner guide
2. Tensioner retaining clip groove
3. Tensioner circlip
4. Tensioner body
5. Tensioner locking clip (Tool #T40267)

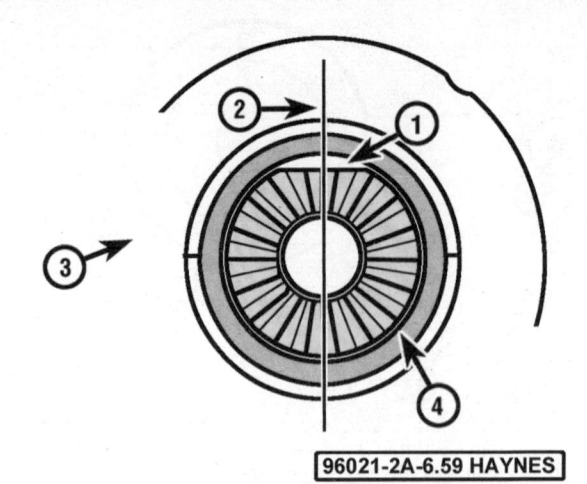

6.59 Mark a line through the center of the crankshaft for TDC identification

1. Crankshaft TDC flat spot 12 o'clock position
2. Permanent marker indexing line
3. Cylinder block
4. Crankshaft locking splines

48 On Version 2 tensioners (with a circlip), use a pair of pliers to compress the ends of the circlip together, then slowly compress the timing chain tensioner piston by prying the timing chain tensioner guide back towards the tensioner until VW special tool #T40267 locking clip **(see illustration)** can be installed into the groove on the outer end of the tensioner to hold it in the compressed position.

49 Install VW special tool #T40271/1 for the exhaust camshaft and #T40271/2 for the intake camshaft to the cylinder head and lock the camshafts in place.

50 The camshaft sprockets must be marked to the special tools #T40271/1 (exhaust camshaft) and #T40271/2 (intake camshaft) before the tools are removed, using paint or a permanent marker. If the camshaft sprockets are to be replaced, the marks must be transferred to the new sprockets.

Note: *To install the special tool #T40271/1 it may be necessary to remove the timing chain tensioner guide top fastener. This may require the help of an assistant to prevent the camshaft from turning while the guide bolt is being removed.*

51 Remove the timing chain upper guide by depressing the lock tab, then slide the top guide off from the top of the chain.

52 In not already done, remove the chain tensioner guide fastener and the chain guide from the left side of the chain.

53 Remove the timing chain tensioner retaining fasteners and tensioner.

Caution: *Do not remove the pin or locking clip holding the tensioner piston in the compressed position once it's removed.*

54 With the oil pump tensioner and guide removed, remove the chain guide fasteners and the chain guide from the right side of the chain.

55 Remove the timing chain.

Crankshaft sprocket

56 Remove the timing chain from the camshaft sprockets and hang it on the camshaft pins, then remove the balance shaft tensioner, guides and chain (see Steps 1 through 54).

57 Loosen the adjusting bolt of the tensioning pin (VW special tool #T10531/2).

6.66 Camshaft timing chain alignment details

1. Exhaust camshaft
2. Intake camshaft
3. Exhaust camshaft timing mark
4. Intake camshaft timing mark
5. Colored link
6. Colored link
7. Crankshaft sprocket timing mark
8. Colored link

Chapter 2 Part A Four-cylinder engines

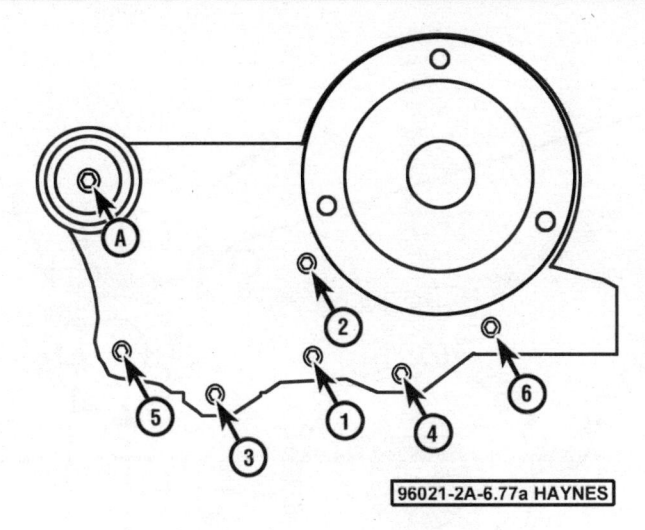

6.77a Bearing bracket tightening sequence - (CXBA, CXBB, CNSA, CNSB, CPRA, CPKA) and 2.0L (CBFA, CCTA, CPLA, CPPA) models

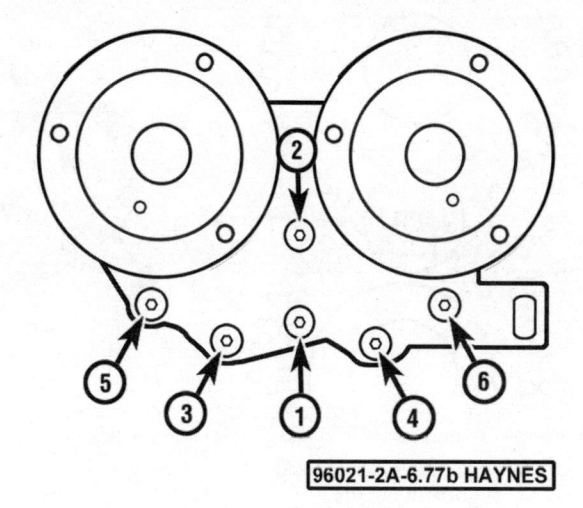

6.77b Bearing bracket tightening sequence - 2.0L (CXCA, CXCB, CNTA, DJJA, CYFB) models

58 If not already removed, remove the oil pump chain and slide the three stage crankshaft sprocket off of the crankshaft.

Installation

Crankshaft sprocket

59 Verify the crankshaft is till at TDC, the flat area on the crankshaft must be at the 12 o'clock position then use a permanent marker and draw a line from the center of the crankshaft straight up **(see illustration)** to aid in installing the three stage crankshaft sprocket.

60 Locate the TDC mark on the three stage crankshaft sprocket, then mark the corresponding tooth with a permanent marker.

61 Tilt the top of three stage chain sprocket towards the engine block to line up the marked tooth of the sprocket with the center mark made in Step 59. Hold it in place and secure it to the crankshaft using VW special tool #T10531/2.

62 Install tensioning pin (special tool #T10531/2) into the threaded end of the crankshaft. Tighten it hand-tight, then install VW special tools #T10531/3 collar and #T10531/4 lock nut, and as the lock nut is tightened, move the crankshaft sprocket back and forth slightly to check if it is seated correctly in the locking splines (tooth contour). Tighten the lock nut until the crankshaft sprocket can no longer be turned.

Camshaft timing chain and tensioner

Caution: *Before starting the engine, carefully rotate the crankshaft by hand through at least two full revolutions (use a socket and breaker bar on the crankshaft pulley center bolt). If you feel any resistance, STOP! There is something wrong - most likely, valves are contacting the pistons. You must find the problem before proceeding. Check your work and see if any updated repair information is available.*

63 Remove all traces of old sealant from the timing chain covers and the mating surfaces of the engine block and cylinder head.

64 Make sure the crankshaft is at TDC and the balance shaft chain marks are aligned **(see illustration 6.84)**.

65 Loop the timing chain around the crankshaft sprocket and align the No.1 colored link with the mark on the crankshaft sprocket. Place the timing chain over the exhaust camshaft and align the colored link with the mark on the exhaust camshaft sprocket.

Note: *There are three bright or colored links on the timing chain. The No.1 colored link is the link farthest away from the two colored links that are closest together.*

66 Once the camshaft sprocket and colored link are aligned **(see illustration)**, install the chain onto the teeth of the camshaft sprocket.

67 Install the right-side chain guides and fasteners, then tighten the fasteners to the torque listed in this Chapter's Specifications.

68 Install the timing chain upper guide to the mounting bracket above the top of the chain.

69 On 2.0L models, rotate the exhaust camshaft sprocket clockwise using VW special tool #T40266 and push VW special holding tool T40271/1 out of the sprocket splines. With the help of an assistant, rotate the camshaft counterclockwise and install the tensioner guide and fastener, then tighten the guide fastener to the torque listed in this Chapter's Specifications.

70 On models without an exhaust camshaft adjuster, use a spanner wrench to rotate the exhaust camshaft sprocket clockwise and push VW special holding tool T40271/1 out of the sprocket splines. With the help of an assistant, rotate the camshaft counterclockwise and install the tensioner guide and fastener. Tighten the guide fastener to the torque listed in this Chapter's Specifications.

71 Install the timing chain tensioner and tighten the retaining fasteners to the torque listed in this Chapter's Specifications.

72 Remove the pin from the tensioner to release the tensioner piston to engage the chain guide, then slowly release the wrench holding the intake camshaft.

Oil pump timing chain

73 Loop the chain over the oil pump drive sprocket and the crankshaft sprocket.

74 Install the oil pump chain tensioner guide against the chain and install the fastener, then tighten the fastener pin to the torque listed in this Chapter's Specifications. Remove the pin from the tensioner to release the tensioner piston to engage the chain guide - make sure the wire spring contacts the tab section of the upper oil pan.

75 Turn the intake camshaft counterclockwise using VW special tool #T40266, then slide the special tool #T40271/2 out from the camshaft sprocket teeth and release the camshaft. Remove both VW special tool #T40271/1 and #T40271/2 camshaft sprocket positioning tools from the engine block.

76 Check that all of the timing marks are aligned **(see illustration 6.66 and 6.85)**.

77 Install the bearing brackets and the retaining fasteners, then tighten the fasteners in sequence **(see illustrations)** to the torque listed in this Chapter's Specifications. On 1.8L (CXBA, CXBB, CNSA, CNSB, CPRA, CPKA) and 2.0L (CBFA, CCTA, CPLA, CPPA) models tighten the bracket-to-exhaust camshaft bolts to the torque listed in this Chapter's Specifications.

Caution: *Carefully rotate the crankshaft by hand through at least two full revolutions (use a socket and breaker bar on the crankshaft pulley center bolt). If you feel any resistance, STOP! There is something wrong - most likely, valves are contacting the pistons. You must find the problem before proceeding.*

2A-18   Chapter 2 Part A Four-cylinder engines

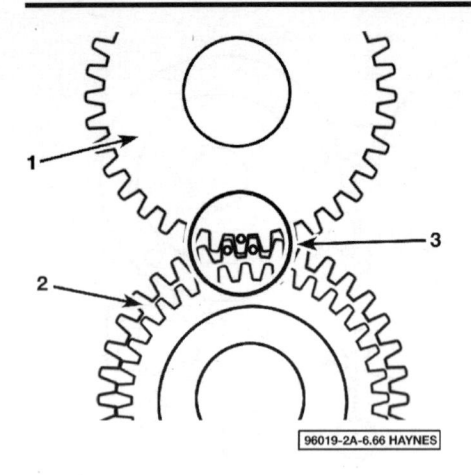

6.84 Intermediate shaft sprocket alignment details

1. Intake side balance shaft
2. Intermediate shaft sprocket
3. Timing marks

78 Install the lower timing chain cover (see Steps 31 through 37), then the upper timing chain cover (see Steps 7 through 10).
79 Install the bearing bracket and tighten the retaining fasteners hand tight.
80 The remainder of installation is the reverse of removal.

Balance shaft timing chain and tensioner

Removal

81 Remove the camshaft timing chain covers, oil pump chain, timing chain and tensioner as previously described in Steps 1 through 58.
82 With the engine set at TDC, unscrew the balance shaft tensioner from the side of the cylinder block.
83 Remove the chain guide retaining pins, the chain guides from each side of the chain and the balance shaft chain.

Installation

Caution: *Before starting the engine, carefully rotate the crankshaft by hand through at least two full revolutions (use a socket and breaker bar on the crankshaft pulley center bolt). If you feel any resistance, STOP! There is something wrong - most likely, valves are contacting the pistons. You must find the problem before proceeding. Check your work and see if any updated repair information is available.*

84 Rotate the intermediate shaft sprocket and intake side balance shaft and align the timing marks on the gears **(see illustration)**.
Note: *The timing marks will only align every seven rotations.*
85 Loop the timing chain around the intermediate shaft sprocket and align one of the colored links of the chain with the mark on the intermediate shaft sprocket. Place the timing chain over the exhaust side balance shaft and crankshaft, then align the colored link with the mark on the exhaust side balance shaft sprocket and crankshaft **(see illustration)**.

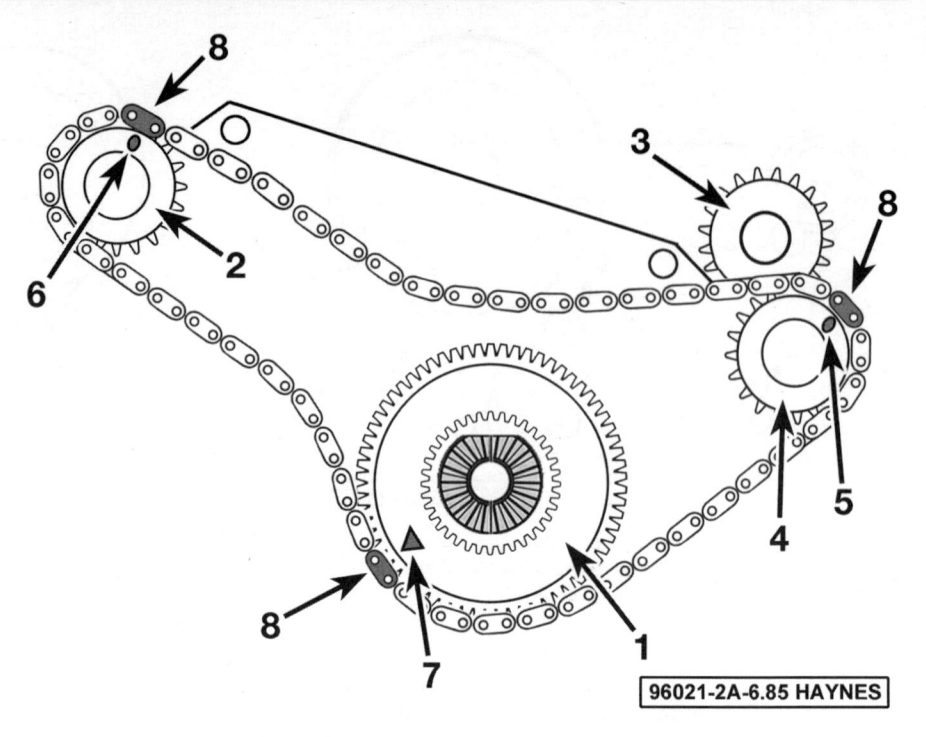

6.85 Balance shaft timing chain alignment details

1. Crankshaft sprocket
2. Exhaust side balance shaft
3. Intake side balance shaft
4. Intermediate sprocket
5. Intermediate sprocket timing mark
6. Exhaust side balance shaft timing mark
7. Crankshaft sprocket timing mark
8. Colored links

Note: *There are three bright or colored links equally spaced on the timing chain.*

86 Install the tensioner side chain guide, then the lower tensioner guide and upper chain guide and pins. Tighten the retaining pins to the torque listed in this Chapter's Specifications.
87 Apply non-hardening thread locking compound to the tensioner, then install the tensioner and tighten it to the torque listed in this Chapter's Specifications.
88 Check that all three balance shaft timing chain marks are correctly aligned.
89 The remainder of installation is the reverse of removal.

7 Intake manifold - removal and installation

Warning: *Wait until the engine is completely cool before beginning this procedure.*
Caution: *If the battery is disconnected, several systems must be re-learned before they will work properly (see Chapter 5, Section 3).*

1 Relieve the fuel system pressure (see Chapter 4).
2 Disconnect the cable from the negative battery terminal (see Chapter 5).
3 Working from above in the engine compartment, remove the engine cover/air filter housing (see Chapter 1, Section 7) or air filter housing (see Chapter 4).

1.4L (CZTA, DGXA) models

Removal

4 Raise the vehicle and support it securely on jackstands then remove the engine splash shield fasteners and shield.
5 Drain the coolant (see Chapter 1).
6 Disconnect the EVAP canister hoses to the intake manifold (see Chapter 6).
7 Remove the throttle body (see Chapter 4, Section 12).
8 Disconnect the quick-connect fitting for the vacuum hose and the coolant hose to the intake manifold then remove the hoses **(see illustration)**.
9 Disconnect the coolant hoses from the intake manifold **(see illustration)**.
10 Disconnect the coolant hose under the intake manifold and remove the pipe fastener **(see illustration)**.
11 Disconnect the electrical connectors from the fuel pressure sensor **(see illustration)**, the oil pressure switch and the canister purge regulator.
12 Remove the intake manifold bolts and remove the intake manifold **(see illustration 7.14)**. If necessary, remove the fuel supply line to the fuel pump and plug the line.

Installation

13 There are individual gaskets for each of the four ports of the intake manifold. Using new manifold gaskets, install the intake mani-

Chapter 2 Part A Four-cylinder engines

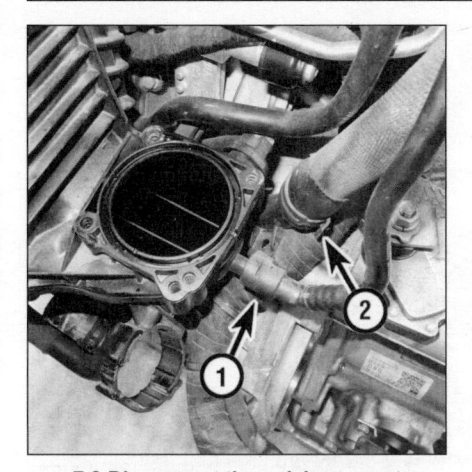

7.8 Disconnect the quick-connect fitting and the coolant hose from the intake manifold

1 Vacuum hose quick-connect fitting
2 Coolant hose clamp and hose

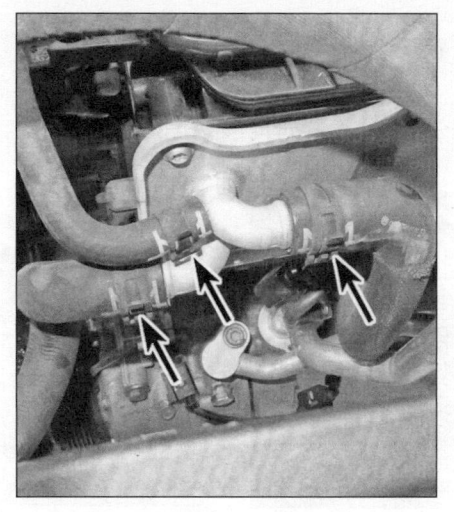

7.9 Disconnect the coolant hoses at the intake manifold

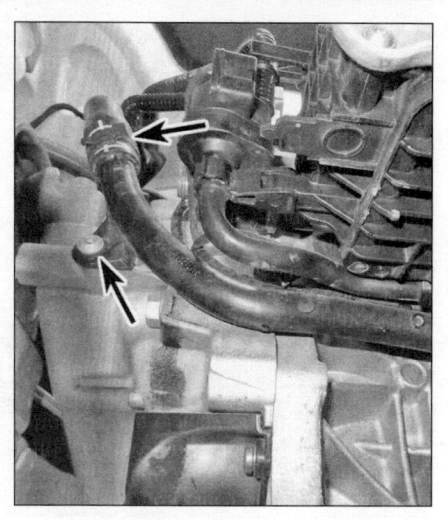

7.10 Disconnect the coolant hose from the pipe then remove the pipe bracket fastener

7.11 Disconnect the electrical connectors from the fuel pressure sensor

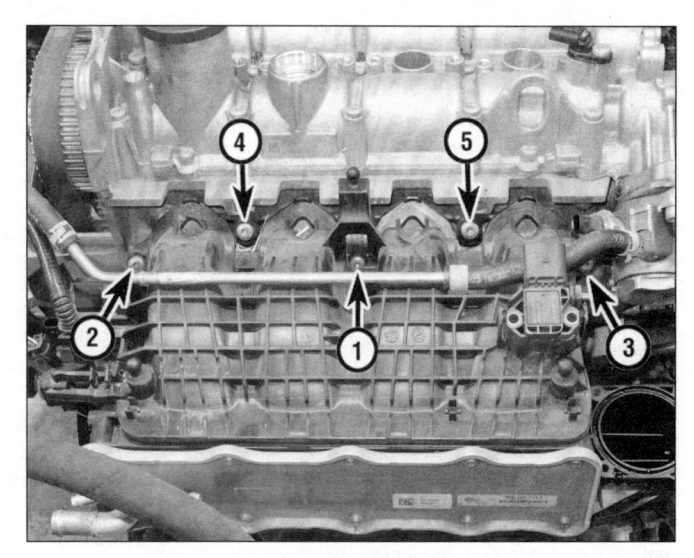

7.13 Intake manifold tightening sequence - 1.4L (CZTA, DGXA) models

fold. Tighten the bolts in sequence **(see illustration)**, to the torque listed in 2Bthis Chapter's Specifications.
14 The remainder of installation is the reverse of removal.

2.0L (CBPA) models

Removal
15 Disconnect the fuel supply line (see Chapter 4). Cover the fuel line openings to prevent debris from entering the fuel system.
16 Clamp off the coolant hoses leading to the throttle body and remove them from the throttle body (see Chapter 4).
17 Disconnect the electrical connectors from the fuel injectors, Camshaft Position (CMP) sensor and the EVAP canister purge valve.
18 Unclip the wiring harness and EVAP hose clamp from the intake manifold, then open the locking ring and remove the EVAP canister purge valve.
19 Remove the oil dipstick tube-to-engine block retaining fastener and rotate the tube out of the way.
20 The intake manifold is secured with eight bolts, four below and four along the top. Unscrew the bolts and pull the intake manifold away from the engine at a slight angle for clearance.

Installation
21 There are individual gaskets for each of the four ports of the intake manifold. Using new manifold gaskets, install the intake manifold. Tighten the bolts in several stages, working from the center out, to the torque listed in 2Bthis Chapter's Specifications.
22 The remainder of installation is the reverse of removal.

1.8L and 2.0L timing chain models

Note: *On models with a Direct Injection fuel injection system, the Teflon combustion chamber seals and injector O-rings must be replaced prior to installation of the intake manifold (see Chapter 4).*

Removal
Note: *If the intake manifold is removed or replaced, the intake manifold runner position sensor must be adapted or relearn to the engine control module (ECM) (see Chapter 6).*
23 Raise the vehicle and support it securely on jackstands, then remove the engine lower splash shield.
24 On 1.8L (CPRA, CPKA) and 2.0L (CPLA, CPPA, CBFA, CBTA) models, remove the cooling fan shroud (see Chapter 3).

2A-20 Chapter 2 Part A Four-cylinder engines

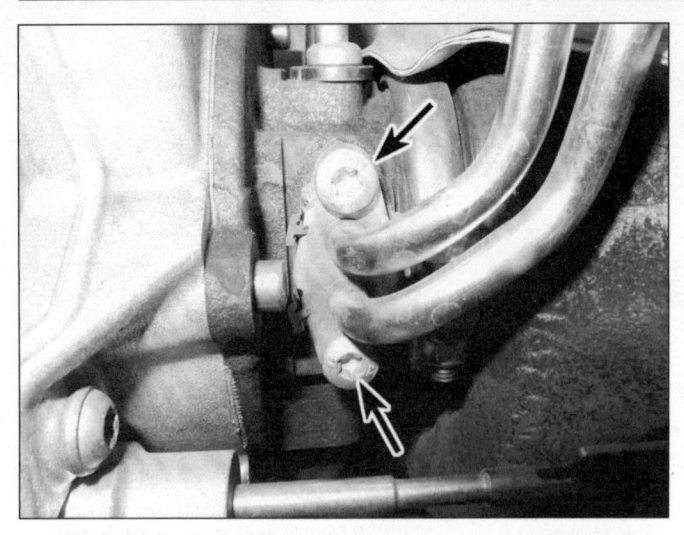

8.8a Remove the coolant pipe fasteners to the turbocharger

8.8b Coolant pipe details
1 Hose connections
2 Tube bracket fasteners

25 Remove the coolant tube bracket bolt and separate the tube from the end of the intake manifold.
26 Remove the charge air cooler inlet and outlet pipes (see Chapter 4).
27 Disconnect the electrical connectors for the cooling system under the intake manifold.
28 Disconnect the fuel supply line (see Chapter 4), then disconnect the electrical connectors from the sensors (see Chapter 6) on the intake manifold.
29 On CBFA, CCTA, models, remove engine oil filter (see Chapter 1) and the throttle body (see Chapter 4).
30 Disconnect the vacuum pump quick-connector, then disconnect the manifold runner control valve from the lower rear side of the intake manifold.
31 Disconnect the fuel line from the high-pressure fuel pump and from the pump to the fuel rail (see Chapter 4, Section 9).
32 Remove the intake manifold support bracket bolts and bracket from under the intake manifold.
33 Disconnect the electrical connector from the intake manifold runner position sensor and the camshaft position sensors (see Chapter 6).
Note: *On CBFA, CCTA models, the intake manifold runner position sensor can only be disconnected once the manifold is pulled back slightly from the cylinder head.*
34 Remove the intake manifold mounting nuts and bolts, then pull the intake manifold back from the cylinder head to locate and access the electrical harness bracket under the manifold. Remove the harness bracket fasteners and remove the harnesses from the intake manifold.
35 Remove the intake manifold from the vehicle and cover all the openings.
36 On 1.8L (CPRA, CPKA) and 2.0L (CPLA, CPPA,CBFA, CBTA) models, remove the fuel rail and replace the injector seals, then reinstall the fuel rail and injectors (see Chapter 4, Section 13) to the intake manifold.

Installation
37 Make sure to seat the gasket around the four ports of the intake manifold using new manifold gasket, then install the intake manifold.
38 Install the harness bracket and fasteners on the bottom of the intake manifold, then tighten the fasteners securely.
39 Tighten the bolts in several stages, working from the center out, to the torque listed in this Chapter's Specifications.
40 The remainder of installation is the reverse of removal.

8 Camshaft housing, camshafts, sprockets, roller rocker arms and lash adjusters - removal, inspection and installation

Warning: *Wait until the engine is completely cool before beginning this procedure.*
Caution: *If the battery is disconnected, several systems must be re-learned before they will work properly (see Chapter 5, Section 3).*
Caution: *Severe engine damage will occur if you make any mistakes. Do not attempt this procedure unless you are highly experienced with this type of repair. If you are at all unsure of your abilities, be sure to consult an expert. Double-check all your work and be sure everything is correct before you attempt to start the engine.*
Note: *The camshafts and lifters should always be thoroughly inspected before installation and camshaft endplay should always be checked prior to camshaft removal. Although the hydraulic lifters are self adjusting and require no periodic service, there is an in-vehicle procedure for checking excessively noisy hydraulic lifters.*

Removal
1.4L (CZTA, DGXA) models
Note: *On 1.4L (CZTA, DGXA) models, the camshafts can not be removed from the camshaft housing, if there is a problem with a camshaft or the camshaft housing the housing and camshafts must be replaced as an assembly.*
1 Disconnect the cable from the negative terminal of the battery (see Chapter 5).
2 Drain the cooling system (see Chapter 1).
3 Remove the air filter housing and resonator (see Chapter 4).
4 Remove the water pump (see Chapter 3, Section 7).
5 Remove the ignition coils (see Chapter 5).
6 Remove the high pressure fuel pump (see Chapter 4).
7 Disconnect the electrical connectors from the intake manifold sensor, the camshaft adjustment valves, and the camshaft position sensors (see Chapter 6).
8 Remove the mounting bolts to the turbocharger coolant pipes **(see illustration)** and disconnect the hoses to the metal lines **(see illustration)** that run across the top of the engine and the pipe retaining bolts, if equipped.
9 Remove the turbocharger heat shield fasteners and support bracket **(see illustration)**.
10 Remove the timing belt upper cover (see Section 2).
11 Set the engine to TDC (see Section 3).
12 Remove the timing belt from the camshaft sprockets (see Section 5).
Caution: *The contact points for the timing belt on the camshaft sprockets, crankshaft sprocket, tensioner and idler pulley must be kept free from oil to prevent damage to the belt and possible premature failure.*
13 Loosen the camshaft housing bolts in the reverse order of the tightening sequence

Chapter 2 Part A Four-cylinder engines

8.9 Turbocharger heat shield and bracket fastener locations

(see illustration 8.93) then carefully separate the camshaft housing from the cylinder head. A gasket sealing material type adhesive is used so it may take some effort to separate the components.

Note: *The camshaft housing bolts have different lengths - be sure to mark the locations of the bolts so new ones can be installed into the original positions.*

14 Remove all traces of gasket sealing material from the camshaft housing cover and cylinder head mating surfaces.

Note: *Be sure to use NEW cover bolts, as the old bolts are stretch-type fasteners that will not provide the correct torque readings if reused.*

15 With the camshaft housing removed, mark and remove the roller rocker and lifters from the cylinder head they must be installed in their original locations.

Camshaft sprockets and seals

16 On models with adjustable camshaft sprockets, place a clean rag under the camshaft adjuster and tensioning roller to catch the engine oil.

17 On the intake camshaft, install VW special tools #T10554 to T10172 to the camshaft sprocket, aligning the four pins of the mounting tool with the holes on the sprockets to prevent the sprockets from turning, then remove the center plug and O-ring using an Allen wrench.

18 On the exhaust camshaft remove the five cap fasteners, then remove the cap and O-ring.

19 Install VW special tools #T10554 to T10172 to the intake camshaft sprocket, aligning the four pins of the mounting tool with the holes on the sprockets. Prevent the sprocket from turning and loosen the center bolt. Install special tools to the exhaust camshaft sprocket, prevent the sprocket from turning and loosen the center bolt, then remove the tool.

20 With the adjustable camshaft sprocket bolts loose, remove the bolt(s) and remove the sprocket(s) from the end of the camshaft(s).

21 On models with non-adjustable camshaft sprockets, hold the sprocket from turning using VW special tools #T10172 to T10172/1 (or equivalent), then remove the mounting bolt and the sprocket from the exhaust camshaft.

22 Using a seal removal tool or VW special tool #T20143, pry the seal out of the camshaft housing.

23 Before installing the camshaft seal, make sure the seal contact surfaces of the camshaft housing cover are perfectly clean.

24 Install the new camshaft seal (with the closed side of the seal facing the seal installer) onto VW special tool #T10478/3 and #T10478/2, or equivalent seal installer then place the seal and tool. Do not apply oil to the new camshaft seal.

Note: *VW special tools use a two stage seal installer to make sure the lips on the seals are in the proper positions.*

25 Slide the seal installer onto the end of the camshaft then install VW special tool #T10478/1A against the seal installer and use the old camshaft sprocket bolt to mount the tool to the end of the camshaft. Tighten the bolt until the camshaft seal and tool are seated flush against the camshaft housing.

26 Working at the rear of the camshaft housing, in place of the camshaft cap **(see illustration 3.16)** attach (VW special tool #T10494) camshaft positioning tool to the intake camshaft with the opposite end of the tool seating on the coolant pump mounting bolt on the exhaust camshaft. With the tool in this position install the camshaft sprocket to the camshaft and hand tighten the new camshaft sprocket center bolt.

27 Install VW special tools #T10554 to T10172 to the camshaft sprocket, aligning the four pins of the mounting tool with the holes on the sprockets to prevent the sprockets from turning, then tighten the camshaft sprocket bolt to the torque listed in this Chapter's Specifications. Remove the tool and repeat this step for the remaining camshaft.

28 On the intake camshaft, install VW special tools #T10554 to T10172 to the camshaft sprocket, aligning the four pins of the mounting tool with the holes on the sprocket to prevent the sprocket from turning. Install the center plug with new O-ring and tighten the it to the torque listed in this Chapter's Specifications. On the exhaust camshaft, install a new O-ring to the sprocket, then install the cap and tighten the five fasteners to the torque listed in this Chapter's Specifications.

Caution: *When tightening the camshaft sprocket bolts, do not use the camshaft positioning tool (VW special tool #T10494) to hold the camshafts from turning or damage to the camshafts may occur.*

2.0L (CBPA) models

29 Set the engine to TDC (see Section 3).

30 Disconnect the cable from the negative terminal of the battery (see Chapter 5).

31 Working from above in the engine compartment, remove the engine cover and air filter housing (see Chapter 4).

32 Remove the valve cover (see Section 4).

33 Clamp the coolant hoses off at the end of the cylinder head, then disconnect and plug the coolant hoses from the coolant tube. Remove the coolant tube retaining fastener from the heat shield and remove the tube.

34 Remove the electrical harness retaining fasteners and pull the harness away from the end of the cylinder head.

35 Verify that the camshaft sprocket mark is aligned with the mark on the rear timing belt cover.

36 Hold the camshaft sprocket from rotating with VW special tool #3415 (or equivalent), then loosen the sprocket center bolt.

37 Remove the timing belt (see Section 5), then side the camshaft sprocket bolt from the end of the camshaft and remove the Woodruff key from the camshaft.

38 Remove the upper rear timing belt guard and camshaft position sensor fastener, then the sensor bracket and the guard from the cylinder head.

39 Loosen both bearing cap nuts for caps 1, 3 and 5, then for caps 2 and 4. Loosen each nut a 1/2 turn at a time, alternating between the caps until all the caps are loose then remove the bearing caps **(see illustration 8.101)**.

Caution: *Keep the caps in order and don't mix them up they must be installed into their original locations.*

40 Remove the camshaft and front seal from the cylinder head.

41 With the camshaft removed, remove the hydraulic lifters and keep them in order they must be installed in their original locations.

Camshaft sprockets and seals

42 If you're trying to replace the camshaft seal without removing the camshaft, remove the timing belt and camshaft sprocket (see Step 36) then use a seal removal tool to pry the seal out of the camshaft bearing cap.

43 To install a new seal, coat the lip of the seal with clean engine oil and drive the the seal into the camshaft bearing cap bore with a seal driver or a large socket slightly smaller in diameter than the seal. The open open side of the seal should face into the engine. Install the camshaft sprocket.

8.81 Measure the outside diameter of each camshaft journal and the inside diameter of each bearing surface on the cylinder head to determine the oil clearance measurement

8.83 Checking camshaft endplay with a dial indicator

1.8L and 2.0L timing chain models

Note: *The camshaft bearings are integrated in the cylinder head and cylinder head cover - the surfaces of the cylinder head cover and the upper portion of the cylinder head must not be machined. If there is a problem, the cylinder head cover and cylinder head must be replaced as an assembly.*

44 Disconnect the cable from the negative terminal of the battery (see Chapter 5).

45 Loosen the right-front wheel bolts, raise the front of the vehicle and support it securely on jackstands. Block the wheels at the opposite end and remove the right-front wheel.

46 Working under the vehicle, remove the lower splash shield below the engine (see Chapter 1, Section 6).

47 Working from above in the engine compartment, remove the engine cover (see Chapter 1, Section 7).

48 Remove the right-front inner fender liner (see Chapter 11).

49 Drain the engine coolant (see Chapter 1).

50 Remove the air filter housing (see Chapter 4).

51 Disconnect the coolant pipe retainers, then disconnect the hoses to the pipes and remove the pipes across the top of the engine.

Caution: *Do not bend the coolant pipes - the internal coating can be cracked and ruined.*

52 Remove the ignition coils (see Chapter 5) and the spark plugs (see Chapter 1).

53 Rotate the engine in the normal direction of rotation (clockwise) until the No. 1 cylinder is located at TDC (see Section 3).

54 Disconnect the electrical connectors to the turbocharger recirculation valve, the camshaft position sensor(s), the fuel pressure regulator and the camshaft adjustment actuators (see Chapter 6) and disengage the wiring harness retainers along the top and sides of the camshaft cover.

55 Disconnect the electrical connector to the canister purge valve, remove the valve (see Chapter 6) and move the harness out of the way.

56 Remove the oil separator from the top of the camshaft cover (see Section 17).

57 Remove the air charge pipe retaining fasteners and the clamps at each end of the air charge pipe.

58 Loosen the hose clamps on the lower air charge hose and remove the hose.

59 Remove the drivebelt and tensioner (see Chapter 1).

60 Remove the high-pressure fuel pump (see Chapter 4).

61 Disconnect the charge air tube from the turbocharger (see Chapter 4).

62 Remove the ground wire fastener below the vacuum pump and the bracket retaining fastener to the left of the ground wire.

63 Disconnect the vacuum hose from the pump, then plug the hose and cap the port on the vacuum pump.

64 Remove the vacuum pump fasteners (see Chapter 9).

65 Remove the upper timing chain cover (see Section 6).

66 Verify the crankshaft is still at TDC mark **(see illustration 3.6b)** and the camshaft sprocket marks match the marks on the camshaft cover at approximately the 11 o'clock position on the exhaust camshaft and the 1 o'clock position on the intake camshaft **(see illustration 8.114)**.

67 Make alignment marks from the camshaft timing chain to the camshaft sprockets using a permanent marker.

Note: *The marks will be used for reinstallation or if the camshaft is being replaced.*

68 Remove the control valve and bearing bracket (see Section 6).

Note: *The control valve(s) is left-hand thread.*

69 With the bearing bracket removed, measure and record the distance between the timing marks on the camshaft sprockets so they can be installed in the exact same position.

70 Remove the lower timing chain cover (see Section 6).

Note: *There are multiple timing chain tensioner versions used and, on some of these versions, the timing chain tensioner can be released without removing the lower timing chain cover. During manufacturing the lower timing chain cover was redesigned and these new style covers prevent access to the tensioner lock, so these steps only show how to proceed with the lower timing chain cover removed to prevent any possible problems.*

71 On Version 1 tensioners (without a circlip), use a small screwdriver or pick to hold the chain tensioner ratchet arm away (UP) from the ratchet stem, then slowly compress the timing chain tensioner piston by prying the timing chain tensioner guide back towards the tensioner until the pin (a drill bit or paper clip will work) can be inserted into the hole to hold it in the compressed position **(see illustration 6.47)**.

72 On Version 2 tensioners (with a circlip), use a pair of pliers to compress the ends of the circlip together, then slowly compress the timing chain tensioner piston by prying the timing chain tensioner guide back towards the tensioner until VW special tool #T40267 locking clip **(see illustration 6.48)** can be installed into the groove on the outer end of the tensioner to hold it in the compressed position.

73 Install VW special tool #T40271/1 (for the exhaust camshaft) and #T40271/2 (for the intake camshaft) camshaft sprocket position tools to the cylinder head and lock the camshafts in place, then remove the timing chain and guides (see Section 6).

74 The camshafts must be marked to the special tools #T40271/1 (exhaust camshaft) and #T40271/2 (intake camshaft) before the tools are removed using paint or a permanent marker. If the camshafts are to be replaced, the marks must be transferred to the new camshaft sprockets.

Chapter 2 Part A Four-cylinder engines   2A-23

8.84 Measuring the camshaft lobe height with a micrometer - make sure you move the micrometer to get the highest reading (top of cam lobe)

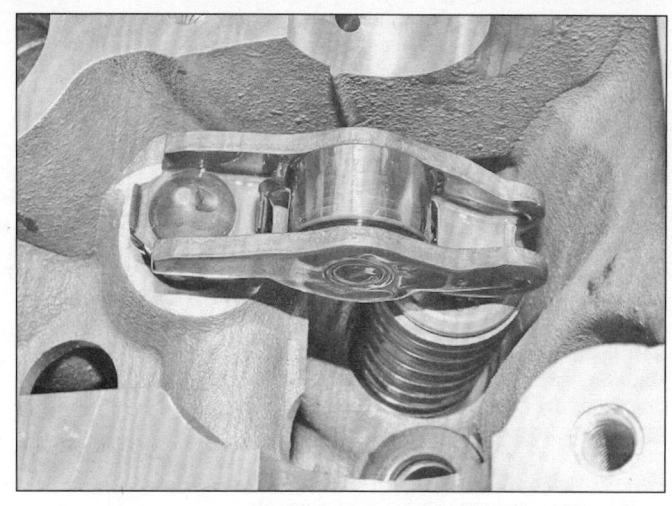

8.85 Typical roller rocker and lash adjuster assembly

75 Install VW special tool #T10531/3 and #T10531/4 to the end of the crankshaft with the flat spot of the tool facing upwards - this is the "TDC point" for the tool - then turn the crankshaft with an open-end wrench counterclockwise out of "TDC".

76 On CBFA and CCTA models, insert a wrench on the flat spot on the back side of the intake camshaft adjuster. On all other models, use VW special tool #T40266 socket on the center of the intake camshaft adjuster, slightly turn the camshaft counterclockwise and slide special tool #T40272 out of the sprocket splines and slowly release the camshaft into the resting position.

77 On CBFA and CCTA models, insert a wrench on the flat spot on the back side of the exhaust camshaft adjuster. On all other models use VW special tool #T40266 socket on the center of the exhaust camshaft adjuster, slightly turn the camshaft clockwise and slide special tool T40271 out of the sprocket splines, then slowly release the camshaft into the resting position.

78 On models without an exhaust camshaft adjuster, use a spanner wrench to firmly hold and slightly turn the camshaft clockwise, slide special tool T40271 out of the sprocket splines, then slowly release the camshaft into the resting position.

79 Remove the cylinder head cover bolts in the reverse of the installation sequence (see illustrations 8.117), then remove the cylinder head cover and camshafts and set them aside in a clean space. Remove all traces of gasket sealing material from the cylinder head cover and cylinder head mating surfaces.
Note: *Be sure to use NEW cover bolts, as the old bolts are stretch-type fasteners that will not provide the correct torque readings if reused.*

Inspection

Note: *On 1.4L (CZTA, DGXA) models, the camshafts cannot be removed from the camshaft housing. If there is a problem with a camshaft or the camshaft housing, the housing and camshafts must be replaced as an assembly.*

80 Check the camshaft cap surfaces for pitting, score marks, galling and abnormal wear. If the bearing surfaces are damaged, the cylinder head and, on timing chain models, the cylinder head cover must be replaced.

81 Measure the outside diameter of each camshaft bearing journal and record your measurements (see illustration). Then measure the inside diameter of each corresponding camshaft bearing and record the measurements. Subtract each cam journal outside diameter from its respective cam bearing bore inside diameter to determine the oil clearance for each bearing. Compare the results to the specified journal-to-bearing clearance. If any of the measurements fall outside the standard specified wear limits in this Chapter, either the camshaft or the cylinder head, or both, must be replaced.

82 Check camshaft runout by placing the camshaft back into the cylinder head and set up a dial indicator on the center journal. Zero the dial indicator. Turn the camshaft slowly and note the dial indicator readings. Record your readings and compare them with the specified runout in this Chapter. If the measured runout exceeds the runout specified in this Chapter, replace the camshaft.

83 Place the camshafts back into the cylinder head and temporarily install the cylinder head cover or guide frame. Check the camshaft endplay by placing a dial indicator with the stem in line with the camshaft and touching the snout (see illustration). Push the camshaft all the way to the rear and zero the dial indicator. Next, pry the camshaft to the front as far as possible and check the reading on the dial indicator. The distance it moves is the endplay. If it's greater than the value listed in this Chapter's Specifications, check the bearing caps/guide frame for wear. If the bearing caps or guide frame are worn, the cylinder head must be replaced.

84 Compare the camshaft lobe height by measuring each lobe with a micrometer (see illustration). Measure each of the intake lobes and record the measurements and relative positions. Then measure each of the exhaust lobes and record the measurements and relative positions also. This will let you compare all of the intake lobes to one another and all of the exhaust lobes to one another. If the difference between the lobes exceeds 0.005 inch, the camshaft should be replaced. Do not compare intake lobe heights to exhaust lobe heights as lobe lift may be different. Only compare intake lobes-to-intake lobes and exhaust lobes-to-exhaust lobes for this comparison.

85 Inspect the contact and sliding surfaces of each rocker arm roller for wear and scratches (see illustration).
Caution: *If the rocker roller is worn, it's a good idea to check the corresponding camshaft. Do not lay the lifters on their side or upside down, or air can become trapped inside and the lifter will have to be bled. The lifters can be laid on their side only if they are submerged in a pan of clean engine oil until reassembly.*

86 Check the contact surfaces of each lash adjuster.

Installation

1.4L (CZTA, DGXA) models

87 Install all the lifters into their original locations, then install the roller rocker arms, making sure they are mounted correctly on the valve shaft ends and are clipped to the lifters.

88 Install a new O-ring seal and screen to the front of the cylinder head.

89 Install the two guide pins to the front side of the cylinder head, then thread the VW special tool #T10288/4 guide pins, or equivalent, into the front-left corner and right rear corner of the cylinder head.

90 Coat the sealing surface of the cylinder head with VW liquid gasket adhesive. Also coat the two guide pins at the front side of the cylinder head.

Chapter 2 Part A Four-cylinder engines

91 Install the camshaft housing over the special tool guide pins, keeping it level and straight while lowering the camshaft housing down onto the cylinder head.

92 Remove the special tool guide pins and install the camshaft housing bolts in their original locations - bolts 1 through 4 (M6 x 52 mm) and bolts 5 through 15 (M6 x 65 mm) - in sequence **(see illustration 8.93)** hand-tight.

93 Tighten the camshaft housing bolts in several stages and in sequence **(see illustration)** to the torque listed in this Chapter's Specifications.

94 Install the timing belt (see Section 5).

Caution: *After 30 minutes, carefully rotate the crankshaft by hand through at least two full revolutions (use a socket and breaker bar on the crankshaft pulley center bolt). If you feel any resistance, STOP! There is something wrong - most likely valves are contacting the pistons. You must find the problem before proceeding. After two revolutions, ensure that both TDC marks still align.*

95 The remainder of installation is the reverse of removal.

96 Refill the cooling system (see Chapter 1), then start the engine check for leaks.

2.0L (CBPA) models

Caution: *On 2.0L (CBPA) models, if new hydraulic valve lifters were installed, wait at least 30 minutes before starting the engine. The lifters must equalize and seat themselves otherwise the valves will hit the pistons.*

97 Clean and lubricate the camshafts, journals and bearing cap surfaces with clean engine oil.

98 Install the camshaft in its original location then install the camshaft oil seal to the front of the camshaft.

99 Install the camshaft bearing caps in their original positions and direction.

100 Install the camshaft nuts and tighten them hand-tight, then tighten bearing caps No. 2 and No. 4 **(see illustration 8.101)** a 1/2-turn at a time, alternating between the two bearing caps, until the camshaft is seated.

8.93 1.4L (CZTA, DGXA) camshaft housing bolt TIGHTENING sequence

101 Once the camshaft is seated, tighten the bearing cap nuts in sequence **(see illustration)** to the torque listed in this Chapter's Specifications.

102 Verify the camshaft seal is seated then install the woodruff key to the camshaft then install the camshaft sprocket and new bolt. Prevent the camshaft sprocket from turning and tighten the sprocket bolt to the torque listed in this Chapter's Specifications.

103 Install the timing belt (see Section 5).

Caution: *If new lifters were installed, don't rotate the engine for 30 minutes. After 30 minutes, carefully rotate the crankshaft by hand through at least two full revolutions (use a socket and breaker bar on the crankshaft pulley center bolt). If you feel any resistance, STOP! There is something wrong - most likely valves are contacting the pistons. You must find the problem before proceeding. After two revolutions, ensure that both TDC marks still align.*

104 The remainder of installation is the reverse of removal.

1.8L and 2.0L timing chain models

105 Clean the camshaft housing mounting surfaces - they must be oil and debris free.

106 Make sure the roller rocker arms make proper contact with the valve stems and are clipped into place on the lifters **(see illustration 8.85)**.

107 Clean and lubricate the camshafts and the cylinder head cover bearing surfaces with clean engine oil.

108 Verify the crankshaft is slightly off of TDC (see Step 75).

109 On all models, place the intake camshaft into the cylinder head with the No.4 cylinder cam lobes point upwards.

110 On 1.8L (CXBA, CXBB, CNSA, CNSB, CPRA, CPKA) and 2.0L (CPLA, CPPA) models, place the exhaust camshaft into the cylinder head with the No. 4 cylinder cam lobes point upwards.

111 On 2.0L (CBFA, CCTA), place the exhaust camshaft into the cylinder head; the recesses on each camshaft should be facing each other.

112 On 2.0L (CNTA, CXCA, CXCB, CYFB, DJJA) models, slide the camshaft lobe assemblies towards each other to line the journals up with the cylinder head cover journals **(see illustration)**, then place the exhaust camshaft into the cylinder head cover.

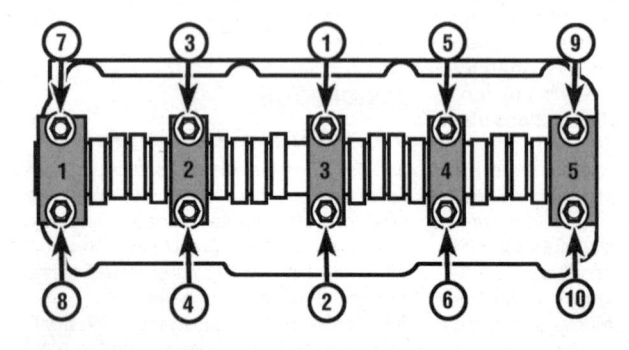

8.101 2.0L (CBPA) seated camshaft bearing cap TIGHTENING sequence

8.112 On 2.0L (CNTA, CXCA, CXCB, CYFB, DJJA) models, slide the exhaust camshaft lobes together until the camshaft journals are matched with the camshaft journals on the cylinder head cover

1 Exhaust camshaft lobes   2 Camshaft journals

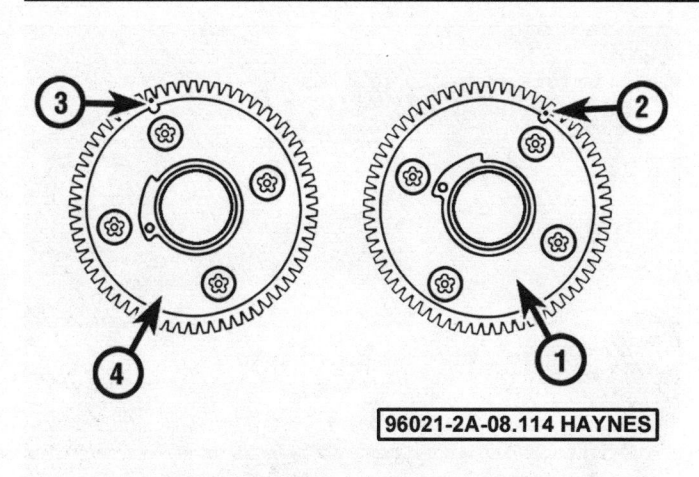

Chapter 2 Part A Four-cylinder engines

8.114 Camshaft timing mark details - adjustable timing sprockets shown, non-adjustable exhaust sprockets similar

1 Intake camshaft sprocket
2 Intake camshaft timing mark
3 Exhaust camshaft timing mark
4 Exhaust camshaft sprocket

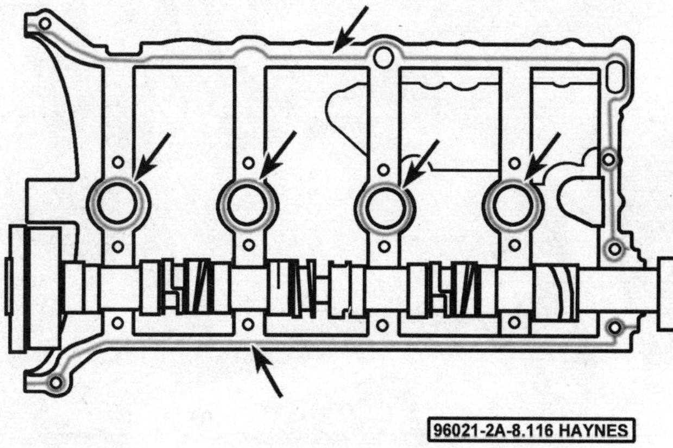

8.116 Apply the sealant to the cylinder head cover as shown - 2.0L (CNTA, CXCA, CXCB, CYFB, DJJA) models shown, other models similar

Note: *On 2.0L (CNTA, CXCA, CXCB, CYFB, DJJA) models, the camshaft lobes can move back in forth on the camshaft when it is removed, the four lobe assemblies are keyed to the camshaft by a spring and check ball for each lobe assembly. Once the camshaft is removed from the cylinder head cover, the lobes must be pushed apart to reinstall them into the original positions.*

113 On 2.0L (CNTA, CXCA, CXCB, CYFB, DJJA) models, rotate the camshaft until the mark on the exhaust sprocket is aligned with the timing mark on the cylinder head cover.

114 On 1.8L (CXBA, CXBB, CNSA, CNSB, CPRA, CPKA) and 2.0L (CBFA, CCTA, CPLA, CPPA) models, verify the timing mark on the intake camshaft sprocket is approximately at the 1 o'clock position and the exhaust camshaft timing mark is approximately at the 11 o'clock position **(see illustration)**.

Note: *On models with a non-adjustable exhaust camshaft sprocket, the timing mark sprocket should be pointing at approximately the 11 o'clock position.*

115 With the camshaft in this position, verify that the intake camshaft lobes for the No. 4 cylinder are now pointing upwards towards 11 o'clock position and the exhaust camshaft lobes for No. 4 cylinder are pointing upwards towards 1 o'clock position.

116 Apply a 1/8-inch (3 mm) bead of RTV sealant to the cylinder head cover mating surfaces **(see illustration)**.

Note: *Once the sealant is applied to the cylinder head cover, the cover must be installed within five minutes.*

117 Install the cover and tighten the new bolts to the torque listed in this Chapter's Specifications, in the order shown **(see illustration)**.

118 On 2.0L (CBFA, CCTA) models, install the seal plug for the intake camshaft into the cylinder head and cylinder head cover.

Note: *The seal plug should not be recessed more than 0.12-inch (2 mm).*

119 If necessary, rotate the camshaft sprocket adjusters until the special tools can be installed on both camshaft sprockets (see Steps 76 and 77).

120 Rotate the crankshaft back to TDC, aligning the crankshaft pulley and install the timing chains (see Section 6).

Caution: *The marks on the chain and sprocket must align or damage will occur.*

Caution: *Don't rotate the crankshaft for 30 minutes. After 30 minutes, carefully rotate the crankshaft by hand through at least two full revolutions (use a socket and breaker bar on the crankshaft pulley center bolt). If you feel any resistance, STOP! There is something wrong - most likely, valves are contacting the pistons. You must find the problem before proceeding.*

121 The remainder of installation is the reverse of removal.

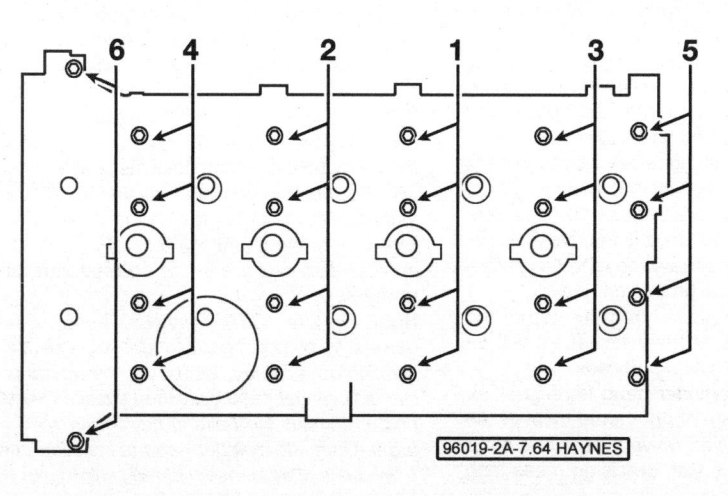

9 Cylinder head - removal and installation

Warning: *Wait until the engine is completely cool before beginning this procedure.*

Caution: *If the battery is disconnected, several systems must be re-learned before they will work properly (see Chapter 5, Section 3).*

Caution: *Performing this procedure may cause a trouble code to be set, and require a scan tool to clear the trouble code (see Chapter 6).*

1 Relieve the fuel pressure (see Chapter 4), then disconnect the cable from the negative terminal of the battery (see Chapter 5).

2 Loosen the right front wheel bolts, raise the front of the vehicle and support it securely on jackstands. Block the wheels at the opposite end and remove the right front wheel.

3 Working under the vehicle, remove the lower splash shield below the engine (see Chapter 1, Section 6).

8.117 Cylinder head cover bolt TIGHTENING sequence

9.9 On 1.4L (CZTA, DGXA) models, remove the engine mount support bracket upper left mounting bolt

9.17 Remove the right driveaxle heat shield bolts

4　Working from above in the engine compartment, remove the engine cover (see Chapter 1, Section 7).
5　Remove the right front inner fender liner (see Chapter 11).
6　Remove the drivebelts (see Chapter 1).
7　Drain the engine coolant (see Chapter 3). Set the engine to TDC (see Section 3).

Timing belt engines

Removal

8　Remove the intake manifold (see Section 7).
9　On 1.4L (CZTA, DGXA) engines, support the engine from below and remove the passenger's side mount (see Section 18), then remove the engine mount support bracket bolt **(see illustration)**.
10　Remove the timing belt (see Section 5).
11　Disconnect the electrical connectors from the engine coolant temperature sensor (ECT), reduced oil pressure switch, fuel pressure sensor, and the fuel injector electrical connectors.
12　Remove the coolant housing fasteners and housing from the cylinder head.
13　On 2.0L (CBPA) models, disconnect the exhaust pipe from the exhaust manifold.
14　On 1.4L (CZTA, DGXA) models, remove the exhaust manifold/turbocharger assembly (see Chapter 4, Section 15).
15　Remove the valve cover on 2.0 L (CBPA) models (see Section 4) and the camshaft housing on 1.4L (CZTA, DGXA) models (see Section 8).
16　Remove the front exhaust pipe/catalytic converter (see Chapter 6).
Note: *Mark the position of the clamp for the catalytic converter to help in installation.*
17　On 1.4L (CZTA, DGXA) models remove the right driveaxle heat shield bolts and shield **(see illustration)**.
18　Remove the oil pressure switch heat shield and oil pressure switch (see Chapter 2C).

19　Remove the intercooler lower connector pipe retaining fasteners and the lower connector pipe.
20　Disconnect all electrical connectors from the turbocharger.
21　Remove the oil supply and coolant supply banjo bolts and plug the lines.
22　Remove the oil return line retaining fasteners and disconnect the return line from the turbocharger.
23　Remove the turbocharger support bracket retaining fasteners and support bracket.
24　Remove coolant reservoir (see Chapter 3).
25　Clamp the coolant hoses off at the end of the cylinder head, then disconnect and plug the coolant hoses from the coolant tube.
26　Remove the coolant tube retaining fastener from the heat shield and remove the tube.
27　On 2.0L (CBPA) models, remove the upper bolt and the rear timing belt guard.
28　Remove the electrical harness retaining fasteners and pull the harness away from the end of the cylinder head.
29　Working in the reverse of the sequence shown in **illustration 9.41a, 9.41b or 9.41c**, progressively loosen the cylinder head bolts, by half a turn at a time, until all bolts can be unscrewed by hand. Discard the bolts - new ones must be installed on reassembly.
30　Check that nothing remains connected to the cylinder head, then lift the head away from the cylinder block; seek assistance if possible, as it is heavy. If resistance is felt, carefully pry the cylinder head upward, beyond the gasket surface, at a casting protrusion.
31　Remove the gasket from the top of the block. Do not discard the gasket - it will be needed for identification purposes.
32　Before the cylinder head is reinstalled, an engine machine shop should inspect the cylinder head, valves, valve guides and other critical dimensions with precision measuring tools.

Installation

33　The mating faces of the cylinder head and cylinder block must be perfectly clean before installing the head. Use a hard plastic or wood scraper to remove all traces of gasket and carbon; also clean the piston crowns. Take particular care during the cleaning operations, as aluminum alloy is easily damaged. Also, make sure that the carbon is not allowed to enter the oil and water passages - this is particularly important for the lubrication system, as carbon could block the oil supply to the engine's components. Using adhesive tape and paper, seal the water, oil and bolt holes in the cylinder block.
34　Check the mating surfaces of the cylinder block and the cylinder head for nicks, deep scratches and other damage. If slight, they may be removed carefully with abrasive paper.
35　If warpage of the cylinder head gasket surface is suspected, use a straight-edge to check it for distortion. At no point should a 0.004-inch feeler gauge fit between the straightedge and the head's gasket surface.
36　Clean out the cylinder head bolt holes using a suitable tap. They must be clean and dry before installation of the head bolts.
37　It is possible for the piston crowns to strike and damage the valve heads if the camshaft(s) is(are) rotated with the timing belt. For this reason, the crankshaft must be set to a position to TDC on No. 1 cylinder, then turn the crankshaft back just a little.
38　Position the new head gasket on the cylinder block, engaging it with the dowel pins. Ensure that the manufacturer's "TOP" or part number markings are facing up **(see illustration 9.95)**.
Note: *On 2.0L (CBPA) models, there are no dowels to center the head gasket, use VW special tool #3450/3, 2A, or cut the heads off two old cylinder head bolts then cut slots in the bolt and install the bolts in the corners a few turns. Once the cylinder head is installed and a few bolts started, use a blade screwdriver to remove the old cut head bolts.*

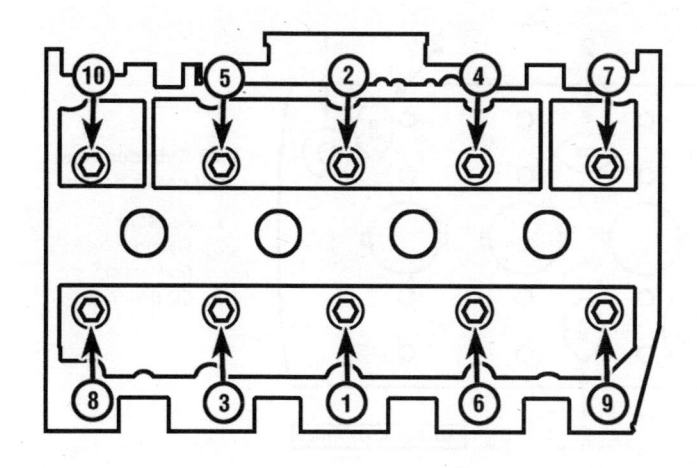

Chapter 2 Part A Four-cylinder engines

9.41a Cylinder head bolt TIGHTENING sequence - 1.4L timing belt (CZTA) engines

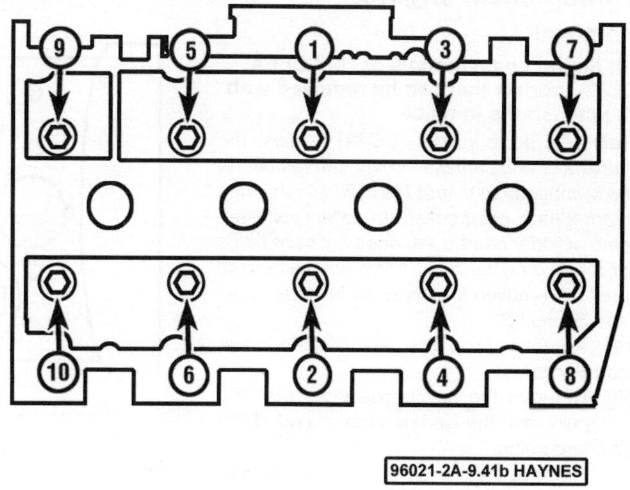

9.41b Cylinder head bolt TIGHTENING sequence - 1.4L timing belt (DGXA) engines

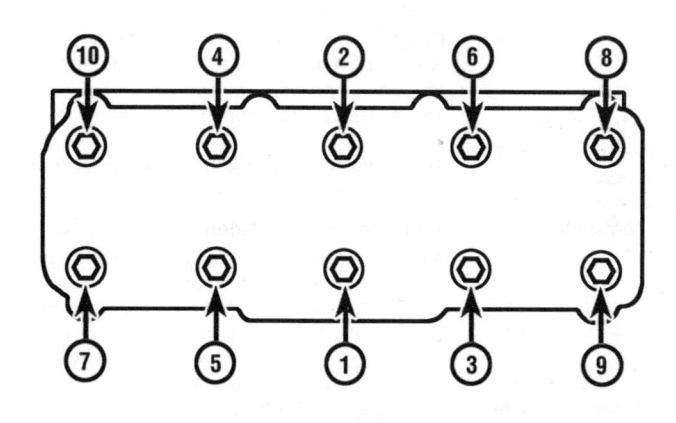

9.41c Cylinder head bolt TIGHTENING sequence - 2.0L timing belt (CBPA) engines

9.41d Using an angle measurement gauge during the final stages of tightening

39 With the help of an assistant, place the cylinder head centrally on the cylinder block, ensuring that the dowel pins engage with the holes in the cylinder head. Check that the head gasket is correctly seated before allowing the full weight of the cylinder head to rest upon it.

Note: *If the cylinder head had been disassembled for repair, be sure the camshafts are reinstalled on the cylinder head with recesses in the camshafts facing each other.*

40 Install the cylinder head bolts and screw them in hand-tight. Use NEW cylinder head bolts, as the old bolts are stretch-type fasteners that will not provide the correct torque readings if reused.

41 Working in the sequence shown **(see illustrations)**, tighten the cylinder head bolts in stages to the torque and angle of rotation listed in this Chapter's Specifications. *It is recommended that an angle-measuring gauge be used during the final stages of the tightening, to ensure accuracy* **(see illustration)**. If a gauge is not available, use white paint to make alignment marks between the bolt head and cylinder head prior to tightening; the marks can then be used to check that the bolt has been rotated through the correct angle during tightening.

42 Rotate the crankshaft clockwise slightly to set the engine back at TDC. Be sure the alignment mark on the crankshaft pulley aligns with the mark on the front cover. Refer to Section 3, if necessary.

43 On 1.4L (CZTA, DGXA) models, install the camshaft housing (see Section 8).

44 Install the intake manifold and the fuel rails (see Section 7).

45 Install the timing belt tensioner and the camshaft sprocket (CBPA models) on the engine if removed.

46 Install and adjust the timing belt as described in Section 5.

Caution: *After the timing belt has been installed, carefully rotate the crankshaft by hand through at least two full revolutions (use a socket and breaker bar on the crankshaft pulley center bolt). If you feel any resistance, STOP! There is something wrong - most likely, valves are contacting the pistons. You must find the problem before proceeding.*

47 The remainder of the installation is the reverse of removal.

48 Change the engine oil and filter, then refill the cooling system (see Chapter 1). Run the engine and check for leaks.

Timing chain engines

Removal

All timing chain models except CBFA, CCTA models that can be removed with the camshafts in place

Note: On some (CBFA, CCTA) models, the camshafts and camshaft cover can remain on the cylinder head unless the cylinder head has internal multi-point bolts with collars installed. If the cylinder head cover does not have plugs for the two center head bolts, ithe camshafts must be removed to access all the head bolts (see Section 8).

49 Remove the upper timing chain cover (see Section 6).

50 Remove the air charge pipe retaining fasteners, and the clamps at each end of the air charge pipe.

51 Remove the charge air cooler lower connector pipe retaining fasteners and lower connector pipe (see Chapter 4).

Note: Place a shop towel into the charge air cooler openings to prevent anything from getting in to the cooler.

52 Support the engine from below and remove the passenger's side engine mount (see Section 18).

53 Remove the timing chain (see Section 6) and camshafts (see Section 8).

54 Remove the coolant hose tube fasteners, then disconnect the hoses along the top of the cylinder head and move the hose assembly out of the way.

55 Loosen both hose clamps on the lower air charge hose and remove the hose.

56 Remove the front exhaust pipe/catalytic converter (see Chapter 6).

Note: Mark the position of the clamp for the catalytic converter to help in installation.

57 Disconnect the electrical connectors from the reduced oil pressure switch and oil pressure switch on the oil filter housing.

58 Remove the intake manifold support bracket fasteners and bracket under the intake manifold.

59 Remove the coolant tube bracket fastener from the intake manifold, then disconnect all the electrical connectors from the intake manifold and the throttle body.

60 Remove the exhaust manifold heat shield retaining fasteners and shield.

61 Disconnect all electrical connectors from the turbocharger.

62 Remove the turbocharger (see Chapter 4).

63 On CBFA and CCTA models, disconnect the vacuum hose from the pump, plug the hose and cap the port on the vacuum pump.

64 Remove the crankcase ventilation hose (see Chapter 6).

65 On CBFA and CCTA models, remove the crankcase ventilation housing (see Section 17), if the camshafts are remaining on the cylinder head.

66 On CBFA models, disconnect the secondary air injection electrical connectors to access the coolant pipe retaining fasteners, then remove the fasteners from the cylinder head.

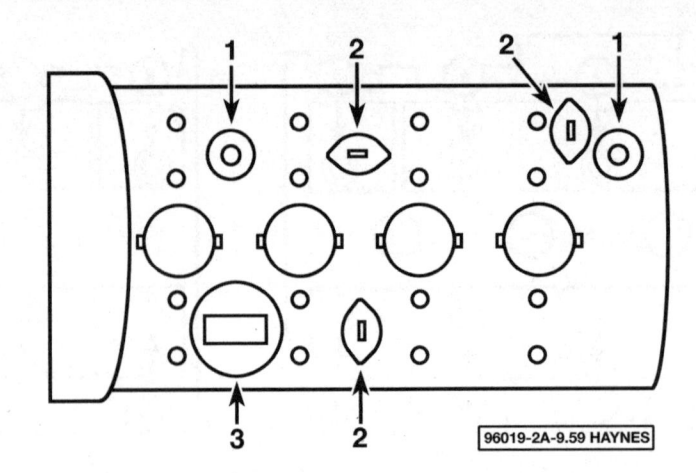

9.80 Cylinder head bolt cover details

1 Ballstud
2 Cylinder head bolt cover
3 Oil filler cap

67 Disconnect the electrical connectors from the intake manifold and EVAP canister.

68 Disconnect the coolant hose to the intake manifold, then remove the coolant tube-to-intake manifold retaining fasteners and tube.

69 Disconnect the heater hoses from the firewall.

CBFA, CCTA models that can be removed with the camshafts in place

70 Insert VW special tool #10352 or equivalent into the two holes on the control valve, then rotate the valve clockwise and pull the valve out from the end of the camshaft.

Note: The control valve has left-hand threads.

71 Remove the bearing bracket retaining fasteners, bearing bracket-to-exhaust camshaft bolt and bearing bracket **(see illustration 6.42)**.

72 Rotate the engine in the normal direction of rotation (clockwise) until the No. 1 cylinder is located at TDC (see Section 3). Verify that the notch on the crankshaft pulley is aligned with the mark on the lower timing chain cover **(see illustration 3.8b)** and the marks on the camshafts are pointing upwards.

73 Make alignment marks from the camshaft timing chain to the camshaft sprockets and upper chain guide to timing chain using a permanent marker.

Note: The marks will be used for reinstallation. If the marks are removed, the lower cover, camshaft timing chain and balance shaft timing chain will have to be removed, reset and reinstalled (see Section 6).

74 Remove the tensioner access hole plug on the left side of the lower timing chain cover. Working through the access hole, use a small screwdriver or pick to hold the chain tensioner ratchet arm away (UP) from the ratchet stem, then slowly compress the timing chain tensioner piston by rotating the crankshaft pulley counterclockwise until the pin (a drill bit or paper clip will work) can be inserted into the hole to hold it in the compressed position **(see illustration 6.45)**.

75 Remove the tensioner-side guide retaining pin from the camshaft timing chain.

76 Using a small screwdriver, release the locking clip and slide the timing chain upper guide rail out and off.

Note: With the lower timing chain cover in place, the loose timing chain cannot fall off.

77 Remove the timing chain from the camshaft sprockets.

Caution: Do not rotate the engine once the timing chain has been removed or damage to the valves or pistons may occur.

78 Remove the ignition coils (see Chapter 5).

79 Disconnect the electrical connector from the camshaft position sensor.

All timing chain models

80 On some CBFA, CCTA models, the cylinder head bolts are covered. To access the head bolts, remove the oil filler cap and unscrew the ballstuds from the cylinder head cover. Rotate the cylinder head bolt covers counterclockwise 90-degrees and remove them **(see illustration)**.

81 Remove the four smaller bolts at the front of the cylinder head first, then remove the cylinder head bolts in reverse order of the installation sequence **(see illustrations 9.90a or 9.90b)**.

82 The cylinder head assembly is heavy; you will have to attach a lifting sling or chain to the cylinder head. Position an engine hoist and connect the sling to it. Fasten the chains or slings to open bolt hole locations on the side of the cylinder head - ones that are strong enough to take the weight, but in locations that will provide good balance. Take up the slack until there is slight tension on the sling or chain. Position the chain on the hoist so it balances the cylinder head as it is lifted.

83 Carefully lift the cylinder head assembly out, making sure the timing chain guide rails are not damaged, if left installed.

84 Lower the cylinder head assembly down onto two blocks of wood, then remove the intake manifold.

Caution: The valves protrude past the bottom of the cylinder head and are easily damaged if the head is placed down on the valves.

Installation

85 Follow Steps 33 through 36.

Chapter 2 Part A Four-cylinder engines

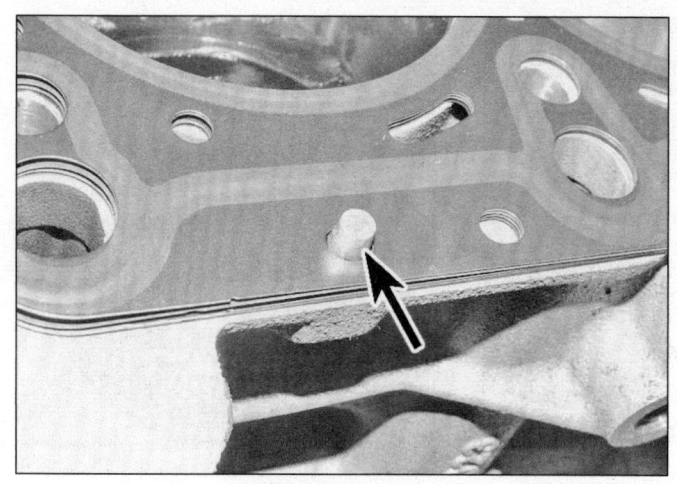

9.87a Center the gasket using the cylinder head dowel pins - typical models shown

9.87b Be sure the "TOP" mark faces upward - typical head gasket shown

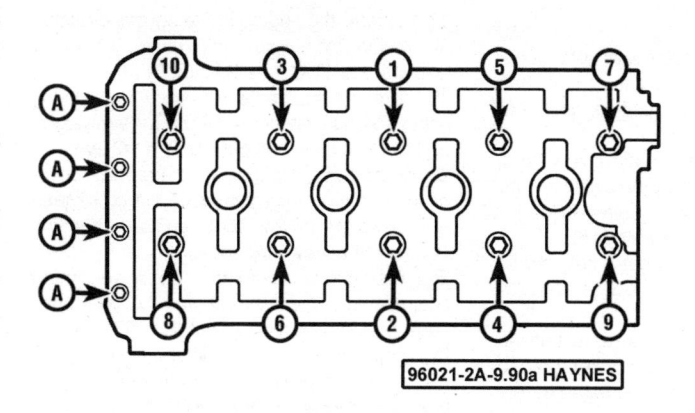

9.90a Cylinder head bolt TIGHTENING sequence, then tighten the front bolts (A) - timing chain engines without camshafts installed

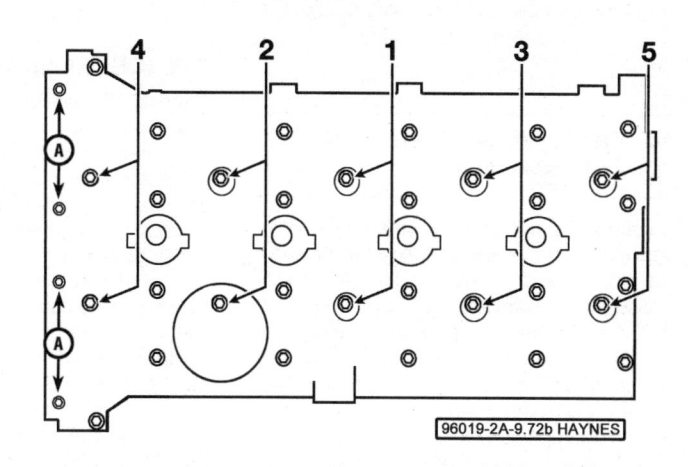

9.90b Tighten cylinder head bolts 1 through 5 (in sequence), then tighten the front bolts (A) - 2.0L timing chain (CCTA, CBFA) engine with camshafts installed

86 It is possible for the piston crowns to strike and damage the valve heads if the camshafts are rotated with the timing chain removed and the crankshaft set to TDC. For this reason, the crankshaft must be set to a position to TDC on No. 1 cylinder and then turn the crankshaft back just a little. Taking care not to damage the timing chain when turning the crankshaft.

87 Position the new head gasket on the cylinder block, engaging it with the dowel pins. Ensure that the manufacturer's "TOP" or part number markings are facing up **(see illustrations)**.

88 With the help of an assistant, place the cylinder head centrally on the cylinder block, ensuring that the dowel pins engage with the recesses in the cylinder head. Check that the head gasket is correctly seated before allowing the full weight of the cylinder head to rest upon it.

Note: *If the cylinder head had been disassembled for repair, be sure the camshafts are reinstalled on the cylinder head with recesses in the camshafts facing each other.*

89 Install the cylinder head bolts and screw them in hand-tight. Use NEW cylinder head bolts, as the old bolts are stretch-type fasteners that will not provide the correct torque readings if reused.

90 Working in the sequence shown **(see illustrations)**, tighten the cylinder head bolts in stages to the torque and angle of rotation listed in this Chapter's Specifications. It is recommended that an angle-measuring gauge be used during the final stages of the tightening, to ensure accuracy **(see illustration)**. If a gauge is not available, use white paint to make alignment marks between the bolt head and cylinder head prior to tightening; the marks can then be used to check that the bolt has been rotated through the correct angle during tightening.

CBFA, CCTA models that can be removed with the camshafts in place

91 Rotate the crankshaft clockwise back to TDC, aligning the crankshaft pulley with the mark on the timing chain cover (see Section 6). Place the timing chain over the camshaft sprocket and align the chain with the marks made in Step 72.

Caution: *The marks on the chain and sprocket must align or damage will occur. If the marks do not align, see Section 3 and follow the procedure for camshaft timing chain removal and installation.*

92 Install the camshaft upper guide and remove the pin holding the tensioner in the compressed position.

Caution: *Carefully rotate the crankshaft by hand through at least two full revolutions (use a socket and breaker bar on the crankshaft pulley center bolt). If you feel any resistance, STOP! There is something wrong - most likely, valves are contacting the pistons. You must find the problem before proceeding.*

All models

93 The remainder of installation is the reverse of removal.

94 Change the engine oil and filter, then refill the cooling system (see Chapter 1). Run the engine and check for leaks.

10.7 A locking chain wrench can be used to prevent the pulley from turning, with a rag wrapped around the pulley to prevent damage to the pulley

10.8 Inspect the teeth on the pulley and crankshaft sprocket for wear

10.11 Crankshaft pulley retaining bolts - timing belt (CBPA) engines

10 Crankshaft pulley - removal and installation

Caution: *On all models, severe engine damage may occur or valve timing may change if the crankshaft is moved out of the TDC position when the crankshaft pulley is removed.*

1 Loosen the right-front wheel bolts, raise the front of the vehicle and support it securely on jackstands. Block the wheels at the opposite end and remove the right-front wheel.
2 Working under the vehicle, remove the splash shield below the engine (see Chapter 1, Section 6).
3 Working from above in the engine compartment, remove the engine cover (see Chapter 1, Section 7).
4 Remove the right-front inner fender liner (see Chapter 11).
5 Remove the drivebelts (see Chapter 1).
6 Rotate the engine in the normal direction of rotation (clockwise) until the No. 1 cylinder is located at TDC (see Section 3).

Timing belt engines

1.4L (CZTA, DGXA) models

7 Prevent the crankshaft pulley from turning using VW special tool #T10475 (or equivalent) and remove the crankshaft pulley bolt **(see illustration)**.

Caution: *Severe engine damage may occur or valve timing may change if the crankshaft is moved out of the TDC position when the crankshaft pulley is removed.*

8 Remove the crankshaft pulley, and inspect the teeth on the pulley and crankshaft sprocket for wear **(see illustration)**.
9 Lightly oil and install the new pulley retaining bolt by hand then tighten the bolt to the torque listed in this Chapter's Specifications.
10 The remainder of installation is the reverse of removal.

2.0L (CBPA) models

Caution: *Severe engine damage may occur or valve timing may change if the crankshaft is moved out of the TDC position when the crankshaft pulley is removed.*

11 Hold the crankshaft stationary with a wrench on the center bolt, then unscrew the pulley bolts and remove the pulley from the crankshaft sprocket **(see illustration)**.

Caution: *If you remove the center bolt, obtain a new one (the manufacturer doesn't recommend re-using it).*

12 If it was removed, be sure to tighten the new crankshaft drive sprocket bolt to the torque and angle of rotation listed in this Chapter's Specifications.
13 Position the crankshaft pulley on the crankshaft drive sprocket and align the mounting holes. Note that the pulley can only go on one way.
14 Install the pulley retaining fasteners and tighten them to the torque listed in this Chapter's Specifications.
15 The remainder of installation is the reverse of removal.

Timing chain engines

Caution: *The crankshaft pulley/vibration damper, the three stage timing chain sprocket and the crankshaft are aligned to each other by teeth - when the crankshaft pulley bolt is tightened the teeth lock together and can't rotate. Once the bolt is removed, if the crankshaft or timing chains are moved the engine will be out of time and severe engine damage may occur.*

16 On 2.0L (CYFB and DJJA) models, remove the electrical connector from the windshield washer reservoir, then remove the fasteners and the reservoir.
17 On CCTA, CBFA, CPRA, CPKA, CPLA, CPPA models, remove the air charge pipe retaining fasteners, the clamps at each end of the air charge pipe, then loosen both hose clamps on the lower air charge hose and remove the hose (see Chapter 4).
18 Rotate the engine in the normal direction of rotation (clockwise) until the No. 1 cylinder is located at TDC (see Section 3).
19 Prevent the crankshaft pulley from turning using VW special tool #T10355 (or equivalent) and loosen crankshaft pulley bolt a 1/2 turn.

Caution: *Severe engine damage may occur or valve timing may change if the crankshaft is moved out of the TDC position when the crankshaft pulley is removed.*

20 Remove two bolts from the lower timing chain cover at the 2 o'clock and 7 o'clock positions. Place VW special tool #T10531/1 on to the crankshaft pulley and timing cover, install two mounting bolts through the tool and into the crankshaft pulley holes (there are four holes on the pulley and only two holes on the tool) then align the knurled knobs on the tool with the bolt holes from the removed bolts on the timing cover and hand-tight knurled bolts. This will keep pressure on the crankshaft pulley/vibration damper, the timing chain sprocket and the crankshaft, preventing any movement.

Note: *Once the cover bolts are removed they must be replaced with new ones.*

21 Remove the crankshaft pulley bolt completely.
22 Before installing VW tensioning pin (special tool #T10531/2), check to see if the outer sleeve of special tool can be easily pushed over the clamping slots and ring of the tool. If it can, push the sleeve down past the locking ring and turn the adjuster bolt on the end of the tool slightly.

Note: *Once the sleeve is down and is locked in place do not turn the adjusting bolt any further, otherwise when you're trying to install the tool into the end of the crankshaft the tool may become jammed in the crankshaft and not seat properly.*

23 Install tensioning pin (special tool #T10531/2) into the threaded end (where the crankshaft pulley bolt was) of the crankshaft and tighten it hand-tight using a wrench.
24 Once the tensioning pin (tool #T10531/2) is tight against the crankshaft sprocket and the adjusting bolt is hand-tight, the chain sprocket

Chapter 2 Part A Four-cylinder engines 2A-31

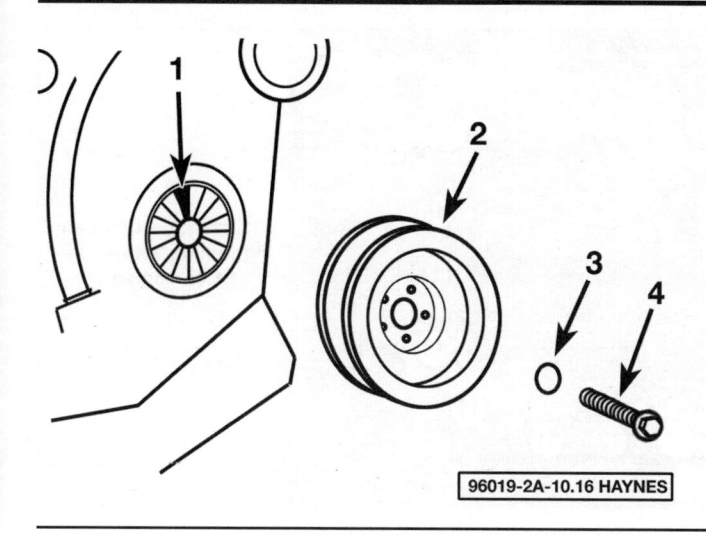

10.25 Crankshaft pulley details - timing chain engines

1. Crankshaft tooth flat index
2. Crankshaft pulley
3. Crankshaft bolt O-ring
4. Crankshaft pulley bolt

11.2 If a seal removal tool is unavailable, the front seal can also be removed with self-tapping screws to pry out the seal - typical seal removal shown

is now secured against the crankshaft. Loosen the knurled knobs on special tool #T10531/1, then pull tool and the crankshaft pulley off of the crankshaft as an assembly.

25 With tensioning pin (special tool #T10531/2) still installed, install the crankshaft pulley to the crankshaft, making sure that both flats on the tooth contours align with the tooth contours on the end of the crankshaft **(see illustration)**.

26 Slide the special tool #T10531/3 on the tensioning pin (tool #T10531/2). Verify the teeth points are still aligned on to the crankshaft pulley. Install special tool #T10531/4 lock nut and as the lock nut is tightened, move the crankshaft pulley back and forth slightly to check if the vibration damper is seated correctly in the tooth contour. Tighten the lock nut until the crankshaft pulley can no longer be turned.

27 Install the special tool #T10531/1 on to the crankshaft pulley and timing cover (see Step 20).

28 Remove the tensioning pin (special tool #T10531/2) from the crankshaft.

29 Install a new pulley retaining bolt and O-ring hand tight.

30 With the crankshaft pulley bolt tight, remove the special tool #T10531/1 from the crankshaft pulley and timing cover.

31 Prevent the crankshaft pulley from turning using VW special tool #T10355 (or equivalent) and tighten the crankshaft bolt to the torque listed in this Chapter's Specifications.

32 Install new lower timing chain cover bolts and tighten them to the torque listed in this Chapter's Specification.

33 The remainder of installation is the reverse of removal.

11 Crankshaft front oil seal and housing - replacement

Timing belt engines

1 Remove the timing belt and crankshaft sprocket (see Section 5).

2 Note how far the seal is recessed in the bore, then carefully pry it out with a screwdriver or seal removal tool. Don't scratch the housing bore or damage the crankshaft in the process (if the crankshaft is damaged, the new seal will end up leaking). If a seal removal tool is unavailable, you can thread two self-tapping screws (180-degrees apart from one another) into the front seal to pry the seal out **(see illustration)**.

3 Clean the bore in the housing and coat the outer edge of the new seal with engine oil or multi-purpose grease. Apply multi-purpose grease to the seal lip.

4 Using a seal driver or a socket with an outside diameter slightly smaller than the outside diameter of the seal, carefully drive the new seal into place with a hammer **(see illustration)**. Make sure it's installed squarely and driven in to the same depth as the original. If a socket isn't available, a short section of large diameter pipe will also work. Check the seal after installation to make sure the spring didn't pop out of place.

5 If the front oil seal housing needs to be removed for access to other components, remove the housing retaining fasteners and remove the housing from the engine while noting the installed position of the fasteners **(see illustration)**. In some instances the front oil seal removal and installation is easier with

11.4 Lubricate the seal lip and drive the new crankshaft seal into place with a seal driver or a large socket and a hammer - typical seal installation shown

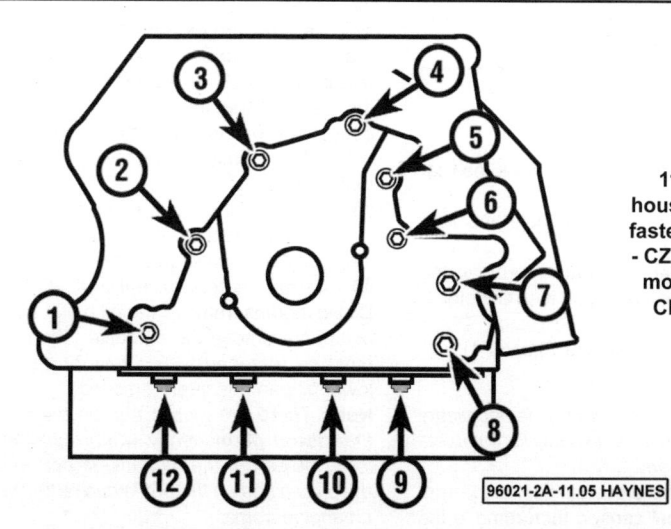

11.5 Oil seal housing retaining fastener locations - CZTA and DGXA models shown, CBPA similar

2A-32   Chapter 2 Part A   Four-cylinder engines

the front housing removed, since the seal can be placed on a workbench and driven straight in and out of the bore with no special tools or adapters.

6   If the front of the oil pan gasket was damaged while removing the housing, use a razor blade or utility knife to cut the pan gasket off flush with front edge of the cylinder block. This part of the oil pan gasket will be replaced with RTV sealant upon installation of the cover.

7   Before installing the front cover, make sure the mating surfaces of the cover, the cylinder block and the oil pan rail are perfectly clean. Use a hard plastic or wood scraper to remove all traces of gasket material. Take particular care when cleaning the front cover, as aluminum alloy is easily damaged.

8   Apply a 3/16-inch (5 mm) bead of RTV sealant to the oil pan flange.

9   Locate the oil seal housing gasket over the dowels on the engine block and install the oil seal housing.

Note: *Be sure to lubricate the oil seal lip before installing the front cover onto the engine. This will aid the installation process and prevent dry start ups, which may damage the seal and lead to future oil leaks.*

10   Tighten the front oil seal housing bolts a little at a time to the torque listed in this Chapter's Specifications.

11   Reinstall the crankshaft sprocket and timing belt (see Section 5).

12   Run the engine and check for oil leaks at the front seal.

Timing chain engines

13   Remove the crankshaft pulley (see Section 10).

14   Refer to Steps 2, 3 and 4 to replace the seal.

Note: *Use VW special tool #T40274 or a hooked tool to work around the tensioning pin (tool #T10531/2) that is installed while the crankshaft pulley is off.*

15   Install the crankshaft pulley (see Section 10).

12 Oil pan(s) - removal and installation

Lower oil pan

Removal

1   Set the parking brake and block the rear wheels.

2   Raise the front of the vehicle and support it securely on jackstands.

3   Working under the vehicle, remove the splash shield below the engine (see Chapter 1, Section 6).

4   Drain the engine oil (see Chapter 1).

1.4L models

Note: *On DGXA models, it may be necessary to remove the fan shroud to allow for complete access to the oil pan.*

5   Disconnect the electrical connector from the oil level thermal sensor, then remove the

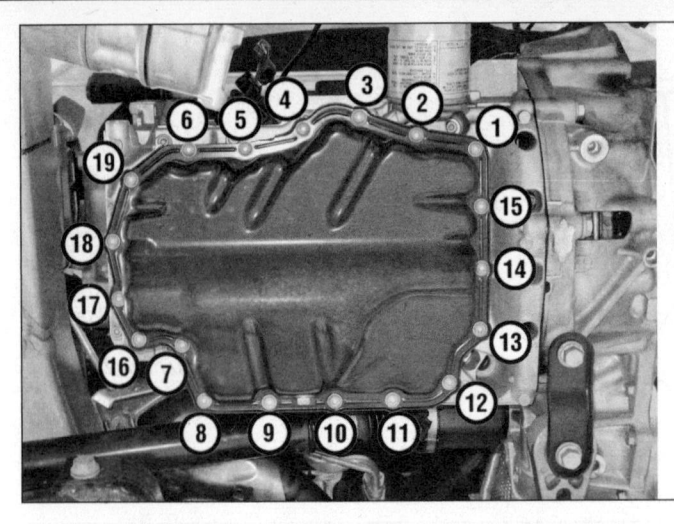

12.21a Lower oil pan tightening sequence - 1.4L engines

mounting nuts and remove the sensor from the lower oil pan. Always replace the sensor seal to the oil pan with a new one.

6   Remove the oil pan bolts in the reverse order of the tightening sequence **(see illustration 12.21a)**, then separate the lower oil pan, inserting special tool #T10561, or equivalent in between the upper and lower oil pans to break the bond of the liquid gasket. Gently tap the special tool with a mallet to free the pan.

7   Remove the lower oil pan.

2.0L (CBPA) engine

Note: *On these models, the oil pan may be removed as one assembly.*

8   Remove the oil pan bolts.

9   Separate the lower oil pan from the upper oil pan. It may be necessary to gently tap the lower oil pan with a mallet to free it.

10   Remove the oil pan and remove the dried liquid sealant from upper oil pan and lower oil pan sealing surfaces.

Timing chain engines

Note: *On some models, two different lower oil pans may be used - a plastic pan or metal pan. If your vehicle has a plastic pan, do not drop it.*

11   Disconnect the electrical connector from the oil level thermal sensor on the bottom of the oil pan.

12   Remove the oil pan bolts in the reverse order of the tightening sequence **(see illustrations 12.21b or 12.21c)**, then separate the lower oil pan from the upper pan using a gasket scraper to break the bond of the liquid gasket. Gently tap the lower oil pan with a mallet to free it.

13   Remove the oil pan and remove the dried liquid sealant from upper oil pan and lower oil pan sealing surfaces.

14   On all models except 2.0L (CYFB and DJJA) models, remove the oil baffle mounting bolts and remove the oil baffle. The oil baffle must be replaced with a new one once the lower oil pan has been removed.

Note: *There are plastic ribs on the oil baffle that distort permanently when tightening the lower oil pan. If the oil baffle is not replaced, the plastic ribs on the baffle can rattle, causing an engine noise.*

Installation

All models

Caution: *When installing the oil pan to an engine that has been removed, make sure that the edge of the oil pan is flush with the cylinder block on the flywheel side before tightening the bolts.*

15   Use a scraper to remove all traces of old sealant from the block and oil pan. Clean the mating surfaces with brake system cleaner.

Note: *Use RTV sealant to seal the oil pan-to-cylinder block mating surface.*

16   Make sure the threaded bolt holes in the block are clean.

17   Check the oil pan flange for distortion, particularly around the bolt holes. Remove any nicks or burrs as necessary.

18   On all timing chain models, except 2.0L (CYFB and DJJA) models, install a new oil baffle and tighten the mounting bolts to the torque listed in this Chapter's Specifications.

19   Apply a 1/16 to 1/8-inch (2 to 3 mm) bead of RTV sealant to the oil pan flange.

Note: *The lower oil pan must be installed within five minutes once the sealant has been applied.*

20   Carefully position the lower oil pan on the upper oil pan, install two mounting fasteners at the front and rear of the oil pan, and tighten them by hand.

21   Install the remaining fasteners into their original locations, then tighten the mounting bolts in sequence **(see illustrations)** to the torque listed in this Chapter's Specifications. On 2.0L (CBPA) engines, tighten the mounting bolts in a diagonal pattern to the torque listed in this Chapter's Specifications.

22   Remaining installation is the reverse of removal.

Upper oil pan

Removal

23   Drain the engine oil and remove the oil filter (see Chapter 1).

24   Remove the lower splash shield from under the vehicle.

25   Remove the lower oil pan (see Steps 1 through 14).

Chapter 2 Part A Four-cylinder engines

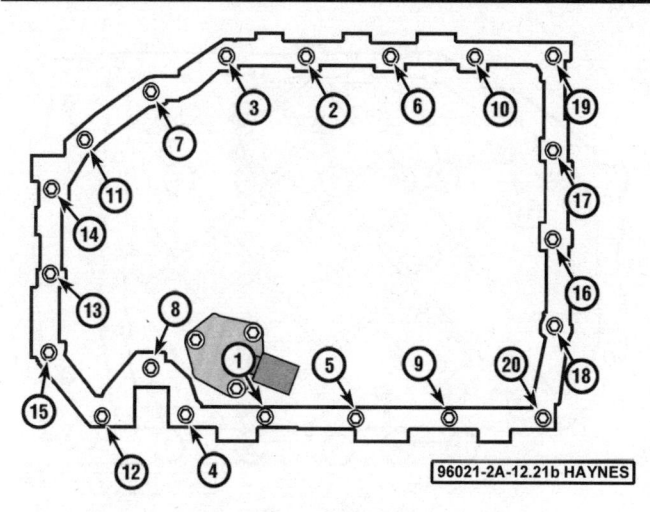

12.21b Lower oil pan tightening sequence - Timing chain models with a metal lower oil pan

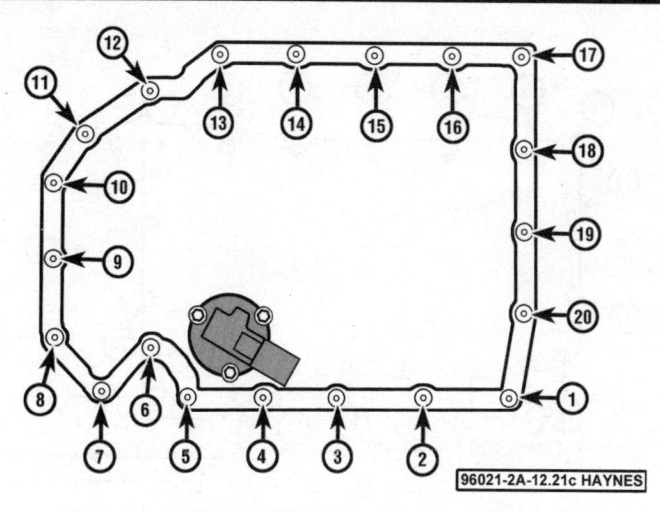

12.21c Lower oil pan tightening sequence - Timing chain models with plastic lower oil pan

1.4L engines

26 Remove the air conditioning compressor (see Chapter 3) without disconnecting the refrigerant lines, then secure the compressor out of the way.

27 Remove the oil pump (see Section 13).

28 Remove the three transaxle-to-oil pan mounting bolts.

29 Loosen the upper oil pan mounting bolts in the reverse order of the tightening sequence **(see illustration 12.51a)**.

30 Insert special tool T10561 or gasket scraper between the cut-out sections around the upper oil pan and the block, then carefully separate the upper oil pan from the block.

Note: *The upper oil pan uses a liquid sealant that has a very strong adhesion and once applied it dries hard and can be difficult to separate the components.*

31 Once the oil pan is removed, remove the oil baffle and clean, if necessary.

2.0L (CBPA) engines

32 Remove the transaxle-to-oil pan mounting bolts.

33 Remove the splash wall/shield fasteners and splash wall from bottom of the pump and pick-up tube.

34 Remove the pick-up tube retaining fasteners and pick-up tube.

35 Remove the upper oil pan mounting fasteners then carefully pry the oil pan from the cylinder block and slide the oil pan down over the oil pump and off the vehicle.

1.8L and 2.0L (CPKA, CPRA, CXBB, CXCA, CXCB, CBFA, CCTA, CPLA, CPPA, CYFB, DJJA and DRLA) engines

36 On CCTA, CBFA, CPRA, CPKA, CPLA and CPPA) models, remove the air charge pipe retaining fasteners, the clamps at each end of the air charge pipe, then loosen both hose clamps on the lower air charge hose and remove the hose (see Chapter 4).

37 On CYFB, DJJA models, disconnect the electrical connectors to the windshield washer reservoir then remove the fasteners and reservoir.

38 Remove the transaxle (see Chapter 7A or Chapter 7B).

39 Remove the oil pan baffle fastener and baffle.

40 Remove the flywheel or driveplate (see Chapter 2A) then remove the rear sealing flange (spacer plate) from the dowels on the end of the engine block.

41 Remove the oil pump (see Chapter 2A).

42 Disconnect the electrical connector from the oil pressure regulator valve on the timing chain cover.

43 Remove the two timing chain cover-to-oil pan bolts at the front of the upper oil pan.

44 Remove the upper oil pan mounting fasteners in reverse of the tightening sequence **(see illustrations 12.51b, 12.51c or 12.51d)** then carefully start prying at the rear of the oil pan, working your way to the front. Make sure not to bend the timing chain cover when prying the oil pan off.

Installation

All models

Caution: *When installing the oil pan to an engine that has been removed, make sure that the edge of the oil pan is flush with the cylinder block on the flywheel side before tightening the bolts.*

45 Use a scraper to remove all traces of old sealant from the block and oil pan. Clean the mating surfaces with brake system cleaner.

Note: *Use RTV sealant to seal the oil pan-to-cylinder block mating surface.*

46 Make sure the threaded bolt holes in the block are clean.

47 Check the oil pan flange for distortion, particularly around the bolt holes. Remove any nicks or burrs as necessary.

Note: *Check to see if the timing chain cover is deformed by mounting the upper oil pan to the engine block without any sealant. Check the gap between the cover and the upper oil pan. If the cover can't be aligned, the cover is deformed and must be replaced after installing the upper oil pan.*

48 On 1.4L engines, install the oil baffle to the engine block.

49 Apply a 1/16 to 1/8-inch (2 to 3 mm) bead of RTV sealant to the oil pan flange **(see illustration)**.

Note: *The upper oil pan must be installed within five minutes once the sealant has been applied.*

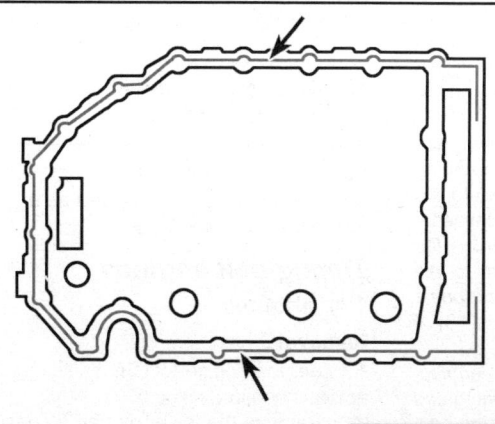

12.49 Apply a 1/16 to 1/8-inch (2 to 3 mm) thick bead of sealant to the upper pan around the sealing surfaces of the upper oil pan (typical for all models)

2A-34   Chapter 2 Part A   Four-cylinder engines

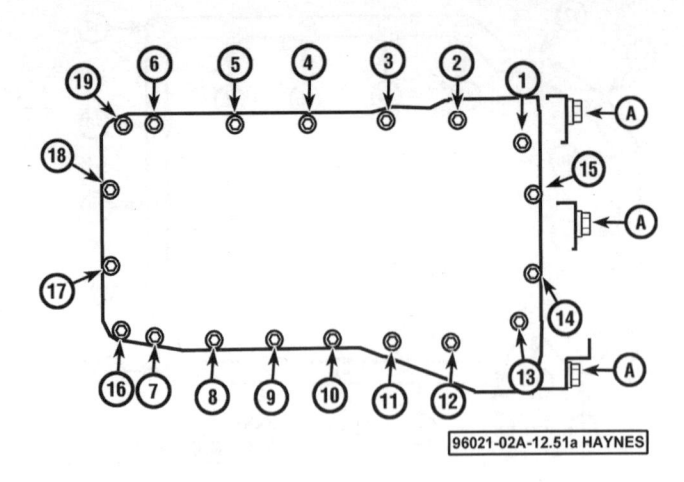

12.51a Upper oil pan tightening sequence - 1.4L engines

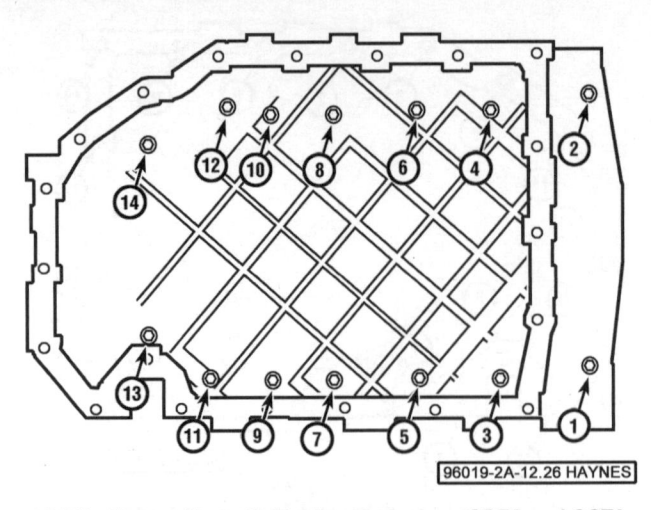

12.51b Upper oil pan tightening sequence - CBFA and CCTA timing chain engines

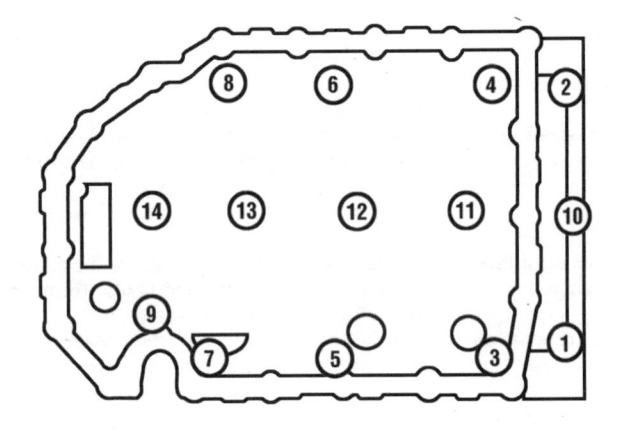

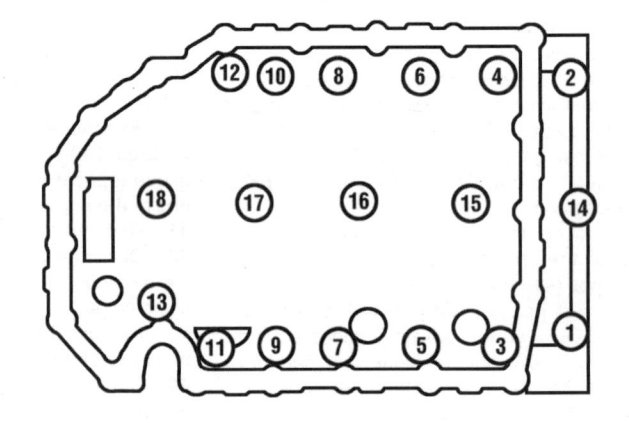

12.51c Upper oil pan tightening sequence - timing chain engines with 14 mounting bolts (except CBFA and CCTA engines)

12.51d Upper oil pan tightening sequence - timing chain engines with 18 mounting bolts (except CBFA and CCTA engines)

50 Carefully position the upper oil pan on the cylinder block, install two mounting fasteners at the front and rear of the oil pan, and tighten them by hand.
Note: On 2.0L (CBPA) models, make sure that the oil pan is positioned flush with the cylinder block on the flywheel side, if the engine or transaxle was removed.
51 Install the remaining fasteners into their original locations, then tighten the mounting bolts in sequence **(see illustrations)** to the torque listed in this Chapter's Specifications.
Note: On 2.0L (CBPA) engines, tighten the mounting bolts in a diagonal pattern to the torque listed in this Chapter's Specifications.
52 Install the oil pump (see Section 13), if removed.
Note: Be sure to follow the sealant manufacturer's recommendations on curing times and allow the sealant to properly cure before adding oil.

53 Remaining installation is the reverse of removal.
54 Run the engine and check for oil pressure and leaks.

13 Oil pump - removal and installation

Note: On all models the oil pump is not serviceable and must be replaced as a unit if a problem is suspected.

Timing belt engines

1.4L engines

Removal

1 Set the engine to TDC on No 1 cylinder as described in Section 3.
2 Remove the lower oil pan as described in Section 12.

3 Using a screwdriver, pry the corners of the oil pump drive chain cover down a little at a time alternating from each corner until the cover can be removed.
4 If you intend to re-use the oil pump drive chain, apply match marks on the chain and sprockets and make an arrow indicating the direction of travel on the chain.
5 Remove the oil pump mounting bolts and lower the rear of the oil pump down, then lift the oil pump drive sprocket out of the chain and remove the pump assembly.
6 Remove the two alignment sleeves from the upper oil pan if they didn't stay on the oil pump, then remove the oil pump intake pipe fastener and remove the intake pipe from the pump.
7 Remove the oil pump seal/strainer from the top of the oil pump.
8 To remove the oil pump drive chain, remove the crankshaft front oil seal housing (see Section 11).

Chapter 2 Part A Four-cylinder engines 2A-35

14.3 Mark the flywheel/driveplate and the crankshaft so they can be reassembled in the same relative positions

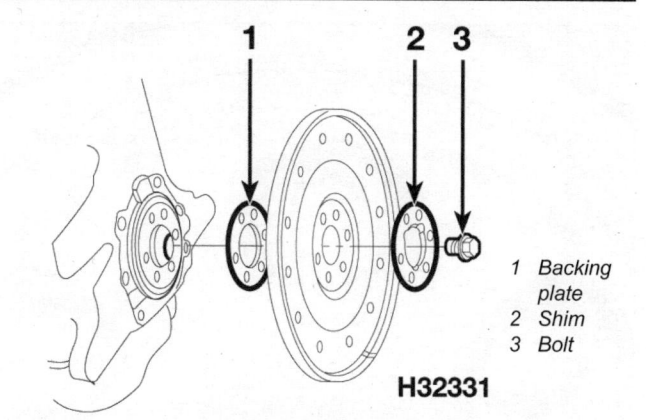

1 Backing plate
2 Shim
3 Bolt

14.5 On vehicles with an automatic transaxle, there is a spacer plate on each side of the driveplate - mark each plate as it is removed so it can be installed back in the same position

Installation

9 If removed, align the marks made on the oil pump drive chain and the crankshaft sprocket, then place the chain onto the sprocket.
10 If removed, install the crankshaft front oil seal housing (see Section 11).
11 Install a new oil pump seal/strainer to the top of the oil pump, then install the alignment sleeves to the top of the oil pump.
12 Place the oil pump drive chain sprocket onto the oil pump drive chain and align the marks, if made, then install the oil pump assembly mounting bolts hand-tight.
13 Tighten the oil pump mounting bolts to the torque listed in this Chapter's Specifications.
14 Install the lower oil pan as described in Section 12 then add oil, start the engine and check for oil pressure and leaks. Recheck the engine oil level (see Chapter 1).

2.0L (CBPA) engines

Removal

15 Set the engine to TDC on No 1 cylinder as described in Section 3.
16 Remove the lower oil pan as described in Section 12.
17 Remove the splash wall mounting bolts and remove the splash wall.
Note: *VW refers to the windage tray on these models as a splash wall.*
18 If you intend to re-use the oil pump drive chain, apply match marks on the chain and sprockets and make an arrow indicating the direction of travel on the chain.
19 While holding the oil pump drive sprocket and chain in place, remove the oil pump sprocket-to-oil pump shaft bolt and slide the chain and sprocket off of oil pump shaft.
20 Remove the oil pump assembly mounting bolts and remove the oil pump from the engine block.
21 Remove the two alignment sleeves from the block if they didn't stay on the oil pump, then remove the oil pump intake pipe fastener and remove the intake pipe from the pump.
22 To remove the oil pump drive chain, remove the crankshaft front oil seal housing (see Section 11).

Installation

23 If removed, align the marks made on the oil pump drive chain, the oil pump sprocket and the crankshaft sprocket then place the chain onto the sprockets.
24 If removed, install the crankshaft front oil seal housing (see Section 11).
25 Install the alignment sleeves to the top of the oil pump, then insert the oil pump shaft into the oil pump drive sprocket and install the oil pump mounting bolts hand-tight.
26 Install the oil pump sprocket-to-oil pump shaft bolt, then tighten the bolt to the torque listed in this Chapter's Specifications while holding the oil pump drive sprocket and chain from moving.
27 Tighten the oil pump assembly mounting bolts to the torque listed in this Chapter's Specifications.
28 Install the lower oil pan as described in Section 12 then add oil, start the engine and check for oil pressure and leaks. Recheck the engine oil level (see Chapter 1).

Timing chain engines (all)

Note: *Two people are required to remove the oil pump.*
29 Remove the lower oil pan (see Section 12).
30 Remove the oil baffle retaining fasteners and oil baffle.
31 This step requires the help of an assistant; pull back on the oil pump chain tensioner using (VW special tool #T10118) or equivalent, then secure the tensioner in place using (VW special tool #T40265) or drill bit.
32 With the tensioner locked in place, remove the oil pump mounting bolts and remove the oil pump from the chain.
33 Remove the oil pick-up tube retaining fastener and pick-up tube from the pump.
Note: *Always replace the O-ring on the pick-up tube.*

Installation

34 Install a new O-ring on the oil pick-up tube, then install the tube to the oil pump and tighten the retaining fastener to the torque listed in this Chapter's Specifications.
35 Have an assistant pull back on the oil pump chain tensioner using VW special tool #T10118 or equivalent and remove special tool #T40265, then install the oil pump to the chain.
Caution: *The chain spring tensioner must be set in the original position or possible engine damage may occur.*
36 Once the oil pump is correctly set onto the chain, install the oil pump retaining fasteners and tighten them to the torque listed in this Chapter's Specifications.
37 Install a new oil baffle and tighten the retaining fasteners to the torque listed in this Chapter's Specifications.
38 Install the lower oil pan (see Section 12).
39 Add oil, start the engine and check for oil pressure and leaks.
40 Recheck the engine oil level.

14 Flywheel/driveplate - removal and installation

Caution: *The manufacturer recommends replacing the flywheel bolts with new ones whenever they are removed.*

Removal

1 On manual transaxle models, remove the transaxle (see Chapter 8) and clutch (see Chapter 7A). Now is a good time to check/replace the clutch components.
2 On automatic transaxle models, remove the automatic transaxle as described in Chapter 7B.

Standard flywheel/driveplate

3 Use a center punch or paint to make alignment marks on the flywheel/driveplate and crankshaft to ensure correct alignment during reinstallation **(see illustration)**.
4 Remove the bolts that secure the flywheel/driveplate to the crankshaft. If the crankshaft turns, wedge a screwdriver in the ring gear teeth to prevent it from turning.
5 Remove the flywheel/driveplate from the crankshaft. Since the flywheel is fairly heavy, be sure to support it while removing the last bolt. Automatic transaxle equipped vehicles have spacers on both sides of the driveplate **(see illustration)**; keep them with the driveplate.

2A-36   Chapter 2 Part A   Four-cylinder engines

14.7 Align the holes with the bolts, then remove the bolts

14.8 Use a locking tool to hold the flywheel from turning

Dual-mass flywheel

6   On engines with the dual-mass flywheel, begin by making alignment marks between the flywheel and the crankshaft.

7   Rotate the outside of the dual-mass flywheel so that the bolts align with the holes **(see illustration)**.

8   Use a locking tool to hold the flywheel **(see illustration)**, then remove the bolts and the flywheel. Discard the bolts; new ones must be installed.

Caution: *In order not to damage the flywheel, do not use an air tool or impact driver to remove the bolts - only use hand tools.*

9   The dual-mass flywheel has a needle-bearing insert that can be removed with a puller and installed with a bearing driver.

Installation

10   Use brake system cleaner on the flywheel to remove any dust, grease and oil. Inspect the surface for cracks, rivet grooves, burned areas and score marks. Light scoring can be removed with emery cloth. Check for cracked and broken ring gear teeth. Lay the flywheel on a flat surface and use a straightedge to check for warpage.

11   Clean and inspect the mating surfaces of the flywheel/driveplate and the crankshaft. If the crankshaft rear seal is leaking, replace it before reinstalling the flywheel/driveplate.

12   Position the flywheel/driveplate and spacer (if used) against the crankshaft. Be sure to align the marks made during removal. Note that some engines have an alignment dowel or staggered bolt holes to ensure correct installation. Before installing the bolts, apply thread locking compound to the threads.

13   Use a locking tool to hold the flywheel **(see illustration 14.8)** from turning and tighten the bolts to the torque listed in this Chapter's Specifications. Follow a criss-cross pattern and work up to the final torque in three or four steps.

14   On vehicles equipped with an automatic transaxle, measure the installed height at three equal places around the driveplate and compare the average measurement to this Chapter's Specifications **(see illustration)**. If the measurement is incorrect, the driveplate must be removed and shimmed to the proper height.

15   The remainder of installation is the reverse of removal.

15 Rear main oil seal - replacement

Note: *The seal is an integral component of the housing, and must be replaced as a unit.*

1   Remove the flywheel/driveplate (see Section 14).

2   On timing belt (CZTA, DGXA and CBPA) engines, the manufacturer recommends removing the oil pan (see Section 12).

Note: *On all timing chain engines, the rear seal and housing can be replaced without removing the oil pan.*

All models except 1.4L engines

Note: *There are two versions of the rear main seal or sealing flange. Once the adapter plate has been removed, check to see which rear main seal is installed. Version 1: The long sealing lip faces the transmission side. Version 2: The rear main seal has two lips - a long sealing lip which faces the engine and a short dust lip that faces the transmission side.*

Caution: *On models equipped with Version 2, the long sealing lip must be pointed towards the engine side. If the seal is installed and is folded or kinked towards the transmission side, it will leak oil.*

3   On 2.0L (CBPA, CCTA, CBFA) engines, remove the rear oil seal housing retaining bolts **(see illustration)**, including the relevant oil pan bolts and, on all other models, remove the bolts in the reverse or of the tightening sequence **(see illustration 15.11)**.

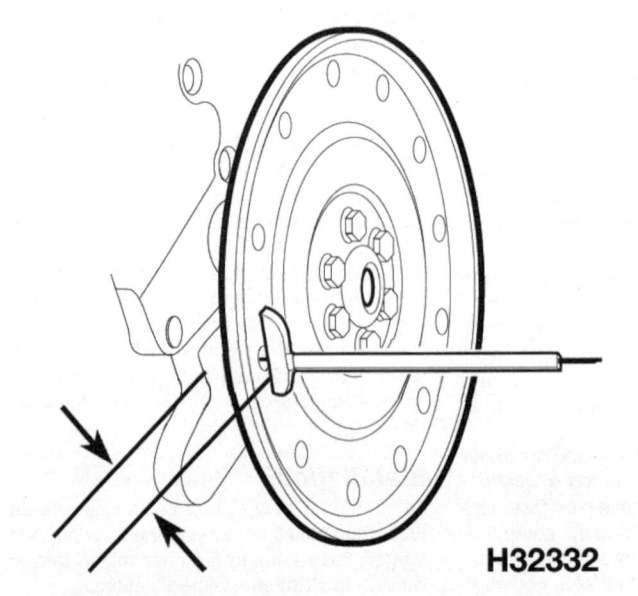

14.14 The driveplate installed height can be measured with a machinist's ruler or depth gauge, from the surface of the driveplate, through one of the holes to the engine block

H32332

Chapter 2 Part A Four-cylinder engines

15.3 Crankshaft rear seal housing mounting bolt locations - timing belt (CBPA) engine shown

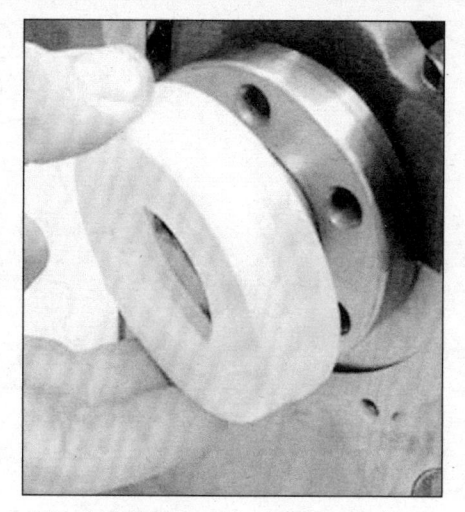

15.7a Install the seal installation tool over the end of the crankshaft...

15.7b ...and slide the oil seal and housing over the tool - 2.0L CBPA engine shown, other models similar

4   Pull the adapter plate from the locating dowels on the rear of the cylinder block, and remove the plate from the engine.

5   Before installing a new rear seal and housing, make sure the mating surfaces of the cover, the intermediate plate, the cylinder block and the oil pan rail are perfectly clean. Use a hard plastic or wood scraper to remove all traces of gasket material. Take particular care when cleaning the rear housing, as aluminum alloy is easily damaged.

6   Apply a 1/8-inch (3 mm) bead of RTV sealant to the new rear oil seal housing flange sealing surfaces.

7   On 2.0L (CBPA) models and models with the Version 1 seal, the new seal is supplied with a guide sleeve installed to the center of the seal. Position the housing and seal over the end of the crankshaft and gently, evenly, push it into position **(see illustrations)**.

8   On 1.8L (CPRA, CPKA, CXBA, CXBB, CNSA, CNSB) and 2.0L (CBFA, CCTA, CPLA, CPPA, CNTA, CXCA, CXCB, CYFB, DJJA, DLRA) engines equipped with Version 2 rear main seal, attach special tool T10122/1 to special tool T10122/6 and slide the real seal housing onto the T10122/1 tool first with the adapter housing facing outwards until the outer side of the seal is on special tool T10122/6, then remove the T10122/1 tool. Make sure the sealing lip is installed correctly **(see illustration)**.

9   On 1.8L (CPRA, CPKA, CXBA, CXBB, CNSA, CNSB) and 2.0L (CBFA, CCTA, CPLA, CPPA, CNTA, CXCA, CXCB, CYFB, DJJA, DLRA) engines, position the special tool T10122/6 and the seal housing over the end of the crankshaft and gently, evenly, push it into position. Once the tool is in position carefully remove the tool.

10   On 2.0L (CBPA) engines, tighten the rear oil seal housing bolts evenly, in several steps to the torque listed in this Chapter's Specifications then remove the tool.

11   On 1.8L (CPRA, CPKA, CXBA, CXBB, CNSA, CNSB) and 2.0L (CBFA, CCTA, CPLA, CPPA, CNTA, CXCA, CXCB, CYFB, DJJA, DLRA) engines, tighten the rear oil seal housing bolts in sequence **(see illustration)** to the torque listed in this Chapter's Specifications then remove the tool.

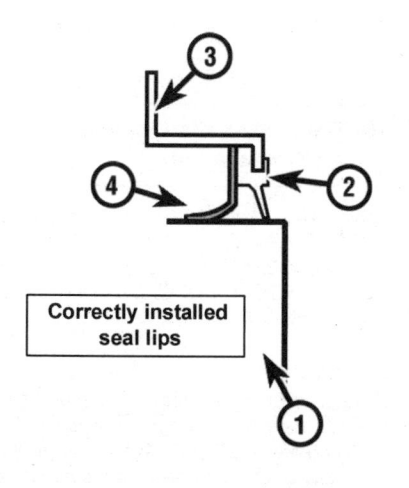

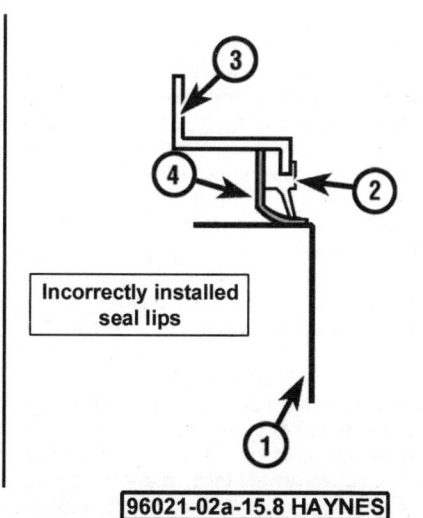

15.8 Version 2 rear main seal installation details

1   Special tool T10122/6 - crankshaft end
2   Rear seal dust lip
3   Rear seal adapter housing
4   Rear seal lip

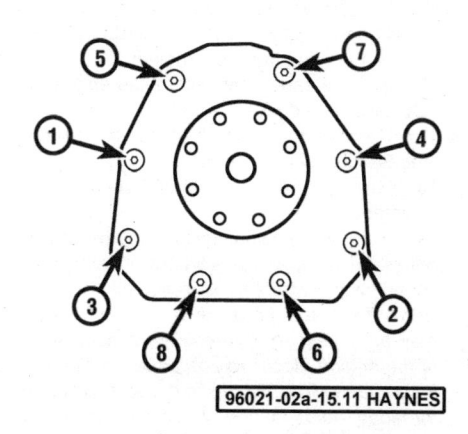

15.11 Rear seal housing tightening sequence

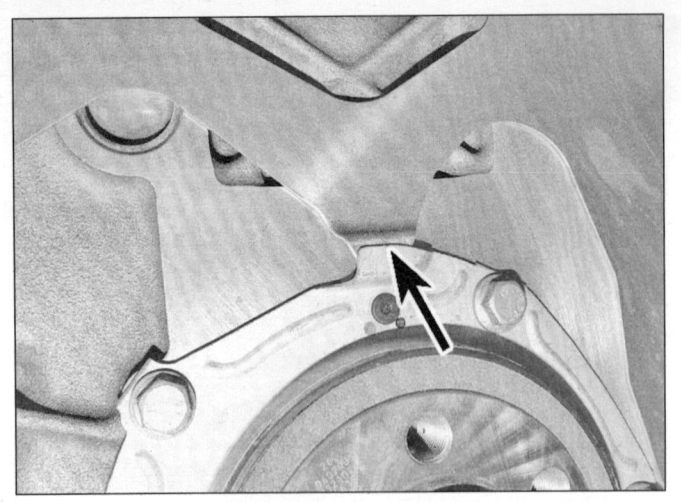

15.13 Install the adapter plate, locating it as shown

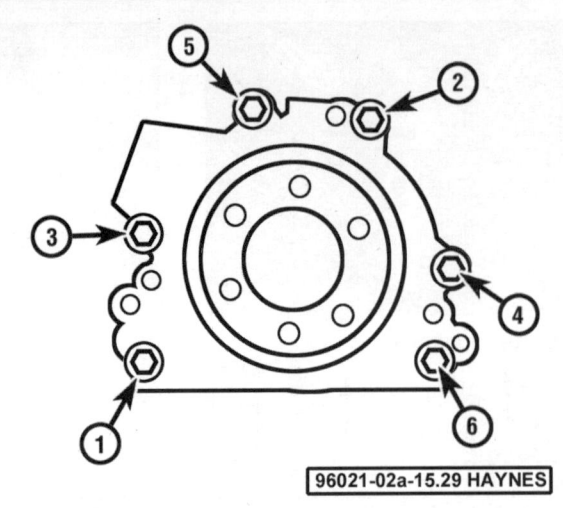

15.29 Rear seal housing tightening sequence - 1.4L engines

12 On all models, be sure to follow the sealant manufacturer's recommendations on curing times and allow the sealant to properly cure before adding oil .
13 Install the adapter plate, locating it over the oil seal housing, and onto the dowels at the back of the cylinder head **(see illustration)**.
14 The remaining steps are the reverse of removal.

1.4L engines

Caution: *Do not attempt this procedure without special tool #T10134, if the rear seal housing and sensor wheel assembly is not installed correctly the engine will not start.*

Note: *The new rear seal housing and sensor wheel may not be separated or rotated after being removed from the packaging and must be installed using special tool #T10134. The special tool must be used to center the sensor wheel guide pin into the hole of the crankshaft and set the correct depth of the housing and sensor wheel.*

15 Pull the adapter plate from the locating dowels on the rear of the cylinder block, and remove the plate from the engine.
16 If not already done, place the engine at TDC (see Section 3).
17 Remove the lower and upper oil pans (see Section 12).
18 Remove the engine speed sensor (see Chapter 6).
19 Remove the sealing housing mounting bolts and remove the sealing housing and speed sensor wheel as an assembly.
Note: *If the seal housing will not come off it may be necessary to press the assembly off of the engine block. To press the housing off, install three M6 x 35 bolts in the sealing flange threaded holes at the 1 o'clock, 5 o'clock and 7 o'clock positions. Slowly tighten the bolts a maximum of 1/2-turn each while alternating between the bolts until the housing and sensor wheel assembly can be removed.*
20 Before installing a new rear seal and housing, make sure the mating surfaces of the cover are clean and oil free.

Caution: *The rear seal housing is equipped with a plastic seal lip support ring which is used as a guide sleeve and must not be removed before installation or the assembly must be replaced with a new one. Also, the rear seal housing and sensor wheel may not be separated or rotated after being removed from the replacement or the assembly must be replaced with a new one.*

21 Place the rear seal housing and sensor ring assembly on a flat surface with sensor ring side down, then push the plastic support guide sleeve down until the edge of the guide sleeve is even or flush with the edge of the sensor ring.
22 Place the front side (transmission side) of the rear seal housing onto the special tool #T10134. It must be flat against the tool. Install the tool to the seal housing following the instructions provided with the tool.
Note: *Once the T10134 is attached to the real seal housing, the locating pin cannot slide out of the sensor ring hole.*
23 Align two guides on VW special tool #T10134 with the holes in the crankshaft, then insert the guide pins on the special tool into the crankshaft flange.
24 Install the two M6x35 mm bolts into the lower two seal housing bolt holes to use as a guide.
25 With the seal housing secured, use a beam-type torque wrench and tighten the tension nut on VW special tool #T10134 to 26 ft-lbs (35 Nm). This will leave a small air gap between the cylinder block and the rear seal housing.
26 Remove VW special tool #T10134, then use a machinist's ruler or depth gauge **(see illustration 14.14)** to measure the difference between the face of the crankshaft flange and the sensor wheel face. There should be a 0.5 mm difference. If there is not, remove the rear housing assembly and repeat Steps 21 through 25 until the proper measurement is achieved.
27 If the sensor wheel measurement is correct, reinstall VW special tool #T10134,

making sure the locating pin is seated in the sensor wheel hole, then tighten the outer hex socket bolts hand-tight and slowly push the tool onto the rear seal assembly.
28 With the seal housing secured, use a beam-type torque wrench and tighten the tension nut on VW special tool #T10134 to 30 ft-lbs (40 Nm). Measure the distance between the sensor wheel and crankshaft face (see Step 26). If the measurement is correct, proceed to the next Step. If the measurement is too small, tighten the tension nut on the special tool to 33 ft-lbs (45 Nm), then proceed to the next Step.
29 Remove VW special tool #T10134 or equivalent and tighten the rear seal housing in sequence **(see illustration)** to the torque listed in this Chapter's Specifications.
30 Install the adapter plate, locating it over the oil seal housing, and onto the dowels at the back of the cylinder head **(see illustration 15.13)**.
31 The remaining steps are the reverse of removal.

16 Balance shafts - removal and installation

Note: *This procedure applies to all 1.8L (CPKA, CPRA, CXBA, CXBB, CNSA, CNSB) and 2.0L (CBFA, CCTA, CPLA, CPPA, CNTA, CXCA, CXCB, CYFB, DJJA, DLRA) timing chain engines only.*

Removal

Note: *The manufacturer requires the balance shaft(s), needle bearings (if equipped) and intermediate shaft sprocket to be replaced with new ones once they are removed from the cylinder block. It may also be necessary to replace the exhaust balance shaft pipe if there is scoring or damage to the bearing riding surfaces of the pipe.*

1 Remove the upper and lower timing chain covers, the timing chain and balance

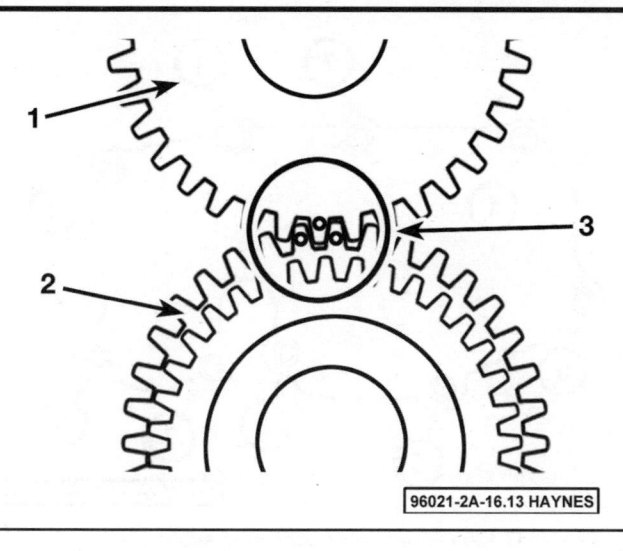

16.13 Intermediate shaft sprocket alignment details

1 Intake balance shaft
2 Intermediate shaft sprocket
3 Timing marks

shaft chain, the timing chain guides and the balance shaft chain guides (see Section 6).
2   Remove the oil pump chain, camshaft and balance shaft timing chains (see Section 6).
3   When removing the intake balance shaft, remove the water pump toothed belt (see Chapter 3), then remove the water pump drive sprocket bolt and the sprocket from the end of the intake balance shaft.
4   On the intake balance shaft, remove the intermediate sprocket retaining fastener and sprocket.
5   Remove the balance shaft(s) retaining bolt and slide the balance shaft(s) out of the cylinder block.
Note: *If the balance shaft(s) will not come out by hand, use (VW special puller tool #T10394) or equivalent.*
6   The exhaust balance shaft has a balance shaft tube installed in the engine block that is removable. Check to make sure the tube is not damaged and that the end of the tube with the slots cut into it is facing towards the outside of the engine.

Intake balance shaft seal - replacement

7   Note how far the intake balance shaft seal is recessed in the bore, then carefully pry it out with a screwdriver. Don't scratch the block bore, it may cause an oil leak.
8   Clean the block bore and coat the outer edge of the new seal with engine oil or multi-purpose grease. Apply multi-purpose grease to the seal lip.
9   Using a seal driver or a socket with an outside diameter slightly smaller than the outside diameter of the seal, carefully drive the new seal into place with a hammer. Make sure it's installed squarely and driven in to the same depth as the original. If a socket isn't available, a short section of large diameter pipe will also work. Check the seal after installation to make sure the spring didn't pop out of place.

Installation

Note: *If the exhaust balance shaft is not level, the balance shaft tube must be removed and installed again.*

10   Place the new balance shaft in a freezer for approximately 30 minutes, then remove and coat the shaft and bearings with engine oil. As quickly as possible, install the balance shaft into the cylinder block.
Note: *The clearance between the balance shaft and the balance shaft tube or cylinder block is very tight. If the shaft will not side in, do not force it; remove the shaft and cool it again until it can be easily installed.*
11   Install the balance shaft retaining fastener and tighten it to the torque listed in this Chapter's Specifications.
12   If removed, replace the O-ring on the intermediate shaft and install the shaft to the cylinder block, aligning the pin on the sprocket shaft with the hole in the cylinder block.
13   Align the timing marks of the intermediate sprocket with the mark on the intake balance shaft. Install the sprocket to the intermediate shaft, making sure the timing marks are aligned **(see illustration)**.
14   Install the intermediate sprocket retaining fastener and tighten the fastener to the torque listed in this Chapter's Specifications.
15   If the intake balance shaft was removed,

17.4 Disconnect the crankcase ventilation hose

install the water pump drive sprocket and bolt, then tighten the bolt to the torque listed in this Chapter's Specifications and install the water pump toothed belt (see Chapter 3).
16   Install the balance shafts, guides, timing and oil pump chains (see Section 6).
17   The remaining steps are the reverse of removal.

17   Oil separator and crankcase ventilator - removal and installation

Note: *This procedure does not apply to 2.0L (CBPA) engines.*

Oil separator

1.4L engines

1   Raise the front of the vehicle and support it securely on jackstands.
2   Working under the vehicle, remove the lower splash shield below the engine (see Chapter 1, Section 6).
3   Remove the charge air cooler pump bolt and move the pump out of the way (see Chapter 3).
4   Disconnect the crankcase ventilation hose from the oil separator **(see illustration)**.
5   Disengage the oil separator cover locking tabs using a screwdriver and remove the cover **(see illustration)**.
6   Loosen the oil separator fasteners in the reverse order of the tightening sequence **(see illustration 17.13a)**.

2.0L (CBFA, CCTA) engines

7   Raise the front of the vehicle and support it securely on jackstands.
8   Working under the vehicle, remove the lower splash shield below the engine (see Chapter 1, Section 6).
9   Remove the oil separator cover retaining fasteners in the reverse of the tightening sequence **(see illustration 17.13b)**.
10   Remove the separator cover and seal from the cylinder block.

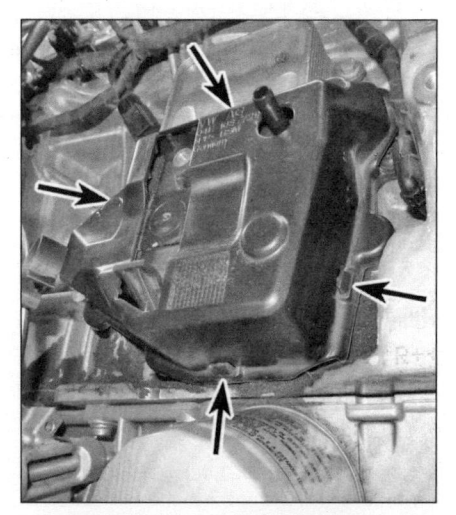

17.5 Oil separator cover locking tab locations - intake manifold removed for clarity

2A-40   Chapter 2 Part A   Four-cylinder engines

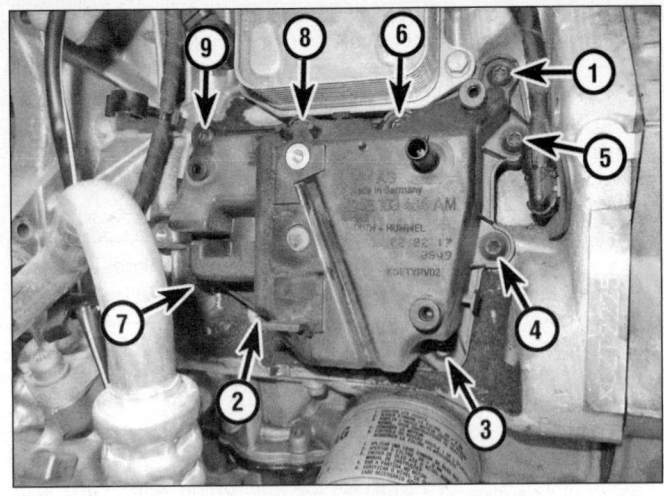

17.13a Oil separator tightening sequence - 1.4L engines

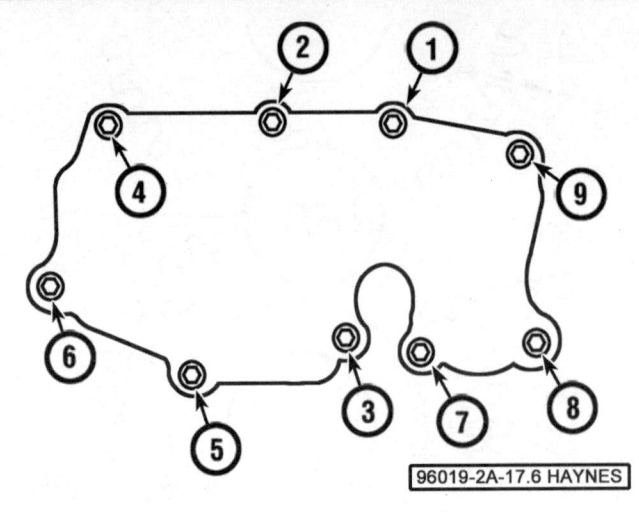

17.13b Oil separator tightening sequence - 2.0L (CCTA, CBFA) engines

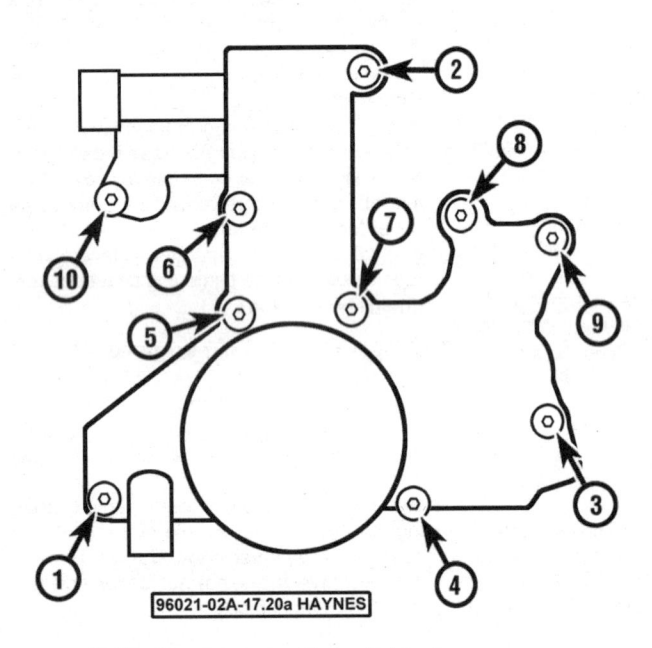

17.20a Crankcase ventilator tightening sequence - 2.0L (CCTA, CBFA) engines

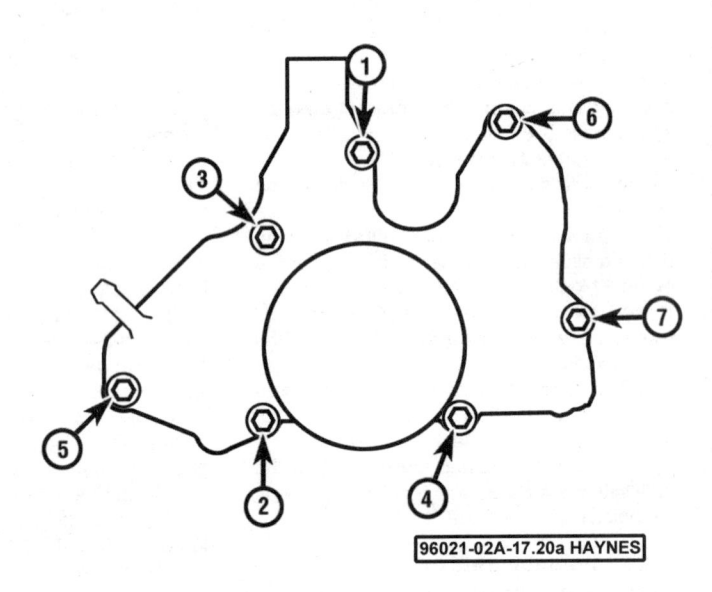

17.20b Crankcase ventilator tightening sequence - 1.8L (CPRA, CPKA, CXBA, CXBB, CNSA, CNSB) and 2.0L (CPLA, CPPA, CXCA, CNTA, CXCB, CYFB, DJJA, DLRA) engines

All models

11   On 1.4L engines, clean the mating surfaces of the cylinder block and cover, then apply a 1/16-inch (2.0 mm) continuous bead of RTV sealant to the sealing surfaces of the separator cover.

Note: *The oil separator must be installed within five minutes of applying the RTV sealant.*

12   On 2.0L (CBFA, CCTA) engines, clean the mating surfaces of the cylinder block and cover, then install a new seal on to the separator cover.

13   On 1.4L and 2.0L (CBFA, CCTA) engines, install the cover to the cylinder block and tighten the fasteners, in sequence **(see illustrations)**, to the torque listed in this Chapter's Specifications.

Crankcase ventilator

1.8L (CPRA, CPKA, CXBA, CXBB, CNSA, CNSB) and 2.0L (CCTA, CBFA, CPLA, CPPA, CXCA, CNTA, CXCB, CYFB, DJJA, DLRA) engines

14   Pull the engine cover up and off of the ballstuds, then remove the cover.

15   Remove the ignition coils from cylinders No. 3 and No. 4 (see Chapter 5).

16   Loosen and disconnect the hose from the canister purge valve (see Chapter 6).

17   Remove the oil separator mounting bolts in the reverse of the tightening sequence **(see illustrations 17.20a or 17.20b)**.

18   Remove the oil separator from the top of the camshaft housing.

19   Clean the mating surfaces of the camshaft housing and sealing surface of the crankcase ventilator, then install a new seal onto the crankcase ventilator, making sure the seal is seated in the grooves.

20   Install the crankcase ventilator to the camshaft housing and tighten the fasteners, in sequence **(see illustrations)**, to the torque listed in this Chapter's Specifications.

21   The remaining steps are the reverse of removal.

18   Engine mounts - check and replacement

1   Engine mounts seldom require attention,

Chapter 2 Part A Four-cylinder engines

18.3 Place a large block of wood between the jack head and the oil pan

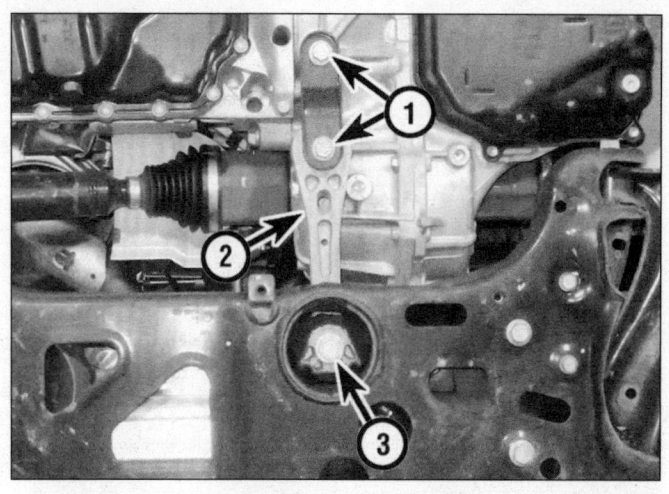

18.8 Pendulum mount details

1. Pendulum-to-transmission bolts
2. Pendulum mount
3. Pendulum mount-to-subframe bolt

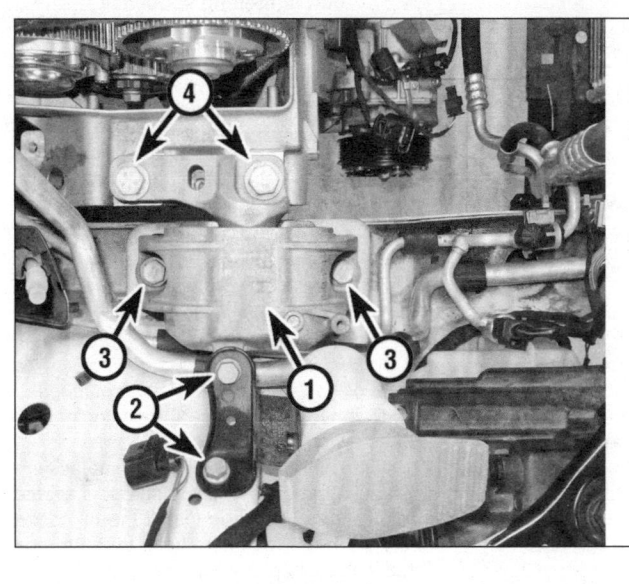

18.18 Passenger side (engine) mount details - 1.4L engine shown, other models similar

1. Passenger side (engine) mount
2. Support bracket-to-body and mount bolts
3. Mount-to-body bolts
4. Mount-to-engine bracket bolts

but broken or deteriorated mounts should be replaced immediately or the added strain placed on the driveline components may cause damage or wear.

Check

2 During the check, the engine must be raised slightly to remove the weight from the mounts.

3 Raise the vehicle and support it securely on jackstands, then position a jack under the engine oil pan. Place a large block of wood between the jack head and the oil pan, then carefully raise the engine just enough to take the weight off the mounts **(see illustration)**. Do not position the wood block under the drain plug.

Warning: *DO NOT place any part of your body under the engine when it's supported only by a jack!*

4 Remove the splash shield under the engine, if equipped. Check the mounts to see if the rubber is cracked, hardened or separated from the metal plates. Sometimes the rubber will split right down the center.

5 Check for relative movement between the mount plates and the engine or frame (use a large screwdriver or pry bar to attempt to move the mounts). If movement is noted, lower the engine and tighten the mount fasteners.

Replacement

Warning: *The weight of the entire engine will be supported by the mounts not being removed during this procedure. Never place any part of your body directly under the engine when performing this procedure. Do not disconnect more than one mount at a time, except during engine removal.*

6 Raise the vehicle and support it securely on jackstands (if not already done). Support the engine as described in Step 3.

7 Remove the under-vehicle splash shield (see Chapter 1, Section 6).

Note: *On some models, it may be necessary to remove the exhaust system bracket from the subframe.*

Pendulum support mount

8 Remove the pendulum mount-to-subframe bolt, then the pendulum-to-transmission bolts **(see illustration)**.

9 Pull the pendulum mount from the subframe.

10 Slide the pendulum mount into the subframe and install the transmission-to-pendulum bolts first, then the pendulum-to-subframe bolt and tighten the bolts to the torque listed in this Chapter's Specifications.

11 Install the splash shield.

Driver's and passenger's side mounts

12 Disconnect the cable from the negative terminal of the battery (see Chapter 5).

13 Working from above in the engine compartment, remove the engine cover (see Chapter 1, Section 7).

Driver's side (transmission) mount

14 See Chapter 7B, Section 13 for the transaxle mount replacement procedure.

Passenger's side (engine) mount

15 Remove the coolant reservoir and position it aside without disconnecting the hoses (see Chapter 3).

16 Remove the EVAP canister (see Chapter 6).

17 Support the engine from above with an engine support fixture or from below **(see illustration 18.3)**, but do not lift the engine.

18 With the engine and transmission supported with a support fixture, slowly remove the engine mount-to-body fasteners **(see illustration)**.

19 Remove the engine mount-to-engine mount bracket fasteners and remove the engine mount.

20 Install the mount and the retaining fasteners, tighten the fasteners hand-tight, then tighten all the fasteners to the torque listed in this Chapter's Specifications.

21 Installation is the reverse of the removal procedure.

19.5 Oil cooler mounting bolt locations - 1.4L models

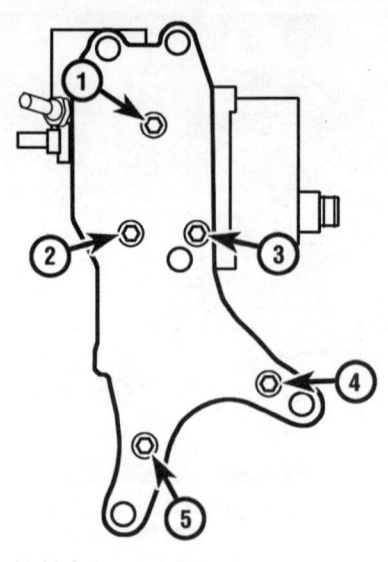

19.22 Sub assembly bracket mounting bolt tightening sequence

19 Engine oil cooler - removal and installation

Warning: *Wait until the engine is completely cool before beginning this procedure.*

1 Raise the front of the vehicle and support it securely on jackstands.
2 Working under the vehicle, remove the lower splash shield below the engine (see Chapter 1, Section 6).
3 Drain the engine coolant (see Chapter 1).

1.4L engines

4 Remove the intake manifold (see Section 7).
5 Remove the oil cooler mounting bolts **(see illustration)** and remove the cooler from the side of the engine.
6 Remove the four O-rings from the engine block and replace them with new ones.
7 Install new O-ring on to the engine block, then install the oil cooler and tighten the mounting bolts to the torque listed in this Chapter's Specifications.
8 The remaining steps are the reverse of removal.
9 Refill the engine oil and coolant (see Chapter 1) then start the engine and check for leaks.

1.8L (CPKA, CPRA, CXBA, CXBB, CNSA, CNSB) and 2.0L (CBFA, CCTA, CPLA, CPPA, CNTA, CXCA, CXCB, CYFB, DJJA, DLRA) engines

10 Remove the oil filter (see Chapter 1) then disconnect the electrical connectors just below the oil filter.
11 Remove the oil dipstick tube mounting bolt, then unclip the tube from the timing chain cover and move the dipstick tube forward.
12 Remove the fan shroud (see Chapter 3).
13 Remove the air inlet duct and charge air cooler inlet pipe (see Chapter 4).
14 Disconnect the electrical connector from the air conditioning compressor, then remove the compressor mounting bolts and secure the compressor away from the sub assembly mounting bracket.
15 Remove the intake manifold support from the sub assembly mounting bracket.
16 Remove the alternator (see Chapter 5) from the sub assembly mounting bracket.
17 Unclip the harness from the sub assembly bracket, then disconnect the electrical connectors from the oil pressure regulator valve and the oil pressure switch. Move the harness out of the way.
18 Remove the sub assembly mounting bracket bolts in reverse order of installation **(see illustration 19.22)** and pull the sub assembly forward until the oil cooler-to-coolant pump housing connector is separated, then remove the sub assembly bracket.
19 Remove the oil cooler mounting bolts and remove the cooler from the sub assembly bracket.

Note: *If large amounts of metal shavings or other debris is found in the engine oil, clean the oil channels thoroughly and replace the engine oil cooler to prevent further damage.*

20 Remove and replace the oil cooler seal with a new one.
21 Install the new seal into the sub assembly bracket grooves, making sure it is seated in all the grooves, then install the oil cooler and tighten the mounting bolts to the torque listed in this Chapter's Specifications.
22 Install the sub assembly bracket to the engine and tighten the mounting bolts in sequence **(see illustration)** to the torque listed in this Chapter's Specifications.
23 The remaining steps are the reverse of removal.
24 Refill the engine oil and coolant (see Chapter 1) then start the engine and check for leaks.

Chapter 2 Part B Five-cylinder engines

Contents

	Section		Section
Camshafts, roller rocker arms and lash adjusters - removal, inspection and installation	11	Oil pan(s) - removal and installation	13
Crankshaft front oil seal flange - replacement	10	Rear main oil seal - replacement	14
Crankshaft pulley - removal and installation	8	Repair operations possible with the engine in the vehicle	2
Cylinder head - removal, inspection and installation	12	Top Dead Center (TDC) for number five piston - locating	3
Engine mounts - check and replacement	15	Upper timing chain cover, cover seal, camshaft timing chain and tensioner - removal and installation	9
Exhaust manifold - removal, inspection and installation	7	Valve cover - removal and installation	4
General Information	1	Valve timing - check and adjustment	5
Intake manifold - removal and installation	6		

Specifications

General
Engine type... Five-cylinder, in-line, DOHC
Firing order... 1-2-4-5-3
Compression pressure................................... See Chapter 2C
Oil pressure... See Chapter 2C

Camshafts
Runout (maximum).. 0.0014 inch (0.035 mm)
Endplay (maximum)....................................... 0.0067 inch (0.17 mm)

Warpage limits
Cylinder head gasket surfaces (head, manifolds and block)................... 0.002 inch (0.05 mm)

2B-2 Chapter 2B Five-cylinder engines

Torque specifications

Note: *One foot-pound (ft-lb) of torque is equivalent to 12 inch-pounds (in-lbs) of torque. Torque values below approximately 15 ft-lbs are expressed in inch-pounds, because most foot-pound torque wrenches are not accurate at these smaller values.*

	Ft-lbs (unless otherwise indicated)	Nm
Accessories bracket (in sequence - **see illustration 12.29**)		
Step 1, bolt 5	Hand-tighten	
Step 2, all bolts	Hand-tighten	
Step 3	15	20
Step 4	Tighten an additional 90 degrees	
Camshaft timing chain cover fasteners	89 in-lbs	10
Camshaft guide frame (in sequence - **see illustration 11.17**)		
Step 1	Hand-tighten all bolts	
Step 2	71 in-lbs	8
Step 3	Tighten an additional 90 degrees	
Exhaust camshaft sprocket bolt		
Step 1	44	60
Step 2	Tighten an additional 90 degrees	
Intake camshaft adjuster bolt		
Step 1	44	60
Step 2	Tighten an additional 90 degrees	
Crankshaft pulley bolts*		
Step 1	37	50
Step 2	Tighten an additional 90 degrees	
Cylinder head bolts (in sequence - **see illustration 12.27a**)		
Step 1	30	40
Step 2	Tighten an additional 90 degrees	
Step 3	Tighten an additional 90 degrees	
Valve cover bolts	89 in-lbs	10
Exhaust manifold nuts*	17	23
Exhaust manifold heat shield bolts	89 in-lbs	10
Engine mount*		
Bracket bolts		
Step 1	15	20
Step 2	Tighten an additional 90 degrees	
Mount-to-engine bracket bolts (passenger's side)		
Step 1	30	40
Step 2	Tighten an additional 90 degrees	
Mount-to-body bracket bolts (passenger's side)		
Step 1	44	60
Step 2	Tighten an additional 90 degrees	
Transmission mount*		
Mount-to-transmission bracket bolts (driver's side)		
Step 1	30	40
Step 2	Tighten an additional 90 degrees	
Mount-to-body bracket bolts (driver's side)		
Step 1	44	60
Step 2	Tighten an additional 90 degrees	
Pendulum mount*		
Mount-to-transmission bolts		
Step 1		
8.8 stamped bolts	30	40
10.9 stamped bolts	37	50
Step 2	Tighten an additional 90 degrees	
Mount-to-subframe bolt		
Step 1	74	100
Step 2	Tighten an additional 90 degrees	
Flywheel/driveplate bolts		
Step 1	44	60
Step 2	Tighten an additional 90 degrees	
Intake manifold-to-cylinder head fasteners	80 in-lbs	9
Intake manifold-to-support bracket fasteners	144 in-lbs	16
Intake manifold support-to-cylinder block fasteners	18	25
Oil dipstick tube-to-cylinder block fastener	18	25
Lower oil pan-to-upper oil pan bolts	89 in-lbs	10
Timing chain tensioner fastener	89 in-lbs	10
Upper oil pan mounting fasteners	18	25

*Replace with new fastener(s)

Chapter 2B Five-cylinder engines

1 General Information

How to use this Chapter

1 This Part of Chapter 2 is devoted to repair procedures possible while the engine is still installed in the vehicle. Since these procedures are based on the assumption that the engine is installed in the vehicle, if the engine has been removed from the vehicle and mounted on a stand, some of the preliminary dismantling steps outlined will not apply.

2 Information concerning engine/transaxle removal, replacement and engine overhaul, can be found in Part C of this Chapter.

Engine description

3 These engines (CBTA and CBUA) are a double overhead camshaft (DOHC), twenty-valve, five-cylinder, in-line type and are transversely mounted at the front of the vehicle, with the transmission on the left-hand end. They incorporate an aluminum cylinder head and an aluminum cylinder block. There are three differences between the CBUA and CBTA engines: The CBUA is a Super Ultra Low Emission Vehicle (SULEV), has a third O2 sensor and has a secondary air injection system (see Chapter 6).

4 The two camshafts are driven by a timing chain, each operating ten valves via roller rocker arms. Valve clearance is maintained automatically at zero lash by hydraulic lash adjusters. Each camshaft rotates in five bearings that are line-bored directly in the cylinder head and the (bolted-on) camshaft guide frame. This means that the guide frame is not available separately from the cylinder head, and must not be interchanged with a guide frame from another engine.

5 These engines incorporate an aluminum timing chain cover, upper and lower oil pans, crankshaft front oil seal housing, rear oil seal housing and six crankshaft main caps. When working on these engines, note that Torx-type (both male and female heads) and hexagon socket (Allen head) fasteners are widely used. A good selection of sockets, with the necessary adapters, will be required so that these can be unscrewed without damage and on reassembly, tightened to the torque wrench settings specified.

Lubrication system

6 The oil pump is driven via a chain from the front of the crankshaft, using the same sprocket that drives the timing chain. The pump forces oil through an externally mounted full-flow cartridge-type filter. From the filter, the oil is pumped into a main gallery in the cylinder block/crankcase, from where it is distributed to the crankshaft (main bearings) and cylinder head.

7 The connecting rod bearings are supplied with oil via internal drillings in the crankshaft. Each piston crown and connecting rod is cooled by a spray of oil.

8 The cylinder head is provided with two oil galleries, one on the intake side and one on the exhaust, to ensure constant oil supply to the camshaft bearings and lifters.

2 Repair operations possible with the engine in the vehicle

1 Many major repair operations can be accomplished without removing the engine from the vehicle.

2 Clean the engine compartment and the exterior of the engine with some type of degreaser before any work is done. It will make the job easier and help keep dirt out of the internal areas of the engine.

3 Depending on the components involved, it may be helpful to remove the hood to improve access to the engine as repairs are performed (refer to Chapter 11 if necessary). Cover the fenders to prevent damage to the paint. Special pads are available, but an old bedspread or blanket will also work.

4 If vacuum, exhaust, oil or coolant leaks develop, indicating a need for gasket or seal replacement, the repairs can generally be made with the engine in the vehicle. The intake and exhaust manifold gaskets, lower oil pan gasket, front crankshaft oil seal and cylinder head gasket are all accessible with the engine in place.

5 Exterior engine components, such as the intake and exhaust manifolds, the lower oil pan, the water pump, the starter motor, the alternator and the fuel system components can be removed for repair with the engine in place.

6 Since the camshaft(s) and cylinder head can be removed without pulling the engine, valve component servicing can also be accomplished with the engine in the vehicle. Replacement of the timing chain and sprockets is also possible with the engine in the vehicle.

7 In extreme cases caused by a lack of necessary equipment, repair or replacement of piston rings, pistons, connecting rods and rod bearings is possible with the engine in the vehicle. However, this practice is not recommended because of the cleaning and preparation work that must be done to the components involved.

3 Top Dead Center (TDC) for number 5 piston - locating and locking

Warning: *Wait until the engine is completely cool before beginning this procedure.*
Caution: *If the battery is disconnected, several systems must be re-learned before they will work properly (see Chapter 5, Section 3).*
Note: *All timing and adjustments are set off of the number 5 cylinder.*
Note: *You will need a crankshaft locking pin (VW special tool #T40069) to perform this procedure.*

1 Disconnect the cable from the negative terminal of the battery (see Chapter 5).

2 Loosen the right front wheel bolts, raise the front of the vehicle and support it securely on jackstands. Block the wheels at the opposite end and remove the right front wheel.

3 Remove the under-vehicle splash shield (see Chapter 1, Section 6).

4 Working from above in the engine compartment, remove the engine cover (see Chapter 1, Section 7).

5 Remove the right front inner fender liner (see Chapter 11).

6 Using a wrench or socket on the crankshaft pulley bolt, rotate the crankshaft clockwise until the mark on the crankshaft pulley is aligned with the mark on the crankshaft front seal flange **(see illustration)**.

Note: *This can be very difficult to see; if the marks can't be seen, you will need the crankshaft adapter tool (VW #T03003, or equivalent). Mount this tool to the crankshaft pulley and rotate the engine until the line on the tool is pointing straight down.*

7 A locking pin hole is located near the lower right front corner of the engine block (on the firewall side) to provide a means of securing the engine on the no. 5 cylinder at TDC. When you locate this hole, remove the locking bolt **(see illustration)**.

8 Using a mirror, look into the bolt hole and make sure the machined hole in the crank-

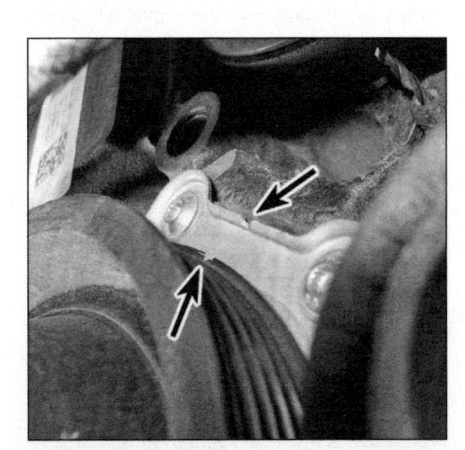

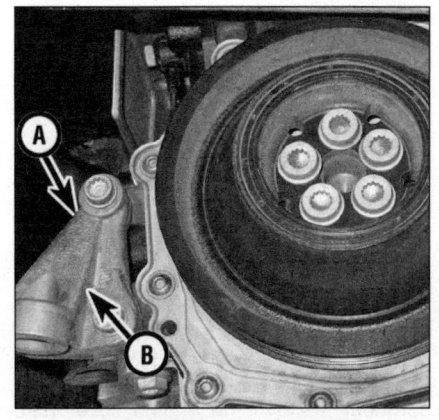

3.6 With the engine at TDC for no. 5 cylinder, the mark on the crankshaft pulley should be aligned with the mark on the crankshaft front seal flange

3.7 The TDC locking hole (A), is located behind the engine support bracket (B)

2B-4 Chapter 2B Five-cylinder engines

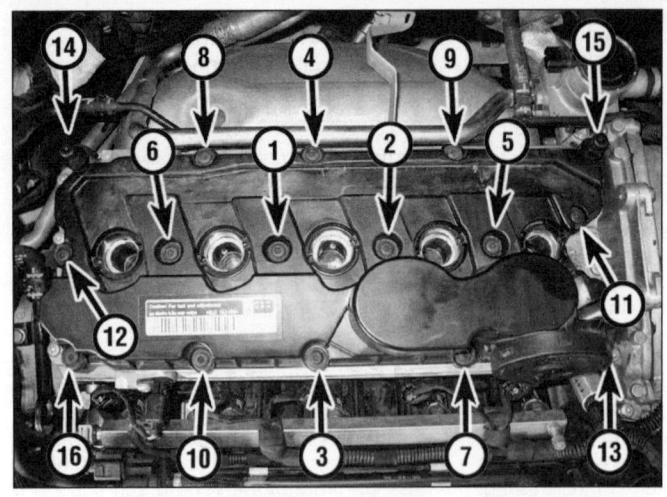

4.9 Valve cover fastener locations and tightening sequence

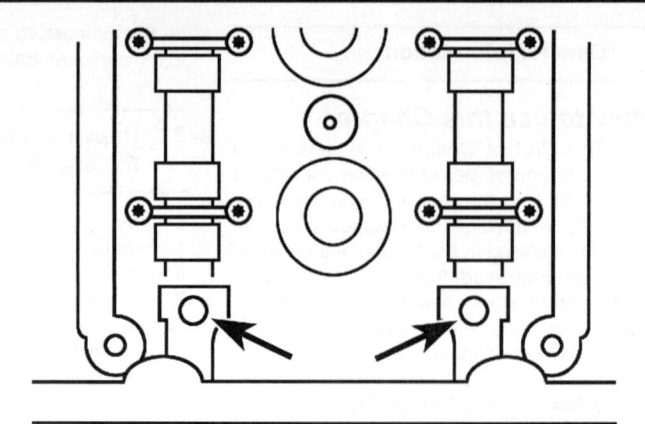

5.10 When the engine is positioned at TDC for cylinder no. 5, the threaded holes in the camshaft should align with the holes in the guide frame

shaft is aligned with the threads of the locking bolt hole, then screw in the locking pin.
Caution: *We don't recommend trying to fabricate a timing pin with a bolt because while you would be able to determine the correct bolt diameter and thread pitch, it is very difficult to determine what the length of the bolt should be. There is no way to determine the correct pin length without comparing it to a factory or aftermarket tool designed to be used with this engine. Using a bolt of the wrong length could damage the engine.*
Note: *It may be necessary to rotate the crankshaft a little up or down to align the two holes.*
9 Once the locking tool is correctly installed and the engine cannot be rotated, the engine is now locked at TDC for the no. 5 cylinder.
10 Before rotating the crankshaft again, make sure that the tool(s) are removed. Do not forget to install the locking bolt and tighten it securely.
11 Once no. 5 cylinder has been positioned at TDC on the compression stroke, TDC for any of the other cylinders can then be located by rotating the crankshaft clockwise 144-degrees at a time and following the firing order (see this Chapter's Specifications).

4 Valve cover - removal and installation

Warning: *Wait until the engine is completely cool before beginning this procedure.*
Caution: *If the battery is disconnected, several systems must be re-learned before they will work properly (see Chapter 5, Section 3).*

Removal

1 Disconnect the cable from the negative battery terminal (see Chapter 5).
2 Working from above in the engine compartment, remove the engine cover (see Chapter 1, Section 7).
3 Remove the individual ignition coil assemblies from the spark plugs (see Chapter 5).

4 Remove the secondary air injection pipe and disconnect the crankcase housing ventilation hose from the corner of the cover.
5 Working progressively, unscrew the valve cover retaining fasteners in reverse order of the tightening sequence **(see illustration 4.9)**, noting the (captive) spacer sleeve and rubber seal. Remove the cover.
6 Discard the cover gasket. This must be replaced if it is leaking or damaged.

Installation

7 Clean the cover and cylinder head gasket faces carefully, then install a new gasket onto the valve cover, ensuring that it is located correctly by the rubber seals and spacer sleeves.
8 Install the cover to the cylinder head, ensuring as the cover is tightened that the gasket remains seated.
9 Working in sequence **(see illustration)**, first tighten the cover bolts by hand only. Once all the bolts are hand-tight, go around once more in sequence, and tighten the bolts to the torque listed in this Chapter's Specifications.
10 The remainder of installation is the reverse of removal. Run the engine and check for oil leaks.

5 Valve timing - check and adjustment

Caution: *The timing system is complex, and severe engine damage will occur if you make any mistakes. Do not attempt this procedure unless you are highly experienced with this type of repair. If you are at all unsure of your abilities, be sure to consult an expert. Double-check all your work and be sure everything is correct before you attempt to start the engine.*
Caution: *If the battery is disconnected, several systems must be re-learned before they will work properly (see Chapter 5, Section 3).*
Note: *The valve timing must be adjusted any time the camshaft sprockets are loosened or removed.*

Check

Note: *You will need a crankshaft locking pin (VW special tool #T40069) and a camshaft holding tool (#T40070).*
1 Disconnect the cable from the negative terminal of the battery (see Chapter 5).
2 Loosen the right front wheel bolts, raise the front of the vehicle and support it securely on jackstands. Block the wheels at the opposite end and remove the right front wheel.
3 Remove the under-vehicle splash shield (see Chapter 1, Section 6).
4 Working from above in the engine compartment, remove the engine cover (see Chapter 1, Section 7).
5 Remove the right front inner fender liner (see Chapter 11).
6 Remove the valve cover (see Section 4).
7 Remove the locking bolt from the side of the engine **(see illustration 3.7)**.
8 Using a wrench or socket on the crankshaft pulley bolt, rotate the crankshaft clockwise only.
Caution: *Do not rotate the engine counterclockwise even a little.*
9 Continue rotating the engine to the point that the machined hole in the crankshaft is 90-percent visible through the locking bolt hole. Use a mirror to make sure the hole has not gone past the opening.
10 Working from the top of the engine, look through the bolt holes of the camshaft guide frame and verify that the threaded holes in the camshafts align with holes of the guide frame **(see illustration)**.
Note: *If the holes in the camshafts and guide frame do not line up, rotate the crankshaft one full turn (360-degrees) and check again.*
11 Confirm that the threaded holes in the camshafts line up with the holes in the guide frame, then insert the locking pin #T40069 in the cylinder block and tighten the pin. The timing will be correct when the locking pin is installed and the threaded holes of the camshaft and camshaft guide frame are aligned.
12 If the camshafts are close but both do not fully align, install the camshaft holding tool

Chapter 2B Five-cylinder engines

6.5 Disconnect the electrical connectors and clamp around the manifold

1. Fuel injector electrical connectors (3 of 5 shown)
2. CMP sensor electrical connector
3. EVAP canister purge electrical connector
4. Harness/hose clamp
5. EVAP canister purge locking ring

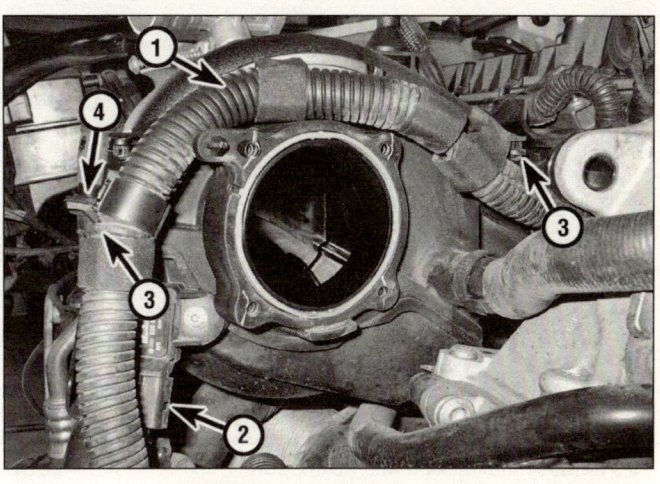

6.8 MAP sensor harness details

1. MAP sensor harness
2. MAP sensor electrical connector
3. Harness clips
4. Oil dipstick grommet/molded bracket location

#T40070 onto the end of the camshaft guide frame and screw the holding tool fasteners into the guide frame holes. Tighten the holding tool fasteners until the tool is flush with the flats on the camshafts. If the bolts cannot be easily tightened, use a wrench on the recess of the exhaust camshaft and rotate the exhaust camshaft slightly clockwise, taking up any slack or wear in the timing chain to align the holes. If the camshaft holding tool can now be easily installed, the timing is correct.

Note: *The engine is not locked correctly in this position for any disassembly.*

13 If the camshaft holding tool cannot be easily installed, the timing is not correct and must be adjusted (see Steps 14 through 25).

Adjustment

14 Remove the upper timing chain cover (see Section 9).
15 With the engine in the timing check position (see Steps 1 through 11), remove the crankshaft locking pin and slightly rotate the crankshaft a few degrees clockwise to the no. 5 cylinder TDC position.
16 Install the camshaft holding tool #T40070 (see Step 12).

Note: *It may be necessary to rotate the crankshaft slightly to help install the camshaft holding tool.*

17 Slightly rotate the crankshaft (if necessary) to the TDC position for the no. 5 cylinder and lock the crankshaft (see Section 3).

Caution: *If the crankshaft is rotated more than a few degrees or is not locked in the TDC position for the no. 5 cylinder, the valves can be damaged.*

18 With a screwdriver placed between the chain guide and the tensioner piston, slowly compress the timing chain tensioner piston and place a pin (a drill bit or paper clip will work) into the hole on the tensioner body to hold it in the fully compressed position (see Section 9).
19 Remove the exhaust camshaft sprocket and intake camshaft adjuster sprocket center retaining fasteners and pull the sprockets off the ends of the camshaft until both sprockets can easily rotate on the ends of the camshafts.
20 Reinstall the sprockets and hand-tighten the new center retaining fasteners (see Section 9, if necessary); the sprockets must still be able to rotate by hand.

Note: *Make sure the timing chain is properly seated against the chain guide.*

21 Remove the pin from the tensioner to release the tensioner piston and allow the piston to engage the chain guide.
22 Install a pin spanner wrench or similar tool to hold the exhaust sprocket in place.
23 With the help of an assistant, use a spanner wrench to carefully rotate (clockwise) and hold the exhaust sprocket to keep the timing chain preloaded.
24 While the assistant holds the pressure against the chain, tighten the intake camshaft adjuster sprocket fastener to step one of the torque listed in this Chapter's Specifications, then tighten the exhaust camshaft sprocket fastener to step one of the torque listed in this Chapter's Specifications.
25 Tighten both the intake and exhaust sprocket fasteners to step two of the torque listed in this Chapter's Specifications.
26 Remove the camshaft holding tool and the crankshaft locking pin.
27 Rotate the engine by hand two full revolutions and check the timing (see Steps 1 through 11). If the timing is incorrect, repeat Steps 14 through 25 until it is correctly adjusted.

Caution: *Before starting the engine, carefully rotate the crankshaft by hand through at least two full revolutions (use a socket and breaker bar on the crankshaft pulley center bolt). If you feel any resistance, STOP! There is something wrong - most likely, valves are contacting the pistons. You must find the problem before proceeding. Check your work and be sure everything is correct before you attempt to start the engine.*

28 The remainder of installation is the reverse of removal.

6 Intake manifold - removal and installation

Warning: *Wait until the engine is completely cool before beginning this procedure.*

Caution: *If the battery is disconnected, several systems must be re-learned before they will work properly (see Chapter 5, Section 3).*

Removal

1 Relieve the fuel system pressure (see Chapter 4).
2 Disconnect the cable from the negative battery terminal (see Chapter 5).
3 Working from above in the engine compartment, remove the engine cover/air filter housing (see Chapter 1, Section 7).
4 To disconnect the fuel supply line, fuel bleeder line (see Chapter 4) and vacuum line (if equipped), press the circlip in at the outer end of the connectors, then lift up on the connector to remove it. Cover the fuel line openings to prevent debris from entering the fuel system.
5 Disconnect the electrical connectors from the fuel injectors, Camshaft Position (CMP) sensor and the EVAP canister purge valve **(see illustration)**.
6 Unclip the wiring harness and EVAP hose clamp from the intake manifold, then open the locking ring and remove the EVAP canister purge valve.
7 Remove the throttle body (see Chapter 4) and the power steering pump and bracket (see Chapter 10).

Note: *On CBUA models the secondary air injection pump must be removed (see Chapter 6) to access the bracket bolts at the bottom on the intake manifold.*

8 Disconnect the Manifold Absolute Pressure (MAP) sensor electrical connector and remove the harness from the manifold **(see illustration)**.

6.9 Remove the intake manifold support bracket retaining fasteners

8.4 Install the crankshaft pulley holding tool

9   Remove the intake manifold support bracket fasteners **(see illustration)**, then push the oil dipstick tube/grommet out from the molded bracket on the manifold.
10   Remove the oil dipstick tube-to-engine block retaining fastener and rotate the tube out of the way.
11   The intake manifold is secured with nine bolts, four below and five along the top. Unscrew the bolts and pull the intake manifold away from the engine at a slight angle for clearance.
Note: *The bolts should remain with the manifold during removal.*

Installation
12   There are individual gaskets for each of the five ports of the intake manifold. Using new manifold gaskets, install the intake manifold. Tighten the bolts in several stages, working from the center out, to the torque listed in this Chapter's Specifications.
13   The remainder of installation is the reverse of removal.

7   Exhaust manifold - removal, inspection and installation

Warning: *Allow the engine to cool completely before beginning this procedure.*

Removal
1   Raise the vehicle and support it securely on jackstands.
2   When the engine has cooled, it will be helpful to soak the manifold heat shield retaining clips and bolts with penetrating oil to loosen any rust. Remove the heat shield. Also apply penetrating oil to the exhaust manifold mounting fasteners.
3   Disconnect the exhaust pipe from the exhaust manifold (see Chapter 4). Remove and discard the old flange gasket.
4   Unplug the electrical connector for the oxygen sensor and remove the oxygen sensor from the exhaust manifold (see Chapter 6).

5   Remove the exhaust manifold mounting nuts, then remove the exhaust manifold and the old manifold gasket.
Note: *On some models it may be necessary to remove the secondary air injection pipe to allow the manifold to be removed.*
6   Use a stud removal tool or two nuts tightened against each other to remove the old studs from the cylinder head.

Inspection
7   Inspect the exhaust manifold for cracks and any other obvious damage. If the manifold is cracked or damaged in any way, replace it.
8   Using a scraper, remove all traces of gasket material from the mating surfaces and inspect them for wear and cracks.
Caution: *When removing gasket material from any surface, especially aluminum, be very careful not to scratch or gouge the gasket surface. Any damage to the surface may cause an exhaust leak. Gasket removal solvents are available from auto parts stores and may prove helpful.*
9   Using a straightedge and feeler gauge, inspect the exhaust manifold mating surface for warpage. Also check the exhaust manifold surface on the cylinder head. If the warpage on any surface exceeds the limits listed in this Chapter's Specifications, the exhaust manifold and/or cylinder head must be replaced or resurfaced at an automotive machine shop.

Installation
10   Install new exhaust studs in the cylinder head and install the manifold with a new gasket and new self-locking nuts. Tighten the nuts in several stages, working from the center out, to the torque listed in this Chapter's Specifications.
Note: *Coat the threads of the exhaust manifold studs with an anti-seize compound.*
11   The remainder of installation is the reverse of removal. Run the engine and check for exhaust leaks.

8   Crankshaft pulley - removal and installation

Removal
1   Remove the drivebelt (see Chapter 1).
2   Set the engine to TDC using the crankshaft locking tools (see Section 3).
3   The crankshaft must be held to prevent its rotation while the pulley bolts are unscrewed. A two-pin type spanner can be used to engage the hub and hold the crankshaft pulley.
Caution: *Use of a chain- or strap-wrench or similar tool can damage the crankshaft pulley.*
4   Insert the holding tool into the spaces in the front face of the pulley **(see illustration)** to hold it in place while removing the crankshaft pulley retaining fasteners.
Caution: *Failure to hold the crankshaft pulley securely while removing the pulley bolts could result in engine damage. Severe engine damage may occur or valve timing may change if the crankshaft is moved out of the TDC position when the crankshaft pulley is removed.*
5   Unscrew all the pulley bolts, then remove the holding tool and pulley.

Installation
6   Lightly coat the crankshaft front seal with clean engine oil, then install the crankshaft pulley. If the seal shows signs of leakage, replace it before installing the crankshaft pulley (see Section 10). Install new crankshaft pulley bolts and hand-tighten only.
7   Insert the holding tool into the pulley and tighten the crankshaft pulley bolts to the torque listed in this Chapter's Specifications.
8   Remove the crankshaft pulley holding tool and the threaded crankshaft alignment bolt from the pulley and front engine cover.
9   Remove the locking pin from the cylinder block.
10   The remainder of installation is the reverse of removal.

9   Upper timing chain cover, cover seal, camshaft timing chain and tensioner - removal and installation

Warning: *Wait until the engine is completely cool before beginning this procedure.*
Caution: *If the battery is disconnected, several systems must be re-learned before they will work properly (see Chapter 5, Section 3).*

Upper timing chain cover
Removal
1   Remove the engine cover/air filter housing (see Chapter 1, Section 7).
2   Remove the battery and battery tray (see Chapter 5).
3   Loosen the right front wheel bolts, then raise the front of the vehicle and support it securely on jackstands. Block the wheels at the opposite end and remove the right front wheel.

Chapter 2B Five-cylinder engines

4 Remove the under-vehicle splash shield (see Chapter 1, Section 6).
5 Drain the coolant from the cooling system (see Chapter 1).
6 Remove coolant reservoir (see Chapter 3).
7 Remove the intake manifold (see Section 6).
8 Disconnect the vacuum hose from the vacuum pump.
9 Disconnect the coolant hoses from the thermostat housing (see Chapter 3).
10 Remove the coolant pipe from the thermostat housing by pulling the retaining clip up from the housing, then pull the pipe out of the housing.
11 Disconnect the engine coolant temperature switch, then remove the thermostat housing retaining fasteners. Remove the thermostat housing (see Chapter 3).
12 Working in a diagonal pattern, remove the upper timing chain cover retaining fasteners and carefully pry the cover from the cylinder head.

Installation

13 Installation is the reverse of removal, noting the following:
 a) Clean the mating surfaces of all sealant. **Note:** *Be careful not to gouge or use any abrasives on the mating surfaces.*
 b) Install the cover within four minutes of applying a 1/8-inch (3 mm) bead of RTV sealant.
 c) Tighten the bolts a little at a time, in a diagonal pattern, to the torque listed in this Chapter's Specifications.
 d) Install a new cover oil seal.
 e) Check and, if necessary, adjust the valve timing as described in Section 5.

Camshaft chain cover seal

Removal

14 Remove the timing chain cover (see Steps 1 through 12).
Note: *The seal can only be removed from the back side of the camshaft chain cover.*
15 Using a socket or tube, drive the seal out of the camshaft cover from the back side.
16 Using a socket or tube, press the seal in to the cover until it is seated.
Caution: *Support the cover on a flat surface when driving the cover seal in, or the housing may be damaged.*
17 Installation is the reverse of removal.

Camshaft timing chain and tensioner

Caution: *The timing system is complex, and severe engine damage will occur if you make any mistakes. Do not attempt this procedure unless you are highly experienced with this type of repair. If you are at all unsure of your abilities, be sure to consult an expert. Double-check all your work and be sure everything is correct before you attempt to start the engine.*
Note: *You will need VW special tools #T40069 (crankshaft locking pin) and #T40070 (camshaft holding tool) for this procedure.*

10.4 Sealing flange retaining fasteners

Removal

18 Remove the upper timing chain cover (see Steps 1 through 12).
19 Remove the valve cover (see Section 4).
20 Lock the crankshaft at TDC for cylinder no. 5 (see Section 3).
21 Install the camshaft holding tool #T40070 (see Section 5).
22 With a screwdriver placed between the chain guide and the tensioner piston, slowly compress the timing chain tensioner piston and place a pin (a drill bit or paper clip will work if special tool #T03006 can't be used) into the hole on the tensioner body to hold it in the fully compressed position.
23 Remove the exhaust camshaft sprocket and intake camshaft adjuster sprocket fasteners and pull the sprockets off the ends of the camshafts.
Caution: *Remove the center mounting fastener on the camshaft sprockets; do not disassemble the camshaft adjuster sprocket.*
24 Remove the tensioner mounting fastener, then slide the tensioner assembly, chain guide and tensioner guide off of the cylinder head dowels.
Caution: *Do not remove the pin holding the tensioner piston in the compressed position.*
25 Remove and discard the tensioner gasket and the strainer from the cylinder head.
26 Remove the timing chain from the housing.
Note: *Mark the original direction of rotation on the chain if it is to be reused.*

Installation

27 Loop the camshaft timing chain around the double sprocket and secure the chain from falling off the sprocket.
28 Install the chain guide (left) and the tensioner guide (right) onto the cylinder head dowels.
29 Insert the tensioner strainer into the cylinder head.
30 Install the tensioner gasket and tensioner assembly onto the cylinder head dowels, then the tensioner fastener. Tighten the fastener to the torque listed in this Chapter's Specifications.

31 Install the intake camshaft adjuster sprocket to the intake camshaft, hand-tightening the sprocket mounting fastener.
32 Loop the timing chain over the sprocket, then the exhaust camshaft sprocket. Install the exhaust camshaft sprocket and chain to the exhaust camshaft, hand-tightening the sprocket mounting fastener.
33 Remove the pin from the tensioner and release the piston to engage the chain guide.
34 Adjust the valve timing (see Section 5).
35 The remainder of installation is the reverse of removal.

10 Crankshaft front oil seal flange - replacement

Note: *The front seal is integrated into the front sealing flange and cannot be replaced separately; if the front oil seal is leaking, the sealing flange must be replaced.*

1 Remove the drivebelt and tensioner (see Chapter 1).
2 Set the engine to TDC using the crankshaft locking tools (see Section 3).
3 Remove the crankshaft pulley (see Section 8).
4 Remove the sealing flange retaining fasteners **(see illustration)**.
5 Starting at the alignment holes, work around the entire flange, then carefully pry the sealing flange from the cylinder block.
Caution: *If the sealing flange is bent or distorted it must be replaced.*
6 Remove the sealing flange from the engine.
7 Before installing the new sealing flange, make sure the mating surfaces of the cylinder block are perfectly clean. Use a hard plastic or wood scraper to remove all traces of gasket material. Take particular care when cleaning as aluminum alloy is easily damaged.
8 Slide VW special tool #T40069 into the crankshaft seal and allow the tool to remain in the seal during installation.
9 Apply a 1/8-inch (3 mm) bead of RTV sealant into the groove of the sealing flange.

2B-8   Chapter 2B Five-cylinder engines

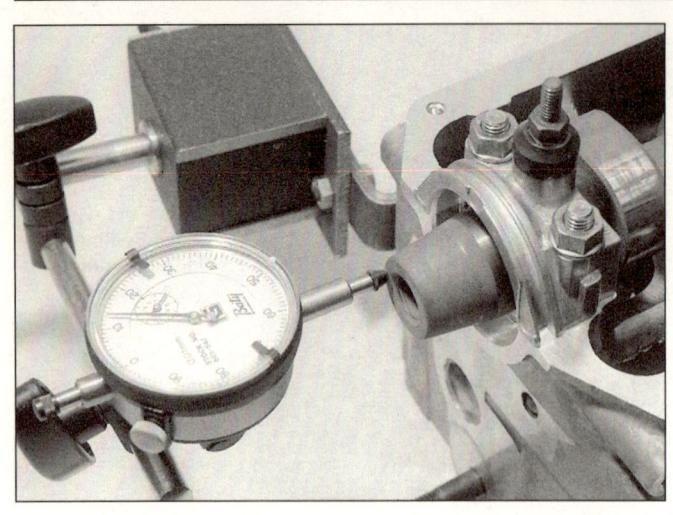

11.9 Checking camshaft endplay with a dial indicator

11.10 Measuring the camshaft lobe height with a micrometer - make sure you move the micrometer to get the highes reading (top of cam lobe)

10   Align special tool #T40069 over the end of the crankshaft, then carefully position the sealing flange on the engine block and install the timing cover retaining fasteners loosely.

11   Working in a diagonal sequence, first tighten the flange retainers by hand only. Once all the fasteners are hand-tight, tighten the retaining fasteners in a diagonal sequence to the torque listed in this Chapter's Specifications.

12   Rotate special tool #T40069 while pulling it off of the crankshaft, making sure the oil seal is not damaged.

13   The remainder of installation is the reverse of removal. Run the engine and check for leaks.

11 Camshafts, roller rocker arms and lash adjusters - removal, inspection and installation

Warning: *Wait until the engine is completely cool before beginning this procedure.*
Caution: *If the battery is disconnected, several systems must be re-learned before they will work properly (see Chapter 5, Section 3).*
Caution: *Severe engine damage will occur if you make any mistakes. Do not attempt this procedure unless you are highly experienced with this type of repair. If you are at all unsure of your abilities, be sure to consult an expert. Double-check all your work and be sure everything is correct before you attempt to start the engine.*
Note: *The camshafts and lifters should always be thoroughly inspected before installation and camshaft endplay should always be checked prior to camshaft removal. Although the hydraulic lifters are self-adjusting and require no periodic service, there is an in-vehicle procedure for checking excessively noisy hydraulic lifters.*

Removal

1   Lock the camshafts in the timing adjustment position as described in Section 5.

2   Remove the camshaft sprocket and the timing chain as described in Section 9.
Note: *When installing the camshafts, the lobes and the slots in the ends of the camshafts must be in the same positions as they were before removal.*

3   This engine does not have individual camshaft bearing caps. All of the caps are part of one assembly called a camshaft guide frame. Working in the reverse of the tightening sequence **(see illustration 11.17)** loosen the camshaft guide frame bolts progressively by half a turn at a time. Work only as described, to gradually release pressure of the valve springs evenly on the guide frame.
Note: *Use NEW guide frame bolts, as the old bolts are stretch-type fasteners that will not provide the correct torque readings if reused.*

4   Remove the guide frame, then remove the camshafts and set them aside in a clean space. Remove all traces of gasket sealing material from the guide frame and cylinder head mating surfaces.

5   Do not reinstall camshaft and valve train components unless a thorough inspection proves they are in perfect condition.

Inspection

6   Visually check the camshaft bearing surfaces for pitting, score marks, galling and abnormal wear. If the bearing surfaces are damaged, the cylinder head and guide frame will have to be replaced.
Note: *If there is scoring on either the guide frame or the camshaft saddles in the cylinder head, both the cylinder head and the guide frame must be replaced.*

7   Measure the outside diameter of each camshaft bearing journal and record your measurements. Measure the inside diameter of each corresponding camshaft bearing and record the measurements. No manufacturer's specifications were available at the time of writing. If the camshafts, camshaft bearing surfaces, lifters or the cylinder head bores are excessively worn, new camshafts, new lifters and/or new camshaft guide frame and cylinder head may be required.

8   To check camshaft runout, place the camshaft back into the cylinder head and set up a dial indicator on the center journal. Zero the dial indicator. Turn the camshaft slowly and note the dial indicator readings. Record your readings and compare them with the specified runout in this Chapter. If the measured runout exceeds the runout specified in this Chapter, replace the camshaft.

9   Place the camshafts back into the cylinder head and temporarily install the cylinder head cover or guide frame. Check the camshaft endplay by placing a dial indicator with the stem in line with the camshaft and touching the snout **(see illustration)**. Push the camshaft all the way to the rear and zero the dial indicator. Next, pry the camshaft to the front as far as possible and check the reading on the dial indicator. The distance it moves is the endplay. If it's greater than the value listed in this Chapter's Specifications, check the guide frame for wear. If the guide frame is worn, the cylinder head must be replaced.

10   Compare the camshaft lobe height by measuring each lobe with a micrometer **(see illustration)**. Measure each of the intake lobes and record the measurements and relative positions. Measure each of the exhaust lobes and record the measurements and relative positions. This will let you compare all of the intake lobes to one another and all of the exhaust lobes to one another. If the difference between the lobes exceeds 0.005 inch, the camshaft should be replaced. Do not compare intake lobe heights to exhaust lobe heights as lobe lift may be different. Only compare intake lobes-to-intake lobes and exhaust lobes-to-exhaust lobes for this comparison.

Chapter 2B Five-cylinder engines   2B-9

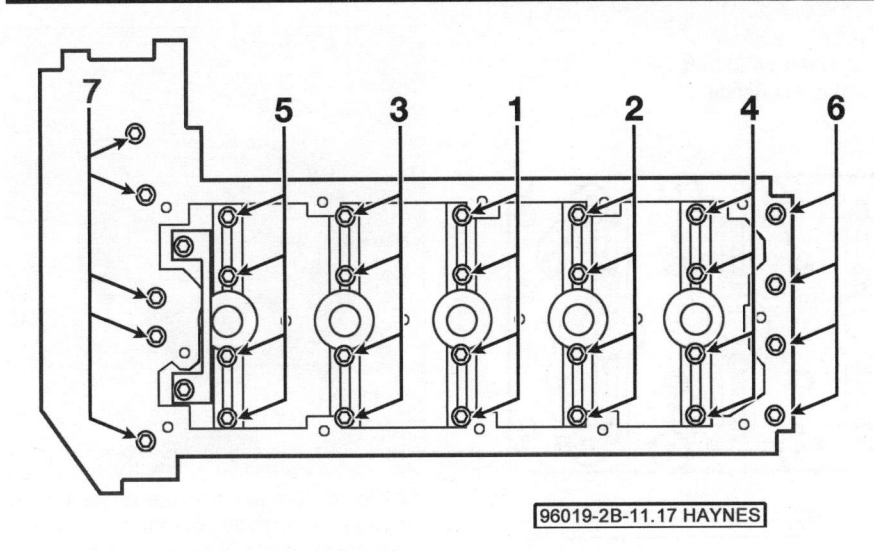

11.17 Camshaft guide frame tightening sequence

11.18 The tapered edge of the sealing plug should be flush with the cylinder head and camshaft guide frame

11 Inspect the contact and sliding surfaces of each lash adjuster and rocker arm roller for wear and scratches.
Note: *If the rocker roller is worn, it's a good idea to check the corresponding camshaft. Do not lay the lash adjusters on their side or upside down, or air can become trapped inside and the lash adjuster will have to be bled. The adjusters can be laid on their side only if they are submerged in a pan of clean engine oil until reassembly.*
12 Check that each lash adjuster moves up and down freely in its bore on the cylinder head. If it doesn't, the valve may stick open and cause internal engine damage.

Installation
13 Lubricate the lash adjusters and roller rocker arms and install them in their original locations in the cylinder head.
14 Place the guide frame upside down on a flat surface, then clean and lubricate the guide frame bearing surfaces with clean engine oil. Clean and lubricate the camshafts with clean engine oil, then place the camshafts into the guide frame with the recesses of the camshaft facing each other. The intake camshaft sensor wheel should be pointing towards the camshaft sensor and the bearing journals of the camshafts should be seated squarely in the guide frame journals.
Caution: *There are two sealing rings on the end of the intake camshaft; the open end of the rings must face up or down. If the rings face side to side, oil will leak past the rings.*
15 With the help of an assistant, turn the guide frame over while keeping the camshafts from falling out. Rotate the camshafts until the threaded holes in the camshafts are aligned with the threaded holes in the camshaft guide frame **(see illustration 5.10)**. Install camshaft holding tool #T40070 and tighten the holding tool to 15 ft-lbs (20 Nm).

16 Turn the guide frame back over and apply a 1/8-inch (3 mm) bead of RTV sealant around the spark plug holes and the outside edges of the guide frame, then immediately install the guide frame assembly to the cylinder head.
17 Install the new guide frame bolts and tighten them in the proper sequence **(see illustration)** to this Chapter's Specifications, wiping off any extra sealant.
Caution: *Follow the sealant manufacturer's recommendations on curing times and allow the sealant to properly cure before adding oil.*
18 Using a socket or pipe, carefully press the sealing plugs in to the end of the cylinder head and camshaft guide frame until the tapered edge of the plug is flush with the cylinder head and camshaft guide **(see illustration)**.
19 Allow the engine to sit for 30 minutes before proceeding.
Caution: *Do not turn the crankshaft during this time.*
20 Adjust the valve timing (see Section 5), then remove the camshaft holding tool and the crankshaft locking pin.
Caution: *Before starting the engine, carefully rotate the crankshaft by hand through at least two full revolutions (use a socket and breaker bar on the crankshaft pulley center bolt). If you feel any resistance, STOP! There is something wrong - most likely, valves are contacting the pistons. You must find the problem before proceeding. Check your work and be sure everything is correct before you attempt to start the engine.*
21 Refill the cooling system (see Chapter 1).
22 The remainder of installation is the reverse of removal.
Note: *Rotate the engine through two complete revolutions and ensure that both TDC marks still align.*

12 Cylinder head - removal, inspection and installation

Warning: *Wait until the engine is completely cool before beginning this procedure.*
Caution: *If the battery is disconnected, several systems must be re-learned before they will work properly (see Chapter 5, Section 3).*

Removal
1 Relieve the fuel system pressure (see Chapter 4)
2 Remove the battery and battery tray (see Chapter 5).
3 Raise the front of the vehicle and support it securely on jackstands. Block the wheels at the opposite end.
4 Remove the under-vehicle splash shield (see Chapter 1, Section 6).
5 Remove the engine cover/air filter housing (see Chapter 1, Section 7).
6 Remove the alternator and ignition coils (see Chapter 5). Remove the power steering pump (see Chapter 10).
7 Drain the cooling system (see Chapter 1).
8 Remove the coolant expansion tank (see Chapter 3).
9 Remove the valve cover (see Section 4).
10 Remove the intake manifold (see Section 6) then remove the accessories support bracket bolts and and bracket **(see illustration 12.29)**. Reinstall the support strap to aid in removing and installing the cylinder head.
11 Remove the camshaft chain cover (see Section 9).
12 Lock the camshafts (see Section 11), then remove the camshaft sprockets from the camshafts (see Section 5) and allow the camshaft timing chain to lie to the side.
13 Disconnect the exhaust pipe from the exhaust manifold.

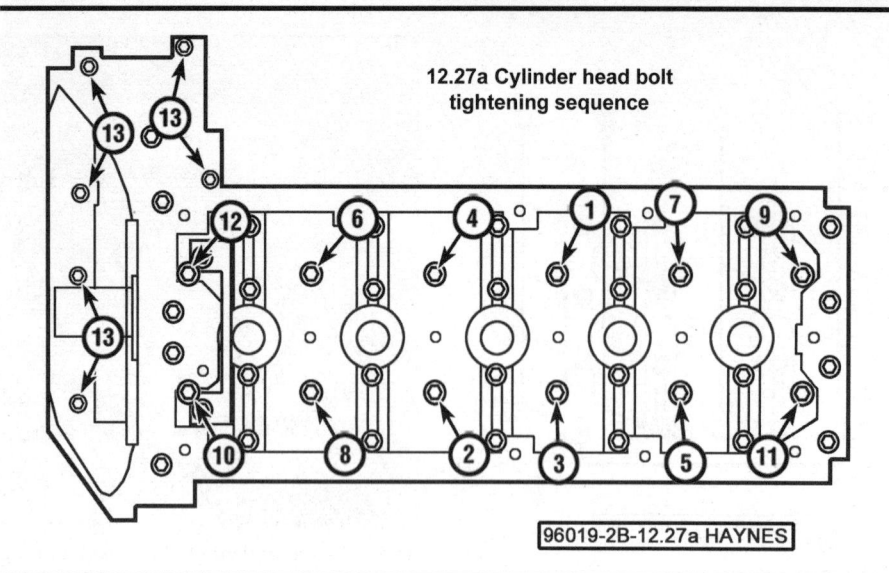

2B-10　Chapter 2B　Five-cylinder engines

12.27a Cylinder head bolt tightening sequence

12.27b You can use a torque angle gauge, or you can carefully note the starting and stopping points of the wrench handle

14　Disconnect the oxygen sensor electrical connectors and move any harness retainers.
15　Remove the cylinder head bolts in the reverse order of installation **(see illustration 12.27a)**.
Note: *If the cylinder head bolt closest to the camshaft holding tool was not able to be removed, loosen the camshaft holding tool one turn and slide the tool away from the bolt. Remove the bolt and retighten the holding tool.*

Inspection

16　The mating faces of the cylinder head and cylinder block must be perfectly clean before replacing the head. Use spray-on gasket remover and a hard plastic or wood scraper to remove all traces of gasket and carbon.
17　Take particular care during the cleaning operations, as aluminum alloy is easily damaged. Also, make sure that the carbon is not allowed to enter the oil and water passages - this is particularly important for the lubrication system, as carbon could block the oil supply to the engine's components.
18　To prevent carbon entering the gap between the pistons and bores, smear a little grease in the gap. After cleaning each piston, use a small brush to remove all traces of grease and carbon from the gap, then wipe away the remainder with a clean rag.
19　Check the mating surfaces of the cylinder block and the cylinder head for nicks, deep scratches and other damage. Also check the cylinder head and cylinder block mating surfaces with a precision straightedge and feeler gauges. If either surface exceeds the warpage limit listed in this Chapter's Specifications, the manufacturer states that the component out of specification must be replaced. If the gasket mating surface of your cylinder head or block is out of specification or is severely nicked or scratched, you may want to consult with an automotive machine shop for advice.

Installation

20　Wipe clean the mating surfaces of the cylinder head and cylinder block. Install the alignment dowels into their original locations.
21　The cylinder head bolt holes must be free from oil or coolant. This is most important, because a hydraulic lock in a cylinder head bolt hole can cause a fracture of the block casting when the bolt is tightened. Note the location of the cylinder head alignment dowels in the block.
22　Apply a 1/8-inch (3 mm) bead of RTV sealant to the clean, sides of the timing chain area of the cylinder block.
Note: *Follow the sealant manufacturer's recommendations on curing times and allow the sealant to properly cure before adding oil.*
23　The cylinder head gasket is easily damaged - only remove it from its packaging just before you're ready to install the gasket. Position a new gasket on the cylinder block surface over the alignment dowels, with the "TOP" mark facing up.
24　Apply a 1/8-inch (3 mm) bead of RTV sealant to the top of the head gasket, at the rear side of the timing chain area (viewed from the driver's side fender looking at the gasket, on the right side, approximately two inches long). Once the sealant is applied to the cylinder head gasket, the cylinder head must be installed within five minutes.
25　As the cylinder head is such a heavy and awkward assembly to install, an assistant should be used. Install the cylinder head, guiding the camshaft chain through the opening at the end of the cylinder head.
26　Coat the new cylinder bolt threads with engine oil - do not apply more than a light film of oil. Install the new cylinder head bolts and screw them in hand-tight.
Note: *New cylinder head bolts must be used.*
27　Working progressively and in the sequence shown **(see illustration)**, tighten all the bolts to the Step 1 torque setting listed in this Chapter's Specifications. There are three tightening stages, the final two using the angle torque method **(see illustration)**.

28　Allow the engine to sit for 30 minutes.
Caution: *Do not turn the crankshaft during this time.*
29　Install the accessories bracket and bolts, then tighten the bolts in sequence **(see illustration)** to the torque listed in this Chapter's Specifications.
30　Adjust the valve timing (see Section 5) and remove the camshaft holding tool and the crankshaft locking pin.
Caution: *Before starting the engine, carefully rotate the crankshaft by hand through at least two full revolutions (use a socket and breaker bar on the crankshaft pulley center bolt). If you feel any resistance, STOP! There is something wrong - most likely, valves are contacting the pistons. You must find the problem before proceeding. Check your work and be sure everything is correct before you attempt to start the engine.*
31　Installation of the other components removed is a reversal of removal.
32　Change the engine oil and filter and refill the cooling system (see Chapter 1).

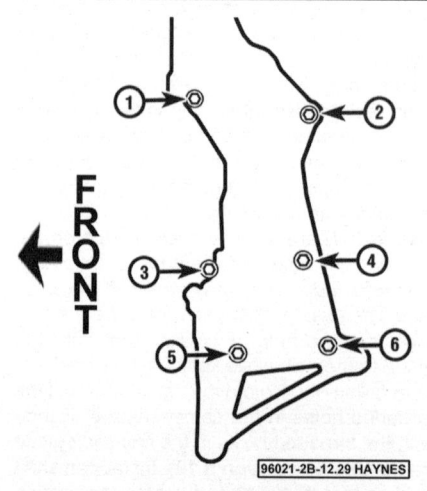

12.29 Accessories bracket bolt tightening sequence

Chapter 2B Five-cylinder engines

13 Oil pan(s) - removal and installation

Note: *This engine is equipped with a two-piece oil pan. The upper pan can only be removed with the engine out of the vehicle. Therefore the following procedure pertains only to removal and installation of the lower oil pan in the vehicle. Refer to Chapter 2C for removal and installation of the upper oil pan.*

Lower oil pan

Removal

1 Set the parking brake and block the rear wheels.
2 Raise the front of the vehicle and support it securely on jackstands.
3 Remove the under-vehicle splash shield (see Chapter 1, Section 6).
4 Drain the engine oil (see Chapter 1).
5 Remove the lower oil pan retaining fasteners in a diagonal sequence.
6 Remove the oil pan, and the gasket if equipped. If it is stuck, tap it gently with a mallet to free it.

Installation

7 Use a scraper to remove all traces of old sealant from the upper oil pan and lower oil pan. Clean the mating surfaces with brake system cleaner.
Note: *Use RTV sealant to seal the oil pan-to-cylinder block mating surface.*
8 Make sure the threaded bolt holes in the block are clean.
9 Check the oil pan flange for distortion, particularly around the bolt holes. Remove any nicks or burrs as necessary.
10 Apply a 1/8-inch (3 mm) bead of RTV sealant to the oil pan flange.
Note: *The oil pan must be installed within five minutes once the sealant has been applied.*
11 Carefully position the lower oil pan on the upper pan. Install the lower oil pan-to-upper oil pan bolts and tighten them by hand. Tighten all bolts in a diagonal sequence, to the torque listed in this Chapter's Specifications.
Note: *Follow the sealant manufacturer's recommendations on curing times and allow the sealant to properly cure before adding oil.*
12 The remainder of installation is the reverse of removal.
13 Run the engine and check for oil pressure and leaks.

Upper oil pan

Removal

14 Remove the lower oil pan (see Steps 1 through 6).
15 Remove the transaxle (see Chapter 7A or Chapter 7B).
16 Remove the crankshaft front oil seal flange (see Section 10).
17 Loosen the camshaft sprockets (see Chapter 2A, Section 5).
18 Remove the control housing (see Chapter 2C, Section 12).

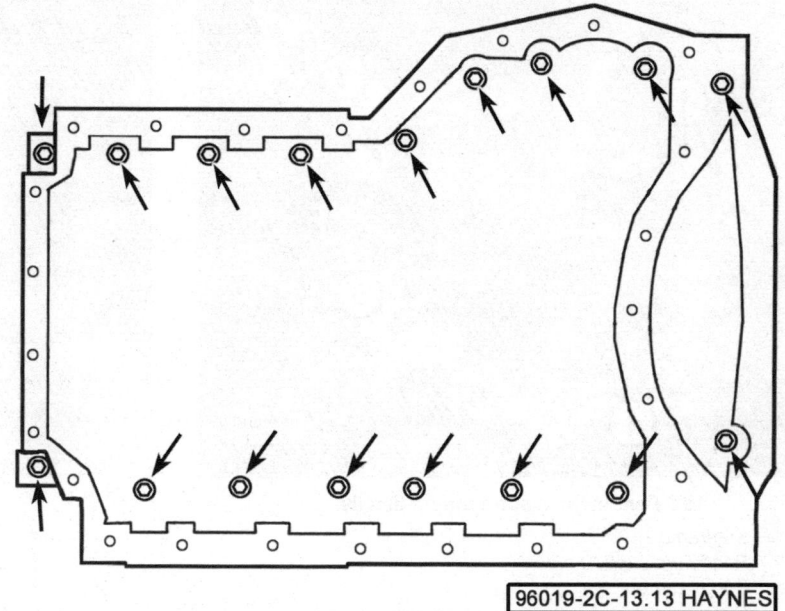

13.22 Upper oil pan mounting fastener locations engines

19 Remove the oil pump chain tensioner and guide (see Chapter 2C, Section 14).
20 Remove the pick-up tube retaining fasteners and pick-up tube.
21 Remove the pick-up tube-to-oil pump seal.
22 Remove the upper oil pan mounting fasteners **(see illustration)** and carefully pry the oil pan off. **Note:** *Pry the oil pan at the number 6 and number 1 crankshaft bearing caps using a flat blade screwdriver or equivalent tool.*

Installation

23 Use a scraper to remove all traces of old sealant from the block and oil pan. Clean the mating surfaces with brake system cleaner.
Note: *Use RTV silicone sealant to seal the oil pan-to-cylinder block mating surface.*
24 Make sure the threaded bolt holes in the block are clean.
25 Check the oil pan flange for distortion, particularly around the bolt holes. Remove any nicks or burrs as necessary.
26 Carefully position the upper oil pan on the cylinder block, install two mounting fasteners at the front and rear of the oil pan, and tighten them by hand.
Note: *The ends of the upper pan and cylinder block must be aligned to create a flat surface for the control housing to be mounted to.*
27 Install VW special tool #2036/1, or equivalent to each side of the engine block.
Note: *This tool has two flat metal bars that are bolted to the cylinder block to keep the oil pan aligned with the cylinder block when the pan is installed.*
28 Align the oil pan with the special tool, check that the edge of the block and the edge of the oil pan are flush, then install the remaining oil pan bolts **(see illustration 13.22)**. Tighten all mounting fasteners in a diagonal sequence, to the torque listed in this Chapter's Specifications.

29 Install the oil pick-up tube to the pump with a new seal, then install the pick-up tube mounting fasteners to the oil pan, tightening all mounting fasteners to the torque listed in this Chapter's Specifications.
30 Install the oil pump timing chain and tensioner (see Chapter 2C, Section 14).
Note: *Be sure to follow the sealant manufacturer's recommendations on curing times and allow the sealant to properly cure before adding oil.*
31 The remainder of installation is the reverse of removal.

14 Rear main oil seal - replacement

1 Remove the transaxle (see Chapter 7A or 7B).
2 Remove the flywheel/driveplate (see Chapter 2A).
3 Pry the oil seal from the rear of the sealing flange. Be careful not to nick or scratch the crankshaft or the flange housing. Note how far it's recessed into the bore before removal, so the new seal can be installed to the same depth. Thoroughly clean the seal bore in the block with a shop towel. Remove all traces of oil and dirt.
4 Lubricate the outside diameter of the seal and install the seal over the end of the crankshaft. Make sure the lip of the seal points toward the engine. Preferably, a seal installation tool (available at some auto parts stores) is needed to press the new seal back into place. If the proper seal installation tool is unavailable, use a large socket, section of pipe or a blunt tool and carefully drive the new seal squarely into the seal bore and flush with the edge of the engine block.
5 Install the flywheel/driveplate (see Chapter 2A).
6 Install the transmission (see Chapter 7A or 7B).

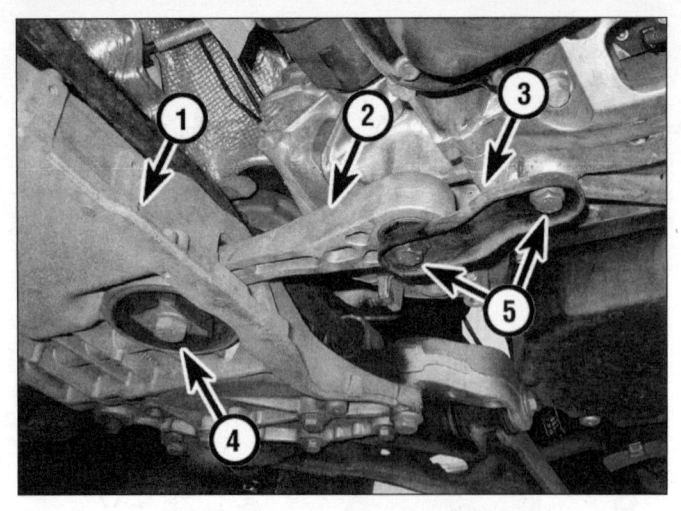

15.8 Pendulum support mount details

1. Subframe
2. Pendulum support mount
3. Pendulum bracket
4. Pendulum mount-to-subframe retaining fastener
5. Pendulum bracket-to-transmission retaining fasteners

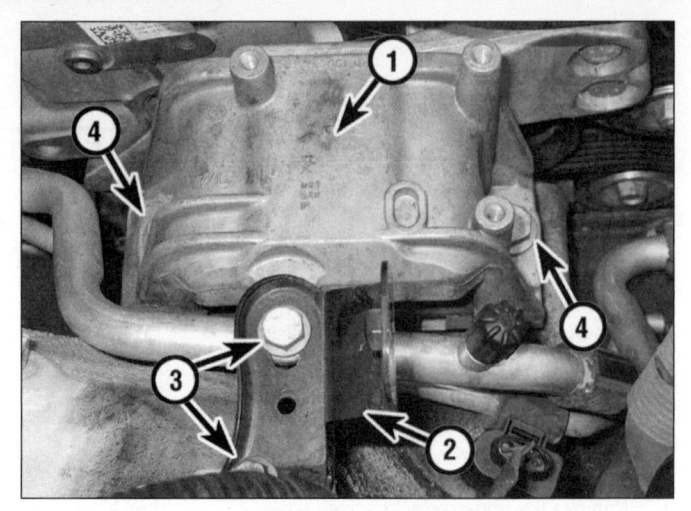

15.15 Engine mount details

1. Engine mount
2. Support bracket
3. Support retaining fasteners
4. Engine mount-to-bracket retaining fasteners

15 Engine mounts - check and replacement

1 Engine mounts seldom require attention, but broken or deteriorated mounts should be replaced immediately or the added strain placed on the driveline components may cause damage or wear.

Check

2 During the check, the engine must be raised slightly to remove the weight from the mounts.
3 Raise the vehicle and support it securely on jackstands, then position a jack under the engine oil pan. Place a large wood block between the jack head and the oil pan to prevent oil pan damage, then carefully raise the engine just enough to take the weight off the mounts.
Warning: *DO NOT place any part of your body under the engine when it's supported only by a jack!*
4 Check the mounts to see if the rubber is cracked, hardened or separated from the bushing in the center of the mount.
5 Check for relative movement between the mount and the engine or chassis. Use a large screwdriver or prybar to attempt to move the mounts. If movement is noted, lower the engine and tighten the mount fasteners.

Replacement

Caution: *If the battery is disconnected, several systems must be re-learned before they will work properly (see Chapter 5, Section 3).*
Note: *Refer to Chapter 7B for information on the transaxle mounts.*
6 Raise the vehicle and support it securely on jackstands (if not already done). Support the engine as described in Step 3.
7 Remove the under-vehicle splash shield (see Chapter 1, Section 6).

Pendulum support mount

8 Remove the pendulum mount-to-subframe bolt, then the pendulum-to-transmission bolts and bracket **(see illustration)**.
9 Pull the pendulum mount from the subframe and remove the pendulum support.
10 Slide the pendulum mount into the subframe and install the bracket and transmission-to-pendulum bolts. Install the pendulum-to-subframe bolt and tighten the bolts to the torque listed in this Chapter's Specifications.
11 Install the splash shield.

Passenger's side (engine) mount

Warning: *The weight of the entire engine will be supported by the mounts not being removed during this procedure. Never place any part of your body directly under the engine when performing this procedure. Do not disconnect more than one mount at a time, except during engine removal.*
Caution: *Remove the heat shield over the inboard joint of the right driveaxle before raising the engine.*
12 Working from above in the engine compartment, remove the engine cover/air filter housing (see Chapter 1, Section 7).
13 Remove the coolant reservoir (see Chapter 3).
Note: *Depending on the model it may be necessary to remove the windshield washer reservoir (see Chapter 12).*
14 Place a floor jack under the engine with a wood block between the jack head and oil pan and raise the engine slightly to relieve the weight from the mounts.
15 Remove the fasteners and detach the support bracket and mount from the body and engine **(see illustration)**.
16 Installation is the reverse of removal. Use thread-locking compound on the mount bolts and tighten them securely.

Chapter 2 Part C
General engine overhaul procedures

Contents

	Section
Control housing cover (five-cylinder engine) - removal and installation	12
Crankshaft - removal and installation	10
Cylinder compression check	3
Engine - removal and installation	7
Engine rebuilding alternatives	5
Engine removal - methods and precautions	6
Engine overhaul - disassembly sequence	8
Engine overhaul - reassembly sequence	11

	Section
General information - engine overhaul	1
Initial start-up and break-in after overhaul	15
Oil pressure check	2
Oil pump, timing chain and tensioner (five-cylinder engine) - removal and installation	14
Pistons and connecting rods - removal and installation	9
Upper oil pan (five-cylinder engine) - removal and installation	13
Vacuum gauge diagnostic checks	4

Specifications

General

Engine designations
  Four-cylinder timing belt engines
    1.4L .... CZTA, DGXA
    2.0L .... CBPA
  Four-cylinder timing chain engines
    1.8L .... CPKA, CPRA, CXBA, CXBB, CNSA, CNSB
    2.0L .... CBFA, CCTA, CPLA, CPPA, CNTA, CXCA, CXCB, CYFB, DJJA
  Five-cylinder engines .... CBTA, CBUA

Displacement
  Four-cylinder engines
    1.4L .... 85.43 cubic inches
    1.8L .... 109.8 cubic inches
    2.0L .... 122.0 cubic inches
  Five-cylinder engines (2.5L) .... 151 cubic inches

Bore and stroke
  Four-cylinder engines
    1.4L .... 2.93 x 3.14 inches (74.5 x 80.0)
    1.8L .... 3.25 x 3.31 inches (82.5 x 84.1)
    2.0L .... 3.25 x 3.65 inches (82.5 x 92.8)
  Five-cylinder engines (2.5L) .... 3.25 x 3.65 inches (82.5 x 92.8)

Compression ratio
  Four-cylinder timing belt engines
    1.4L 10.0: 1
    2.0L 10.3: 1
  Four-cylinder timing chain engines
    1.8L 9.6: 1
    2.0L 9.6: 1
  Five-cylinder engines 9.5: 1

General (continued)

Cylinder compression
  Jetta
    1.4L
      Standard .... 145 to 218 psi (1000 to 1500 kPa)
      Maximum variation between cylinders .... 44 psi (300 kPa)
    1.8L
      Standard .... 160 to 203 psi (1100 to 1400 kPa)
      Maximum variation between cylinders .... 44 psi (300 kPa)
    2.0L, 2014 and earlier models
      Standard .... 145 to 188 psi (1000 to 1300 kPa)
      Maximum variation between cylinders .... 44 psi (300 kPa)
    2.0L, 2015 and later models
      Standard .... 160 to 188 psi (1100 to 1300 kPa)
      Maximum variation between cylinders .... 44 psi (300 kPa)
    2.5L
      Standard .... 130 to 188 psi (896 to 1300 kPa)
      Maximum variation between cylinders .... 44 psi (300 kPa)
  GLI
    Standard .... 160 to 203 psi (1100 to 1400 kPa)
    Maximum variation between cylinders .... 44 psi (300 kPa)
  Golf, Golf R, Golf Sportwagen, Golf Alltrack, GTI
    1.4L
      Standard .... 145 to 218 psi (1000 to 1500 kPa)
      Maximum variation between cylinders .... 44 psi (300 kPa)
    1.8L
      Standard .... 160 to 203 psi (1100 to 1400 kPa)
      Maximum variation between cylinders .... 44 psi (300 kPa)
Oil pressure
  1.4L (CZTA, DGXA)
    At idle @ 176-degrees F (80-degrees C) minimum .... 8.7 psi (60 kPa)
    2000 rpm @ 176-degrees F (80-degrees C) minimum .... 21.75 psi (150 kPa)
    Oil pressure regulator valve disconnected
      3800 rpm @ 176-degrees F (80-degrees C) minimum .... 40.6 psi (2.8 bar)
  2.0L (CBPA)
    At idle @ 176-degrees F (80-degrees C) minimum .... 17.4 to 23.2 psi (120 to 161 kPa)
    2000 rpm @ 176-degrees F (80-degrees C) minimum .... 39.1 to 65 psi (269 to 448 kPa)
  2.0L (CBFA, CCTA)
    At idle @ 176-degrees F (80-degrees C) minimum .... 12.3 to 30.5 psi (85 to 210 kPa)
    2000 rpm @ 176-degrees F (80-degrees C) minimum .... 23.2 to 30.45 psi (160 to 210 kPa)
    3500 rpm @ 176-degrees F (80-degrees C) minimum .... 43.51 to 58.01 psi (300 to 400 kPa)
  1.8L (CXBA, CXBB, CNSA) and 2.0L (CXCA, CNTA, CXCB)
    At idle @ 176-degrees F (80-degrees C) minimum .... 12.3 to 23.2 psi (85 to 161 kPa)
    2000 rpm @ 176-degrees F (80-degrees C) minimum .... 17.4 to 23.2 psi (120 to 161 kPa)
    3700 rpm @ 176-degrees F (80-degrees C) minimum .... 17.4 to 23.2 psi (120 to 161 kPa)
    Oil pressure regulator valve disconnected
      At idle @ 176-degrees F (80-degrees C) minimum .... 12.3 to 43 psi (85 to 296 kPa)
      2000 rpm @ 176-degrees F (80-degrees C) minimum .... 29 to 58 psi (200 to 400 kPa)
      3700 rpm @ 176-degrees F (80-degrees C) minimum .... 43 to 58 psi (300 to 400 kPa)
  1.8L (CPRA, CPKA) and 2.0L (CPLA, CPPA)
    At idle @ 176-degrees F (80-degrees C) minimum .... 13.05 to 34.80 psi (93 to 240 kPa)
    2000 rpm @ 176-degrees F (80-degrees C) minimum .... 29 to 34.8 psi (200 to 239 kPa)
    3700 rpm @ 176-degrees F (80-degrees C) minimum .... 43.5 to 62.3 psi (300 to 430 kPa)
    Oil pressure regulator valve disconnected
      At idle @ 176-degrees F (80-degrees C) minimum .... 13.05 to 62.3 psi (93 to 430 kPa)
      2000 rpm @ 176-degrees F (80-degrees C) minimum .... 43.05 to 62.3 psi (300 to 430 kPa)
      3700 rpm @ 176-degrees F (80-degrees C) minimum .... 43.05 to 62.3 psi (300 to 430 kPa)
  2.0L (CYFB, DJJA) Golf R 4-Motion models
    At idle @ 176-degrees F (80-degrees C) minimum .... 12.3 to 23.2 psi (85 to 161 kPa)
    2000 rpm @ 176-degrees F (80-degrees C) minimum .... 17.4 to 23.2 psi (120 to 161 kPa)
    3700 rpm @ 176-degrees F (80-degrees C) minimum .... 17.4 to 23.2 psi (120 to 161 kPa)
    Oil pressure regulator valve disconnected
      At idle @ 176-degrees F (80-degrees C) minimum .... 12.3 to 43 psi (85 to 296 kPa)
      2000 rpm @ 176-degrees F (80-degrees C) minimum .... 29 to 43 psi (200 to 300 kPa)
      3700 rpm @ 176-degrees F (80-degrees C) minimum .... 43.5 to 58 psi (300 to 400 kPa)
  2.5L (CBTA, CBUA)
    2000 rpm @ 176-degrees F (80-degrees C) .... 39 to 65 psi (269 to 448 kPa)
    Maximum pressure .... 102 psi (703 kPa)

Chapter 2 Part C General engine overhaul procedures

Torque specifications — Ft-lbs (unless otherwise indicated)   Nm

Note: *One foot-pound (ft-lb) of torque is equivalent to 12 inch-pounds (in-lbs) of torque. Torque values below approximately 15 ft-lbs are expressed in inch-pounds, because most foot-pound torque wrenches are not accurate at these smaller values.*

Connecting rod bolts (new)
    Four-cylinder timing belt engines
        1.4L (CZTA, DGXA) models
            Step 1 .... 22 — 60
            Step 2 .... Tighten an additional 90-degrees
        2.0L (CBPA) models
            Step 1 .... 22 — 30
            Step 2 .... Tighten an additional 90 degrees
    Four-cylinder timing chain engines
        1.8L (CPKA, CPRA, CXBA, CXBB, CNSA, CNSB) models
            Step 1 .... 33 — 45
            Step 2 .... Tighten an additional 90-degrees
        2.0L (CBFA, CCTA, CPLA, CPPA, CXCA, CNTA, CXCB, CYFB, DJJA)
            Step 1 .... 33 — 45
            Step 2 .... Tighten an additional 90-degrees
    Five-cylinder engines
        2.5L five-cylinder (CBTA, CBUA) models
            Step 1 .... 22 — 30
            Step 2 .... Tighten an additional 90-degrees
Control housing cover fasteners (CBTA, CBUA) models
    Step 1 .... 89 in-lbs — 10
    Step 2 .... 18.5 — 25
Main bearing bolts (new) (tighten in sequence, **see illustrations 10.19a, 10.19b, 10.19c or 10.19d**)
    Four-cylinder timing belt engines
        2.0L (CBPA) models
            Step 1 .... 48 — 65
            Step 2 .... Tighten an additional 90-degrees
    Four-cylinder timing chain engines
        1.8L and 2.0L models
            Step 1 .... Tighten all bolts by hand
            Step 2, cap bolts (1 through 10) .... 48 — 65
            Step 3, cap bolts (1 through 10) .... Tighten an additional 90-degrees
            Step 4, jack bolts (side "A" bolts) .... 15 — 20
            Step 5, jack bolts (side "A" bolts) .... Tighten an additional 90-degrees
    Five-cylinder (CBTA, CBUA) engines
        Step 1 .... 30 — 40
        Step 2 .... Tighten an additional 90-degrees
Oil pump fasteners
    2.5L five-cylinder (CBTA, CBUA) .... 15 — 25
Oil pump pick-up tube fasteners
    2.5L five-cylinder (CBTA, CBUA) .... 89 in-lbs — 10
Oil pump sprocket retaining bolt (new)
    2.5L five-cylinder (CBTA, CBUA)
        Step 1 .... 15 — 20
        Step 2 .... Tighten an additional 90-degrees
Oil pump tensioner fastener .... 89 in-lbs — 10
Piston oil spray nozzle relief valve fastener .... 20 — 27
Upper oil pan mounting fasteners
    2.5L (CBTA, CBUA) .... 18.5 — 25

Chapter 2 Part C General engine overhaul procedures

1.10a An engine block being bored. An engine rebuilder will use special machinery to recondition the cylinder bores

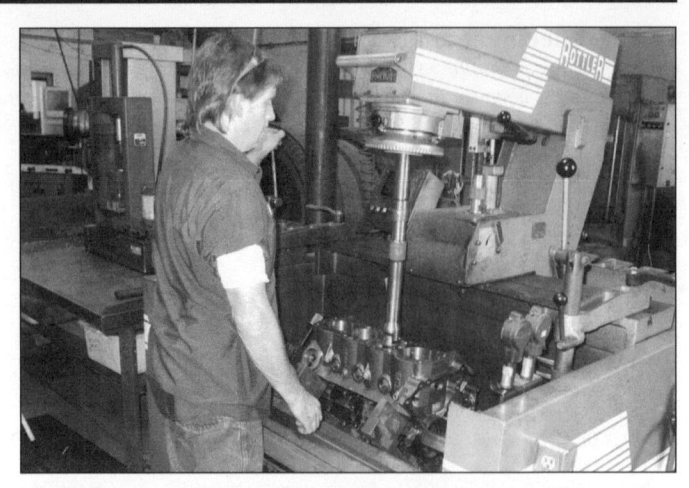

1.10b If the cylinders are bored, the machine shop will normally hone the engine on a machine like this

1 General information - engine overhaul

1   Included in this Chapter are general information and diagnostic testing procedures for determining the overall mechanical condition of your engine.

2   The information ranges from advice concerning preparation for an overhaul and the purchase of replacement parts and/or components to detailed, step-by-step procedures covering removal and installation.

3   The following Sections have been written to help you determine whether your engine needs to be overhauled and how to remove and install it once you've determined it needs to be rebuilt. For information concerning in-vehicle engine repair, see Chapter 2A or 2B.

4   The Specifications included in this Part are general in nature and include only those necessary for testing the oil pressure and checking the engine compression. Refer to Chapter 2A or 2B for additional engine Specifications.

5   It's not always easy to determine when, or if, an engine should be completely overhauled, because a number of factors must be considered.

6   High mileage is not necessarily an indication that an overhaul is needed, while low mileage doesn't preclude the need for an overhaul. Frequency of servicing is probably the most important consideration. An engine that's had regular and frequent oil and filter changes, as well as other required maintenance, will most likely give many thousands of miles of reliable service. Conversely, a neglected engine may require an overhaul very early in its service life.

7   Excessive oil consumption is an indication that piston rings, valve seals and/or valve guides are in need of attention. Make sure that oil leaks aren't responsible before deciding that the rings and/or guides are bad. Perform a cylinder compression check to determine the extent of the work required (see Section 4). Also check the vacuum readings under various conditions (see Section 3).

8   Check the oil pressure with a gauge installed in place of the oil pressure sending unit and compare it to this Chapter's Specifications (see Section 2). If it's extremely low, the bearings and/or oil pump are probably worn out.

Note: *On 1.4L, 1.8L and 2.0L engines, the oil pump is regulated and has two different pressure stages which must be checked in sequence, one after the other. The test must start at the reduced oil pressure switch then at the oil pressure regulation valve, to have an accurate reading of the oil pressure system. On 1.8L and 2.0L engines, the oil pressure regulator valve must be removed and a special fitting installed to perform the test. On 1.4L engines, the electrical connector to the regulator must be disconnected to perform the test.*

9   Loss of power, rough running, knocking or metallic engine noises, excessive valve train noise and high fuel consumption rates may also point to the need for an overhaul, especially if they're all present at the same time. If a complete tune-up doesn't remedy the situation, major mechanical work is the only solution.

10   An engine overhaul involves restoring the internal parts to the specifications of a new engine. During an overhaul, the piston rings are replaced and the cylinder walls are reconditioned (rebored and/or honed) **(see illustrations 1.10a and 1.10b)**. If a rebore is done by an automotive machine shop, new oversize pistons will also be installed. The main bearings, connecting rod bearings and camshaft bearings are generally replaced with new ones and, if necessary, the crankshaft may be reground to restore the journals **(see illustration 1.10c)**. Generally, the valves are serviced as well, since they're usually in less-than-perfect condition at this point. While the engine is being overhauled, other components, such as the starter and alternator, can be rebuilt or replaced as well. The end result should be similar to a new engine that will give many trouble free miles.

Note: *Critical cooling system components such as the hoses, drivebelts, thermostat and water pump should be replaced with new parts when an engine is overhauled. The radiator should be checked carefully to ensure that it isn't clogged or leaking (see Chapter 3). If you purchase a rebuilt engine or short block, some rebuilders will not warranty their engines unless the radiator has been professionally flushed. Also, we don't recommend overhauling the oil pump - always install a new one when an engine is rebuilt.*

11   Overhauling the internal components on today's engines is a difficult and time-consuming task which requires a significant amount of specialty tools and is best left to a professional engine rebuilder **(see illustrations 1.11a, 1.11b and 1.11c)**. A competent

1.10c A crankshaft having a main bearing journal ground

Chapter 2 Part C General engine overhaul procedures

1.11a A machinist checks for a bent connecting rod, using specialized equipment

1.11b A bore gauge being used to check a cylinder bore

1.11c Uneven piston wear like this indicates a bent connecting rod

engine rebuilder will handle the inspection of your old parts and offer advice concerning the reconditioning or replacement of the original engine. Never purchase parts or have machine work done on other components until the block has been thoroughly inspected by a professional machine shop. As a general rule, time is the primary cost of an overhaul, especially since the vehicle may be tied up for a minimum of two weeks or more. Be aware that some engine builders only have the capability to rebuild the engine you bring them while other rebuilders have a large inventory of rebuilt exchange engines in stock. Also be aware that many machine shops could take as much as two weeks time to completely rebuild your engine depending on shop workload. Sometimes it makes more sense to simply exchange your engine for another engine that's already rebuilt to save time.

2 Oil pressure check

1   Low engine oil pressure can be a sign of an engine in need of rebuilding. A low oil pressure indicator (often called an "idiot light") is not a test of the oiling system. Such indicators only come on when the oil pressure is dangerously low. Even a factory oil pressure gauge in the instrument panel is only a relative indication, although much better for driver information than a warning light. A better test is with a mechanical (not electrical) oil pressure gauge.

2   Locate the oil pressure sending unit:

   a) *On 1.4L (CZTA, DGXA) engines, there are two oil pressure sending units; an oil pressure switch located on the exhaust side of the engine block behind the driveaxle boot heat shield, and a reduced oil pressure switch located on the front side of the engine just below the intake manifold* **(see illustration)**. *When checking the oil pressure on a 1.4L engine, remove the reduced oil pressure sending unit and install the gauge adapter in its place.*

   b) *On 2.0L (CBPA) engines, the oil pressure switch is located on the rear of the oil filter housing.*

   c) *On 1.8L (CPKA, CPRA, CXBA, CXBB, CNSA and CNSB) and 2.0L (CBFA, CCTA, CPLA, CPPA, CNTA, CXCA and CXCB) engines, there are two oil pressure sending units, both located on the oil filter housing at top of the engine; the reduced oil pressure switch is the first (or top) switch and the oil pressure switch is the second switch on the housing just below the reduced pressure switch. When checking the oil pressure on one of these engines, remove the reduced oil pressure sending unit and install the gauge adapter in its place.*

   d) *On 2.5L (CBPA) engines, the oil pressure sending unit is located on the oil filter housing* **(see illustration)**, *threaded into the oil filter adapter.*

Note: *On 1.8L (CPKA, CPRA, CXBA, CXBB, CNSA and CNSB) and 2.0L (CBFA, CCTA, CPLA, CPPA, CNTA, CXCA and CXCB) engines, the oil pressure is electronically regulated and has two different pressure stages that must be checked one after the other to get an accurate oil pressure reading.*

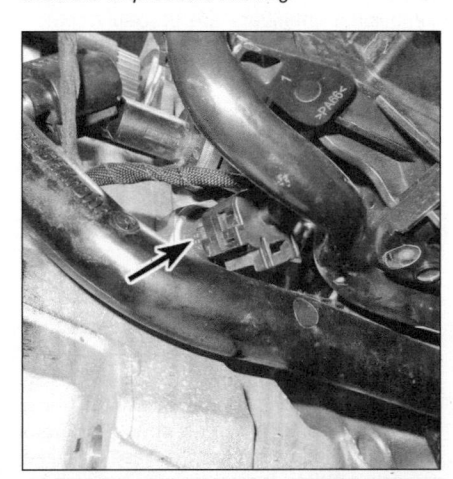

3   Unscrew the oil pressure sending unit or reduced oil pressure sending unit and screw in the hose for your oil pressure gauge. If necessary, install an adapter fitting. Use Teflon tape or thread sealant on the threads of the adapter and/or the fitting on the end of your gauge's hose.

Note: *Once the sending unit or pressure switch is removed, reconnect the switch or sensor and ground the outside of switch or sensor to the engine.*

4   Connect an accurate tachometer to the engine, according to the tachometer manufacturer's instructions.

5   Check the oil pressure with the engine running (normal operating temperature) at the specified engine speed, and compare it to this Chapter's Specifications. If it's extremely low, the bearings and/or oil pump are probably worn out.

6   On all models except 2.0L (CBPA, CBFA, CCTA) and 2.5L (CBTA, CBUA) engines, after the oil pressure is checked, disconnect the electrical connector from the oil pressure regulator valve, located at the lower front corner of the timing chain cover and test the oil pressure (see Steps 3 through 5), compare it to this Chapter's Specifications.

2.2a Reduced oil pressure sending unit location - 1.4L (CZTA and DGXA) engines

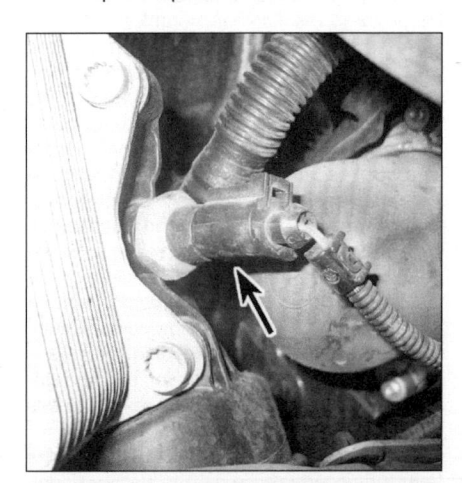

2.2b The oil pressure sending unit is located on the side of the oil filter housing - five-cylinder engine shown

3.5 Use a compression gauge with a threaded fitting for the spark plug hole, not the type that requires hand pressure to maintain the seal.

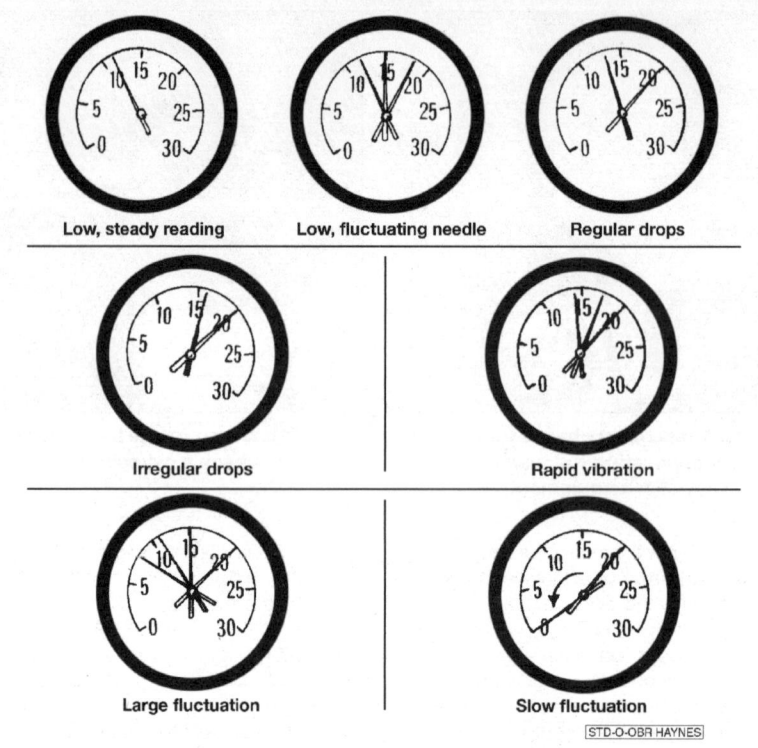

4.6 Typical vacuum gauge readings

3 Cylinder compression check

1 A compression check will tell you what mechanical condition the upper end of your engine (pistons, rings, valves, head gaskets) is in. Specifically, it can tell you if the compression is down due to leakage caused by worn piston rings, defective valves and seats or a blown head gasket.

Note: *The engine must be at normal operating temperature and the battery must be fully charged for this check.*

2 Begin by cleaning the area around the ignition coils before you remove them (compressed air should be used, if available).

3 Remove all of the spark plugs from the engine (see Chapter 1).

4 Remove the fuel pump fuse (see Chapter 4, Section 2) or disconnect the electrical connector from the fuel pump (see Chapter 4, Section 7).

5 Install a compression gauge in the spark plug hole **(see illustration)**.

6 Have an assistant depress the accelerator pedal to the floor and crank the engine over at least seven compression strokes while you watch the gauge. The compression should build up quickly in a healthy engine. Low compression on the first stroke, followed by gradually increasing pressure on successive strokes, indicates worn piston rings. A low compression reading on the first stroke, which doesn't build up during successive strokes, indicates leaking valves or a blown head gasket (a cracked head could also be the cause). Deposits on the undersides of the valve heads can also cause low compression. Record the highest gauge reading obtained.

7 Repeat the procedure for the remaining cylinders and compare the results to this Chapter's Specifications.

8 Add some engine oil (about three squirts from a plunger-type oil can) to each cylinder, through the spark plug hole, and repeat the test.

9 If the compression increases after the oil is added, the piston rings are definitely worn. If the compression doesn't increase significantly, the leakage is occurring at the valves or head gasket. Leakage past the valves may be caused by burned valve seats and/or faces or warped, cracked or bent valves.

10 If two adjacent cylinders have equally low compression, there's a strong possibility that the head gasket between them is blown. The appearance of coolant in the combustion chambers or the crankcase would verify this condition.

11 If one cylinder is slightly lower than the others, and the engine has a slightly rough idle, a worn lobe on the camshaft could be the cause.

12 If the compression is unusually high, the combustion chambers are probably coated with carbon deposits. If that's the case, the cylinder head(s) should be removed and decarbonized.

13 If compression is way down or varies greatly between cylinders, it would be a good idea to have a leak-down test performed by an automotive repair shop. This test will pinpoint exactly where the leakage is occurring and how severe it is.

4 Vacuum gauge diagnostic checks

1 A vacuum gauge provides inexpensive but valuable information about what is going on in the engine. You can check for worn rings or cylinder walls, leaking head or intake manifold gaskets, incorrect carburetor adjustments, restricted exhaust, stuck or burned valves, weak valve springs, improper ignition or valve timing and ignition problems.

2 Unfortunately, vacuum gauge readings are easy to misinterpret, so they should be used in conjunction with other tests to confirm the diagnosis.

3 Both the absolute readings and the rate of needle movement are important for accurate interpretation. Most gauges measure vacuum in inches of mercury (in-Hg). The following references to vacuum assume the diagnosis is being performed at sea level. As elevation increases (or atmospheric pressure decreases), the reading will decrease. For every 1,000 foot increase in elevation above approximately 2,000 feet, the gauge readings will decrease about one inch of mercury.

4 Connect the vacuum gauge directly to the intake manifold vacuum, not to ported (throttle body) vacuum. Be sure no hoses are left disconnected during the test or false readings will result.

5 Before you begin the test, allow the engine to warm up completely. Block the wheels and set the parking brake. With the transaxle in Park, start the engine and allow it to run at normal idle speed.

Warning: *Keep your hands and the vacuum gauge clear of the fans.*

6 Read the vacuum gauge; an average, healthy engine should normally produce about 17 to 22 in-Hg with a fairly steady needle **(see illustration)**. Refer to the following vacuum gauge readings and what they indicate about the engine's condition:

7 A low steady reading usually indicates a leaking gasket between the intake manifold and cylinder head(s) or throttle body, a leaky vacuum hose, late ignition timing or incorrect camshaft timing. Check ignition timing with a

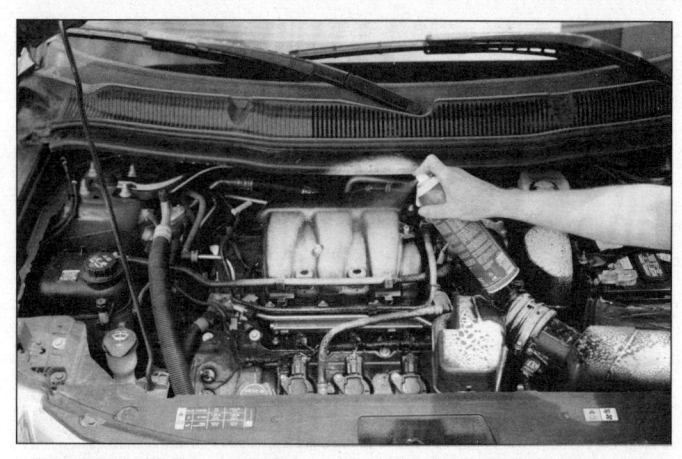

Chapter 2 Part C General engine overhaul procedures 2C-7

6.3a After tightly wrapping water-vulnerable components, use a spray cleaner on everything, with particular concentration on the greasiest areas, usually around the valve cover and lower edges of the block. If one section dries out, apply more cleaner

6.3b Depending on how dirty the engine is, let the cleaner soak in according to the directions and then hose off the grime and cleaner. Get the rinse water down into every area you can get at; then dry important components with a hair dryer or paper towels

timing light and eliminate all other possible causes, utilizing the tests provided in this Chapter before you remove the timing chain cover to check the timing marks.

8 If the reading is three to eight inches below normal and it fluctuates at that low reading, suspect an intake manifold gasket leak at an intake port or a faulty fuel injector.

9 If the needle has regular drops of about two-to-four inches at a steady rate, the valves are probably leaking. Perform a compression check or leak-down test to confirm this.

10 An irregular drop or down-flick of the needle can be caused by a sticking valve or an ignition misfire. Perform a compression check or leak-down test and read the spark plugs.

11 A rapid vibration of about four in-Hg vibration at idle combined with exhaust smoke indicates worn valve guides. Perform a leakdown test to confirm this. If the rapid vibration occurs with an increase in engine speed, check for a leaking intake manifold gasket or head gasket, weak valve springs, burned valves or ignition misfire.

12 A slight fluctuation, say one inch up and down, may mean ignition problems. Check all the usual tune-up items and, if necessary, run the engine on an ignition analyzer.

13 If there is a large fluctuation, perform a compression or leak-down test to look for a weak or dead cylinder or a blown head gasket.

14 If the needle moves slowly through a wide range, check for a clogged PCV system, incorrect idle fuel mixture, throttle body or intake manifold gasket leaks.

15 Check for a slow return after revving the engine by quickly snapping the throttle open until the engine reaches about 2,500 rpm and let it shut. Normally the reading should drop to near zero, rise above normal idle reading (about 5 in-Hg over) and then return to the previous idle reading. If the vacuum returns slowly and doesn't peak when the throttle is snapped shut, the rings may be worn. If there is a long delay, look for a restricted exhaust system (often the muffler or catalytic converter). An easy way to check this is to temporarily disconnect the exhaust ahead of the suspected part and redo the test.

5 Engine rebuilding alternatives

1 The do-it-yourselfer is faced with a number of options when purchasing a rebuilt engine. The major considerations are cost, warranty, parts availability and the time required for the rebuilder to complete the project. The decision to replace the engine block, piston/connecting rod assemblies and crankshaft depends on the final inspection results of your engine. Only then can you make a cost effective decision whether to have your engine overhauled or simply purchase an exchange engine for your vehicle.

2 Some of the rebuilding alternatives include:

3 **Individual parts** - If the inspection procedures reveal that the engine block and most engine components are in reusable condition, purchasing individual parts and having a rebuilder rebuild your engine may be the most economical alternative. The block, crankshaft and piston/connecting rod assemblies should all be inspected carefully by a machine shop first.

4 **Short block** - A short block consists of an engine block with a crankshaft and piston/connecting rod assemblies already installed. All new bearings are incorporated and all clearances will be correct. The existing camshafts, valve train components, cylinder head and external parts can be bolted to the short block with little or no machine shop work necessary.

5 **Long block** - A long block consists of a short block plus an oil pump, oil pan, cylinder head, valve cover, camshaft and valve train components, timing sprockets and chain or gears and timing cover. All components are installed with new bearings, seals and gaskets incorporated throughout. The installation of manifolds and external parts is all that's necessary.

6 **Low mileage used engines** - Some companies now offer low mileage used engines which are a very cost effective way to get your vehicle up and running again. These engines often come from vehicles which have been totaled in accidents or come from other countries which have a higher vehicle turnover rate. A low mileage used engine also usually has a similar warranty like the newly remanufactured engines.

7 Give careful thought to which alternative is best for you and discuss the situation with local automotive machine shops, auto parts dealers and experienced rebuilders before ordering or purchasing replacement parts.

6 Engine removal - methods and precautions

1 If you've decided that an engine must be removed for overhaul or major repair work, several preliminary steps should be taken. Read all removal and installation procedures carefully prior to committing to this job. These engines are removed by lowering the engine to the floor, along with the transmission, and then raising the vehicle sufficiently to slide the assembly out; this will require a vehicle hoist as well as an engine hoist. Make sure the engine hoist is rated in excess of the combined weight of the engine and transmission. A transmission jack is also very helpful. Safety is of primary importance, considering the potential hazards involved in removing the engine from the vehicle.

2 Locating a suitable place to work is extremely important. Adequate work space, along with storage space for the vehicle, will be needed. If a shop or garage isn't available, at the very least a flat, level, clean work surface made of concrete or asphalt is required.

3 Cleaning the engine compartment and engine before beginning the removal procedure will help keep tools clean and organized **(see illustrations)**.

2C-8   Chapter 2 Part C General engine overhaul procedures

6.5a Get an engine stand sturdy enough to firmly support the engine while you're working on it. Stay away from three-wheeled models: they have a tendency to tip over more easily, so get a four-wheeled unit

6.5b A clutch alignment tool is necessary if you plan to install a rebuilt engine mated to a manual transaxle

4   If you're a novice at engine removal, get at least one helper. One person cannot easily do all the things you need to do to remove a big heavy engine and transmission assembly from the engine compartment. Also helpful is to seek advice and assistance from someone who's experienced in engine removal.

5   Plan the operation ahead of time. Arrange for or obtain all of the tools and equipment you'll need prior to beginning the job **(see illustrations)**. Some of the equipment necessary to perform engine removal and installation safely and with relative ease are (in addition to a vehicle hoist and an engine hoist) a heavy duty floor jack (preferably fitted with a transmission jack head adapter), complete sets of wrenches and sockets as described in the front of this manual, wooden blocks, plenty of rags and cleaning solvent for mopping up spilled oil, coolant and gasoline.

6   Plan for the vehicle to be out of use for quite a while. A machine shop can do the work that is beyond the scope of the home mechanic. Machine shops often have a busy schedule, so before removing the engine, consult the shop for an estimate of how long it will take to rebuild or repair the components that may need work.

7   Engine - removal and installation

Warning: *The models covered by this manual are equipped with a Supplemental Restraint System (SRS), more commonly known as airbags. Always disable the airbag system before working in the vicinity of airbag system components to avoid the possibility of accidental deployment of the airbag, which could cause personal injury (see Chapter 12).*
Warning: *Gasoline is extremely flammable, so take extra precautions when you work on any part of the fuel system. Don't smoke or allow open flames or bare light bulbs near the work area, and don't work in a garage where a gas-type appliance (such as a water heater or clothes dryer) is present. Since gasoline is carcinogenic, wear fuel-resistant gloves when there's a possibility of being exposed to fuel, and, if you spill any fuel on your skin, rinse it off immediately with soap and water. Mop up any spills immediately and do not store fuel-soaked rags where they could ignite. The fuel system is under constant pressure, so, if any fuel lines are to be disconnected, the fuel pressure in the system must be relieved first (see Chapter 4 for more information). When you perform any kind of work on the fuel system, wear safety glasses and have a Class B type fire extinguisher on hand.*
Warning: *The engine must be completely cool before beginning this procedure.*
Caution: *If the battery is disconnected, several systems must be re-learned before they will work properly (see Chapter 5, Section 3).*
Note: *Engine removal on these models is a difficult job, especially for the do-it-yourself mechanic working at home. Because of the vehicle's design, the manufacturer states that the engine and transaxle have to be removed as a unit from the bottom of the vehicle, not the top. With a floor jack and jackstands the vehicle can't be raised high enough and supported safely enough for the engine/transaxle assembly to slide out from underneath. The manufacturer recommends that removal of the engine/transaxle assembly only be performed on a vehicle hoist.*

Removal

1   Park the vehicle on a frame-contact type vehicle hoist. The pads of the hoist arms must contact the body welt along each side of the vehicle.
2   Remove the engine cover (see Chapter 1, Section 7).
3   Relieve the fuel system pressure (see Chapter 4). Remove the battery cover and disconnect the cable from the negative battery terminal (see Chapter 5).
4   Remove the air filter housing and inlet and outlet ducts (see Chapter 4).
5   Detach the wires from the engine compartment fuse and relay panel.
6   Place protective covers on the fenders and cowl and remove the hood (see Chapter 11).
7   Remove the battery and battery tray (see Chapter 5).
8   Loosen the front wheel bolts and the driveaxle/hub bolts, then raise the vehicle on the hoist. Remove the under-vehicle splash shield (see Chapter 1, Section 6).
9   Drain the cooling system, engine oil, power steering fluid and transmission fluid (see Chapter 1).
10   Remove the air conditioning compressor drivebelt (see Chapter 3), then, without disconnecting the refrigerant lines from the compressor, unbolt the compressor and secure it out of the way using rope or wire (see Chapter 1).
Note: *Do not disconnect or stretch the lines to the compressor when securing it out of the way.*
11   Remove the windshield wiper motor (see Chapter 12) and cowl assembly (see Chapter 11).
12   Clearly label, then disconnect all vacuum lines, coolant and emissions hoses, wiring harness connectors **(see illustration)**, ground straps and fuel lines. Masking tape and/or a touch up paint applicator work well for marking items. Take photos or sketch the locations of components and brackets. It's also a good idea when removing brackets or larger components to label and bag the bolts for the bracket or component.
13   On models equipped with a noise generator, open the locking ring on the charge air pipe, pull the fuel lines from the retaining clip, then remove the EVAP canister retaining bolts and canister (see Chapter 6).
14   On turbocharged models, remove the charge air cooler inlet, outlet pipes and cooler (see Chapter 4) then cover the openings to prevent debris from falling into the cooler.
15   Disconnect the quick-connect fitting from the vacuum pump and secure the lines out of the way.
16   Remove the cover and disconnect the electrical connectors from the Engine Control Module (ECM) (see Chapter 6).
17   Unclip the automatic transaxle control module from the bracket on the front side

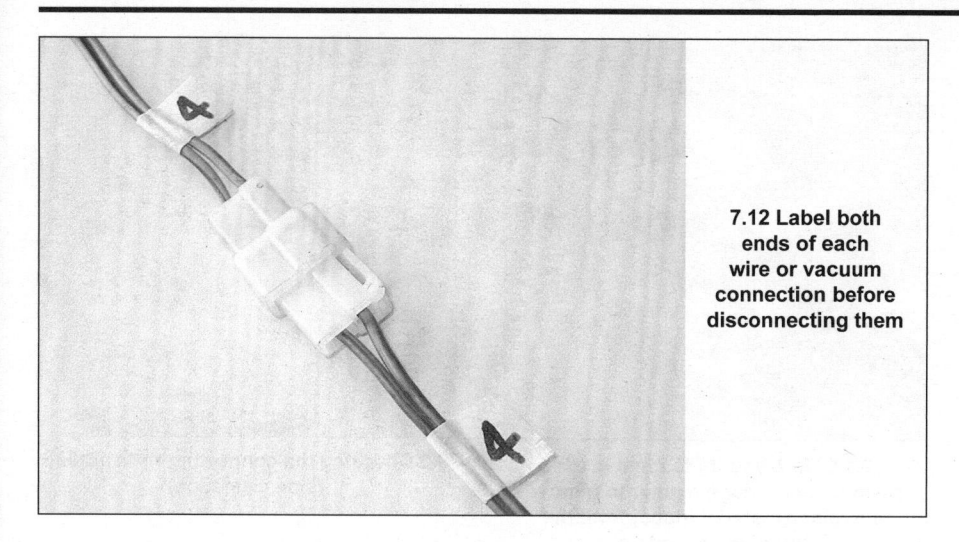

Chapter 2 Part C General engine overhaul procedures 2C-9

7.12 Label both ends of each wire or vacuum connection before disconnecting them

of the transaxle and remove the control module.
18 Remove the alternator and the starter (see Chapter 5).
19 Remove the power steering pump, if equipped (see Chapter 10).
Note: *On some models it will be necessary to remove the power steering pump pressure line retainers mounting the line around the transaxle.*
20 Remove the coolant reservoir tank and set it aside (see Chapter 3).
21 On models equipped with secondary air injection, disconnect the lines from the air injection pump and remove the lines (see Chapter 6).
22 Remove the engine cooling fan (see Chapter 3).
Note: *If the radiator is not taken out, it will be necessary to secure the radiator and condenser assembly to the front of the vehicle using special tool #VAS 531003 or a couple of moving straps. It will also be necessary to detach the transmission oil cooler lines from the bottom of the radiator on automatic transaxle equipped vehicles.*
23 Disconnect the shift cable(s) from the transaxle (see Chapter 7A or Chapter 7B). Also disconnect any wiring harness connectors from the transaxle.
24 Disconnect the radiator hoses, heater hoses and coolant recirculation pump, if equipped (see Chapter 3).
Note: *On models equipped with a coolant recirculation pump, it may be necessary to remove the pump fasteners and pump after the hoses have been disconnected.*
25 If equipped with a manual transaxle, disconnect the clutch release cylinder from the transaxle (see Chapter 8).
26 Detach the exhaust pipe(s) from the exhaust manifold(s) (see Chapter 4).
27 Detach all remaining wiring harnesses and hoses from between the engine/transaxle and the chassis. Be sure to mark all connectors to facilitate reassembly.
28 Remove the driveaxle heat shield and driveaxles (see Chapter 8).
29 Remove the pendulum mount (see Chapter 2A, Section 17 or Chapter 2B, Section 15) and, on models so equipped, disconnect the power steering lines from the mounting clips.
30 Attach a lifting sling or chain to the engine. Position an engine hoist and connect the sling to it. If no lifting hooks or brackets are present, you'll have to fasten the chains or slings to some substantial part of the engine - ones that are strong enough to take the weight, but in locations that will provide good balance. Take up the slack until there is slight tension on the sling or chain. Position the chain on the hoist so it balances the engine and the transaxle level with the vehicle.
31 Remove the front subframe (see Chapter 10).
32 Remove the transaxle mount and the engine mount.
33 Recheck to be sure nothing except the mounts are still connecting the engine to the vehicle or to the transaxle. Disconnect and label anything still remaining.
34 Slowly lower the engine/transaxle from the vehicle.
Note: *Placing a sheet of cardboard or paneling between the engine and the floor makes moving the powertrain easier.*
35 Once the powertrain is on the floor, disconnect the engine lifting hoist and raise the vehicle hoist until the vehicle clears the powertrain.
36 Reconnect the chain or sling and raise the engine/transaxle with the hoist. Support the transaxle with a jack (preferably one with a transmission jack head adapter). Separate the engine from the transaxle (see Chapter 7A).
37 Remove the flywheel/driveplate and mount the engine on a stand.

Installation
Note: *The manufacturer recommends replacing all subframe and suspension fasteners with new ones whenever they are loosened or removed.*

38 Installation is the reverse of removal, noting the following points:

a) Check the engine/transaxle mounts. If they're worn or damaged, replace them.
b) Attach the transaxle to the engine following the procedure described in Chapter 7A.
c) Add coolant, oil, power steering and transaxle fluids as needed (see Chapter 1).
d) Align the subframe reference marks before tightening the bolts.
e) Tighten the subframe, suspension and steering fasteners to the torque listed in the Specifications in Chapter 10.
f) Reconnect the negative battery cable (see Chapter 5).
g) Run the engine and check for proper operation and leaks. Shut off the engine and recheck fluid levels.

8 Engine overhaul - disassembly sequence

1 It's much easier to remove the external components if it's mounted on a portable engine stand. A stand can often be rented quite cheaply from an equipment rental yard. Before the engine is mounted on a stand, the flywheel/driveplate should be removed from the engine.
2 If a stand isn't available, it's possible to remove the external engine components with it blocked up on the floor. Be extra careful not to tip or drop the engine when working without a stand.
3 If you're going to obtain a rebuilt engine, all external components must come off first, to be transferred to the replacement engine. These components include:

Clutch and flywheel (models with manual transaxle)
Driveplate (models with automatic transaxle)
Control housing cover
Emissions-related components
Engine mounts and mount brackets
Fuel injection components
Intake/exhaust manifolds
Oil filter
Ignition coils and spark plugs
Thermostat and housing assembly
Water pump

Note: *When removing the external components from the engine, pay close attention to details that may be helpful or important during installation. Note the installed position of gaskets, seals, spacers, pins, brackets, washers, bolts and other small items.*

4 If you're going to obtain a short block (assembled engine block, crankshaft, pistons and connecting rods), remove the timing belt, cylinder head, oil pan, oil pump pick-up tube, oil pump and water pump from your engine so that you can turn in your old short block to the rebuilder as a core. See *Engine rebuilding alternatives* for additional information regarding the different possibilities to be considered.

9 Pistons and connecting rods - removal and installation

Removal

Note: *Prior to removing the piston/connecting rod assemblies, remove the cylinder head, oil pans, oil pump and windage tray if equipped.*

1 Use your fingernail to feel if a ridge has formed at the upper limit of ring travel (about 1/4-inch down from the top of each cylinder). If carbon deposits or cylinder wear have produced ridges, they must be completely removed with a special tool **(see illustration)**. Follow the manufacturer's instructions provided with the tool. Failure to remove the ridges before attempting to remove the piston/connecting rod assemblies may result in piston breakage.

2 After the cylinder ridges have been removed, turn the engine so the crankshaft is facing up.

3 Before the main bearing cap assembly and connecting rods are removed, check the connecting rod endplay with feeler gauges. Slide them between the first connecting rod and the crankshaft throw until the play is removed **(see illustration)**. Repeat this procedure for each connecting rod. The endplay is equal to the thickness of the feeler gauge(s). Check with an automotive machine shop for the endplay service limit (a typical end play limit should measure between 0.005 to 0.015 inch [0.127 to 0.381 mm]). If the play exceeds the service limit, new connecting rods will be required. If new rods (or a new crankshaft) are installed, the endplay may fall under the minimum allowable. If it does, the rods will have to be machined to restore it. If necessary, consult an automotive machine shop for advice.

4 Check the connecting rods and caps for identification marks. The marks on the face of the rods and caps should always point towards the drivebelt end of the engine and the marks on the ends of the rods and caps are for cylinder designation. If they aren't plainly marked, use paint or marker to clearly identify each rod and cap (1, 2, 3, etc., depending on the cylinder they're associated with) **(see illustration)**.

5 Loosen each of the connecting rod cap bolts 1/2-turn at a time until they can be removed by hand.

Note: *New connecting rod cap bolts must be used when reassembling the engine, but save the old bolts for use when checking the connecting rod bearing oil clearance.*

6 Remove the number one connecting rod cap and bearing insert. Don't drop the bearing insert out of the cap.

7 Remove the bearing insert and push the connecting rod/piston assembly out through the top of the engine. Use a wooden or plastic hammer handle to push on the upper bearing surface in the connecting rod. If resistance is felt, double-check to make sure that all of the ridge was removed from the cylinder.

8 Repeat the procedure for the remaining cylinders.

9 After removal, reassemble the connecting rod caps and bearing inserts in their respective connecting rods and install the cap bolts finger-tight. Leaving the old bearing inserts in place until reassembly will help prevent the connecting rod bearing surfaces from being accidentally nicked or gouged.

10 The pistons and connecting rods are now ready for inspection and overhaul at an automotive machine shop.

Piston ring installation

11 Before installing the new piston rings, the ring end gaps must be checked. It's assumed that the piston ring side clearance has been checked and verified correct.

12 Lay out the piston/connecting rod assemblies and the new ring sets so the ring sets will be matched with the same piston and cylinder during the end gap measurement and engine assembly.

13 Insert the top (number one) ring into the first cylinder and square it up with the cylinder walls by pushing it in with the top of the piston **(see illustration)**. The ring should be near the bottom of the cylinder, at the lower limit of ring travel.

9.1 Before you try to remove the pistons, use a ridge reamer to remove the raised material (ridge) from the top of the cylinders

9.3 Checking the connecting rod endplay (side clearance)

9.4 If the connecting rods and caps are not marked, use permanent ink or paint to mark the caps to the rods by cylinder number (for example, this would be the No. 4 connecting rod)

9.13 Install the piston ring into the cylinder then push it down into position using a piston so the ring will be square in the cylinder

Chapter 2 Part C General engine overhaul procedures

9.14 With the ring square in the cylinder, measure the ring end gap with a feeler gauge

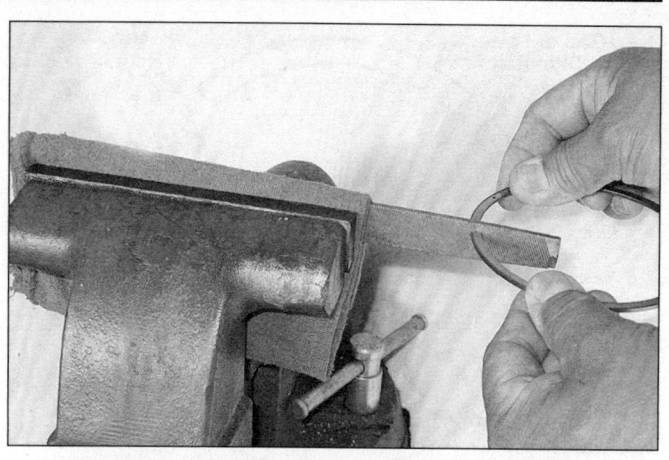

9.15 If the ring end gap is too small, clamp a file in a vise as shown and file the piston ring ends - be sure to remove all raised material

14 To measure the end gap, slip feeler gauges between the ends of the ring until a gauge equal to the gap width is found **(see illustration)**. The feeler gauge should slide between the ring ends with a slight amount of drag. A typical ring gap should fall between 0.010 and 0.020 inch (0.25 to 0.50 mm) for compression rings and up to 0.030 inch (0.76 mm) for the oil ring steel rails. If the gap is larger or smaller than specified, double-check to make sure you have the correct rings before proceeding.

15 If the gap is too small, it must be enlarged or the ring ends may come in contact with each other during engine operation, which can cause serious damage to the engine. If necessary, increase the end gaps by filing the ring ends very carefully with a fine file. Mount the file in a vise equipped with soft jaws, slip the ring over the file with the ends contacting the file face and slowly move the ring to remove material from the ends. When performing this operation, file only by pushing the ring from the outside end of the file towards the vise **(see illustration)**.

16 Excess end gap isn't critical unless it's greater than 0.0315 inch (0.8 mm). Again, double-check to make sure you have the correct ring type.

17 Repeat the procedure for each ring that will be installed in the first cylinder and for each ring in the remaining cylinders. Remember to keep rings, pistons and cylinders matched up.

18 Once the ring end gaps have been checked/corrected, the rings can be installed on the pistons.

19 The oil control ring (lowest one on the piston) is usually installed first. It's composed of three separate components. Slip the spacer/expander into the groove **(see illustration)**. If an anti-rotation tang is used, make sure it's inserted in the drilled hole in the ring groove. Next, install the upper side rail in the same manner **(see illustration)**. Don't use a piston ring installation tool on the oil ring side rails, as they may be damaged. Instead, place one end of the side rail into the groove between the spacer/expander and the ring land, hold it firmly in place and slide a finger around the piston while pushing the rail into the groove. Finally, install the lower side rail.

20 After the three oil ring components have been installed, check to make sure that both the upper and lower side rails can be rotated smoothly inside the ring grooves.

21 The number two (middle) ring is installed next. It's usually stamped with a mark which must face up, toward the top of the piston. Do not mix up the top and middle rings, as they have different cross-sections.

Note: *Always follow the instructions printed on the ring package or box - different manufacturers may require different approaches.*

22 Use a piston ring installation tool and make sure the identification mark is facing the top of the piston, then slip the ring into the middle groove on the piston **(see illustration)**. Don't expand the ring any more than necessary to slide it over the piston.

23 Install the number one (top) ring in the same manner. Make sure the mark is facing up. Be careful not to confuse the number one and number two rings.

24 Repeat the procedure for the remaining pistons and rings.

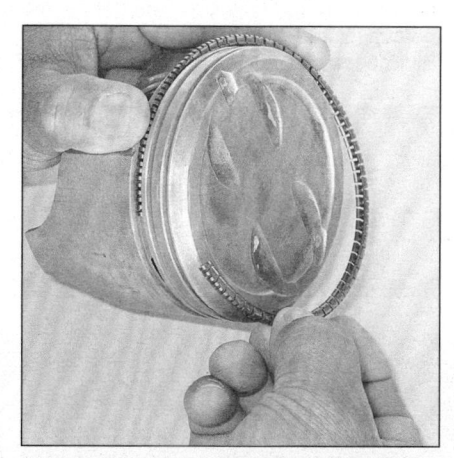

9.19a Installing the spacer/expander in the oil ring groove

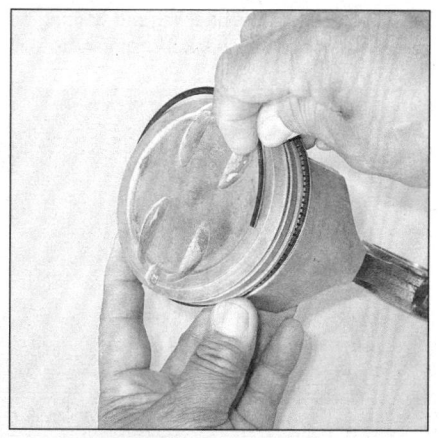

9.19b DO NOT use a piston ring installation tool when installing the oil control side rails

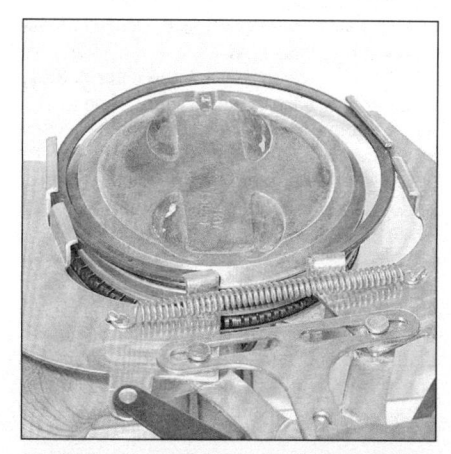

9.22 Use a piston ring installation tool to install the number 2 and the number 1 (top) rings - be sure the directional mark on the piston ring(s) is facing toward the top of the piston

2C-11

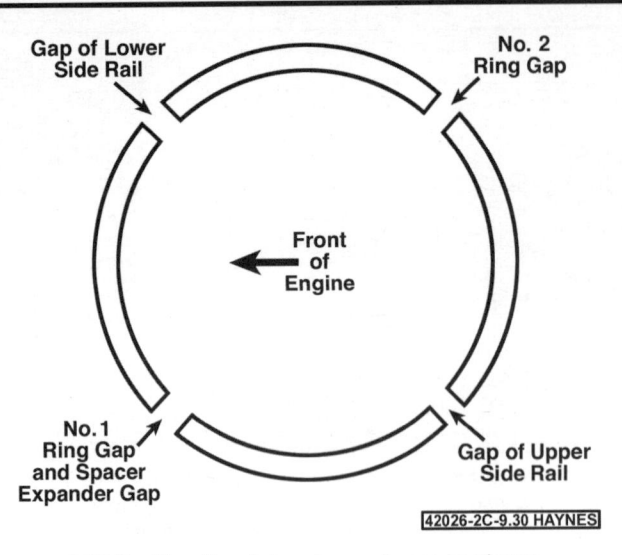

9.30 Position the piston ring end gaps as shown

9.35 Use a plastic or wooden hammer handle to push the piston into the cylinder

Installation

25 Before installing the piston/connecting rod assemblies, the cylinder walls must be perfectly clean, the top edge of each cylinder bore must be chamfered, and the crankshaft must be in place.

26 Remove the cap from the end of the number one connecting rod (refer to the marks made during removal). Remove the original bearing inserts and wipe the bearing surfaces of the connecting rod and cap with a clean, lint-free cloth. They must be kept spotlessly clean.

Connecting rod bearing oil clearance check

27 Clean the back side of the new upper bearing insert, then lay it in place in the connecting rod.

28 Make sure the tab on the bearing fits into the recess in the rod. Don't hammer the bearing insert into place and be very careful not to nick or gouge the bearing face. Don't lubricate the bearing at this time.

29 Clean the back side of the other bearing insert and install it in the rod cap. Again, make sure the tab on the bearing fits into the recess in the cap, and don't apply any lubricant. It's critically important that the mating surfaces of the bearing and connecting rod are perfectly clean and oil free when they're assembled.

30 Position the piston ring gaps at 90-degree intervals around the piston as shown **(see illustration)**.

31 Lubricate the piston and rings with clean engine oil and attach a piston ring compressor to the piston. Leave the skirt protruding about 1/4-inch to guide the piston into the cylinder. The rings must be compressed until they're flush with the piston.

32 Rotate the crankshaft until the number one connecting rod journal is at BDC (bottom dead center) and apply a liberal coat of engine oil to the cylinder walls.

33 With the mark on top of the piston facing the front (timing belt or chain end) of the engine, gently insert the piston/connecting rod assembly into the number one cylinder bore and rest the bottom edge of the ring compressor on the engine block.

34 Tap the top edge of the ring compressor to make sure it's contacting the block around its entire circumference.

35 Gently tap on the top of the piston with the end of a wooden or plastic hammer handle **(see illustration)** while guiding the end of the connecting rod into place on the crankshaft journal. The piston rings may try to pop out of the ring compressor just before entering the cylinder bore, so keep some downward pressure on the ring compressor. Work slowly, and if any resistance is felt as the piston enters the cylinder, stop immediately. Find out what's hanging up and fix it before proceeding. Do not, for any reason, force the piston into the cylinder - you might break a ring and/or the piston.

36 Once the piston/connecting rod assembly is installed, the connecting rod bearing oil clearance must be checked before the rod cap is permanently installed.

37 Cut a piece of the appropriate size Plastigage slightly shorter than the width of the connecting rod bearing and lay it in place on the number one connecting rod journal, parallel with the journal axis **(see illustration)**.

38 Clean the connecting rod cap bearing face and install the rod cap. Make sure the mating mark on the cap is on the same side as the mark on the connecting rod **(see illustration 9.4)**.

39 Install the old rod bolts, at this time, and tighten them to the torque listed in this Chapter's Specifications.

Note: *Use a thin-wall socket to avoid erroneous torque readings that can result if the socket is wedged between the rod cap and the bolt head. If the socket tends to wedge itself between the fastener and the cap, lift up on it slightly until it no longer contacts the cap. DO NOT rotate the crankshaft at any time during this operation.*

40 Remove the fasteners and detach the rod cap, being very careful not to disturb the Plastigage.

9.37 Place Plastigage on each connecting rod bearing journal parallel to the crankshaft centerline

Chapter 2 Part C General engine overhaul procedures 2C-13

9.41 Use the scale on the Plastigage package to determine the bearing oil clearance - be sure to measure the widest part of the Plastigage and use the correct scale; it comes with both standard and metric scales

10.1 Checking crankshaft endplay with a dial indicator

10.3 Checking the crankshaft endplay with feeler gauges at the thrust bearing journal

41 Compare the width of the crushed Plastigage to the scale printed on the Plastigage envelope to obtain the oil clearance **(see illustration)**. The connecting rod oil clearance is usually about 0.001 to 0.002 inch (0.025 to 0.05 mm). Consult an automotive machine shop for the clearance specified for the rod bearings on your engine.

42 If the clearance is not as specified, the bearing inserts may be the wrong size (which means different ones will be required). Before deciding that different inserts are needed, make sure that no dirt or oil was between the bearing inserts and the connecting rod or cap when the clearance was measured. Also, recheck the journal diameter. If the Plastigage was wider at one end than the other, the journal may be tapered. If the clearance still exceeds the limit specified, the bearing will have to be replaced with an undersize bearing.

Caution: *When installing a new crankshaft always use a standard size bearing.*

Final installation

43 Carefully scrape all traces of the Plastigage material off the rod journal and/or bearing face. Be very careful not to scratch the bearing - use your fingernail or the edge of a plastic card.

44 Make sure the bearing faces are perfectly clean, then apply a uniform layer of clean moly-base grease or engine assembly lube to both of them. You'll have to push the piston into the cylinder to expose the face of the bearing insert in the connecting rod.

Note: *If there was a large amount metal shavings or debris, replace the oil piston nozzle fastener and nozzle.*

45 Slide the connecting rod back into place on the journal, install the rod cap, install the *new* bolts and tighten them to the torque listed in this Chapter's Specifications. Again, work up to the torque in three steps.

Caution: *Install new connecting rod cap bolts.*

Do NOT reuse old bolts - they have stretched and cannot be reused.

46 Repeat the entire procedure for the remaining pistons/connecting rods.

47 The important points to remember are:
 a) *Keep the back sides of the bearing inserts and the insides of the connecting rods and caps perfectly clean when assembling them.*
 b) *Make sure you have the correct piston/rod assembly for each cylinder.*
 c) *The mark on the piston must face the front (timing chain end) of the engine.*
 d) *Lubricate the cylinder walls liberally with clean oil.*
 e) *Lubricate the bearing faces when installing the rod caps after the oil clearance has been checked.*
 f) *Make sure the piston oil spray nozzles were not bent or damaged.*

48 After all the piston/connecting rod assemblies have been correctly installed, rotate the crankshaft a number of times by hand to check for any obvious binding.

49 As a final step, check the connecting rod endplay, as described in Step 3. If it was correct before disassembly and the original crankshaft and rods were reinstalled, it should still be correct. If new rods or a new crankshaft were installed, the endplay may be inadequate. If so, the rods will have to be removed and taken to an automotive machine shop for resizing.

10 Crankshaft - removal and installation

Removal

Caution: *On 1.4L engines, there is a risk of permanently damaging the bearing block when removing the crankshaft. The loosening of the crankshaft bearing cap bolts deforms the cylinder bearing block and causes bearing damage. The manufacturer recommends not removing the crankshaft.*

Note: *The crankshaft can be removed only after the engine has been removed from the vehicle. It's assumed that the flywheel or driveplate, crankshaft pulley, crankshaft front oil seal flange, timing belt or timing chains, oil pans, oil pump body, control housing cover (five-cylinder engine), oil filter and piston/connecting rod assemblies have already been removed. On four-cylinder engines, the rear main oil seal retainer must be unbolted and separated from the block before proceeding with crankshaft removal.*

1 Before the crankshaft is removed, measure the endplay. Mount a dial indicator with the indicator in line with the crankshaft and just touching the end of the crankshaft as shown **(see illustration)**.

2 Pry the crankshaft all the way to the rear and zero the dial indicator. Next, pry the crankshaft to the front as far as possible and check the reading on the dial indicator. The distance traveled is the endplay. A crankshaft endplay will fall between 0.003 to 0.009 inch (0.07 to 0.23 mm) for four-cylinder engines and 0.003 to 0.008 inch (0.07 to 0.20 mm) on five-cylinder engines. If it is greater than that, check the crankshaft thrust surfaces for wear after it's removed. If no wear is evident, new main bearings should correct the endplay.

3 If a dial indicator isn't available, feeler gauges can be used. Gently pry the crankshaft all the way to the front of the engine. Slip feeler gauges between the crankshaft and the front face of the thrust bearing or washer to determine the clearance **(see illustration)**.

4 Loosen the main bearing cap bolts 1/4-turn at a time each, until they can be removed by hand. Loosen the bolts in the reverse of the tightening sequence **(see illustration 10.19a, 10.19b, 10.19c or 10.19d)**. On all four-cylinder timing chain engines, remove the jack bolts (side bolts) before removing the bearing cap bolts.

5 Remove the main bearing caps, gently tap the caps with a soft-face hammer around its perimeter and pull the cap straight up and off the cylinder block.

2C-14   Chapter 2 Part C   General engine overhaul procedures

6 Carefully lift the crankshaft out of the engine. It may be a good idea to have an assistant available, since the crankshaft is quite heavy and awkward to handle. With the bearing inserts in place inside the engine block and main bearing caps, reinstall the main bearing caps onto the engine block and tighten the bolts finger-tight.

Installation

7 Crankshaft installation is the first step in engine reassembly. It's assumed at this point that the engine block and crankshaft have been cleaned, inspected and repaired or reconditioned.
8 Position the engine block with the bottom facing up.
9 Remove the bolts and lift off the main bearing caps.
10 If they're still in place, remove the original bearing inserts. Wipe the bearing surfaces of the block and main bearing cap assembly with a clean, lint-free cloth. They must be kept spotlessly clean. This is critical for determining the correct bearing oil clearance.

Main bearing oil clearance check

11 Without mixing them up, clean the back sides of the new upper main bearing inserts (with grooves and oil holes) and lay one in each main bearing saddle in the engine block. Each upper bearing (engine block) has an oil groove and oil hole in it. **Caution:** *The oil holes in the block must line up with the oil holes in the upper bearing inserts.* **Note:** *The thrust bearing is located on the engine block number 3 (center) journal.* Clean the back sides of the lower main bearing inserts and lay them in the corresponding location in the main bearing caps or the lower cylinder block. Make sure the tab on the bearing insert fits into its corresponding recess.
Caution: *Do not hammer the bearing insert into place and don't nick or gouge the bearing faces. DO NOT apply any lubrication at this time.*

12 Clean the faces of the bearing inserts in the block and the crankshaft main bearing journals with a clean, lint-free cloth.
13 Check or clean the oil holes in the crankshaft, as any dirt here can go only one way - straight through the new bearings.
14 Once you're certain the crankshaft is clean, carefully lay it in position in the cylinder block. Lube and insert the thrust washer on either side of journal no. 3 for all engines.
Note: *Install the thrust washers with the groove in the thrust washer facing the crankshaft and the smooth sides facing the main bearing saddle.*
15 Before the crankshaft can be permanently installed, the main bearing oil clearance must be checked.
16 Cut several strips of the appropriate size of Plastigage. They must be slightly shorter than the width of the main bearing journal.
17 Place one piece on each crankshaft main bearing journal, parallel with the journal axis as shown **(see illustration)**.
18 Clean the faces of the bearing inserts in the main bearing caps or the lower cylinder block. Hold the bearing inserts in place and install the caps or the lower cylinder block onto the crankshaft and cylinder block. DO NOT disturb the Plastigage.
19 Apply clean engine oil to all bolt threads prior to installation, then install all bolts finger-tight. Tighten the bolts in the sequence shown **(see illustrations)** progressing in steps, to the torque listed in this Chapter's Specifications. DO NOT rotate the crankshaft at any time during this operation. **Note:** *Use the old bolts at this time, not the new ones. It is not necessary to tighten the jack bolts (side bolts) for this Step.*
20 Remove the bolts in the reverse order of the tightening sequence and carefully lift the caps straight up and off the block. Do not disturb the Plastigage or rotate the crankshaft.

10.17 Place the Plastigage onto the crankshaft bearing journal as shown

21 Compare the width of the crushed Plastigage on each journal to the scale printed on the Plastigage envelope to determine the main bearing oil clearance **(see illustration)**. Main bearing oil clearance will fall between 0.007 to 0.0145 inch (0.17 to 0.37 mm) for four-cylinder engines and 0.009 to 0.017 inch (0.23 to 0.43 mm) on five-cylinder engines.
22 If the clearance is not as specified, the bearing inserts may be the wrong size (which means different ones will be required). Before deciding if different inserts are needed, make sure that no dirt or oil was between the bearing inserts and the cap assembly or block when the clearance was measured. If the Plastigage was wider at one end than the other, the crankshaft journal may be tapered. If the clearance still exceeds the limit specified, the bearing insert(s) will have to be replaced with an undersize bearing insert(s).
23 The upper bearing inserts are installed with different thicknesses. Colored dots serve to identify the bearing thicknesses; S-black, R-red, G-yellow, B-blue and W-white. On four-

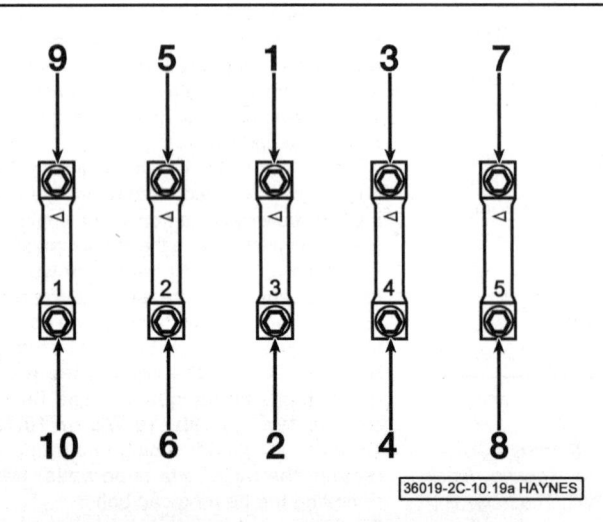

10.19a Main bearing cap beam bolt tightening sequence - four-cylinder timing belt (CBPA) engines

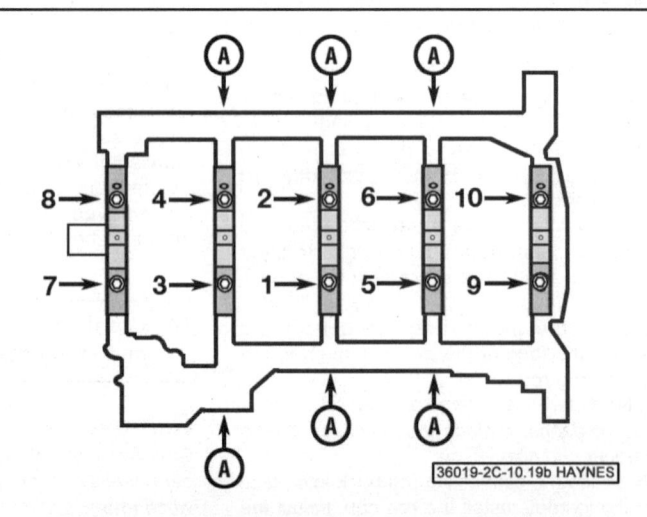

10.19b Main bearing cap beam bolt tightening sequence - 1.8L and 2.0L four-cylinder timing chain engines (except CBFA and CCTA models). (A) are the jack (side) bolts

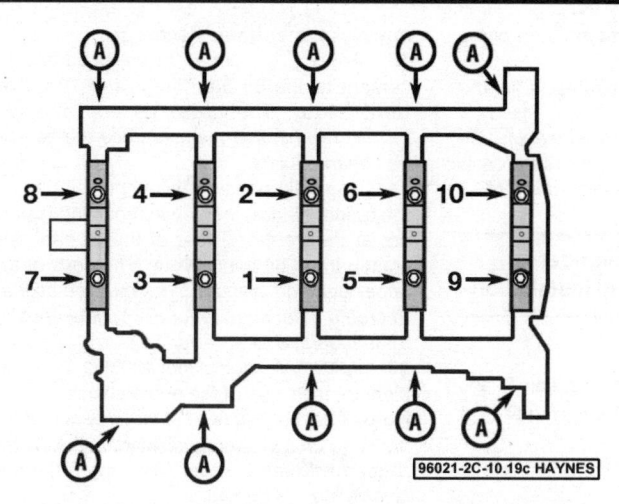

Chapter 2 Part C General engine overhaul procedures   2C-15

10.19c Main bearing cap beam bolt tightening sequence - CBFA and CCTA 2.0L four-cylinder timing chain engines. (A) are the jack (side) bolts

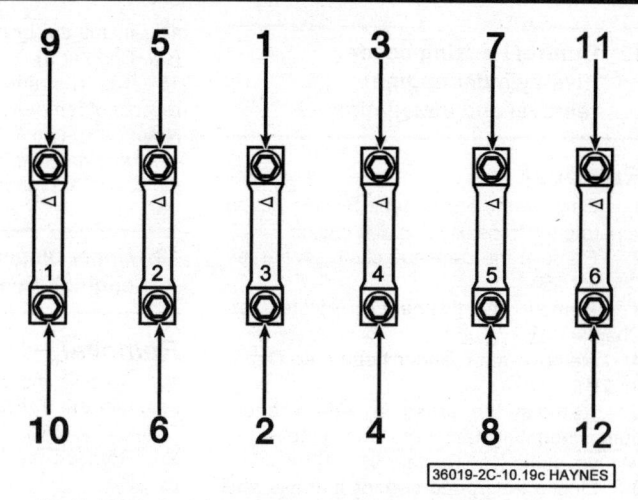

10.19d Main bearing cap beam bolt tightening sequence - five-cylinder engines

cylinder timing belt (CBPA) engines, these letters are marked on the lower sealing surface of the cylinder block where the oil pan is mounted. The first letter, starting from the belt side of the engine, and reading right to the left is for bearing cap no. 1, the second letter is for bearing cap no. 2, through all five caps to identify which bearing thickness must be installed in which location. On four-cylinder timing chain engines except for CBFA and CCTA, these letters are marked on the end of the crankshaft and on CBFA and CCTA engines, the letters are marked on a machined pad at the upper rear of the engine block. The first letter, starting from the left and reading to the right is for bearing cap no.1, the second letter is for bearing cap no. 2, through all five caps. On five-cylinder engines, these letters are marked on the lower sealing surface of the cylinder block where the oil pan is mounted. The first letter, starting from the belt side of the engine, and reading left to right is for bearing cap no. 1, the second letter if for bearing cap no. 2, through all six. The lower bearing inserts are always installed with Y-yellow dotted inserts, for all engines.

10.21 Use the scale on the Plastigage package to determine the bearing oil clearance - be sure to measure the widest part of the Plastigage and use the correct scale; it comes with both standard and metric scales

Note: *If the colored dots cannot be properly identified, always replace the inserts with blue dotted inserts.*

24 Carefully scrape all traces of the Plastigage material off the main bearing journals and/or the bearing insert faces. Be sure to remove all residue from the oil holes. Use your fingernail or the edge of a plastic card - don't nick or scratch the bearing faces.
25 Carefully lift the crankshaft out of the cylinder block.
26 Clean the bearing insert faces in the cylinder block, then apply a thin, uniform layer of moly-base grease or engine assembly lube to each of the bearing surfaces. Be sure to coat the thrust faces as well as the journal face of the thrust bearing.
27 Make sure the crankshaft journals are clean, then lay the crankshaft back in place in the cylinder block.
28 Clean the bearing insert faces and apply the same lubricant to them. Clean the engine block and the bearing caps/lower cylinder block thoroughly. The surfaces must be free of oil residue.
29 Install the main bearing caps with the arrows on the caps facing the front of the engine.
30 Prior to installation, apply clean engine oil to all bolt threads wiping off any excess, then install all bolts finger-tight.
Note: *The manufacturer requires using NEW bolts for the main caps and jack bolts (side bolts)onfour-cylinder timing chain engines.*
31 Tighten the **new** main bearing cap bolts, following the correct torque sequence **(see illustrations 10.19a, 10.19b, 10.91c or 10.19d)**. Torque the bolts to the Specifications listed in this Chapter. On four-cylinder timing chain engines, tighten the jack bolts (side bolts) at the same time as the bearing cap bolts.
32 Recheck the crankshaft endplay with a feeler gauge or a dial indicator. The endplay should be correct if the crankshaft thrust faces aren't worn or damaged and if new bearings have been installed.
33 Rotate the crankshaft a number of times by hand to check for any obvious binding. It should rotate with a running torque of 50 in-lbs or less. If the running torque is too high, correct the problem at this time.
34 Install a new rear main oil seal (see Chapter 2A or Chapter 2B).

11 Engine overhaul - reassembly sequence

1 Before beginning engine reassembly, make sure you have all the necessary new parts, gaskets and seals as well as the following items on hand:

Common hand tools
A 1/2-inch drive torque wrench
New engine oil
Gasket sealant
Thread locking compound

2 If you obtained a short block it will be necessary to install the cylinder head, the control housing cover, the oil pump and pick-up tube, upper and lower the oil pans, the water pump, the timing belt/chain and timing cover, and the valve cover (see Chapter 2A or 2B). In order to save time and avoid problems, the external components must be installed in the following general order:

Thermostat and housing cover
Water pump
Intake and exhaust manifolds
Fuel injection components
Emission control components
Spark plugs
Ignition coils
Oil filter
Engine mounts and mount brackets
Clutch and flywheel (manual transaxle)
Driveplate (automatic transaxle)

12 Control housing cover (five-cylinder engine) - removal and installation

Removal

1 Remove the engine (see Section 7) and separate the transaxle from the engine.
2 Remove the camshaft chain cover (see Chapter 2B).
3 Remove the flywheel/driveplate (see Chapter 2A)
4 Remove the cylinder head (see Chapter 2B).
5 Remove the brake booster vacuum pump mounting fasteners and remove the vacuum pump (see Chapter 9).
6 Unclip the speed sensor harness and remove the speed sensor (see Chapter 6).
7 Remove the control housing cover retaining fasteners **(see illustration)**, then carefully pry the housing from the cylinder block and upper oil pan, starting at the dowels, on each side of the cover.

Installation

8 Use a scraper to remove all traces of old sealant from the block and upper oil pan. Clean the mating surfaces with brake system cleaner.
Note: *Before removing all the traces of the old sealant from the control housing cover, follow the outline of the sealant and copy it, to apply the new sealant along the same line.*
9 Make sure the threaded bolt holes in the block are clean.
10 Check the cover housing for distortion, particularly around the bolt holes. Remove any nicks or burrs as necessary.
11 Apply a 1/16-inch (2 mm) bead of RTV sealant to the control housing flange.
12 Carefully position the housing cover on to the alignment dowels, install the retaining fasteners and tighten them by hand.
13 Tighten all control housing retaining fasteners to the torque listed in this Chapter's Specifications.
14 Install the brake booster vacuum pump (see Chapter 9).
15 The remainder of installation is the reverse of removal.
Note: *Be sure to follow the sealant manufacturer's recommendations on curing times and allow the sealant to properly cure before adding oil.*

13 Upper oil pan (five-cylinder engine) - removal and installation

Removal

1 Remove the engine (see Section 7) and separate the transaxle from the engine (see Chapter 7A or Chapter 7B).
2 Remove the lower oil pan (see Chapter 2B).
3 Remove the crankshaft front oil seal flange (see Chapter 2B).
4 Loosen the camshaft sprockets (see Chapter 2B, Section 5).
5 Remove the control housing (see Section 12).
6 Remove the oil pump chain tensioner and guide (see Section 14).
7 Remove the pick-up tube retaining fasteners and pick-up tube.
8 Remove the pick-up tube-to-oil pump seal.
9 Remove the upper oil pan mounting fasteners **(see illustration)** and carefully pry the oil pan off.
Note: *Pry the oil pan at the number 6 and number 1 crankshaft bearing caps using a flat-bladed screwdriver or equivalent tool.*

Installation

10 Use a scraper to remove all traces of old sealant from the block and oil pan. Clean the mating surfaces with brake system cleaner.
Note: *Use RTV sealant to seal the oil pan-to-cylinder block mating surface.*
11 Make sure the threaded bolt holes in the block are clean.
12 Check the oil pan flange for distortion, particularly around the bolt holes. Remove any nicks or burrs as necessary.
13 Apply a 1/16-inch (2 mm) bead of RTV sealant to the oil pan flange **(see illustration)**. **Note:** *The upper oil pan must be installed within five minutes once the sealant has been applied.*
14 Carefully position the upper oil pan on the cylinder block, install two mounting fasteners at the front and rear of the oil pan, and tighten them by hand. **Note:** *The ends of the upper pan and cylinder block must be aligned to create a flat surface for the control housing to be mounted to.*
15 Install VW special tool #2036/1, or equivalent to each side of the engine block.
Note: *This tool has two flat metal bars that are bolted to the cylinder block to keep the oil pan aligned with the cylinder block when the pan is installed.*
16 Align the oil pan with the special tool, check that the edge of the block and the edge of the oil pan are flush, then install the remaining oil pan bolts **(see illustration 13.9)**. Tighten all mounting fasteners in a diagonal sequence, to the torque listed in this Chapter's Specifications.
17 Install the oil pick-up tube to the pump with a new seal, then install the pick-up tube mounting fasteners to the oil pan, tightening all mounting fasteners to the torque listed in this Chapter's Specifications.
18 Install the oil pump timing chain and tensioner (see Section 14).
19 The remainder of installation is the reverse of removal.
Note: *Be sure to follow the sealant manufacturer's recommendations on curing times and allow the sealant to properly cure before adding oil.*

14 Oil pump, timing chain and tensioner (five-cylinder engine) - removal and installation

Note: *The oil pump is not serviceable; if there is a problem, it must be replaced.*

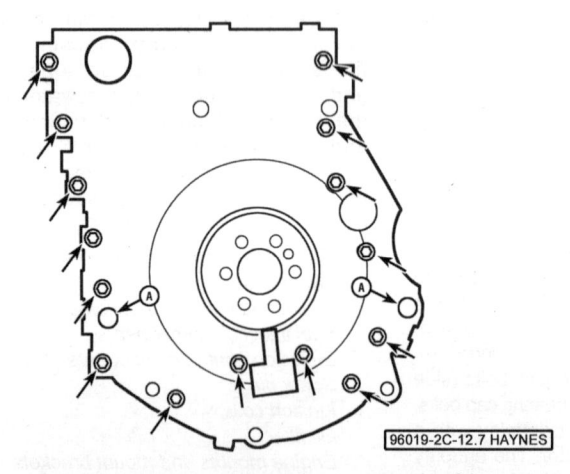

12.7 Remove the control housing cover retaining bolts, then pry the cover off, starting at the dowel (A) locations

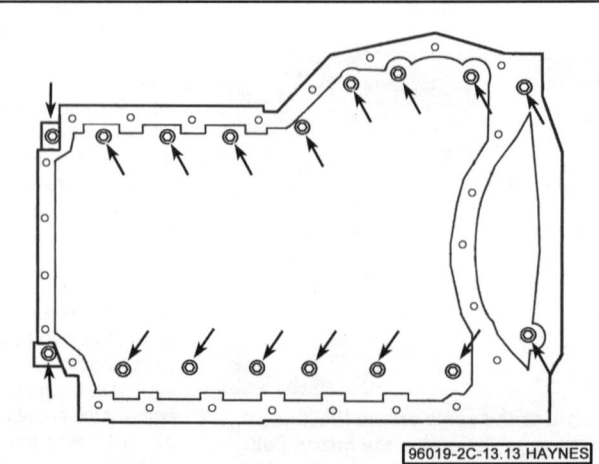

13.9 Upper oil pan mounting fastener locations

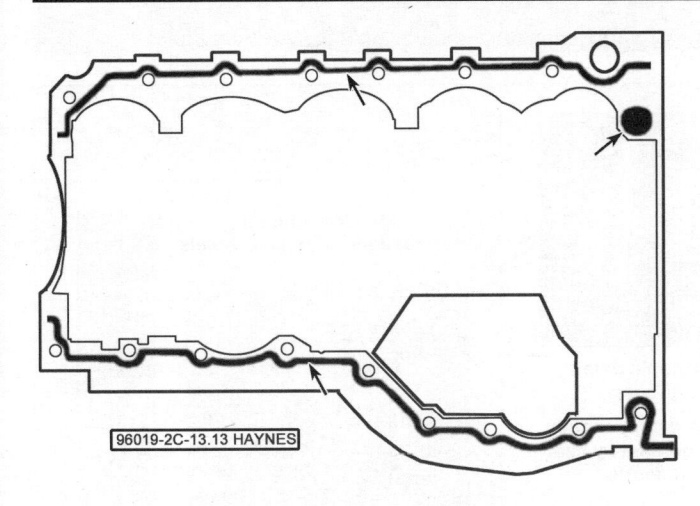

Chapter 2 Part C General engine overhaul procedures 2C-17

13.13 Apply a 1/16-inch (2.0 mm) thick bead of sealant to the upper pan at these points

14.8 Use a holding tool on the oil pump drive sprocket to remove the retaining bolt

Removal

1 Remove the engine (see Section 7), then separate the transmission from the engine (see Chapter 7A or 7B).
2 Remove the upper oil pan and oil pick-up tube (see Section 13).
3 Remove the crankshaft front oil seal flange (see Chapter 2B).
4 Loosen the camshaft sprockets (see Chapter 2A, Section 5).
5 Remove the control housing (see Section 12).
6 With a screwdriver placed between the chain guide and the tensioner piston, slowly compress the oil pump timing chain tensioner piston and place a pin (a drill bit or paper clip will work) or special tool #T10172 into the hole on the tensioner body to hold it in the fully compressed position.
7 Remove the tensioner mounting fastener, then slide the tensioner assembly, chain guide and tensioner guide off of the cylinder block dowels.
Caution: *Do not remove the pin holding the tensioner piston in the compressed position.*
8 While holding the oil pump drive sprocket with a suitable tool, remove the sprocket retaining bolt from the oil pump, then remove the sprocket and chain **(see illustration)**.
9 Remove oil pump retaining fasteners, then the oil pump and O-ring.
Note: *Always replace the O-ring with a new one.*

Installation

10 Lock the crankshaft in the TDC position (see Chapter 2B, Section 3).
11 Install a new O-ring onto the pump, then install the oil pump and hand-tighten the retaining fasteners.
12 Place the oil pump sprocket onto the pump, with the lettering on the sprocket facing outwards. Install the sprocket retaining bolt and tighten it to the torque listed in this Chapter's Specifications.
13 Loosen the oil pump bolts so that the oil pump can slide, then use VW special alignment tool #T03005 to align the oil pump sprocket with the crankshaft sprocket.
Note: *Special shim #T03005/1 and T03005 must be installed between the alignment tool and crankshaft.*
14 Mount the tool to the end of the crankshaft and tighten the tool mounting bolts to 22 ft-lbs (30 Nm).
15 Slide the oil pump sprocket towards the tool, which has magnets attached, and allow the sprocket to contact to the magnets.
Note: *The magnets and sprocket must be clean and free from any metal shavings.*
16 Remove the crankshaft locking pin (T40069).
17 Press the crankshaft towards the front (drivebelt side) and hold it in place using a shim or screwdriver between the last main bearing cap and the crankshaft counter weight.
18 Press the oil pump body lightly towards the oil pump chain side. Tighten the two oil pump retaining bolts at the back of the pump, then the bolt at the front of the pump, to the torque listed in this Chapter's Specifications.
19 Lock the crankshaft in the TDC position again (see Chapter 2B, Section 3), and remove the special alignment tool.
20 If a new pump is being installed, fill the oil pump through the pick-up tube opening with oil, while rotating the oil pump sprocket several times.
21 Place the oil pump chain onto the sprockets and check to make sure the chain runs in a straight line between three sprockets.
22 Install the upper oil pan (see Section 13).
23 Install the chain guide, the tensioner guide and the tensioner assembly onto the cylinder block dowels, then install the tensioner fastener and tighten it to the torque listed in this Chapter's Specifications.
24 Remove the pin from the tensioner and release the piston to engage the chain guide.
25 The remainder of installation is the reverse of removal.

15 Initial start-up and break-in after overhaul

Warning: *Have a fire extinguisher handy when starting the engine for the first time.*
1 Once the engine has been installed in the vehicle, double-check the engine oil and coolant levels.
2 With the spark plugs out of the engine and the fuel pump disabled (see Chapter 4, Section 3), crank the engine until oil pressure registers on the gauge or the light goes out.
3 Install the spark plugs, hook up the plug wires and restore the ignition system and fuel pump functions.
4 Start the engine. It may take a few moments for the fuel system to build up pressure, but the engine should start without a great deal of effort.
5 After the engine starts, it should be allowed to warm up to normal operating temperature. While the engine is warming up, make a thorough check for fuel, oil and coolant leaks.
6 Shut the engine off and recheck the engine oil and coolant levels.
7 Drive the vehicle to an area with minimum traffic, accelerate from 30 to 50 mph, then allow the vehicle to slow to 30 mph with the throttle closed. Repeat the procedure 10 or 12 times. This will load the piston rings and cause them to seat properly against the cylinder walls. Check again for oil and coolant leaks.
8 Drive the vehicle gently for the first 500 miles (no sustained high speeds) and keep a constant check on the oil level. It is not unusual for an engine to use oil during the break-in period.
9 At approximately 500 to 600 miles, change the oil and filter.
10 For the next few hundred miles, drive the vehicle normally. Do not pamper it or abuse it.
11 After 2,000 miles, change the oil and filter again and consider the engine broken in.

ENGINE BEARING ANALYSIS

Debris

Babbitt bearing embedded with debris from machinings

Microscopic detail of debris

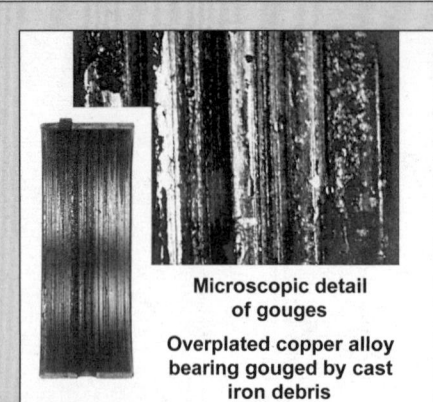

Microscopic detail of gouges

Overplated copper alloy bearing gouged by cast iron debris

Aluminum bearing embedded with glass beads

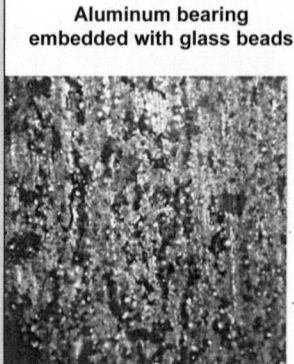

Microscopic detail of glass beads

Damaged lining caused by dirt left on the bearing back

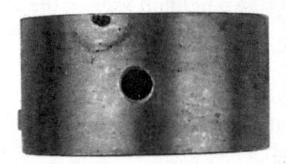

Misassembly

Result of a lower half assembled as an upper - blocking the oil flow

Excessive oil clearance is indicated by a short contact arc

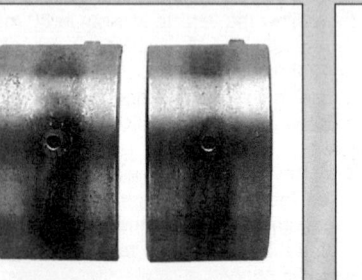

Polished and oil-stained backs are a result of a poor fit in the housing bore

Result of a wrong, reversed, or shifted cap

Overloading

Damage from excessive idling which resulted in an oil film unable to support the load imposed

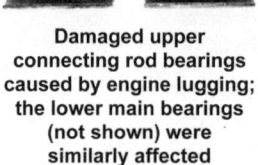

Damaged upper connecting rod bearings caused by engine lugging; the lower main bearings (not shown) were similarly affected

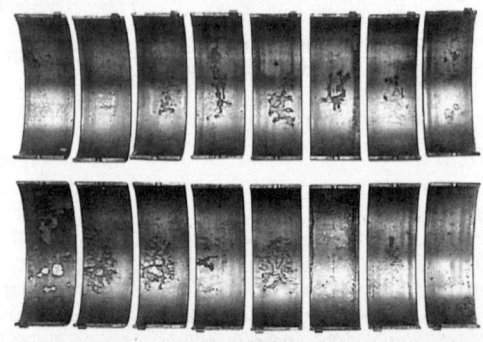

The damage shown in these upper and lower connecting rod bearings was caused by engine operation at a higher-than-rated speed under load

Misalignment

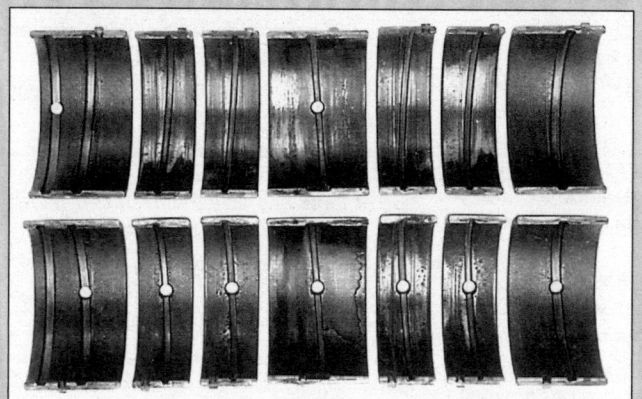

A warped crankshaft caused this pattern of severe wear in the center, diminishing toward the ends

A poorly finished crankshaft caused the equally spaced scoring shown

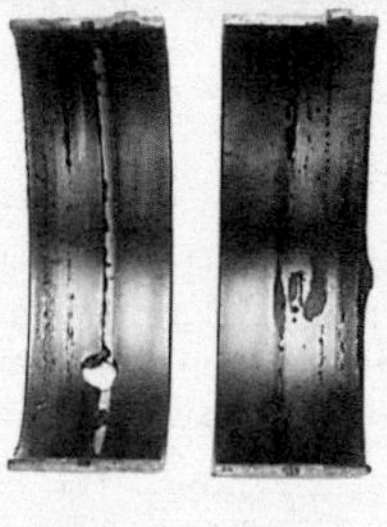

A bent connecting rod led to the damage in the "V" pattern

A tapered housing bore caused the damage along one edge of this pair

Lubrication

Result of dry start: The bearings on the left, farthest from the oil pump, show more damage

Result of a low oil supply or oil starvation

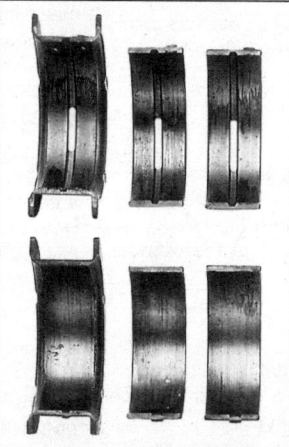

Severe wear as a result of inadequate oil clearance

Corrosion

Microscopic detail of corrosion

Corrosion is an acid attack on the bearing lining generally caused by inadequate maintenance, extremely hot or cold operation, or inferior oils or fuels

Microscopic detail of cavitation

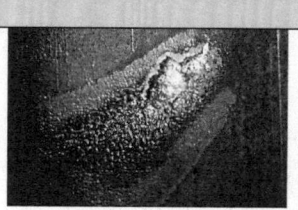

Example of cavitation - a surface erosion caused by pressure changes in the oil film

Damage from excessive thrust or insufficient axial clearance

Bearing affected by oil dilution caused by excessive blow-by or a rich mixture

© 1986 Federal-Mogul Corporation
Copy and photographs courtesy of Federal Mogul Corporation

COMMON ENGINE OVERHAUL TERMS

B

Backlash - The amount of play between two parts. Usually refers to how much one gear can be moved back and forth without moving the gear with which it's meshed.

Bearing Caps - The caps held in place by nuts or bolts which, in turn, hold the bearing surface. This space is for lubricating oil to enter.

Bearing clearance - The amount of space left between shaft and bearing surface. This space is for lubricating oil to enter.

Bearing crush - The additional height which is purposely manufactured into each bearing half to ensure complete contact of the bearing back with the housing bore when the engine is assembled.

Bearing knock - The noise created by movement of a part in a loose or worn bearing.

Blueprinting - Dismantling an engine and reassembling it to EXACT specifications.

Bore - An engine cylinder, or any cylindrical hole; also used to describe the process of enlarging or accurately refinishing a hole with a cutting tool, as to bore an engine cylinder. The bore size is the diameter of the hole.

Boring - Renewing the cylinders by cutting them out to a specified size. A boring bar is used to make the cut.

Bottom end - A term which refers collectively to the engine block, crankshaft, main bearings and the big ends of the connecting rods.

Break-in - The period of operation between installation of new or rebuilt parts and time in which parts are worn to the correct fit. Driving at reduced and varying speed for a specified mileage to permit parts to wear to the correct fit.

Bushing - A one-piece sleeve placed in a bore to serve as a bearing surface for shaft, piston pin, etc. Usually replaceable.

C

Camshaft - The shaft in the engine, on which a series of lobes are located for operating the valve mechanisms. The camshaft is driven by gears or sprockets and a timing chain. Usually referred to simply as the cam.

Carbon - Hard, or soft, black deposits found in combustion chamber, on plugs, under rings, on and under valve heads.

Cast iron - An alloy of iron and more than two percent carbon, used for engine blocks and heads because it's relatively inexpensive and easy to mold into complex shapes.

Chamfer - To bevel across (or a bevel on) the sharp edge of an object.

Chase - To repair damaged threads with a tap or die.

Combustion chamber - The space between the piston and the cylinder head, with the piston at top dead center, in which air-fuel mixture is burned.

Compression ratio - The relationship between cylinder volume (clearance volume) when the piston is at top dead center and cylinder volume when the piston is at bottom dead center.

Connecting rod - The rod that connects the crank on the crankshaft with the piston. Sometimes called a con rod.

Connecting rod cap - The part of the connecting rod assembly that attaches the rod to the crankpin.

Core plug - Soft metal plug used to plug the casting holes for the coolant passages in the block.

Crankcase - The lower part of the engine in which the crankshaft rotates; includes the lower section of the cylinder block and the oil pan.

Crank kit - A reground or reconditioned crankshaft and new main and connecting rod bearings.

Crankpin - The part of a crankshaft to which a connecting rod is attached.

Crankshaft - The main rotating member, or shaft, running the length of the crankcase, with offset throws to which the connecting rods are attached; changes the reciprocating motion of the pistons into rotating motion.

Cylinder sleeve - A replaceable sleeve, or liner, pressed into the cylinder block to form the cylinder bore.

D

Deburring - Removing the burrs (rough edges or areas) from a bearing.

Deglazer - A tool, rotated by an electric motor, used to remove glaze from cylinder walls so a new set of rings will seat.

E

Endplay - The amount of lengthwise movement between two parts. As applied to a crankshaft, the distance that the crankshaft can move forward and back in the cylinder block.

F

Face - A machinist's term that refers to removing metal from the end of a shaft or the face of a larger part, such as a flywheel.

Fatigue - A breakdown of material through a large number of loading and unloading cycles. The first signs are cracks followed shortly by breaks.

Feeler gauge - A thin strip of hardened steel, ground to an exact thickness, used to check clearances between parts.

Free height - The unloaded length or height of a spring.

Freeplay - The looseness in a linkage, or an assembly of parts, between the initial application of force and actual movement. Usually perceived as slop or slight delay.

Freeze plug - See Core plug.

G

Gallery - A large passage in the block that forms a reservoir for engine oil pressure.

Glaze - The very smooth, glassy finish that develops on cylinder walls while an engine is in service.

H

Heli-Coil - A rethreading device used when threads are worn or damaged. The device is installed in a retapped hole to reduce the thread size to the original size.

I

Installed height - The spring's measured length or height, as installed on the cylinder head. Installed height is measured from the spring seat to the underside of the spring retainer.

J

Journal - The surface of a rotating shaft which turns in a bearing.

K

Keeper - The split lock that holds the valve spring retainer in position on the valve stem.

Key - A small piece of metal inserted into matching grooves machined into two parts fitted together - such as a gear pressed onto a shaft - which prevents slippage between the two parts.

Knock - The heavy metallic engine sound, produced in the combustion chamber as a result of abnormal combustion - usually detonation. Knock is usually caused by a loose or worn bearing. Also referred to as detonation, pinging and spark knock. Connecting rod or main bearing knocks are created by too much oil clearance or insufficient lubrication.

L

Lands - The portions of metal between the piston ring grooves.

Lapping the valves - Grinding a valve face and its seat together with lapping compound.

Lash - The amount of free motion in a gear train, between gears, or in a mechanical assembly, that occurs before movement can begin. Usually refers to the lash in a valve train.

Lifter - The part that rides against the cam to transfer motion to the rest of the valve train.

M

Machining - The process of using a machine to remove metal from a metal part.

Main bearings - The plain, or babbit, bearings that support the crankshaft.

Main bearing caps - The cast iron caps, bolted to the bottom of the block, that support the main bearings.

O

O.D. - Outside diameter.

Oil gallery - A pipe or drilled passageway in the engine used to carry engine oil from one area to another.

Oil ring - The lower ring, or rings, of a piston; designed to prevent excessive amounts of oil from working up the cylinder walls and into the combustion chamber. Also called an oil-control ring.

Oil seal - A seal which keeps oil from leaking out of a compartment. Usually refers to a dynamic seal around a rotating shaft or other moving part.

O-ring - A type of sealing ring made of a special rubberlike material; in use, the O-ring is compressed into a groove to provide the sealing action.

Overhaul - To completely disassemble a unit, clean and inspect all parts, reassemble it with the original or new parts and make all adjustments necessary for proper operation.

P

Pilot bearing - A small bearing installed in the center of the flywheel (or the rear end of the crankshaft) to support the front end of the input shaft of the transmission.

Pip mark - A little dot or indentation which indicates the top side of a compression ring.

Piston - The cylindrical part, attached to the connecting rod, that moves up and down in the cylinder as the crankshaft rotates. When the fuel charge is fired, the piston transfers the force of the explosion to the connecting rod, then to the crankshaft.

Piston pin (or wrist pin) - The cylindrical and usually hollow steel pin that passes through the piston. The piston pin fastens the piston to the upper end of the connecting rod.

Piston ring - The split ring fitted to the groove in a piston. The ring contacts the sides of the ring groove and also rubs against the cylinder wall, thus sealing space between piston and wall. There are two types of rings: Compression rings seal the compression pressure in the combustion chamber; oil rings scrape excessive oil off the cylinder wall.

Piston ring groove - The slots or grooves cut in piston heads to hold piston rings in position.

Piston skirt - The portion of the piston below the rings and the piston pin hole.

Plastigage - A thin strip of plastic thread, available in different sizes, used for measuring clearances. For example, a strip of plastigage is laid across a bearing journal and mashed as parts are assembled. Then parts are disassembled and the width of the strip is measured to determine clearance between journal and bearing. Commonly used to measure crankshaft main-bearing and connecting rod bearing clearances.

Press-fit - A tight fit between two parts that requires pressure to force the parts together. Also referred to as drive, or force, fit.

Prussian blue - A blue pigment; in solution, useful in determining the area of contact between two surfaces. Prussian blue is commonly used to determine the width and location of the contact area between the valve face and the valve seat.

R

Race (bearing) - The inner or outer ring that provides a contact surface for balls or rollers in bearing.

Ream - To size, enlarge or smooth a hole by using a round cutting tool with fluted edges.

Ring job - The process of reconditioning the cylinders and installing new rings.

Runout - Wobble. The amount a shaft rotates out-of-true.

S

Saddle - The upper main bearing seat.

Scored - Scratched or grooved, as a cylinder wall may be scored by abrasive particles moved up and down by the piston rings.

Scuffing - A type of wear in which there's a transfer of material between parts moving against each other; shows up as pits or grooves in the mating surfaces.

Seat - The surface upon which another part rests or seats. For example, the valve seat is the matched surface upon which the valve face rests. Also used to refer to wearing into a good fit; for example, piston rings seat after a few miles of driving.

Short block - An engine block complete with crankshaft and piston and, usually, camshaft assemblies.

Static balance - The balance of an object while it's stationary.

Step - The wear on the lower portion of a ring land caused by excessive side and back-clearance. The height of the step indicates the ring's extra side clearance and the length of the step projecting from the back wall of the groove represents the ring's back clearance.

Stroke - The distance the piston moves when traveling from top dead center to bottom dead center, or from bottom dead center to top dead center.

Stud - A metal rod with threads on both ends.

T

Tang - A lip on the end of a plain bearing used to align the bearing during assembly.

Tap - To cut threads in a hole. Also refers to the fluted tool used to cut threads.

Taper - A gradual reduction in the width of a shaft or hole; in an engine cylinder, taper usually takes the form of uneven wear, more pronounced at the top than at the bottom.

Throws - The offset portions of the crankshaft to which the connecting rods are affixed.

Thrust bearing - The main bearing that has thrust faces to prevent excessive endplay, or forward and backward movement of the crankshaft.

Thrust washer - A bronze or hardened steel washer placed between two moving parts. The washer prevents longitudinal movement and provides a bearing surface for thrust surfaces of parts.

Tolerance - The amount of variation permitted from an exact size of measurement. Actual amount from smallest acceptable dimension to largest acceptable dimension.

U

Umbrella - An oil deflector placed near the valve tip to throw oil from the valve stem area.

Undercut - A machined groove below the normal surface.

Undersize bearings - Smaller diameter bearings used with re-ground crankshaft journals.

V

Valve grinding - Refacing a valve in a valve-refacing machine.

Valve train - The valve-operating mechanism of an engine; includes all components from the camshaft to the valve.

Vibration damper - A cylindrical weight attached to the front of the crankshaft to minimize torsional vibration (the twist-untwist actions of the crankshaft caused by the cylinder firing impulses). Also called a harmonic balancer.

W

Water jacket - The spaces around the cylinders, between the inner and outer shells of the cylinder block or head, through which coolant circulates.

Web - A supporting structure across a cavity.

Woodruff key - A key with a radiused backside (viewed from the side).

Notes

Chapter 3
Cooling, heating and air conditioning systems

Contents

	Section		Section
Air conditioning and heating system - check and maintenance.....	3	Engine cooling fans - removal and installation	5
Air conditioning compressor - removal and installation	13	General Information...	1
Air conditioning condenser - removal and installation	15	Heater and air conditioning control assembly and	
Air conditioning evaporator core - removal and installation...........	16	flexible shafts - removal and installation	10
Air conditioning expansion valve - general information	17	Heater and air conditioning housing - removal and installation.....	12
Air conditioning high-pressure sensor - replacement	18	Heater core - removal and installation..	11
Air conditioning refrigerant desiccant cartridge - replacement	14	Radiator and expansion tank - removal and installation................	6
Blower motor and blower motor resistor -		Thermostat/engine temperature control actuator -	
removal and installation	9	replacement	4
Coolant temperature gauge sending unit -		Troubleshooting	2
check and replacement	8	Water pump and electric coolant pump - replacement	7

Specifications

General
Coolant capacity	See Chapter 1
Refrigerant type*...	R-134a or R-1234yf

* Refer to the underhood HVAC sticker for the type of refrigerant used in your vehicle.

Torque specifications

Ft-lbs (unless otherwise indicated)   **Nm**

Note: *One foot-pound (ft-lb) of torque is equivalent to 12 inch-pounds (in-lbs) of torque. Torque values below approximately 15 ft-lbs are expressed in inch-pounds, because most foot-pound torque wrenches are not accurate at these smaller values.*

Cooling fan-to-fan shroud fasteners	44 in-lbs	5
Cooling fan shroud-to-radiator fasteners...	44 in-lbs	5
Thermostat housing fasteners		
Four-cylinder engines		
1.4L (in sequence, **see illustration 4.22**)....................................	72 in-lbs	8
2.0L (CBPA)...	133 in-lbs	15
1.8L and 2.0L (all except CBPA)	79 in-lbs	9
Five-cylinder engines		
Thermostat housing-to-cylinder block...	18	25
Thermostat housing cover	44 in-lbs	5
Water pump (toothed) belt cover fasteners	79 in-lbs	9
Water pump (toothed) belt drive gear bolt		
Step 1	89 in-lbs	10
Step 2	Tighten an additional 90-degrees	
Water pump fasteners		
Four-cylinder engines		
1.4L (in sequence, **see illustration 7.22**)		
Step 1	Hand-tighten	
Step 2	89 in-lbs	10
Step 3	Loosen all bolts one turn	
Step 4, Bolts 2, 1 and 5	89 in-lbs	10
Step 5, Bolts 3, 4, 5, 1 and 2	108 in-lbs	12
2.0L (CBPA)...	133 in-lbs	15
1.8L and 2.0L (all except CBPA)	79 in-lbs	9
Five-cylinder engines	88 in-lbs	10
Air conditioning refrigerant lines-to-compressor fasteners	16	22
Air conditioning compressor mounting bolts...	18	25

2.2 The cooling system pressure tester is connected in place of the pressure cap, then pumped up to pressurize the system

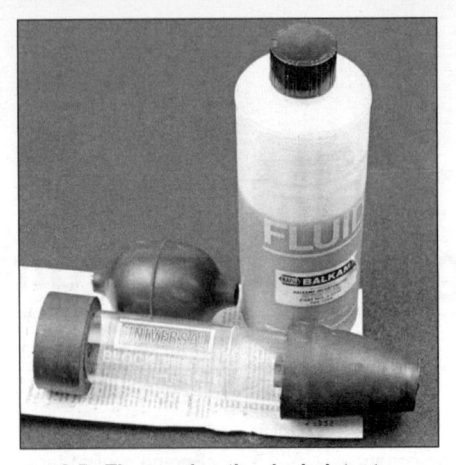

2.5a The combustion leak detector consists of a bulb, syringe and test fluid

2.5b Place the tester over the cooling system filler neck and use the bulb to draw a sample into the tester

1 General information

Warning: *Do not allow antifreeze to come in contact with your skin or painted surfaces of the vehicle. Rinse off spills immediately with plenty of water. Antifreeze is highly toxic if ingested. Never leave antifreeze lying around in an open container or in puddles on the floor; children and pets are attracted by its sweet smell and may drink it. Check with local authorities about disposing of used antifreeze. Many communities have collection centers which will see that antifreeze is disposed of safely. Never dump used antifreeze on the ground or pour it into drains.*

Engine cooling system

1 All modern vehicles employ a pressurized engine cooling system with thermostatically controlled coolant circulation. The cooling system consists of a radiator, an expansion tank or coolant reservoir, a pressure cap (located on the expansion tank or radiator), a thermostat, a cooling fan, and a water pump.

2 The water pump circulates coolant through the engine. The coolant flows around each cylinder and around the intake and exhaust ports, near the spark plug areas and in close proximity to the exhaust valve guides.

3 A thermostat controls engine coolant temperature. During warm up, the closed thermostat prevents coolant from circulating through the radiator. As the engine nears normal operating temperature, the thermostat opens and allows hot coolant to travel through the radiator, where it's cooled before returning to the engine.

Heating system

4 The heating system consists of a blower fan and heater core located in a housing under the dash, the hoses connecting the heater core to the engine cooling system and the heater/air conditioning control head on the dashboard. Hot engine coolant is circulated through the heater core. When the heater mode is activated, a flap door in the housing opens to expose the heater core to the passenger compartment through air ducts. A fan switch on the control head activates the blower motor, which forces air through the core, heating the air.

Air conditioning system

5 The air conditioning system consists of a condenser mounted in front of the radiator, an evaporator mounted adjacent to the heater core, a compressor mounted on the engine, a receiver-drier or accumulator and the plumbing connecting all of the above components.

6 A blower fan forces the warmer air of the passenger compartment through the evaporator core (sort of a radiator-in-reverse), transferring the heat from the air to the refrigerant. The liquid refrigerant boils off into low pressure vapor, taking the heat with it when it leaves the evaporator.

7 There can be two types of refrigerant and systems used depending on the year and model of the vehicle. A sticker is located at the front of the engine compartment detailing what was installed in the vehicle's air conditioning system, either R134a or R1234yf.

Caution: *The refrigerants used in R134a or R1234yf and the components are not compatible - do not mix the refrigerants or oil used with either system or damage to the system will occur.*

2 Troubleshooting

Coolant leaks

1 A coolant leak can develop anywhere in the cooling system, but the most common causes are:

a) A loose or weak hose clamp
b) A defective hose
c) A faulty pressure cap
d) A damaged radiator
e) A bad heater core
f) A faulty water pump
g) A leaking gasket at any joint that carries coolant

2 Coolant leaks aren't always easy to find. Sometimes they can only be detected when the cooling system is under pressure. Here's where a cooling system pressure tester comes in handy. After the engine has cooled completely, the tester is attached in place of the pressure cap, then pumped up to the pressure value equal to that of the pressure cap rating **(see illustration)**. Now, leaks that only exist when the engine is fully warmed up will become apparent. The tester can be left connected to locate a nagging slow leak.

Coolant level drops, but no external leaks

3 If you find it necessary to keep adding coolant, but there are no external leaks, the probable causes include:

a) A blown head gasket
b) A leaking intake manifold gasket (only on engines that have coolant passages in the manifold)
c) Cracked cylinder head or cylinder block

4 Any of the above problems will also usually result in contamination of the engine oil, which will cause it to take on a milkshake-like appearance. A bad head gasket or cracked head or block can also result in engine oil contaminating the closed cooling system.

5 Combustion leak detectors (also known as block testers) are available at most auto parts stores. These work by detecting exhaust gases in the cooling system, which indicates a compression leak from a cylinder into the coolant. The tester consists of a large bulb-type syringe and bottle of test fluid **(see illustration)**. A measured amount of the fluid is added to the syringe. The syringe is placed over the cooling system filler neck and, with the engine running, the bulb is squeezed and a sample of the gases present in the cooling system are drawn up through the test fluid **(see illustration)**. If any combustion gases are present in the sample taken, the test fluid will change color.

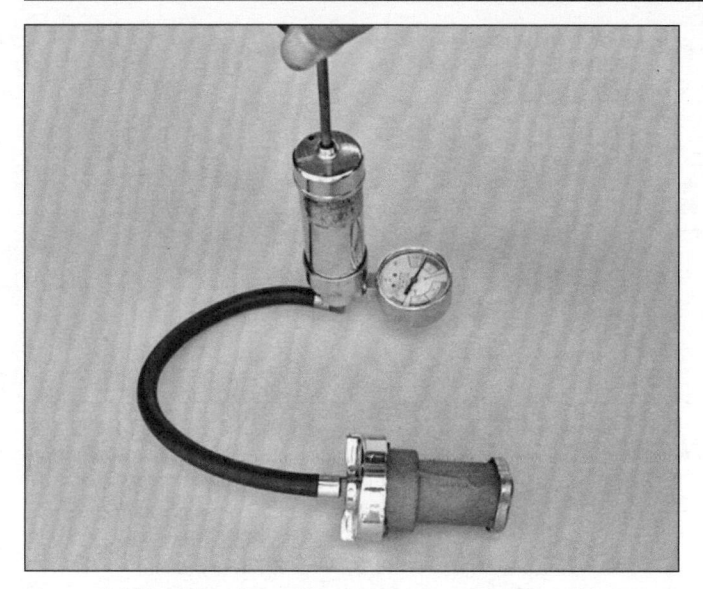

Chapter 3 Cooling, heating and air conditioning systems 3-3

2.8 Checking the cooling system pressure cap with a cooling system pressure tester

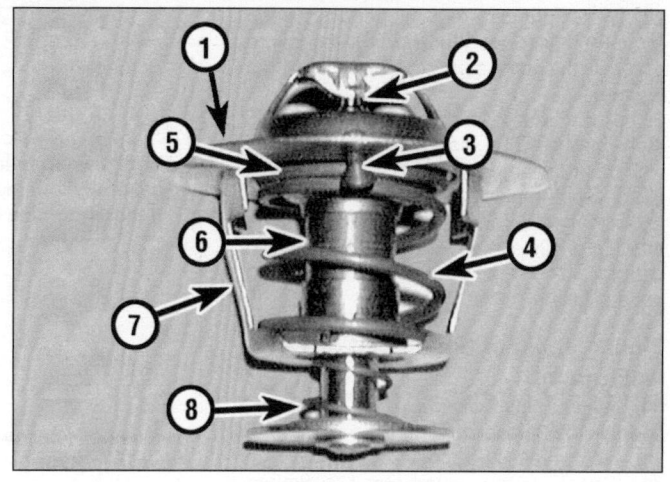

2.10 Typical thermostat

1	Flange	5	Valve seat
2	Piston	6	Valve
3	Jiggle valve	7	Frame
4	Main coil spring	8	Secondary coil spring

6   If the test indicates combustion gas is present in the cooling system, you can be sure that the engine has a blown head gasket or a crack in the cylinder head or block, and will require disassembly to repair.

Pressure cap

Warning: *Wait until the engine is completely cool before beginning this check.*

7   The cooling system is sealed by a spring-loaded cap, which raises the boiling point of the coolant. If the cap's seal or spring are worn out, the coolant can boil and escape past the cap. With the engine completely cool, remove the cap and check the seal; if it's cracked, hardened or deteriorated in any way, replace it with a new one.

8   Even if the seal is good, the spring might not be; this can be checked with a cooling system pressure tester **(see illustration)**. If the cap can't hold a pressure within approximately 1-1/2 lbs of its rated pressure (which is marked on the cap), replace it with a new one.

9   The cap is also equipped with a vacuum relief spring. When the engine cools off, a vacuum is created in the cooling system. The vacuum relief spring allows air back into the system, which will equalize the pressure and prevent damage to the radiator (the radiator tanks could collapse if the vacuum is great enough). If, after turning the engine off and allowing it to cool down you notice any of the cooling system hoses collapsing, replace the pressure cap with a new one.

Thermostat

10   Before assuming the thermostat **(see illustration)** is responsible for a cooling system problem, check the coolant level and drivebelt tension (see Chapter 1), and temperature gauge (or light) operation.

11   If the engine takes a long time to warm up (as indicated by the temperature gauge or heater operation), the thermostat is probably stuck open. Replace the thermostat with a new one.

12   If the engine runs hot or overheats, a thorough test of the thermostat should be performed.

13   Definitive testing of the thermostat can only be made when it is removed from the vehicle. If the thermostat is stuck in the open position at room temperature, it is faulty and must be replaced.

Caution: *Do not drive the vehicle without a thermostat. The computer may stay in open loop and emissions and fuel economy will suffer.*

14   To test a thermostat, suspend the (closed) thermostat on a length of string or wire in a pot of cold water.

15   Heat the water on a stove while observing thermostat. The thermostat should fully open before the water boils.

16   If the thermostat doesn't open and close as specified, or sticks in any position, replace it.

Cooling fan

Electric cooling fan

17   If the engine is overheating and the cooling fan is not coming on when the engine temperature rises to an excessive level, unplug the fan motor electrical connector(s) and connect the motor directly to the battery with fused jumper wires. If the fan motor doesn't come on, replace the motor.

18   If the radiator fan motor is okay, but it isn't coming on when the engine gets hot, the fan relay might be defective. A relay is used to control a circuit by turning it on and off in response to a control decision by the Powertrain Control Module (PCM). These control circuits are fairly complex, and checking them should be left to a qualified automotive technician. Sometimes, the control system can be fixed by simply identifying and replacing a bad relay.

19   Locate the fan relays in the engine compartment fuse/relay box.

20   Test the relay (see Chapter 12).

21   If the relay is okay, check all wiring and connections to the fan motor. Refer to the wiring diagrams at the end of Chapter 12. If no obvious problems are found, the problem could be the Engine Coolant Temperature (ECT) sensor or the Powertrain Control Module (PCM). Have the cooling fan system and circuit diagnosed by a dealer service department or repair shop with the proper diagnostic equipment.

Belt-driven cooling fan

22   Disconnect the cable from the negative terminal of the battery and rock the fan back and forth by hand to check for excessive bearing play.

23   With the engine cold (and not running), turn the fan blades by hand. The fan should turn freely.

24   Visually inspect for substantial fluid leakage from the clutch assembly. If problems are noted, replace the clutch assembly.

25   With the engine completely warmed up, turn off the ignition switch and disconnect the negative battery cable from the battery. Turn the fan by hand. Some drag should be evident. If the fan turns easily, replace the fan clutch.

Water pump

26   A failure in the water pump can cause serious engine damage due to overheating.

Drivebelt-driven water pump

27   There are two ways to check the operation of the water pump while it's installed on the engine. If the pump is found to be defective, it should be replaced with a new or rebuilt unit.

28 Water pumps are equipped with weep (or vent) holes (see illustration). If a failure occurs in the pump seal, coolant will leak from the hole.

29 If the water pump shaft bearings fail, there may be a howling sound at the pump while it's running. Shaft wear can be felt with the drivebelt removed if the water pump pulley is rocked up and down (with the engine off). Don't mistake drivebelt slippage, which causes a squealing sound, for water pump bearing failure.

Timing chain or timing belt-driven water pump

30 Water pumps driven by the timing chain or timing belt are located underneath the timing chain or timing belt cover.

31 Checking the water pump is limited because of where it is located. However, some basic checks can be made before deciding to remove the water pump. If the pump is found to be defective, it should be replaced with a new or rebuilt unit.

32 One sign that the water pump may be failing is that the heater (climate control) may not work well. Warm the engine to normal operating temperature, confirm that the coolant level is correct, then run the heater and check for hot air coming from the ducts.

33 Check for noises coming from the water pump area. If the water pump impeller shaft or bearings are failing, there may be a howling sound at the pump while the engine is running.

34 It you suspect water pump failure due to noise, wear can be confirmed by feeling for play at the pump shaft. This can be done by rocking the drive sprocket on the pump shaft up and down. To do this you will need to remove the tension on the timing chain or belt as well as access the water pump.

Note: *Be careful not to mistake drivebelt noise (squealing) for water pump bearing or shaft failure.*

All water pumps

35 In rare cases or on high-mileage vehicles, another sign of water pump failure may be the presence of coolant in the engine oil. This condition will adversely affect the engine in varying degrees.

Note: *Finding coolant in the engine oil could indicate other serious issues besides a failed water pump, such as a blown head gasket or a cracked cylinder head or block.*

36 Even a pump that exhibits no outward signs of a problem, such as noise or leakage, can still be due for replacement. Removal for close examination is the only sure way to tell. Sometimes the fins on the back of the impeller can corrode to the point that cooling efficiency is diminished significantly.

Heater system

37 Little can go wrong with a heater. If the fan motor will run at all speeds, the electrical part of the system is okay. The three

2.28 The water pump weep hole is generally located on the underside of the pump

basic heater problems fall into the following general categories:

a) *Not enough heat*
b) *Heat all the time*
c) *No heat*

38 If there's not enough heat, the control valve or door is stuck in a partially open position, the coolant coming from the engine isn't hot enough, or the heater core is restricted. If the coolant isn't hot enough, the thermostat in the engine cooling system is stuck open, allowing coolant to pass through the engine so rapidly that it doesn't heat up quickly enough. If the vehicle is equipped with a temperature gauge instead of a warning light, watch to see if the engine temperature rises to the normal operating range after driving for a reasonable distance.

39 If there's heat all the time, the control valve or the door is stuck wide open.

40 If there's no heat, coolant is probably not reaching the heater core, or the heater core is plugged. The likely cause is a collapsed or plugged hose, core, or a frozen heater control valve. If the heater is the type that flows coolant all the time, the cause is a stuck door or a broken or kinked control cable.

Air conditioning system

41 If the cool air output is inadequate:
a) *Inspect the condenser coils and fins to make sure they're clear.*
b) *Check the compressor clutch for slippage.*
c) *Check the blower motor for proper operation.*
d) *Inspect the blower discharge passage for obstructions.*
e) *Check the system air intake filter for clogging.*

42 If the system provides intermittent cooling air:
a) *Check the circuit breaker, blower switch and blower motor for a malfunction.*
b) *Make sure the compressor clutch isn't slipping.*
c) *Inspect the plenum door to make sure it's operating properly.*
d) *Inspect the evaporator to make sure it isn't clogged.*
e) *If the unit is icing up, it may be caused by excessive moisture in the system,*

incorrect super heat switch adjustment or low thermostat adjustment.

43 If the system provides no cooling air:
a) *Inspect the compressor drivebelt. Make sure it's not loose or broken.*
b) *Make sure the compressor clutch engages. If it doesn't, check for a blown fuse.*
c) *Inspect the wire harness for broken or disconnected wires.*
d) *If the compressor clutch doesn't engage, bridge the terminals of the A/C pressure switch(es) with a jumper wire; if the clutch now engages, and the system is properly charged, the pressure switch is bad.*
e) *Make sure the blower motor is not disconnected or burned out.*
f) *Make sure the compressor isn't partially or completely seized.*
g) *Inspect the refrigerant lines for leaks.*
h) *Check the components for leaks.*
i) *Inspect the receiver-drier/accumulator or expansion valve/tube for clogged screens.*

44 If the system is noisy:
a) *Look for loose panels in the passenger compartment.*
b) *Inspect the compressor drivebelt. It may be loose or worn.*
c) *Check the compressor mounting bolts. They should be tight.*
d) *Listen carefully to the compressor. It may be worn out.*
e) *Listen to the idler pulley and bearing and the clutch. Either may be defective.*
f) *The winding in the compressor clutch coil or solenoid may be defective.*
g) *The compressor oil level may be low.*
h) *The blower motor fan bushing or the motor itself may be worn out.*
i) *If there is an excessive charge in the system, you'll hear a rumbling noise in the high pressure line, a thumping noise in the compressor, or see bubbles or cloudiness in the sight glass.*
j) *If there's a low charge in the system, you might hear hissing in the evaporator case at the expansion valve, or see bubbles or cloudiness in the sight glass.*

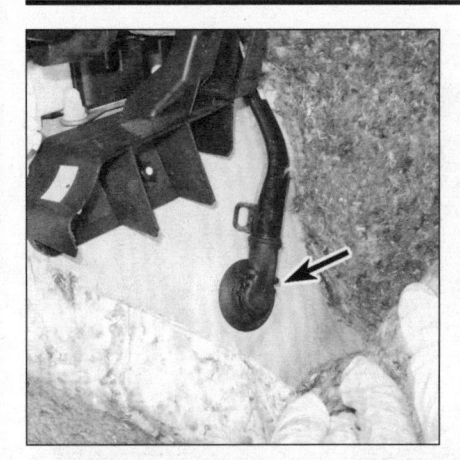

Chapter 3 Cooling, heating and air conditioning systems   3-5

3.1 Evaporator drain hose, is located under the front section of the center console

3.9 Insert a thermometer in the center vent, turn on the air conditioning system and wait for it to cool down; depending on the humidity, the output air should be 30 to 40 degrees cooler than the ambient air temperature

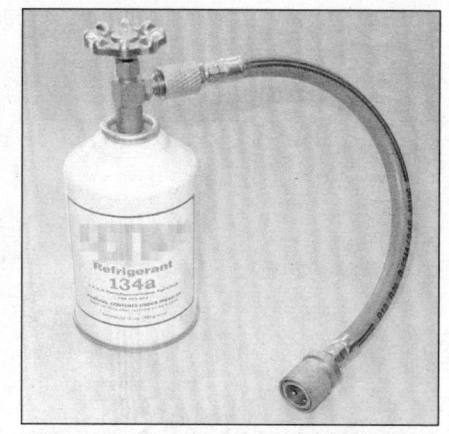

3.11 R-134a automotive air conditioning charging kit

3 Air conditioning and heating system - check and maintenance

Air conditioning system

Warning: *The air conditioning system is under high pressure. Do not loosen any hose fittings or remove any components until after the system has been discharged. Air conditioning refrigerant should be properly discharged into an EPA-approved recovery/recycling unit at a dealer service department or an automotive air conditioning repair facility. Always wear eye protection when disconnecting air conditioning system fittings.*

Caution: *The models covered by this manual use environmentally friendly R-134a or R1234yf. The refrigerants (and the appropriate refrigerant oils) are not compatible with each other or their system components and must never be mixed or the components will be damaged.*

Caution: *When replacing entire components, additional refrigerant oil should be added equal to the amount that is removed with the component being replaced. Be sure to read the can before adding any oil to the system, to make sure it is compatible with the R-134a or R1234yf systems.*

1 The following maintenance checks should be performed on a regular basis to ensure that the air conditioning continues to operate at peak efficiency.

a) *Inspect the condition of the compressor drivebelt. If it is worn or deteriorated, replace it (see Chapter 1).*
b) *Check the drivebelt tension (see Chapter 1).*
c) *Inspect the system hoses. Look for cracks, bubbles, hardening and deterioration. Inspect the hoses and all fittings for oil bubbles or seepage. If there is any evidence of wear, damage or leakage, replace the hose(s).*
d) *Inspect the condenser fins for leaves, bugs and any other foreign material that may have embedded itself in the fins. Use a fin comb or compressed air to remove debris from the condenser.*
e) *Make sure the system has the correct refrigerant charge.*
f) *If you hear water sloshing around in the dash area or have water dripping on the carpet, check the evaporator housing drain tube* **(see illustration)** *and insert a piece of wire into the opening to check for blockage.*

2 It's a good idea to operate the system for about ten minutes at least once a month. This is particularly important during the winter months because long term non-use can cause hardening, and subsequent failure, of the seals. Note that using the Defrost function operates the compressor.

3 If the air conditioning system is not working properly, proceed to Step 6 and perform the general checks outlined below.

4 Because of the complexity of the air conditioning system and the special equipment necessary to service it, in-depth troubleshooting and repairs beyond checking the refrigerant charge and the compressor clutch operation are not included in this manual. However, simple checks and component replacement procedures are provided in this Chapter. For more complete information on the air conditioning system, refer to the *Haynes Automotive Heating and Air Conditioning Manual*.

5 The most common cause of poor cooling is simply a low system refrigerant charge. If a noticeable drop in system cooling ability occurs, one of the following quick checks will help you determine if the refrigerant level is low.

Checking the refrigerant charge

6 Warm the engine up to normal operating temperature.

7 Place the air conditioning temperature selector at the coldest setting and put the blower at the highest setting.

8 After the system reaches operating temperature, feel the larger pipe exiting the evaporator at the firewall. The outlet pipe should be cold (the tubing that leads back to the compressor). If the evaporator outlet pipe is warm, the system probably needs a charge.

9 Insert a thermometer in the center air distribution duct **(see illustration)** while operating the air conditioning system at its maximum setting - the temperature of the output air should be 30 to 40 degrees F below the ambient air temperature (down to approximately 40 degrees F). If the ambient (outside) air temperature is very high, say 110 degrees F, the duct air temperature may be as high as 60 degrees F, but generally the air conditioning is 30 to 40 degrees F cooler than the ambient air.

10 Further inspection or testing of the system requires special tools and techniques and is beyond the scope of the home mechanic.

Adding refrigerant

Caution: *Make sure any refrigerant, refrigerant oil or replacement component you purchase is designated as compatible with R-134a systems.*

11 Purchase an R-134a automotive charging kit at an auto parts store **(see illustration)**. A charging kit includes a can of refrigerant, a tap valve and a short section of hose that can be attached between the tap valve and the system low side service valve.

Caution: *Never add more than one can of refrigerant to the system. If more refrigerant than that is required, the system should be evacuated and leak tested.*

12 Back off the valve handle on the charging kit and screw the kit onto the refrigerant can, making sure first that the O-ring or rubber seal inside the threaded portion of the kit is in place.

Warning: *Wear protective eyewear when dealing with pressurized refrigerant cans.*

3-6 Chapter 3 Cooling, heating and air conditioning systems

3.13 Location of the low-side charging port

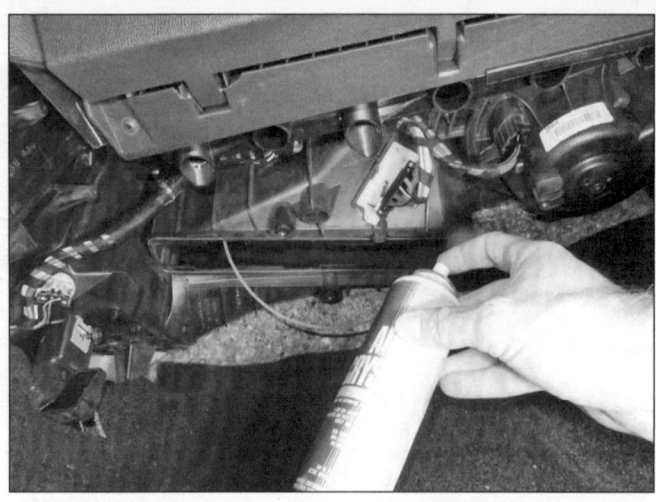

3.24 Remove the interior ventilation filter and spray the disinfectant into the blower housing with the blower on high (be sure to wear safety goggles)

13   Remove the dust cap from the low-side charging port and attach the hose's quick-connect fitting to the port **(see illustration)**. **Warning:** *DO NOT hook the charging kit hose to the system high side! The fittings on the charging kit are designed to fit only on the low side of the system.*

14   Warm up the engine and turn On the air conditioning. Keep the charging kit hose away from the fan and other moving parts. **Note:** *The charging process requires the compressor to be running. If the clutch cycles off, you can put the air conditioning switch on High and leave the car doors open to keep the clutch on and compressor working. The compressor can be kept on during the charging by removing the connector from the pressure switch and bridging it with a paper clip or jumper wire during the procedure.*

15   Turn the valve handle on the kit until the stem pierces the can, then back the handle out to release the refrigerant. You should be able to hear the rush of gas. Keep the can upright at all times, but shake it occasionally. Allow stabilization time between each addition. **Note:** *The charging process will go faster if you wrap the can with a hot-water-soaked rag to keep the can from freezing up.*

16   If you have an accurate thermometer, you can place it in the center air conditioning duct inside the vehicle and keep track of the output air temperature. A charged system that is working properly should cool down to approximately 40 degrees F. If the ambient (outside) air temperature is very high, say 110 degrees F, the duct air temperature may be as high as 60 degrees F, but generally the air conditioning is 30 to 40 degrees F cooler than the ambient air.

17   When the can is empty, turn the valve handle to the closed position and release the connection from the low-side port. Reinstall the dust cap.

18   Remove the charging kit from the can and store the kit for future use with the piercing valve in the UP position, to prevent inadvertently piercing the can on the next use.

Heating systems

19   If the carpet under the heater core is damp, or if antifreeze vapor or steam is coming through the vents, the heater core is leaking. Remove it (see Section 11) and install a new unit (most radiator shops will not repair a leaking heater core).

20   If the air coming out of the heater vents isn't hot, the problem could stem from any of the following causes:

a) *The thermostat is stuck open, preventing the engine coolant from warming up enough to carry heat to the heater core. Replace the thermostat (see Section 4).*
b) *There is a blockage in the system, preventing the flow of coolant through the heater core. Feel both heater hoses at the firewall. They should be hot. If one of them is cold, there is an obstruction in one of the hoses or in the heater core, or the heater control valve is shut. Detach the hoses and back flush the heater core with a water hose. If the heater core is clear but circulation is impeded, remove the two hoses and flush them out with a water hose.*
c) *If flushing fails to remove the blockage from the heater core, the core must be replaced (see Section 11).*

Eliminating air conditioning odors

21   Unpleasant odors that often develop in air conditioning systems are caused by the growth of a fungus, usually on the surface of the evaporator core. The warm, humid environment there is a perfect breeding ground for mildew to develop.

22   The evaporator core on most vehicles is difficult to access, and factory dealerships have a lengthy, expensive process for eliminating the fungus by opening up the evaporator case and using a powerful disinfectant and rinse on the core until the fungus is gone. You can service your own system at home, but it takes something much stronger than basic household germ-killers or deodorizers.

23   Aerosol disinfectants for automotive air conditioning systems are available in most auto parts stores, but remember when shopping for them that the most effective treatments are also the most expensive. The basic procedure for using these sprays is to start by running the system in the RECIRC mode for ten minutes with the blower on its highest speed. Use the highest heat mode to dry out the system and keep the compressor from engaging by disconnecting the wiring connector at the compressor.

24   The disinfectant can usually comes with a long spray hose. Insert the nozzle into an intake port inside the cabin, and spray according to the manufacturer's recommendations **(see illustration)**.

25   Once the evaporator has been cleaned, the best way to prevent the mildew from coming back again is to make sure your evaporator housing drain tube is clear **(see illustration 3.1)**.

Automatic heating and air conditioning systems

26   Some vehicles are equipped with an optional automatic climate control system. This system has its own computer that receives inputs from various sensors in the heating and air conditioning system. This computer, like the PCM, has self-diagnostic capabilities to help pinpoint problems or faults within the system. Vehicles equipped with automatic heating and air conditioning systems are very complex and considered beyond the scope of the home mechanic. Vehicles equipped with automatic heating and air conditioning systems should be taken to dealer service department or other qualified facility for repair.

Chapter 3 Cooling, heating and air conditioning systems   3-7

4.7 Remove the two hose clamps and separate the hoses from the thermostat cover - 1.4L engines

4.8a Remove the thermostat cover fasteners...

4.8b... and remove the cover - 1.4L engines

4.9 Using snap-ring pliers rotate, the thermostat counterclockwise and remove it from the housing - 1.4L engines

4.11 Make sure the centering pin fits into the guide hole in the bottom of the housing - 1.4L (CZTA, DGXA) engines

4 Thermostat/engine temperature control actuator - replacement

Warning: *The engine must be completely cool when this procedure is performed.*
Caution: *Don't drive the vehicle without a thermostat! The computer may stay in open loop mode and emissions and fuel economy will suffer.*
Caution: *If the battery is disconnected, several systems must be re-learned before they will work properly (see Chapter 5, Section 3).*
Note: *2.0L (CBPA) and 2.5L (CBTA, CBUA) models are equipped with a bypass thermostat located in the top line of the bypass hose.*

1   Remove the engine cover (see Chapter 1, Section 7).
2   Disconnect the cable from the negative terminal of the battery (see Chapter 5).
3   Raise the front of the vehicle and support it securely on jackstands.
4   Remove the under-vehicle splash shield (see Chapter 1, Section 6).

5   Drain the coolant from the radiator (see Chapter 1). If the coolant is new, save it and reuse it. If it is to be replaced, see Section 1 for cautions about proper handling of used antifreeze.

1.4L four-cylinder engines

Note: *There are two thermostats used on these models: a thermostat used primarily for the main (radiator) cooling circuit and a thermostat used for the engine block cooling circuit. The thermostats are installed in the thermostat housing and water pump housing.*

Main cooling system circuit thermostat

6   Remove the intake air resonator, the air filter housing and charge air cooler air pipe (see Chapter 4).
7   Disengage the coolant hose clamps, then remove the hoses from the thermostat cover **(see illustration)**.
8   Remove the thermostat housing fasteners **(see illustrations)**, then detach the cover. Be prepared for some coolant to spill as the gasket seal is broken. Clean the mating surfaces of the engine block and the thermostat.
9   Using special thermostat removal tool #T10508 or a pair of snap-ring pliers **(see illustration)**, rotate the thermostat counterclockwise to remove it from the thermostat housing.
Note: *It may be necessary to push the thermostat downwards while rotating it at first before the thermostat will start to come out.*
10   Remove the cover gasket and clean the mating surfaces of the thermostat housing and the cover.
11   Install a new thermostat into the housing making sure the centering pin on the thermostat **(see illustration)** is centered into the guide hole in the housing.
12   Install a new gasket to the housing then reattach the thermostat cover and tighten the bolts to the torque listed in this Chapter's Specifications.
13   The remainder of installation is the reverse of the removal procedure.

3-8   **Chapter 3 Cooling, heating and air conditioning systems**

4.17 Separate the water pump from the thermostat housing - 1.4L engines

4.18 Remove the thermostat from the thermostat housing - 1.4L engines

4.20 Remove and replace the gasket from the thermostat housing

14 If you're not replacing the small thermostat refer to Chapter 1 and refill and bleed the system, then run the engine and check carefully for leaks.

Cylinder block circuit thermostat

15 Remove the water pump (see Section 7).
16 Remove the thermostat housing-to-water pump fasteners in reverse of the tightening sequence **(see illustration 4.22)**.
17 Separate the water pump from the thermostat housing **(see illustration)**.
18 Remove the thermostat from the housing or water pump **(see illustration)**.
19 Install a new thermostat into the water pump, making sure the centering pin on the thermostat is centered into the guide hole in the water pump.
20 Remove the gasket from the thermostat housing, then clean the mating surfaces of the thermostat housing and the water pump **(see illustration)**.
21 Coat the new gasket with coolant then install the gasket to the thermostat housing.
22 Reattach the thermostat cover to the water pump and tighten the bolts in sequence **(see illustration)** to the torque listed in this Chapter's Specifications.

23 The remainder of installation is the reverse of the removal procedure.
24 Refer to Chapter 1 and refill and bleed the system, then run the engine and check carefully for leaks.

2.0L (CBPA) four-cylinder engine

25 Locate the thermostat housing under the intake manifold on the left side of the engine block, then disconnect the radiator hose clamp and remove the radiator hose from the thermostat housing.
26 Remove the thermostat housing fasteners, then detach the housing. Be prepared for some coolant to spill as the gasket seal is broken. Clean the mating surfaces of the engine block and the thermostat.
27 Install a new O-ring and thermostat to the housing, then reattach the thermostat housing to the engine block. Tighten the bolts to the torque listed in this Chapter's Specifications.
Note: *The jiggle valve must be at the top or point upwards when the thermostat is installed.*
28 The remainder of installation is the reverse of the removal procedure.

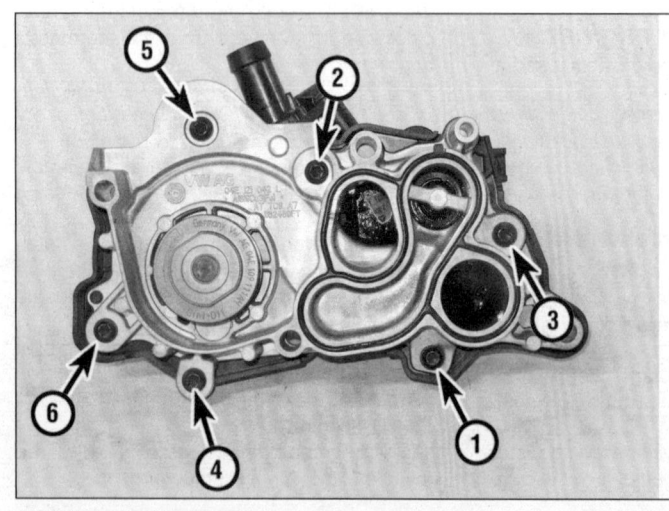

4.22 Water pump-to-thermostat housing tightening sequence - 1.4L engines

29 Refer to Chapter 1 and refill and bleed the system, then run the engine and check carefully for leaks.

1.8L and 2.0L (all except CBPA) four-cylinder engines

2.0L (CBFA, CCTA) four-cylinder engines

30 On models equipped with a noise generator, open the locking ring on the charge air pipe, disconnect the electrical connector to the charge air pipe, then remove the mounting fastener and the charge air pipe.
31 Remove the wiring harness retaining strap fastener and strap from under the radiator hose.
32 Remove the after-run coolant pump bracket fastener and move the after-run coolant pump assembly out of the way.
33 Disconnect the coolant hoses to the thermostat housing and secure them out of the way. Be prepared for some coolant to spill.
34 Remove the intake manifold support bracket fasteners and move the bracket to the right and secure it out of the way.
35 Remove the oil separator (see Chapter 2A).
36 Remove the thermostat cover retaining fasteners and cover. Be prepared for some coolant to spill as the cover is removed.
Note: *A centering pin is mounted in the thermostat cover; if the cover is turned over, the pin may fall out of the cover.*
Note: *Depending on the version, a thermostat can be installed without a centering pin.*
37 Note how the thermostat is installed (spring end inside the water pump housing), then remove the thermostat from the water pump housing.
38 Clean the mating surfaces of the thermostat housing and the thermostat cover.
39 Install the thermostat and make sure the correct end faces out; the spring end is installed into the thermostat housing assembly.
40 Install a new O-ring to the thermostat cover and reattach the cover to the pump

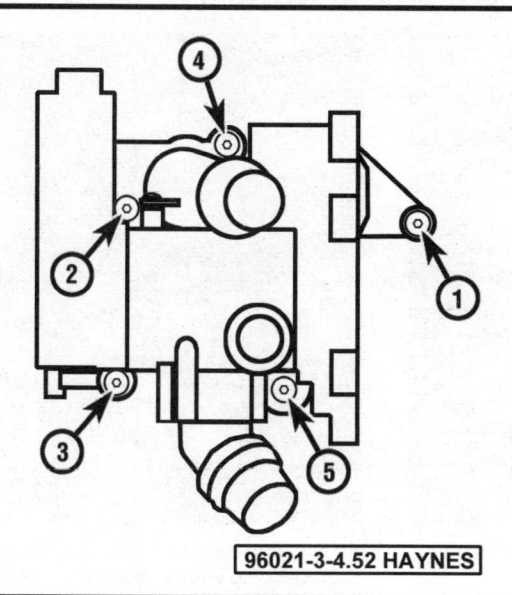

Chapter 3 Cooling, heating and air conditioning systems

4.52 Engine temperature control actuator assembly tightening sequence - 2.0L (CXCA, CXCB, CNTA) engine shown, other models similar

housing. Tighten the retaining fasteners to the torque listed in this Chapter's Specifications.

41 The remainder of installation is the reverse of the removal procedure.

42 Refer to Chapter 1 and refill and bleed the system, then run the engine and check carefully for leaks.

1.8L (CXBA, CXBB, CNSA, CNSB) and 2.0L (CXCA, CXCB, CYFB, DJJA) four-cylinder engines

Note: On 1.8L engines, the thermostat is an electronic assembly called the engine temperature control actuator and is replaced as an assembly.

43 Remove the water pump (see Section 7).

44 Remove the throttle body/control module (see Chapter 4).

45 Remove the coolant pipe bracket fasteners from the intake manifold.

46 Disconnect the coolant hose clips and hose connections from the engine temperature control actuator assembly and secure them out of the way. Be prepared for some coolant to spill.

47 Disconnect the electrical connector from the engine control actuator assembly.

48 Remove the engine temperature control actuator assembly mounting bolts, then pull the assembly off of the center pins on the engine block. Be prepared for some coolant to spill.

49 Separate the engine temperature control actuator assembly from the engine oil cooler, then remove the connector pipe from the oil cooler if it has remained in the cooler.

50 Clean the mating surfaces of the engine temperature control actuator and the water pump.

51 Install new O-rings on to the connector pipe then install the connector pipe into the engine oil cooler.

52 Install the alignment pins into the cylinder block, then install the engine temperature control actuator assembly to the oil cooler connector pipe and align the actuator assembly with the alignment pins. Install the assembly fasteners and tighten the retaining fasteners in sequence **(see illustration)** to the torque listed in this Chapter's Specifications.

53 The remainder of installation is the reverse of the removal procedure.

54 Refer to Chapter 1 and refill and bleed the system, then run the engine and check carefully for leaks.

1.8L (CPKA, CPRA) and 2.0L (CPLA, CPPA) four-cylinder engines

Note: Volkswagen requires the thermostat housing to be removed to access both thermostat cover fasteners.

55 Remove the water pump (see Section 7).

56 Remove the throttle body/control module (see Chapter 4).

57 Remove the coolant pipe bracket fasteners from the intake manifold.

58 Disconnect the coolant hose clips and hose connections from the thermostat housing and secure them out of the way. Be prepared for some coolant to spill.

59 Disconnect the harness electrical connectors under the intake manifold then remove the connector bracket fasteners and bracket.

60 Disconnect the electrical connector from the after-run coolant pump, then remove the pump bracket nut and move the after-run coolant pump out of the way.

61 Remove the thermostat housing mounting bolts, then pull the assembly off of the center pins on the engine block. Be prepared for some coolant to spill.

62 Separate the thermostat housing from the engine oil cooler, then remove the connector pipe from the oil cooler if it has remained in the cooler.

63 Remove the thermostat cover fasteners and separate the cover from the housing, then remove the thermostat.

64 Clean the mating surfaces of the thermostat housing, the cover, the engine block and the water pump.

65 Install the O-ring and thermostat, in the same order it was removed. Make sure the correct end faces out; the spring end is installed into the thermostat housing.

66 Tighten the thermostat cover fastener to the torque listed in this Chapter's Specifications.

67 Install new O-rings on to the connector pipe then install the connector pipe into the engine oil cooler.

68 Install the alignment pins into the cylinder block, then install the thermostat housing to the oil cooler connector pipe and align the housing with the alignment pins. Install the housing fasteners and tighten the retaining fasteners in sequence **(see illustration 4.52)** to the torque listed in this Chapter's Specifications.

69 The remainder of installation is the reverse of the removal procedure.

70 Refer to Chapter 1 and refill and bleed the system, then run the engine and check carefully for leaks.

2.5L five-cylinder engines

71 Remove the intake manifold (see Chapter 2B).

Note: Reinstall the oil dipstick tube to prevent coolant from running into the oil pan.

72 Remove the thermostat cover retaining fasteners and cover.

73 Note how the thermostat is installed (jiggle valve on top, and the spring end inside the thermostat housing), then remove the thermostat from the housing.

74 Remove the seal from the thermostat housing and clean the mating surfaces of the thermostat housing and the thermostat cover.

75 Install the thermostat seal to the housing. Install the thermostat, making sure the correct end faces out - with the jiggle valve up and the spring end installed into the housing.

Note: The jiggle valve must be at the top or point upwards when the thermostat is installed.

76 Install a new O-ring to the thermostat cover and reattach the cover to the thermostat housing. Tighten the retaining fasteners to the torque listed in this Chapter's Specifications.

77 The remainder of installation is the reverse of the removal procedure.

78 Refer to Chapter 1 and refill and bleed the system, then run the engine and check carefully for leaks.

5 Engine cooling fans - removal and installation

Warning: Keep hands, tools and clothing away from the fan. To avoid injury or damage, DO NOT operate the engine with a damaged fan. Do not attempt to repair fan blades - always replace a damaged fan with a new one.

Caution: If the battery is disconnected, sev-

3-10 Chapter 3 Cooling, heating and air conditioning systems

5.6 Disengage the red locking tab, then disconnect the engine harness connector from the cooling fan connector

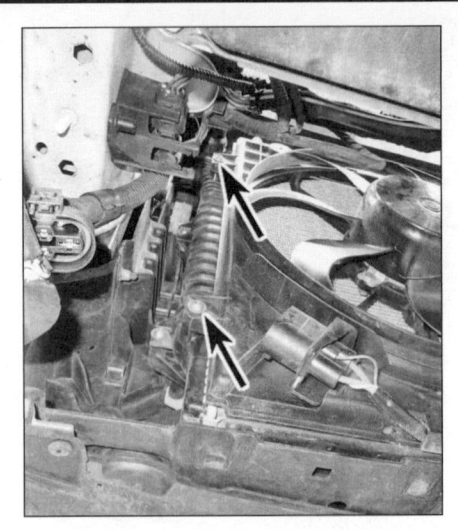

5.7a Cooling fan shroud left side retaining fasteners. . .

5.7b . . . and right side retaining fasteners - CZTA engine shown, other models similar

5.8 Fan shroud and cooling fan details

1. Fan shroud
2. Cooling fan motor connection
3. Cooling fan retaining fasteners

eral systems must be re-learned before they will work properly (see Chapter 5, Section 3).
1   Remove the engine cover (see Chapter 1, Section 7). Disconnect the cable from the negative terminal of the battery (see Chapter 5).
2   Raise the front of the vehicle and support it securely on jackstands.
3   Remove the under-vehicle splash shield (see Chapter 1, Section 6).
4   On four-cylinder engine, remove the air inlet duct (see Chapter 4).
5   On four-cylinder engines equipped with a noise generator, open the locking ring on the charge air pipe, disconnect the electrical connector from the charge air pipe, then remove the mounting fastener and the left-side charge air pipe (see Chapter 4).
6   Using a small screwdriver, disengage the red locking tab and disconnect the electrical connector from the cooling fan(s) harness **(see illustration)**.
7   Remove the cooling fan shroud fasteners **(see illustrations)** and remove the assembly from the bottom of the vehicle.
8   Disconnect and separate the electrical connector for the cooling fans, then remove the cooling fan motor fasteners and fans from the shroud **(see illustration)**.
9   If the fan blades or the fan motor are damaged, they can be replaced by removing the fan blade from the fan motor.
10 The remainder of installation is the reverse of the removal procedure.

6   Radiator and expansion tank - removal and installation

Warning: *The engine must be completely cool when this procedure is performed.*
Caution: *If the battery is disconnected, several systems must be re-learned before they will work properly (see Chapter 5, Section 3).*
Note: *For the replacement of the radiator shutter assembly (if equipped), see Chapter 11, Section 13.*

Radiator

1   Raise the front of the vehicle and support it securely on jackstands. Disconnect the cable from the negative terminal of the battery (see Chapter 5).
2   Remove the under-vehicle splash shield (see Chapter 1, Section 6) and drain the cooling system as described in Chapter 1, Section 26. Refer to the coolant **Warning** in Section 1.
3   Remove the engine cover (see Chapter 1, Section 7).
4   Remove the cooling fan shroud (see Section 5).

2.0L (CBPA) four-cylinder engines and 2.5L (CBTA, CBUA) five-cylinder engines

5   Disconnect the hoses from the radiator. Be prepared for some coolant to spill as the hoses are removed.
6   Remove the front bumper cover (see Chapter 11).
7   Remove the air guide fasteners and remove the air guides from each side of the condenser, then remove the condenser to radiator bolts (see Section 15).
8   Remove the radiator mount bolts and push the radiator towards the rear until the upper mounts can be removed from the sides of the radiator.
9   Lift the radiator up and out of the lower mounts, then remove the radiator from the vehicle.
10 Installation is the reverse of removal.
11 Refill and bleed the system, then run the engine and check carefully for leaks (see Chapter 1).

1.4L, 1.8L (CPRA, CPKA) and 2.0L (CPLA, CPPA, CBFA, CCTA) four-cylinder engines

12 Remove the clip securing the upper left hose connector to the radiator **(see illustration)**. Be prepared for some coolant to spill as the hoses are removed.

Chapter 3 Cooling, heating and air conditioning systems

6.12 Use a small screwdriver to lift the locking clip and disconnect the upper hose connector from the radiator - 1.4L CZTA engine shown, other models similar

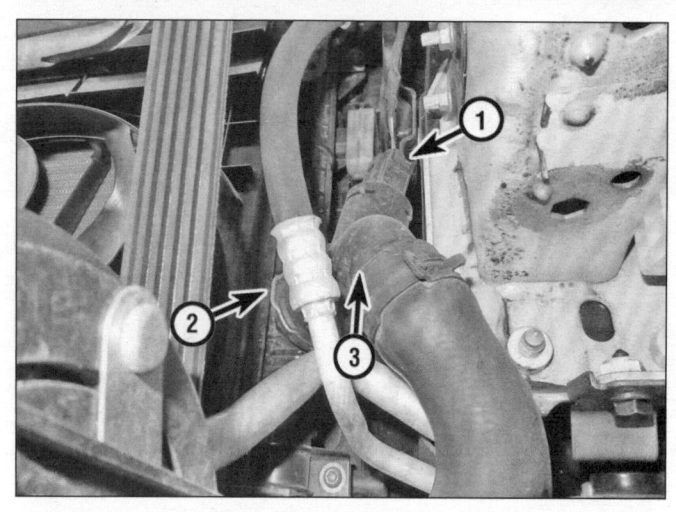

6.13 Lower right radiator hose details - 1.4L CZTA engine shown, other models similar

1. Coolant temperature sensor electrical connector
2. Hose connector retaining clip
3. Lower hose-to-radiator connector

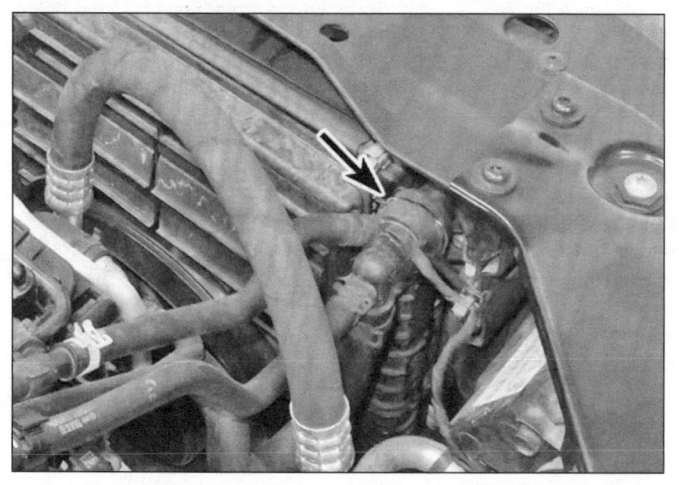

6.14 Lift the lock clip up and disconnect the hose connector from the charge air system circuit cooler hose - 1.4L CZTA engine shown, other models similar

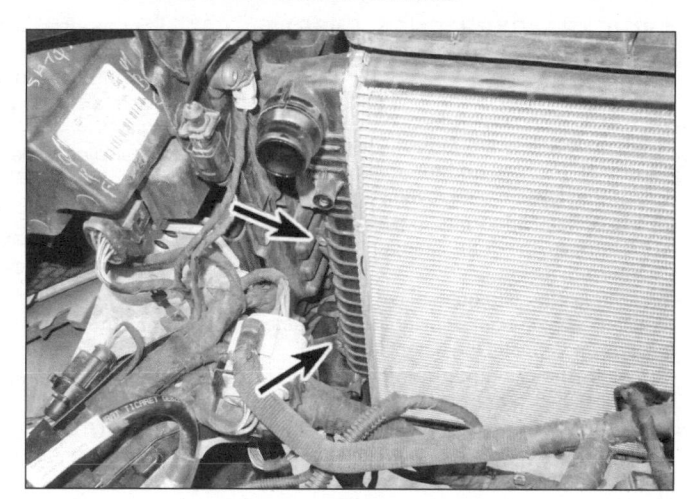

6.15a Radiator left side mounting bolt locations...

6.15b... and radiator right side mounting bolt locations - 1.4L CZTA engine shown, other models similar

13 Disconnect the coolant temperature sensor electrical connector at the radiator lower right hose connector, then remove the clip and disconnect the hose connector from the radiator **(see illustration)**. Be prepared for some coolant to spill as the hoses are removed.

14 Lift the clip securing the hose connector to the charge air system circuit cooler (upper right connector) and disconnect the connector from the cooler **(see illustration)**.

15 Remove the radiator mounting bolts **(see illustrations)** and slide the radiator towards the engine until the upper mounts are free, then lift the radiator up and out of the vehicle.

16 Prior to installation of the radiator, replace hose connector O-rings, any damaged hose clips and/or radiator hoses.

17 Installation is the reverse of removal.

18 Refill and bleed the system, then run the engine and check carefully for leaks (see Chapter 1).

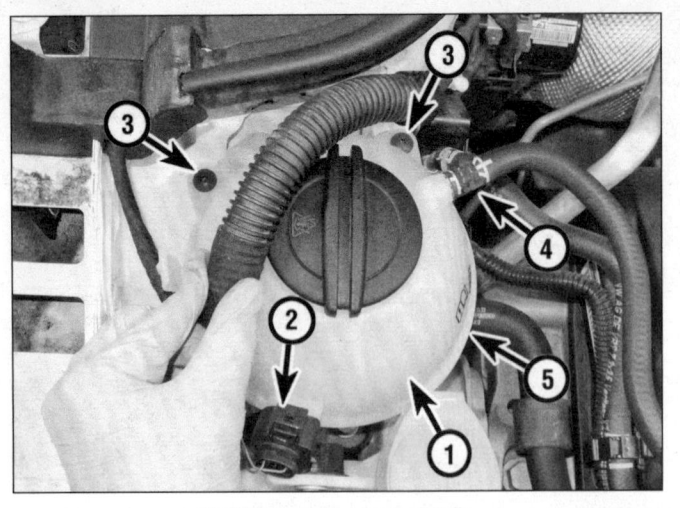

3-12 Chapter 3 Cooling, heating and air conditioning systems

6.29 Expansion tank details
1. Expansion tank
2. Coolant level sensor
3. Mounting screws
4. Upper coolant hose
5. Lower coolant hose

7.8 Remove crankcase ventilation tube fasteners and the tube

1.8L (CXBA, CXBB, CNSA, CNSB) and 2.0L (CNTA, CXCA, CXCB, CYFB, DJJA) four-cylinder engines

19 Remove the clip securing the upper right hose connector to the radiator (see illustration 6.12). Be prepared for some coolant to spill as the hoses are removed.
20 Remove the fan shroud (see Section 5).
21 Remove the front bumper cover (see Chapter 11).
22 Unclip the air guides by pushing the retaining tabs on both sides of the lock carrier inwards and remove the air guide.
23 With the air guide removed, insert a blade screwdriver into the openings on each side of the lock carrier and depress the radiator locking tabs down and pull the radiator back and off of the charge air cooler.
24 Slide the radiator back, then slightly rotate the radiator and remove it from the vehicle.
25 Prior to installation of the radiator, replace the hose connector O-rings, any damaged hose clips and/or radiator hoses.
26 Installation is the reverse of removal.
27 Refill and bleed the system, then run the engine and check carefully for leaks (see Chapter 1).

Expansion tank

28 Drain the cooling system as described in Chapter 1 until the expansion tank is empty. Refer to the coolant **Warning** in Section 1.
29 Detach the upper coolant hoses from the expansion tank (see illustration).
30 Disconnect the coolant level sensor connector. Unclip the wiring harness from the retainers and lift it up to access the expansion tank mounting screws. Remove the mounting screws, then lift the expansion tank up and disconnect the lower hose.
31 Remove the expansion tank from the engine compartment.

32 Prior to installation, make sure the reservoir is clean and free of debris which could be drawn into the radiator (wash it with soapy water and a brush if necessary, then rinse thoroughly).
33 Installation is the reverse of removal.

7 Water pump and electric coolant pump - replacement

Caution: *If the battery is disconnected, several systems must be re-learned before they will work properly (see Chapter 5, Section 3).*

Water pump

Warning: *Wait until the engine is completely cool before starting this procedure.*
1 Remove the engine cover (see Chapter 1, Section 7).
2 Disconnect the cable from the negative terminal of the battery (see Chapter 5).
3 Raise the front of the vehicle and support it securely on jackstands.

4 Remove the lower splash shield below the engine (see Chapter 1, Section 6).
5 Drain the cooling system (see Chapter 1). If the coolant is new, save it and reuse it. If it is to be replaced, see Section 1 for cautions about proper handling of used antifreeze.

1.4L (CZTA, DGXA) four-cylinder engines

6 Remove the battery and battery tray (see Chapter 5).
7 Remove the air inlet resonator (see Chapter 4).
8 Disconnect the EVAP hose quick-connect fitting and remove the hose from the ventilation tube. Remove the fasteners and the crankcase ventilation tube (see illustration). Once the breather tube is removed, cover the openings to prevent debris from falling to the openings.
Note: *When removing the tube take care not to damage the ground wires that are attached to the tube on each end.*
9 Locate the radiator hoses at the thermostat housing, then disconnect the clamps and remove the hoses (see illustration).

7.9 Disconnect the hose clamps and separate the radiator hoses from the thermostat housing

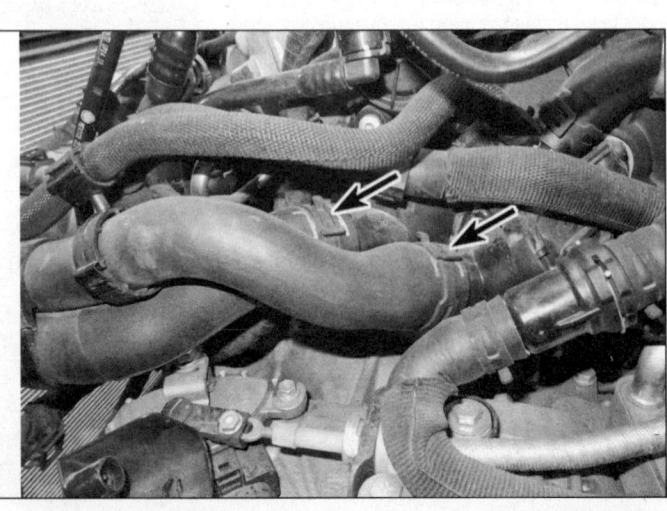

Chapter 3 Cooling, heating and air conditioning systems   3-13

7.11 Remove the connector fasteners, then remove the connector from the turbocharger

7.12a Use a small screwdriver to release the locking tab...

7.12b... then slide the harness retainer off of the bracket

10   Disconnect the coolant hoses from the thermostat housing **(see illustration 4.7)**.

11   Remove the connector fasteners and separate the connector from the turbocharger **(see illustration)**. Carefully remove the connector and plastic line as an assembly.

12   Disengage the wiring harness retainer using a small screwdriver, then slide the harness and retainer off of the bracket on the end of the cylinder head **(see illustrations)**.

13   Disengage the wiring harness from the retainers **(see illustration)** in front of water pump toothed belt cover and move the harness out of the way.

14   Remove the water pump toothed belt cover fasteners and cover **(see illustrations)**.

15   Disconnect the coolant hose to the rear of the thermostat housing **(see illustration)**.

16   Remove the water pump mounting fasteners in reverse order of installation **(see illustration 7.22)** then lift the pump up slightly to remove the toothed belt from the water pump pulley **(see illustration)** and remove the water pump assembly.

7.13 Disengage the wiring harness retainers and move the harness out of the way

7.14a Remove the water pump toothed belt cover fasteners...

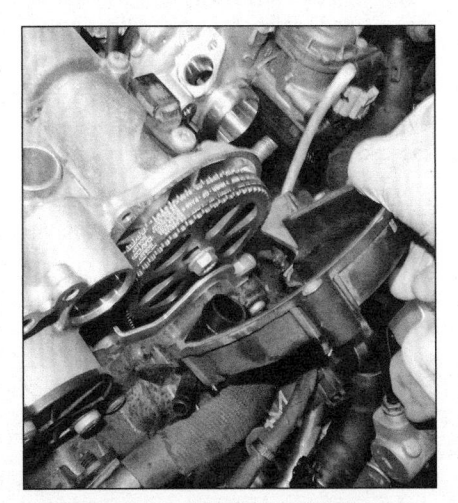

7.14b... and remove the cover

7.15 Disengage the hose clamp and remove the hose from the thermostat housing

7.16 Lift the water pump assembly up and unhook the toothed belt from the water pump pulley

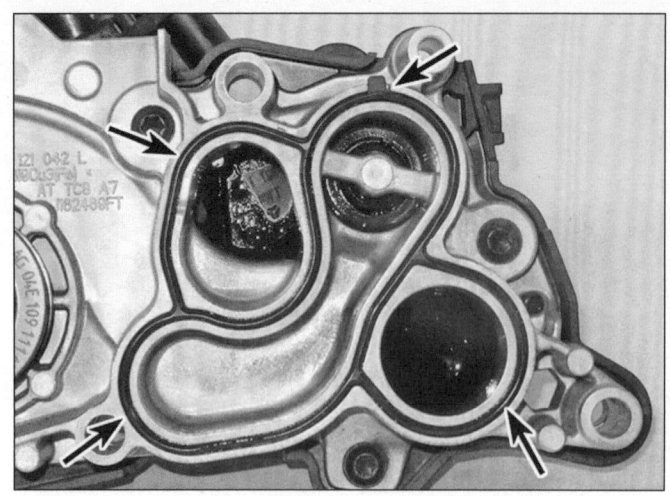

3-14   Chapter 3   Cooling, heating and air conditioning systems

7.19 Install a new seal to the water pump, making sure the seal fits correctly into the grooves on the pump

7.21 Install the water pump pulley into the belt

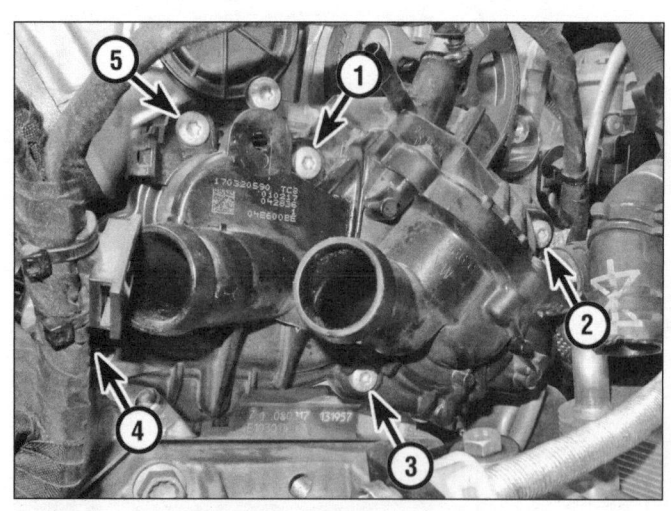

7.22 Water pump bolt tightening sequence

7.24 Location of the water pump 10 mm hex opening

17   Separate the thermostat housing from the water pump and discard the seal (see Section 4).

18   Clean the surface of the cylinder head, the thermostat housing and new water pump, then install the new seal to the water pump housing and thermostat housing and tighten the bolts in sequence (see Section 4) to the torque listed in this Chapter's Specifications.

19   Replace the water pump toothed belt with a new one and install a new seal to the water pump **(see illustration)** then coat the new seal with clean coolant.

20   Rotate the engine by hand until number 1 cylinder is at TDC (see Chapter 2A, Section 3).

21   Place the new toothed belt onto the drive pulley, then install the pump pulley onto the toothed belt **(see illustration)**.

22   Install the water pump onto the cylinder head and tighten the mounting bolts in sequence **(see illustration)** using Step 1 and Step 2 of the tightening specifications to the torque listed in this Chapter's Specifications.

23   The water pump toothed belt must be correctly tensioned before the final torque is completed. Loosen the all the water pump bolts in reverse of the tightening sequence **(see illustration 7.22)** one full turn.

Note: *This is Step 3 of the tightening sequence listed in this Chapter's Specifications.*

Note: *Step 24 and 25 require the help of an assistant when adjusting the water pump toothed belt tension to insure accurate tension is maintained during the tightening process.*

24   Insert a 10 mm Allen head socket through the thermostat housing and into the water pump 10 mm hex opening **(see illustration)**.

25   Using special tool #VAS6583 or a torque wrench attached to the 10 mm Allen head socket, insert the socket into the hex opening in the water pump **(see illustration 7.24)**, rotate the pump housing clockwise using 22 ft-lbs (30 Nm) of force and tighten the water pump bolts, in the sequence shown in illustration 7.22, using Steps 4 and 5 listed in this Chapter's Specifications.

Note: *This will require two torque wrenches - one for rotating the water pump and tensioning the belt, and another for tightening the bolts.*

26   Installation is the reverse of removal.

27   Refill and bleed the cooling system (see Chapter 1). Start the engine and check for the proper coolant level and the water pump and hoses for leaks.

2.0L (CBPA) four-cylinder engine

28   Remove the timing belt and timing belt tensioner (see Chapter 2A).

29   Remove the timing belt rear guard fasteners and guard.

30   Remove the water pump mounting bolts and remove the water pump.

31   Remove the O-ring located between the housing and block and discard it; a new one should be used on installation.

32   Remove the gasket from the thermostat housing and discard it; a new one should be used on installation **(see illustration 4.20)**.

33   Installation is the reverse of removal. Be sure to install a new O-ring between the

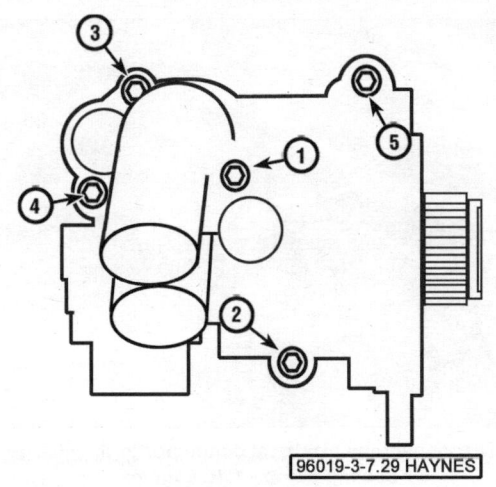

Chapter 3 Cooling, heating and air conditioning systems   3-15

7.55a Water pump mounting fastener tightening sequence - 2.0L (CBFA, CCTA) four-cylinder engines

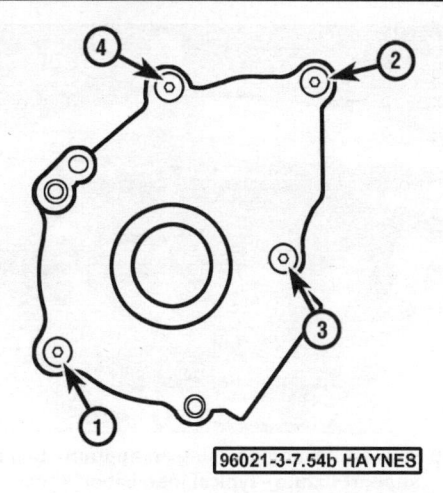

7.55b Water pump mounting fastener tightening sequence - 1.8L and 2.0L (CPLA, CPPA, CNTA, CXCA, CXCB, CYFB, DJJA) four-cylinder four-cylinder engines

water pump housing and the engine block, and tighten the fasteners to the torque listed in this Chapter's Specifications. Refer to Chapter 2A, Section 4 and install the timing belt.

34 Refill and bleed the cooling system (see Chapter 1). Start the engine and check for the proper coolant level and the water pump and hoses for leaks.

1.8L and 2.0L (all except CBPA) four-cylinder engines

35 Remove the right front inner fender liner (see Chapter 11).

36 On models equipped with a noise generator, open the locking ring on the charge air pipe, disconnect the electrical connector to the charge air pipe, then remove the mounting fasteners and the charge air pipes.

37 On 2.0L (CBFA, CCTA) engines, remove the intake manifold (see Chapter 2A).

38 Disconnect the electrical connectors above the small coolant pipe and pull the harness back enough to access the pipe.

39 Remove the small coolant pipe-to-cylinder block mounting fasteners. Disconnect the heater hose clamp from the top end of the coolant pipe and the clamp from the connection at the bottom of the pipe (where it meets the turbocharger coolant pipe).

40 Remove the coolant pipe from the hoses and from the water pump housing.

Note: *Always replace the small coolant pipe-to-water pump housing O-ring.*

41 On 1.8L and 2.0L (all except CBPA) engines, remove the throttle body assembly (see Chapter 4).

42 Remove the wiring harness retaining strap fastener and strap from under the radiator hose.

43 Remove the after-run coolant pump bracket fastener.

44 Remove the radiator hoses located behind the air conditioning compressor (see Chapter 1).

45 Remove the water pump (toothed) belt cover fasteners and cover.

46 On 1.8L (CPKA, CPRA) and 2.0L (CBFA, CBTA, CPLA, CPPA) engines, use a wrench or socket to hold the crankshaft pulley from turning, then remove the bolt that holds the water pump (toothed) belt and drive gear to the end of the balance shaft. Remove the belt.

Caution: *The water pump belt drive gear mounting bolt is left-hand thread.*

47 On 1.8L (CXBA, CXBB, CNSA, CNSB) and 2.0L (CXCA, CXCB, CNTA, CYFB, DJJA) engines, loosen the water pump mounting bolts. Allow the pump to rotate slightly and remove the water pump (toothed) belt.

48 On 2.0L (CBFA and CCTA) engines, remove the engine coolant temperature sensor (see Chapter 6).

49 Remove the water pump mounting fasteners in the reverse order of the tightening sequence **(see illustration 7.54a or 7.54b)**, then lift the water pump off of the center pins.

Note: *The centering pins can stick in the water pump when the pump is removed. Remove the pins from the pump and install them back in the cylinder block.*

50 On 2.0L (CBFA and CCTA) engines, as the water pump assembly is removed, the connector between the oil cooler and water pump should remain in the oil cooler.

51 On 2.0L (CBFA and CCTA) engines, remove the connector from the oil cooler and replace both O-rings on the connector, then install the connector back into the oil cooler.

52 Clean the surface of the cylinder block and new water pump, then install the seal to the water pump housing.

53 On 2.0L (CBFA and CCTA) engines, slide the water pump onto the connector piece in the oil cooler then onto the centering pins.

Note: *Make sure the water pump housing is seated on the centering pins; use a small mirror if necessary.*

54 On 2.0L (CBFA and CCTA) engines, the manufacturer recommends replacing the drive gear seal whenever the drive gear is removed, as follows:

a) Note how the seal is installed - the new one must be installed to the same depth and facing the same way. Carefully pry the oil seal out of the cylinder block with a seal puller or a small screwdriver. Be very careful not to scratch the balance shaft! Wrap electrician's tape around the tip of the screwdriver to avoid damage to the balance shaft.

b) Apply clean engine oil or multi-purpose grease to the outer edge of the new seal and balance shaft, then install it over the balance shaft and in the cylinder block with the lip (spring side) facing IN. Drive the seal into place with a seal driver or a large socket and a hammer. Make sure the seal enters the bore squarely and stop when the front face is at the proper depth.

55 Install the water pump and fasteners. Then tighten in sequence **(see illustrations)** to the torque listed in this Chapter's Specifications.

Note: *On 2.0L (CBFA, CCTA) engines, install the water pump without the engine coolant temperature sensor. The engine coolant temperature sensor is slightly in the way of the water pump making it difficult to install the water pump.*

56 Place the water pump (toothed) belt over the water pump pulley, then place the drive gear at the bottom of the belt and install a new drive gear bolt. Counter hold the crankshaft sprocket and tighten it to the torque listed in this Chapter's Specifications.

Note: *The collar on the new drive gear bolt faces the transaxle.*

57 Install the water pump (toothed) belt cover and fasteners, then tighten the fasteners to the torque listed in this Chapter's Specifications.

58 Installation is the reverse of removal.

59 Refill and bleed the cooling system (see Chapter 1). Start the engine and check for the proper coolant level and the water pump and hoses for leaks.

3-16   **Chapter 3   Cooling, heating and air conditioning systems**

7.70 Support the engine using an appropriate engine support fixture - typical installation shown

7.82 Disconnect the electrical connector to the charge air cooling pump - 1.4L engines

2.5L (CBTA, CBUA) five-cylinder engines

60   Remove the battery and battery tray (see Chapter 5).
61   Remove the drivebelt (see Chapter 1).
62   Remove the exhaust pipe-to-exhaust manifold fasteners and secure the exhaust pipe to the side.
63   Remove the pendulum support (see Chapter 2B, Section 15).
64   Remove the air inlet pipe between the throttle body and the air filter housing (see Chapter 4).
65   On manual transaxles, remove the shift levers (see Chapter 7A).
66   On automatic transaxles, disconnect the selector lever cable (see Chapter 7B) and remove the Transmission Control Module (TCM) (see Chapter 7B).
67   On automatic transaxles, remove the TCM bracket fasteners and secure the bracket assembly out of the way. It's not necessary to remove the power steering reservoir from the bracket. Be careful that the power steering lines are not kinked.
68   Remove the cowl cower and grille (see Chapter 11).
69   Remove the transport strap fasteners and remove the strap from the cylinder head.

70   Install an engine support fixture **(see illustration)** and slightly raise the engine.
71   Remove the windshield wiper reservoir and coolant expansion tank fasteners, then move both tanks to the side.
72   Remove the engine mount (see Chapter 2B).
73   Loosen the transmission mount bolts and slide the engine as far back as possible.
74   Remove the water pump mounting bolts and carefully maneuver the pump out of the vehicle (at an angle).
75   Install the water pump into the engine block with the sealed plug facing downwards.
76   Install the water pump bolts and tighten them to the torque listed in this Chapter's Specifications.
77   The remainder of the installation procedure is the reverse of removal. Refill and bleed the cooling system (see Chapter 1).
78   Start the engine and check for the proper coolant level, and check the water pump and hoses for leaks.

Electric coolant pump

Warning: *Wait until the engine is completely cool before starting this procedure.*
Note: *The electric coolant pump on 1.4L engines is called the charge air cooling pump. On 1.8L (CPKA, CPRA) and 2.0L (CBFA, CCTA, CPLA, CPPA) engines it is called*

an after-run pump. On 1.8L (CXBA, CXBB, CNSA, CNSB) and 2.0L (CXCA, CXCB, CNTA, CYFB, DJJA) engines it is called a coolant heater support pump.

79   Raise the vehicle and support it securely on jackstands.
80   Remove the lower splash shield below the engine (see Chapter 1, Section 6).
81   Drain the cooling system (see Chapter 1). If the coolant is new, save it and reuse it. If it is to be replaced, see Section 1 for cautions about proper handling of used antifreeze.

1.4L (CZTA, DGXA) engines

82   Working under the vehicle, disconnect the electrical connector to the charge air cooling pump **(see illustration)**.
83   Remove the inlet and outlet hoses from the charge air coolant pump, then remove the mounting bolt from the top of the pump **(see illustration)**.
84   Remove the insulator mounting bracket and coolant pump as an assembly **(see illustration)**.
85   Remove the charge air coolant pump-to-insulator mounting bracket fasteners **(see illustration)** and separate the pump from the bracket.
86   Installation is the reverse of removal. Refill and bleed the cooling system (see Chapter 1).

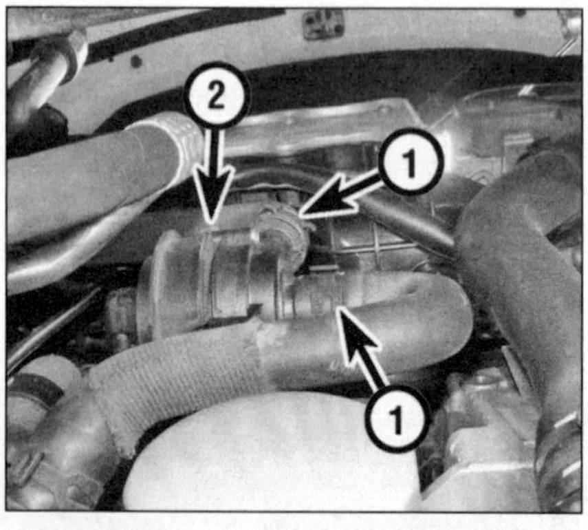

7.83 Charge air coolant pump details - 1.4L engines

1   Inlet and outlet hoses
2   Insulator mounting bracket bolt

7.84 Lift the insulator mounting bracket up from the bottom, then slide it off the post

7.85 Remove the charge air cooler pump fasteners - 1.4L engines

Chapter 3 Cooling, heating and air conditioning systems   3-17

2.0L (CBFA, CCTA) engines

87 If equipped with a noise generator, open the locking ring on the charge air pipe, disconnect the electrical connector from the charge air pipe, then remove the mounting fastener and the charge air pipe.
88 Remove the inlet and outlet hoses from the after-run coolant pump.
89 Disconnect the electrical connector from the pump, then remove the retaining bracket bolt and detach the after-run coolant pump from the engine.
90 Installation is the reverse of removal. Refill and bleed the cooling system (see Chapter 1).

1.8L (CPKA, CPRA) and 2.0L (CPLA, CPPA) engines

91 Remove the engine cover (see Chapter 1).
92 Remove the air filter housing (see Chapter 4).
93 Disconnect the charge air pressure sensor connector (see Chapter 6), then remove the charge air cooler air inlet pipe fasteners and pipe.
94 Lift the clip securing the hose connector to the thermostat housing and disconnect the hose connector. Be prepared for some coolant to spill as the hoses are removed.
95 Disconnect the electrical connector from the pump, then remove the retaining bracket nut and detach the after-run coolant pump from the engine.
96 Installation is the reverse of removal. Refill and bleed the cooling system (see Chapter 1).

1.8L (CNSA, CNSB) and 2.0L CXCA, CXCB, CNTA, CYFB, DJJA) engines

97 Loosen the charge air pipe clamps and remove the pipe.
98 Open the heat protective sleeve around the pump.
99 Disconnect the electrical connector from the pump, then remove the retaining bracket bolts and detach the coolant heater support pump from the engine.
100 Installation is the reverse of removal. Refill and bleed the cooling system (see Chapter 1).

8 Coolant temperature gauge sending unit - check and replacement

Check

Warning: *Wait until the engine is completely cool before beginning this procedure.*

1 The coolant temperature indicator system consists of the temperature gauge, a sensor mounted on the engine or lower radiator hose and the vehicle's main computer. The Engine Coolant Temperature (ECT) sensor provides a signal to the Powertrain Control Module (PCM) (see Chapter 6). The PCM uses this signal to control the temperature gauge.
2 If the temperature gauge goes above normal and begins to read hot, check the

9.3 Blower motor details
1 Blower motor
2 Blower motor electrical connector
3 Locking tab
4 Locking tab stop

9.6 Push the locking tab inwards towards the connector to release the retaining tab

coolant level in the system (see Chapter 1). Also, refer to the *Troubleshooting* Section before assuming that the temperature indicator is faulty.
3 Start the engine and monitor it while it warms up for 10 minutes. If the gauge has not moved from the cold position, check the wiring harness connections going to the instrument cluster and the sensor.
4 If there is a problem with the ECT sensor, it is likely that the CHECK ENGINE light will come on and the sensor will have to be replaced or the circuit will have to be repaired (see Chapter 6).

Replacement

5 Refer to Chapter 6 for the engine coolant temperature sensor replacement procedure.

9 Blower motor and blower motor resistor - removal and installation

Warning: *These models have airbags. Always disable the airbag system before working in the vicinity of any airbag system component to avoid the possibility of accidental deployment of the airbag(s), which could cause personal injury (see Chapter 12).*

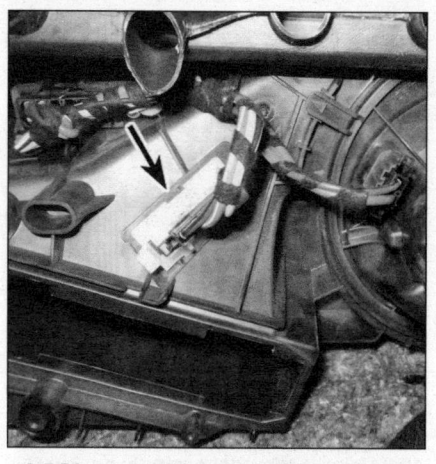

9.5 Disconnect the electrical connector from the resistor

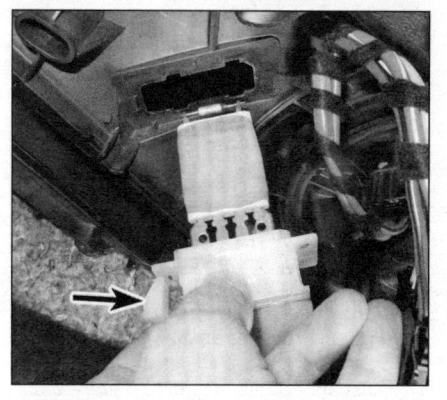

9.7 Depress the tab inward and remove the blower motor resistor from the housing

Note: *The blower control module is an integral part of the blower motor and can't be replaced separately.*

1 Remove the trim panel fasteners and trim panel from under the glove box (see Chapter 1, Section 15).

Blower motor

2 Disconnect the electrical connector from the blower motor **(see illustration 9.3)**.
3 Pull the locking tab down on the blower motor downwards, rotate the blower motor assembly counterclockwise and withdraw the blower motor **(see illustration)**.
Note: *Some models may have a retaining screw that will have to be removed before the housing will rotate.*
4 Installation is the reverse of removal; rotate the blower housing until it contacts the stop.

Blower motor resistor

5 Disconnect the electrical connector from the blower motor resistor **(see illustration)**.
6 Push the locking tab towards the connector terminals to release the resistor **(see illustration)**.
7 Remove the blower motor resistor from the heater and air conditioning housing **(see illustration)**.
8 Installation is the reverse of removal.

3-18   **Chapter 3   Cooling, heating and air conditioning systems**

10.2 Using a trim tool carefully pry out the control assembly trim panel

10.3 Heater and air conditioning control assembly screw locations

10.4 Unplug all the electrical connectors from the control assembly

10   Heater and air conditioning control assembly and flexible shafts - removal and installation

Warning: *These models have airbags. Always disable the airbag system before working in the vicinity of any airbag system component to avoid the possibility of accidental deployment of the airbag(s), which could cause personal injury (see Chapter 12).*

Caution: *If the battery is disconnected, several systems must be re-learned before they will work properly (see Chapter 5, Section 3).*

1   Disconnect the negative battery cable (see Chapter 5).

Heater and air conditioning control assembly

2   Using a trim tool, carefully pry out the control assembly trim panel starting along the bottom edge of the panel, then the sides, until the panel can be removed **(see illustration)**.
3   Remove the control assembly screws and pull the control assembly out from the instrument panel **(see illustration)**.
4   Disconnect the electrical connectors from the rear of the control head **(see illustration)**.
5   The remainder of installation is the reverse of removal.

Flexible shafts

6   Remove the glove box (see Chapter 11).
7   Remove the instrument panel support fasteners **(see illustration)**.
8   Depress the large retaining tab and remove the instrument panel support from behind the instrument panel **(see illustration)**.
Note: *The retaining tab must be depressed before the support can be installed.*
9   Looking through the right side of the instrument panel, where the instrument panel was removed, rotate the air distribution knob until the retainer for the flexible shaft is visible in the set-up gear **(see illustration)**.
10   Remove the heater and air conditioning control assembly (see Steps 2 through 4).

11   Depress the cable retaining clip and separate the cable from the control housing **(see illustration)**.
12   Carefully pry each side of the control housing out of the instrument panel **(see illustration)**.
13   Use a small screwdriver to lift the retainer and disconnect the flexible shaft from the gear **(see illustration)**.
14   Depress the retainer and disconnect the flexible shaft from the control housing **(see illustration)**.
15   The remainder of installation is the reverse of removal.

11   Heater core - removal and installation

Warning: *Wait until the engine is completely cool before beginning this procedure.*
Warning: *These models have airbags. Always disable the airbag system before working in the vicinity of any airbag system component to avoid the possibility of accidental deployment*

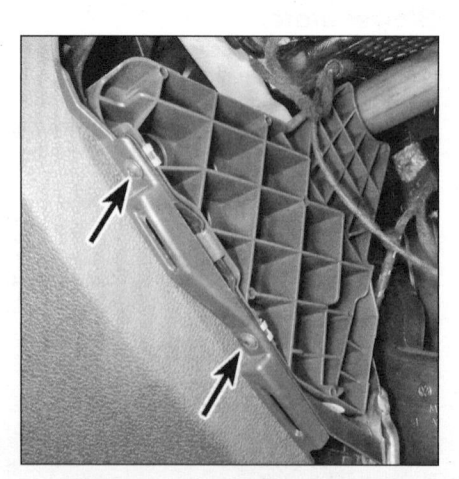

10.7 Remove the instrument panel support fasteners

10.8 Slide the instrument panel support out from behind the instrument panel

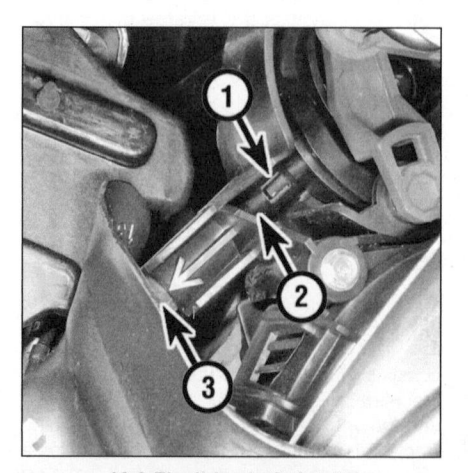

10.9 Flexible shaft details

1   *Flexible shaft retainer*
2   *Set-up gear*
3   *Flexible shaft*

Chapter 3 Cooling, heating and air conditioning systems   3-19

10.11 Depress the cable retaining clip and separate the cable from the control housing

10.12 Carefully pry each side of the control housing out of the instrument panel

of the airbag(s), which could cause personal injury (see Chapter 12).

Caution: *If the battery is disconnected, several systems must be re-learned before they will work properly (see Chapter 5, Section 3).*

1   Remove the engine cover (see Chapter 1, Section 7).
2   Disconnect the cable from the negative terminal of the battery (see Chapter 5).
3   Drain the cooling system (see Chapter 1). If the coolant is new, save it and reuse it. If it is to be replaced, see Section 1 for cautions about proper handling of used antifreeze.
4   Remove the battery and battery tray (see Chapter 5).
5   Mark the heater hoses, then remove the locking clips and disconnect the hoses at the heater core inlet and outlet on the engine side of the firewall **(see illustration)**.

Note: *The heater core is designed for coolant flow in one specific direction. The coolant hoses must be connected to their original locations or the system will not work properly.*

6   Using a suction gun, remove as much of the coolant as possible from the heater core tubes and plug the open tubes.
7   Remove the center console lower side trim panels (see Chapter 11).
8   Remove the driver's knee bolster and bolster bracket (see Chapter 11).
9   On models equipped with an auxiliary heating element, disconnect the ground wire retaining nut and move the ground wire out of the way. Locate the heating element connector, press the connector inwards and slide the retainer up to disconnect the connector from the heating element. Once the connector is disconnected remove the heating element mounting bolts and remove the element.

Caution: *If the heating element is hot it can burn skin or heat metal enough to burn skin.*

10   Place plastic over the carpeting, then lay rags over the plastic to prevent any coolant from getting on the carpet.
11   Working on the upper line, lift the tab on the center of the U-shaped clamp, then slide the clamp down and off and remove the heater core coolant tube from the heater core. Be prepared for coolant to spill out of the line.
12   Working on the lower line, loosen the screw-type clamp and remove the heater core coolant tube from the heater core. Be prepared for coolant to spill out of the line.
13   Remove the heater core cover fasteners, then push the locking tabs apart and remove the cover.
14   Once the remaining coolant has drained, pull the heater core from the housing.
15   Installation is the reverse of removal.

Note: *When reinstalling the heater core, make sure any original insulating/sealing materials are in place around the heater core pipes and around the core.*

16   Refill and bleed the cooling system (see Chapter 1).
17   Start the engine and check for proper operation.

10.13 Use a small screwdriver to lift the retainer and disconnect the flexible shaft from the gear

10.14 Depress the retainer and disconnect the flexible shaft from the control housing

11.5 Remove the heater hose locking clips

3-20   **Chapter 3   Cooling, heating and air conditioning systems**

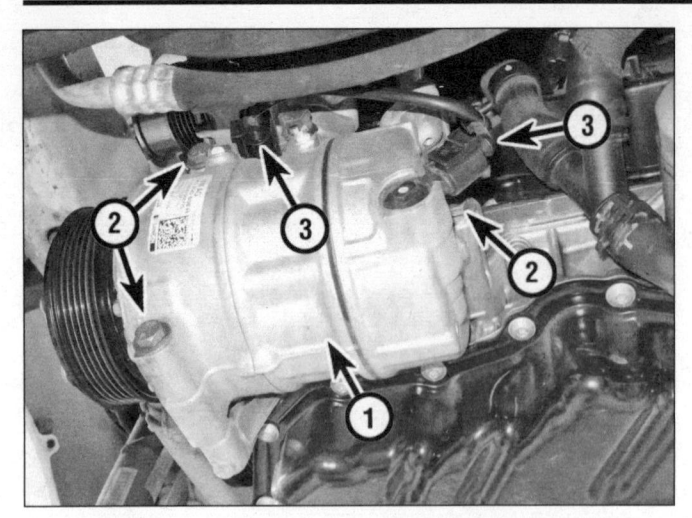

13.11 Air conditioning compressor mounting details - 1.4L (CZTA) four-cylinder engine shown, others similar

1   Air conditioning compressor   3   Electrical connectors
2   Mounting bolts

13.12 Air conditioning compressor suction and discharge line bolt locations

12   Heater and air conditioning housing - removal and installation

Warning: *Wait until the engine is completely cool before beginning this procedure.*
Warning: *The air conditioning system is under high pressure. DO NOT loosen any fittings or remove any components until after the system has been discharged. Air conditioning refrigerant must be properly discharged into an EPA-approved container at a dealership service department or an automotive air conditioning repair facility. Always wear eye protection when disconnecting air conditioning system fittings.*
Warning: *These models have airbags. Always disable the airbag system before working in the vicinity of any airbag system component to avoid the possibility of accidental deployment of the airbag(s), which could cause personal injury (see Chapter 12).*
Caution: *If the battery is disconnected, several systems must be re-learned before they will work properly (see Chapter 5, Section 3).*
Note: *This is a very difficult procedure for the home mechanic, involving numerous hard-to-find fasteners, clips and electrical connectors.*

1   Have the air conditioning system discharged (see **Warning** above).
2   Remove the engine covers. Disconnect and remove the battery (see Chapter 5).
3   Drain the cooling system (see Chapter 1). Refer to the coolant **Warning** in Section 1.
4   Mark, then disconnect the hoses at the heater core inlet and outlet on the engine side of the firewall **(see illustration 11.5)**.
Note: *If the hoses are stuck to the pipes, cut them off and replace them with new ones upon installation.*
5   On models equipped with an auxiliary heating element, disconnect the ground wire retaining nut and move the ground wire out of the way. Locate the heating element connector, press the connector inwards and slide the retainer up to disconnect the connector from the heating element.
6   Disconnect the heater core tubes from the heater core (see Section 11).
7   Remove the air conditioning line fastener from the evaporator core fitting at the firewall.
8   Remove the instrument panel (see Chapter 11).
9   Remove the passenger's side airbag (see Chapter 12).
10   Remove the center console support brace. Remove the upper and lower air ducts from the heating/air conditioning unit.
11   From the engine compartment side, disconnect the electrical connection fasteners and heater HVAC housing fastener.
12   Disconnect the electrical connectors at the right side of the instrument panel.
13   Remove the fasteners from the right side of the cross beam support, unclip the main harness from the back side of the cross beam support, then pull the beam forward.
14   With the help of an assistant, remove the cross beam support from the vehicle.
15   Carefully remove the housing assembly from the vehicle.
16   Installation is the reverse of removal.
17   Refill and bleed the cooling system (see Chapter 1).
18   Have the system evacuated, recharged and leak tested by the shop that discharged it.
19   Check for proper operation.

13   Air conditioning compressor - removal and installation

Warning: *The air conditioning system is under high pressure. DO NOT loosen any fittings or remove any components until after the system has been discharged. Air conditioning refrigerant must be properly discharged into an EPA-approved container at a dealership service department or an automotive air conditioning repair facility. Always wear eye protection when disconnecting air conditioning system fittings.*
Caution: *If the battery is disconnected, several systems must be re-learned before they will work properly (see Chapter 5, Section 3).*

1   The refrigerant desiccant cartridge (see Section 14) should be replaced whenever the compressor is replaced.
2   Have the air conditioning system discharged (see **Warning** above).
3   Remove the engine cover (see Chapter 1, Section 7). Disconnect the negative battery cable (see Chapter 5).
4   Loosen the right front wheel bolts, then raise the vehicle on the hoist. Remove the under-vehicle splash shield (see Chapter 1, Section 6).
5   Depending on the engine you have it may be easier to remove the right inner fender splash shield (see Chapter 11, Section 15).
6   Remove the drivebelt (see Chapter 1).
7   Remove the air filter housing (see Chapter 4).
8   On turbocharged models, it may be necessary to remove the charge air cooler pipes (see Chapter 4, Section 15).
9   Clean the compressor thoroughly around the refrigerant line fittings.
10   Disconnect the electrical connectors from the top and back of the air conditioning compressor.
11   Remove the compressor mounting bolts **(see illustration)**. Detach the compressor from the mounting bracket and lower the compressor from the engine compartment.
12   Disconnect the suction and discharge lines from the compressor **(see illustration)**. Plug the open fittings to prevent the entry of dirt and moisture, and discard the seals between the plates and compressor

Chapter 3 Cooling, heating and air conditioning systems   3-21

14.4a Depress the air duct upper retaining tabs . . .

14.4b . . . the lower retaining tabs . . .

14.4c . . . then remove the air duct

13 If a new compressor is being installed, pour the oil from the old compressor into a graduated container and add that exact amount of new refrigerant oil to the new compressor. Also follow any directions included with the new compressor.
Note: *Some replacement compressors come with refrigerant oil in them. Follow the directions with the compressor regarding the draining of excess oil prior to installation.*
Warning: *The oil used must be labeled as compatible with R-134a or R1234yf refrigerant systems.*
14 Installation is the reverse of removal. When installing the line fitting bolt to the compressor, use new seals lubricated with clean refrigerant oil, and tighten the bolt securely.
Note: *Make sure the alignment sleeves in the compressor bolt holes are seated correctly and the contact surfaces are clean and flat.*
15 Have the system evacuated, recharged and leak tested by the shop that discharged it.

14 Air conditioning refrigerant desiccant cartridge - replacement

Warning: *The air conditioning system is under high pressure. DO NOT loosen any fittings or remove any components until after the system has been discharged. Air conditioning refrigerant must be properly discharged into an EPA-approved container at a dealer service department or an automotive air conditioning repair facility. Always wear eye protection when servicing the air conditioning system.*
Caution: *If the battery is disconnected, several systems must be re-learned before they will work properly (see Chapter 5, Section 3).*
Note: *The receiver-dryer cartridge can't be disassembled to replace the desiccant bag.*
1 Have the air conditioning system discharged by a dealer service department or by an automotive air conditioning shop before proceeding (see **Warning** above).

2 Raise the front of the vehicle and support it securely on jackstands.
3 Remove the front bumper cover (see Chapter 11).
4 Depress the air duct retaining tabs, then remove the air duct from the side of the condenser (**see illustrations**).
5 Remove the cartridge upper fastener (**see illustration**).
6 Remove the cartridge lower fasteners (**see illustrations**).
7 Slide the dryer cartridge up and out of the vehicle.
8 Slide the cartridge down in the channel on the condenser, then install the fasteners and tighten them securely.
9 Have the system evacuated, recharged and leak tested by the shop that discharged it.

14.5 Remove the cartridge upper fastener

14.6a Remove the cartridge lower fastener (1) then remove the lower mounting bracket fastener (2)

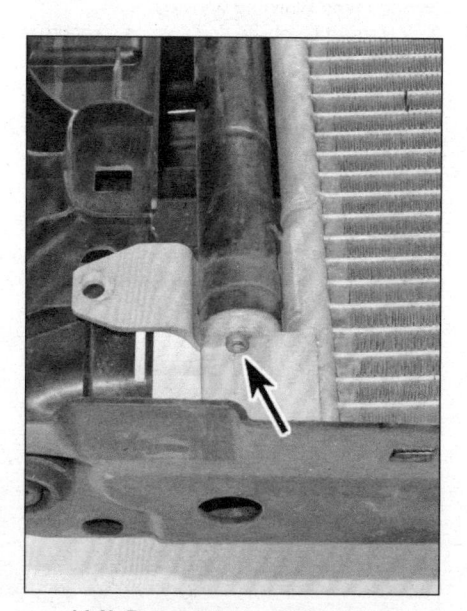

14.6b Remove the cartridge lower fastener from the engine compartment side

3-22   **Chapter 3 Cooling, heating and air conditioning systems**

15.9 Location of the condenser suction and discharge line fasteners

15.10 Slide the charge air system circuit cooler back until the cooler upper mounts (1) are free from the mounting slots (2) - left side shown, right side identical

15 Air conditioning condenser - removal and installation

Warning: *The air conditioning system is under high pressure. DO NOT loosen any fittings or remove any components until after the system has been discharged. Air conditioning refrigerant must be properly discharged into an EPA-approved container at a dealership service department or an automotive air conditioning repair facility. Always wear eye protection when disconnecting air conditioning system fittings.*

Warning: *The engine must be completely cool before beginning this procedure.*

1   Have the air conditioning system discharged (see **Warning** above).
2   Raise the vehicle and support it securely on jackstands.
3   Remove the under-vehicle splash shield (if equipped).
4   Drain the coolant from the radiator (see Chapter 1). If the coolant is new, save it and reuse it. If it is to be replaced, see Section 1 for cautions about proper handling of used antifreeze.
5   Remove the front bumper assembly (see Chapter 11).
6   Remove the cooling fan shroud (see Section 5).
7   Remove the radiator (see Section 6).
8   Remove the air duct retaining tabs, then remove the air ducts from the side of the condenser (see Section 14).
9   Remove the suction and discharge line fasteners, then disconnect the lines from the right side of the condenser **(see illustration)**. Plug the open fittings to prevent the entry of dirt and moisture, and discard the seals between the plates and condenser.
10   If equipped with a charge air system circuit cooler, slide the charge air system circuit cooler back until the cooler upper mounts are free **(see illustration)**, then lift the cooler out of the lower mounts and move it towards the engine and secure it out of the way.
11   Remove the condenser fasteners **(see illustration)**, then tilt the top of the condenser to the rear and remove the condenser.

12   Installation is the reverse of removal. Always use new O-rings on air conditioning system fittings. If you are replacing the condenser with a new one, add fresh refrigerant oil to the new unit following the directions included with the new condenser (the oil must be R-134a or R1234yf compatible, depending on the system your vehicle uses).
13   Refill and bleed the system, then run the engine and check carefully for leaks (see Chapter 1).
14   Have the system evacuated, recharged and leak tested by the shop that discharged it.

16 Air conditioning evaporator core - removal and installation

Warning: *The air conditioning system is under high pressure. DO NOT loosen any fittings or remove any components until after the system has been discharged. Air conditioning refrigerant must be properly discharged into an EPA-approved container at a dealership service department or an automotive air conditioning repair facility. Always wear eye protection when disconnecting air conditioning system fittings.*

1   Have the air conditioning system discharged by a dealership service department or an automotive air conditioning facility.
2   Remove the heater and air conditioner housing (see Section 12).
3   Set the housing on a clean working surface.
4   Remove the evaporator core fasteners, separate the housing and carefully remove the evaporator core from the housing.
5   Installation is the reverse of removal.

Note: *When reinstalling the evaporator core, make sure any original insulating/sealing materials are in place around the evaporator core pipes and around the core.*

6   Installation is the reverse of removal. Always use new O-rings on air conditioning system fittings. If you are replacing the

15.11 Condenser mounting bolt locations

Chapter 3 Cooling, heating and air conditioning systems

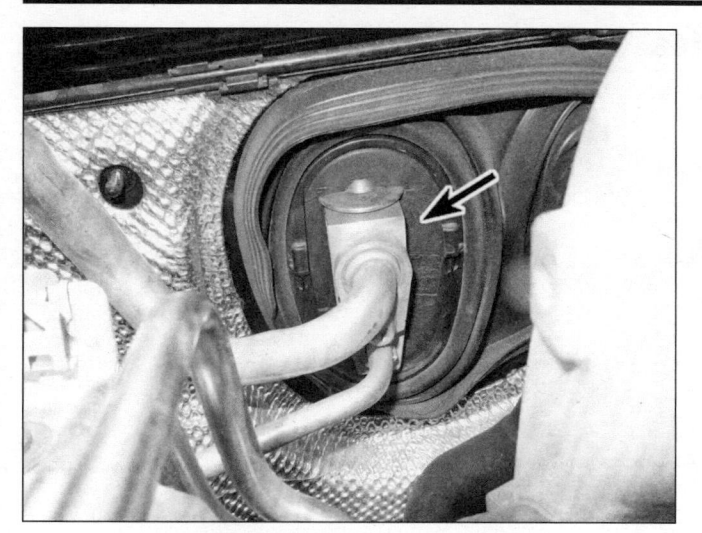

17.2 Typical expansion valve location

18.2 Disconnect the electrical connector from the sensor

evaporator core with a new one, add fresh refrigerant oil to the new unit following the directions included with the new evaporator (the oil must be R-134a or R1234yf compatible, depending on which refrigerant your vehicle uses).

7 Have the system evacuated, recharged and leak tested by the shop that discharged it.

17 Air conditioning expansion valve - general information

Warning: *The air conditioning system is under high pressure. DO NOT loosen any hose fittings or remove any components until the system has been discharged. Air conditioning refrigerant must be properly discharged into an EPA-approved recovery/recycling unit by a dealer service department or an automotive air conditioning repair facility. Always wear eye protection when disconnecting air conditioning system fittings.*

1 There are several ways that air conditioning systems convert the hot high-pressure liquid refrigerant from the compressor to cold lower-pressure vapor. The conversion takes place at the air conditioning evaporator; the evaporator is chilled as the refrigerant passes through, cooling the airflow through the evaporator for delivery to the vents. The conversion is accomplished by a sudden change in pressure. Many vehicles have a removable controlled orifice in one of the air conditioning pipes at the firewall.

2 The models covered by this manual use an expansion valve that accomplishes the same purpose as a controlled orifice **(see illustration)**. To remove the expansion valve, have the air conditioning system discharged by a licensed air conditioning technician. Remove the heat shield fasteners and heat shield, if equipped, then disconnect the refrigerant lines from the expansion valve at the firewall. Remove the bolt(s) securing the valve, then remove the valve.

3 Installation is the reverse of removal. Make sure to replace all the O-ring with new ones.

4 Have the system evacuated, recharged and leak tested by the shop that discharged it.

18 Air conditioning high-pressure sensor - replacement

Warning: *There is a check valve installed in the hard line that the sensor threads into - there is no need to discharge the refrigerant to replace the sensor. However, a small amount of refrigerant might escape when the sensor is unscrewed. Wear eye protection when performing this procedure.*

1 The pressure sensor is located on the high-pressure hard line from the condenser to the air conditioning thermostatic expansion valve (TXV).
Note: *On some models, the high pressure sensor is installed near the condenser but is removed in the same way.*

2 Unplug the electrical connector from the sensor **(see illustration)**.

3 Unscrew the pressure sensor.

4 Replace the O-ring with a new one. Before installing it, lubricate it with clean refrigerant oil of the correct type.

5 Screw the new sensor into place until hand-tight, then tighten it securely.

6 Reconnect the electrical connector.

Notes

Chapter 4
Fuel and exhaust systems

Contents

	Section		Section
Air filter housing - removal and installation	11	General Information	1
Exhaust system servicing - general information	14	High-pressure fuel pump (Direct Injection models) - removal and installation	9
Fuel lines and fittings - general information and disconnection	6	Throttle body/control module - removal and installation	12
Fuel pressure - check	4	Transfer pump/fuel level sender - replacement	8
Fuel pressure relief procedure	3	Troubleshooting	2
Fuel pump module/fuel level sender - removal and installation	7	Turbocharger/exhaust manifold and charge air cooler - check and replacement	15
Fuel rail and injectors - removal and installation	13		
Fuel system - bleeding (five-cylinder models only)	5		
Fuel tank - removal and installation	10		

Specifications

Fuel system pressure
At idle
- 2.5L engines .... 50 to 65 psi (3.5 to 4.5 bar)
- 2015 and earlier non-turbocharged 2.0L models .... 55 to 60 psi (3.8 to 4.2 bar)
- All other models .... 58 to 101.5 psi (4.0 to 7.0 bar)*

Holding pressure, after 10 minutes .... No more than 14.5 psi (1 bar) less than idle low end specification

On vehicles with direct injection, this specification is for the in tank fuel supply pump, not the high pressure pump.
Warning: *Never attempt to measure fuel pressure on the high-pressure side of a Direct Injection system.*

Torque specifications   Ft-lbs (unless otherwise indicated)   Nm
Note: *One foot-pound (ft-lb) of torque is equivalent to 12 inch-pounds (in-lbs) of torque. Torque values below approximately 15 ft-lbs are expressed in inch-pounds, because most foot-pound torque wrenches are not accurate at these smaller values.*

Turbocharger assembly
Turbocharger-to-cylinder head nuts
- 1.4L engines .... 120 in-lbs   14
- 1.8L and 2.0L engines (except 2.0L CBFA an CCTA) 18.5   25

Exhaust manifold-to-cylinder head nuts (2.0L CBFA an CCTA)
- Step 1 .... 44 in-lbs   5
- Step 2 .... 106 in-lbs   12
- Step 3 .... 142 in-lbs   16
- Step 4 .... 18.5   25

Oil supply line banjo fasteners 24   33
Oil supply and return line flange fasteners (both ends) .... 80 in-lbs   9
Coolant supply and return line banjo fasteners (both ends) ... 28   38
Coolant supply and return line flange fasteners (both ends) .. 80 in-lbs   9

Throttle body fasteners
- Except 2.0L CCTA, CBFA, CBPA engines .... 58 in-lbs   7
- 2.0L CBPA (non-turbo) engines .... 80 in-lbs   9
- 2.0L CCTA, CBFA engines .... 44 in-lbs   5
- 2.5L Five-cylinder models .... 57 in-lbs   6.5

Direct-injection system
High-pressure fuel pump mounting fasteners
- Step 1
  - M6 bolt .... 78 in-lbs   8
  - M8 bolt 15   20
- Step 2 .... Tighten an additional 90 degrees

High-pressure fuel line fittings
- 1.4L engines
  - Step 1 .... 142 in-lbs   16
  - Step 2 .... Tighten an additional 45 degrees
- 2.0L CCTA, CBFA engines 13   18
- Except 2.0L CCTA, CBFA, engines 18   27

4-1

Chapter 4 Fuel and exhaust systems

1 General Information

Fuel system warnings

Warning: *Gasoline is extremely flammable and repairing fuel system components can be dangerous. Consider your automotive repair knowledge and experience before attempting repairs which may be better suited for a professional mechanic.*

a) Don't smoke or allow open flames or bare light bulbs near the work area
b) Don't work in a garage with a gas-type appliance (water heater, clothes dryer)
c) Use fuel-resistant gloves. If any fuel spills on your skin, wash it off immediately with soap and water
d) Clean up spills immediately
e) Do not store fuel-soaked rags where they could ignite
f) Prior to disconnecting any fuel line, you must relieve the fuel pressure (see Section 3)
g) Wear safety glasses
h) Have a proper fire extinguisher on hand

Fuel system

Warning: *The fuel delivery systems on direct-injection models are equipped with both a low-pressure system and a second, high-pressure system that operates at extremely high pressure. At idle, the fuel pressure in the fuel rail is about 725 psi; under certain operating conditions such as hard acceleration, the system pressure can reach 1700 psi.*

1   The multi-port fuel system consists of the fuel tank, electric fuel pump/fuel level sending unit (located in the fuel tank), fuel rail and fuel injectors. Multi-port fuel injection uses timed impulses to inject the fuel directly into the intake port of each cylinder. The Powertrain Control Module (PCM) controls the fuel pump and injectors. The PCM monitors various engine parameters and delivers the exact amount of fuel required into the intake ports.

2   The direct-injection fuel system consists of the fuel tank, electric fuel pump/fuel level sending unit (located in the fuel tank), the high-pressure fuel pump, fuel rail and fuel injectors. Direct-injection uses high-pressure timed impulses to inject the fuel directly into the combustion chamber of each cylinder. The Powertrain Control Module (PCM) controls the in-tank fuel pump and injectors, the high-pressure fuel pump is mounted to the end of the cylinder head and driven by the intake camshaft. The PCM monitors various engine parameters and delivers the exact amount of fuel required into the intake ports.

3   Fuel is circulated from the in-tank fuel pump to the fuel rail (or high-pressure fuel) pump through fuel lines running along the underside of the vehicle. Various sections of the fuel line are either rigid metal or nylon, or flexible fuel hose. The various sections of the fuel hose are connected either by quick-connect fittings or threaded metal fittings. Direct-injection models use a steel fuel line between the high-pressure fuel pump and the fuel rail.

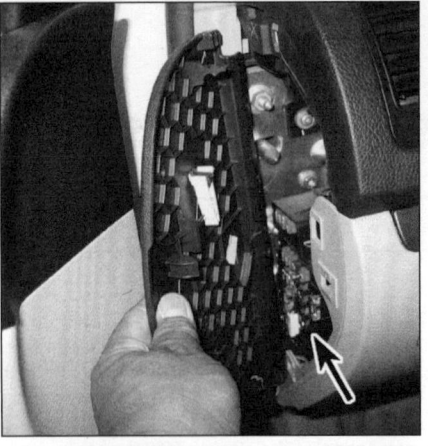

2.2a The fuel pump fuse is located in the fuse box at the left end of the instrument panel (be sure to check the guide on the end panel or in your owner's manual to locate the fuse for your particular model)

4   On AWD models, a separate transfer pump and fuel level sender are part of the fuel system. The transfer pump keeps the fuel level even between the two sumps of the fuel tank. The fuel level sender monitors the fuel level to ensure the sumps of the fuel tank are filled equally.

Exhaust system

5   The exhaust system consists of the exhaust manifold(s), catalytic converter(s), muffler(s), tailpipe and all connecting pipes, flanges and clamps. The catalytic converters are an emission control device added to the exhaust system to reduce pollutants.

2 Troubleshooting

Fuel pump

Note: *The following information applies to the in-tank fuel pump only.*

Warning: *The fuel delivery systems on direct-injection models are equipped with both a low-pressure system and a second, high-pressure system that operates at extremely high pressure. At idle, the fuel pressure in the fuel rail is about 725 psi; under certain operating conditions such as hard acceleration, the system pressure can reach 1700 psi.*

1   The fuel pump is located inside the fuel tank. Sit inside the vehicle with the windows closed, turn the ignition key to On (not Start) and listen for the sound of the fuel pump as it's briefly activated. You will only hear the sound for a second or two, but that sound tells you that the pump is working. Alternatively, have an assistant listen at the fuel filler cap.

2   If the pump does not come on, check the fuel pump fuse **(see illustration)** and on five-cylinder models, the fuel pump relay **(see illustration)**. All models are equipped with a fuel pump fuse, but there is no fuel pump relay on four-cylinder models; the fuel pump circuit

2.2b On five-cylinder models, the fuel pump relay is located at the far left end of the relay panel, which is located inside the left end of the instrument panel, to the left of the steering column. Four-cylinder models do not use a fuel pump relay. Instead, they use a control module located on the cover of the fuel level sensor

on these models is controlled by the fuel pump control module (FPCM), which cannot be diagnosed without special diagnostic equipment.

Fuel injection system

Note: *The following procedure is based on the assumption that the fuel pump is working and the fuel pressure is adequate (see Section 4).*

3   Check all electrical connectors that are related to the system. Check the ground wire connections for tightness.
4   Verify that the battery is fully charged (see Chapter 5).
5   Inspect the air filter element (see Chapter 1).
6   Check all fuses related to the fuel system (see Chapter 12).
7   Check the air induction system between the throttle body and the intake manifold for air leaks. Also inspect the condition of all vacuum hoses connected to the intake manifold and to the throttle body.
8   Remove the air intake duct from the throttle body and look for dirt, carbon, varnish, or other residue in the throttle body, particularly around the throttle plate. If it's dirty, clean it with carb cleaner, a toothbrush and a clean shop towel.

Note: *For the next step, on direct-injection models, the injectors may be located under sound insulation that must be removed to access the injectors.*

9   With the engine running, place an automotive stethoscope against each injector, one at a time, and listen for a clicking sound that indicates operation **(see illustration)**.

Warning: *Stay clear of the drivebelt and any rotating or hot components.*

10   If you can hear the injectors operating, but the engine is misfiring, the electrical circuits are functioning correctly, but the injectors might be dirty or clogged. Try a com-

Chapter 4 Fuel and exhaust systems

2.9 With the engine running, place an automotive stethoscope against each injector, one at a time, and listen for a clicking sound that indicates operation

3.3 Electrical connector on the mechanical high-pressure pump - 1.4L engine shown

mercial injector cleaning product (available at auto parts stores). If cleaning the injectors doesn't help, replace the injectors.

11 If an injector is not operating (it makes no sound), disconnect the injector electrical connector and measure the resistance across the injector terminals with an ohmmeter. Compare this measurement to the other injectors. If the resistance of the non-operational injector is quite different from the other injectors, replace it.

12 If the injector is not operating, but the resistance reading is within the range of resistance of the other injectors, the PCM or the circuit between the PCM and the injector might be faulty.

3 Fuel pressure relief procedure

Warning: *Gasoline is extremely flammable. See Fuel system warnings in Section 1.*
Note: *Whenever the battery is disconnected, several systems must be re-learned before they will work properly (see Chapter 5, Section 3).*

1 Remove the fuel filler cap to relieve any pressure built-up in the fuel tank.

Direct-injection models

Warning: *The fuel delivery systems on direct-injection models are equipped with both a low-pressure system and a second, high-pressure system that operates at extremely high pressure. At idle, the fuel pressure in the fuel rail is about 725 psi; under certain operating conditions such as hard acceleration, the system pressure can reach 1700 psi. So before you remove or replace any component (mechanical high-pressure pump, high-pressure fuel lines, fuel rail and/or fuel injectors) in the high-pressure part of the fuel injection system, be sure to use the following procedure to relieve fuel pressure.*

2 Remove the fuel pump fuse. This opens the circuit to the Fuel Pump Control Module (FPCM) so that the fuel pump inside the fuel tank will not operate.

a) *Golf - Fuse No. 27 located in the interior fuse block (fuse panel c).*
b) *Jetta - Fuse No. 47 located in the interior fuse block (fuse panel c).*

3 Disconnect the electrical connector from the fuel pressure regulator valve on the mechanical high-pressure pump **(see illustration)**.

4 Start the engine and allow it to idle until it stalls, then turn the ignition key to the Off position and disconnect the cable from the negative terminal of the battery (see Chapter 5). Even though the low-pressure pump inside the fuel tank is already disabled, the engine might run briefly on the residual pressure inside the high-pressure part of the system.

5 At this point, the residual fuel pressure in the high side of the system might still be about 87 psi (6 bar). So when you crack open the first fitting, make sure that you are wearing safety goggles and that you completely surround the fitting with plenty of shop rags to catch any fuel that might spill out.

Warning: *The high-pressure line must be opened shortly after the pressure has been released or the pressure could build back up due to warming of the fuel.*

Multi-port fuel injection models

6 Remove the fuel pump fuse as described in the direct-injection procedure or the fuel pump relay **(see illustrations 2.2a and 2.2b)**.

7 Attempt to start the engine; it should immediately stall. Crank the engine several more times to ensure the fuel system has been completely relieved. Disconnect the cable from the negative terminal of the battery before working on the fuel system.

8 Cover any fuel connection to be disassembled with rags to absorb the residual fuel that may leak out. Properly dispose of the rags.

9 Remove the cap from the fuel rail bleeder and surround the fitting with rags, then slowly depress the Schrader valve with a small screwdriver and allow the residual fuel pressure to bleed off.

10 The fuel system is now depressurized. Properly dispose of the rags.

4 Fuel pressure - check

Warning: *Gasoline is extremely flammable. See Fuel system warnings in Section 1.*
Warning: *Before performing this check, review the fuel pressure values in this Chapter's Specifications and make sure your gauge is capable of reading the pressures that the system on your vehicle can produce.*
Warning: *This fuel pressure check procedure does not include the high-pressure side of turbocharged models. The fuel pressure between the mechanical high-pressure pump and the fuel rail on those models is between 725 psi (idle) to almost 1700 psi. Do not attempt to measure it.*
Caution: *On five-cylinder engine models, whenever the fuel system has been opened it must be bled (see Section 5). If the system is not bled, the catalytic converter can be damaged.*
Note: *The following procedure assumes that the fuel pump is receiving voltage and runs.*
Note: *Whenever the battery is disconnected, several systems must be re-learned before they will work properly (see Chapter 5, Section 3).*

1 Raise the vehicle and support it securely on jackstands (unless you will be using Method 2 or 3 in the next Step).

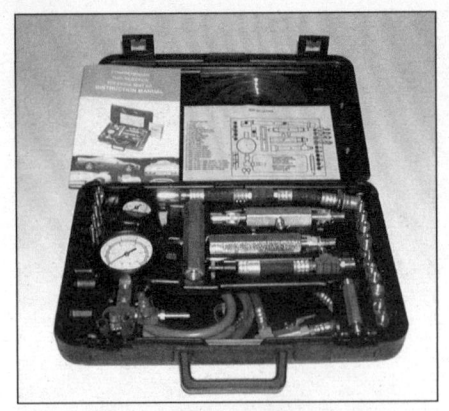

Chapter 4 Fuel and exhaust systems

4.2a This fuel pressure testing kit contains all the necessary fittings and adapters, along with the fuel pressure gauge, to test most automotive systems

4.2b Using the proper adapters, the fuel pressure gauge can be installed between the fuel filter and the fuel line

5.1 Schrader valve location (five-cylinder models only)

2 Relieve the fuel system pressure (see Section 3), then disconnect the fuel line from the outlet side of the fuel filter (see Section 6). Use an adapter to connect the fuel pressure gauge between the fuel line and the fuel filter **(see illustrations)** (Method 1), or, if equipped, at the fuel feed line connection at the fuel rail or high-pressure fuel pump (Method 2). Alternatively, if the fuel rail is equipped with a Schrader valve, the pressure gauge can be connected to it (Method 3).

3 Start the engine and allow it to idle. Note the gauge reading as soon as the pressure stabilizes, and compare it with the pressure listed in this Chapter's Specifications.

4 If the fuel pressure is not within specifications, check the following:
If the pressure is lower than specified, check for a restriction in the fuel system (kinked fuel line, plugged fuel pump inlet strainer or clogged fuel filter). If no restrictions are found, check the fuel pressure sensor.
If the pressure is higher than specified, check the fuel pressure sensor.

5 Turn off the engine. Fuel pressure should not fall more than 14.5 psi over ten (10) minutes. If it does, the problem could be a leaky fuel injector, fuel line leak, or faulty fuel pump.

6 Relieve the fuel system pressure (see Section 3), then disconnect the fuel pressure gauge.

7 Reconnect the fuel line to the fuel filter.

5 Fuel system - bleeding (five-cylinder models only)

Warning: *Gasoline is extremely flammable. See Fuel system warnings in Section 1.*
Note: *Whenever the battery is disconnected, several systems must be re-learned before they will work properly (see Chapter 5, Section 3).*

1 Remove the bleeder valve cap on the fuel rail **(see illustration)**.

2 Connect an adapter from the fuel pressure test kit **(see illustration 4.2a)** to the Schrader valve with a drain hose inserted in an appropriate container.

3 Have an assistant turn the ignition key to the "On" position for 5 to 10 seconds - fuel will flow through the adapter and drain hose into the container.

4 Once the fuel flow has little to no air bubbles clamp the drain hose off and turn the ignition key to the "Off" position.

5 Surround the Schrader valve fitting with plenty of shop rags to catch any fuel that might spill out and remove the adapter and drain hose.

6 The fuel system is now bled. Reinstall the cap and properly dispose of the rags.
Note: *This procedure might set a trouble code that will require a scan tool to clear (see Chapter 6).*

6 Fuel lines and fittings - general information and disconnection

Warning: *Gasoline is extremely flammable. See Fuel system warnings in Section 1.*
Warning: *The fuel delivery systems on direct-injection models are equipped with both a low-pressure system and a second, high-pressure system that operates at extremely high pressure. At idle, the fuel pressure in the fuel rail is about 725 psi; under certain operating conditions such as hard acceleration, the system pressure can reach 1700 psi.*
Caution: *On five-cylinder models, whenever the fuel system has been opened it must be bled (see Section 5). If the system is not bled, the catalytic converter can be damaged.*
Note: *Whenever the battery is disconnected, several systems must be re-learned before they will work properly (see Chapter 5, Section 3).*

1 Relieve the fuel pressure before servicing fuel lines or fittings (see Section 3), then disconnect the cable from the negative battery terminal (see Chapter 5) before proceeding.

2 On multi-port fuel injection models, the fuel supply line connects the fuel pump in the fuel tank to the fuel rail on the engine. On direct-injection models, the fuel supply line connects the fuel pump in the fuel tank to the high-pressure fuel pump on the engine.

3 The Evaporative Emission (EVAP) system lines connect the fuel tank to the EVAP canister and connect the canister to the intake manifold.

4 Whenever you're working under the vehicle, be sure to inspect all fuel and evaporative emission lines for leaks, kinks, dents and other damage. Always replace a damaged fuel or EVAP line immediately.

5 If you find signs of dirt in the lines during disassembly, disconnect all lines and blow them out with compressed air. Inspect the fuel strainer on the fuel pump pick-up unit for damage and deterioration.

Steel tubing

Note: *Direct-injection models use a steel fuel line between the high pressure fuel pump and the fuel rail.*

6 It is critical that the fuel lines be replaced with lines of equivalent type and specification.

7 Some steel fuel lines have threaded fittings. When loosening these fittings, hold the stationary fitting with a wrench while turning the tube nut.

Plastic tubing

8 When replacing fuel system plastic tubing, use only original equipment replacement plastic tubing.
Caution: *When removing or installing plastic fuel line tubing, be careful not to bend or twist it too much, which can damage it. Also, plastic fuel tubing is NOT heat resistant, so keep it away from excessive heat.*

Flexible hoses

9 When replacing fuel system flexible hoses, use only original equipment replacements.

10 Don't route fuel hoses (or metal lines) within four inches of the exhaust system or within ten inches of the catalytic converter. Make sure that no rubber hoses are installed directly against the vehicle, particularly in places where there is any vibration. If allowed to touch some vibrating part of the vehicle, a hose can easily become chafed and it might start leaking. A good rule of thumb is to maintain a minimum of 1/4-inch clearance around a hose (or metal line) to prevent contact with the vehicle underbody.

Chapter 4 Fuel and exhaust systems

4-5

Disconnecting Fuel Line Fittings

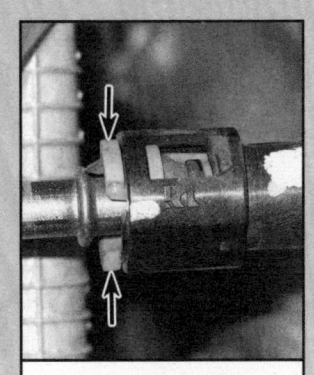

Two-tab type fitting; depress both tabs with your fingers, then pull the fuel line and the fitting apart

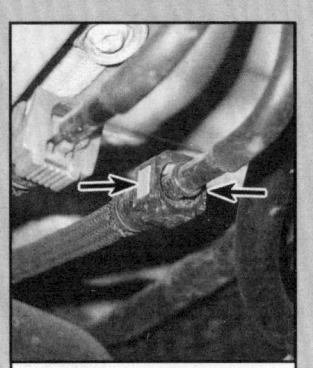

On this type of fitting, depress the two buttons on opposite sides of the fitting, then pull it off the fuel line

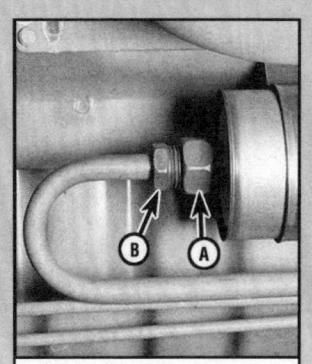

Threaded fuel line fitting; hold the stationary portion of the line or component (A) while loosening the tube nut (B) with a flare-nut wrench

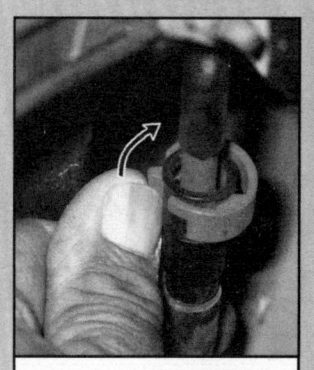

Plastic collar-type fitting; rotate the outer part of the fitting

Metal collar quick-connect fitting; pull the end of the retainer off the fuel line and disengage the other end from the female side of the fitting . . .

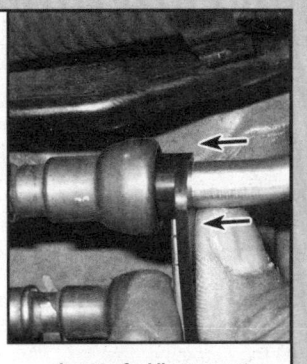

. . . insert a fuel line separator tool into the female side of the fitting, push it into the fitting and pull the fuel line off the pipe

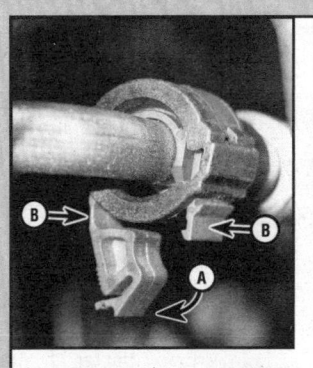

Some fittings are secured by lock tabs. Release the lock tab (A) and rotate it to the fully-opened position, squeeze the two smaller lock tabs (B) . . .

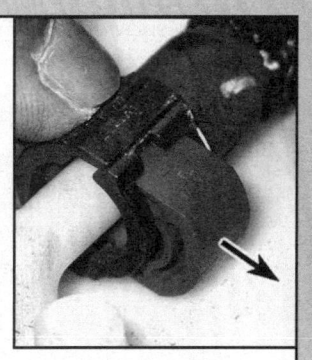

. . . then push the retainer out and pull the fuel line off the pipe

Spring-lock coupling; remove the safety cover, install a coupling release tool and close the tool around the coupling . . .

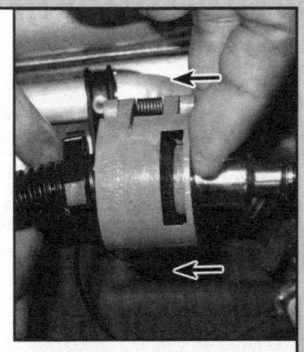

. . . push the tool into the fitting, then pull the two lines apart

Hairpin clip type fitting: push the legs of the retainer clip together, then push the clip down all the way until it stops and pull the fuel line off the pipe

4-6 Chapter 4 Fuel and exhaust systems

7.3 Fuel pump/fuel level sensor cover

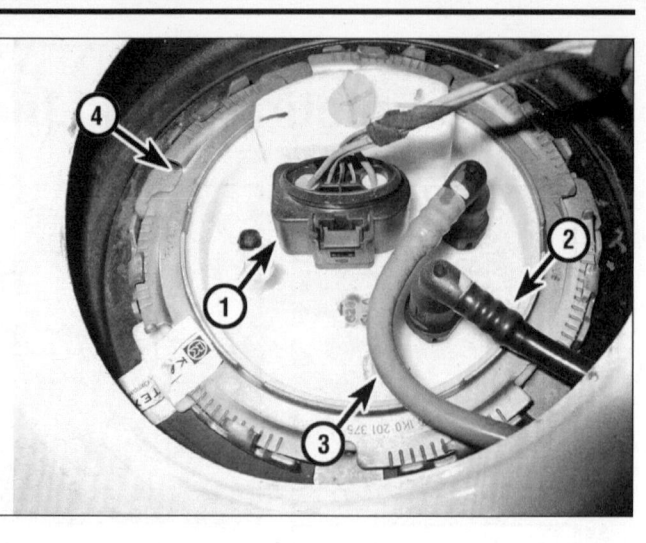

7.5 Fuel pump module/fuel level sender connections

1. Fuel pump/fuel level sender electrical connector
2. Fuel pump supply line
3. Fuel return line
4. Module lock ring

7 Fuel pump module/fuel level sender - removal and installation

Warning: *Gasoline is extremely flammable, so take extra precautions when you work on any part of the fuel system. See the Fuel system warnings in Section 1.*

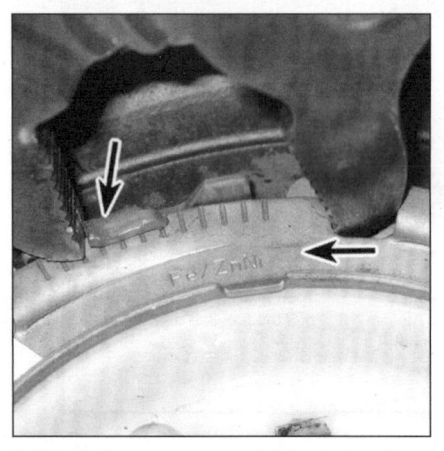

7.6 Use pliers to loosen the lock ring if the special lock ring tool isn't available

Caution: *On five-cylinder models, whenever the fuel system has been opened it must be bled (see Section 5). If the system is not bled, the catalytic converter can be damaged.*
Note: *Whenever the battery is disconnected, several systems must be re-learned before they will work properly (see Chapter 5, Section 3).*
Note: *The following procedure applies to the in-tank fuel pump for all models.*

Removal

1 Relieve the fuel system pressure (see Section 3).
2 Disconnect the cable from the negative battery terminal (see Chapter 5).
3 To access the fuel pump module inspection hole cover, remove the rear seat cushion (see Chapter 11), then lift the carpeting back and remove the access hole cover **(see illustration)**.
4 On four-cylinder models, the Fuel Pump Control Module (FPCM) is attached to the top of the cover, so handle the cover carefully. It's not necessary to disconnect the electrical connector from the FPCM unless you're replacing the FPCM.

5 Disconnect the fuel pump module electrical connector and lines **(see illustration)**. Mark the lines to prevent mix-ups when reconnecting them.
6 Remove the fuel pump module lock ring. A special tool (available at most auto parts stores) is available to unscrew the lock ring, but a pair of pliers will work **(see illustration)**. Lock one end of the pliers on the lock ring and the other end on one of the tank tabs, then slowly close the pliers to move the ring counterclockwise.
7 Lift the fuel pump/fuel level sensor module out of the tank as a single assembly **(see illustration)**. Angle the module as necessary to protect the fuel level sensor float arm.
Caution: *On some models, there may be a suction tube attached to the fuel pump module. When removing the fuel pump module, use care, and disconnect the suction tube before attempting to remove the module from the vehicle.*
Note: *At the time this manual was written, the fuel pump module was not serviceable, except for the fuel level sending unit.*
8 Inspect the flange seal **(see illustration)**. If the seal is damaged, replace it.

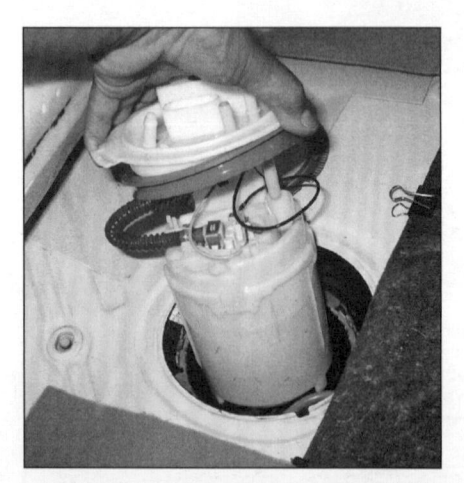

7.7 Carefully remove the fuel pump/fuel level sensor module from the tank

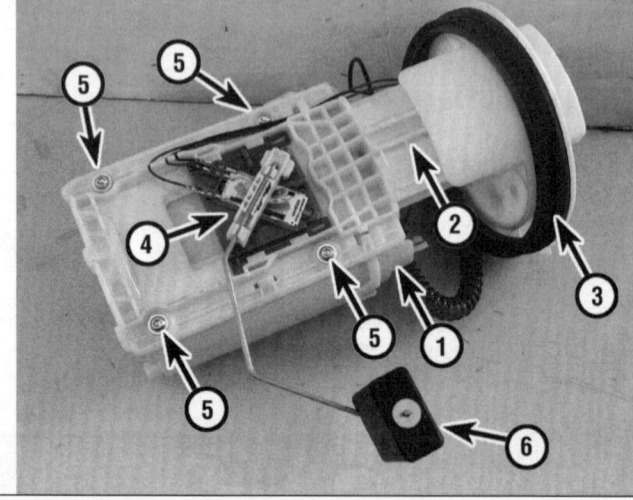

7.8 Typical fuel pump/fuel level sensor module details

1. Fuel delivery pump
2. Fuel pump module assembly
3. Flange seal
4. Fuel level sender location
5. Fuel pump-to-module fasteners
6. Fuel level sensor arm and float

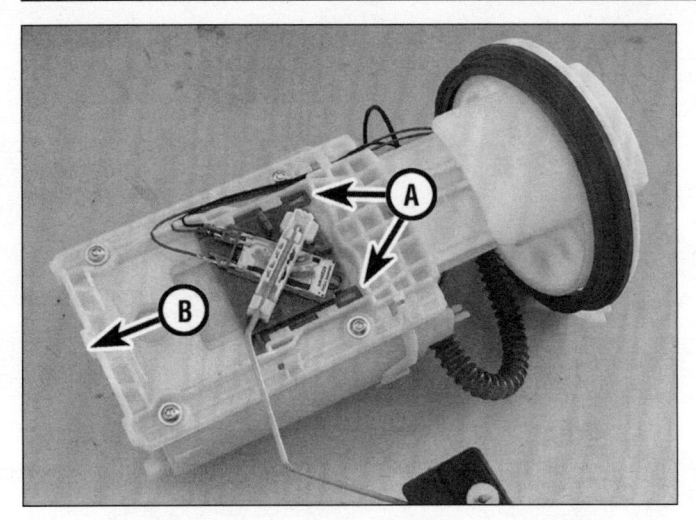

Chapter 4 Fuel and exhaust systems

7.9 Lift up on the locking tabs (A), then slide the fuel level sender down (B) and out - early style sensor shown

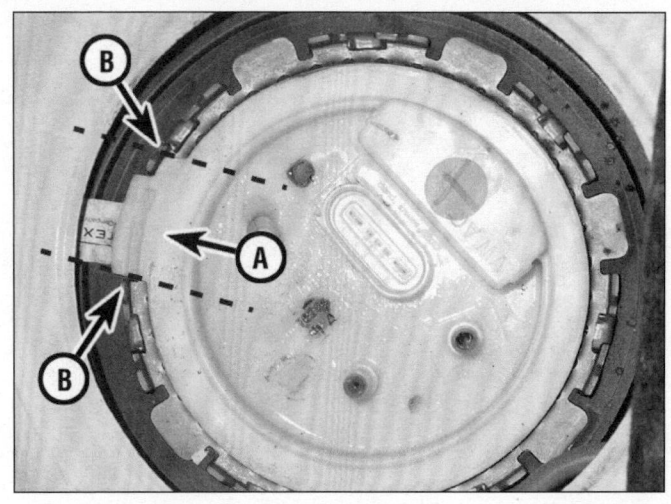

7.17 Make sure the tab (A) on the fuel delivery unit is placed in between the tabs (B) of the tank before trying to install the lock ring

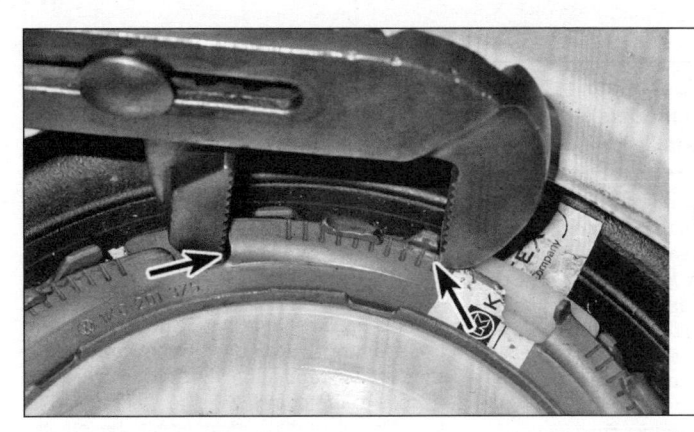

7.18 Rotate the lock ring clockwise to lock the module in place

Fuel level sender

Early version

9   To remove the fuel level sender, disconnect the sender electrical connectors, then use a small screwdriver to disengage the two sensor lock tabs that secure the sender to the fuel pump module **(see illustration)**. Pull the sensor straight up out of the pump module.
Note: *Mark the color and location of the wires before you disconnect them.*
10   When installing the fuel level sender on the fuel pump module, make sure that the locking tabs snap into place.
11   No further disassembly of the fuel pump module/fuel level sender is possible.
12   Installation is the reverse of removal.

Later version

13   Holding the fuel level sensor arm, pull the sensor in and lift the sensor up to release it from the retaining tab.
14   Disconnect the electrical wires from the sensor, noting the wire locations and colors for installation, and remove the sensor from the top of the module.
15   Install the fuel level sensor module on the fuel pump module, making sure that the locking tabs snap into place.
16   Installation is the reverse of removal.

Installation

17   Install the fuel pump/fuel level sensor module in to the tank and align the tab on the module with the tank **(see illustration)**.
Note: *Use a new seal for the locking ring if the old seal is damaged.*
18   Place the locking ring over the module and lock the ring into place using a pair of pliers **(see illustration)**.
19   The remainder of installation is the reverse of removal.
20   On five-cylinder models, bleed the fuel system (see Section 5).

8   Transfer pump/fuel level sender - replacement

Warning: *Gasoline is extremely flammable, so take extra precautions when you work on any part of the fuel system. See the Fuel system warnings in Section 1.*
Note: *The following procedure applies to AWD models.*
Note: *Whenever the battery is disconnected, several systems must be re-learned before they will work properly (see Chapter 5, Section 3).*
1   Disconnect the cable from the negative battery terminal (see Chapter 5).
2   To access the fuel transfer pump/fuel level sender inspection hole cover, remove the rear seat cushion (see Chapter 11), then lift the carpeting back and remove the access hole cover **(see illustration 7.3)**.
3   Disconnect the electrical connector.
4   Remove the inspection cover lock ring. A special tool (available at most auto parts stores) is available to unscrew the lock ring, but a pair of pliers will work **(see illustration 7.6)**. Lock one end of the pliers on the lock ring and the other end on one of the tank tabs, then slowly close the pliers to move the ring counterclockwise.

Fuel level sender

Note: *It is recommended to drain the gas from the tank for this procedure.*
5   Reach inside of the fuel tank and locate the fuel level sender harness connector and disconnect.
6   Press in on the tabs and release the fuel level sender harness connector from the bracket. Press the connector through the bracket.
7   To remove the fuel level sender, release the two tabs at the top of the sender while sliding the sender upwards. The arm remains attached to the sender.
8   Installation is reverse of removal. Ensure the sender harness connector is secured in the bracket before reconnecting the vehicle harness.

Transfer pump

9   The transfer pump is part of the fuel tank and is not serviceable separately. The fuel tank must be replaced if the transfer pump is defective. See Section 10.

4-8 Chapter 4 Fuel and exhaust systems

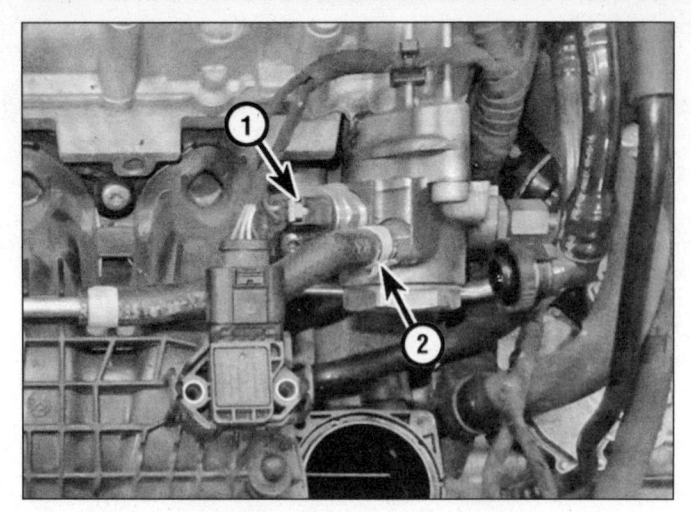

9.6 High-pressure fuel pump details (1.4L engine with Hitachi pump shown, others similar)
1 Pressure regulator electrical connector
2 Fuel supply line

9.8 Remove the high-pressure fuel tube (A). Hold the high-pressure fuel pump fitting (1) with a wrench while loosening the tube nut (2) to prevent it from turning (1.4L engine shown, others similar)

9 High-pressure fuel pump (Direct Injection models) - removal and installation

Warning: *Gasoline is extremely flammable, so take extra precautions when you work on any part of the fuel system. See the Fuel system warnings in Section 1.*

Warning: *The fuel delivery systems on direct-injection models are equipped with both a low-pressure system and a second, high-pressure system that operates at extremely high pressure. At idle, the fuel pressure in the fuel rail is about 725 psi; under certain operating conditions such as hard acceleration, the system pressure can reach 1700 psi.*

Warning: *Wait until the engine is completely cold before performing this procedure.*

Note: *Whenever the battery is disconnected, several systems must be re-learned before they will work properly (see Chapter 5, Section 3).*

1 Relieve the fuel system pressure (see Section 3).

2 Remove the engine cover.
3 Disconnect the cable from the negative battery terminal (see Chapter 5).
4 On 1.8L and 2.0L engines (except CBFA and CCTA), remove the air filter housing assembly (see Section 11).
5 On Golf models with 1.8L and 2.0L engines (except CBFA and CCTA), remove the charge air pipe that runs over the bellhousing and to the intercooler. The charge air pipe is attached with bolts and hose clamps. Remove the harness fasteners and reposition the harness to gain access to the pump.
6 On all models, disconnect the electrical connector from the fuel pressure regulator valve and disconnect the fuel supply line from the high-pressure fuel pump **(see illustration)**.
7 On 1.8L and 2.0L engines (except CBFA and CCTA), remove the fastener securing the bracket of the high pressure fuel line tube.
8 On all models, disconnect the high pressure fuel line tube from the high-pressure pump and fuel rail **(see illustration)**.

Warning: *Surround the fitting and wrenches with a rag before loosening the high-pressure fuel line fitting.*

9 Remove the high-pressure fuel pump mounting fasteners **(see illustration)** and remove the pump. Unscrew the bolts evenly, a little at a time, until all pressure is released from the spring.

Note: *The plunger (cam follower) might remain in the cylinder head.*

10 Remove and discard the old pump O-ring. Always use a new O-ring when installing the pump.
11 Before installing the pump, coat the plunger with clean engine oil and insert the plunger into the cylinder head. Then rotate the crankshaft with a socket and breaker bar on the center bolt of the crankshaft pulley until the plunger reaches its lowest point.

Note: *Hold the plunger down in its bore with your finger as you rotate the engine.*

12 Install the pump with a new O-ring. Install the pump mounting fasteners and tighten them alternately, a little at a time until the pump is seated, then to the torque listed in this Chapter's Specifications.
13 Reconnect the high-pressure fuel tube and tighten the fittings securely. Reconnect the fuel supply line.

Note: *Hold the high-pressure fuel pump fitting with a wrench while tightening the tube nut to prevent it from turning and distorting the fuel tube.*

14 Connect the electrical connector.
15 The remainder of installation is the reverse of the removal procedure.
16 Start the engine and check for leaks.

10 Fuel tank - removal and installation

Warning: *Gasoline is extremely flammable. See Fuel system warnings in Section 1.*
Caution: *On five-cylinder models, whenever the fuel system has been opened it must be*

9.9 High-pressure fuel pump mounting fasteners (1.4L engine shown, others similar)

Chapter 4 Fuel and exhaust systems

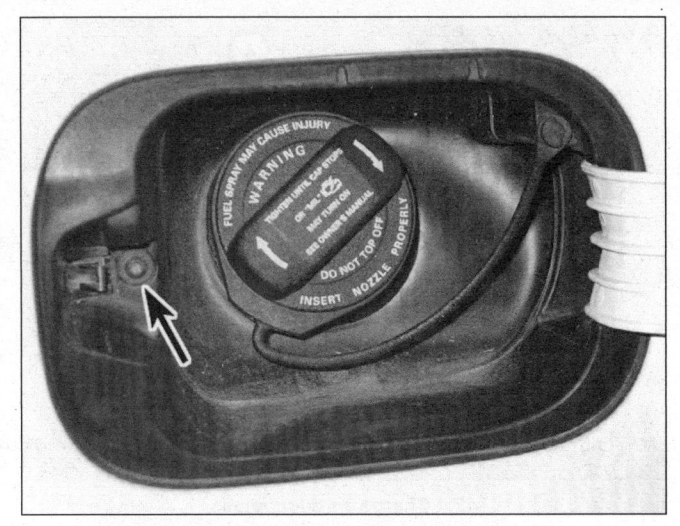

10.7a Remove the fuel filler liner mounting fastener...

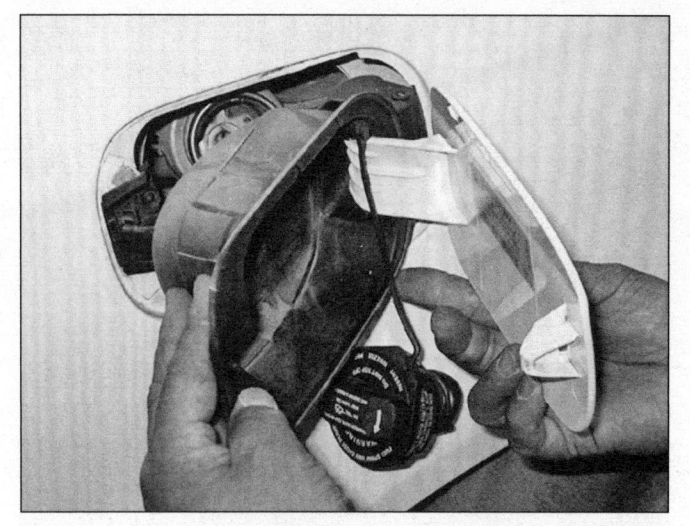

10.7b ...then rotate the liner out

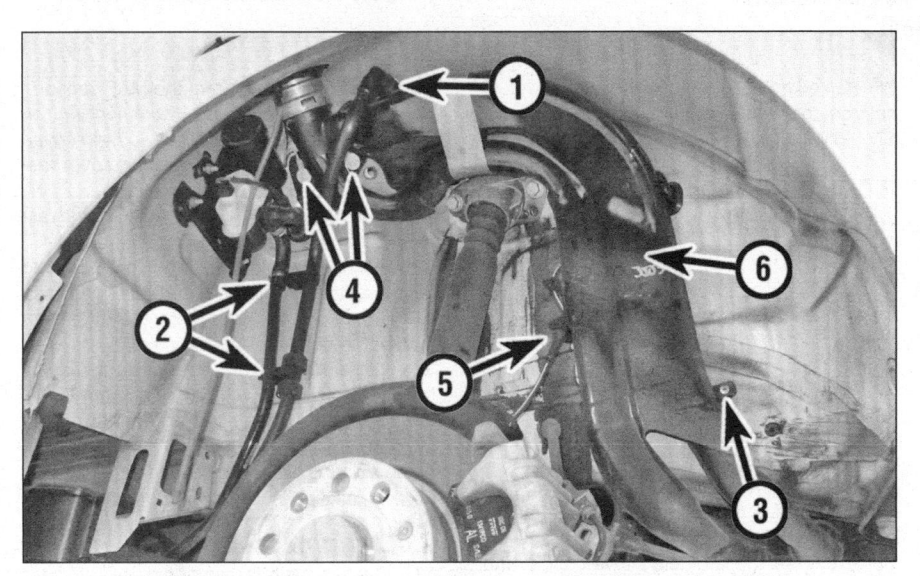

10.8 Fuel tank filler tube details (all Jetta and 2014 and earlier Golf)

1. EVAP hose connection
2. Clips
3. Rivet
4. Fuel filler tube mounting fasteners
5. Connector for the ABS sensor wire harness
6. Protective shield

10.11 Fuel tank mounting fastener locations

bled (see Section 5). If the system is not bled, the catalytic converter can be damaged.

Note: *Whenever the battery is disconnected, several systems must be re-learned before they will work properly (see Chapter 5, Section 3).*

Note: *The following procedure is much easier to perform if the fuel tank is empty.*

1 Remove the fuel tank filler cap to relieve fuel tank pressure.

2 Relieve the fuel system pressure (see Section 3).

3 Disconnect the cable from the negative battery terminal (see Chapter 5).

4 Remove the rear seat cushion, disconnect the fuel pump control module (four-cylinder models), then disconnect the electrical connector and fuel supply line quick-connect fittings from the fuel pump module (see Section 7).

Note: *AWD models have a transfer pump/fuel level sender under an additional access cover. Be sure to remove the cover and disconnect this connector as well.*

5 Raise the rear of the vehicle and support it securely on jackstands.

6 Remove the right-rear wheel, then remove the rear inner fender splash shield (see Chapter 11).

7 On Golf models, open the fuel tank filler flap and remove the mounting fastener and plastic liner **(see illustrations)**.

8 Remove the fuel filler tube mounting fasteners and disconnect the hoses. On all Jetta models and 2014 and earlier Golf models, unclip the ABS sensor wire from the bracket, then remove the rivet and protective shield **(see illustration)**.

9 Disconnect the exhaust system hangers, then remove the center and rear mufflers from the vehicle (see Section 14).

10 If equipped, remove the fuel filter (see Chapter 1) and cap the fuel lines.

Note: *During the next step, note that the fuel filler tubes remain attached to the fuel tank and are removed as a unit with the fuel tank.*

11 Support the fuel tank securely, then remove the fuel tank mounting fasteners, and the retaining strap mounting fastener **(see illustration)**. Remove the strap and carefully lower the fuel tank.

12 Installation is the reverse of removal. Tighten the fuel tank mounting fasteners to the torque listed in this Chapter's Specifications. On five-cylinder models, bleed the fuel system (see Section 5).

13 Reconnect the cable to the negative battery terminal (see Chapter 5), then start the engine and check for fuel leaks.

4-10   Chapter 4 Fuel and exhaust systems

11.1 Disconnect the air hoses (A) and the breather hose (B) (Jetta 1.4L CZTA engine shown)

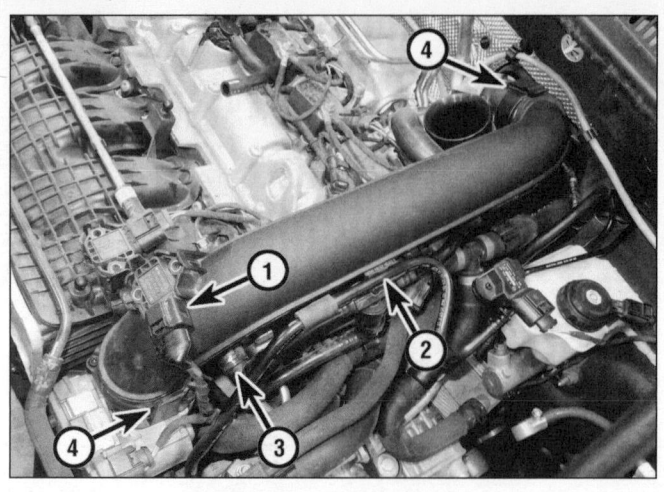

12.4 Charge air duct details
1. Charge air pressure sensor electrical connector
2. Emission hose clips
3. Hose clamp
4. Charge air duct clamps

11 Air filter housing - removal and installation

1.4L engines

1 Loosen the clamps that secure the air intake duct(s) to the air filter housing, detach the intake duct(s) **(see illustration)**.
2 If equipped, detach the crankcase breather hose.
3 On DGXA engines, disconnect the electrical connector from the Mass Air Flow (MAF) sensor.
4 On all engines, firmly pull the air filter housing upwards to detach the rubber grommets under the housing.
5 On DGXA engines, rotate the housing forward to disconnect from the duct.
6 On all engines, remove from the vehicle.
7 Installation is reverse of removal.

1.8L and 2.0L turbocharged engines

8 On Jetta models, release the air guide top cover clips and remove the top cover.
9 Release the air guide lower clips and remove the guide with the air duct hose.
10 On all models, loosen the clamps that secure the air intake duct to the air filter housing and throttle body, then detach the intake duct.
11 On models with air injection, disconnect the air injection hose.
12 On all models, disconnect the vacuum line from the air filter housing lid.
13 On Golf models, firmly pull the air filter housing upwards to detach the rubber grommets under the housing.
14 Rotate the housing upwards at the rear to disconnect from the duct.
15 On Jetta models, remove the bolt at the rear attaching the housing to the vehicle.

16 On all models, remove the air filter housing from the vehicle.
17 Installation is reverse of removal.

2.0L non-turbocharged (CBPA) engines

Note: *This engine is available in Jetta models only.*

18 Loosen the clamp that secures the air intake duct to the air filter housing, then detach the intake duct.
19 Firmly pull the air filter housing upwards to detach the rubber grommets under the housing.
20 Lift the air filter housing up and disconnect the housing from the intake air duct at the bottom of the housing.
21 To remove the air guide, release the air guide top cover clips and remove the top cover.
22 Release the air guide lower clips and remove the guide from the air duct hose.
23 To remove the lower housing assembly, remove the left front inner fender liner.
24 Remove the bolt attaching the guide air intake duct to the bracket.
25 Loosen the clamp that secures the air intake duct to the lower housing, then detach the duct.
26 Remove the fasteners securing the lower housing to the vehicle, and carefully remove the lower housing. The duct that goes to the air filter housing is attached to the battery tray.
27 Installation is reverse of removal.

2.0L turbocharged (CBFA and CCTA) engines

28 On Jetta models, release the air guide top cover clips and remove the top cover.
29 Release the air guide lower clips and remove the guide with the air duct hose.
30 On all models, disconnect the electrical connector from the Mass Air Flow (MAF) sensor (see Chapter 6).
31 Loosen the clamps that secure the air intake duct to the air filter housing, then

detach the intake duct from the MAF sensor.
Note: *On CBFA engines, disconnect the air injection hose from the air filter housing cover.*
32 Remove the air filter housing cover fasteners and cover.
Note: *Do not try to completely remove the fasteners from the cover.*
33 Remove the air filter housing mounting fastener, then lift the air filter housing up and disconnect the housing from the intake air duct at the bottom of the housing.
34 While the air filter housing is out, inspect the condition of the water drain and hose.
35 Before reassembling and installing the air filter housing, thoroughly blow out all parts of the housing with compressed air.
36 Installation is the reverse of removal.

2.5L engines

37 On five-cylinder engines, the air filter housing is incorporated into the engine cover (see Chapter 1, Section 7 for removal and installation).

12 Throttle body/control module - removal and installation

Note: *Whenever the battery is disconnected, several systems must be re-learned before they will work properly (see Chapter 5, Section 3).*
Note: *If a new throttle body control module is installed, the Engine Control Module (ECM) will need to be programmed using the factory scan tool.*

1 Disconnect the cable from the negative battery terminal (see Chapter 5).
2 Remove the engine cover (see Chapter 1, Section 7).

1.4L engines

3 Remove the air filter housing (see Section 11).
4 Disconnect the air charge pressure sensor connector **(see illustration)**.

Chapter 4 Fuel and exhaust systems

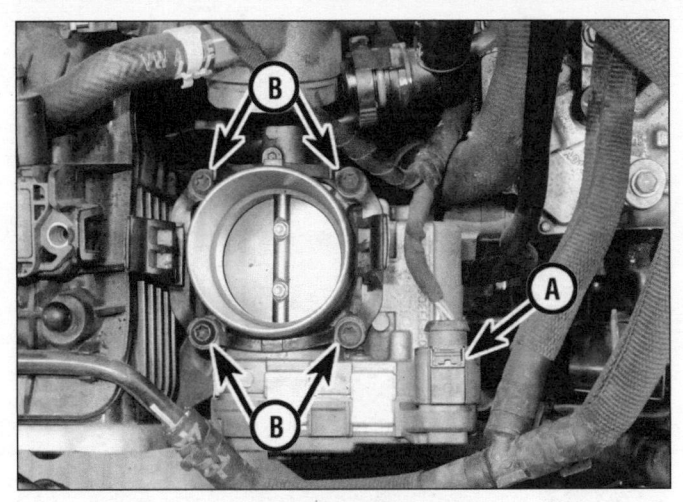

4-11

12.8 Throttle body details
A Electrical connector
B Mounting bolts

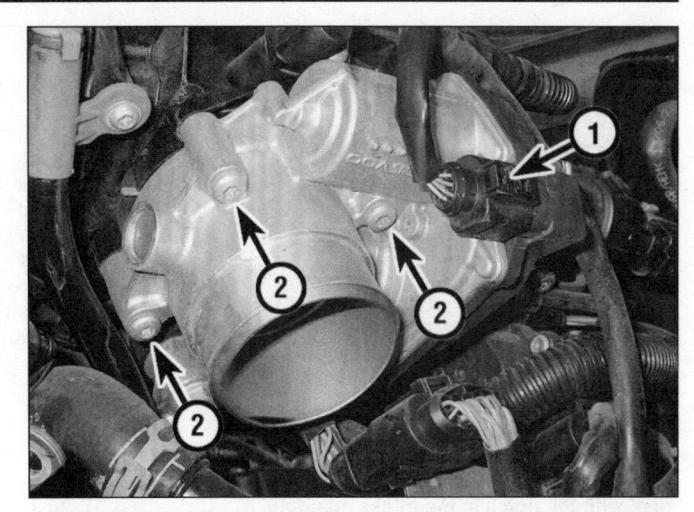

12.13 Throttle body details (1.8L and 2.0L turbocharged engines)
1 Electrical connector
2 Throttle body mounting fasteners (fourth fastener not visible)

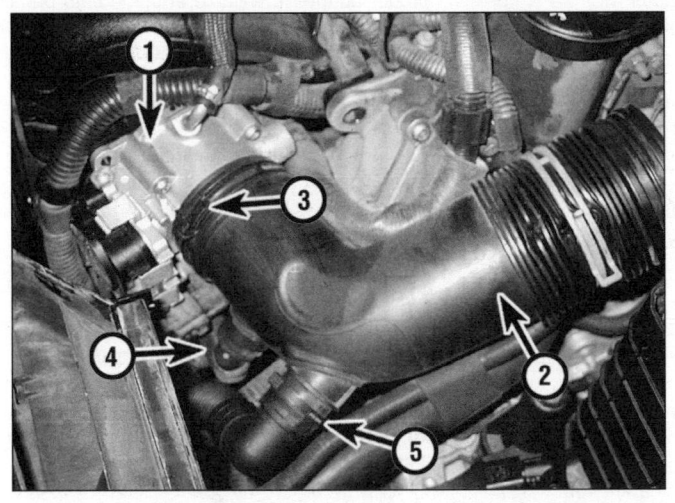

12.23 Throttle body air inlet hose details - five-cylinder models
1 Throttle body
2 Air inlet hose
3 Spring clamp
4 Secondary air injection pump hose (if equipped)
5 Crankcase vent hose

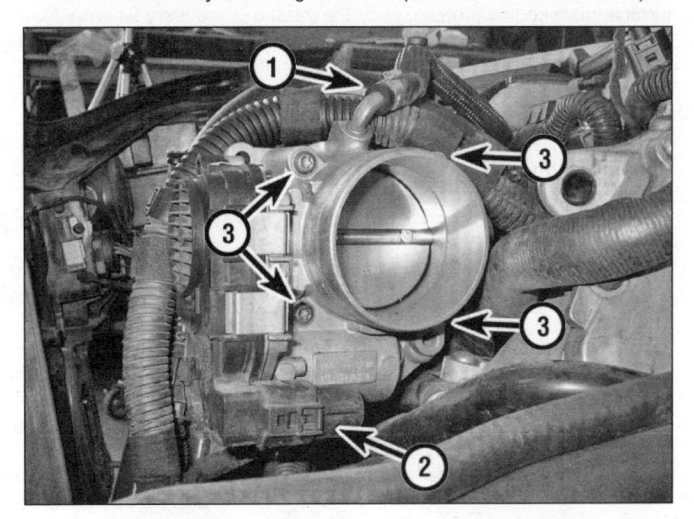

12.25 Throttle body details - five-cylinder models
1 Vent hose
2 Throttle motor electrical connector
3 Mounting fasteners

5   Detach the emission hoses from the clips on the intake air duct.
6   Disconnect the hose attached to the duct.
7   Release the clamps securing the duct to the throttle body and the turbocharger outlet.
8   Disconnect the throttle body electrical connector **(see illustration)**.
9   Remove the four bolts and bracket and remove the throttle body from the intake manifold.
10 Installation is reverse of removal.

1.8L and 2.0L turbocharged engines

11 Raise the vehicle and support it securely on jackstands.
12 Remove the charge air pipe fasteners, then disconnect the electrical connector at the pipe. Loosen the clamps at each end of the charge air pipe, and remove the pipe.

13 Disconnect the electrical connector from the throttle body **(see illustration)**.
14 Remove the throttle body mounting fasteners and carefully remove the throttle body from the intake manifold.
15 Remove and inspect the throttle body gasket. If it's flattened, hardened or cracked, replace it. If it's in good condition, it can be reused.
16 Cover the intake manifold opening with a clean shop towel to prevent anything from entering.
17 Installation is the reverse of removal. Tighten the throttle body fasteners to the torque listed in this Chapter's Specifications.

2.0L non-turbocharged (CBPA) engines

Note: *This engine is available in Jetta models.*
18 Loosen the intake air hose clamp at the throttle body and pull the hose off of the throttle body.

19 Disconnect the throttle body electrical connector.
20 Remove the four bolts and detach the throttle body.
21 Inspect the throttle body seal and replace if damaged.
22 Installation is reverse of removal.

2.5L engines

23 Disconnect the vent hose, secondary air injection pump hose and air inlet pipe clamp. Remove the air inlet pipe between the throttle body and the air filter housing **(see illustration)**.
24 On some models, it may be necessary to remove the air guide connector-to-radiator support fasteners and the connector.
25 Disconnect the electrical connector and the vent hose, remove the throttle body mounting fasteners and carefully remove the throttle body from the intake manifold **(see illustration)**.

4-12   Chapter 4 Fuel and exhaust systems

12.29 Align the tab of the throttle body seal with the notch in the intake manifold

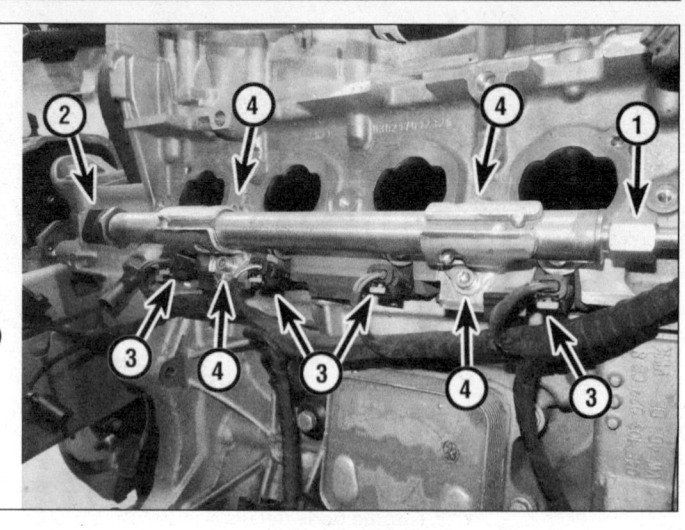

13.6 Disconnect the high pressure fuel tube (1), fuel pressure sensor (2), injectors (3) and fuel rail bolts (4) to remove the fuel rail (1.4L engine shown)

26   On models that use coolant to heat the throttle valve control module, once the throttle body is removed, clamp off the coolant lines to the throttle body, disconnect the clamps and remove the coolant lines.
Note: *Always replace the hose clamps to the coolant lines.*
27   Remove and inspect the throttle body gasket. If it's flattened, hardened or cracked, replace it. If it's in good condition, it can be reused.
28   Cover the intake manifold opening with a clean shop towel to prevent anything from entering.
29   Install the gasket to the intake manifold, making sure the gasket is properly seated **(see illustration)**.
30   Installation is the reverse of removal. Tighten the throttle body fasteners to the torque listed in this Chapter's Specifications.

13 Fuel rail and injectors - removal and installation

Warning: *Gasoline is extremely flammable, so take extra precautions when you work on any part of the fuel system. See the Fuel system warnings in Section 1.*
Warning: *The fuel delivery systems on direct-injection models are equipped with both a low-pressure system and a second, high-pressure system that operates at extremely high pressure. At idle, the fuel pressure in the fuel rail is about 725 psi; under certain operating conditions such as hard acceleration, the system pressure can reach 1700 psi.*
Warning: *Wait until the engine is completely cool before beginning this procedure.*
Caution: *On five-cylinder models, whenever the fuel system has been opened it must be bled (see Section 5). If the system is not bled, the catalytic converter can be damaged.*
Note: *Whenever the battery is disconnected, several systems must be re-learned before they will work properly (see Chapter 5, Section 3).*
1   Relieve the fuel system pressure (see Section 3).

2   Disconnect the cable from the negative battery terminal (see Chapter 5).
3   Remove the engine cover (see Chapter 1, Section 7).
4   Wrap a shop rag around the fuel line connectors at the front of the engine and disconnect the fittings (see Section 6). Plug the supply line and the fuel inlet fitting.

1.4L, 1.8L and 2.0L direct-injection engines

5   Remove the intake manifold (see Chapter 2A, Section 7).
6   Remove the high-pressure fuel tube from the high-pressure pump and fuel rail **(see illustration)**.
7   Disconnect the fuel pressure sensor electrical connector.
8   Clean the fuel rail and injector base to prevent debris from entering the engine.
9   Remove the bolts and pull the fuel rail off of the fuel injectors.
10   To remove the injectors from the cylinder head, use a pair of screwdrivers or a pair of angled needle-nose pliers to pry them out of their bores **(see illustrations)**. If they are stuck in the head, a slide hammer with the proper adapter (available at some auto parts stores and most specialty tool suppliers) will have to be used **(see illustrations)**.
11   To replace the injector seals, refer to the *Fuel injector seal replacement* procedure in this section.
12   Thoroughly clean the injector bores with a small nylon brush. If any of the valves are in the way, carefully rotate the engine just enough to provide clearance to reach all of the bore **(see illustration)**.
13   Install the fuel injectors in the cylinder head (NOT in the fuel rail). You should be able to push each assembled injector into its bore in the cylinder head. The bore is tapered, so you will encounter some resistance as the Teflon seal nears the bottom of the bore. Press the injector into its bore until it stops. If necessary, gently tap it into place using a hammer and a small block of wood **(see illustration)**.
Caution: *Do NOT oil the Teflon seals.*
14   Install the fuel rail onto the injectors and install the bolts. Tighten the fuel rail bolts to the torque listed in this Chapter's Specifications.

13.10a Sometimes the injectors can be pried out with a pair of screwdrivers . . .

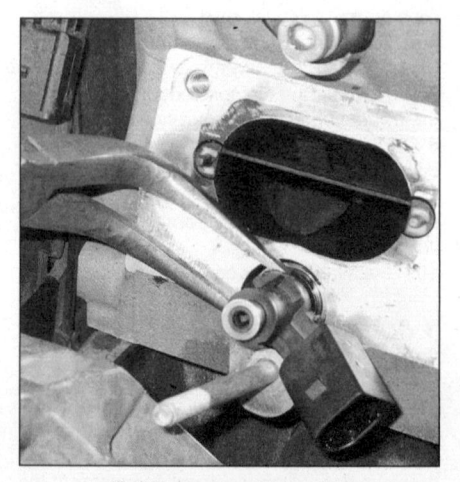

13.10b . . . or with angled needle-nose pliers

Chapter 4 Fuel and exhaust systems

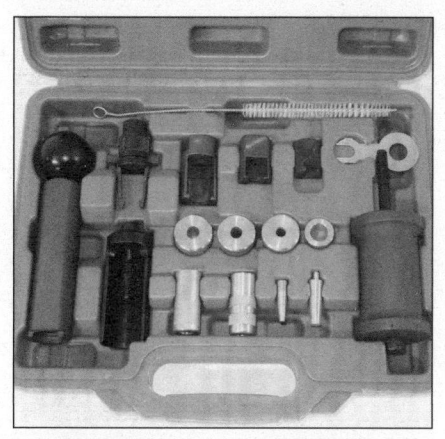

4-13

13.10c Special injector tool kits like this are available at many auto parts stores and specialty tool distributors. It includes everything for removing DI injectors and installing/resizing the Teflon seals

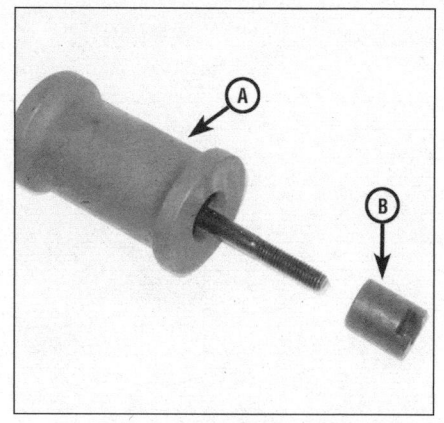

13.10d If the injectors are stuck, a slide hammer (A) and the proper adapter (B) will have to be used

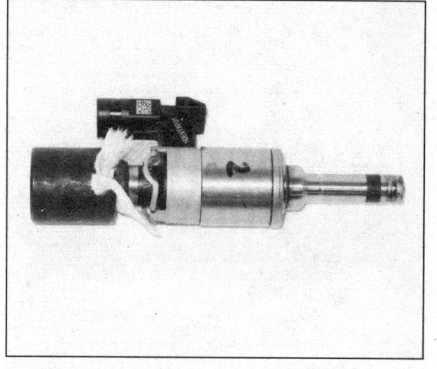

13.10e Assemble the adapter to the top of the injector (shown removed for clarity), then thread the slide hammer shaft into the adapter and pull the injector from the head. A piece of paper towel is being used here to provide a tighter fit on the injector

Fuel injector seal replacement

Caution: *Whenever the injectors have been removed, all of the seals and support rings must be replaced with new ones.*

15 Remove and inspect the support ring **(see illustration)**. If it's damaged, replace it.

16 Remove the old combustion chamber Teflon sealing ring and the upper O-ring from each injector **(see illustration)**.

Caution: *Be extremely careful not to damage the groove for the seal or the rib in the floor of the groove. If you damage the groove or the rib, you must replace the injector.*

17 Before installing the new Teflon seal on each injector, thoroughly clean the groove for the seal and the injector shaft. Remove all combustion residue and varnish with a clean shop rag.

Teflon seal installation

18 The manufacturer recommends that you use the tools included in the special injector tool set to install the Teflon lower seals on the injectors **(see illustration 13.10c)**.

13.12 Clean the injector bores with a nylon brush

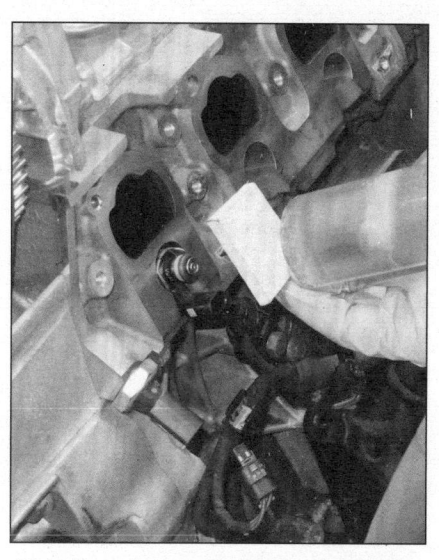

13.13 Use a hammer and piece of wood to tap the injector into its bore, if necessary

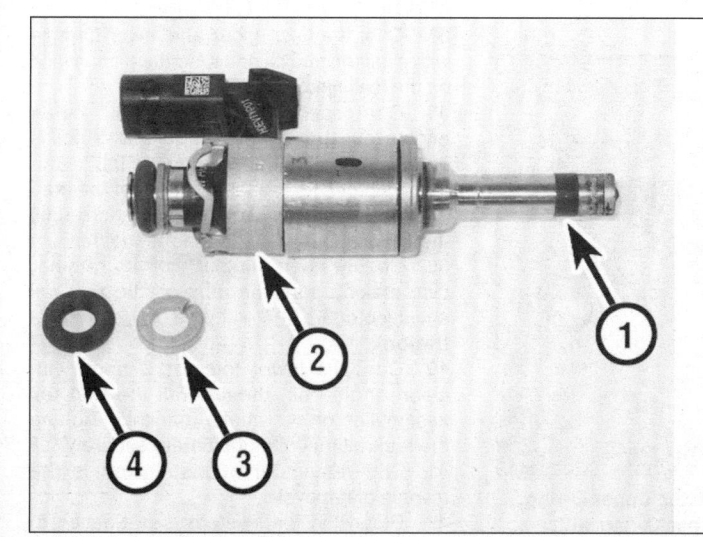

13.15 Fuel injector details - direct injection models

1. Combustion chamber Teflon sealing ring
2. Support ring
3. Spacer ring
4. Injector upper O-ring

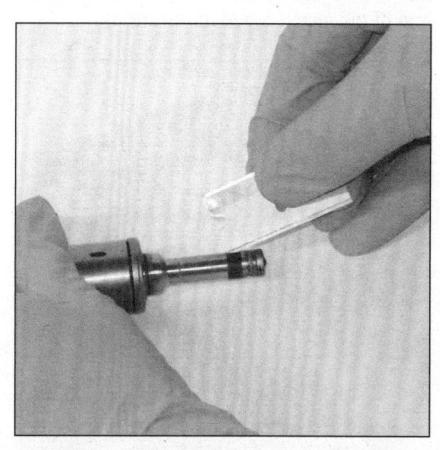

13.16 To remove the Teflon sealing ring, cut it off with a hobby knife (be careful not to scratch the injector groove)

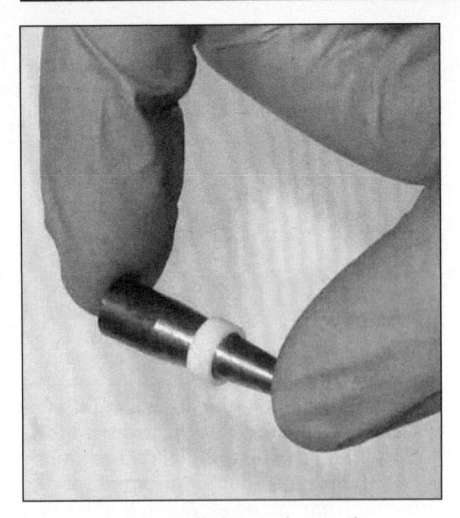

4-14   **Chapter 4   Fuel and exhaust systems**

13.19 Install the seal onto the installation cone

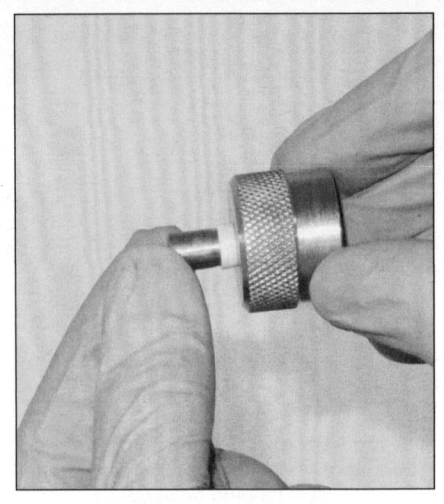

13.20 Insert the cone, with the seal installed, into the spacer

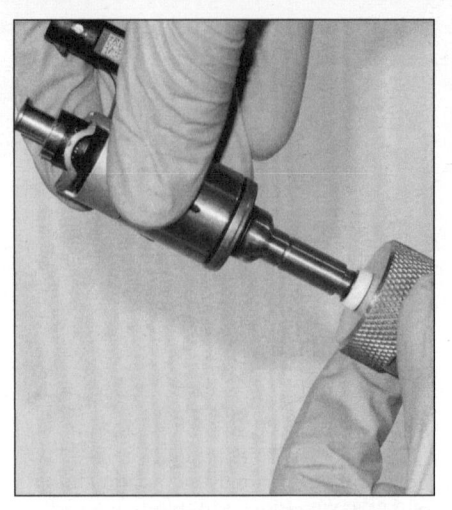

13.21 Press the seal over the cone and onto the injector using the sizing sleeve

19   Install the special seal assembly cone on the injector **(see illustration)**.

20   Install the special sleeve on the injector and use the sleeve to push on the assembly cone, which pushes the Teflon seal into place on its groove. Do NOT use any lubricants to do so **(see illustration)**.

21   Pushing the Teflon seal into place in its groove, the cone expands it slightly **(see illustration)**. There are two sizing sleeves in the special tool set with progressively smaller inside diameters. Using a clockwise rotating motion of about 180 degrees, install the slightly larger sleeve onto the injector and over the Teflon seal until the sleeve hits its stop, then carefully turn the sleeve counter-clockwise as you pull it off the injector. Use the slightly smaller sizing sleeve the same way. The seal is now sized. Repeat this step for each injector.

Caution: *Do NOT oil the new Teflon seal.*

Spacer ring and upper O-ring installation

22   Install the new spacer ring at the upper end of the injector **(see illustration)**.

23   Lubricate the new upper O-ring with clean engine oil and install it on the injector **(see illustration)**. Note that the seal is installed *above* the spacer.

24   Repeat this procedure until all four injectors are reassembled.

2.0L non-turbocharged engines

25   Detach the harness from the fuel rail.

26   Disconnect the fuel injector connectors and the Camshaft Position Sensor (CMP) connector.

27   Clean the fuel rail and injector base to prevent debris from entering the engine.

28   Remove the bolts securing the fuel rail.

29   Carefully but firmly pull the fuel rail and injectors out of the engine.

30   Remove the fuel injector retaining clip and pull the injector from the fuel rail (see Step 35).

31   Replace the injector O-rings (see Step 36).

32   When installing the injectors and fuel rail, coat the O-rings with clean engine oil.

2.5L engines

Removal

33   Disconnect the electrical connectors to the EVAP canister purge valve and fuel injectors, release the locking ring securing the valve to the transport strap then lift the hose out of the clips (see Chapter 2B, illustration 6.5).

34   Remove the fuel rail mounting fasteners **(see illustration)** and lift the fuel rail and injectors out from the intake manifold **(see illustration)**.

35   Remove the retaining clip from each fuel injector **(see illustration)**, then pull the injectors out of the fuel rail.

36   Remove and discard the old injector O-rings and install new ones **(see illustration)**.

Caution: *Even if you only removed the fuel rail assembly to replace a single injector or a leaking O-ring, it is recommended to remove all of the injectors from the fuel rails and replace all of the O-rings at the same time.*

Installation

37   Coat the new upper and lower O-rings with clean engine oil and slide them into place on the fuel injectors.

38   Coat each upper O-ring with clean engine oil, then insert the injector into its bore in the fuel rail, then align the tab on the fuel injector with the tab on the fuel rail **(see illustration)** and push it into the bore until the injector is fully seated in the fuel rail.

39   Secure each injector with its retainer clip, making sure the tabs on the fuel rail and injector are seated in the clip **(see illustration)**.

40   Coat the lower injector O-rings with clean engine oil, then install the fuel rail assemblies on the intake manifold. Tighten the fuel rail mounting fasteners securely.

41   The remainder of installation is the reverse of removal.

42   Bleed the fuel system (see Section 5) and inspect the injectors for fuel leaks.

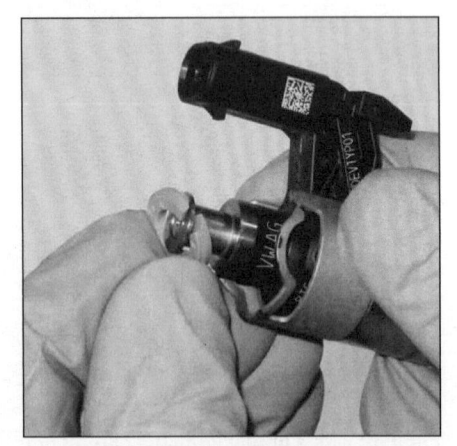

13.22 Install the spacer ring to the top of the injector...

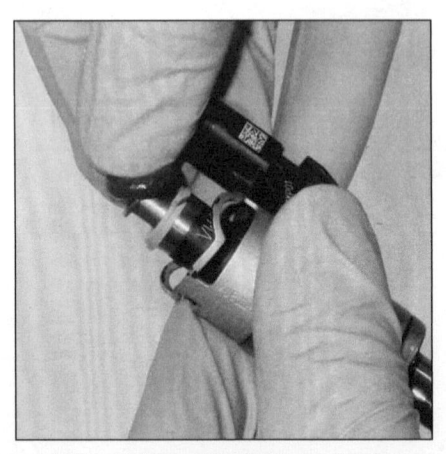

13.23 ...followed by the upper O-ring. Lubricate the upper O-ring with clean engine oil

Chapter 4 Fuel and exhaust systems

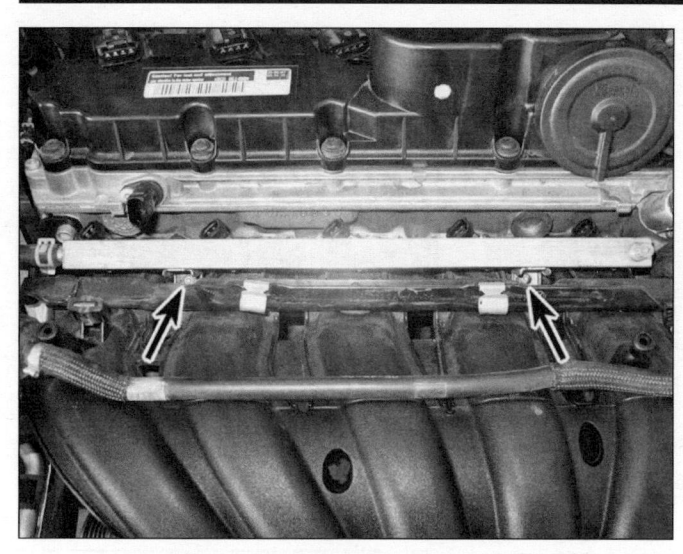

4-15

13.34a Fuel rail mounting fastener locations - five-cylinder models

13.34b Remove the fuel rail and injectors as an assembly from the intake manifold and cylinder head - five-cylinder models

13.35 Remove each injector retaining clip, then pull the injector out of the fuel rail - five-cylinder models

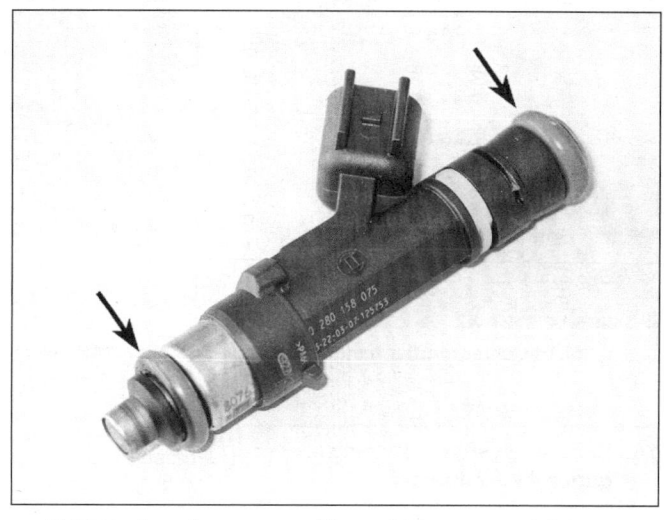

13.36 Replace the upper and lower O-rings on each injector

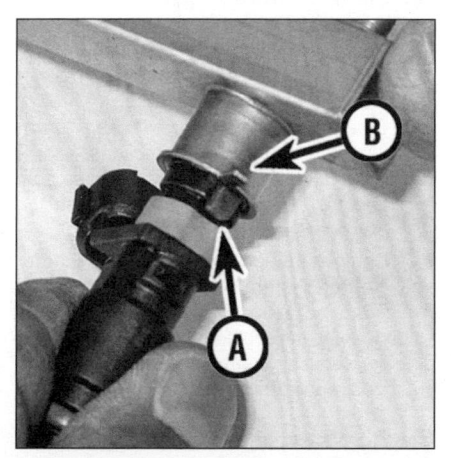

13.38 Insert the injector into the fuel rail and align the tab of the injector (A), with tab on the fuel rail (B) - five-cylinder models

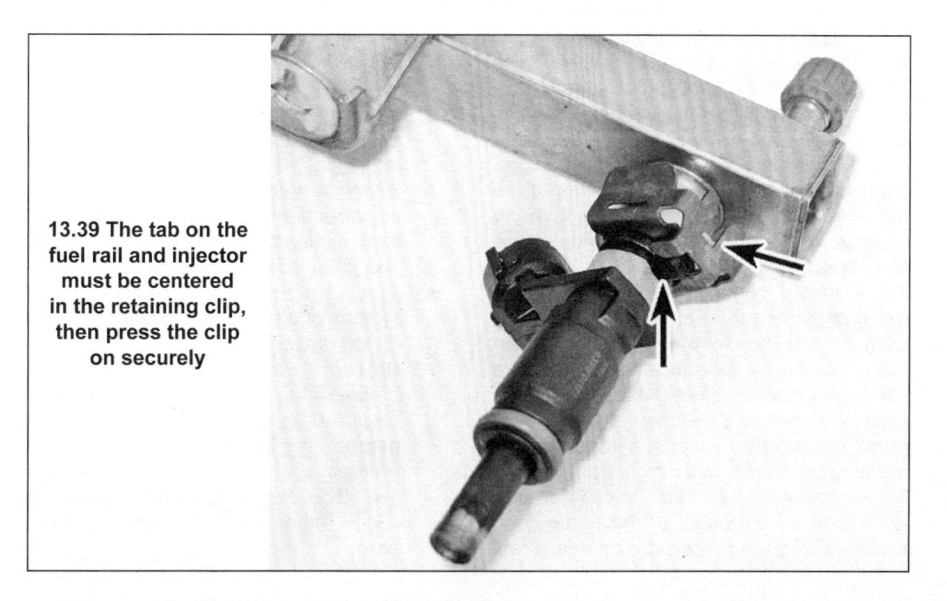

13.39 The tab on the fuel rail and injector must be centered in the retaining clip, then press the clip on securely

4-16   Chapter 4 Fuel and exhaust systems

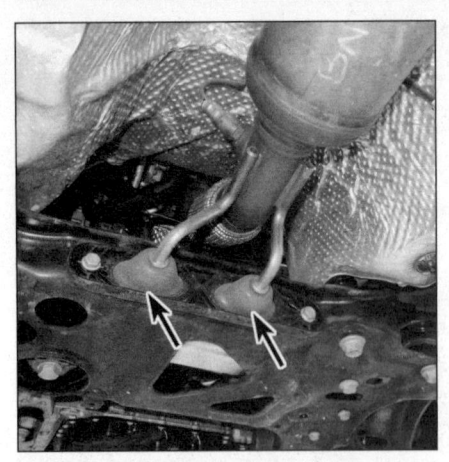

14.1a Front exhaust hangers

14.1b Exhaust coupler

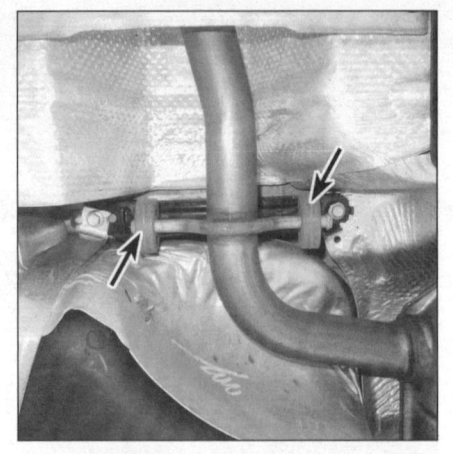

14.1c Rear exhaust hangers

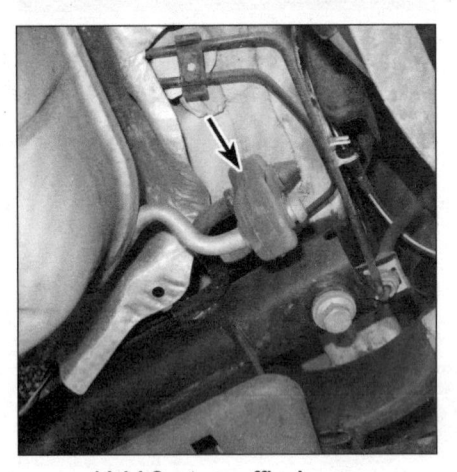

14.1d Center muffler hanger

14.1e Rear muffler hanger

14 Exhaust system servicing - general information

Warning: *Allow exhaust system components to cool before inspection or repair. Also, when working under the vehicle, make sure it is securely supported on jackstands.*

1   The exhaust system consists of the exhaust manifolds, catalytic converter, muffler, tailpipe and all connecting pipes, flanges and clamps. The exhaust system is isolated from the vehicle body and from chassis components by a series of rubber hangers **(see illustrations)**. Periodically inspect these hangers for cracks or other signs of deterioration, replacing them as necessary.

2   Conduct regular inspections of the exhaust system to keep it safe and quiet. Look for any damaged or bent parts, open seams, holes, loose connections, excessive corrosion or other defects which could allow exhaust fumes to enter the vehicle. Do not repair deteriorated exhaust system components; replace them with new parts.

3   If the exhaust system components are extremely corroded, or rusted together, a cutting torch is the most convenient tool for removal. Consult a properly-equipped repair shop. If a cutting torch is not available, you can use a hacksaw, or if you have compressed air, there are special pneumatic cutting chisels that can also be used. Wear safety goggles to protect your eyes from metal chips and wear work gloves to protect your hands.

4   Here are some simple guidelines to follow when repairing the exhaust system:
 a) *Work from the back to the front when removing exhaust system components.*
 b) *Apply penetrating oil to the exhaust system component fasteners to make them easier to remove.*
 c) *Use new gaskets, hangers and clamps.*
 d) *Apply anti-seize compound to the threads of all exhaust system fasteners during reassembly.*
 e) *Be sure to allow sufficient clearance between newly installed parts and all points on the underbody to avoid overheating the floor pan and possibly damaging the interior carpet and insulation. Pay particularly close attention to the catalytic converter and heat shield.*

15 Turbocharger/exhaust manifold and charge air cooler - check and replacement

Turbocharger system check

1   The turbocharger is a precision component which can be severely damaged by a lack of lubrication or from foreign material entering the air intake duct. Turbocharger failure may be indicated by poor engine performance, blue/gray exhaust smoke or unusual noises from the turbocharger. If a turbocharger failure is suspected, check the following areas:
 a) *Check the intake air duct for looseness or damage. Make sure there are no restrictions in the air intake system. Check for a dirty air filter element or damaged intercooler.*
 b) *Check the system vacuum hoses for restrictions or damage.*
 c) *Check the system wiring for damage and electrical connectors for looseness or corrosion.*
 d) *Make sure the wastegate actuator linkage is not binding.*
 e) *Check the exhaust system for damage or restrictions.*
 f) *Check the lubricating oil supply and drainback lines for damage or restrictions.*
 g) *Check the coolant supply and return lines for damage and restrictions.*
 h) *If the turbocharger requires replacement due to failure, be sure to change the engine oil and filter (see Chapter 1).*

2   Complete diagnosis of the turbocharger and control system requires special techniques and equipment. If the previous checks fail to identify the problem, take the vehicle to a dealer service department or other properly equipped repair facility for diagnosis.

Turbocharger
Removal
Warning: *Wait until the engine is completely cool before beginning this procedure.*

Chapter 4 Fuel and exhaust systems

4-17

15.8 Remove the driveaxle boot heat shield (1.4L engine)

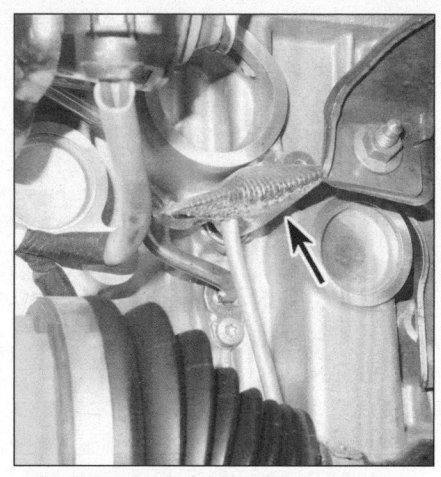

15.9 Locate the oil pressure switch and remove the protective cover to disconnect the electrical connector (1.4L engine)

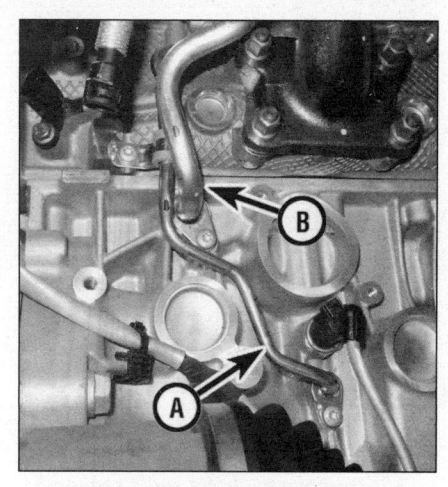

15.12 Identifying the turbocharger oil supply (A) and oil return (B) lines (1.4L engine)

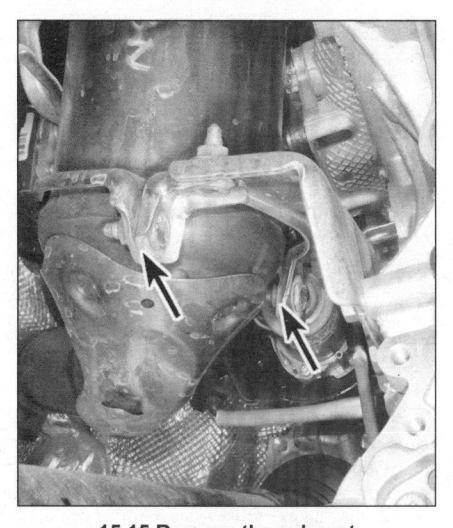

15.15 Remove the exhaust pipe bracket nuts (1.4L engines)

15.16 Remove the exhaust clamp and support the exhaust (heat shield shown removed) (1.4L engine)

15.23 Remove the bolts and the turbocharger inlet (1.4L engine)

Note: *Whenever the battery is disconnected, several systems must be re-learned before they will work properly (see Chapter 5, Section 3).*

Note: *The turbocharger can only be replaced as a unit with the exhaust manifold.*

3   Warm up the engine and drain the engine oil, then allow the engine to cool off and drain the cooling system (see Chapter 1).
4   Disconnect the cable from the negative terminal of the battery (see Chapter 5).
5   Loosen the right-front wheel bolts, raise the front of the vehicle and support it securely on jackstands. Block the wheels at the opposite end and remove the right-front wheel.
6   Remove the under-vehicle splash shield (see Chapter 1, Section 6).
7   Remove the engine cover (see Chapter 1, Section 7). Remove the right-side inner fender liner (see Chapter 11).

1.4L engines
8   Remove the right side driveaxle heat shield mounting fasteners and shield **(see illustration)**.
9   Remove the protective heat shield and disconnect the oil pressure switch connector **(see illustration)**.
10   On Golf models, remove the secondary air injection pump (see Chapter 6, Section 18).
11   On Jetta models, if necessary, remove the electric brake vacuum pump (see Chapter 9, Section 17).
12   On all models, remove the bolts and detach the oil supply line from the turbocharger and engine block **(see illustration)**.
13   Remove the bolts and detach the oil return pipe from the turbocharger and engine block.
14   On Jetta models, disconnect the coolant temp sensor and charge air pressure sensor connectors.
15   On all models, remove the exhaust pipe bracket nuts **(see illustration)**.

16   Remove the exhaust clamp at the catalytic converter and secure the exhaust using a length of wire **(see illustration)**.
17   On Jetta models, release the air guide top cover clips and remove the top cover.
18   Release the air guide lower clips and remove the guide with the air duct hose.
19   On Golf models, disconnect the charge air pressure sensor, release the retaining tabs from both ends of the air duct pipe and remove the pipe from the engine compartment.
20   On all models, disconnect the EVAP canister hose from the crankcase vent hose. Remove the bolts and remove the crankcase vent hose from the engine.
21   Detach the coolant hose from the bracket clamp on the rear of the engine.
22   Disconnect the electrical connector from the turbocharger.
23   Remove the bolts attaching the air inlet to the turbocharger and remove the air inlet **(see illustration)**.

4-18 Chapter 4 Fuel and exhaust systems

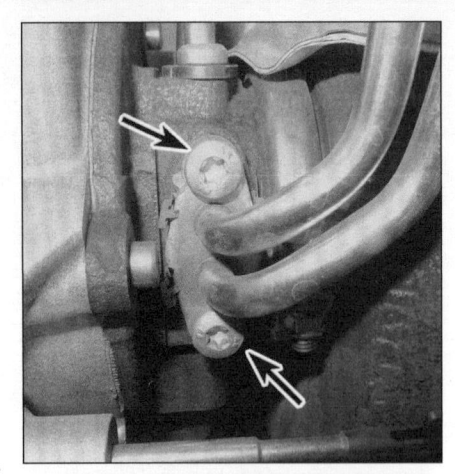

15.24 Remove the bolts and disconnect the coolant lines from the turbocharger (1.4L engine)

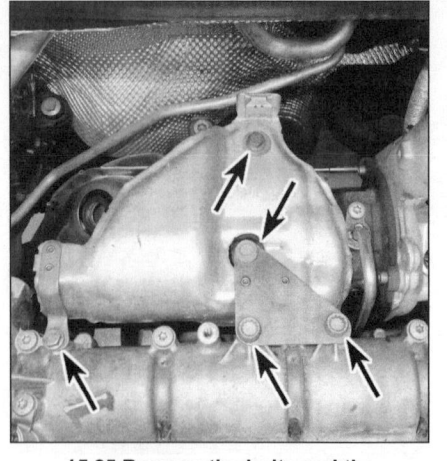

15.25 Remove the bolts and the turbocharger heat shield (1.4L engine)

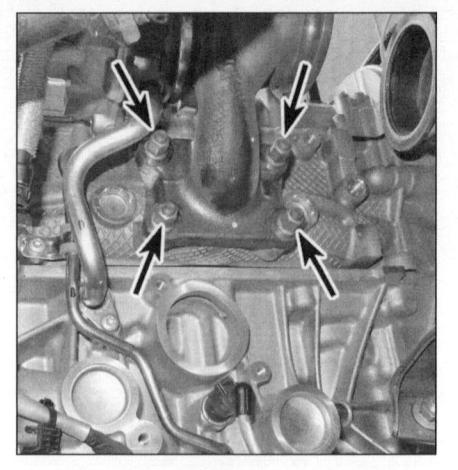

15.26 Turbocharger mounting nuts (1.4L engines)

24 Remove the bracket bolt securing the coolant lines to the top of the engine. Remove the bolts and disconnect the coolant lines from the turbocharger (see illustration). Disconnect the hose at the rear of the engine compartment (if equipped) and move the coolant lines to the side.
25 Remove the bolts attaching the heat shield to the turbocharger (see illustration) and remove the heat shield.
26 Remove the nuts attaching the turbocharger to the cylinder head (see illustration). Remove the turbocharger from the vehicle.

Golf and GTI 1.8L and 2.0L engines (except CBFA and CBTA)

27 Remove the air filter housing with the duct to the turbocharger (see Section 11).
28 Remove the air guide pipe bracket bolt and disconnect the pipe from the turbocharger.
29 Disconnect the coolant hose near the oil fill.
30 Disconnect the ignition coil electrical connectors and the two additional connectors on the left side of the engine that share the same harness. Detach the harness from the valve cover bracket and position the harness to the side.
31 On CNTA, CXCA, CXCB and DKFA engines, disconnect the camshaft adjustment actuator connectors.
32 On all engines, disconnect any related connectors at the rear of the engine to allow clearance to remove the turbocharger and exhaust manifold assembly. Remove the ground cable bolt from the rear of the engine.
33 Disconnect the coolant hose from the pipe above the turbocharger. Remove the bracket bolts from the pipe and move the pipe to the side.
34 Remove the oxygen sensor located before the catalytic converter.
35 On CNTA, CXCA, CXCB and DKFA engines, remove the fasteners and the camshaft adjustment actuators from the cylinder head.

36 On all engines, remove the ignition coils.
37 Remove the four bolts and single nut attaching the heat shield to the cylinder head and remove the heat shield.
38 Remove the turbocharger support bracket nut and bolt from the turbo, loosen the lower bolt but do not remove, and remove the bracket from the vehicle.
39 Disconnect the electrical connectors from the wastegate bypass regulator valve and the turbocharger recirculating valve and detach the harness clips.
40 On DKFA engines, disconnect the crankcase vent hose from the turbo inlet.
41 On CNTA, CXCA, and CXCB engines, remove the crankcase vent from the valve cover.
42 On all engines, remove the bolts and detach the oil supply line from the turbocharger and engine block.
43 Remove the bolts and detach the coolant pipe from the turbocharger and engine block.
44 Remove the bolts and detach the oil return pipe from the turbocharger and engine block.
45 Disconnect the coolant hose near the cylinder head on the rear of the engine.
46 Working under the vehicle, remove the strut rod engine mount between the transmission and the subframe.
47 Using a ratchet strap or similar, rotate the engine so as the top of the engine moves towards the front of the engine compartment, this provides room to remove the turbocharger.
48 Remove the nuts attaching the turbocharger to the cylinder head. Remove the turbocharger from the vehicle.

Jetta 1.8L and 2.0L engines (except CBFA and CBTA)

49 Remove the turbocharger duct elbow from the engine compartment.
50 Remove the turbocharger duct bracket bolt at the rear of the cylinder head and disconnect the duct from the turbocharger. The

duct may be connected to the turbocharger using hose clamps or two locking tabs.
51 Remove the bolt attaching the crankcase vent to the valve cover and disconnect the hose. Remove the turbocharger duct from the vehicle.
52 Remove the bolts securing the air guide pipe across the front of the engine, below the balancer pulley. Disconnect the ends of the pipe from the turbocharger and the charge air cooler, and remove from the vehicle.
53 Disconnect the oil level sensor connector and detach the harness bracket from the subframe.
54 Remove the right-side driveaxle heat shield mounting fasteners and shield (see Chapter 8).
55 Remove the exhaust system and catalytic converter mount support fasteners and mount, then lower the exhaust out of the way (see Section 14).
56 Disconnect the electrical connectors from the wastegate bypass regulator valve and the turbocharger recirculating valve and detach the harness clips.
57 Disconnect the ignition coil connectors. Detach the harness from the harness clips and position the harness to the side.
58 Remove the bolts attaching the small coolant pipe around the rear of the valve cover.
59 Remove the exhaust manifold heat shield fasteners and remove the heat shield.
60 Remove the bolts and detach the coolant pipe from the turbocharger and the engine block.
61 Remove the bolts and detach the oil return pipe from the turbocharger and the engine block.
62 Remove the turbocharger support bracket nut and bolt from the turbo, loosen the lower bolt but do not remove, and remove the bracket from the vehicle.
63 Remove the bolts and detach the oil supply line and heat shield from the turbocharger.
64 Remove the nuts attaching the turbocharger to the cylinder head. Remove the turbocharger from the vehicle.

Chapter 4 Fuel and exhaust systems

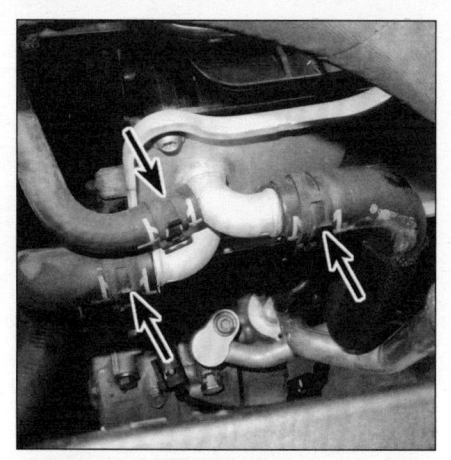

15.92 Disconnect the coolant hoses from the charge air cooler (1.4L engines)

15.93 Remove the bolts around the perimeter of the charge air cooler (1.4L engines)

2.0L CBFA and CBTA engines

65 Disconnect the electrical connectors and harness from the ignition coils (see Chapter 5).
66 Disconnect the coolant hose to the coolant expansion tank (see Chapter 3).
67 Disconnect the coolant hoses from the coolant pipe that is at the right of the cylinder head, over the bell housing.
68 Disconnect the ground wire at the right side of the cylinder head.
69 Disconnect the vacuum line union at the rear of the engine, above the exhaust heat shield.
70 Remove the turbocharger duct bracket bolt at the rear of the cylinder head and disconnect the duct from the turbocharger. Position the duct to the side.
Note: *For the next step, a 6 mm hex fitting socket at least 2" (5 cm) long is needed.*
71 Remove the exhaust manifold heat shield fasteners and remove the heat shield with the coolant pipe.
72 Disconnect the oil supply line from the turbocharger.
73 Remove the bolts securing the air guide pipe across the front of the engine, below the balancer pulley. Disconnect the ends of the pipe from the turbocharger and the charge air cooler, and remove from the vehicle.
74 Remove the right side driveaxle heat shield mounting fasteners and shield (see Chapter 8).
75 Remove the exhaust system and catalytic converter mount support fasteners and mount, then lower the exhaust out of the way (see Section 14).
76 Disconnect the electrical connectors from the wastegate bypass regulator valve and the turbocharger recirculating valve and detach the harness clips.
77 Remove the bolts and the turbocharger support bracket from the engine block.
78 Remove the banjo bolt attaching the coolant pipe to the turbocharger and detach the pipe.
79 Remove the bolts attaching the oil return line to the turbocharger and detach the line. Remove the bracket bolt for the oil supply line.
80 Disconnect the lower coolant hose from the coolant pipe near the AC receiver drier.
81 Remove the exhaust manifold nuts and remove the turbocharger and exhaust manifold from the vehicle.
82 If you're replacing the exhaust manifold/turbocharger assembly, place the assembly on a clean work bench and remove the components that don't come with the new or remanufactured exhaust manifold/turbocharger assembly and install them on the new unit.

Installation - All models

83 When installing the turbocharger assembly, be sure to:
 a) *Coat the threads of all exhaust manifold and turbocharger fasteners and all banjo fasteners with anti-seize compound*
 b) *Replace all seals, gaskets and self-locking nuts*
 c) *Use new sealing washers for all banjo fittings*
 d) *Tighten all turbocharger fasteners to the torque listed in this Chapter's Specifications (if no torque specification is listed, tighten the fastener securely)*
 e) *Add oil to the turbocharger through the oil supply line*
 f) *Make sure that all hoses are clean inside before installing them*
 g) *If you're replacing any hose clamps, the replacement clamps must be the same types as the old ones*
 h) *Make sure that the exhaust system is properly aligned before reconnecting it to the turbocharger*
 i) *Refill the engine oil and coolant (see Chapter 1) before starting the engine*
 j) *After starting the engine, let it idle for about one minute to ensure an adequate oil supply to the turbocharger*

84 Installation is otherwise the reverse of removal.

Charge air cooler - replacement

85 Raise the vehicle and support it securely on jackstands.
86 Remove the under-vehicle splash shield (see Chapter 1, Section 6).
87 Remove the engine cover (see Chapter 1, Section 7).
88 Drain the cooling system (see Chapter 1).

1.4L engines

Note: *The charge air cooler is mounted to the intake manifold on 1.4L engines.*
89 Remove the cooling fan shroud (see Chapter 3).
90 Detach the air conditioning refrigerant line from the bracket and clamp on the radiator core support.
91 Unbolt the air conditioning compressor from the bracket and secure it out of the way towards the passenger's side of the vehicle (see Chapter 3, Section 13).
92 Disconnect the coolant hoses from the charge air cooler **(see illustration)**.
93 Remove the bolts securing the charge air cooler to the intake manifold **(see illustration)**.
94 Slide the charge air cooler forward and upwards to remove from the intake manifold and the vehicle.
95 Installation is reverse of removal. Be sure to install the seal on the cooler.
96 Tighten the bolts in a diagonal pattern, starting at the center and working outwards.
97 Refill the cooling system (see Chapter 1).

1.8L and 2.0L engines

98 Remove the front bumper cover (see Chapter 11).
99 Loosen the clamps (Jetta - disconnect the charge air pressure sensor) and remove the charge air hoses from the charge pipes and charge air cooler.
100 Remove the cooling fan shroud and radiator (see Chapter 3).
101 On Golf models, remove the air duct attached to the radiator core support. Remove the three fasteners and the center guide from the core support.
102 On all models, remove the charge air cooler bracket bolts.
103 Detach the condenser from the charge air cooler. Secure the condenser to the radiator core support.
104 Remove the charge air cooler from under the vehicle.
105 Installation is the reverse of removal. Replace any damaged ducts, hoses and/or hose clamps.
106 Refill the cooling system (see Chapter 1).

Notes

Chapter 5
Engine electrical systems

Contents

	Section		Section
Alternator - removal and installation	7	General information and precautions	1
Battery and battery tray - removal and installation	4	Ignition coils - removal and installation	6
Battery cables - replacement	5	Starter motor - removal and installation	8
Battery - disconnection and reconnection	3	Troubleshooting	2

Specifications

General
Firing order.. See Chapter 1
Battery voltage
    Engine off.. 12.0 to 12.6 volts
    Engine running.. Approximately 13.5 volts

Torque specifications
Ft-lbs (unless otherwise indicated) **Nm**

Note: One foot-pound (ft-lb) of torque is equivalent to 12 inch-pounds (in-lbs) of torque. Torque values below approximately 15 ft-lbs are expressed in inch-pounds, because most foot-pound torque wrenches are not accurate at these smaller values.

Alternator mounting bolts	18	25
Starter motor mounting bolts		
2014 and earlier models		
M10 bolts (if equipped)	30	40
M12 bolts	55	75
2015 and later models	59	80
Ignition coil bolts	84 in-lbs	10
Spark plugs	See Chapter 1	

1 General information and precautions

General information

Ignition system

1 The electronic ignition system consists of the Camshaft Position (CMP) sensor, the Knock Sensor (KS), the Powertrain Control Module (PCM), the ignition switch, the battery, the individual ignition coils, and the spark plugs. For more information on the CMP and KS sensors, as well as the PCM, refer to Chapter 6.

Charging system

2 The charging system includes the alternator (with an integral voltage regulator), the Powertrain Control Module (PCM), the Body Control Module (BCM), a charge indicator light on the dash, the battery, the Battery Monitoring Control Module (vehicles equipped with active start/stop system), a fuse or fusible link and the wiring connecting all of these components. The charging system supplies electrical power for the ignition system, the lights, the radio, etc. The alternator is driven by a drivebelt.

Starting system

3 The starting system consists of the battery, the ignition switch, the starter relay, the Powertrain Control Module (PCM), the Body Control Module (BCM), the Transmission Range (TR) switch or clutch start switch, the starter motor and solenoid assembly, and the wiring connecting all of the components.

4 Some models may also be equipped with an active start/stop system. This system turns the engine off when coming to a stop (such as a stoplight) to conserve fuel. When the driver is ready to drive again, the engine is started. When using the start/stop system, this places additional wear on the starter due to the increased use of the starter to start the engine.

Precautions

5 Always observe the following precautions when working on the electrical system:

a) *Be extremely careful when servicing engine electrical components. They are easily damaged if checked, connected or handled improperly.*
b) *Never leave the ignition switched on for long periods of time when the engine is not running.*
c) *Never disconnect the battery cables while the engine is running.*
d) *Maintain correct polarity when connecting battery cables from another vehicle during jump starting - see the "Booster battery (jump) starting" Section at the front of this manual.*
e) *Always disconnect the cable from the negative battery terminal before working on the electrical system, but read the battery disconnection procedure first (see Section 3).*

6 It's also a good idea to review the safety-related information regarding the engine electrical systems located in the *Safety first!* Section at the front of this manual before beginning any operation included in this Chapter.

Start/stop technology

Warning: *Vehicles equipped with the Start/Stop System can restart at any time if the key has been left in the On position during any maintenance or repairs. Be sure to deactivate the Start/Stop system before performing any repairs. A message will be displayed in the instrument cluster indicating the Start/Stop has been disabled.*

7 The Start/Stop System, on models so equipped, reduces fuel consumption by turning off the engine when the vehicle comes to a complete stop, and restarting the engine when the driver takes their foot off of the brake. The system works automatically when the vehicle is driven for approximately four seconds at a minimum speed of 2 mph.

8 The system should be disabled when doing any type of service work on the vehicle that requires raising the vehicle or working under the hood.

9 The switch for the start/stop is located in the console just in front of the shifter lever.

10 Vehicles equipped with a start/stop system use an AGM (Absorbent Glass Mat) battery instead of a traditional lead acid battery (acid flooded type). The AGM battery appearance is very similar to the standard flooded type battery. When replacing the battery, make sure the replacement battery is of the proper type.

2 Troubleshooting

Ignition system

1   If a malfunction occurs in the ignition system, do not immediately assume that any particular part is causing the problem. First, check the following items:

a) Make sure that the cable clamps at the battery terminals are clean and tight.
b) Test the condition of the battery (see Steps 16 through 19). If it doesn't pass all the tests, replace it.
c) Check the ignition coil connections.
d) Check any relevant fuses in the engine compartment fuse and relay box (see Chapter 12). If they're burned, determine the cause and repair the circuit.

Check

Warning: *Because of the high voltage generated by the ignition system, use extreme care when performing a procedure involving ignition components.*
Note: *The ignition system components on these vehicles are difficult to diagnose. In the event of ignition system failure that you can't diagnose, have the vehicle tested at a dealer service department or other qualified auto repair facility.*
Note: *You'll need a spark tester for the following test. Spark testers are available at most auto supply stores.*

2   If the engine turns over but won't start, verify that there is sufficient ignition voltage to fire the spark plugs as follows.
3   On models with a coil-over-plug type ignition system, remove a coil and install the tester between the boot at the lower end of the coil and the spark plug **(see illustration)**. On models with spark plug wires, disconnect a spark plug wire from a spark plug and install the tester between the spark plug wire boot and the spark plug.
4   Crank the engine and note whether or not the tester flashes.
Caution: *Do NOT crank the engine or allow it to run for more than five seconds; running the engine for more than five seconds may set a Diagnostic Trouble Code (DTC) for a cylinder misfire.*
5   If the tester flashes during cranking, the coil is delivering sufficient voltage to the spark plug to fire it. On coil-over-plug models, repeat this test for each cylinder to verify that the other coils are OK.
6   If the tester doesn't flash on coil-over-plug models, remove a coil from another cylinder and swap it for the one being tested. If the tester now flashes, you know that the original coil is bad. If the tester doesn't flash on models with spark plug wires, check the spark plug wire and coil electrical connections. If the tester still doesn't flash, a sensor, the PCM or wiring harness may be defective. Check for store DTCs related to the ignition system (such as the camshaft position sensor). Have the PCM checked out by a dealer service department or other qualified repair shop

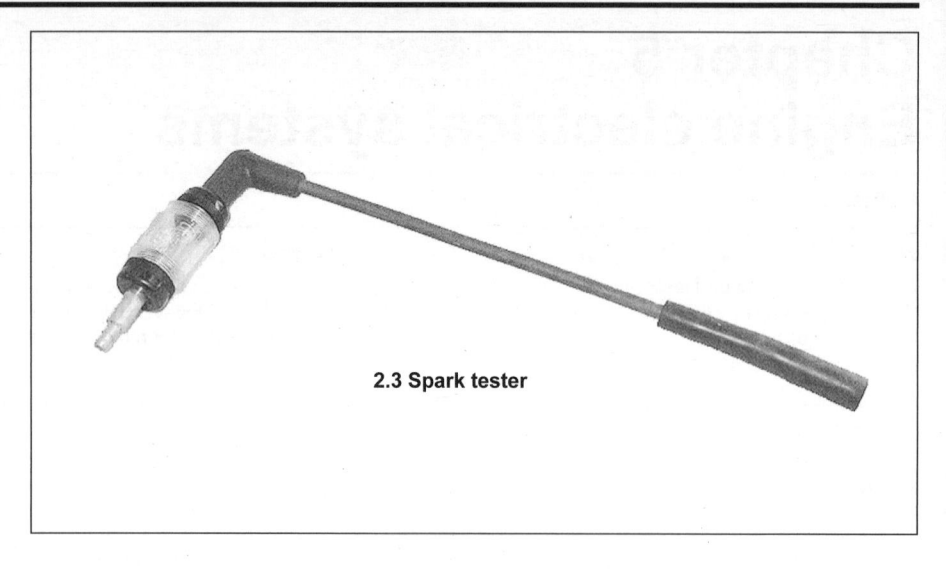

2.3 Spark tester

(testing the PCM is beyond the scope of the do-it-yourselfer because it requires expensive special tools).
7   If the tester flashes during cranking but a misfire code (related to the cylinder being tested) has been stored, the spark plug could be fouled or defective.
8   If all coils are firing correctly, but the engine misfires, check that an ignition coil boot isn't leaking spark from the side and grounding to the cylinder head (coil-over-plug), or that a spark plug wire is not grounding against the engine and leaking spark. If everything checks out ok, check for a bad or fouled spark plug or faulty fuel injector. Additional diagnostics may be necessary by a qualified repair facility with specialized equipment.

Charging system

9   If a malfunction occurs in the charging system, do not automatically assume the alternator is causing the problem. First check the following items:

a) Check the drivebelt tension and condition, as described in Chapter 1. Replace it if it's worn or deteriorated.
b) Make sure the alternator mounting bolts are tight.
c) Inspect the alternator wiring harness and the connectors at the alternator and voltage regulator. They must be in good condition, tight and have no corrosion.
d) Check the fusible link (if equipped) or main fuse in the underhood fuse/relay box. If it is burned, determine the cause, repair the circuit and replace the link or fuse (the vehicle will not start and/or the accessories will not work if the fusible link or main fuse is blown).
e) Start the engine and check the alternator for abnormal noises (a shrieking or squealing sound indicates a bad bearing).
f) Check the battery. Make sure it's fully charged and in good condition (one bad cell in a battery can cause overcharging by the alternator).

g) Disconnect the battery cables (negative first, then positive). Inspect the battery posts and the cable clamps for corrosion. Clean them thoroughly if necessary (see Chapter 1). Reconnect the cables (positive first, negative last).

Alternator - check

10   Use a voltmeter to check the battery voltage with the engine off. It should be at least 12.6 volts **(see illustration 2.16)**.
11   Start the engine and check the battery voltage again. It should now be approximately 13.5 to 15 volts.
12   If the voltage reading is more or less than the specified charging voltage, the voltage regulator is probably defective, which will require replacement of the alternator (the voltage regulator is not replaceable separately). Remove the alternator and have it bench tested (most auto parts stores will do this for you).
13   The charging system (battery) light on the instrument cluster lights up when the ignition key is turned to On, but it should go out when the engine starts.
14   If the charging system light stays on after the engine has been started, there is a problem with the charging system. Before replacing the alternator, check the battery condition, alternator belt tension and electrical cable connections.
15   If replacing the alternator doesn't restore voltage to the specified range, have the charging system tested by a dealer service department or other qualified repair shop.

Battery - check

16   Check the battery state of charge. Visually inspect the indicator eye on the top of the battery (if equipped with one); if the indicator eye is black in color, charge the battery as described in Chapter 1. Next perform an open circuit voltage test using a digital voltmeter. With the engine and all accessories Off, touch the negative probe of the voltmeter to the negative terminal of the battery and the positive probe to the positive terminal of the battery **(see illustration)**. The battery volt-

Chapter 5 Engine electrical systems

2.16 To test the open circuit voltage of the battery, touch the black probe of the voltmeter to the negative terminal and the red probe to the positive terminal of the battery; a fully charged battery should be at least 12.6 volts

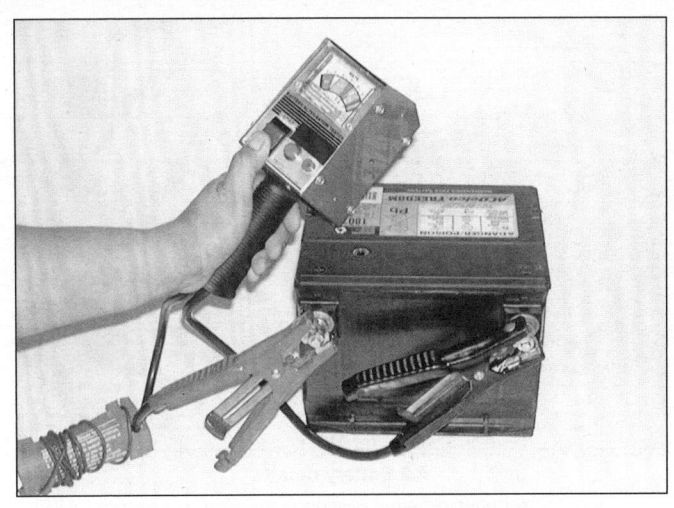

2.18 Connect a battery load tester to the battery and check the battery condition under load following the tool manufacturer's instructions

age should be 12.6 volts or slightly above. If the battery is less than the specified voltage, charge the battery before proceeding to the next test. Do not proceed with the battery load test unless the battery charge is correct.

Note: *The battery's surface charge must be removed before accurate voltage measurements can be made. Turn on the high beams for ten seconds, then turn them off and let the vehicle stand for two minutes.*

17 Disconnect the negative battery cable, then the positive cable from the battery.

18 Perform a battery load test. An accurate check of the battery condition can only be performed with a load tester **(see illustration)**. This test evaluates the ability of the battery to operate the starter and other accessories during periods of high current draw. Connect the load tester to the battery terminals. Load test the battery according to the tool manufacturer's instructions. This tool increases the load demand (current draw) on the battery.

19 Maintain the load on the battery for 15 seconds and observe that the battery voltage does not drop below 9.6 volts. If the battery condition is weak or defective, the tool will indicate this condition immediately.

Note: *Cold temperatures will cause the minimum voltage reading to drop slightly. Follow the chart given in the manufacturer's instructions to compensate for cold climates. Minimum load voltage for freezing temperatures (32 degrees F) should be approximately 9.1 volts.*

Starting system

The starter rotates, but the engine doesn't

20 Remove the starter (see Section 8). Check the overrunning clutch and bench test the starter to make sure the drive mechanism extends fully for proper engagement with the flywheel ring gear. If it doesn't, replace the starter.

21 Check the flywheel ring gear for missing teeth and other damage. With the ignition turned off, rotate the flywheel so you can check the entire ring gear.

The starter is noisy

22 If the solenoid is making a chattering noise, first check the battery (see Steps 16 through 19). If the battery is okay, check the cables and connections.

23 If you hear a grinding, crashing metallic sound when you turn the key to Start, check for loose starter mounting bolts. If they're tight, remove the starter and inspect the teeth on the starter pinion gear and flywheel ring gear. Look for missing or damaged teeth.

24 If the starter sounds fine when you first turn the key to Start, but then stops rotating the engine and emits a zinging sound, the problem is probably a defective starter drive that's not staying engaged with the ring gear. Replace the starter.

The starter rotates slowly

25 Check the battery (see Steps 16 through 19).

26 If the battery is okay, verify all connections (at the battery, the starter solenoid and motor) are clean, corrosion-free and tight. Make sure the cables aren't frayed or damaged.

27 Check that the starter mounting bolts are tight so it grounds properly. Also check the pinion gear and flywheel ring gear for evidence of a mechanical bind (galling, deformed gear teeth or other damage).

The starter does not rotate at all

28 Check the battery (see Steps 16 through 19).

29 If the battery is okay, verify all connections (at the battery, the starter solenoid and motor) are clean, corrosion-free and tight. Make sure the cables aren't frayed or damaged.

30 Check all of the fuses in the underhood fuse/relay box.

31 Check that the starter mounting bolts are tight so it grounds properly.

32 Check for voltage at the starter solenoid "S" terminal when the ignition key is turned to the start position. If voltage is present, replace the starter/solenoid assembly. If no voltage is present, the problem could be the starter relay, the Transmission Range (TR) switch (see Chapter 7B) or clutch pedal switch (see Chapter 8), or with an electrical connector somewhere in the circuit (see the wiring diagrams at the end of this manual). Also, on many modern vehicles, the Powertrain Control Module (PCM) and the Body Control Module (BCM) control the voltage signal to the starter solenoid; on such vehicles a special scan tool is required for diagnosis.

3 Battery - disconnection and reconnection

Caution: *Always disconnect the cable from the negative battery terminal FIRST and hook it up LAST or the battery may be shorted by the tool being used to loosen the cable clamps.*

Caution: *If the audio system is equipped with an anti-theft system, make sure you have the correct activation code before disconnecting the battery.*

1 Some systems on the vehicle require battery power to be available at all times, either to maintain continuous operation (alarm system, power door locks, etc.), or to maintain control unit memory (radio station presets, Powertrain Control Module and other control units). When the battery is disconnected, the power that maintains these systems is cut, and the systems must be re-learned before they will work properly. So, before you disconnect the battery, please note that on a vehicle with power door locks, it's a wise precaution

5-4   Chapter 5   Engine electrical systems

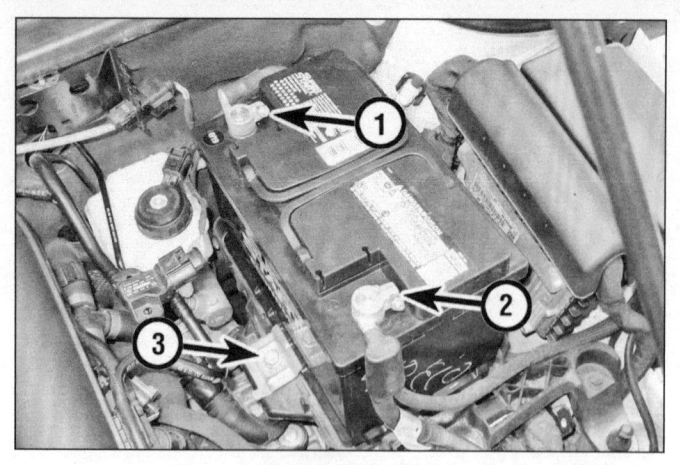

4.2 Battery details
1 Battery negative terminal (disconnect FIRST, reconnect LAST)
2 Battery positive terminal
3 Battery hold-down clamp

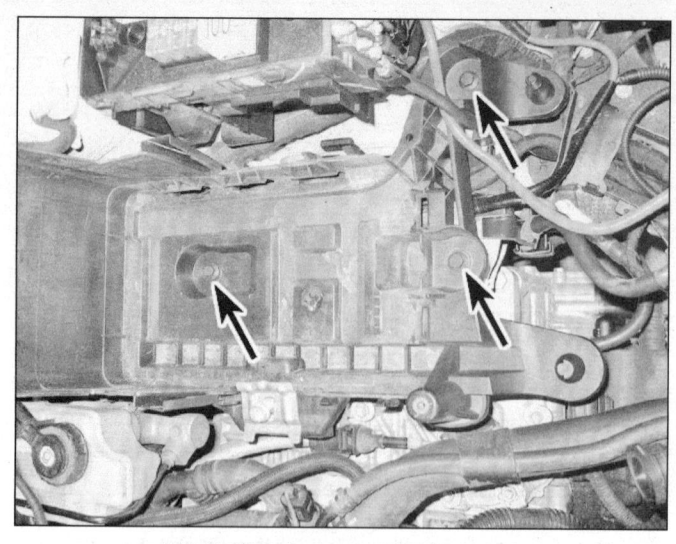

4.10 Battery tray mounting fasteners

to remove the key from the ignition and to keep it with you, so that it does not get locked inside if the power door locks should engage accidentally when the battery is reconnected!
2 Devices known as "memory-savers" can be used to avoid some of these problems. Precise details vary according to the device used. The typical memory saver is plugged into the cigarette lighter and is connected to a spare battery. Then the vehicle battery can be disconnected from the electrical system. The memory saver will provide sufficient current to maintain audio unit security codes, PCM memory, etc., and will provide power to always hot circuits such as the clock and radio memory circuits.
Warning: *Some memory savers deliver a considerable amount of current in order to keep vehicle systems operational after the main battery is disconnected. If you're using a memory saver, make sure that the circuit concerned is actually open before servicing it.*
Warning: *If you're going to work near any of the airbag system components, the battery MUST be disconnected and a memory saver must NOT be used. If a memory saver is used, power will be supplied to the airbag, which means that it could accidentally deploy and cause serious personal injury.*
3 To disconnect the battery for service procedures requiring power to be cut from the vehicle:
  a) Turn the ignition off and remove the key.
  b) Remove the battery cover and/or negative battery terminal cover.
  c) Loosen the cable end bolt and disconnect the cable from the negative battery terminal. Isolate the cable end to prevent it from coming into accidental contact with the battery terminal.

Battery initialization (re-learning)

4 Reconnect the negative battery terminal.
5 Insert the key into the ignition switch and turn the key to On position, then the Off position.
6 Connect a scan tool and check for any trouble codes (see Chapter 6).
7 Turn the ignition on and reset the clock. Fully open and close all windows and the sunroof to reset the auto stop functions. When the windows are fully closed, continue holding switch in the closed position until a relay click is heard.
8 Start the vehicle; the electro-mechanical steering warning light (ASR/ESP) should now be on. Drive the vehicle in a straight direction for a short distance to a maximum speed of 10 to 13 mph - the steering angle sensor should now be re-initialized and the warning light turned off.

4   Battery and battery tray - removal and installation

Caution: *If the battery is disconnected, several systems must be re-learned before they will work properly (see Section 3).*

Battery

1 Open the battery jacket flap.
2 Disconnect the cable from the negative battery terminal first, then disconnect the cable from the positive battery terminal **(see illustration)**.
3 Remove the battery hold-down clamp bolt and clamp at the base of the battery.
4 Lift out the battery. Be careful - it's heavy.
Note: *Battery straps and handlersare available at most auto parts stores for reasonable prices. They make it easier to remove and carry the battery.*
5 If you are replacing the battery, make sure you get one that's identical, with the same dimensions, amperage rating, cold cranking rating, etc. Also, be sure to remove the heat shield (if equipped) from the old battery and install it on the new battery.

6 Installation is the reverse of removal. Be sure to connect the positive cable first and the negative cable last.

Battery tray

7 Remove the battery (see Steps 1 through 5).
8 If needed for access, remove the air filter housing. See Chapter 4, Section 11.
9 Disconnect any harness retainers, or components attached to the battery tray. On 2015 and later Jetta models, on automatic transmission models this may include the Transmission Control module (TCM), and on vehicles equipped with 5-cylinder engines, may include the power steering reservoir.
10 Remove the battery tray mounting fasteners **(see illustration)** and lift out the battery tray.
11 Installation is the reverse of removal. Be sure to connect the positive cable first and the negative cable last.

5   Battery cables - replacement

Caution: *On models equipped with the active start/stop system, a Battery Motioning Control Module is part of the negative battery cable.*

1 When removing the cables, always disconnect the cable from the negative battery terminal first and hook it up last, or you might accidentally short out the battery with the tool you're using to loosen the cable clamps. Even if you're only replacing the cable for the positive terminal, be sure to disconnect the negative cable from the battery first.
2 Disconnect the old cables from the battery, then trace each of them to their opposite ends and disconnect them. Be sure to note the routing of each cable before disconnecting it to ensure correct installation.
3 If you are replacing any of the old cables, take them with you when buying new cables. It is vitally important that you replace the cables with identical parts.

Chapter 5 Engine electrical systems

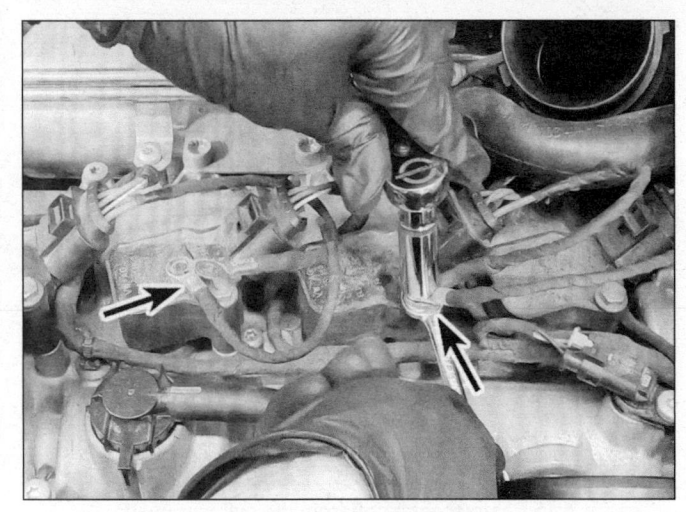

6.4 Remove the nut and ground wires from the coil fasteners (1.4L shown, others similar)

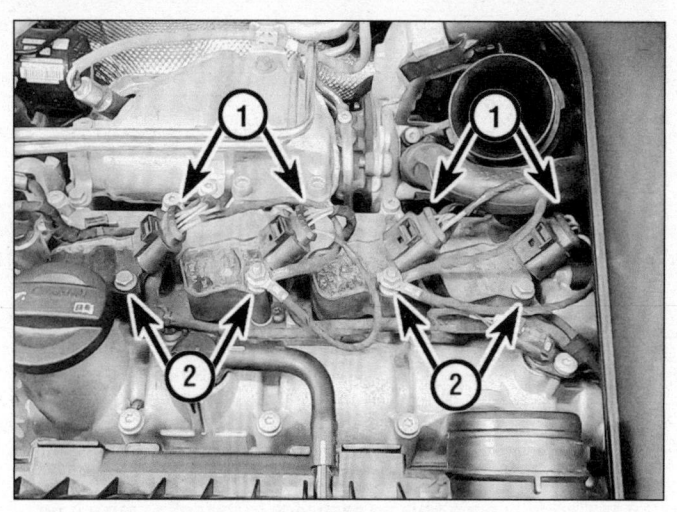

6.5 Disconnect the coil electrical connectors and remove the bolts (1.4L shown, others similar)

4 Clean the threads of the solenoid or ground connection with a wire brush to remove rust and corrosion. Apply a light coat of battery terminal corrosion inhibitor or petroleum jelly to the threads to prevent future corrosion.

5 Attach the cable to the solenoid or ground connection and tighten the mounting nut/bolt securely.

6 Before connecting a new cable to the battery, make sure that it reaches the battery post without having to be stretched.

7 Connect the cable to the positive battery terminal first, *then* connect the ground cable to the negative battery terminal.

Battery Monitoring Control Module

8 On models equipped with the active start/stop system, the Battery Motioning Control Module is attached to the negative battery cable. The module is part of the battery cable and cannot be replaced separately - it must be replaced with the cable. When replacing the negative battery cable on a vehicle with active start/stop, if the old negative cable has a Battery Motioning Control Module, be sure the new battery cable is equipped with the Battery Motioning Control Module.

6 Ignition coils - removal and installation

Caution: *If the battery is disconnected, several systems must be re-learned before they will work properly (see Section 3).*

All engines

1 Disconnect the cable from the negative battery terminal (see Section 3).

2 Remove the engine cover/air filter housing (see Chapter 1, Section 7).

1.4L turbocharged engines, 1.8L turbocharged engines and 2.0L turbocharged (CNTA, CPLA, CPPA, CXCA, CXCB) engines

3 Remove the air filter housing lid or any air intake duct work to allow access for removing the coils. The ducts may be attached to the engine with brackets, secured by bolts.

4 If equipped, remove the nuts and ground cables from the coil mounting fasteners **(see illustration)**.

5 Disconnect the coil electrical connectors and remove the the coil fasteners **(see illustration)**.

6 Using a prybar or similar tool, carefully pry the ignition coil from the valve cover and spark plug to remove **(see illustration)**.

7 Apply a little silicone dielectric compound to the inside of the spark plug boot before installing the coil. Installation is otherwise the reverse of removal.

6.6 Use a pry bar or similar tool to carefully release the coils from the spark plugs and valve cover (1.4L shown, others similar)

2.0L non-turbocharged engine

Note: *This model uses a single ignition coil and spark plug wires.*

8 Mark the location of the spark plug wires and disconnect from the coil terminals.

9 Disconnect the electrical connector from the coil.

10 Remove the bolts securing the coil and remove the ignition coil.

11 Apply a little silicone dielectric compound to the inside of the spark plug wire boot before installing to the coil. Installation is otherwise the reverse of removal.

All other engines

12 On four-cylinder engines, remove the ignition coil harness fasteners.

13 On five-cylinder engines, disconnect the electrical connector from the ignition coils **(see illustration)**. On four-cylinder engines, the ignition coil harness is rigid, so you must pull the coils up about an inch (see Step 5), disconnect all electrical connectors from the coils and pull the harness and connectors away as a unit.

6.13 Depress the tabs to release the connectors

5-6   Chapter 5   Engine electrical systems

6.14 Carefully pry a little on each side of the coil to raise it up

6.15 Remove each coil from the cylinder head

14   Using special tool #T40039, slide the tool in to the top slot of the coil and pull the coil up approximately 1-1/4 inches. If the special tool is not available, use two flat blade screwdrivers or a trim panel tool to carefully pry the ignition coil up **(see illustration)**.

15   Grasp the ignition coil firmly and pull it straight up **(see illustration)**.

16   Apply a little silicone dielectric compound to the inside of the spark plug boot before installing the coil. Installation is otherwise the reverse of removal.

7   Alternator - removal and installation

Warning: *Wait until the engine is completely cool before performing this procedure.*

Note: *Some alternators may be equipped with a pulley with a one-way clutch. The pulley may need to be transferred over to the new alternator.*

1   Disconnect the cable from the negative battery terminal (see Section 3).

2   Remove the engine cover.

Four-cylinder models

2014 and earlier models

3   On 2.0L turbocharged models, raise and support the front of the vehicle on jackstands.

4   On all turbocharged models, remove the air inlet hoses (see Chapter 1, Section 7).

5   Remove the turbocharger charge air pipe (see Chapter 4).

6   On non-turbocharged models, remove the air filter housing (see Chapter 4, Section 11).

7   On all models, remove the drivebelt (see Chapter 1).

8   On turbocharged models, loosen the coolant pipe mounting bolts from the intake manifold, and position the coolant pipe upwards to gain additional clearance.

9   On 2.0L turbocharged models, remove the engine undercover and detach the air conditioning compressor and position to the side (see Chapter 3, Section 13).

10   On all models, disconnect the electrical connectors from the alternator. On some models, it may be necessary to remove the mounting bolts and reposition the alternator to remove the connectors.

11   Remove the alternator mounting fasteners and remove the alternator.

2015 and later models

Note: *On models with the EVAP canister mounted in the engine compartment, detach the canister from the bracket and position it to the side with the lines connected.*

Caution: *Before installing the alternator, ensure that the alternator mounting bushings move freely in the mounts. Otherwise the alternator may not be secured to the bracket properly.*

12   Raise and support the front of the vehicle on jackstands.

13   Remove the engine undercover.

1.4L engines

14   On 2015 through 2017 Jetta models, detach the lower coolant hose from the clip attached to the fan shroud and secure the hose towards the engine for working room. **(see illustrations)**

15   On 2015 through 2017 Jetta models, carefully pry the auxiliary coolant pump rub-

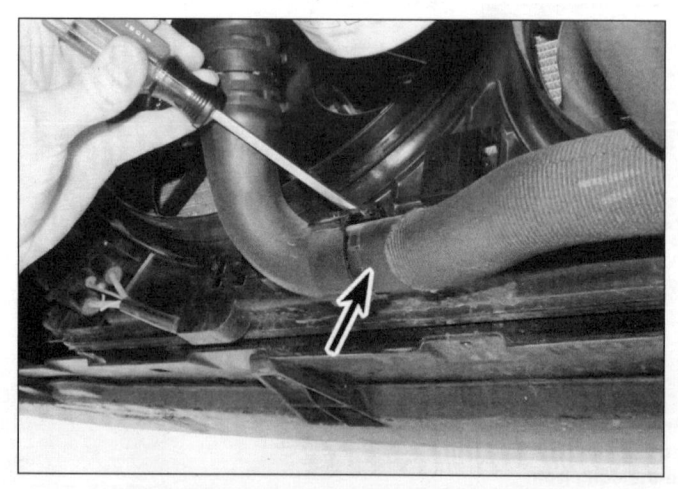

7.14a Detach the lower hose from the clip . . .

7.14b . . . and secure the hose towards the engine for working room (1.4L shown)

Chapter 5 Engine electrical systems

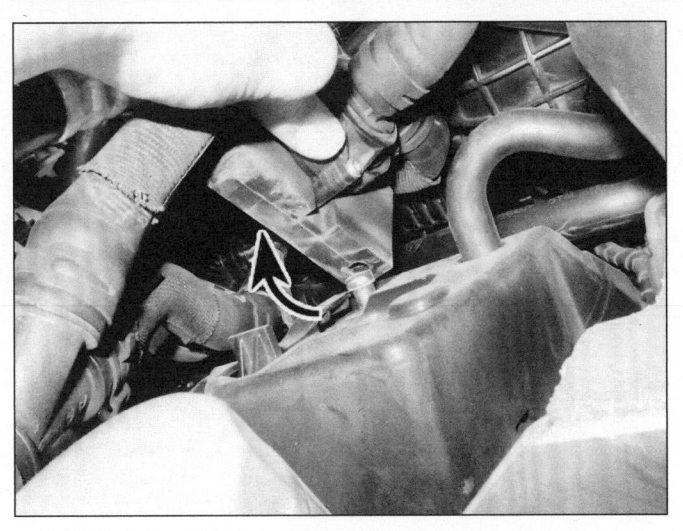

7.15a Pull the rubber auxiliary water pump mounting support from the stud on the engine . . .

7.15b . . . and disconnect the pump electrical connector

ber mount from the mounting studs on the engine. Re-position the pump and disconnect the electrical connector. **(see illustrations)**

16 On all models, remove the drivebelt (see Chapter 1, Section 19).

17 Remove the air conditioning compressor mounting bolts and support the compressor in a position to the side (see Chapter 3, Section 13) **(see illustration)**.
Warning: *Do not disconnect the refrigerant lines.*

18 Remove the alternator mounting bolts and pry the alternator from the support **(see illustrations)**.

19 Rotate the alternator to allow access to the electrical connectors from under the vehicle **(see illustration)**.

20 Disconnect the electrical connectors and remove the alternator from the vehicle.

Jetta 1.8L and 2.0L engines (2015 through 2017)

21 Remove the bolts attaching the coolant pipe above the alternator. Secure the pipe to the side.

7.17 Remove the air conditioning compressor mounting bolts and secure the compressor so as to not damage the refrigerant lines (1.4L shown)

7.18a Remove the alternator bolts . . .

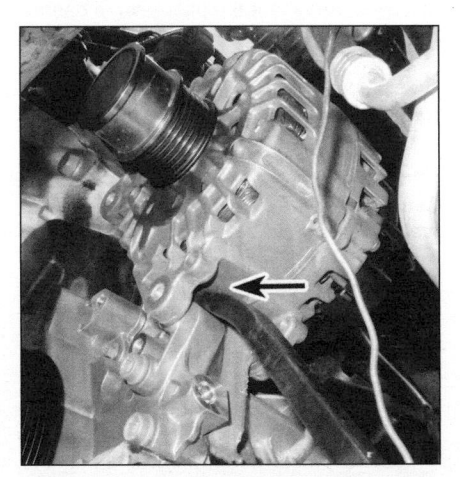

7.18b . . . pry the bottom of the alternator from the bracket . . .

7.18c . . . then pry the top of the alternator from the bracket (1.4L shown)

7.19 After re-positioning the alternator to access the electrical connectors, disconnect them (1.4L shown)

5-8   **Chapter 5   Engine electrical systems**

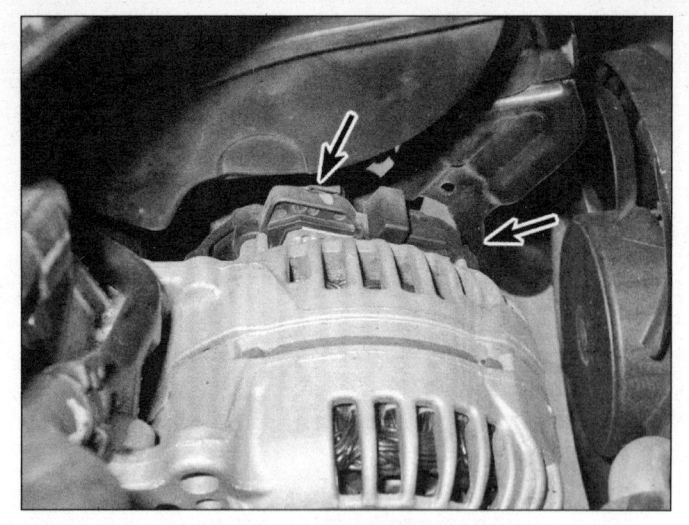

7.43 Alternator electrical connectors

7.44a Alternator upper mounting bolt location . . .

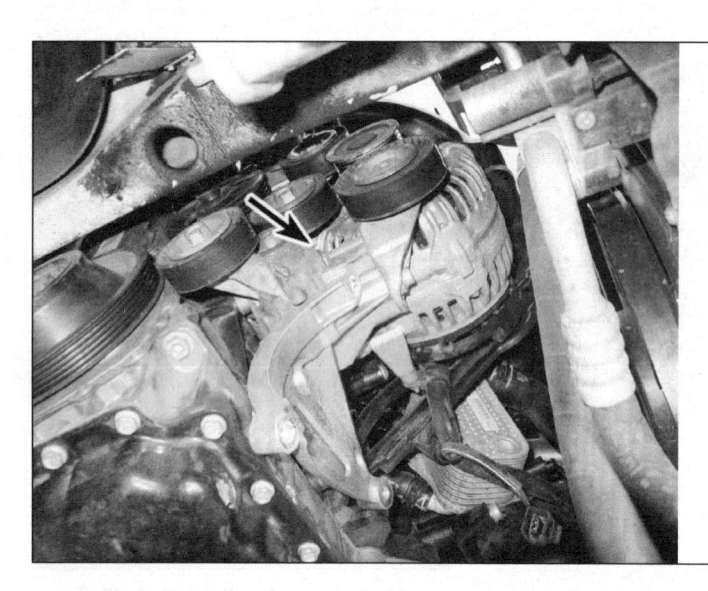

7.44b . . . and lower mounting bolt location

22   Disconnect and if necessary, remove the oil pressure switch near the alternator to prevent the switch from getting damaged during alternator replacement (see Chapter 6, Section 20).
23   Remove the drivebelt (see Chapter 1, Section 19).
24   Working from under the vehicle, remove the lower alternator bolts.
25   Working from the top, remove the upper alternator bolts.
26   Rotate the alternator to allow access to the alternator electrical connectors.
27   Disconnect the electrical connectors and remove the alternator from the vehicle.

Golf 1.8L and 2.0L engines/Jetta 2.0L engines (2018 and later)

28   Remove the engine cooling fans (see Chapter 3, Section 5).
29   Remove the drivebelt (see Chapter 1, Section 19).
30   Remove the upper alternator bracket bolts.
31   Loosen the hose clamps and remove the turbocharger piping from under the front of the engine.
32   Detach the coolant hose that runs along the front of the engine, from the clip.
33   Remove the bolts securing the turbocharger piping located behind the alternator. Loosen the upper hose clamp and detach the piping. This piping attaches to the piping removed from under the engine previously.
34   Disconnect the electrical connector for the turbocharger piping and remove from the vehicle.
35   Remove the AC compressor mounting bolts and support the compressor in a position to the side (see Chapter 3, Section 13).
36   Remove the lower alternator bolts.
37   Rotate the alternator to allow access to the alternator electrical connectors.

38   Disconnect the electrical connectors and remove the alternator from the vehicle.

Five-cylinder models

39   Raise and support the front of the vehicle on jackstands.
40   Remove the air inlet from the radiator support (see Chapter 1, Section 7).
41   Place the radiator support panel in the service position (see Chapter 11).
42   Remove the drivebelts and interfering tensioners/idlers (see Chapter 1).
43   Disconnect the electrical connectors from the alternator **(see illustration)**.
44   Remove the alternator mounting fasteners **(see illustrations)** and remove the alternator.

All models

45   Installation is the reverse of removal. Be sure to tighten the alternator mounting fasteners securely.
46   Reconnect the cable to the negative terminal of the battery (see Section 3).
47   Check the charging voltage (see Section 2) to verify that the alternator is operating correctly.

8   Starter motor - removal and installation

Caution: *If the battery is disconnected, several systems must be re-learned before they will work properly (see Section 3).*

1   Disconnect the cable from the negative battery terminal (see Section 3).
2   Remove the engine cover/air filter housing (see Chapter 1, Section 7).
3   Raise the vehicle and place it on jackstands. Remove the engine under-covers (see Chapter 1, Section 6).
4   Slide the protective cover off of the

Chapter 5 Engine electrical systems

8.4a Slide the solenoid protective cover (A) off of the solenoid ...

8.4b ... then disconnect the electrical connectors

starter solenoid **(see illustration)**, then disconnect the starter motor electrical connectors **(see illustration)** and harness retainer, if equipped.

5   Remove the plastic harness retainer fastener and retainer to access the starter motor upper mounting bolt. Remove the starter motor upper mounting bolt. On automatic transaxle models, remove the bracket nut and bracket from the lower bolt **(see illustration)**.

6   Remove the starter lower mounting bolt and remove the starter from below.

Note: *On models equipped with a DSG transaxle, the starter motor is removed out from the top of the engine compartment.*

7   Installation is the reverse of removal. Be sure to tighten the starter motor mounting fasteners securely.

8.5 Remove the bracket retaining nut (A), from the starter lower mounting bolt, then remove the harness bracket (B) from the starter

Notes

Chapter 6
Emissions and engine control systems

Contents

	Section		Section
Accelerator Pedal Position (APP) sensor - replacement	4	Mass Airflow (MAF) sensor - replacement	11
Camshaft Position (CMP) sensor - replacement	5	Obtaining and clearing Diagnostic Trouble Codes (DTCs)	3
Catalytic converter - replacement	15	Oil pressure switch - replacement	20
Charge Air Pressure sensor - replacement	23	On Board Diagnosis (OBD) system	2
Crankcase Ventilation (CCV) system - component replacement	17	Oxygen sensors - replacement	12
Crankshaft Position (CKP)/engine speed sensor - replacement	6	Powertrain Control Module (PCM) - removal and installation	14
Engine Coolant Temperature (ECT) sensor - replacement	7	Secondary Air Injection (AIR) system - component replacement	18
Evaporative Emissions Control (EVAP) system - component replacement	16	Transmission speed sensors - replacement	13
Fuel Pressure Sensor - replacement	22	Variable camshaft adjustment solenoid valve - replacement	19
General Information	1	Variable intake manifold actuator (1.8L and 2.0L turbocharged engines) - replacement	21
Intake Air Temperature (IAT) sensor - replacement	10		
Knock sensor - replacement	8		
Manifold Absolute Pressure (MAP)/Intake Air Temperature (IAT) sensor - replacement	9		

Specifications

Torque specifications   Ft-lbs (unless otherwise indicated)   Nm

Note: *One foot-pound (ft-lb) of torque is equivalent to 12 inch-pounds (in-lbs) of torque. Torque values below approximately 15 ft-lbs are expressed in inch-pounds, because most foot-pound torque wrenches are not accurate at these smaller values.*

Engine oil cooler	18.5	25
Engine oil filter bracket housing		
Step 1	133 in-lbs	15
Step 2	Tighten an additional 90-degrees	
Knock sensor mounting bolt		
1.4L turbocharged and 2.0L non-turbocharged engines	15	20
1.8L and 2.0L (except CBFA, CBTA) turbocharged engines		
Step 1	70 in-lbs	8
Step 2	Tighten an additional 90-degrees	
2.0L (CBFA, CBTA) turbocharged engines	16	22
Oil pressure switch	15	20
Oxygen sensors	41	55
Fuel pressure sensor		
1.4L engines	16	22
1.8L and 2.0L turbocharged engines	20	27

1 General Information

1 To prevent pollution of the atmosphere from incompletely burned and evaporating gases, and to maintain good driveability and fuel economy, a number of emission control systems are incorporated. They include the:

Catalytic converter

2 A catalytic converter is an emission control device in the exhaust system that reduces certain pollutants in the exhaust gas stream. There are two types of converters: oxidation converters and reduction converters.

3 Oxidation converters contain a monolithic substrate (a ceramic honeycomb) coated with the semi-precious metals platinum and palladium. An oxidation catalyst reduces unburned hydrocarbons (HC) and carbon monoxide (CO) by adding oxygen to the exhaust stream as it passes through the substrate, which, in the presence of high temperature and the catalyst materials, converts the HC and CO to water vapor (H_2O) and carbon dioxide (CO_2).

4 Reduction converters contain a monolithic substrate coated with platinum and rhodium. A reduction catalyst reduces oxides of nitrogen (NOx) by removing oxygen, which in the presence of high temperature and the catalyst material produces nitrogen (N) and carbon dioxide (CO_2).

5 Catalytic converters that combine both types of catalysts in one assembly are known as "three-way catalysts" or TWCs. A TWC can reduce all three pollutants.

Evaporative Emissions Control (EVAP) system

6 The Evaporative Emissions Control (EVAP) system prevents fuel system vapors (which contain unburned hydrocarbons) from escaping into the atmosphere. On warm days, vapors trapped inside the fuel tank expand until the pressure reaches a certain threshold. Then the fuel vapors are routed from the fuel tank through the fuel vapor vent valve and the fuel vapor control valve to the EVAP canister, where they're stored temporarily until the next time the vehicle is operated. When the conditions are right (engine warmed up, vehicle up to speed, moderate or heavy load on the engine, etc.) the PCM opens the canister purge valve, which allows fuel vapors to be drawn from the canister into the intake manifold. Once in the intake manifold, the fuel vapors mix with incoming air before being drawn through the intake ports into the combustion chambers where they're burned up with the rest of the air/fuel mixture. The EVAP system is complex and virtually impossible to troubleshoot without the right tools and training.

Secondary Air Injection (AIR) system

7 Some vehicles are equipped with a secondary air injection (AIR) system. The secondary air injection system is used to reduce tailpipe emissions on initial engine start-up. The system uses an electric motor/pump assembly, relay, vacuum valve/solenoid, air shut-off valve, check valves and tubing to inject fresh air directly into the exhaust manifolds. The fresh air (oxygen) reacts with the exhaust gas in the catalytic converter to reduce HC and CO levels. The air pump and solenoid are controlled by the PCM through the AIR relay. During initial start-up, the PCM energizes the AIR relay, the relay supplies battery voltage to the air pump and the vacuum valve/solenoid, engine vacuum is applied to the air shut-off valve which opens and allows air to flow through the tubing into the exhaust manifolds. The PCM will operate the air pump until closed loop operation is reached (approximately four minutes). During normal operation, the check valves prevent exhaust backflow into the system.

Powertrain Control Module (PCM)

8 The Powertrain Control Module (PCM) is the brain of the engine management system. It also controls a wide variety of other vehicle systems. In order to program the new PCM, the dealer needs the vehicle as well as the new PCM. If you're planning to replace the PCM with a new one, there is no point in trying to do so at home because you won't be able to program it yourself.

Positive Crankcase Ventilation (PCV) system

9 The Positive Crankcase Ventilation (PCV) system reduces hydrocarbon emissions by scavenging crankcase vapors, which are rich in unburned hydrocarbons. A PCV valve or orifice regulates the flow of gases into the intake manifold in proportion to the amount of intake vacuum available.

10 The PCV system generally consists of the fresh air inlet hose, the PCV valve or orifice and the crankcase ventilation hose (or PCV hose). The fresh air inlet hose connects the air intake duct to a pipe on the valve cover. The crankcase ventilation hose (or PCV hose) connects the PCV valve or orifice in the valve cover to the intake manifold.

2 On Board Diagnosis (OBD) system

General description

1 All models are equipped with the second generation OBD-II system. This system consists of an on-board computer known as the Powertrain Control Module (PCM), and information sensors, which monitor various functions of the engine and send data to the PCM. This system incorporates a series of diagnostic monitors that detect and identify fuel injection and emissions control system faults and store the information in the computer memory. This system also tests sensors and output actuators, diagnoses drive cycles, freezes data and clears codes.

2 The PCM is the brain of the electronically controlled fuel and emissions system. It receives data from a number of sensors and other electronic components (switches, relays, etc.). Based on the information it receives, the PCM generates output signals to control various relays, solenoids (fuel injectors) and other actuators. The PCM is specifically calibrated to optimize the emissions, fuel economy and driveability of the vehicle.

3 It isn't a good idea to attempt diagnosis or replacement of the PCM or emission control components at home while the vehicle is under warranty. Because of a federally-mandated warranty which covers the emissions system components and because any owner-induced damage to the PCM, the sensors and/or the control devices may void this warranty, take the vehicle to a dealer service department if the PCM or a system component malfunctions.

Scan tool information

4 Because extracting the Diagnostic Trouble Codes (DTCs) from an engine management system is now the first step in troubleshooting many computer-controlled systems and components, a code reader, at the very least, will be required (see illustration). More powerful scan tools can also perform many of the diagnostics once associated with expensive factory scan tools (see illustration). If you're planning to obtain a generic scan tool for your vehicle, make sure that it's compatible with OBD-II systems. If you don't plan to purchase a code reader or scan tool and don't have access to one, you can have the codes extracted by a dealer service department or an independent repair shop.

Note: *Some auto parts stores even provide this service.*

2.4a Simple code readers are an economical way to extract trouble codes when the CHECK ENGINE light comes on

2.4b Hand-held scan tools like these can extract computer codes and also perform diagnostics

Chapter 6 Emissions and engine control systems

Information Sensors

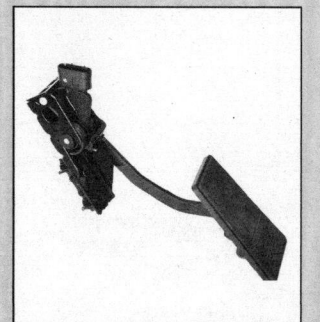

Accelerator Pedal Position (APP) sensor - as you press the accelerator pedal, the APP sensor alters its voltage signal to the PCM in proportion to the angle of the pedal, and the PCM commands a motor inside the throttle body to open or close the throttle plate accordingly

Camshaft Position (CMP) sensor - produces a signal that the PCM uses to identify the number 1 cylinder and to time the firing sequence of the fuel injectors

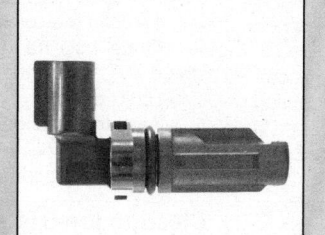

Crankshaft Position (CKP) sensor - produces a signal that the PCM uses to calculate engine speed and crankshaft position, which enables it to synchronize ignition timing with fuel injector timing, and to detect misfires

Engine Coolant Temperature (ECT) sensor - a thermistor (temperature-sensitive variable resistor) that sends a voltage signal to the PCM, which uses this data to determine the temperature of the engine coolant

Fuel tank pressure sensor - measures the fuel tank pressure and controls fuel tank pressure by signaling the EVAP system to purge the fuel tank vapors when the pressure becomes excessive

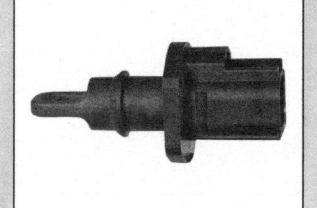

Intake Air Temperature (IAT) sensor - monitors the temperature of the air entering the engine and sends a signal to the PCM to determine injector pulse-width (the duration of each injector's on-time) and to adjust spark timing (to prevent spark knock)

Knock sensor - a piezoelectric crystal that oscillates in proportion to engine vibration which produces a voltage output that is monitored by the PCM. This retards the ignition timing when the oscillation exceeds a certain threshold

Manifold Absolute Pressure (MAP) sensor - monitors the pressure or vacuum inside the intake manifold. The PCM uses this data to determine engine load so that it can alter the ignition advance and fuel enrichment

Mass Air Flow (MAF) sensor - measures the amount of intake air drawn into the engine. It uses a hot-wire sensing element to measure the amount of air entering the engine

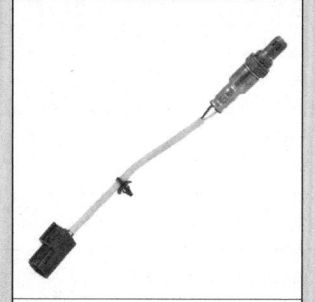

Oxygen sensors - generates a small variable voltage signal in proportion to the difference between the oxygen content in the exhaust stream and the oxygen content in the ambient air. The PCM uses this information to maintain the proper air/fuel ratio. A second oxygen sensor monitors the efficiency of the catalytic converter

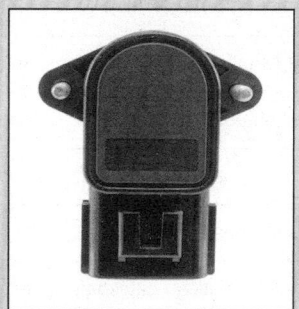

Throttle Position (TP) sensor - a potentiometer that generates a voltage signal that varies in relation to the opening angle of the throttle plate inside the throttle body. Works with the PCM and other sensors to calculate injector pulse width (the duration of each injector's on-time)

Photos courtesy of Wells Manufacturing, except APP and MAF sensors.

6-4 Chapter 6 Emissions and engine control systems

3 Obtaining and clearing Diagnostic Trouble Codes (DTCs)

1 All models covered by this manual are equipped with on-board diagnostics. When the PCM recognizes a malfunction in a monitored emission or engine control system, component or circuit, it turns on the Malfunction Indicator Light (MIL) on the dash. The PCM will continue to display the MIL until the problem is fixed and the Diagnostic Trouble Code (DTC) is cleared from the PCM's memory. You'll need a scan tool to access any DTCs stored in the PCM.

2 Before outputting any DTCs stored in the PCM, thoroughly inspect ALL electrical connectors and hoses. Make sure that all electrical connections are tight, clean and free of corrosion. And make sure that all hoses are correctly connected, fit tightly and are in good condition (no cracks or tears).

Accessing the DTCs

3 The Diagnostic Trouble Codes (DTCs) can only be accessed with a code reader or scan tool. Professional scan tools are expensive, but relatively inexpensive generic code readers or scan tools (see illustrations 2.4a and 2.4b) are available at most auto parts stores. Simply plug the connector of the scan tool into the diagnostic connector (see illustration). Then follow the instructions included with the scan tool to extract the DTCs.

4 Once you have outputted all of the stored DTCs, look them up on the accompanying DTC chart.

5 After troubleshooting the source of each DTC, make any necessary repairs or replace the defective component(s).

Clearing the DTCs

6 Clear the DTCs with the code reader or scan tool in accordance with the instructions provided by the tool's manufacturer.

Diagnostic Trouble Codes

7 The accompanying tables are a list of the Diagnostic Trouble Codes (DTCs) that can be accessed by a do-it-yourselfer working at home (there are many, many more DTCs available to professional mechanics with proprietary scan tools and software, but those codes cannot be accessed by a generic scan tool). If, after you have checked and repaired the connectors, wire harness and vacuum hoses (if applicable) for an emission-related system, component or circuit, the problem persists, have the vehicle checked by a dealer service department or other qualified repair shop.

3.3 The Data Link Connector (DLC) is located under the left end of the instrument panel

Diagnostic Trouble Codes

Code	Probable cause
P0010	Intake camshaft position actuator circuit open (bank 1)
P0011	"A" Camshaft position - timing over-advanced (bank 1)
P0012	"A" Camshaft position - timing over-retarded (bank 1)
P0013	"B" Camshaft position - actuator circuit malfunction (bank 1)
P0014	"B" Camshaft position - timing over-advanced or system performance problem (bank 1)
P0015	"B" Camshaft position - timing over-retarded (bank 1)
P0016	Crankshaft position - camshaft position correlation problem (bank 1 sensor a)
P0017	Crankshaft position - camshaft position correlation problem (bank 1 sensor b)
P0018	Crankshaft position - camshaft position correlation problem (bank 2 sensor a)
P0019	Crankshaft position - camshaft position correlation problem (bank 2 sensor b)
P0020	Intake camshaft position actuator circuit open (bank 2)
P0021	Intake camshaft position-timing over-advanced (bank 2)
P0022	Intake camshaft position-timing over-retarded (bank 2)
P0023	"B" Camshaft position - actuator circuit (bank 2)
P0024	"B" Camshaft position - timing over-advanced or system performance problem (bank 2)

Chapter 6 Emissions and engine control systems

Code	Probable cause
P0025	"B" Camshaft position - timing over-retarded (bank 2)
P0030	HO2S heater control circuit (bank 1, sensor 1)
P0031	HO2S heater control circuit low (bank 1, sensor 1)
P0032	HO2S heater control circuit high (bank 1, sensor 1)
P0036	HO2S heater control circuit (bank 1 sensor 2)
P0037	HO2S heater control circuit low (bank 1, sensor 2)
P0038	HO2S heater control circuit high (bank 1, sensor 2)
P0040	Upstream oxygen sensors swapped from bank to bank (HO2S - bank 1, sensor 1/bank 2, sensor 1)
P0041	Downstream oxygen sensors swapped from bank to bank (HO2S - bank 1, sensor 2/bank 2, sensor 2)
P0042	HO2S heater control circuit (bank 1, sensor 3
P0043	HO2S heater control circuit low (bank 1, sensor 3
P0044	HO2S heater control circuit high (bank 1 sensor 3
P0050	HO2S heater control circuit (bank 2, sensor 1)
P0051	HO2S heater control circuit low (bank 2, sensor 1)
P0052	HO2S heater control circuit high (bank 2, sensor 1)
P0056	HO2S heater control circuit malfunction (bank 2, sensor 2)
P0057	HO2S heater control circuit low (bank 2, sensor 2)
P0058	HO2S heater control circuit high (bank 2, sensor 2)
P0068	MAP/MAF - throttle position correlation problem
P0087	Fuel rail/system pressure - too low
P0088	Fuel rail/system pressure - too high
P0089	Fuel pressure regulator 1 performance problem
P0097	Intake air temperature sensor 2 circuit low
P0098	Intake air temperature sensor 2 circuit high
P0100	Mass or volume air flow "a" circuit
P0101	Mass air flow or volume air flow circuit, range or performance problem
P0102	Mass air flow or volume air flow circuit, low input
P0103	Mass air flow or volume air flow circuit, high input
P0105	Manifold absolute pressure/barometric pressure circuit
P0106	Manifold absolute pressure or barometric pressure circuit, range or performance problem
P0107	Manifold absolute pressure/barometric pressure circuit low

Diagnostic Trouble Codes (continued)

Code	Probable cause
P0108	Manifold absolute pressure/barometric pressure circuit high
P0111	Intake air temperature circuit, range or performance problem
P0112	Intake air temperature circuit, low input
P0113	Intake air temperature circuit, high input
P0116	Engine coolant temperature circuit range/performance problem
P0117	Engine coolant temperature circuit, low input
P0118	Engine coolant temperature circuit, high input
P0120	Throttle/pedal position sensor/switch "a" circuit
P0121	Throttle position or pedal position sensor/switch circuit, range or performance problem
P0122	Throttle position or pedal position sensor/switch circuit, low input
P0123	Throttle position or pedal position sensor/switch circuit, high input
P0125	Insufficient coolant temperature for closed loop fuel control
P0128	Coolant Thermostat (coolant temperature below normal range)
P0130	O2 sensor circuit malfunction (bank 1, sensor 1)
P0131	O2 sensor circuit low voltage (bank 1, sensor 1)
P0132	O2 sensor circuit high voltage (bank 1, sensor 1)
P0133	O2 sensor circuit, slow response (bank 1, sensor 1)
P0134	O2 sensor circuit no activity detected (bank 1, sensor 1
P0135	O2 sensor heater circuit malfunction (bank 1, sensor 1)
P0136	O2 sensor circuit malfunction (bank 1, sensor 2)
P0137	O2 sensor circuit, low voltage (bank 1, sensor 2)
P0138	O2 sensor circuit, high voltage (bank 1, sensor 2)
P0139	O2 sensor circuit, slow response (bank 1, sensor 2)
P0140	O2 sensor circuit - no activity detected (bank 1, sensor 2)
P0141	O2 sensor heater circuit malfunction (bank 1, sensor 2)
P0142	O2 Sensor Circuit, Bank 1 Sensor 3
P0143	O2 Sensor Circuit Low Voltage, Bank 1 Sensor 3
P0144	O2 Sensor Circuit High Voltage, Bank 1 Sensor 3
P0145	O2 Sensor Circuit Slow Response, Bank 1 Sensor 3
P0146	O2 Sensor Circuit No Activity Detected, Bank 1 Sensor 3

Chapter 6 Emissions and engine control systems

Code	Probable cause
P0147	O2 Sensor Heater Circuit, Bank 1 Sensor 3
P0150	O2 sensor circuit malfunction (bank 2, sensor 1)
P0151	O2 sensor circuit low voltage (bank 2 Sensor 1)
P0152	O2 sensor circuit high voltage (bank 2 Sensor 1)
P0153	O2 sensor circuit, slow response (bank 2, sensor 1)
P0154	O2 sensor circuit no activity detected, bank 2 sensor 1
P0155	O2 sensor heater circuit malfunction (bank 2, sensor 1)
P0156	O2 sensor circuit malfunction (bank 2, sensor 2)
P0157	O2 sensor circuit, low voltage (bank 2, sensor 2)
P0158	O2 sensor circuit, high voltage (bank 2, sensor 2)
P0159	O2 sensor circuit, slow response (bank 2, sensor 2)
P0160	O2 sensor circuit - no activity detected (bank 2, sensor 2)
P0161	O2 sensor heater circuit malfunction (bank 2, sensor 2)
P0169	Incorrect fuel composition
P0171	System too lean (bank 1)
P0172	System too rich (bank 1)
P0174	System too lean (bank 2)
P0175	System too rich (bank 2)
P0181	Fuel temperature sensor "a" circuit range/performance problem
P0182	Fuel temperature sensor "a" circuit low
P0183	Fuel temperature sensor "a" circuit high
P0188	Fuel temperature sensor "b" circuit high
P0190	Fuel rail pressure sensor "a" circuit
P0192	Fuel rail pressure sensor "a" circuit low
P0200	Injector Circuit/Open
P0201	Injector circuit malfunction - cylinder no. 1
P0202	Injector circuit malfunction - cylinder no. 2
P0203	Injector circuit malfunction - cylinder no. 3
P0204	Injector circuit malfunction - cylinder no. 4
P0205	Injector circuit malfunction - cylinder no. 5
P0216	Injector/injection timing control circuit

Diagnostic Trouble Codes (continued)

Code	Probable cause
P0219	Engine overspeed condition
P0221	Throttle position or pedal position sensor/switch b, range or performance problem
P0222	Throttle position or pedal position sensor/switch b circuit, low input
P0223	Throttle position or pedal position sensor/switch b circuit, high input
P0225	Throttle/pedal position sensor/switch "c" circuit
P0226	Throttle/pedal position sensor/switch "c" circuit range/performance problem
P0227	Throttle/pedal position sensor/switch "c" circuit low
P0228	Throttle/pedal position sensor/switch "c" circuit high
P0230	System too rich at idle
P0234	Turbocharger overboost condition
P0236	Turbocharger boost sensor a circuit, range or performance problem
P0237	Turbocharger boost sensor a circuit, low
P0238	Turbocharger boost sensor a circuit, high
P0240	Turbocharger boost sensor "b" circuit range/performance problem
P0241	Turbocharger boost sensor "b" circuit low
P0242	Turbocharger boost sensor "b" circuit high
P0243	Turbocharger wastegate solenoid "a" malfunction
P0245	Turbocharger wastegate solenoid a, low
P0246	Turbocharger wastegate solenoid a, high
P0247	Turbocharger wastegate solenoid "b"
P0249	Turbocharger wastegate solenoid "b" low
P0250	Turbocharger wastegate solenoid "b" high
P0251	Injection pump fuel metering control "a"
P0252	Injection pump fuel metering control "a" range/performance problem
P0261	Cylinder no. 1 injector circuit, low
P0262	Cylinder no. 1 injector circuit, high
P0263	Cylinder no. 1 contribution/balance
P0264	Cylinder no. 2 injector circuit, low
P0265	Cylinder no. 2 injector circuit, high
P0266	Cylinder no. 2 contribution/balance

Chapter 6 Emissions and engine control systems

Code	Probable cause
P0267	Cylinder no. 3 injector circuit, low
P0268	Cylinder no. 3 injector circuit, high
P0269	Cylinder no. 3 contribution/balance
P0270	Cylinder no. 4 injector circuit, low
P0271	Cylinder no. 4 injector circuit, high
P0272	Cylinder no. 4 contribution/balance
P0273	Cylinder no. 5 injector circuit, low
P0274	Cylinder no. 5 injector circuit, high
P0275	Cylinder no. 5 contribution/balance
P0299	Turbocharger underboost condition
P0300	Random/multiple cylinder misfire detected
P0301	Cylinder no. 1 misfire detected
P0302	Cylinder no. 2 misfire detected
P0303	Cylinder no. 3 misfire detected
P0304	Cylinder no. 4 misfire detected
P0305	Cylinder no. 5 misfire detected
P0321	Crankshaft position (CKP) sensor/engine speed (RPM) sensor - range or performance problem
P0322	Crankshaft position (CKP) sensor/engine speed (RPM) sensor - no signal
P0324	Knock control system error
P0327	Knock sensor no. 1 circuit, low input (bank 1 or single sensor)
P0328	Knock sensor no. 1 circuit, high input (bank 1 or single sensor)
P0332	Knock sensor no. 2 circuit, low input (bank 2)
P0333	Knock sensor no. 2 circuit, high input (bank 2)
P0340	Camshaft position sensor "A" - circuit malfunction (bank 1)
P0341	Camshaft position sensor "A" - range or performance problem (bank 1)
P0342	Camshaft position sensor "A" - low input (bank 1)
P0343	Camshaft position sensor "A" - high input (bank 1)
P0345	Camshaft position sensor "A" - circuit malfunction (bank 2)
P0346	Camshaft position sensor "A" - range/performance problem (bank 2)
P0347	Camshaft position sensor "A" - low input (bank 2)
P0348	Camshaft position sensor "A" - range/performance problem (bank 2)

Diagnostic Trouble Codes (continued)

Code	Probable cause
P0351	Ignition coil 1 primary or secondary circuit malfunction
P0352	Ignition coil 2 primary or secondary circuit malfunction
P0353	Ignition coil 3 primary or secondary circuit malfunction
P0354	Ignition coil 4 primary or secondary circuit malfunction
P0355	Ignition coil 5 primary or secondary circuit malfunction
P0365	Camshaft position sensor "B" - circuit malfunction (bank 1)
P0366	Camshaft position sensor "B" - range/performance problem (bank 1)
P0367	Camshaft position sensor "B" - low input (bank 1)
P0368	Camshaft position sensor "B" circuit high input (bank 1)
P0390	Camshaft position sensor "B" - circuit malfunction
P0391	Camshaft position sensor "B" - range/performance problem (bank 2)
P0392	Camshaft position sensor "B" - low input (bank 2)
P0393	Camshaft position sensor "B" - high input (bank 2)
P0400	Exhaust gas recirculation flow
P0401	Exhaust gas recirculation flow insufficient detected
P0402	Exhaust gas recirculation flow excessive detected
P0403	Exhaust gas recirculation control circuit
P0404	Exhaust gas recirculation control circuit range/performance
P0411	Secondary air injection system, incorrect flow detected
P0412	Secondary air injection system switching valve A - circuit malfunction
P0413	Secondary air injection system switching valve A - open circuit
P0414	Secondary air injection system switching valve A - shorted circuit
P0416	Secondary air injection system switching valve "B" circuit open
P0417	Secondary air injection system switching valve "B" circuit shorted
P0418	Secondary air injection system, pump relay A - circuit malfunction
P0420	Catalyst system efficiency below threshold (bank 1)
P0421	Warm-up catalyst efficiency below threshold (bank 1)
P0422	Main catalyst efficiency below threshold (bank 1)
P0423	Heated catalyst efficiency below threshold (bank 1)
P0424	Heated catalyst temperature below threshold (bank 1)

Chapter 6 Emissions and engine control systems

Code	Probable cause
P0425	Catalyst temperature sensor circuit (bank 1, sensor 1)
P0426	Catalyst temperature sensor circuit range/performance (bank 1, sensor 1)
P0427	Catalyst temperature sensor circuit low (bank 1, sensor 1)
P0428	Catalyst temperature sensor circuit high (bank 1, sensor 1)
P0429	Catalyst heater control circuit (bank 1)
P0430	Catalyst system efficiency below threshold (bank 2)
P0431	Warm-up catalyst efficiency below threshold (bank 2)
P0432	Main catalyst efficiency below threshold (bank 2)
P0440	Evaporative emission control system malfunction
P0441	Evaporative emission control system, incorrect purge flow
P0442	Evaporative emission control system, small leak detected
P0443	Evaporative emission control system, purge control valve circuit malfunction
P0444	Evaporative emission control system, open purge control valve circuit
P0445	Evaporative emission control system, short in purge control valve circuit
P0449	Evaporative emission system vent valve/solenoid circuit
P0455	Evaporative emission (EVAP) control system leak detected (no purge flow or large leak)
P0456	Evaporative emission (EVAP) control system leak detected (very small leak)
P0458	Evaporative emission system purge control valve circuit low
P0459	Evaporative emission system purge control valve circuit high
P0480	Fan 1 control circuit
P0481	Fan 2 control circuit
P0489	Exhaust gas recirculation control circuit low
P0490	Exhaust gas recirculation control circuit high
P0491	Secondary air injection system (bank 1)
P0492	Secondary air injection system (bank 2)
P0498	Evaporative emission system vent valve control circuit low
P0499	Evaporative emission system vent valve control circuit high
P0501	Vehicle speed sensor, range or performance problem
P0505	Idle air control system
P0506	Idle control system, rpm lower than expected
P0507	Idle control system, rpm higher than expected

Chapter 6 Emissions and engine control systems

Diagnostic Trouble Codes (continued)

Code	Probable cause
P0510	Closed throttle position switch
P0544	Exhaust gas temperature sensor circuit (bank 1 sensor 1)
P0545	Exhaust gas temperature sensor circuit low (bank 1 sensor 1)
P0546	Exhaust gas temperature sensor circuit high (bank 1 sensor 1)
P0560	System voltage malfunction
P0562	System voltage low
P0563	System voltage high
P0564	Cruise control multi-function input "a" circuit
P0571	Cruise control/brake switch a, circuit malfunction
P0597	Thermostat heater control circuit/open
P0598	Thermostat heater control circuit low
P0599	Thermostat heater control circuit high
P0600	Serial communication link malfunction
P0601	Internal control module, memory check sum error
P0602	Control module programming error
P0603	Internal control module keep alive memory (KAM) error
P0604	Internal control module, random access memory (RAM) error
P0605	Internal control module, read only memory (ROM) error
P0606	PCM processor fault
P0607	Control module performance
P0608	Control module VSS output "a"
P0609	Control module VSS output "b"
P0613	TCM processor
P0614	ECM / TCM incompatible
P0627	Fuel pump "a" control circuit/open
P0629	Fuel pump "a" control circuit high
P0638	Throttle actuator control range/performance problem (bank 1)
P0639	Throttle actuator control range/performance problem (bank 2)
P0641	Sensor reference voltage "a" circuit/open
P0642	Sensor reference voltage "a" circuit low

Chapter 6 Emissions and engine control systems

Code	Probable cause
P0643	Sensor reference voltage "a" circuit high
P0651	Sensor reference voltage "b" circuit/open
P0652	Sensor reference voltage "b" circuit low
P0653	Sensor reference voltage "b" circuit high
P0657	Actuator supply voltage "a" circuit/open
P0658	Actuator supply voltage "a" circuit low
P0659	Actuator supply voltage "a" circuit high
P0685	ECM power relay, control - circuit open
P0686	ECM power relay control - circuit low
P0687	Engine, control relay - short to ground
P0688	Engine, control relay - short to positive
P0691	Fan 1 control circuit low
P0692	Fan 1 control circuit high
P0693	Fan 2 control circuit low
P0694	Fan 2 control circuit high
P0697	Sensor reference voltage "c" circuit/open
P0698	Sensor reference voltage "c" circuit low
P0699	Sensor reference voltage "c" circuit high
P0704	Clutch switch input circuit malfunction

4 Accelerator Pedal Position (APP) sensor - replacement

1   The APP sensor is located at the top of (and is an integral component of) the accelerator pedal (see illustration). There is no reason to remove the APP sensor except to replace it. But if the APP sensor is replaced, the new unit must be programmed with a factory scan tool. If you get a Diagnostic Trouble Code indicating a problem with the APP sensor, have the sensor replaced by a dealer service department or other qualified repair shop equipped with the necessary scan tool.

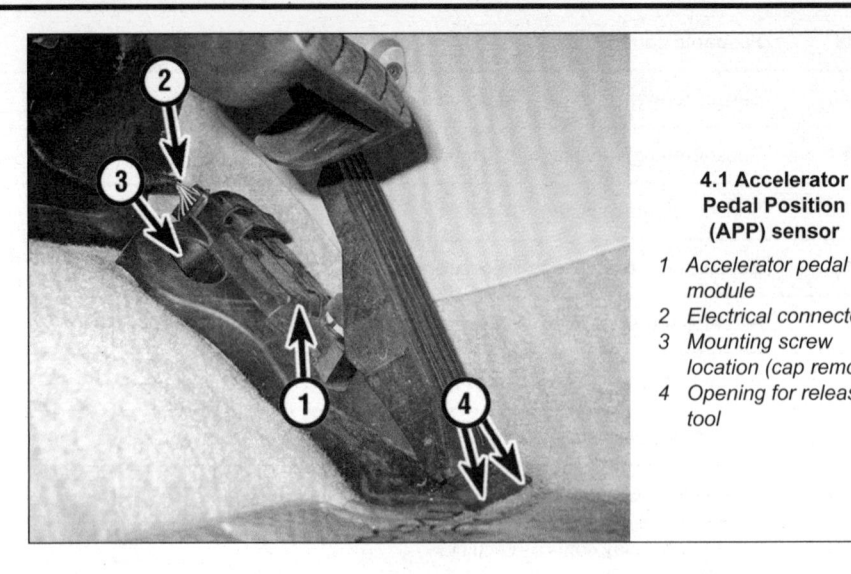

5 Camshaft Position (CMP) sensor - replacement

Note: *If the battery is disconnected, several systems must be re-learned before they will work properly (see Chapter 5, Section 3).*
Note: *1.4L engines and Golf 2.0L turbocharged engines (CXCA, CXCB, DKFA) are equipped with two CMP sensors.*

1   Disconnect the cable from the negative battery terminal (see Chapter 5).
2   Remove the engine cover (see Chapter 1, Section 7).
3   The CMP sensor is located at the rear end of the cylinder head cover on 1.4L engines (see illustration), under the camshaft sprocket on 2.0L non-turbocharged engines, on the front side of the cylinder head cover just below the oil filler cap on 1.8L an 2.0L turbocharged engines, or on the camshaft guide frame on 2.5L five-cylinder engines (see illustration). 2.0L turbocharged engines CXCA, CXCB, DKFA sensor 2 is attached to the cylinder head cover.
4   On 2.0L non-turbocharged engines, remove the timing belt cover (see Chapter 2A, Section 5).

4.1 Accelerator Pedal Position (APP) sensor

1 Accelerator pedal module
2 Electrical connector
3 Mounting screw location (cap removed)
4 Opening for release tool

5   On 1.8L and 2.0L turbocharged engines (and Golf CXCA, CXCB, DKFA, for sensor 1) remove the intake manifold (see Chapter 2A Section 7) to access the sensor.
6   On all engines, disconnect the CMP sensor electrical connector.
7   Remove the CMP sensor mounting fastener and remove the CMP sensor.
8   Installation is the reverse of removal.

6 Crankshaft Position (CKP)/engine speed sensor - replacement

Note: *The CKP sensor is also referred to as the Engine Speed Sensor.*

1.4L engines
Note: *The CKP sensor is located on the front left (driver's) end of the engine block, near the transmission bellhousing.*
1   Raise the vehicle and support it securely on jackstands.

2   Remove the engine undercover.
3   If equipped, remove the bolt attaching the bracket of the charge air cooler pump and position the pump to the side.
4   Locate the CKP sensor and disconnect the electrical connector (see illustration).
5   Remove the rubber seal from the engine block (see illustration).
6   Remove the bolt (see illustration) and pull the sensor out of the block.
7   Inspect the O-ring and replace if damaged.
8   Installation is reverse of removal.

1.8L and 2.0L engines
Note: *The CKP sensor is located on the front left (driver's) end of the engine block, near the transmission bellhousing.*
9   Raise the vehicle and support it securely on jackstands.
10 Remove the engine under cover.
11 On turbocharged models, for access to the CKP sensor, remove the charge air pipe bolts, detach the ends, and remove the pipe.

5.3a The CMP sensors are located on the rear end the cylinder head cover (1.4L engines)

5.3b The CMP sensor is located on the front side of the camshaft guide frame (2.5L five-cylinder engines)

6.4 Disconnect the CKP sensor connector (1.4L engines)

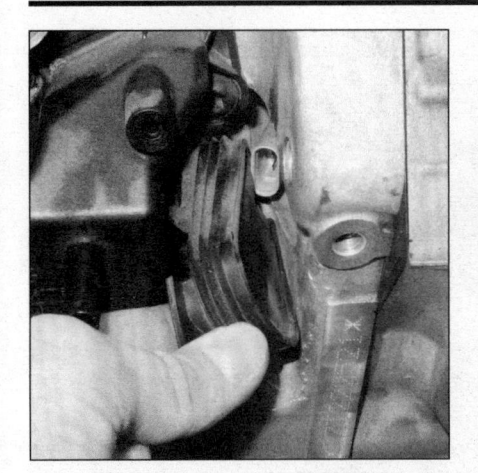

Chapter 6 Emissions and engine control systems 6-15

6.5 Remove the CKP sensor seal from the engine block (1.4L engines)

6.6 Remove the bolt and remove the sensor (1.4L engines)

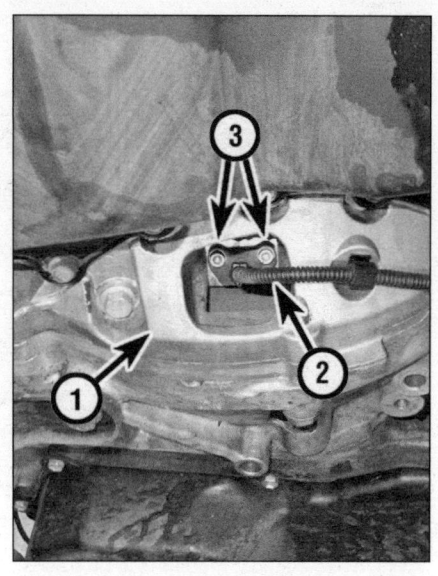

6.18 The CKP sensor is located on the left bottom end of the engine block, near the flywheel (five-cylinder engine)

1. Control housing
2. CKP sensor and electrical connector
3. Mounting fasteners

12 On all models, disconnect the CKP sensor electrical connector.
13 Remove the CKP sensor mounting fastener and remove the CKP sensor.
14 Inspect the O-ring and replace if damaged.
15 Installation is the reverse of removal.

2.5L engines

16 Loosen the right-front wheel bolts. Raise the front of the vehicle and support it securely on jackstands. Block the wheels at the opposite end and remove the right-front wheel.
17 Remove the under-vehicle splash shield (see Chapter 1, Section 6).
18 Locate the CKP sensor on the control housing, near the flywheel **(see illustration)**.
19 Trace the CKP sensor electrical lead to the sensor electrical connector and disconnect the connector.
20 Remove the engine speed sensor mounting fasteners and remove the sensor.
21 Installation is the reverse of removal.

7 Engine Coolant Temperature (ECT) sensor - replacement

Warning: *Wait until the engine has cooled completely before beginning this procedure.*
1 Remove the engine cover (see Chapter 1, Section 7).
2 Release the pressure from the cooling system and reinstall the cap (see Chapter 1). **Note:** *This will help minimize coolant loss. However, remove the old sensor and install the new one as quickly as possible.*
3 Place a container under the ECT sensor location to catch any coolant.

1.4L, 1.8L and 2.0L (except CBFA, CCTA) engines

Note: *Turbocharged engines are equipped with two ECT sensors. The ECT sensors are located on the rear of the cylinder head and on the radiator inlet.*

Engine-mounted sensor

Note: *Non-turbocharged 2.0L engines are only equipped with this ECT sensor.*

4 On turbocharged models, remove the charge air duct (see Chapter 4, **illustration 12.4**).
5 On all models, locate the ECT sensor and disconnect the electrical connector **(see illustration)**.
6 Depending on the engine, remove the bolt or clip securing the sensor to the engine **(see illustration)**, then pull the sensor from the engine and install the new one.

7.5 Locate and disconnect the ECT sensor connector (1.4L engine shown; the ECT sensor is located at the left-rear [driver's side] corner of the cylinder head, below the turbocharger inlet)

7.6 Remove the fastener and remove the ECT sensor (1.4L engine, shown from below)

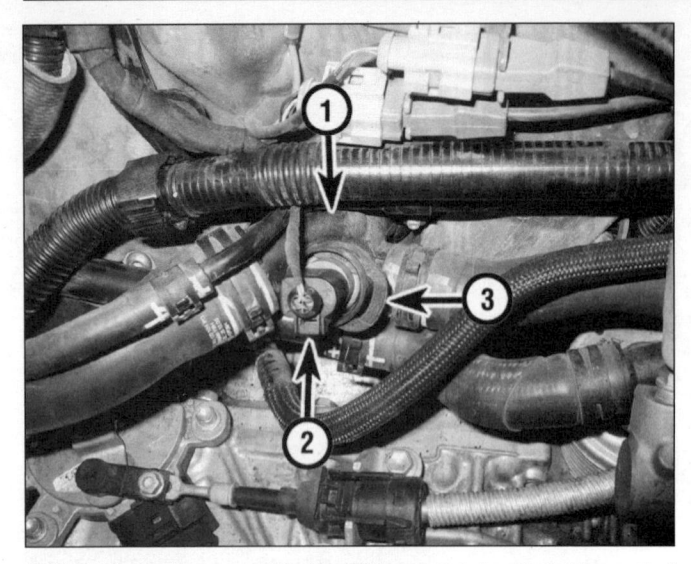

7.17 On five-cylinder models, the ECT sensor is located at the end of the timing chain cover (which is on the left end of the cylinder head)
1 Coolant housing
2 ECT sensor electrical connector
3 Retaining clip

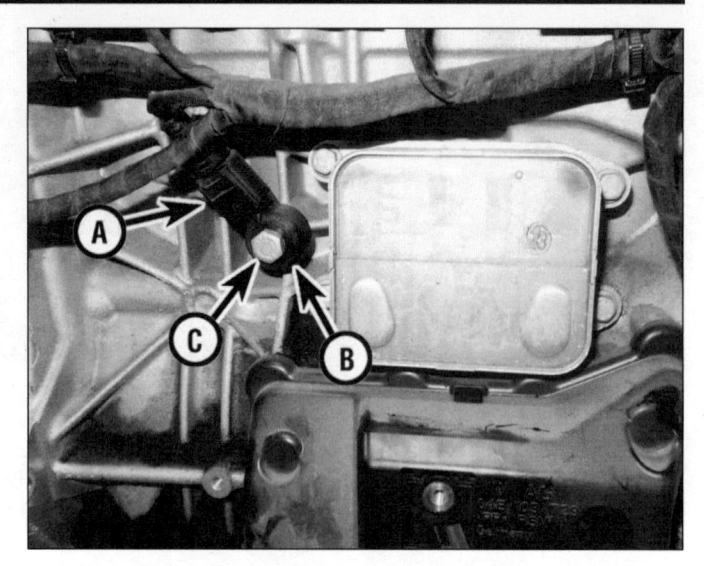

8.5 Knock sensor details (1.4L engines; the sensor is located on the front side of the engine block)
A Knock sensor connector
B Knock sensor
C Knock sensor mounting bolt

Radiator-mounted sensor

7   Raise the front of the vehicle and support it securely on jackstands.
8   Remove the engine under cover.
9   Locate the sensor on the lower radiator outlet and disconnect the electrical connector.
10 Remove the retaining clip, then pull the sensor out of the radiator outlet.

2.0L turbocharged engines (CBFA, CCTA)

11 Remove the throttle body (see Chapter 4, Section 12).
12 Remove the intake manifold support bracket fasteners and support.
13 Disconnect the electrical connector from the ECT sensor.
14 On engines with a snap-in ECT sensor, press the locking tabs on the sensor inwards, then pull the sensor out of the housing and install the new one.
15 On engines with a bolt-in ECT sensor, remove the bolts and the retaining plate, then pull the sensor from the engine.

2.5L engines

16 Disconnect the inlet air hose from the throttle body (see Chapter 4).
17 Disconnect the electrical connector from the ECT sensor **(see illustration)**.
18 Pull out the retaining clip and remove the sensor.

All engines

19 Remove and discard the sensor O-ring.
20 Before installing the ECT sensor, install a new O-ring and apply some engine coolant to the O-ring.
21 Installation is the reverse of removal.
22 Refill the cooling system (see Chapter 1).

8   Knock sensor - replacement

Warning: *Don't attempt to remove a knock sensor when the engine is hot. If the engine has just been operated, allow sufficient time for it to cool down.*
Caution: *Incorrectly tightening a knock sensor fastener can affect the performance of the knock sensor.*
Note: *If the battery is disconnected, several systems must be re-learned before they will work properly (see Chapter 5, Section 3).*

1.4L engines and 2.0L non-turbocharged engines

1   Raise the front of the vehicle and support it securely on jackstands.
2   Remove the under-vehicle splash shield (see Chapter 1, Section 6).
3   Remove the drivebelt (see Chapter 1, Section 19).
4   On 1.4L engines, remove the air conditioning compressor bolts and secure the compressor to the side (see Chapter 3, Section 13).
Warning: *Don't disconnect the refrigerant lines.*
5   On all engines, locate and disconnect the knock sensor electrical connector **(see illustration)**.
6   Remove the bolt and detach the knock sensor from the block.
7   Installation is reverse of removal.
Caution: *Incorrectly tightening a knock sensor fastener can affect the performance of the knock sensor.*
8   Tighten the knock sensor bolt to the torque listed in this Chapter's Specifications.

1.8L and 2.0L turbocharged engines

9   Raise the front of the vehicle and support it securely on jackstands.
10 Remove the under-vehicle splash shield (see Chapter 1, Section 6).
11 Disconnect the cable from the negative battery terminal (see Chapter 5).
12 Drain the cooling system (see Chapter 1).
13 Remove the intake manifold (see Chapter 4).
14 Remove the water pump (see Chapter 3, Section 7).
15 Disconnect the hoses from the Engine Temperature Control Actuator (water pump housing). Disconnect the electrical connector from the bottom of the actuator.
16 Remove the five bolts attaching the actuator to the engine.
17 Remove the actuator from the engine by disconnecting from the oil cooler.
18 Discard the O-rings from the oil cooler connection and use new o-rings for installation.
19 Disconnect the knock sensor harness connector.
20 Remove the knock sensor bolt and remove the knock sensor from the vehicle.
21 Installation is reverse of removal. Tighten the knock sensor bolt to the torque listed in this Chapter's Specifications.
Caution: *Incorrectly tightening a knock sensor fastener can affect the performance of the knock sensor.*
22 Tighten the knock sensor bolt to the specifications located in this Chapter's Specifications.

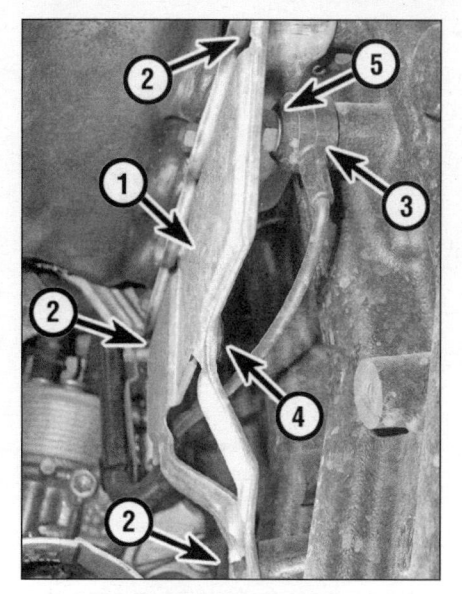

Chapter 6 Emissions and engine control systems

8.23 Knock sensor details - five-cylinder models

1. Knock sensor heat shield
2. Heat shield mounting fasteners
3. Knock sensor 1
4. Knock sensor 2
5. Knock sensor 1 mounting bolt

2.5L engines

Note: *There are two knock sensors - both are located under the exhaust manifold. Sensor 1 is located near the drivebelt end of the engine and sensor 2 is located near the transmission end of the engine. This procedure applies to either knock sensor. The green electrical connector is for sensor 1 and the gray connector is for sensor 2.*

23 Locate the knock sensors under the exhaust manifold **(see illustration)**, then trace the electrical lead for each sensor back to the sensor electrical connector and disconnect it.
24 Remove the knock sensor heat shield fasteners and the heat shield.
25 Remove the knock sensor mounting bolt and remove the sensor.
Caution: *Incorrectly tightening a knock sensor fastener can affect the performance of the knock sensor.*
26 When installing a knock sensor, be sure to tighten the sensor mounting fastener to the torque listed in this Chapter's Specifications.
27 Installation is otherwise the reverse of removal.

9 Manifold Absolute Pressure (MAP)/Intake Air Temperature (IAT) sensor - replacement

Note: *This procedure applies to all engines except 2.0L turbocharged CBFA and CCTA engines. The MAP sensor is located on the intake manifold near the throttle body. The Intake Air Temperature (IAT) sensor is integrated with the Manifold Absolute Pressure (MAP) sensor and is referred to as the Intake Manifold Sensor.*

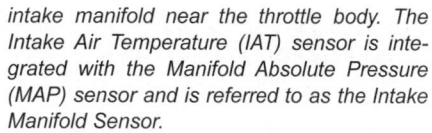

1 Remove the engine cover (see Chapter 1, Section 7).
2 On 1.4L engines, remove the air filter housing (see Chapter 4, Section 11).
3 On all engines, locate and disconnect the electrical connector from the MAP/IAT sensor **(see illustrations)**.
4 Unclip the MAP sensor from the manifold or remove the MAP sensor mounting fasteners, as applicable, and remove the sensor.
Note: *Do not discard the O-ring; the O-ring does not have a replacement part (it only comes with a new sensor).*
5 Installation is the reverse of removal.

9.3a MAP/IAT sensor location (1.4L engines)

9.3b MAP/IAT sensor location (2.5L engines)

10 Intake Air Temperature (IAT) sensor - replacement

Note: *This procedure applies to 2.0L turbocharged CBFA and CCTA engines. On all other engines except 2018 and later 1.4L, the IAT is part of the MAP sensor and the Charge Air Pressure sensor (see Section 9). The IAT sensor is located at the back side of the intake manifold. On 2018 and later 1.4L engines, the IAT sensor is part of the MAF sensor located on the air filter housing lid (see Section 11).*

1 Remove the engine cover (see Chapter 1, Section 7).
2 Locate and disconnect the electrical connector from the IAT sensor.
3 Remove the IAT sensor mounting fastener and remove the sensor from the intake manifold.
Note: *Do not discard the O-ring; the O-ring does not have a replacement part (it only comes with a new sensor).*
4 Installation is the reverse of removal.

11 Mass Airflow (MAF) sensor - replacement

Note: *Some 2018 and later models with 1.4L engines are equipped with a MAF sensor.*
1 Remove the engine cover.
2 Locate the MAF sensor on the air filter housing lid intake tube.
3 Disconnect the MAF sensor electrical connector.
4 Remove the screws attaching the MAF to the air filter housing lid, and carefully pull the MAF sensor from the lid.
5 Installation is reverse of removal.

12 Oxygen sensors - replacement

Note: *Because the oxygen sensors are installed in the exhaust system, which contracts when cool, they can be difficult to loosen when the engine is cold. Rather than risk damage to an oxygen sensor or its mounting threads, start and run the engine for a minute or two, then shut it off. Be careful not to burn yourself during the following procedure.*

1 Be particularly careful when servicing an oxygen sensor:
a) Oxygen sensors have a permanently attached pigtail and an electrical connector that cannot be removed. Damaging or removing the pigtail or electrical connector will render the sensor useless.
b) Keep grease, dirt and other contaminants away from the electrical connector and the louvered end of the sensor.
c) Do not use cleaning solvents of any kind on an oxygen sensor.
d) Oxygen sensors are extremely delicate. Do not drop a sensor, throw it around or handle it roughly.
e) Make sure that the silicone boot on the sensor is installed in the correct position. Otherwise, the boot might melt and prevent the sensor from operating correctly.

6-18   Chapter 6   Emissions and engine control systems

12.5a The upstream oxygen sensor is located at the upper end of the catalytic converter (1.4L engine shown, other four-cylinder models are similar)

A   Upstream O2 sensor electrical connector
B   Upstream O2 sensor

12.5b The upstream oxygen sensor is located at the end of the exhaust manifold (2.5L engines)

2   Remove the engine cover (see Chapter 1, Section 7).

3   If you're removing a downstream sensor, or an upstream sensor on a five-cylinder model, raise the front of the vehicle and support it securely on jackstands.

4   If you're removing a downstream sensor, or an upstream sensor on a five-cylinder model, remove the lower splash shield below the engine (see Chapter 1, Section 6).

Upstream oxygen sensors

5   Locate the upstream oxygen sensor **(see illustrations)**, then trace the wiring harness to its electrical connector and disconnect the connector. Disengage the sensor harness from any harness clips.

6   Unscrew the upstream oxygen sensor **(see illustration)**.

7   If you're going to install the old sensor, apply anti-seize compound to the threads of the sensor to facilitate future removal. If you're going to install a new oxygen sensor, it's not necessary to apply anti-seize compound to the threads; the threads on new sensors already have anti-seize compound on them.

8   Installation is the reverse of removal. Be sure to tighten the oxygen sensor to the torque listed in this Chapter's Specifications.

Downstream oxygen sensors

Note: *On some models, the O2 sensor harness may disappear under a plastic shield. remove the shield to access the downstream O2 sensor connector.*

9   Locate the downstream oxygen sensor(s) **(see illustrations)**, then trace the lead up to the electrical connector and disconnect the connector.

10   Unscrew the downstream oxygen sensor.

11   If you're going to install the old sensor, apply anti-seize compound to the threads of the sensor to facilitate future removal. If you're going to install a new oxygen sensor, it's not necessary to apply anti-seize compound to the threads. The threads on new sensors already have anti-seize compound on them.

12   Installation is the reverse of removal. Be sure to tighten the oxygen sensor to the torque listed in this Chapter's Specifications.

13   Transmission speed sensors - replacement

1   On automatic transaxles, speed sensors are integral components of the valve body/Transmission Control Module (TCM) assembly, which is located inside the transaxle. It is recommended that these sensors be replaced by a qualified automotive repair facility.

12.6 An oxygen sensor socket will allow you to work in tight quarters where a wrench would be difficult to use (and might round-off the corners of the sensor's hex)

12.9a Downstream oxygen sensor location (1.4L engine shown, other four-cylinder models are similar)

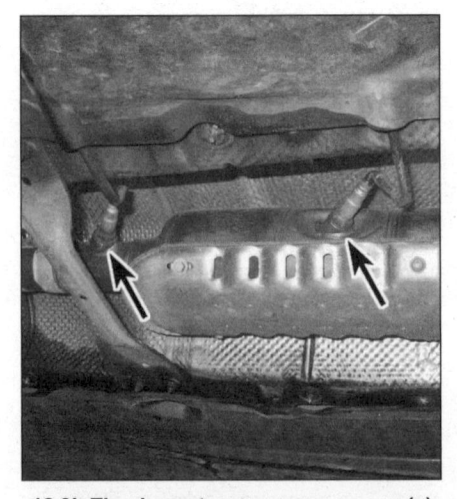

12.9b The downstream oxygen sensor(s) are located in the middle and after the catalyst (2.5L engines)

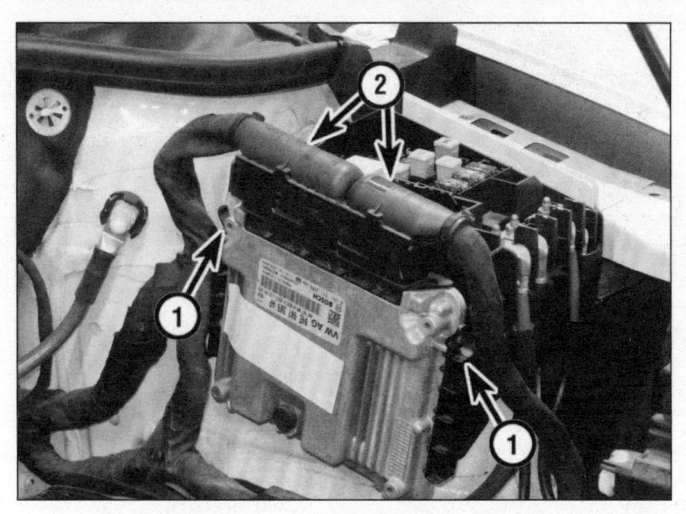

Chapter 6 Emissions and engine control systems

14.5 Details of the Powertrain Control Module (PCM) on model not equipped with anti-theft equipment:

1. Powertrain Control Module (PCM) bracket retaining tabs
2. Electrical connectors

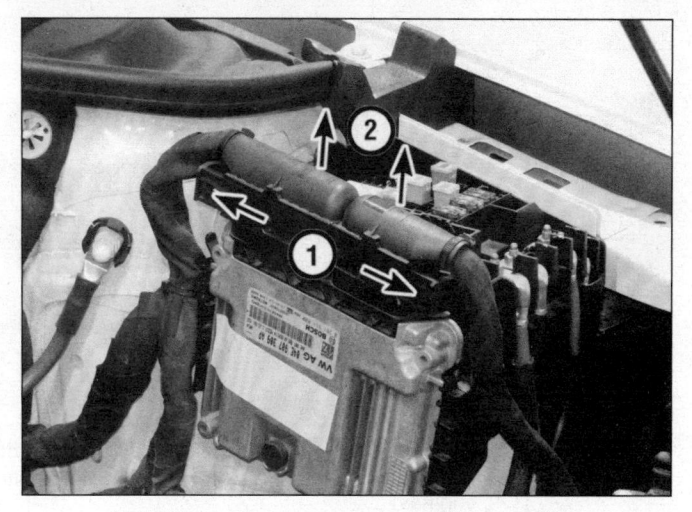

14.6 Release the connector locks (1) and pull the connectors from the PCM (2)

14 Powertrain Control Module (PCM) - removal and installation

Note: *If the battery is disconnected, several systems must be re-learned before they will work properly (see Chapter 5, Section 3).*

Note: *The Powertrain Control Module (PCM) cannot be replaced at home because the new unit must be reprogrammed with a proprietary scan tool. The PCM in many models is also housed inside a special metal anti-theft box and is extremely difficult to remove.*

Basic removal

1 Disconnect the cable from the negative battery terminal (see Chapter 5).
2 Remove the cover from the electronics box.
3 Remove the air filter housing if needed (see Chapter 4, Section 11).
4 Release the tabs and remove the PCM cover if equipped.
5 Pull back the tabs using your fingers and pull the PCM up and out of the bracket with the connector still connected and position to the side **(see illustration)**.
6 On models without anti-theft equipment, slide the connector retainers out and pull the connectors form the PCM **(see illustration)**.
Caution: *On models with the anti-theft equipment, if the PCM only needs to be moved to the side for servicing of other components, stop here.*
7 Installation is the reverse of the removal procedure.

Anti-theft equipment removal

Note: *Some models are equipped with an anti-theft housing or cover plate in which the PCM is mounted to prevent the connectors from being disconnected.*

8 Perform the steps outlined above to release the PCM from the bracket.

9 The anti-theft equipment is secured with shear-head bolts. Cut slots in the shear-head bolts using a small cutting wheel or grinder and use a flat blade screwdriver to unscrew the bolts. Alternatively, drill through the bolts and use a screw extractor to remove them.
10 Disconnect the electrical connectors and remove the PCM.
11 Installation is the reverse of the removal procedure.

15 Catalytic converter - replacement

Note: *The front exhaust pipe and catalytic converter are a single assembly.*

1 Remove the engine cover (see Chapter 1, Section 7).
2 Loosen the right-front wheel bolts, then raise the front of the vehicle and support it securely on jackstands.
3 Remove the under-vehicle splash shield (see Chapter 1, Section 6).
4 If needed, remove the right-side inner fender liner (see Chapter 11).
5 Remove the upstream and downstream oxygen sensors from the catalytic converter (see Section 12).
6 Remove the heat shield at the turbocharger (if equipped).
7 Remove the fasteners or clamps that secure the front exhaust pipe with catalytic converter to the manifold, cylinder head, or turbocharger.
8 Remove the front exhaust pipe support bracket bolts **(see illustration)** and remove the support bracket (see Chapter 4, Section 14).
9 Disconnect the clamps that hold the catalytic converter to the rear exhaust.
10 Remove the catalytic converter with front exhaust pipe.
11 Remove and discard the old flange gasket and clamps.
12 Installation is the reverse of removal. Be sure to use new gaskets and self-locking nuts and tighten all fasteners securely.

15.8 Typical catalytic converter details (2.5L engines)

1. Catalytic converter
2. Front exhaust pipe
3. Exhaust pipe support bracket
4. Exhaust pipe support bracket mounting fasteners
5. Clamps

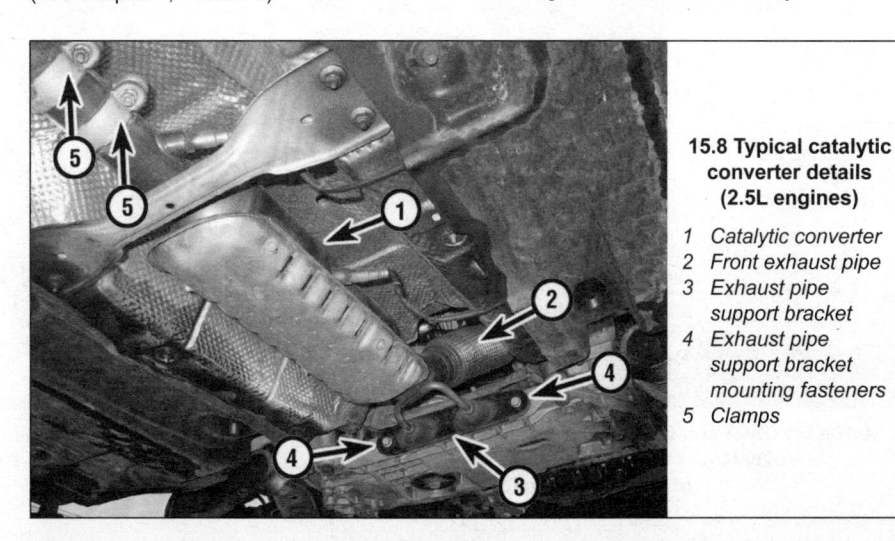

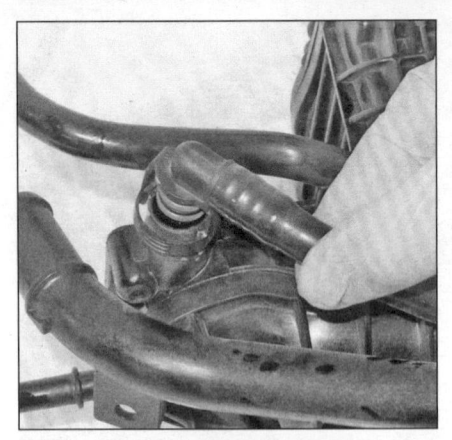

6-20 Chapter 6 Emissions and engine control systems

16.3 Disconnect the EVAP fitting from the intake (1.4L engines)

16.4 Detach the solenoid and tube from the intake (1.4L engines)

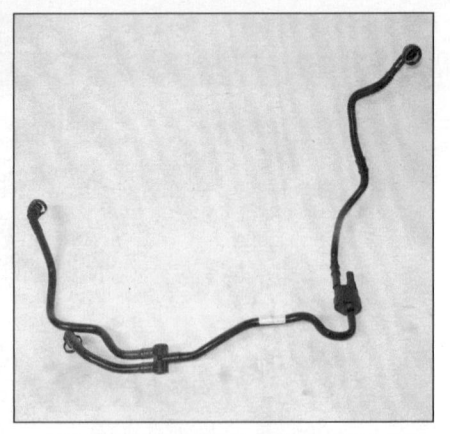

16.5 EVAP purge solenoid and hoses shown removed from the vehicle (1.4L engines)

16 Evaporative Emissions Control (EVAP) system - component replacement

EVAP purge control solenoid valve

Note: The purge control solenoid valve is located on the left end of intake manifold (1.4L engines), on the intake manifold next to the throttle body (1.8L and 2.0L CCTA, CBFA engines), or next to fuel rail (2.5L engines).

1   Remove the engine cover (see Chapter 1, Section 7).

1.4L engines

Note: On 1.4L engines, the purge control solenoid and hoses are one unit. The solenoid cannot be replaced separately.

2   Remove the intake manifold (see Chapter 2A, Section 7).
3   Disconnect the EVAP hose quick-disconnect fitting from the intake manifold (see illustration).
4   Detach the purge solenoid and hose from the intake manifold (see illustration).
5   Remove the EVAP purge solenoid and hose assembly (see illustration).
6   Installation is the reverse of removal.

2.0L turbocharged engines (CBFA and CCTA)

7   Remove the intake manifold (see Chapter 2A, Section 7).
8   Locate the EVAP purge solenoid and remove it from the intake manifold.
9   Installation is the reverse of removal.

All except 1.4L and CBFA/CCTA 2.0L turbocharged engines

10 Locate and disconnect the electrical connector from the purge control solenoid valve **(see illustrations)**.
11 Disconnect the hoses from the purge control solenoid valve.
12 Remove the purge control valve from its mounting bracket.
13 Installation is the reverse of removal.

EVAP canister

2017 and earlier Jetta models - engine compartment mounted

Note: On some 2017 and earlier Jetta models, the EVAP canister is located in the passenger front corner of the engine compartment

14 Locate the EVAP canister and disconnect the two lines.
15 Release the retaining tab at the rear of the canister and pull the canister upwards and out of the vehicle.
16 Installation is the reverse of removal.

2017 and earlier Jetta models - under-vehicle mounted

Note: On some 2017 and earlier Jetta models, the EVAP canister is located underneath the vehicle, next to the spare tire well.

16.10a On CBFA and CCTA engines, the EVAP canister purge control solenoid valve is located at the left rear part of the intake manifold

16.10b On 2.5L engines, the EVAP canister purge control solenoid valve is located next to the fuel rail

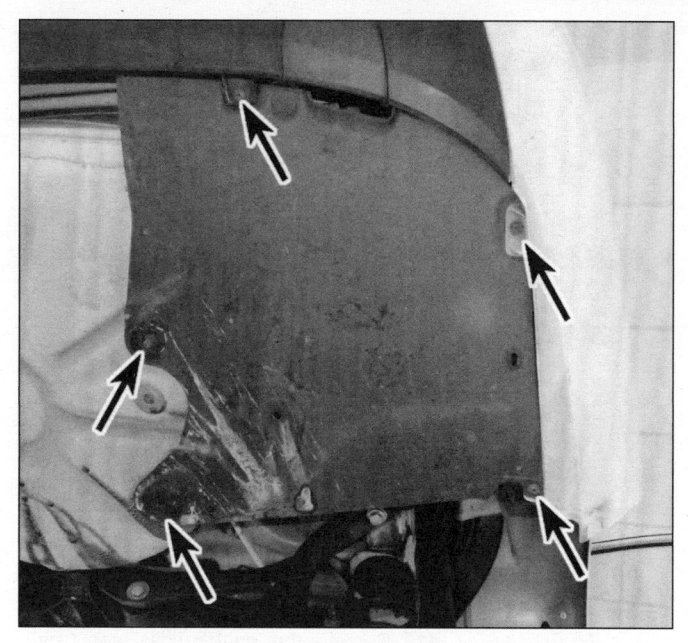

Chapter 6 Emissions and engine control systems

16.18 EVAP canister cover fasteners (Jetta models)

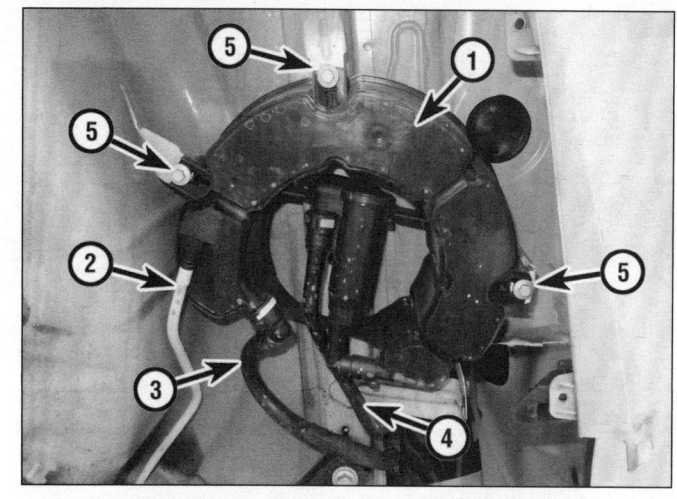

16.19 EVAP canister details (Jetta models)

1. EVAP canister
2. EVAP purge line (to purge control solenoid valve in engine compartment)
3. EVAP vent hose (from fuel tank)
4. EVAP hose (to leak detection pump)
5. Mounting bolts

17 Raise the vehicle and place it securely on jackstands.
18 Remove the EVAP canister cover **(see illustration)**.
19 Disconnect the hoses **(see illustration)** from the canister. To disconnect each hose quick-connect fitting, press the release ring toward the canister (away from the hose) and simultaneously pull off the hose.
20 Unscrew the mounting bolts and remove the canister.
21 Installation is the reverse of removal.

Golf models and 2018 and later Jetta models

Note: *The EVAP canister is located inside the left rear wheel well.*

Note: *On some models, the leak detection pump is part of the EVAP canister.*

22 Loosen the left rear wheel bolts, raise the vehicle and place it securely on jackstands. Remove the left rear wheel. Remove the left rear fenderwell splash shield (see Chapter 11).
23 On models with integrated leak detection pump, disconnect the electrical connector from the leak detection pump.
24 On all models, disconnect the two quick-connect EVAP hoses attached to the EVAP canister.
25 To remove the EVAP canister, remove the bolt/screw at the bottom of the canister, and depress the retaining tab at the top. Pull the EVAP canister (and leak detection pump if equipped) downwards and out of the bracket.
26 On models with integrated leak detection pump, the air filter and leak detection pump can be removed from the canister and transferred over to another canister.
27 Installation is the reverse of removal.

Leak detection pump (2017 and earlier Jetta models)

Note: *On Golf and 2018 Jetta models, the leak detection pump is part of the EVAP canister. For removal and installation procedures, see the EVAP canister removal for Golf models.*

Note: *The leak detection pump is located inside the left rear wheel well.*

28 Loosen the left rear wheel bolts, raise the vehicle and place it securely on jackstands. Remove the left rear wheel. Remove the left rear fenderwell splash shield (see Chapter 11).

16.29 EVAP leak detection pump details (2017 and earlier Jetta models)

1. Mounting bracket nuts (three of four visible)
2. Electrical connector
3. Air filter hose
4. Air filter
5. Hose from EVAP canister
6. Leak detection pump

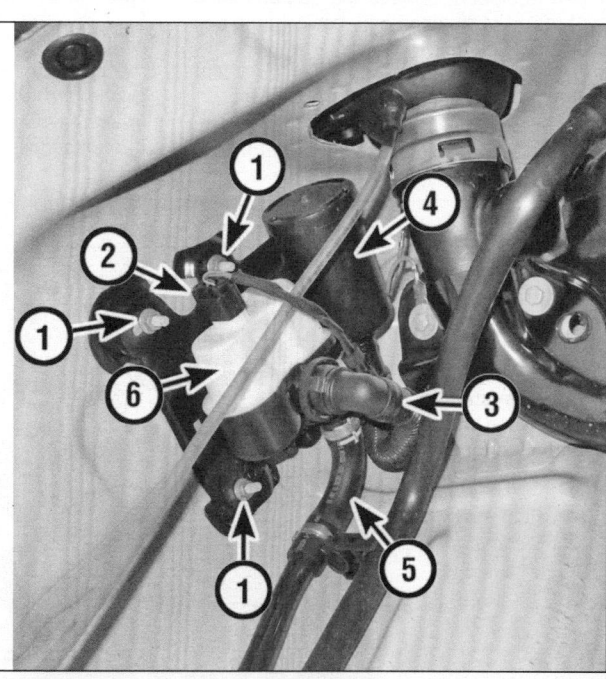

29 Remove the leak detection pump mounting bracket fasteners **(see illustration)**, pull down the pump and mounting bracket, then disconnect the electrical connector from the leak detection pump.
30 Cut the hose clamp that secures the EVAP canister hose to the leak detection pump and disconnect the hose from the leak detection pump.
31 Disconnect the filter and hose from the leak detection pump.
32 Remove the fasteners that secure the leak detection pump to its mounting bracket.
33 Installation is the reverse of removal.

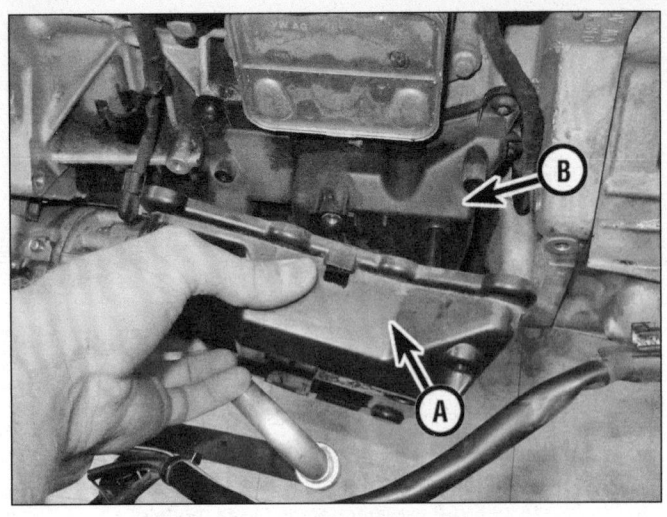

6-22   Chapter 6 Emissions and engine control systems

17.5 Remove the cover (A) that protects the CCV assembly (B) (1.4L engines)

17.6 Remove the fasteners and detach the CCV assembly from the engine block (1.4L engines)

17 Crankcase Ventilation (CCV) system - component replacement

Note: *Over time, the plastic hoses for the CCV system become brittle and breaks easily when attempting to disconnect. When replacing the CCV assembly, it is a good idea to replace the hoses as well.*

1.4L engines

Note: *On these engines, the CCV assembly (oil separator) is bolted to the front side of the engine block.*

1 Raise and support the front of the vehicle on jackstands.
2 Remove the engine under cover.
3 If equipped, remove the bolt attaching the bracket of the charge air cooler pump and position the pump to the side.
4 Disconnect the hose from the CCV assembly.
5 Release the clips and remove the cover from the CCV assembly **(see illustration)**.

Caution: *Clean the area around the CCV assembly to prevent debris from entering the engine block before removing the CCV assembly.*

6 Remove the bolts attaching the CCV assembly to the engine block **(see illustration)**.
7 Carefully pull the CCV assembly from the engine block. The CCV assembly is sealed to the block using adhesive and may require effort to remove.
8 Clean the adhesive from the block (and the CCV assembly if it is to be reinstalled).
9 Apply sealant to the CCV before installing.
10 Installation is reverse of removal.

1.8L and 2.0L engines

Note: *The following procedure applies to 1.8L and 2.0L turbocharged engines, where the CCV assembly is attached to the cylinder head cover. On 2.0L non-turbocharged engines, the CCV is part of the intake air duct. The orifice can be cleaned, but not serviced separately.*

11 Remove the ignition coils (see Chapter 5, Section 6) an any additional items to access the CCV assembly.
12 Disconnect the hoses from the CCV assembly on top of the cylinder head cover.
13 Remove the CCV assembly mounting fasteners and remove the housing from the top of the cylinder head cover.
14 Install a new gasket to the CCV assembly, then install the housing to the cylinder head cover.
15 Install the mounting fasteners and tighten them securely.
16 Reconnect the CCV system hoses.
17 The remainder of the installation is reverse of removal.

2.5L engines

18 The crankcase ventilation assembly is integrated within the valve cover and can't be replaced separately **(see illustration)**.
19 If there is a problem with the assembly, the valve cover will need to be replaced (see Chapter 2B).

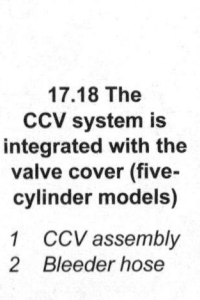

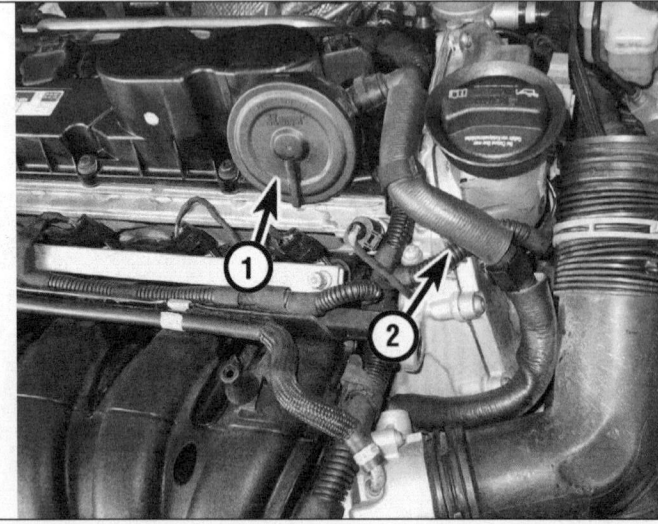

18 Secondary Air Injection (AIR) system - component replacement

Note: *2.0L non-turbocharged engines are not equipped with a secondary air injection system.*

Secondary air injection pump

1.4L engines (2018 and later models)

1 Remove the air filter housing (see Chapter 4, Section 11).
2 Locate the air injection pump and disconnect the two quick-connect hoses.
3 Raise the front of the vehicle and support it securely on jackstands.
4 Remove the lower splash shield below the engine (see Chapter 1, Section 6).
5 Remove the bolts and the driveaxle heat shield.

17.18 The CCV system is integrated with the valve cover (five-cylinder models)

1 CCV assembly
2 Bleeder hose

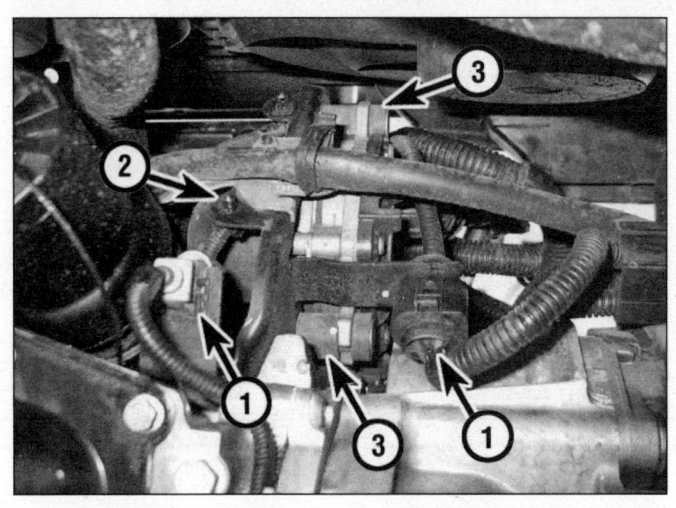

Chapter 6 Emissions and engine control systems   6-23

18.13 Secondary air injection pump details (except 1.4L engines)

1. Electrical connectors
2. Harness retainer
3. AIR pump mounts (two of three visible)

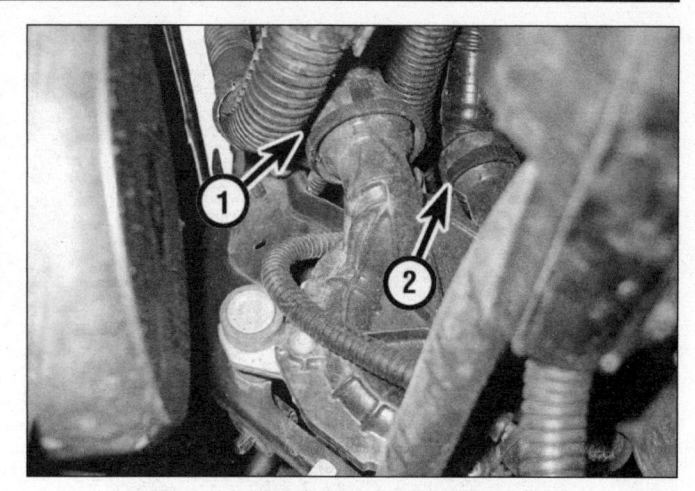

18.14 Secondary air injection pump tube locations (except 1.4L engines)

1. Vent tube-to-throttle body inlet tube
2. Pressure tube-to-secondary air injection solenoid valve

6   Disconnect the air injection pump electrical connector.
7   Detach the coolant pipe from the air pump bracket.
8   Remove the air pump fasteners and detach the air pump from the bracket.
9   Installation is reverse of removal.

All except 1.4L engines

Note: *The secondary air injection pump motor is located in the lower right front corner of the engine compartment.*

10   Raise the front of the vehicle and support it securely on jackstands.
11   Remove the lower splash shield below the engine (see Chapter 1, Section 6).
12   If necessary, remove the driver's inner wheel well.
13   Remove the harness retainer and disconnect the electrical connectors **(see illustration)**.
14   Disconnect the two air tubes from the secondary AIR pump **(see illustration)**.
15   Remove the secondary AIR pump mounting nuts.
Note: *The AIR pump is mounted on three rubber mounts with studs through the middle.*
16   Slightly press the lower rubber mount downwards while removing the secondary air pump to clear the transaxle.
17   Installation is the reverse of removal.

Secondary air injection solenoid valve

1.4L engines (2018 and later)

18   Raise the front of the vehicle and support it securely on jackstands.
19   Remove the lower splash shield below the engine (see Chapter 1, Section 6).
20   Remove the band clamp for the catalytic converter at the turbocharger housing, remove the two bracket bolts and secure the catalytic converter to the side.
21   Remove the bolts and the driveaxle heat shield.
22   Remove the catalytic converter bracket from the engine.
23   Disconnect the air injection hose quick-disconnect from the solenoid valve.
24   Locate the heat shield and remove the three bolts and the heat shield.
25   Disconnect the solenoid valve electrical connector.
26   Remove the bolts securing the solenoid valve base to the engine and remove the solenoid valve assembly.
27   Discard the gasket and use a new one for installation.
28   Installation is reverse of removal.

1.8L and 2.0L turbocharged engines

29   Remove the air filter housing (see Chapter 4, Section 11).
30   Remove the battery and battery tray (see Chapter 5, Section 4).
31   If necessary, remove the air intake duct to access the solenoid valve.
32   Locate the solenoid valve and disconnect the electrical connector.
33   Disconnect the air injection hose quick-connect fitting.
34   Remove the bracket bolt and position the coolant hose and pipe to the side.
35   Remove the bolts attaching the solenoid valve base to the cylinder head and remove the solenoid valve assembly.
36   Discard the seal for the solenoid valve base to the cylinder head and use a new one.
37   Installation is reverse of removal.

2.5L engines

38   Remove the engine cover (see Chapter 1, Section 7).
39   Disconnect the two air tubes from the secondary air injection solenoid valve **(see illustration)**.
40   Disconnect the electrical connector to the solenoid valve.
41   Remove the wire bracket from the base of the secondary air injection solenoid valve. Remove the mounting fasteners and remove the solenoid valve from the cylinder head.
42   Installation is the reverse of removal. Be sure to use new seals and tighten the valve mounting fasteners securely.

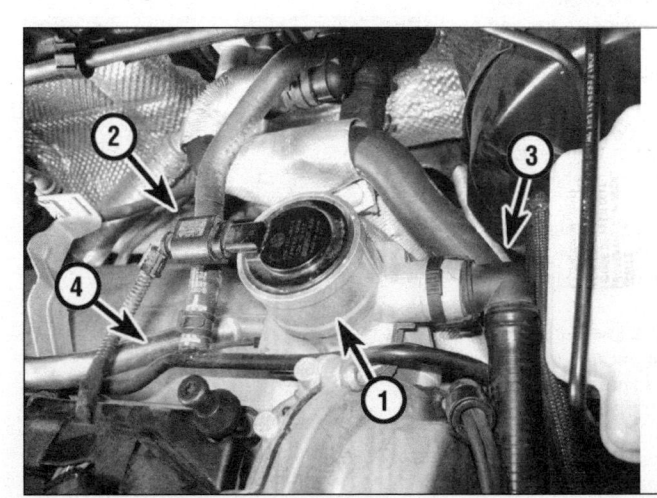

18.39 Secondary air injection solenoid valve details (2.5L engines)

1. Secondary air injection solenoid valve
2. Electrical connector
3. Pressure tube from the air injection pump
4. Secondary air injection pipe to cylinder head

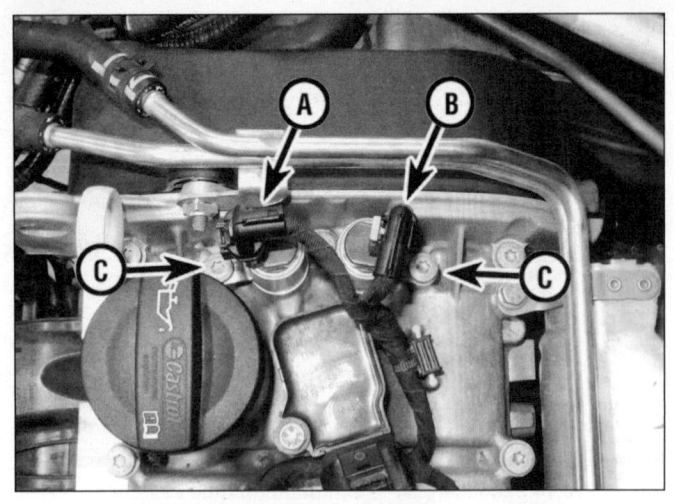

19.2 Variable camshaft adjustment solenoid valve details (1.4L engines)

- A Intake camshaft solenoid valve electrical connector
- B Exhaust camshaft solenoid valve electrical connector
- C Solenoid valve mounting bolt

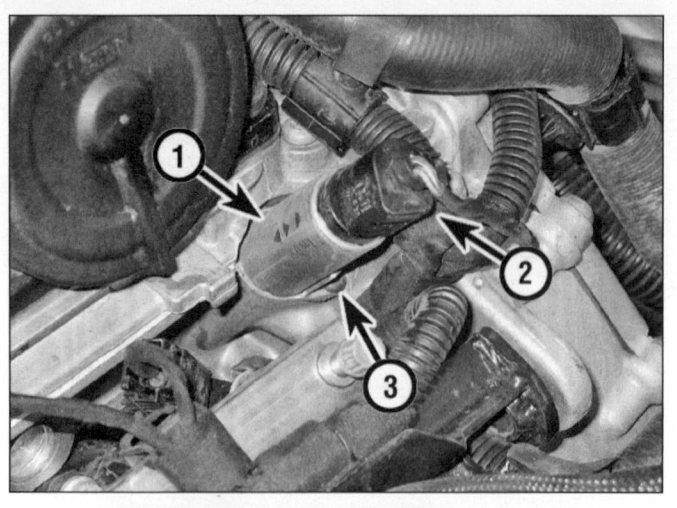

19.13 Variable camshaft adjustment solenoid valve details (2.5L engines)

1. Variable camshaft adjustment solenoid valve
2. Camshaft adjustment solenoid valve electrical connector
3. Solenoid valve mounting fastener

19 Variable camshaft adjustment solenoid valve - replacement

1.4L engines

Note: *The variable camshaft adjustment solenoids are located on the valve cover near the cam cover.*

1. Remove the engine cover.
2. Disconnect the electrical connector from the camshaft adjustment solenoid valve (**see illustration**).
3. Remove the camshaft adjustment solenoid valve mounting bolt and pull the valve out of the valve cover.
4. Remove and discard the solenoid valve O-ring.
5. Installation is the reverse of removal. Use a new O-ring.

1.8L and 2.0L turbocharged engines

Note: *The variable camshaft adjustment solenoid is located on the front of the upper timing chain cover. Some models may be equipped with two solenoids.*

6. Remove the engine cover (see Chapter 1, Section 7).
7. Disconnect the electrical connector to the camshaft adjustment valve, then remove the adjustment valve retaining fasteners, the valve and seal.

Note: *Always replace the adjustment valve seal and O-ring on the adjustment valve.*

8. Disconnect the electrical connector from the camshaft adjustment valve.
9. Remove the camshaft adjustment solenoid valve mounting fasteners and pull the valve out of the upper timing chain cover.
10. Remove and discard the solenoid valve O-ring.
11. Installation is the reverse of removal. Lubricate the new O-ring with engine oil.

2.5L engines

12. Remove the engine cover (see Chapter 1, Section 7).
13. Disconnect the electrical connector from the camshaft adjustment solenoid valve (**see illustration**).
14. Remove the camshaft adjustment solenoid valve mounting fastener and pull the valve out of the cylinder head.
15. Remove and discard the solenoid valve O-ring.
16. Installation is the reverse of removal. Use a new O-ring.

20 Oil pressure switch - replacement

1.4L engines

Oil pressure switch

Note: *The oil pressure switch on 1.4L engines is located on the rear of the engine block, behind the driveaxle heat shield.*

1. Raise and support the front of the vehicle on jackstands.
2. Remove the engine undercover.
3. Remove the passenger's driveaxle heat shield bolts and remove the heat shield.
4. Locate the switch and disconnect the electrical connector (**see illustration**).
5. Unscrew and remove the switch from the engine block.
6. Discard the oil pressure switch seal and use a new one for installation if the same switch will be used.
7. Installation is reverse of removal.

Reduced oil pressure switch

Note: *The reduced oil pressure switch on 1.4L engines is located on the front of the engine block, near the alternator.*

8. Detach the EVAP canister purge valve from the intake manifold (see Section 16).
9. Locate the switch and disconnect the electrical connector (**see illustration**).
10. Unscrew and remove the switch from the engine block.

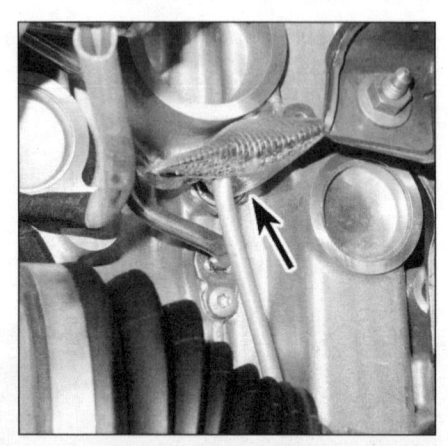

20.4 Identifying the oil pressure switch (1.4L engines)

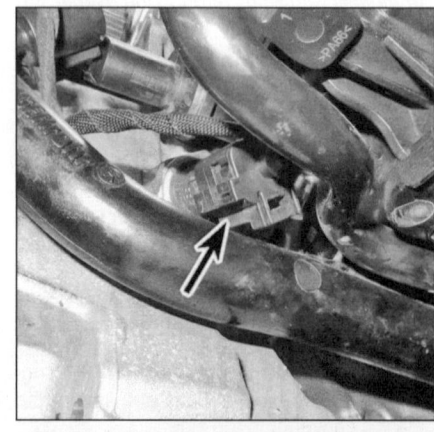

20.9 Identifying the reduced oil pressure switch (1.4L engines)

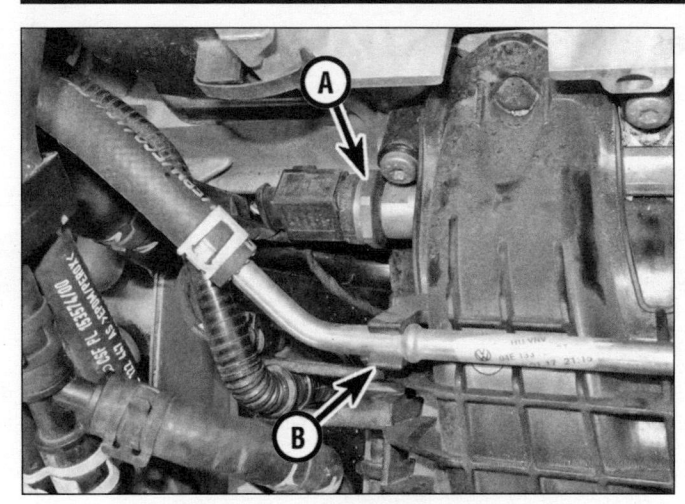

Chapter 6 Emissions and engine control systems

22.4 Fuel pressure sensor (A) and fuel feed line clip (B) (1.4L engines)

23.3 Identifying the charge air pressure sensor (1.4L models)

11 Discard the oil pressure switch seal and use a new one for installation if the same switch will be used.
12 Installation is reverse of removal.

1.8L, 2.0L and 2.5L engines

Note: *On 2.0L CBPA engines, the oil pressure and reduced oil pressure switches are located on the oil filter housing. On all other 1.8L and 2.0L engines, the oil pressure switch is located on the rear of the engine block and the reduced oil pressure switch is located on the oil filter housing assembly (2.0L CBPA engines are not equipped with a reduced oil pressure switch). On 2.5L engines the oil pressure and reduced oil pressure switches are located on the engine block near the oil filter housing assembly.*

13 On 2.0L CBPA engines, remove the EVAP canister (see Section 16).
14 To remove the oil pressure switch on 1.8L and 2.0L engines (except CBPA), raise and support the front of the vehicle.
15 On all engines, locate the switches and disconnect the electrical connector.
16 Unscrew the switch to remove from the engine block or oil filter housing.
17 Discard the oil pressure switch seal and use a new one for installation if the same switch will be used.
18 Installation is reverse of removal.

21 Variable intake manifold actuator (1.8L and 2.0L turbocharged engines) - replacement

Note: *The intake flap is controlled by a vacuum actuator mounted to the end of the intake manifold below the high pressure fuel pump.*
Note: *On some models, the actuator may not be serviceable separately.*

1 Remove the engine cover (see Chapter 1, Section 7).
2 Disconnect the vacuum hose to the vacuum actuator.
3 Disconnect the vacuum actuator-to-crank arm linkage.
4 Remove the actuator fasteners and actuator from the manifold.
5 Installation is the reverse of removal.

22 Fuel Pressure Sensor - replacement

1 Release the fuel system pressure (see Chapter 4, Section 3).
2 Disconnect the negative battery cable (see Chapter 5, Section 3).

1.4L engines
Note: *The fuel pressure sensor is located at the end of the fuel rail.*

3 Remove the air filter housing if necessary to access the sensor (see Chapter 4, Section 11).
4 Locate the fuel pressure sensor and disconnect the electrical connector **(see illustration)**.
5 Unscrew the sensor from the fuel rail.
Caution: *Do not apply any sealant to the threads of the sensor during installation.*
6 Installation is reverse of removal.

1.8L and 2.0L (except CBFA and CCTA) turbocharged engines

7 Remove the engine cover.
8 Remove the coolant pipe bracket bolts near the alternator.
9 Disconnect the charge air pressure sensor connector.
10 Remove the bracket bolts, loosen the clamps and remove the intake duct between the throttle body and the intercooler.
11 Remove the intake manifold bracket.
12 Locate the fuel pressure sensor and disconnect the electrical connector.
Note: *If needed, disconnect the alternator connector for additional clearance for removal of the sensor.*

13 Unscrew the sensor to remove from the vehicle.
14 Installation is reverse of removal. Apply clean engine oil to the threads before installing.

2.0L (CBFA and CCTA) turbocharged engines

15 If needed, remove the passenger's side intake air duct between the intercooler and the turbocharger.
16 Remove the fasteners attaching the bracket to the alternator.
17 Remove the bolts securing the coolant line to the intake manifold and position the coolant line out of the way.
18 Disconnect the fuel pressure sensor electrical connector.
19 Unscrew and remove the fuel pressure sensor from the fuel rail.
20 Installation is reverse of removal. Apply clean engine oil to the threads before installing.

23 Charge Air Pressure sensor - replacement

Note: *The sensor is located in the charge air intake duct.*

1 On 1.8L and 2.0L CPLA, CPPA, CPKA and CPRA engines, remove the air filter housing assembly to access the sensor.
2 On all other 1.8L and 2.0L engines, raise and support the front of the vehicle on jackstands and remove the engine undercover.
3 Locate the sensor and disconnect the electrical connector **(see illustration)**.
4 Remove the sensor fasteners or release the tabs, as applicable, then remove the sensor from the duct.
5 Discard the O-ring and use a new one when installing the same sensor.
6 Installation is reverse of removal.

Notes

Chapter 7 Part A
Manual transaxle

Contents

	Section		Section
Back-up light switch - removal and installation	7	Oil seals - replacement	8
General Information	1	Shift cables - removal, installation and adjustment	3
Manual transaxle - removal and installation	4	Shift lever assembly - removal and installation	2
Manual transaxle overhaul - general information	5	Transaxle mounts - check and replacement	6

Specifications

General

Designations
- 5-speed manual 0AF, 0A4
- 6-speed manual .... 02Q (A, H, K, S), 0BB
- Lubricant type and capacity .... See Chapter 1

Torque specifications   Ft-lbs (unless otherwise indicated)   Nm

Note: *One foot-pound (ft-lb) of torque is equivalent to 12 inch-pounds (in-lbs) of torque. Torque values below approximately 15 ft-lbs are expressed in inch-pounds, because most foot-pound torque wrenches are not accurate at these smaller values.*

Driveaxle flange mounting bolt	24	33
Driveaxle heat shield-to-engine bolts	26	35
Transaxle-to-engine bolts **(see illustrations 4.29a and 4.29b)**		
5-speed models		
Bolts A (M12 x 65mm)	59	80
Bolts B (M12 x 150mm)	59	80
Bolts C (M12 x 165mm)	59	80
Bolt D (M10 x 50mm)	30	40
Bolt E (M12 x 85mm)	59	80
Bolt F (M6 x 8mm)	88 in-lbs	10
6-speed models		
Bolts A (M12 x 55mm short thread)	59	80
Bolts B (M12 x 55mm long thread)	59	80
Bolts C (M12 x 70mm)	59	80
Bolt D (M10 x 50mm)	30	40
Bolt E (M10 x 105mm)	30	40
Bolt F (M12 x 165mm)	59	80
Bolt G (M12 x 165mm)	59	80
Transaxle mount-to-body bolts*		
Step 1	44	60
Step 2	Tighten an additional 90-degrees	
Lubricant level filler/check plug and drain plug	See Chapter 1	

* Do not reuse bolts - always replace bolts with new ones.

Chapter 7 Part A Manual transaxle

1.3a Typical 5-speed shift linkage details - 6-speed models similar

1. Selector lever
2. Damper
3. Selector lever housing
4. Bushing
5. Bolt
6. Fulcrum pin
7. Bushing
8. Spring
9. Selector lever gate
10. Retaining screw
11. Housing seal
12. Securing clip
13. Bushing
14. Spring
15. Cover plate
16. Damper collar
17. Bearing
18. Selector lever ball/guide
19. Damping washer
20. Shift cable
21. Baseplate
22. Locking nut

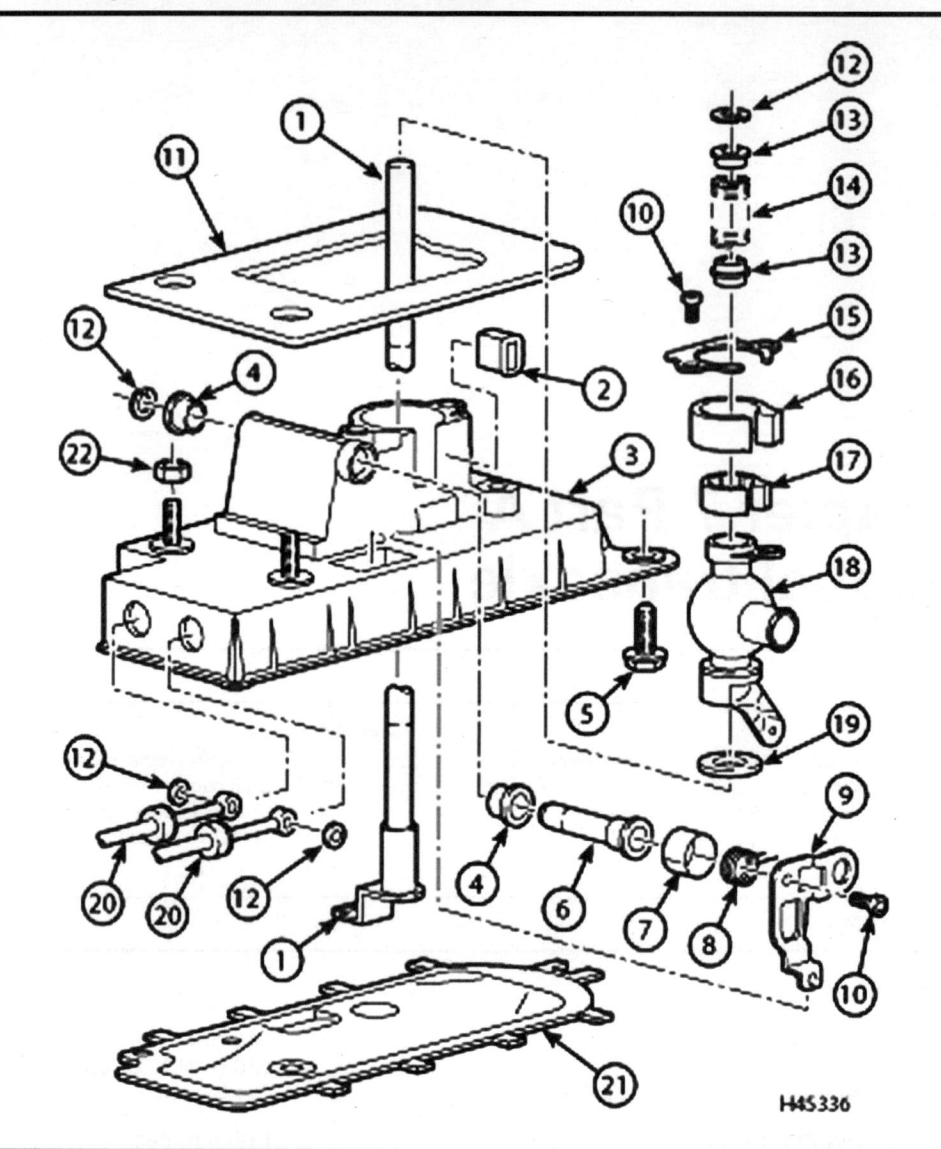

1.3b Transaxle end of the shift cables

1 General Information

1 The manual transaxle is bolted to the left end of the engine. The transaxle case is aluminum alloy.

2 Drive from the crankshaft is transmitted through the clutch to the transaxle input shaft, which is splined to the clutch friction disc.

3 All gears including reverse incorporate a synchromesh engagement. The floor-mounted shift lever is connected to the transaxle by shift cables (see illustrations).

2 Shift lever assembly - removal and installation

Note: *On later Golf models with 6-speed transaxles, the shift cables are part of the shift lever assembly and cannot be replaced separately. They can still be adjusted, but not replaced without replacing the shift lever assembly with them.*

Shift knob and boot

Note: *The badge on leather and plastic shift knobs can be removed by carefully prying out with a plastic trim tool.*

1 Carefully pry the shift boot from the center console using a plastic trim tool or similar.

2 Pull the shift boot up to expose the shift lever.

3 Open the retaining clamp attaching the shift knob to the lever and remove the shift knob and boot together.

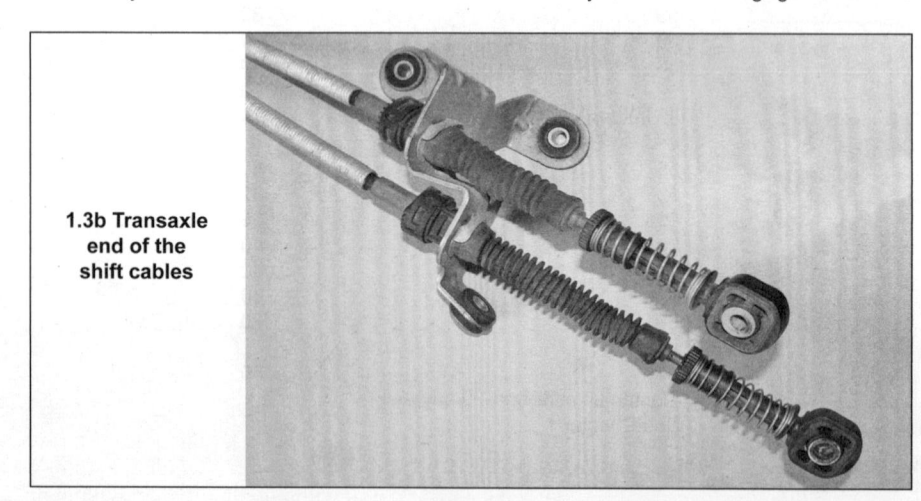

Chapter 7 Part A Manual transaxle

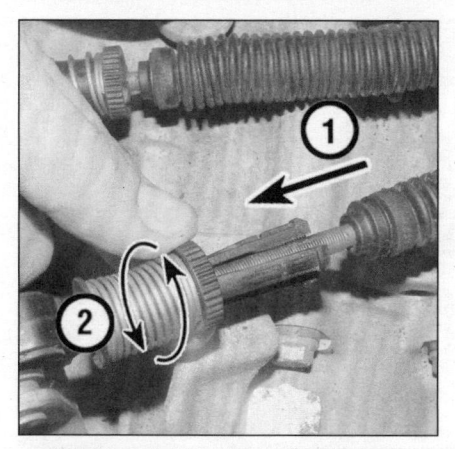

3.8 Push the locking collar forward to compress the spring (1), then turn clockwise to lock (2) - 5-speed model shown, 6-speed models similar

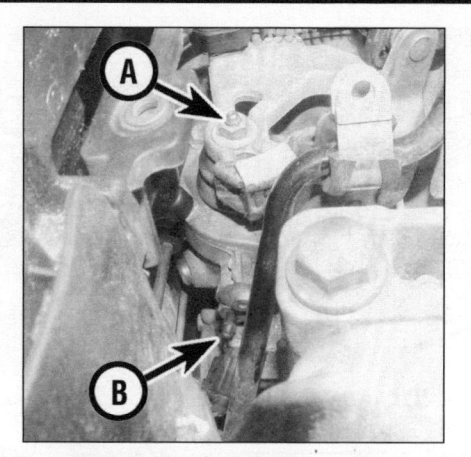

3.19 Press down on the selector shaft (A), then rotate the locking lever (B) while pushing the lever in to lock the shaft - 5-speed model shown, 6-speed models similar

3.20 Lock the gear lever in position with a drill bit - 5-speed model shown, 6-speed models similar

4 After removing the shift knob and boot, the lower insulation inside the shift lever assembly (if equipped) can be removed by disengaging the retainers and pulling up and out.
5 Installation is reverse of removal.
Note: It is recommended that a new clamp be used during installation.

Shift lever assembly

6 Disconnect the cable from the negative battery terminal (see Chapter 5, Section 3).
7 Raise the vehicle and support it securely on jackstands.
8 Working inside of the vehicle, remove the shift knob and boot as described in the previous procedure.
9 Remove the center console (see Chapter 11, Section 27).
10 If equipped, remove the nuts securing the shift lever assembly cover, and remove the cover.
11 Remove the nuts securing the shift lever assembly to the floor of the vehicle.
12 On later Golf models with 6-speed transaxles, working in the engine compartment, remove the air filter housing (see Chapter 4, Section 11).
13 Disconnect the cables from the shift levers and bracket at the transaxle (see Section 3).
14 On all models, working under the vehicle, remove the under body covers and center tunnel cross-member fasteners and the crossmember.
15 Disconnect the exhaust at the front coupling and lower it to access the shift lever assembly.
16 Remove the fasteners and the panel and heat shield from the tunnel.
17 Lower the shift lever assembly.
18 On later Golf models with 6-speed transaxles, remove the shift lever assembly with the cables. Note the routing of the cables for installation.
19 On all other models, disconnect the shift cables from the shift lever assembly (see Section 3).
20 Remove the shift lever assembly.
21 Installation is reverse of removal.
22 On later Golf models with 6-speed transaxles, route the shift cables in the original path.
23 Adjust the shift cables (see Section 3).

3 Shift cables - removal, installation and adjustment

Caution: If the battery is disconnected, several systems must be re-learned before they will work properly (see Chapter 5, Section 3).
Note: On later Golf models with 6-speed transaxles, the shift cables are part of the shift lever assembly and cannot be replaced separately. They can still be adjusted, but not replaced without replacing the shift lever assembly with them.

Removal

1 Place the shift lever into neutral position.
2 Remove the shift lever assembly (see Section 2).
3 At the bottom of the shift lever assembly housing, bend the base plate tabs up and remove the base plate and seal.
4 Pry the shift cables off of the shift lever pins.
5 Pull up and remove the shift cable housing retainers at the housing and pull the cables through the shift lever assembly housing.
6 Working in the engine compartment, remove the engine cover/air filter housing if necessary to access the transaxle shift levers and cables (see Chapter 1, Section 7).
7 Remove the battery and battery tray, if necessary, to access the transaxle shift levers and cables (see Chapter 5).
8 With the shift lever in the neutral position, push the two locking collars on the cable retainers (one on each cable) forward to compress the springs, then turn them clockwise to lock into position **(see illustration)**.
9 Pull the cables from the retainers.
10 Pull up and remove the shift cable housing retainers at the bracket and pull the cables through the shift cable bracket.
11 To remove the retainers, use a screwdriver to pry from the shift levers.

Installation

12 Installation is reverse of removal.
13 When installing the cables, secure the cable housing with the retainers before attaching the cable ends.
14 To attach the cable ends at the transaxle, insert the cables into the retainers, then rotate the two locking collars counterclockwise to allow the springs to expand and lock the cables into the retainers.
15 Adjust the shift cables as instructed in this section and ensure the shifter operates properly.

Adjustment

16 Remove the engine cover/air filter housing if necessary to access the transaxle shift levers and cables (see Chapter 1, Section 7).
17 Remove the battery and battery tray, if necessary, to access the transaxle shift levers and cables (see Chapter 5).
18 With the shift lever in the neutral position, push the two locking collars (one on each cable) forwards to compress the springs, then turn them clockwise to lock into position **(see illustration 2.9)**.
19 Press down on the selector shaft on the top of the transaxle. Push the locking pin (special tool T10027) into the transaxle while turning it clockwise until it engages and locks the selector shaft **(see illustration)**.
20 Working inside the vehicle, unclip the shift lever boot from the center console. Still in the neutral position, move the shift lever as far to the left as possible and insert a screwdriver (or drill bit) through the hole in the base of the gear lever and into the hole in the housing **(see illustration)**.

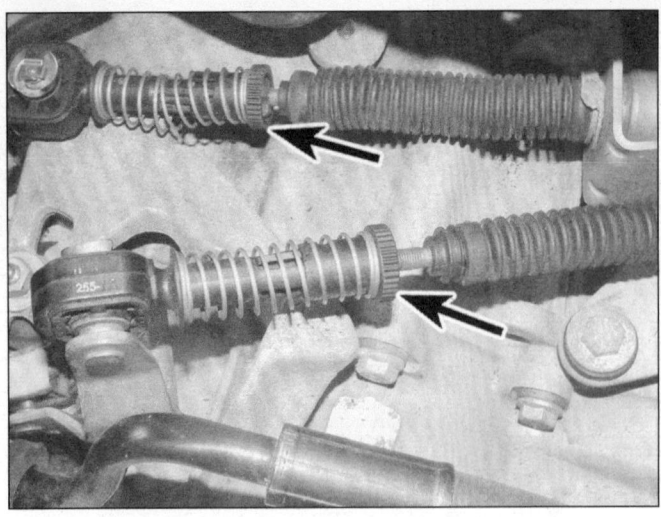

3.21 Release the two locking collars back into position - 5-speed model shown, 6-speed models similar

4.5 On models with metal relay levers, use a small screwdriver to disconnect the retaining clip

21 Working back in the engine compartment, turn the two locking collars on the cables counterclockwise so that the springs will release them back into position and lock the cables **(see illustration)**.
22 With the cable adjustment set, the locking pin can now be turned counterclockwise to its original position (pointing upwards) and the selector shaft can be moved.
23 Inside the vehicle, remove the drill bit from the shift lever, then check the operation of the selector mechanism. When the shift lever is at rest in neutral, it should be centered, ready to select 3rd or 4th. The shift lever boot can now be installed to the center console.
24 The remaining installation is the reverse of removal.

4 Manual transaxle - removal and installation

Caution: *If the battery is disconnected, several systems must be re-learned before they will work properly (see Chapter 5, Section 3).*

Removal

1 Park the vehicle on a solid, level surface. Give yourself enough space to move around it easily. Apply the parking brake and chock the rear wheels.
2 Remove the battery and battery tray (see Chapter 5). Also remove the engine cover/air filter housing (see Chapter 1, Section 7).
3 Loosen the front wheel bolts and the driveaxle/hub bolts.
4 Raise the front of the vehicle and support it securely on jackstands. Remove the splash guard from under the engine compartment (see Chapter 1, Section 6).
5 On models equipped with a metal relay lever, remove the clips from the shift lever cables and relay lever **(see illustration)**. Remove the cables and lever from the top of the transaxle.
6 On models equipped with a plastic relay lever, push the two locking collars (one on each cable) forwards to compress the springs, then turn them clockwise to lock into position **(see illustration 3.9)**. Using a small screwdriver, press the retaining tab down at the relay lever and slide the cables and relay off as a unit.
7 Remove the shift cable mounting bracket bolts and bracket **(see illustration)**.
8 On AWD Golf models, if necessary, remove the windshield washer fluid reservoir.
9 On all models, unbolt the release cylinder hydraulic line bracket, the remove the transaxle support bracket bolts and bracket.
10 Unbolt the release cylinder from the transaxle, and tie it to one side without disconnecting the hydraulic line.
Caution: *Do not depress the clutch pedal with the release cylinder removed.*
11 Disconnect the electrical connector from the back-up light switch on the transaxle. On models with start/stop system, disconnect the transaxle neutral position sensor.
12 Disconnect the starter solenoid electrical connector. Check that all wiring and ground straps have been disconnected from the transaxle.
13 Remove the starter motor (see Chapter 5).
14 Support the engine from above with an engine support fixture (see Chapter 2C).
15 Remove the transaxle-to-engine bolts accessible from above.
16 Remove the inner fender liner (see Chapter 10). Remove the suspension level control sensor, if equipped (see Chapter 11).
17 Remove the engine cooling fans from the radiator (see Chapter 3, Section 5).
18 On turbocharged models, remove the lower intercooler charge hose.
19 On models with hydraulic power steering, remove the power steering lines and brackets from the transaxle.
20 On models with level control, remove the front level control sensor from the vehicle.
21 On all models, remove the front portion of the exhaust system (see Chapter 4). Take care not to excessively bend the flexible section of the front pipe.
Note: *On some models, it may not be nec-*

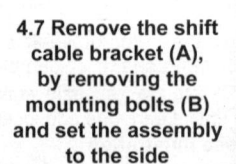

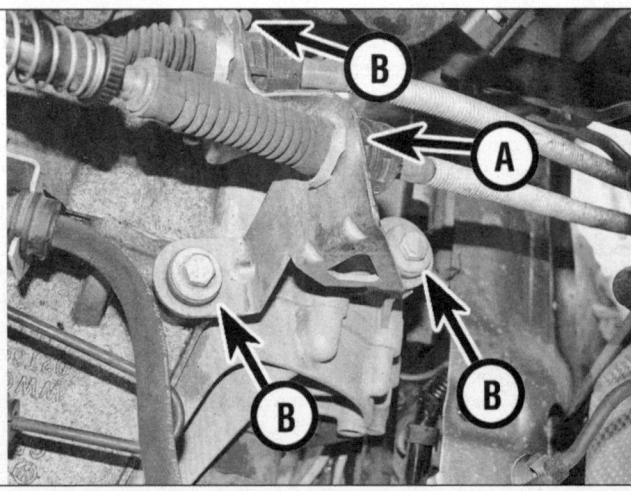

4.7 Remove the shift cable bracket (A), by removing the mounting bolts (B) and set the assembly to the side

Chapter 7 Part A Manual transaxle

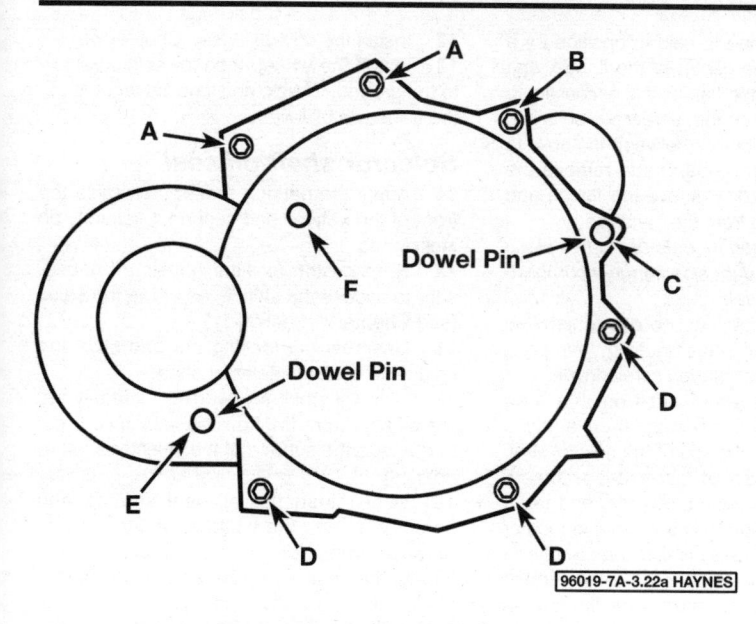

4.29a 5-speed mounting bolt identification
(refer to letters for torque specifications when installing)

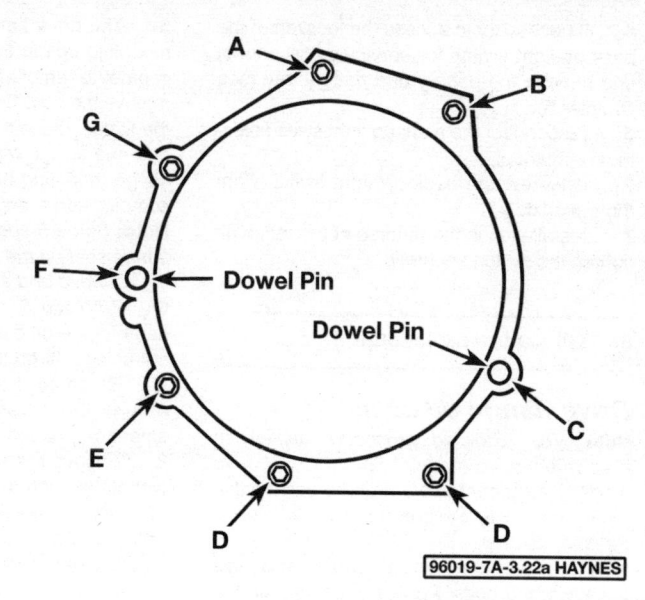

4.29b 6-speed mounting bolt identification
(refer to letters for torque specifications when installing)

---

essary to remove the exhaust for transaxle removal.

22 Remove the driveaxles (see Chapter 8).

23 On AWD models, disconnect the driveshaft (see Chapter 8, Section 16) and support out of the way, and disconnect any AWD components electrical connectors.

24 On all models, working under the vehicle, remove the pendulum support mounting bolt and support (see Chapter 2A).

25 Disconnect any remaining connectors and detach harnesses from routing clips to prevent damage to the harness or components.

26 Support the transaxle on a jack, preferably one made for this purpose. Secure the transaxle to the jack with a safety chain. Unbolt the transaxle mounts, and if necessary, remove the mounts and the brackets.

27 On 5-speed models, if equipped, remove the flywheel cover plate.

28 On all models, using the support fixture, lower the engine and transaxle until any remaining mount bolts can be accessed and removed.

29 Make sure that the transaxle is adequately supported, then unscrew the remaining bolts securing the transaxle to the engine **(see illustrations)**.

30 With the help of an assistant, withdraw the transaxle from the locating dowels on the rear of the engine, making sure that the input shaft does not hang on the clutch. **Caution:** *Make sure that the transaxle remains steady on the jack head. Keep the transaxle level until the input shaft is fully withdrawn from the clutch friction plate.* **Note:** *On 6-speed models, the transaxle is pressed to the engine dowels and will have to be slowly and carefully separated.*

31 Lower the transaxle to the ground.

32 Service the clutch as needed (see Chapter 8).

Installation

33 Before installing the transaxle, make sure that the location dowels are correctly positioned in the engine cylinder block rear face.

34 Installation of the transaxle is a reversal of the removal procedure, but note the following points:

a) *Check the rubber mounts and replace them if necessary.*
b) *Apply a little high-melting-point grease to the splines of the transaxle input shaft.*
c) *Tighten all fasteners* **(see illustration 4.29a or 4.29b)** *to the specified torque.*
d) *On completion, refer to Section 3 and check the shift cable adjustment.*

5 Manual transaxle overhaul - general information

1 Overhauling a manual transaxle unit is a difficult and involved job for the home mechanic. In addition to disassembling and reassembling many small parts, clearances must be precisely measured and, if necessary, changed by selecting shims and spacers. Internal transaxle components are also often difficult to obtain and in many instances, extremely expensive. Because of this, if the transaxle develops a fault or becomes noisy, the best course of action is to have the unit overhauled by a transaxle specialist or to obtain an exchange reconditioned unit.

2 Nevertheless, it is not impossible for the more experienced mechanic to overhaul the transaxle if the special tools are available and the job is carried out in a deliberate step-by-step manner, to ensure that nothing is overlooked.

3 The tools necessary for an overhaul include internal and external snap-ring pliers, bearing pullers, a slide hammer, a set of pin punches, a dial test indicator and possibly a hydraulic press. In addition, a large, sturdy workbench and a vise will be required.

4 During disassembly of the transaxle, make careful notes of how each component is fitted to make reassembly easier and accurate.

5 Before disassembling the transaxle, it will help if you have some idea of where the problem lies. Certain problems can be closely related to specific areas in the transaxle which can make component examination and replacement easier. Refer to the *Troubleshooting* Section in this manual for more information.

6 Transaxle mounts - check and replacement

1 Manual transaxle mounts are similar to automatic transaxle mounts. For checking and replacing transaxle mounts, see Chapter 7B, Section 13.

7 Back-up light switch - removal and installation

Note: *The back-up light switch is located near the transaxle shift lever.*

1 Raise the vehicle and support it securely on jackstands.

2 Remove the splash shield under the engine (see Chapter 1, Section 6).

3 Remove the engine cover/air filter housing (see Chapter 1, Section 7).

4 If necessary to access the location of the back-up light switch for servicing, disconnect and remove the battery and battery tray (see Chapter 5).
5 Disconnect the back-up light switch electrical connector.
6 Unscrew the back-up light switch from the transaxle.
7 Installation is the reverse of the removal; tighten the switch securely.

8 Oil seals - replacement

Drive flange oil seals

Note: *The following procedure applies to front-wheel-drive models. For all wheel drive models, additional seals may be serviceable (see Chapter 8, Section 14 and Chapter 8, Section 15).*

1 Loosen the wheel bolts and the driveaxle/hub bolts. Apply the parking brake, then raise the front of the vehicle and support it securely on jackstands. Remove the wheel.
2 Refer to Chapter 8 remove the driveaxle.
3 Unbolt and remove the heat shield from the right side driveaxle.
Note: *On 6-speed models, it may be necessary to remove the inner fender liner for better access to the flange (see Chapter 11).*
4 Position a suitable container beneath the transaxle to catch spilled oil.
5 The drive flange is held in position by a mounting bolt in the center of the flange, and in order to remove the flange, it is necessary to remove the bolt. To do this, screw two bolts into the flange, place a prybar between the bolts to hold the flange from turning and remove the flange mounting bolt. Remove the flange and compression spring from the transaxle.
Note: *Some 5-speed models use a two piece flange seal on the right side flange incorporating a sleeve and seal.*
6 With the flange out, note the installed depth of the oil seal in the housing, then pry it out using a large flat-bladed screwdriver.
7 Clean all traces of dirt from the area around the oil seal opening, then apply a smear of grease to the lips of the new oil seal.
8 Ensure the seal is correctly positioned, with its sealing lip facing inwards, and tap it squarely into position, using a suitable tubular drift (such as a socket) which bears only on the hard outer edge of the seal. If the surface of the flange is good, make sure the seal is installed at the same depth in its housing as originally noted.
9 Clean the oil seal and apply a smear of multi-purpose grease to its lips.
10 Insert the drive flange with compression spring through the oil seal, engaging it with the differential gear.
11 Install the flange mounting bolt and tighten the bolt to the torque listed in this Chapter's Specifications.
12 Install the driveaxle (see Chapter 8).
13 Install the wheel, then lower the vehicle to the ground. Check, and if necessary top up, the transaxle oil level.

Selector shaft oil seal

14 Apply the parking brake, then raise the front of the vehicle and support it securely on jackstands.
15 Remove the air filter housing if necessary to access the shift levee at the transaxle (see Chapter 4, Section 11).
16 Unscrew the locking nut and slide the shift lever from the selector shaft.
17 Using a small screwdriver, carefully pry the oil seal from the housing, taking care not to damage the surface of the selector shaft or housing.
18 Wipe clean the oil seal seating and selector shaft, then smear a little multi-purpose grease on the new oil seal lips and locate the seal over the end of the shaft. Make sure the closed side of the seal faces outwards. To prevent damage to the oil seal, temporarily wrap some adhesive tape around the end of the shaft.
19 Tap the oil seal squarely into position, using a suitable tubular drift which bears only on the hard outer edge of the seal.
20 Install the gearshift coupling and tighten the locking bolt.
21 Lower the vehicle to the ground.

Chapter 7 Part B
Automatic transaxle

Contents

	Section		Section
Automatic transaxle overhaul - general information	12	Shift knob - removal and installation	5
Automatic transaxle - removal and installation	11	Shift lever assembly - removal and installation	6
Diagnosis - general	2	Transaxle fluid cooler - removal and installation	9
Electronic control system	3	Transaxle mounts - check and replacement	13
General Information	1	Transaxle range switch - replacement and adjustment	10
Shift cable - removal, installation and adjustment	7	Transmission Control Module (TCM) - removal and installation	4
Shift interlock system - description and check	8		

Specifications

General
Designations
6-speed automatic	09G, 09GL
8-speed automatic	09S
6-speed automatic direct shift transaxle	02E, 0D9
7-speed automatic direct shift transaxle	0CW, 0AM, 0GC
Automatic transaxle fluid type and capacity	See Chapter 1

Torque specifications

Ft-lbs (unless otherwise indicated)   Nm

Note: *One foot-pound (ft-lb) of torque is equivalent to 12 inch-pounds (in-lbs) of torque. Torque values below approximately 15 ft-lbs are expressed in inch-pounds, because most foot-pound torque wrenches are not accurate at these smaller values.*

	Ft-lbs	Nm
Automatic selector cable bracket bolt	80 in-lbs	9
Transaxle oil cooler fasteners		
6-speed transaxle		
Round oil cooler bolt	27	36
Rectangle oil cooler bolts	15	20
6-speed DSG transaxle		
Step 1	15	20
Step 2	Tighten an additional 90 degrees	
Driveplate-to-torque converter (new bolts)	44	60
6-speed auto transaxle-to-engine mounting fasteners **(see illustration 11.28a)**		
Five-cylinder models		
Bolts 1 (M10 x 65mm)	30	40
Bolt 2 (M10 x 76mm)	30	40
Bolts 3 (M12 x 65mm)	59	80
Bolt 4 (M12 x 95mm)	59	80
Bolt 5 (M12 x 170mm)	59	80
Bolt 6 (M12 x 170mm)	59	80
All 2011 four-cylinder models		
Bolts 1 (M10 x 55mm)	30	40
Bolt 2 (M10 x 65mm)	30	40
Bolts 3 (M12 x 55mm)	59	80
Bolt 4 (M12 x 70mm)	59	80
Bolt 5 (M12 x 165mm)	59	80
Bolt 6 (M12 x 165mm)	59	80

Torque specifications (continued)

Note: One foot-pound (ft-lb) of torque is equivalent to 12 inch-pounds (in-lbs) of torque. Torque values below approximately 15 ft-lbs are expressed in inch-pounds, because most foot-pound torque wrenches are not accurate at these smaller values.

	Ft-lbs (unless otherwise indicated)	Nm
6 and 7-speed DSG transaxle-to-engine mounting fasteners (see illustration 11.28b)		
Bolts 1 (M10 x 50mm)	30	40
Bolts 2 (M12 x 70mm)	59	80
Bolt 3 (M10 x 40mm)	30	40
Bolts 4 (M12 x 55mm)	59	80
Bolt 5 (M10 x 45mm)	30	40
8-speed transaxle-to-engine mounting fasteners	Not available	
Transaxle mount*		
Mount-to-transaxle bracket bolts (driver side)		
Step 1	29.5	40
Step 2	Tighten an additional 90 degrees	
Mount-to-body bracket bolts (driver side)		
Step 1	44	60
Step 2	Tighten an additional 90 degrees	
Roll restrictor bolts		
Subframe		
Step 1	96	130
Step 2	Tighten an additional 90 degrees	
Transaxle bracket		
Step 1	36	50
Step 2	Tighten an additional 90 degrees	

* Replace all mount bolts with new ones.

1 General Information

1   The automatic transaxles on the covered models include either a 6-speed automatic, 8-speed automatic, 6-speed DSG (Direct Shift Transaxle) or 7-speed DSG.

2   The DSG transaxle uses two multi-disc clutch packs instead of a conventional torque converter. This allows the driver to shift manually through the gears without a clutch pedal.

3   The identification of the transaxle designation is the first three digits of the transaxle number. On all transaxles, the numbers are located on side of the transaxle on a pad **(see illustration)**.

4   The overall operation of the transaxle is managed by the Engine Control Module (ECM) and the Transaxle Control Module (TCM). Comprehensive troubleshooting can therefore only be carried out using dedicated electronic test equipment such as a factory scan tool.

5   Due to the complexity of the transaxle and its control system, major repairs and overhaul operations should be left to a dealer service department or other qualified repair facility, who will be equipped to carry out troubleshooting and repair. The information in this Chapter is therefore limited to a description of the removal and installation of the transaxle as a complete unit. The removal, installation and adjustment of the shift cable and key interlock cable is also described.

2 Diagnosis - general

1   Automatic transaxle malfunctions may be caused by five general conditions:
 a) Poor engine performance
 b) Improper adjustments
 c) Hydraulic malfunctions
 d) Mechanical malfunctions
 e) Malfunctions in the computer or its signal network

2   Diagnosis of these problems should always begin with a check of the easily repaired items: fluid level and condition (see Chapter 1) and shift cable adjustment. Next, perform a road test to determine if the problem has been corrected or if more diagnosis is necessary. If the problem persists after the preliminary tests and corrections are completed, additional diagnosis should be done by a dealer service department or transaxle repair shop. Refer to the *Troubleshooting* Section at the front of this manual for information on symptoms of transaxle problems.

1.3 Transaxle identification number location

Preliminary checks

3   Drive the vehicle to warm the transaxle to normal operating temperature.
4   Check the fluid level as described in Chapter 1.
  a) If the fluid level is unusually low, add enough fluid to bring it up to the proper level, then check for external leaks (see below).
  b) If the fluid level is abnormally high, drain off the excess, then check the drained fluid for contamination by coolant. The presence of engine coolant in the automatic transaxle fluid indicates that a failure has occurred in the transaxle fluid cooler.
  c) If the fluid is foaming, drain it and refill the transaxle, then check for coolant in the fluid.

5   Look for a CHECK ENGINE light glowing on the instrument panel (see Chapter 6 for information).
Note: *If the engine or its control network is malfunctioning, do not proceed with the preliminary checks until it has been repaired and runs normally.*
6   Inspect the shift cable (see Section 7). Make sure that it's properly adjusted and that it operates smoothly.

Fluid leak diagnosis

7   Most fluid leaks are easy to locate visually. Repair usually consists of replacing a seal or gasket. If a leak is difficult to find, the following procedure may help.
8   Identify the fluid. Make sure it's transaxle fluid and not engine oil or brake fluid.
9   Try to pinpoint the source of the leak. Drive the vehicle several miles, then park it over a large sheet of cardboard. After a minute or two, you should be able to locate the leak by determining the source of the fluid dripping onto the cardboard.
10   Make a careful visual inspection of the suspected component and the area immediately around it. Pay particular attention to gasket mating surfaces. A mirror is often helpful for finding leaks in areas that are hard to see.
11   If the leak still cannot be found, clean the suspected area thoroughly with a degreaser or solvent, then dry it.
12   Drive the vehicle for several miles at normal operating temperature and varying speeds. After driving the vehicle, visually inspect the suspected component again.
13   Once the leak has been located, the cause must be determined before it can be properly repaired. If a gasket is replaced but the sealing flange is bent, the new gasket will not stop the leak. The bent flange must be straightened.
14   Before attempting to repair a leak, check to make sure that the following conditions are corrected or they may cause another leak.
Note: *Some of the following conditions cannot be fixed without highly specialized tools and expertise. Such problems must be referred to a transaxle shop or a dealer service department.*

Gasket leaks

15   Check the pan periodically. Make sure the bolts are tight, no bolts are missing, the gasket is in good condition and the pan is flat (dents in the pan may indicate damage to the valve body inside).
16   If the pan gasket is leaking, the fluid level may be too high, the vent may be plugged, the pan bolts may be too tight, the pan sealing flange may be warped, the sealing surface of the transaxle housing may be damaged, the gasket may be damaged or the transaxle casting may be cracked or porous. If sealant instead of gasket material has been used to form a seal between the pan and the transaxle housing, it may be the wrong sealant.

Seal leaks

17   If a transaxle seal is leaking, the fluid level may be too high, the vent may be plugged, the seal bore may be damaged, the seal itself may be damaged or improperly installed, the surface of the shaft protruding through the seal may be damaged or a loose bearing may be causing excessive shaft movement.
18   Make sure the fluid drain/check plug in the transaxle fluid pan is in good condition. If leaking transaxle fluid is evident, replace the O-ring on the drain plug.

Case leaks

19   If the case itself appears to be leaking, the casting is porous and will have to be repaired or replaced.
20   Make sure the oil cooler hose fittings are tight and in good condition.

Fluid comes out vent pipe

21   If this condition occurs, the transaxle is overfilled, there is coolant in the fluid, the vent is plugged or the drain-back holes are plugged.

3   Electronic control system

Trouble codes

1   The electronic control system for the transaxle has some self-diagnostic capabilities. If certain kinds of system malfunctions occur, the PCM stores the appropriate diagnostic trouble code in its memory and the CHECK ENGINE indicator light illuminates to inform the driver. The diagnostic trouble codes can only be extracted from the PCM using a SCAN tool that can be linked to the On Board Diagnostic (OBD II) computer via the Data Link connector (DLC). Codes are listed below for reference, but can only be extracted with the correct scan tool connected to the Data Link connector (DLC) under the left side of the instrument panel (refer to Chapter 6 for additional information).

Trouble codes

Code	Code identification
P0700	Transaxle control system (mil request)
P0701	Transaxle control system range/performance problem
P0702	Transaxle control system electrical
P0704	Clutch switch input circuit malfunction
P0705	Transaxle range sensor "a" circuit (PRNDL input)
P0706	Transaxle range sensor "a" circuit range/performance problem
P0710	Transaxle fluid temperature sensor "a" circuit
P0711	Transaxle fluid temperature sensor "a" circuit range/performance problem
P0712	Transaxle fluid temperature sensor "a" circuit low

Trouble codes (continued)

Code	Code identification
P0713	Transaxle fluid temperature sensor "a" circuit high
P0714	Transaxle fluid temperature sensor "a" circuit intermittent
P0715	Input/turbine speed sensor "a" circuit
P0716	Input/turbine speed sensor "a" circuit range/performance problem
P0717	Input/turbine speed sensor "a" circuit no signal
P0720	Generator control circuit
P0721	Output speed sensor circuit range/performance problem
P0722	Output speed sensor circuit no signal
P0725	Engine speed input circuit
P0726	Engine speed input circuit range/performance problem
P0727	Engine speed input circuit no signal
P0729	Gear 6 incorrect ratio
P0730	Incorrect gear ratio
P0731	Gear 1 incorrect ratio
P0732	Gear 2 incorrect ratio
P0733	Gear 3 incorrect ratio
P0734	Gear 4 incorrect ratio
P0735	Gear 5 incorrect ratio
P0736	Reverse incorrect ratio
P0740	Torque converter clutch circuit/open
P0741	Torque converter clutch circuit performance problem/stuck off
P0743	Torque converter clutch circuit electrical problem
P0746	Pressure control solenoid "a" performance problem/stuck off
P0747	Pressure control solenoid "a" stuck on
P0748	Pressure control solenoid "a" electrical problem
P0749	Pressure control solenoid "a" intermittent
P0751	Shift solenoid "a" performance problem/stuck off
P0752	Shift solenoid "a" stuck on
P0753	Shift solenoid "a" electrical problem
P0756	Shift solenoid "b" performance problem/stuck off

Code	Code identification
P0757	Shift solenoid "b" stuck on
P0758	Shift solenoid "b" electrical problem
P0761	Shift solenoid "c" performance problem/stuck off
P0762	Shift solenoid "c" stuck on
P0763	Shift solenoid "c" electrical
P0766	Shift solenoid "d" performance problem/stuck off
P0768	Shift solenoid "d" electrical
P0771	Shift solenoid "e" performance problem/stuck off
P0773	Shift solenoid "e" electrical
P0776	Pressure control solenoid "b" performance problem/stuck off
P0777	Pressure control solenoid "b" stuck on
P0778	Pressure control solenoid "b" electrical problem
P0779	Pressure control solenoid "b" intermittent
P0780	Shift error
P0781	1-2 shift
P0782	2-3 shift
P0783	3-4 shift
P0784	4-5 shift
P0785	Shift timing solenoid "a"
P0791	Intermediate shaft speed sensor "a" circuit
P0796	Pressure control solenoid "c" performance problem/stuck off
P0797	Pressure control solenoid "c" stuck on
P0798	Pressure control solenoid "c" electrical problem
P0799	Pressure control solenoid "c" intermittent
P0811	Excessive clutch "a" slippage
P0820	Gear lever x-y position sensor circuit
P0821	Gear lever x position circuit
P0822	Gear lever y position circuit
P0829	5-6 shift
P0840	Transaxle fluid pressure sensor/switch "a" circuit
P0841	Transaxle fluid pressure sensor/switch "a" circuit range/performance problem

Chapter 7 Part B Automatic Transaxle

Trouble codes (continued)

Code	Code identification
P0842	Transaxle fluid pressure sensor/switch "a" circuit low
P0845	Transaxle fluid pressure sensor/switch "b" circuit
P0846	Transaxle fluid pressure sensor/switch "b" circuit range/performance problem
P0863	TCM communication circuit
P0864	TCM communication circuit range/performance problem
P0865	TCM communication circuit low
P0884	TCM power input signal intermittent
P0886	TCM power relay control circuit low
P0887	TCM power relay control circuit high
P0889	TCM power relay sense circuit range/performance problem
P0890	TCM power relay sense circuit low
P0891	TCM power relay sense circuit high
P0892	TCM power relay sense circuit intermittent
P0914	Gear shift position circuit
P0915	Gear shift position circuit range/performance problem
P0918	Gear shift position circuit intermittent
P0919	Gear shift position control error
P0929	Gear shift lock solenoid control circuit range/performance problem

4 Transmission Control Module (TCM) - removal and installation

Caution: *The following procedures only describe removing and installing the Transmission Control Module (TCM). If a TCM requires replacement, the procedure should be performed by a qualified technician or repair facility with the proper equipment to transfer the data from the old module to the new module.*
Note: *Only automatic transaxle vehicles are equipped with a TCM. On DSG model transaxles, the control module is built into the transaxle and should by serviced by a qualified technician or repair facility.*

Golf models

Note: *On Golf models, the TCM is located behind the driver's front inner fender's splash shield.*

1   Loosen the left-front wheel bolts, then raise the front of the vehicle and support it securely on jackstands.
2   Remove the left-front wheel.
3   Remove the inner fender splash shield.
4   Locate the TCM and disengage the TCM bracket retaining clips. Remove the TCM from the bracket.
5   Disconnect the TCM connector latch and unplug the connector.
6   Installation is reverse of removal.

Jetta models

Note: *On 1.4L, 1.8L and 2.0L Jetta models, the TCM is located under the wiper cowl, on the passenger's side. On 2.5L Jetta models, the TCM is located next to the battery.*

7   On 1.4L, 1.8L and 2.0L models, remove the wiper cowl (see Chapter 11, Section 16) to access the TCM **(see illustration)**.
8   On all models, disconnect the negative battery cable (see Chapter 5, Section 3).
9   Locate the TCM, disconnect the TCM connector latch and unplug the connector **(see illustration)**.
10 Disengage the TCM bracket retaining clips and remove the TCM from the vehicle.
11 Installation is reverse of removal.

5 Shift knob - removal and installation

Removal

1   Put the shift lever into the Drive (D) position.
2   On shift knobs with a side button, pull the button away from the handle until gap appears between the knob and the button.
3   Hold the button in this position using a zip-tie or similar.
Note: *On shift knobs with a front button, the shift knob locks in place when the shift knob is removed. DO NOT press the shift knob button while removed.*
4   On all models, detach the shift boot from the center console **(see illustration)**.

Chapter 7 Part B Automatic Transaxle

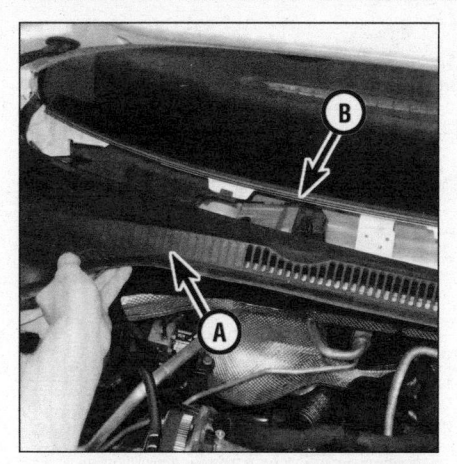

4.7 Remove the wiper cowl (A) to access the TCM (B) (1.4L Jetta model shown)

4.9 Disconnect the TCM connector (1) and release the tabs on the bracket (2) to remove the module (1.4L Jetta model shown)

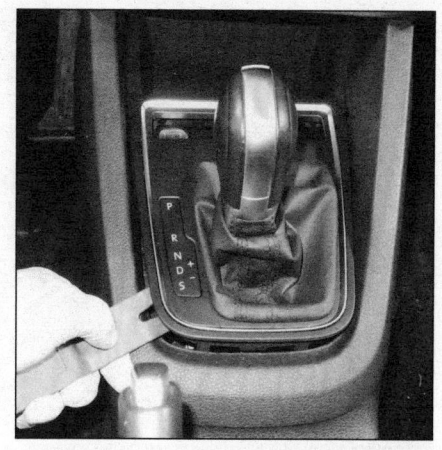

5.4 Carefully pry the shift boot away from the center console

5 On models with a sleeve securing the shift knob, pull the sleeve upwards towards the knob to release the knob.

6 On models that use a clamp to secure the shift knob, cut the clamp off **(see illustration)**.

7 On all models, remove the shift knob from the shift lever assembly without pressing in on the shift knob button.

Installation

8 Installation is reverse of removal.

Note: *On shift knobs with a side button, if the shift knob is pressed in by mistake, pull the button away from the handle until a gap appears between the knob and the button, then secure in this position for installation.*

9 On shift knobs with a front button, if the knob is pressed in by mistake, remove the knob trim by prying it upwards, then push the small rod into the groove of the shaft hole **(see illustrations)**.

5.6 Cut the clamp securing the shift knob to the shift lever assembly (if equipped)

5.9a Pry up the knob trim cap . . .

5.9b . . . move the small rod downward (against spring pressure) with a screwdriver . . .

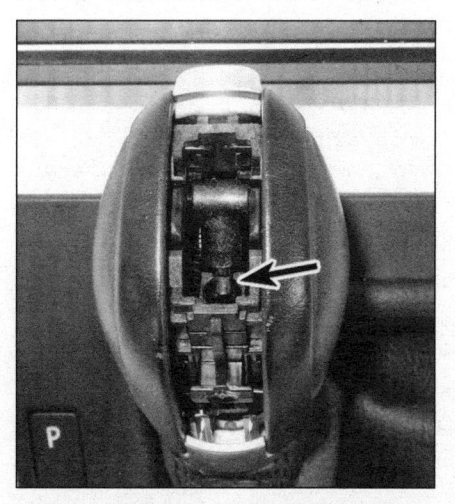

5.9c . . . until it locks into the groove . . .

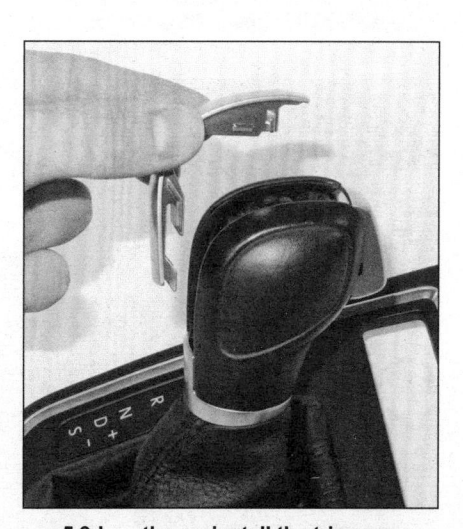

5.9d . . . then reinstall the trim cap, observing the location of the tabs

7B-8   Chapter 7 Part B   Automatic Transaxle

5.12 Install a new screw-type clamp (available at hardware stores) onto the knob base

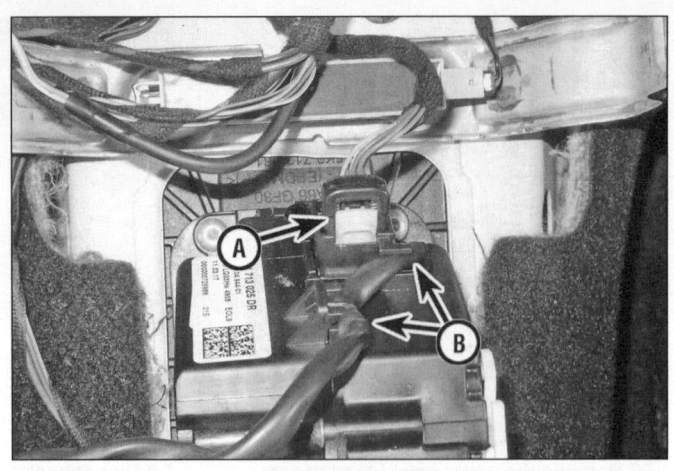

6.5 Disconnect the connectors (A) and detach the harnesses (B) from the shift lever housing

10 Install the shift knob and boot onto the shift lever assembly.

11 On models with a sleeve securing the shift knob, push the sleeve down to lock the shift knob.

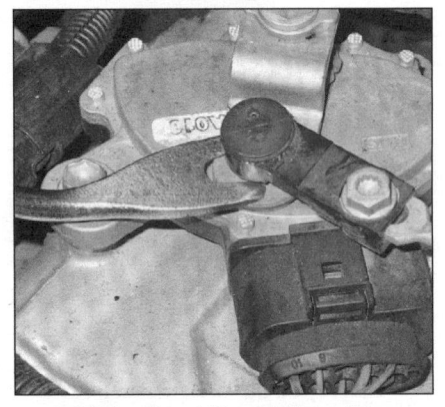

12 On models that use a clamp to secure the shift knob, install a new screw-type clamp **(see illustration)**.

6   Shift lever assembly - removal and installation

1 Raise the vehicle and support it securely on jackstands.

2 Remove the shift knob (see Section 5).

3 Remove the center console (see Chapter 11, Section 27).

4 Remove the fasteners and the shift lever assembly insulation.

5 Disconnect the shift lever assembly electrical connectors and harness clips **(see illustration)**.

6 Remove the air filter housing (see Chapter 4, Section 11), battery and battery tray (see Chapter 5, Section 4) as necessary to access the shift lever on the transaxle.

7 Place the shift lever assembly in neutral (N) and disconnect the shift cable from the lever and bracket at the transaxle **(see illustration)**. Detach the cable from any brackets.

Golf models

Note: The following procedure instructs on removing the shift lever assembly from under the vehicle. An alternative removal may be possible through the passenger compartment as described in the instructions for Jetta models.

8 Remove the cross braces from under the vehicle. Disconnect and and lower the exhaust.

9 Remove the heat shield and tunnel cover.

10 Remove the fasteners from inside of the passenger compartment attaching the shift lever assembly floor bracket **(see illustration)**.

11 Lower the shift lever assembly and remove with the cable from the vehicle.

12 Installation is reverse of removal.

13 Adjust the shift cable as necessary (see Section 7).

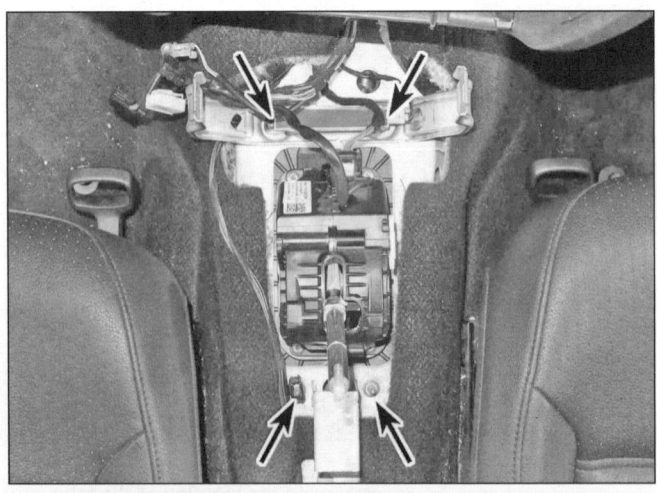

6.7 Pry the shift cable from the shift lever at the transaxle (6-speed automatic shown, DSG similar)

6.10 Remove the nuts securing the shifter to the floor (Golf models)

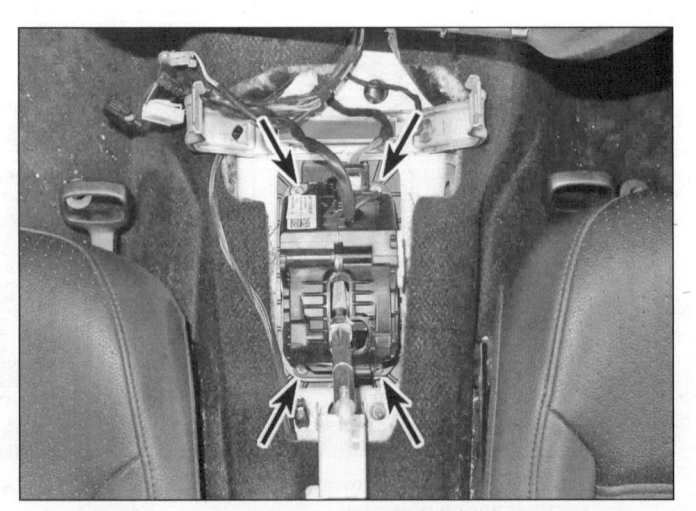

6.15 Remove the bolts attaching the shifter to the floor (Jetta models)

Chapter 7 Part B Automatic Transaxle   7B-9

7.7 Pry up the clip to remove the shift cable from the bracket

7.21 Shift cable adjustment details (6-speed automatic shown, others similar)
1. Selector lever adjustment bolt
2. Selector shaft lever
3. Shift cable

Jetta models

Note: *The following procedure instructs on removing the shift lever assembly through the passenger compartment. An alternative removal is from under the vehicle as described in the instructions for Golf models.*

14 Attach a piece of string or cord to the end of the shift cable that is long enough to reach from the transaxle to the shift lever assembly opening. This is used to install the cable when installing the shifter.
15 Working in the passenger compartment, remove the bolts attaching the shift lever assembly to the floor of the vehicle **(see illustration)**.
16 Lift the shift lever assembly out of the floor and carefully pull the cable from the opening in the floor.
17 When the string or cord is seen in the opening, detach the string from the shift cable end.
18 Installation is reverse of removal.
19 Attach the string or cord to the end of the shift cable and pull the string and cable through the vehicle while installing the shift lever assembly.
20 Adjust the shift cable as necessary (see Section 7).

7 Shift cable - removal, installation and adjustment

Note: *The following procedure applies to Golf models. On Jetta models, the shift cable is part of the shift lever assembly and not serviceable separately.*

Removal

1 Block the wheels and apply the parking brake.
2 Raise the front of the vehicle and support it securely on jackstands.
3 Place the shift lever in the neutral (N) position.
4 Remove the engine cover (see Chapter 1, Section 7).
5 Remove the air filter housing (see Chapter 4, Section 11).
6 At the transaxle end of the cable, use a screwdriver to pry the end of the shift cable up from the transaxle lever **(see illustration 6.7)**.
7 Remove the clip holding the shift lever cable to the bracket and remove the cable **(see illustration)**.
8 Unscrew the bolt and detach the support bracket and cable from the side of the transaxle.
9 Working under the vehicle, lower the exhaust system (see Chapter 4) and remove the heat shield fasteners and heat shield.
10 On ZF model shift lever assemblies, remove the plug from the bottom of the shifter and remove the cable housing clip from the front of the shift lever assembly.
11 Using a larger flat-tipped screwdriver, rotate the cable lock (under the plastic plug) 90 degrees in a clockwise direction.
12 On ECS model shift lever assemblies, use a screwdriver to pry off the cover from the bottom of the shift lever assembly.
13 Release the shift cable from the shift lever assembly.
14 Remove the lock from the shift cable.
15 On all models, remove the shift cable from the shift lever assembly.

Installation

16 On ZF shifters with a plug and lock mechanism, insert the cable fully and rotate the cable lock 90 degrees in a counter-clockwise direction, then insert the cable housing clip.
17 Remaining installation is a reversal of removal.
18 Before lowering the vehicle to the ground and before reconnecting the cable to the transaxle lever, adjust the cable as follows.

Adjustment

19 Apply the parking brake, then raise the front of the vehicle and support it securely on jackstands (if not already on stands).
20 Place the shift lever in the P position.
21 Loosen the selector lever adjustment bolt **(see illustration)**.
22 Move the selector shaft lever into the P position. Tighten the selector lever bolt.
23 Verify that the transaxle is in park by rotating both front wheels by hand until the park lock actuator engages in the transaxle, and the front wheels can no longer be rotated in the same direction at the same time. If the wheels can still be rotated, move the selector shaft lever forward until the park lock engages.
24 Lightly press the shift handle button on the shifter handle and make sure the handle is still in the P position.
25 Tighten the selector lever adjustment bolt to the torque listed in this Chapter's Specifications.
26 Move the shift lever through the gear ranges and verify that the shifter operates correctly.

8 Shift interlock system - description and check

Warning: *These models are equipped with airbags. Always disable the airbag system before working in the vicinity of any airbag system component to avoid the possibility of accidental deployment of the airbag(s), which could cause personal injury (see Chapter 12).*
Warning: *Do not use a memory saving device to preserve the ECM's memory when working on or near airbag system components.*

Description

1 The shift lock system prevents the shift lever from being shifted out of Park or Neutral until the brake pedal is applied and the button on the lever is pushed in. It also prevents the ignition key from being turned to the Lock position until the shift lever has been placed in the Park position.

Chapter 7 Part B Automatic Transaxle

Solenoid check

Note: *The shift solenoid is not removable from the shift lever and must be replaced as a complete assembly.*

2 Remove the center console (see Chapter 11).

3 Follow the wiring harness to the shift lock solenoid electrical connector, then unplug the connector. Using a pair of jumper wires, momentarily apply battery voltage and ground to the solenoid terminals and verify that there's an audible click.

Caution: *Don't apply battery voltage any longer than necessary to perform this check.*

4 If the shift lock solenoid doesn't click when energized, the solenoid is defective.

Shift interlock manual release

Note: *The shift interlock manual release allows shifting out of park and into other gears when the battery is dead or the shift interlock system is defective.*

5 Carefully pry the shift boot base from the center console.

6 Press the brake pedal or apply the parking brake.

7 Locate the shift interlock manual release on the right side of the shift lever assembly **(see illustration)**.

8 Press down and hold the manual release to shift the transaxle out of park and into other gears **(see illustration)**.

9 Release the manual release to resume normal operation of the shift interlock system.

9 Transaxle fluid cooler - removal and installation

1 Remove the engine cover.
2 Remove the air filter housing if needed for access (see Chapter 4, Section 11).
3 Remove the battery and battery tray if needed for access (see Chapter 5, Section 4).
4 On models equipped with an electric water pump, remove the pump (see Chapter 3, Section 7).
5 On all models, pinch or clamp off the oil cooler coolant hoses to prevent coolant loss.
6 Remove the hose clamps and pull the coolant hoses off of the oil cooler fittings **(see illustration)**.

Note: *If needed, detach the shift cable from the housing clamp near the oil cooler and position the cable to the side.*

7 On round-style oil coolers, remove the single bolt and washers going through the center of the oil cooler.
8 On rectangular style oil coolers, remove the bolts securing the oil cooler to the transaxle case **(see illustration)**.
9 On all oil coolers, remove the cooler from the transaxle and discard the O-rings.
10 Installation is reverse of removal, noting the following items:
  a) Use NEW O-rings for installation.
  b) Tighten the fasteners to the torque listed in this Chapter's Specifications.
  c) Top-off the engine coolant level as necessary.
  d) Top off the transaxle fluid level as necessary.

10 Transaxle range switch - replacement and adjustment

Note: *Only automatic transaxle vehicles are equipped with a range switch. DSG model transaxles, are not equipped with a range switch.*

Replacement

1 Block the wheels to prevent the vehicle from rolling and place the shift lever in Neutral (N) position.
2 Remove the air filter housing (see Chapter 4, Section 11), battery and battery tray (see Chapter 5, Section 4) as necessary to access the range switch below the shift lever on the transaxle.

8.7 Locate the shift interlock manual release (6-speed automatic shown)

8.8 Press and hold the shift interlock manual release to shift out of park (6-speed automatic shown)

9.6 Remove the clamps and coolant hoses (A) and re-position the shift cable (B) (Jetta with 1.4L and 6-speed auto shown)

9.8 Remove the bolts securing the cooler to the transaxle (Jetta with 1.4L and 6-speed auto shown)

Chapter 7 Part B Automatic Transaxle

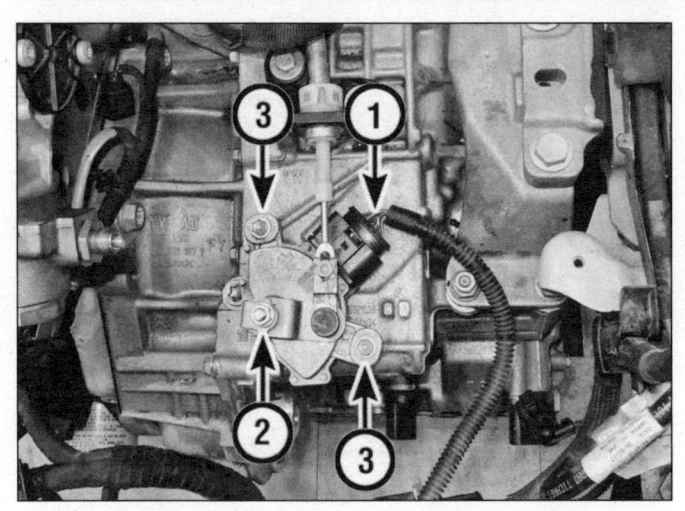

10.4 Disconnect the range switch electrical connector (1) then remove the nut and shift lever (2) and the mounting bolts (3) to remove the range switch (1.4L model shown)

11.3 Unbolt the shift cable support bracket (6-speed automatic shown)

3 Disconnect the shift cable from the shift lever and bracket at the transaxle (see Section 7).

4 Disconnect the range switch electrical connector **(see illustration)**.

5 Secure the shift lever on the transaxle from rotating and remove the retaining nut. Remove the shift lever from the shift shaft.

6 Remove the bolts and remove the range switch from the transaxle.

7 Installation is reverse of removal. Adjust the range switch as instructed.

Adjustment

Note: *Adjustment of the range switch requires use of special tool T10173 (Gauge, Multi Function Switch).*

8 With the wheels blocked to prevent the vehicle from rolling, ensure the shift lever in Neutral (N) position.

9 With the shift cable and shift lever removed, and the bolts attaching the transaxle range switch loose, place the gauge over the shift shaft and adjust the range switch so the gauge slides over it.

Note: *If the gauge does not fit over the range switch no matter the adjustment, the shift shaft is not in neutral position.*

10 Tighten the bolts on the range switch securely.

11 Automatic transaxle - removal and installation

Caution: *If the battery is disconnected, several systems must be re-learned before they will work properly (see Chapter 5, Section 3).*

Removal

1 Park the vehicle on a solid, level surface. Give yourself enough space to move around it easily. Apply the parking brake and chock the rear wheels.

2 Remove the engine cover (see Chapter 1, Section 7). Remove the battery and battery tray (see Chapter 5).

3 With the selector lever in position P, carefully disconnect the inner cable from the shift lever, then unbolt the support bracket **(see illustration)**. Position the cable to one side (see Section 7).

4 Support the engine with a three-bar engine support located on the front fender inner channels **(see illustration)**. The engine should be supported using the lifting eyes at each end. Depending on the engine, temporarily remove components as necessary to attach the support fixture.

5 Unscrew and remove the transaxle-to-engine mounting bolts accessible from the top of the engine compartment **(see illustration)**.

6 Loosen the front wheel bolts and driveaxle/hub bolts, then raise the front of the vehicle and support it securely on jackstands. Remove both front wheels.

11.4 An engine support fixture can be obtained at most equipment rental yards

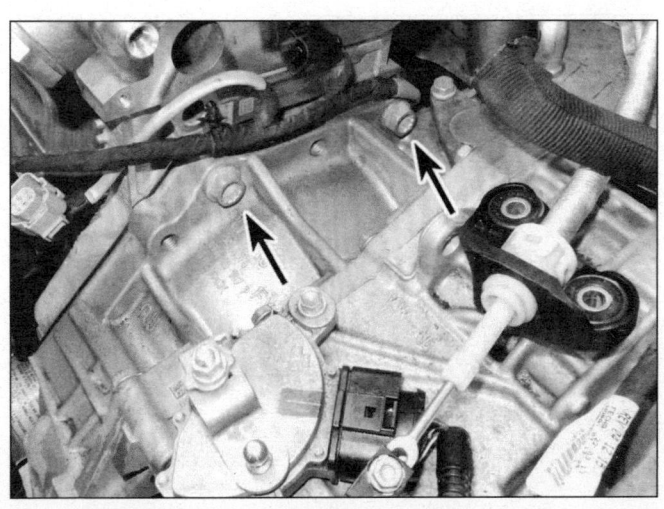

11.5 Remove the upper transaxle-to-engine bolts (1.4L 6-speed automatic shown)

7B-12 Chapter 7 Part B Automatic Transaxle

11.11 Driveaxle heat shield mounting fastener locations

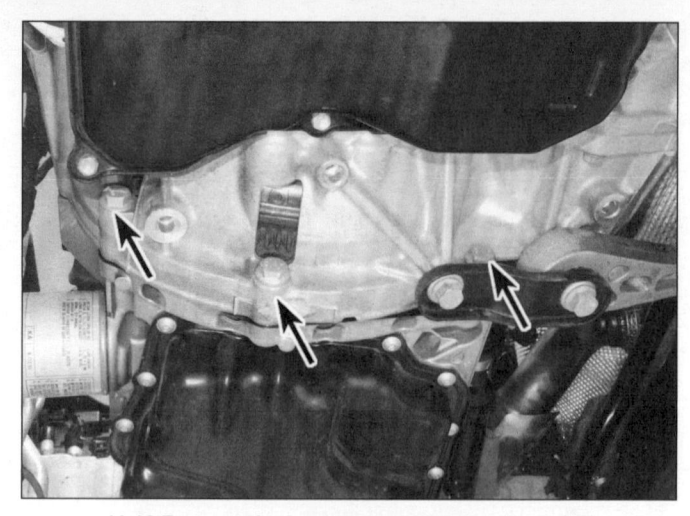

11.18 Remove the lower transaxle-to-engine bolts (1.4L 6-speed automatic shown)

7 Remove the under-vehicle splash shield (see Chapter 1, Section 6).
8 Remove the exhaust system (see Chapter 4).
9 Disconnect the wiring from the sensors on the transaxle (see Chapter 6).
10 Identify the electrical connectors on the transaxle, then unplug them. Loosen and detach the wiring support, and position the wiring to one side.
11 Remove the heat shield fasteners and shield from over the inner end of the driveaxle **(see illustration)**.
12 Remove the driveaxles (see Chapter 8).
13 On AWD models, remove the transfer case support and disconnect the driveshaft from the transfer case (see Chapter 8, Section 16).
14 On all models, position a suitable container beneath the transaxle to collect spilled fluid.
15 Unbolt the transaxle cooler from the transaxle and secure to the side with the coolant hoses attached (see Section 9). Plug the openings in the transaxle housing to prevent entry of dust and dirt.
16 Support the transaxle on a jack, preferably one made for this purpose. Secure the transaxle to the jack with a safety chain. Unbolt the right-hand transaxle mount complete with rubber bushing.
17 Remove the left and right-hand transaxle mounts, keeping track of the location of various-length bolts (see Section 13). At the rear of the transaxle, remove the nuts securing the rear mount to the crossmember, then remove the rear crossmember.
18 Unscrew the transaxle-to-engine mounting bolts accessible from under the car **(see illustration)**.
19 Remove the starter motor (see Chapter 5).
Note: *On some models, a single transaxle-to-engine bolt is located behind the starter.*
20 Remove the converter nut access plug, then turn the engine to locate one of the torque converter-to-driveplate nuts in the opening **(see illustrations)**. Unscrew and remove the nuts while preventing the engine from turning using a wide-bladed screwdriver engaged with the ring gear teeth on the driveplate. Unscrew the remaining five fasteners, turning the engine a third of a turn at a time to locate them. Mark the relationship of the torque converter to the driveplate so balance will be preserved when reinstalling the transaxle.
21 Remove the bolts and the transaxle roll-restrictor **(see illustration)**.
22 Unbolt the front exhaust mount from the subframe (see Chapter 4, Section 14).
23 Remove the balljoint nuts and separate the balljoints from the control arms (see Chapter 10, Section 8). On DSG transaxles, remove the subframe and left control arm (see Chapter 10, Section 17).
24 On DSG transaxles, remove the small cover plate (above the right driveaxle flange).
25 With the help of an assistant, withdraw the transaxle from the locating dowels on the rear of the engine, making sure that the

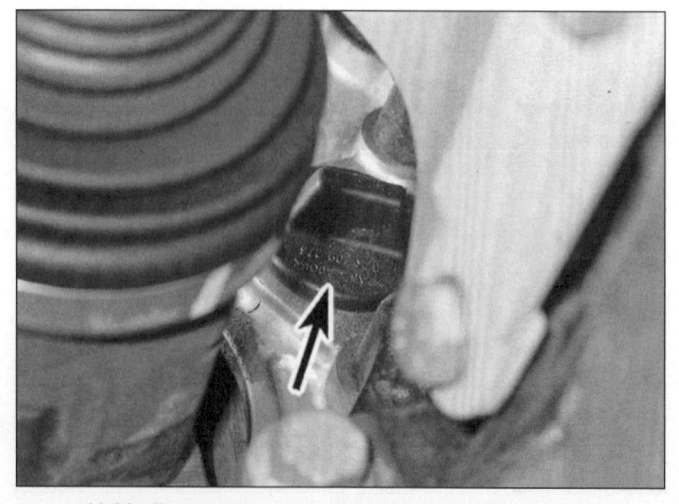

11.20a Torque converter fastener cover plug location

11.20b Rotate the engine to access and remove the torque converter nuts through the opening

Chapter 7 Part B Automatic Transaxle   7B-13

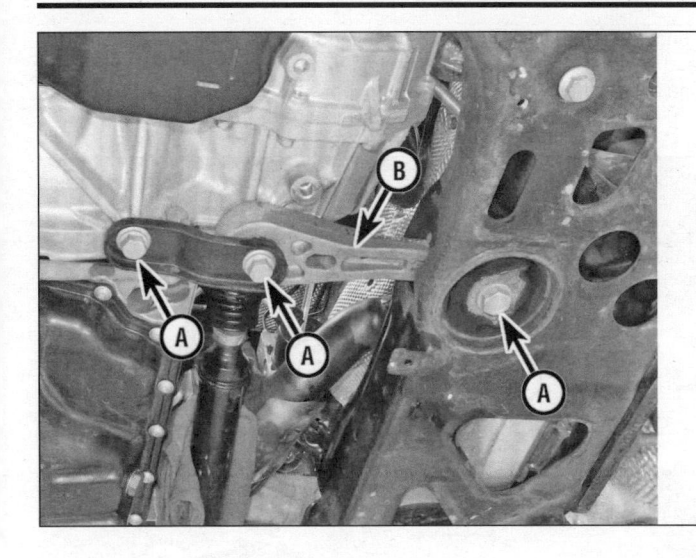

11.21 Remove the roll-restrictor bolts (A) and the roll restrictor (B)

torque converter remains fully engaged with the transaxle input shaft. If necessary, use a prybar to release the torque converter from the driveplate.

26 When the locating dowels are clear of their mounting holes, lower the transaxle to the ground using the jack. Strap a restraining bar across the front of the bellhousing to keep the torque converter in position.

Warning: *Make sure that the transaxle remains steady on the jack head.*

27 Where necessary, remove the intermediate plate from the locating dowels.

Installation

28 Installation of the transaxle is a reversal of the removal procedure, but note the following points:
a) As the torque converter is reinstalled, ensure that the drive pins at the center of the torque converter hub engage with the recesses in the automatic transaxle fluid pump inner wheel.
b) When installing the transaxle, make sure the marks on the torque converter and driveplate are in alignment.
c) Install the bellhousing bolts in the specific location **(see illustrations)**, then tighten the bolts and torque converter-to-driveplate nuts to the specified torque. Always replace self-locking nuts and bolts.
d) Replace the O-ring seals on the fluid pipes and filler tube attached to the transaxle casing.
e) Tighten the transaxle mounting bolts to the correct torque.
f) Tighten the driveshaft and driveaxle inner CV joint bolts to the torque values listed in the Chapter 8 Specifications.
g) Check the final drive oil level and transaxle fluid level as described in Chapter 1.
h) Check the shift cable adjustment (see Section 7).

12 Automatic transaxle overhaul - general information

1 In the event of a fault occurring, it will be necessary to establish whether the fault is electrical, mechanical or hydraulic in nature, before repair work can be contemplated. Diagnosis requires detailed knowledge of the transaxle's operation and construction, as well as access to specialized test equipment, and so is deemed to be beyond the scope of this manual. It is therefore essential that problems with the automatic transaxle are referred to a dealer service department or other qualified repair facility for assessment.

2 Note that a faulty transaxle should not be removed before the vehicle has been assessed by a knowledgeable technician equipped with the proper tools, as troubleshooting must be performed with the transaxle installed in the vehicle.

13 Transaxle mounts - check and replacement

Warning: *The weight of the entire engine and transaxle will be supported by the mounts not being removed during this procedure. Never place any part of your body directly under the engine or transaxle when performing this procedure.*

Check

1 Raise the vehicle and support it securely on jackstands.

2 Insert a large screwdriver or prybar between the transaxle and the subframe and try to pry the transaxle up slightly.

3 The transaxle should not move much at all - if the mount is cracked or torn, replace it.

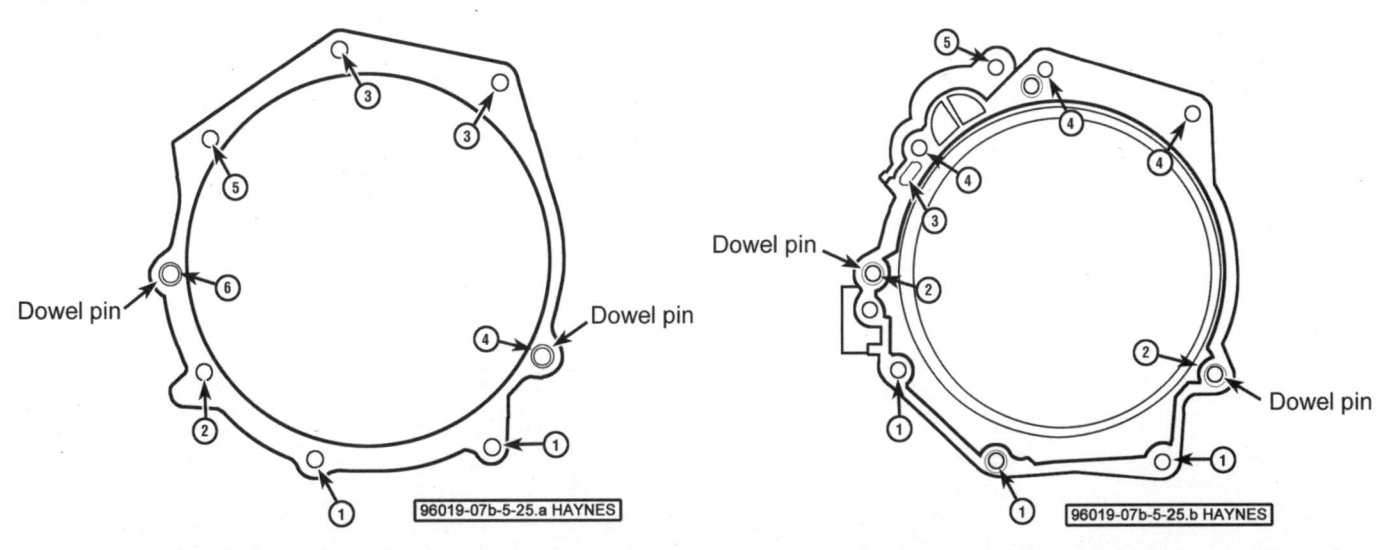

11.28a Transaxle bolt locations - 6-speed automatic transaxles

11.28b Transaxle bolt locations - DSG transaxles

Replacement

Mount

4   Working from above in the engine compartment, remove the engine cover (see Chapter 1, Section 7).
5   Remove the battery and battery tray (see Chapter 5, Section 4).
6   Remove the air filter housing (see Chapter 4, Section 11).
7   Disassemble and remove the electrical harness retainer from the top of the mount (if equipped).
8   Support the transaxle with a floor jack. Place a block of wood on the jack head to act as a cushion.
9   Remove the bolts and nuts attaching the mount to the subframe and transaxle **(see illustration)**.
Warning: *Do not disconnect more than one mount at a time, except during engine or transaxle removal.*
10   Raise the transaxle slightly with the jack and remove the mount.
11   Installation is the reverse of removal. Use thread-locking compound on the mount bolts and be sure to tighten them securely.

Roll restrictor

Note: *The roll restrictor bushing is pressed in the sub frame and must be pressed out and back in for replacement. The following procedure describes removing the bracket to access the bushing.*
12   Raise the vehicle and support it securely on jackstands.
13   Remove the vehicle undercover.
14   Remove the bolts and bracket and roll restrictor from the sub frame and the transaxle **(see illustration 11.21)**.
15   Installation is the reverse of removal. Use thread-locking compound on the mount bolts and be sure to tighten to the torque listed in this Chapter's Specifications.

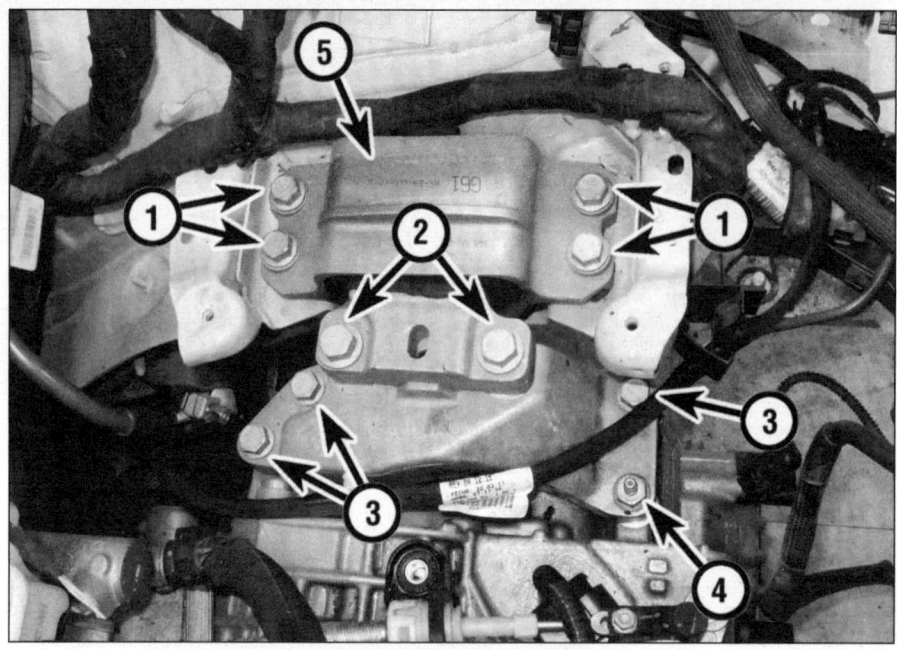

13.9 Transaxle mount details
1   Transaxle mount-to-body mounting fasteners
2   Transaxle mount-to-transaxle bracket mounting fasteners
3   Transaxle bracket-to-transaxle mounting fasteners
4   Transaxle bracket-to-transaxle mounting studded fastener
5   Transaxle mount

Chapter 8
Clutch and driveline

Contents

	Section		Section
All-wheel drive clutch pump - removal and installation	20	Clutch release cylinder - removal and installation	4
All-wheel drive clutch - removal and installation	19	Driveaxle (front) - removal and installation	12
All-wheel drive control module - replacement	18	Driveaxle (rear) - removal and installation	13
Clutch - description and check	2	Driveaxles - general information and inspection	11
Clutch components - removal, inspection and installation	6	Driveshaft - removal and installation	16
Clutch hydraulic system - bleeding	5	Front axle differential lock and seals - replacement	14
Clutch master cylinder - removal and installation	3	General Information	1
Clutch pedal bracket assembly - removal and installation	8	Rear differential - removal and installation	17
Clutch pedal springs - removal and installation	10	Transfer case and seals - replacement	15
Clutch pedal switch - removal and installation	9		
Clutch release bearing and lever - removal, inspection and installation	7		

Specifications

Torque specifications

	Ft-lbs (unless otherwise indicated)	Nm

Note: *One foot-pound (ft-lb) of torque is equivalent to 12 inch-pounds (in-lbs) of torque. Torque values below approximately 15 ft-lbs are expressed in inch-pounds, since most foot-pound torque wrenches are not accurate at these smaller values.*

Clutch pedal bracket retaining nuts*	18.5	25
Clutch pedal-to-bracket mounting bolt/nut*	18.5	25
Clutch release cylinder retaining bolts		
5-speed models	15	20
6-speed models		
Without thread lock (metal housing)	106 in-lbs	12
With thread lock	133 in-lbs	15
Clutch pressure plate bolts		
Four cylinder engines	15	20
Five-cylinder engines	120 in-lbs	13
Flywheel bolts	See Chapter 2A or 2B	
Driveaxle flange bolts		
Step 1	88 in-lbs	10
Step 2		
5-speed models (M8)	30	40
6-speed models (M10)	52	70
Driveaxle flange-to-shaft center bolt	22	30
Driveaxle/hub bolt*		
Hex bolt		
Step 1	148	200
Step 2	Tighten an additional 180-degrees (1/2-turn)	
12 point (ribbed**) bolt		
Step 1	52	70
Step 2	Tighten an additional 90-degrees (1/4-turn)	
12 point (non-ribbed) bolt		
Step 1	148	200
Step 2	Tighten an additional 180-degrees (1/2-turn)	

* Replace with new bolt(s)
** The ribbed section of the bolt is the contact surface under the bolt head.

Torque specifications (continued)

Note: *One foot-pound (ft-lb) of torque is equivalent to 12 inch-pounds (in-lbs) of torque. Torque values below approximately 15 ft-lbs are expressed in inch-pounds, since most foot-pound torque wrenches are not accurate at these smaller values.*

	Ft-lbs (unless otherwise indicated)	Nm
Driveshaft center support fasteners	18.5	25
Driveshaft rubber disc-to-flange bolts		
Except Golf R models		
Step 1	37	50
Step 2	Tighten an additional 90-degrees (1/4-turn)	
Golf R models	45	60
Transfer case-to-transaxle bolts	30	40
Transfer case output shaft flange nut	354	480
Front axle differential lock-to-transaxle bolts		
Step 1	30	40
Step 2	Tighten an additional 90-degrees (1/4-turn)	
Differential mounting bolts*		
Step 1	44	60
Step 2	Tighten an additional 180-degrees (1/2-turn)	
Differential pinion flange nut	155	210
AWD clutch housing mounting bolts	37	50

* Replace with new bolt(s)

1 General Information

1 The information in this Chapter deals with the components from the rear of the engine to the drive wheels, except for the transaxle, which is dealt with in Chapters 7A and 7B. Included in this Chapter is service information on the clutch and its release system, the transfer case (AWD models) and the driveaxles.

2 Since nearly all the procedures covered in this Chapter involve working under the vehicle, make sure it's securely supported on sturdy jackstands or a hoist where the vehicle can be easily raised and lowered.

2 Clutch - description and check

1 All models with a manual transaxle use either a "Sachs" or "LuK" single dry plate, diaphragm spring type clutch. The clutch disc has a splined hub which allows it to slide along the splines of the transaxle input shaft. The clutch and pressure plate are held in contact by spring pressure exerted by the diaphragm in the pressure plate.

2 The clutch release system is hydraulically operated. The release system consists of the clutch pedal, the clutch master cylinder, the clutch release cylinder and the hydraulic line between the master cylinder and release cylinder. On 6-speed transaxles, the release cylinder and release bearing are one unit.

3 When pressure is applied to the clutch pedal to release the clutch, the clutch master cylinder transmits this movement to the clutch release cylinder, which moves the clutch release lever. As the lever pivots, the release bearing pushes against the fingers of the diaphragm spring of the pressure plate assembly, which in turn releases the clutch plate.

4 Terminology can be a problem regarding the clutch components because common names have in some cases changed from that used by the manufacturer. For example, the clutch release cylinder is sometimes referred to as a slave cylinder, the driven plate is also called the clutch plate or disc, the pressure plate assembly is also known as the clutch cover, and the clutch release bearing is sometimes called a throw-out bearing.

5 Other than replacing components that have obvious damage, some preliminary checks should be performed to diagnose a clutch system failure.

  a) *To check clutch spin down time, run the engine at normal idle speed with the transaxle in Neutral (clutch pedal up, engaged). Disengage the clutch (pedal down), wait several seconds and shift the transaxle into Reverse. No grinding noise should be heard. A grinding noise would most likely indicate a problem in the pressure plate or the clutch disc.*

  b) *To check for complete clutch release, run the engine (with the parking brake applied to prevent movement) and hold the clutch pedal approximately 1/2-inch from the floor. Shift the transaxle between 1st gear and Reverse several times. If the shift is not smooth, component failure is indicated.*

  c) *Visually inspect the clutch pedal pivot at the top of the clutch pedal to make sure there is no sticking or excessive wear.*

  d) *Make sure that the hydraulic lines aren't leaking at either the master cylinder or the release cylinder (see Sections 3 and 4).*

  e) *Bleed the system if necessary (see Section 5).*

3 Clutch master cylinder - removal and installation

Warning: *The models covered by this manual are equipped with a Supplemental Restraint System (SRS), more commonly known as airbags. Always disable the airbag system before working in the vicinity of any airbag system components to avoid the possibility of accidental deployment of the airbags, which could cause personal injury (see Chapter 12).*
Caution: *If the battery is disconnected, several systems must be re-learned before they will work properly (see Chapter 5, Section 3).*

Removal

1 Disconnect the cable from the negative terminal of the battery (see Chapter 5).

2 Remove the clutch pedal assembly (see Section 8).

Golf models

3 Remove the clutch pedal switch (see Section 9).

4 Press in on the sides of the three retainers that attach the master cylinder to the brake pedal assembly, and remove the mounting pin.

5 Remove the master cylinder assembly from the brake pedal assembly.

Jetta models

6 On Jetta, using a pin punch or small screwdriver, disconnect the clip tabs on each side of the clutch pedal that hold the clutch master cylinder push rod to the pedal.

7 Place a spacer approximately 1-1/2 inches (40 mm) long between the clutch pedal and clutch pedal stop.

Note: *A standard half-inch drive socket could be used.*

Chapter 8 Clutch and driveline 8-3

4.7 Typical clutch release (slave) cylinder details - early model 5-speed transaxle shown

1 Mounting bolts
2 Fluid pipe and retaining clip

5.4 Typical bleeder valve location - early model 5-speed transaxle shown

8   Release the lock from the clutch master cylinder to the clutch pedal assembly, then rotate the clutch master cylinder counter-clockwise and pull the clutch master cylinder from the pedal bracket.

Installation
Golf models
9   Installation is reverse of removal, replace the mounting pin with a new part when installing.

Jetta models
10   Remove the spacer and insert the clutch master cylinder into the pedal bracket, then insert the clip on to the end of the pushrod.
11   Place the spacer between the clutch pedal and clutch pedal stop, then push the master cylinder into the clutch pedal bracket. Rotate the cylinder clockwise to lock it in place.
12   Push the master cylinder push rod into the pedal until the clip on the end of the pushrod is heard or felt locking into the pedal.

All models
13   Install the clutch pedal assembly (see Section 8).
14   Installation is otherwise the reverse of removal, with the following points:
Tighten all fasteners to the torque values listed in this Chapter's Specifications.
Bleed the clutch hydraulic system (see Section 5).
Check the brake fluid level in the brake fluid reservoir, adding as necessary to bring it to the appropriate level (see Chapter 1).

4 Clutch release cylinder - removal and installation

Caution: *If the battery is disconnected, several systems must be re-learned before they will work properly (see Chapter 5, Section 3).*

Note: *The clutch release cylinder may also be referred to as the slave cylinder.*

Removal
Note: *On 5-speed models, the release cylinder is located on the top of the transaxle housing and access is from the engine compartment. On 6-speed models, the release cylinder and release bearing are a single unit, located in the bellhousing of the transaxle; see Section 7 for removal and installation.*

1   Disconnect the cable from the negative terminal of the battery (see Chapter 5).
2   Remove the engine top cover/air filter assembly (see Chapter 1, Section 7).
3   Remove the battery and battery tray (see Chapter 5).
4   If needed, disconnect the gear selector cables from the gear selector levers, as described in Chapter 7A. Remove the clip and remove the relay lever. Unscrew the nut and remove the selector lever from the top of the transaxle. Remove the bolts and the shift cable bracket.
5   Remove the transaxle support bracket mounting fasteners and bracket.
6   Have some rags handy, as some fluid will be spilled when the line is disconnected.
7   Pull the fluid pipe retaining clip from the union on the release cylinder, then pull the pipe from the union (see illustration).
8   Remove the release cylinder mounting bolts and remove the release cylinder from the transaxle.

Installation
9   Lubricate the end of the pushrod with copper grease. Apply lithium-base grease to the area of the boot that seats in the bore in the transaxle. Install the release cylinder, inserting it straight into its bore (otherwise the pushrod may not seat in its pocket in the release lever).
10   Connect the pressure line to the cylinder. Insert the clip and make sure the line is completely attached and won't pull off.
11   Install the release cylinder mounting bolt and tighten it to the torque listed in this Chapter's Specifications.
12   The remainder of installation is the reverse of removal, with the additional following points:
   a) Bleed the system (see Section 5).
   b) Check the fluid level in the brake fluid reservoir, adding as necessary to bring it to the appropriate level (see Chapter 1).
   c) Wash off any spilled brake fluid with water.

5 Clutch hydraulic system - bleeding

1   The hydraulic system should be bled of all air whenever any part of the system has been removed or if the fluid level has been allowed to fall so low that air has been drawn into the master cylinder. The procedure is similar to bleeding a brake system.
2   Fill the brake master cylinder with new brake fluid conforming to DOT 4 specifications.

Note: *Do not re-use any of the fluid coming from the system during the bleeding operation or use fluid which has been inside an open container for an extended period of time.*

3   Remove the engine cover (see Chapter 1, Section 7). Remove the air filter housing if needed to gain access (see Chapter 4, Section 11).
4   Locate the bleeder screw on the clutch release cylinder (see illustration). Remove the dust cap from the bleeder screw and push a length of snug-fitting (preferably clear) hose over the screw. Place the other end of the hose into a clear container with about two inches of brake fluid in it. The hose end must be submerged in the fluid.
5   Have an assistant depress the clutch pedal and hold it. Open the bleeder screw on the release cylinder, allowing fluid to flow through the hose. Close the bleeder screw

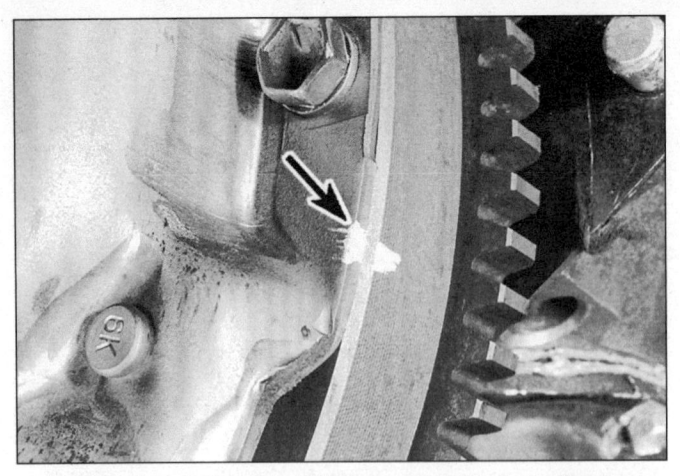

6.4 Mark the relationship of the pressure plate to the flywheel (in case you're going to re-use the same pressure plate)

6.6 When removing the pressure plate, be careful not to let the clutch disc fall out

when fluid stops flowing from the hose. Once closed, have your assistant release the pedal.

6 Continue this process until all air is evacuated from the system, indicated by a full, solid stream of fluid being ejected from the bleeder screw each time and no air bubbles in the hose or container. Keep a close watch on the fluid level inside the brake master cylinder reservoir; if the level drops too low, air will be sucked back into the system and the process will have to be started over again.

7 Install the dust cap on the bleeder screw. Check carefully for proper operation before placing the vehicle in normal service.

8 Lower the vehicle.

9 Recheck the brake fluid level.

6 Clutch components - removal, inspection and installation

Warning: *Dust produced by clutch wear and deposited on clutch components is hazardous to your health. DO NOT blow it out with compressed air and DO NOT inhale it. DO NOT use gasoline or petroleum-based solvents to remove the dust. Brake system cleaner should be used to flush the dust into a drain pan. After the clutch components are wiped clean with a rag, dispose of the contaminated rags and cleaner in a labeled, covered container.*

Removal

1 Access to the clutch components is normally accomplished by removing the transaxle, leaving the engine in the vehicle. If, of course, the engine is being removed for major overhaul, then the opportunity should always be taken to check the clutch for wear and replace worn components as necessary. However, the relatively low cost of the clutch components compared to the time and labor involved in gaining access to them warrants their replacement any time the engine or transaxle is removed, unless they are new or in near-perfect condition. The following procedures assume that the engine will stay in place.

2 Remove the transaxle from the vehicle (see Chapter 7A). Support the engine while the transaxle is out. Preferably, an engine hoist or support fixture should be used to support it from above. However, if a jack is used underneath the engine, make sure a piece of wood is used between the jack and oil pan to spread the load.

3 The release lever and release bearing (on 5-speed models) or release bearing and release cylinder (on 6-speed models) can remain attached to the transaxle; however, you should inspect them (see Section 7) while the transaxle is removed.

4 Carefully inspect the flywheel and pressure plate for indexing marks. If they cannot be found, scribe marks yourself so the pressure plate and the flywheel will be in the same alignment during installation **(see illustration).** Of course, this won't be necessary if you're planning on replacing the pressure plate with a new one.

Note: *Two types of pressure plates are used - a "Sachs" or a "LuK." The components are not interchangeable so make sure you replace them with the same type.*

5 Slowly loosen the pressure plate-to-flywheel bolts. Work in a criss-cross pattern and loosen each bolt a little at a time until all spring pressure is relieved.

6 Hold the pressure plate securely and completely remove the bolts, followed by the pressure plate and clutch disc **(see illustration).**

7 Inspect the pilot bearing recessed into the end of the crankshaft for wear or damage. If the needle bearings are good, clean the area and apply new grease to the needle bearings. If the bearing is damaged, insert a puller (available at most auto parts stores) and pull the bearing out from the crankshaft.

8 Using a bearing driver, drive the pilot bearing into the crankshaft with the letters on the needle bearing facing the outside. The bearing should be inserted 1/8-inch below the edge of the crankshaft. Lubricate the bearing with high-temperature grease.

Inspection

9 Ordinarily, when a problem occurs in the clutch, it can be attributed to wear of the clutch driven plate assembly (clutch disc). However, all components should be inspected at this time.

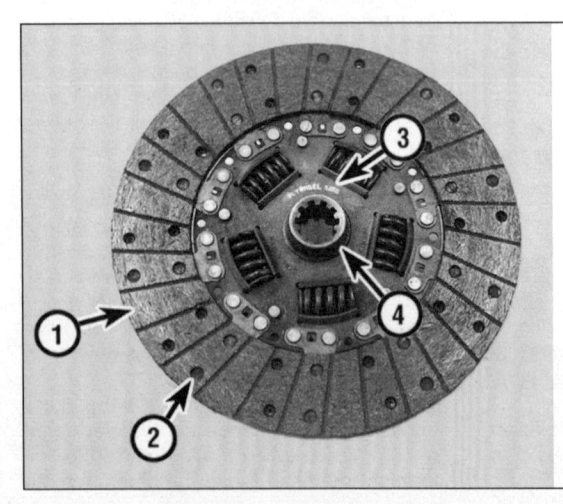

6.11 The clutch disc

1 **Lining** - this will wear down in use
2 **Springs or dampers** - check for cracking and deformation
3 **Splined hub** - the splines must not be worn and should slide smoothly on the transmission input shaft splines
4 **Rivets** - these secure the lining and will damage the flywheel or pressure plate if allowed to contact the surfaces

Chapter 8 Clutch and driveline

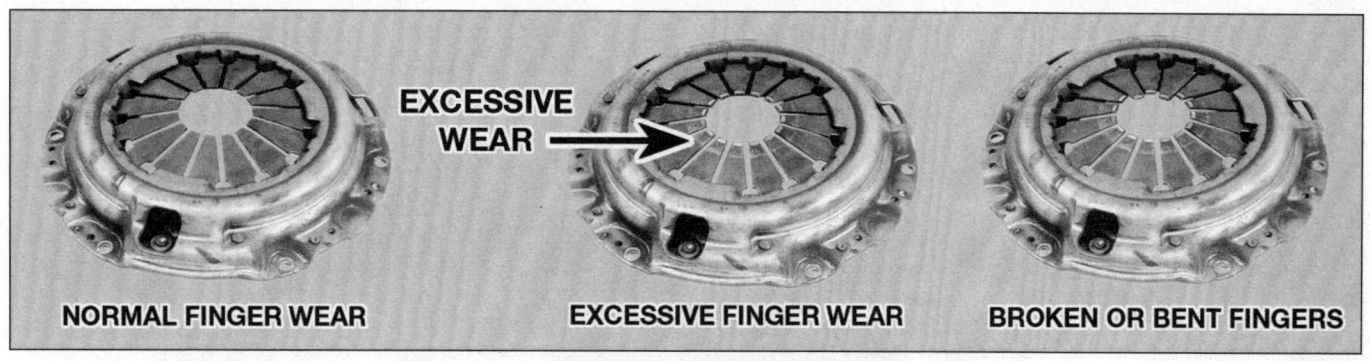

6.13a Replace the pressure plate if excessive wear is noted

6.13b Examine the pressure plate friction surface for score marks, cracks and evidence of overheating

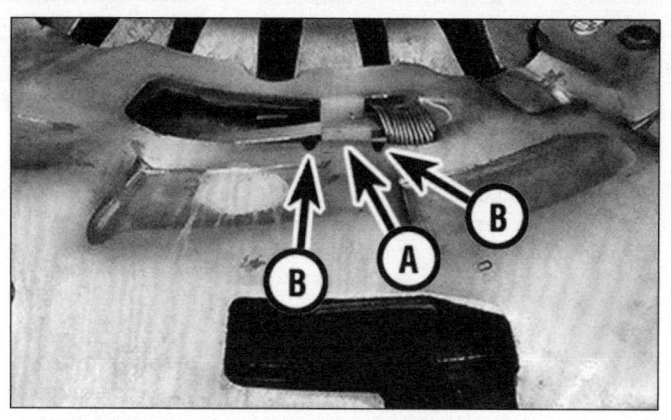

6.15 If installing a new LuK pressure plate, make sure the edges of the adjusting ring (A) are located between the notches in the clutch cover cut-outs (B) (one of three cut-outs shown)

10 Inspect the flywheel for cracks, heat checking, score marks and other damage. If the imperfections are slight, a machine shop can resurface it to make it flat and smooth. Refer to Chapter 2A or 2B for the flywheel removal procedure. **Note:** *On models equipped with a dual-mass flywheel, the flywheels may not be able to be resurfaced, and will need to be replaced.*

11 Inspect the lining on the clutch disc. There should be at least 1/16-inch of lining above the rivet heads. Check for loose rivets, distortion, cracks, broken springs and other obvious damage **(see illustration)**. As mentioned above, ordinarily the clutch disc is replaced as a matter of course, so if in doubt about the condition, replace it with a new one.

12 The release bearing should be replaced along with the clutch disc (see Section 7).

13 Check the machined surface and the diaphragm spring fingers of the pressure plate **(see illustrations)**. If the surface is grooved or otherwise damaged, replace the pressure plate assembly. Also check for obvious damage, distortion, cracking, etc. Light glazing can be removed with emery cloth or sandpaper. If a new pressure plate is indicated, new or factory rebuilt units are available.

Installation

14 Position the clutch disc and pressure plate with the clutch held in place with an alignment tool. Make sure the disc is installed properly. Find the word "Getriebeseite" which means "transmission side" stamped on the hub of the disc; this side must face the pressure plate.

15 Tighten the pressure plate-to-flywheel bolts only finger-tight, working around the pressure plate. On models equipped with a new "LuK" clutch assembly (only), make sure both edges of the adjusting ring are located between the notches in the clutch cover cut-outs **(see illustration)**. If the adjusting ring is in a different position with a new pressure plate, do not install the pressure plate and replace it with a new one and check again.

16 Center the clutch disc by ensuring the alignment tool is through the splined hub and into the recess in the crankshaft **(see illustration)**. Wiggle the tool up, down or side-to-side as needed to bottom the tool. Tighten the pressure plate-to-flywheel bolts a little at a time, working in a criss-cross pattern to prevent distortion of the cover. After all of the bolts are snug, tighten them to the torque listed in this Chapter's Specifications. Remove the alignment tool.

17 Using moly-base grease, lubricate the inner surface of the release bearing and the face of the bearing where it contacts the fingers of the pressure plate diaphragm spring. Also place grease on the release lever contact areas and the transaxle input shaft. **Caution:** *Don't use too much grease.*

18 Install the clutch release bearing (see Section 7).

19 Install the transaxle and all components removed previously, tightening all fasteners to the proper torque specifications.

6.16 Center the clutch disc with a clutch alignment tool, then tighten the pressure plate bolts a little at a time, in a criss-cross pattern, to the torque listed in this Chapter's Specifications

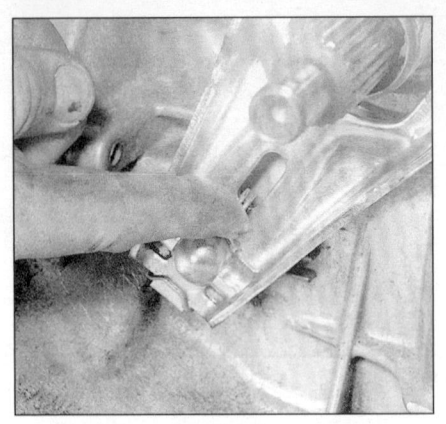

7.3 Push the detent back to release the lever - 5-speed models

7.4 Pry inward on the plastic tabs retaining the bearing to the lever, then separate the bearing from the lever - 5-speed models

7.9 To check the bearing, hold it by the outer race and rotate the inner race while applying pressure; if the bearing doesn't turn smoothly or if it's noisy, replace it - 5-speed models

7 Clutch release bearing and lever - removal, inspection and installation

Warning: *Dust produced by clutch wear and deposited on clutch components is hazardous to your health. DO NOT blow it out with compressed air and DO NOT inhale it. DO NOT use gasoline or petroleum-based solvents to remove the dust. Brake system cleaner should be used to flush it into a drain pan. After the clutch components are wiped clean with a rag, dispose of the contaminated rags and cleaner in a labeled, covered container.*

Removal

1 Remove the transaxle as described in Chapter 7A.

5-speed models

2 Remove the leaf spring mounting bolt at the end of the clutch release lever and remove the spring.
3 Push the detent spring back at the pivot end to release lever by pushing it through the hole (see illustration). This will release the pivot end of the lever. Withdraw the lever together with the release bearing from the guide sleeve.

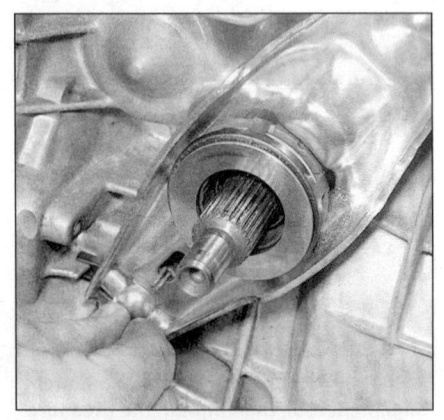

7.12 Press the release lever onto the ballstud until the spring clip holds it in position - 5-speed models

4 Use a screwdriver to depress the tabs and separate the bearing from the lever (see illustration).
5 Remove the plastic pivot from the ballstud. The release lever locates on the plastic pivot.

6-speed models

6 Remove the clip securing the breather assembly to the end of the release unit on the outside of the transaxle.
7 Separate the breather assembly from the release cylinder. Have a plug and cap ready and immediately plug the breather assembly and cap the release cylinder.
8 From inside the transaxle housing, remove the release cylinder/release bearing unit mounting fasteners and unit.

Inspection

5-speed models

9 Hold the bearing by the outer race and rotate the inner race while applying pressure (see illustration). If the bearing doesn't turn smoothly, or if it's noisy, replace the bearing with a new one. Wipe the bearing with a clean rag and inspect it for damage, wear and cracks. Don't immerse the bearing in solvent; it's sealed for life and to do so would ruin it. Also check the release lever for cracks and bends.

6-speed models

10 Check for any sign of leaks, then hold the release unit and rotate the release bearing while applying pressure. If there are signs of leaks or the bearing doesn't turn smoothly, or if it's noisy, replace the release cylinder/release bearing unit with a new one. Wipe the bearing with a clean rag and inspect it for damage, wear and cracks. Don't immerse the assembly in solvent; to do so would ruin it. Also check the slave cylinder housing for cracks and bends.
Note: *The release cylinder and bearing are one piece and cannot be separated. If either component is bad the entire assembly must be replaced.*

Installation

5-speed models

11 Begin installation by lubricating the ballstud and plastic pivot with a little copper grease. Smear a little grease on the release bearing surface which contacts the diaphragm spring fingers and the release lever, and also on the guide sleeve.
12 Install the spring onto the release lever, making sure the plastic pivot is in place on the ball stud. Position the lever and bearing and press the release lever onto the ballstud until the spring holds it in position (see illustration).
13 Install the transaxle (see Chapter 7A).

6-speed models

Note: *Some models are equipped with a plastic washer on the contact surface of the bearing. This is for pressure plates with a lower installed height.*

14 Slide the release cylinder/bearing onto the transmission input shaft, install and tighten the mounting fasteners (in small amounts) to the torque listed in this Chapter's Specifications.
Caution: *If the mounting fasteners are not tightened in small, even amounts the ears on the slave cylinder/release bearing could break off.*

15 Replace the O-ring on the end of the release cylinder.
16 Install the transaxle (see Chapter 7A).
17 Connect the breather assembly to the slave cylinder and push the clip in to lock the line.
18 The remainder of installation is the reverse of removal, with the additional following points:
   a) Bleed the system (see Section 5).
   b) Check the fluid level in the brake fluid reservoir, adding as necessary to bring it to the appropriate level (see Chapter 1).
   c) Wash off any spilled brake fluid with water.

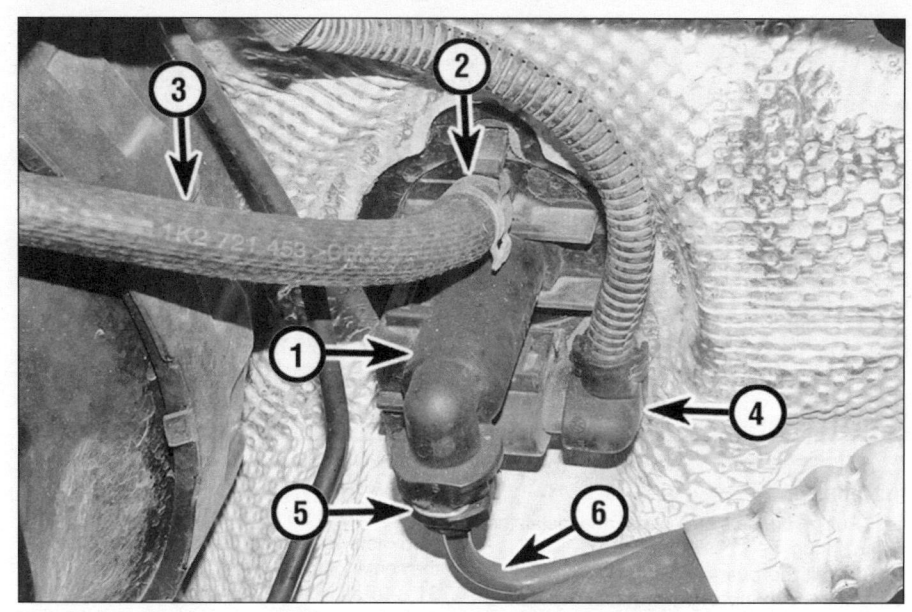

Chapter 8 Clutch and driveline

8.18 Jetta clutch master cylinder details

1. Clutch master cylinder
2. Supply hose clamp
3. Supply hose
4. Clutch position sensor
5. Retaining clip
6. Pressure line (to release cylinder)

8 Clutch pedal bracket assembly - removal and installation

Caution: *If the battery is disconnected, several systems must be re-learned before they will work properly (see Chapter 5, Section 3).*

Removal

1 Disconnect the cable from the negative terminal of the battery (see Chapter 5).
2 Remove the engine cover (see Chapter 1, Section 7).
3 Remove the air filter housing assembly (see Chapter 4, Section 11).
4 Remove the battery and the battery tray (see Chapter 5, Section 4).
5 Place enough rags on the floor under the clutch pedal to absorb any brake fluid that may spill.

Golf models

6 Move the driver's seat to the full rearward position.
7 Tilt the steering wheel as far up as possible.
8 Disconnect the cable from the negative terminal of the battery (see Chapter 5).
9 Remove the driver's side under-dash cover, if equipped, and knee bolster (see Chapter 11).
Note: *Some models may be equipped with a knee airbag (see Chapter 11, Section 28).*
10 Remove the impact bolster support fasteners and support from the instrument panel.
11 Remove the driver's footwell vent assembly.
12 Disconnect and remove the control module (Data Bus and On-Board Diagnostic Interface) above the lower part of the steering column.
13 Remove the single bolt attaching the bolster bracket above the upper part of the steering column and move the bracket to the side.
14 If equipped, disconnect and remove the parking aid control module and bracket, next to the clutch pedal assembly.
15 Disconnect the clutch position switch connector.
16 Disconnect the fluid reservoir hose and clutch pressure line from the clutch master cylinder.
17 Remove the three bolts and remove the pedal assembly and bracket from the vehicle.

Jetta models

18 Working in the engine compartment, clamp the fluid supply hose leading from the brake fluid reservoir to the clutch master cylinder using a brake hose clamp, then remove the clamp and detach the hose from the clutch master cylinder **(see illustration)**. Have a plug ready and immediately plug the line.
19 Pull out the clip and separate the pressure line from the clutch master cylinder. Plug the line to prevent excessive fluid loss and the entry of contaminants.
Note: *There is a seal on the end of pressure line which can stay in the clutch master when the line is removed. Use a small pick or screwdriver to remove the seal from the master cylinder and always replace the seal with a new one.*
20 Disconnect the clutch pedal switch electrical connector from the switch and leave the switch connected to the master cylinder.
Note: *If the clutch master is going to be replaced, remove the clutch pedal switch (see Section 9).*
21 Remove the driver's side under cover, if equipped, and knee bolster (see Chapter 11).
22 Remove the impact bolster support fasteners and support the instrument panel.
23 Remove the clutch pedal cover push fasteners from the pedal bracket and remove the cover (if equipped).
24 Remove the pedal bracket fasteners and remove the pedal assembly from the firewall.
25 If necessary, disconnect the clutch master cylinder push rod from the pedal (see Section 3).

Installation

26 Install the pedal assembly and tighten the fasteners to the torque listed in this Chapter's Specifications.
27 The remainder of installation is the reverse of removal, with the additional following points:
 a) Bleed the system (see Section 5).
 b) Check the fluid level in the brake fluid reservoir, adding as necessary to bring it to the appropriate level (see Chapter 1).
 c) Wash off any spilled brake fluid with water.

9 Clutch pedal switch - removal and installation

Caution: *If the battery is disconnected, several systems must be re-learned before they will work properly (see Chapter 5, Section 3).*
Note: *The clutch pedal switch is also called the clutch position sensor in the scan tool diagnostic code references.*

Golf models

1 Move the driver's seat to the furthest position away from the instrument panel.
2 Locate and disconnect the clutch switch connector on the clutch master cylinder (between the clutch and brake pedal assemblies).
3 Press the retaining clip on the switch and remove from the master cylinder.
4 Installation is reverse of removal.

Jetta models
Removal

5 Disconnect the cable from the negative terminal of the battery (see Chapter 5).
6 Remove the engine cover (see Chapter 1, Section 7).
7 Remove the battery and battery tray (see Chapter 5).
8 On models equipped with a pressure line that has a round dial type component, remove the pressure line retaining clip, then disconnect and plug the line.
Note: *There is a seal on the end of pressure line which can stay in the clutch master when the line is removed. Use a small pick or screwdriver to remove the seal from the master cylinder and always replace the seal with a new one.*

8-8 Chapter 8 Clutch and driveline

9 Using a small screwdriver, depress the tab on clutch pedal switch (see illustration), then pull the switch down, disconnect the switch electrical connector and remove the switch.

Installation
10 Connect the pedal switch electrical connector to the switch, then place the clutch pedal switch under the clutch master cylinder and press the switch up, making sure the switch engages with the clutch master cylinder until the switch locks in place.
11 Bleed the clutch system if the pressure line was removed (see Section 5).
12 The remainder of installation is the reverse of removal.

10 Clutch pedal springs - removal and installation

Caution: *If the battery is disconnected, several systems must be re-learned before they will work properly (see Chapter 5, Section 3).*

Golf models
Removal
1 Move the driver's seat to the full rearward position.
2 Tilt the steering wheel as far up as possible.
3 Disconnect the cable from the negative terminal of the battery (see Chapter 5).
4 Remove the driver's side under-dash cover, if equipped, and knee bolster (see Chapter 11).
Caution: *Some models may be equipped with a knee airbag (see Chapter 11, Section 28).*
5 Remove the impact bolster support fasteners and support from the instrument panel.
6 Remove the driver's footwell vent assembly.
7 Disconnect and remove the control module (Data Bus and On-Board Diagnostic Interface) above the lower part of the steering column.
8 Remove the single bolt attaching the bolster bracket above the upper part of the steering column and move the bracket to the side.
9 If equipped, disconnect and remove the parking aid control module and bracket, next to the clutch pedal assembly.
10 Press the retainer clip together and pull the clutch master cylinder actuator rod from the clip.

Tension spring
11 Press the clutch pedal toward the floor to access and remove the tension spring from the pedal assembly.

Over-center spring
12 Pull the clutch pedal up and detach and remove the over-center spring assembly from the clutch pedal assembly. Note the mounting pins for installation.

Return spring
13 Pull the clutch pedal up to release tension and remove the return spring from the clutch pedal assembly. Note the direction of the spring for installation.

9.9 Jetta clutch (pedal) position sensor details
1 Clutch position sensor
2 Mounting tab
3 Electrical connector

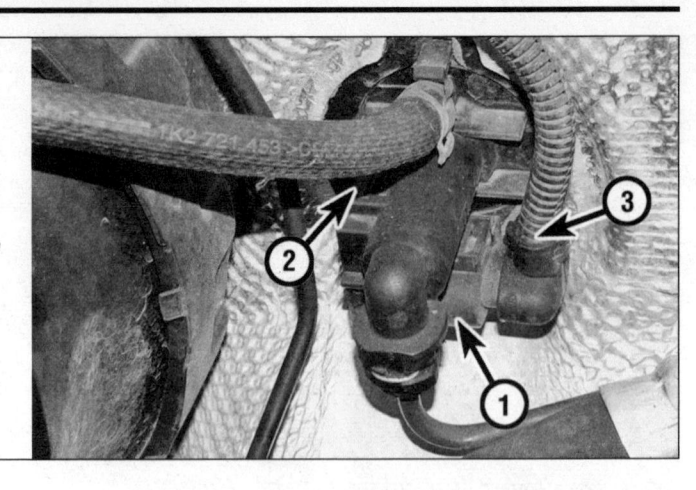

Installation
14 Pull up on the clutch pedal and install the return spring.
15 Install the over-center spring by pulling the clutch pedal up slightly, attach the spring to the pedal, align the pin and bracket mounts, and release the pedal.
16 Press the pedal down and install the tension spring.
17 Push the master cylinder push rod into the pedal until the clip on the end of the push-rod is heard or felt locking into the pedal.
18 The remainder of installation is reverse of removal.

Jetta models
Note: *Jetta models are only equipped with an over-center spring.*

Removal
19 Move the driver's seat to the full rearward position.
20 Tilt the steering wheel as far up as possible.
21 Disconnect the cable from the negative terminal of the battery (see Chapter 5).
22 Remove the driver's side under-dash cover, if equipped, and knee bolster (see Chapter 11).
Caution: *Some models may be equipped with a knee airbag (see Chapter 11, Section 28).*
23 Detach the harness retainer cover from under the dash.
24 Remove the driver's foot well vent assembly.
25 Remove the impact bolster support fasteners and support from the instrument panel.
26 Remove the clutch pedal cover push fasteners from the pedal bracket and remove the cover (if equipped).
27 Remove the clutch pedal through-bolts and nut from the pedal bracket.
28 Maneuver the clutch pedal downwards slightly, then remove the over-center spring from the slot in the pedal bracket.
Note: *The clutch master cylinder push rod remains attached to the pedal.*

Installation
29 Install the over-center spring into the pedal bracket.
30 Using a flat-blade screwdriver or VW special tool #T10178, push the over-spring back and install the clutch pedal to the bracket. Make sure the pivot end of the pedal sits squarely in the center of the over-center spring.
31 To align the holes for the through-bolt, push the spring in with the pedal and install the through-bolt, then tighten the nut to the torque listed in this Chapter's Specifications.
32 The remaining installation steps are reverse of removal.

11 Driveaxles - general information and inspection

1 Power is transmitted from the transaxle to the front wheels through a pair of driveaxles. The inner end of the driveaxle is either bolted to a drive flange protruding from the differential or splined to the differential in the transaxle; the outer end of each driveaxle has a stub shaft that is splined to the hub and bearing assembly and locked in place with a large bolt.
2 The inner ends of the driveaxles are equipped with sliding constant velocity (CV) joints, which are capable of both angular and axial motion. Each inner CV joint assembly consists of either a triple rotor-type bearing or a ball-and-cage type bearing and a housing in which the joint is free to slide in and out as the driveaxle moves up and down with the wheel.
3 The outer ends of the driveaxles are equipped with ball-and-cage type CV joints, which are capable of angular but not axial movement. Each outer CV joint consists of six ball bearings running between an inner race and an outer cage.
4 The boots should be inspected periodically for damage and leaking lubricant. Torn CV joint boots must be replaced immediately or the joints will be damaged. If either boot of a driveaxle is damaged, that driveaxle must be replaced.
5 Should a boot be damaged, the CV joint can be disassembled and cleaned, but if any parts are damaged, the entire driveaxle assembly may have to be replaced as a unit - check with your local auto parts store regarding the availability of replacement parts and CV joints (see Section 12).

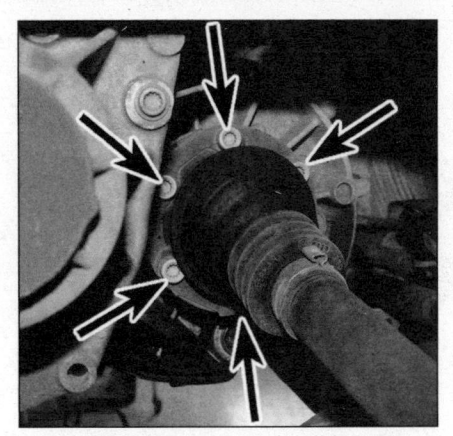

Chapter 8 Clutch and driveline

12.4 Driveaxle inner CV joint fastener locations, five of six shown - bolted flange models

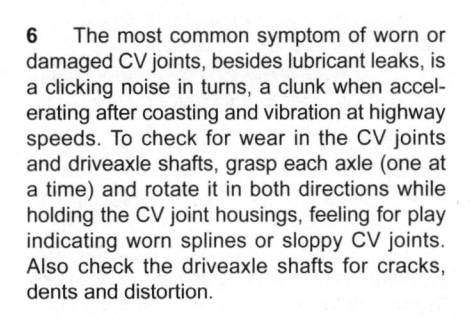

12.5 Carefully pry the inner end of the driveaxle from the transmission at opposite points - splined shaft models

12.7 Loosen and remove the driveaxle hub bolt (Jetta shown)

6   The most common symptom of worn or damaged CV joints, besides lubricant leaks, is a clicking noise in turns, a clunk when accelerating after coasting and vibration at highway speeds. To check for wear in the CV joints and driveaxle shafts, grasp each axle (one at a time) and rotate it in both directions while holding the CV joint housings, feeling for play indicating worn splines or sloppy CV joints. Also check the driveaxle shafts for cracks, dents and distortion.

12   Driveaxle (front) - removal and installation

Warning: *The manufacturer recommends replacing the driveaxle/hub bolt with a new one whenever it is removed.*

Removal

1   Remove the wheel trim/hub cap (as applicable) and loosen the driveaxle/hub bolt 1/4-turn with the vehicle resting on its wheels. Also loosen the wheel bolts.
Caution: *Don't loosen the driveaxle/hub bolt any more than 1/4-turn with the wheel on the*

ground - the wheel bearing could be damaged.
2   Raise the front of the vehicle and support it securely on jackstands. Block the rear wheels to prevent the vehicle from rolling off the stands. Remove the wheel.
3   Remove the lower splash shield below the engine (see Chapter 1, Section 6).

With bolt-on flange

4   Unscrew the bolts securing the inner CV joint to the transaxle drive flange or half shaft and remove the retaining plates from underneath the bolts **(see illustration)**. Support the driveaxle by suspending it with a strap - do not allow it to hang under its weight, or the joint may be damaged.

With splined shaft

5   Place a drain pan under the CV joint to be removed, then pry the inner CV joint out of the transaxle slightly, using a large screwdriver or prybar positioned between the transaxle and the CV joint housing **(see illustration)**. Be careful not to damage the differential seal. Support the driveaxle by suspending it with a strap - do not allow it to hang under its weight, or the joint may be damaged.

Note: *When the driveaxle is removed, transmission fluid will drain out of the transaxle.*

All models

6   Pull the ABS wheel speed sensor partially out of its mounting hole in the steering knuckle (see Chapter 9). This will prevent it from becoming damaged as the driveaxle is withdrawn from the hub.
7   Remove the driveaxle/hub bolt **(see illustration)**.
8   Remove the balljoint mounting nuts and separate the steering knuckle and balljoint from the control arm (see Chapter 10).
Note: *Disconnect the front level control sensor arm-to-control arm nut and separate the sensor arm from the control arm, if equipped.*
9   If the splines of the outer joint are stuck in the hub, tap the joint out of the hub using a brass drift or dead blow hammer **(see illustration)**. If this fails to free it from the hub, the joint will have to be pressed out using a puller **(see illustration)**. Carefully swing the steering knuckle outwards and withdraw the driveaxle outer constant velocity joint from the hub **(see illustration)**.
10   Maneuver the driveaxle out from underneath the vehicle. Remove the gasket from

12.9a Strike the driveaxle bolt with a deadblow hammer to remove

12.9b If necessary, press the joint out of the hub using a puller

12.9c Carefully swing the knuckle assembly away from the driveaxle and secure out of the way

the end of the inner constant velocity joint, if present. Discard the gasket; a new one should be used on installation.

11 Don't allow the vehicle to rest on its wheels with one (or both) driveaxle(s) removed, as damage to the wheel bearing(s) may result. If moving the vehicle is unavoidable, temporarily insert the outer end of the driveaxle(s) in the hub(s) and tighten the driveaxle/hub bolt(s); in this case, the inner end(s) of the driveaxle(s) must be supported, for example by suspending with string from the vehicle underbody. Do not allow the driveaxle to hang down, as the joint may be damaged.

Installation

With bolt-on flange

12 Ensure that the transaxle drive flange and inner joint mating surfaces are clean and dry.
13 Ensure that the outer joint and hub splines and threads are clean, then lubricate the splines with a light coat of multi-purpose grease.
14 Maneuver the driveaxle into position and engage the outer joint with the hub. Install a new bolt and tighten it to draw the joint fully into position. Don't tighten it completely yet (wait until the wheel is installed and the vehicle has been lowered to the ground).
15 Connect the steering knuckle and balljoint to the control arm, tightening the balljoint nuts to the torque listed in the Chapter 10 Specifications.
16 Align the driveaxle inner joint with the transaxle flange and install the retaining bolts and plates. Tighten the bolts to the torque listed in this Chapter's Specifications.

With splined shaft

17 Pry the old spring clip from the inner end of the driveaxle and install a new one **(see illustration)**. Lubricate the seal with multi-purpose grease and raise the driveaxle into position while supporting the CV joints.
Note: *Position the spring clip with the opening facing down; this will ease insertion of the driveaxle and prevent damage to the clip.*
18 Push the splined end of the inner CV joint into the differential side gear and make sure the spring clip locks in its groove.
19 Apply a light coat of multi-purpose grease to the outer CV joint splines, pull out on the steering knuckle assembly and install the stub axle into the hub.
20 Insert the balljoint studs into the control arm and tighten the nuts to the torque listed in the Chapter 10 Specifications.
21 Install a new bolt and tighten it to draw the joint fully into position. Don't tighten it completely yet (wait until the wheel is installed and the vehicle has been lowered to the ground).
22 Grasp the inner CV joint housing (not the driveaxle) and pull out to make sure the driveaxle has seated securely in the transaxle.

All models

23 Ensure that the outer joint is drawn fully into position, then install the wheel and lower the vehicle to the ground.
24 The remainder of installation is the reverse of removal, with the following points:
  a) Tighten the driveaxle/hub bolt to the torque and angle of rotation listed in this Chapter's Specifications.
  b) Once the driveaxle/hub bolt is correctly tightened, tighten the wheel bolts to the torque listed in the Chapter 1 Specifications and install the wheel trim/hub cap.
  c) Refill the transmission fluid (see Chapter 1) if any fluid leaked out.

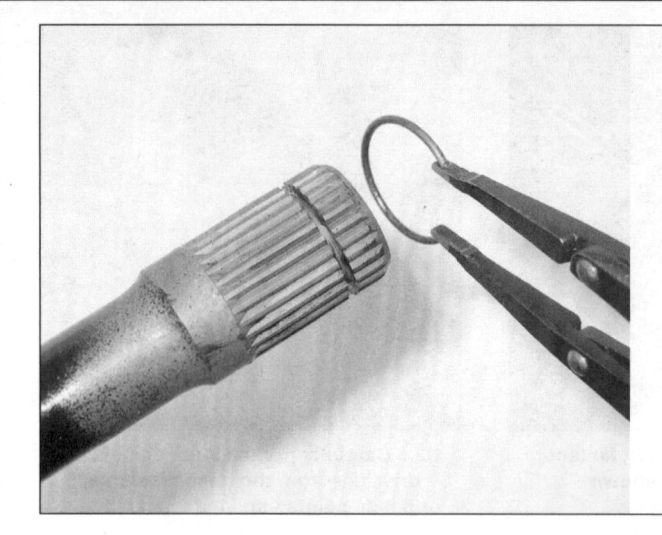

12.17 Carefully pry the spring clip from the driveaxle and install a new one - With splined shaft (AAR2600i models)

13 Driveaxle (rear) - removal and installation

Warning: *The manufacturer recommends replacing the driveaxle/hub bolt with a new one whenever it is removed.*
Note: *The following procedure applies to AWD Golf models only.*

Removal

1 Remove the wheel trim/hub cap (as applicable) and loosen the driveaxle/hub bolt 1/4-turn with the vehicle resting on its wheels. Also loosen the wheel bolts.
Caution: *Don't loosen the driveaxle/hub bolt any more than 1/4-turn with the wheel on the ground - the wheel bearing could be damaged.*
2 Raise the rear of the vehicle and support it securely on jackstands. Block the front wheels to prevent the vehicle from rolling off the stands. Remove the wheel.
3 On models equipped with level control, detach the level control sensor from the control arm.
4 On models equipped with Adaptive Chassis DCC, disconnect the electrical connector from the shock absorber.
5 Drill out the rivets and remove the stone shield from the control arm, on models so equipped).
6 Detach the stabilizer bar link from the stabilizer bar, then remove the stabilizer bar bracket bolts. Detach the stabilizer bar from the subframe and allow it to pivot down.
7 Detach the shock absorber, lower control arm and tie-rod from the knuckle (see Chapter 10).
8 Remove the coil spring (see Chapter 10).
9 Unscrew the bolts securing the inner CV joint to the differential drive flange and remove the retaining plates from underneath the bolts **(see illustration 12.4)**. Support the driveaxle by suspending it with a strap - do not allow it to hang under its weight, or the joint may be damaged.
10 Remove the driveaxle/hub bolt.
11 If the splines of the outer joint are stuck in the hub, tap the joint out of the hub using a dead-blow hammer or a brass drift. If this fails to free it from the hub, the joint will have to be pressed out using a puller **(see illustration 12.9b)**. Carefully swing the knuckle outwards and withdraw the driveaxle outer constant velocity joint from the hub.
12 Maneuver the driveaxle out from underneath the vehicle. Remove the gasket from the end of the inner constant velocity joint, if present. Discard the gasket; a new one should be used on installation.
13 Don't allow the vehicle to rest on its wheels with one (or both) driveaxle(s) removed, as damage to the wheel bearing(s) may result. If moving the vehicle is unavoidable, temporarily insert the outer end of the driveaxle(s) in the hub(s) and tighten the driveaxle/hub bolt(s); in this case, the inner end(s) of the driveaxle(s) must be supported, for example by suspending with string from the vehicle underbody. Do not allow the driveaxle to hang down, as the joint may be damaged.

Installation

14 Ensure that the differential drive flange and inner joint mating surfaces are clean and dry.
15 Ensure that the outer joint and hub splines and threads are clean, then lubricate the splines with a light coat of multi-purpose grease.

Chapter 8 Clutch and driveline 8-11

16 Maneuver the driveaxle into position and engage the outer joint with the hub. Install a new bolt and tighten it to draw the joint fully into position. Don't tighten it completely yet (wait until the wheel is installed and the vehicle has been lowered to the ground).
17 Install the coil spring (see Chapter 10).
18 Connect the lower control arm, tie-rod and shock absorber to the control arm, tightening the fasteners to the torque listed in the Chapter 10 Specifications.
19 Align the driveaxle inner joint with the differential flange and install the retaining bolts and plates. Tighten the bolts to the torque listed in this Chapter's Specifications.
20 Ensure that the outer joint is drawn fully into position, then install the wheel and lower the vehicle to the ground.
21 The remainder of installation is the reverse of removal, with the following points:
 a) *Tighten the driveaxle/hub bolt to the torque and angle of rotation listed in this Chapter's Specifications.*
 b) *Once the driveaxle/hub bolt is correctly tightened, tighten the wheel bolts to the torque listed in the Chapter 1 Specifications and install the wheel trim/hub cap.*

14 Front axle differential lock and seals - replacement

Note: *The following procedures apply to Golf GTI models with DSG 09D transaxles only.*

Differential lock assembly

1 Raise the front of the vehicle and support it securely on jackstands.
2 Place the transaxle in the Park position.
3 Remove the passenger's side driveaxle heat shield and driveaxle (see Section 12).
4 Disconnect the front exhaust pipe from the engine and support under the vehicle (see Chapter 4, Section 14).
5 Locate the differential lock and disconnect the electrical connector from the module and the oil level thermal sensor.
6 Remove the lower transaxle mount bracket bolts and mount bracket.
7 Remove the bolt securing the passenger's driveaxle flange on the differential lock.
8 Remove the bolts attaching the differential lock housing to the transaxle.
Note: *Place a drain pan under the transaxle to catch any fluid during the next steps.*
9 Carefully remove the differential lock assembly from the transaxle.
10 Installation is reverse of removal. Tighten the fasteners to the torque listed in this Chapter's Specifications.
11 Refill the differential lock fluid to the proper levels (see Chapter 1 Specifications).

Input shaft seals

12 With the differential lock removed, pry out the seal that seals the differential lock to the transaxle.
13 Use a small slide hammer to remove the shaft seal from inside of the differential lock housing.
14 Coat the new seal contact surfaces with grease.
15 Drive the new seals into the differential lock, with the open sides of the seals facing the differential lock housing.

Output shaft seal

Note: *The output shaft seal can be replaced with the differential lock in the vehicle.*
16 Unbolt the passenger's side driveaxle from the flange (see Section 12).
17 Remove the bolt from the center of the flange.
18 Use a slide hammer to remove the flange and shaft assembly.
19 Pry the seal out of the differential lock housing.
20 Coat the new seal contact surfaces with grease.
21 Drive the new seal into the differential lock, with the open side of the seal facing the differential lock housing.
Note: *The next step is similar to installing a driveaxle into a transaxle.*
22 Install the flange and shaft squarely into the differential lock housing as far as it will go, and use a dead-blow hammer to secure the locking ring inside of the differential lock.
23 The remainder of the installation is reverse of removal.

15 Transfer case and seals - replacement

Note: *The following procedures apply to Golf AWD models only. The transfer case is also known as the Bevel Box.*

Transfer case

Note: *On DSG transaxles, place the transaxle in the Park position.*
1 Raise the vehicle and support it securely on jackstands.
2 Drain the transaxle and transfer case fluid (see Chapter 1).
3 Disconnect the driveshaft from the transfer case flange and support out of the way (see Section 16).
4 If necessary, remove the driveshaft heat shield.
5 Remove the passenger's driveaxle heat shield and driveaxle (see Section 12).
6 Remove the front exhaust bracket from the sub-frame (see Chapter 4, Section 14).
7 On models with DSG (02E) transaxles, disconnect the exhaust at the connection after the bracket, and secure the exhaust up under the vehicle.
8 On all models except DSG (02E) transaxles, remove the lower transaxle mount bolts and mount from the subframe and transaxle (see Chapter 7A).
9 Rotate the engine forward and secure in that position using a block of wood.
10 On all models, remove the bolt securing the passenger's driveaxle flange on the transfer case.
11 Disconnect any electrical connectors or harnesses that may prevent removal of the transfer case.
12 Remove the transaxle support bracket from the engine and transmission.
13 On models with DSG (02E) transaxles, remove the front subframe (see Chapter 10, Section 17).
14 On all models, remove the bolts attaching the transfer case to the transaxle.
Note: *Place a drain pan under the transaxle to catch any fluid during the next steps.*
15 Carefully remove the transfer case from the transaxle.
16 Installation is reverse of removal. Tighten the fasteners to the torque listed in this Chapter's Specifications.
17 Refill the transaxle and transfer case fluid to the proper levels (see Chapter 1 Specifcations).

Input shaft seals

18 With the transfer case removed, the procedure is the same as the Front Axle Differential Lock input seals (see Section 14). The small input shaft seal described in this procedure is on the output side and replaced after removing the output shaft and flange.

Output shaft seals

19 The driveaxle output shaft seal replacement procedure is the same as the Front Axle Differential Lock output seal and can be performed with the transfer case in the vehicle (see Section 14). On the transfer case, there is also a small output shaft inside that is removed in the same way as the mall input shaft in the input shaft procedures.

Removal

20 The driveshaft output shaft seal is replaced with the transfer case removed.
21 Remove the output shaft flange nut and use a puller to remove the flange.
22 Use a press to remove the bearing from the flange.
23 Pry the seal from the transfer case housing.

Installation

24 Place the bearing into the race in the transfer case.
25 Coat the new seal contact surfaces with grease.
26 Drive the new seal into the transfer case, with the open side of the seal facing the transfer case housing.
27 Using a puller, hook the puller over the edge pinion housing and tighten to press the flange onto the shaft.
28 Tighten the flange nut to the torque listed in this Chapter's Specifications.
29 The remainder of the installation is revere of removal.

16 Driveshaft - removal and installation

Note: *The following procedure applies to Golf AWD models only.*

Removal

1 Raise the front of the vehicle and support it securely on jackstands.
2 Remove the under body panels to allow access to the driveshaft.
3 Remove the muffler and exhaust pipes as necessary to provide clearance to removal of the driveshaft (see Chapter 4, Section 14).
4 Remove the bolts attaching the lower transaxle mount bracket to the transaxle.
5 Rotate the engine and transaxle forward and hold it in the rotated position using a block of wood.
6 Remove the bolts and the driveshaft heat shield from the transfer case.
7 Remove the bolts and the center heat shield from the body.
8 Remove the mounting nuts and remove the rear tunnel brace from the vehicle.
9 Mark the driveshaft rubber discs in relation to the flanges on the differential and transfer case for alignment during installation.
10 While holding the driveshaft in place, remove the driveshaft flange bolts.
Caution: *Support the ends of the driveshaft and do not let the ends of the driveshaft hang down to prevent damage to the driveshaft. The remainder of the procedure requires two people.*
11 With the assistance of another person, support the center of the driveshaft and remove the center support bolts. Some models may have two supports requiring fastener removal.
12 Remove the driveshaft from the vehicle.
Caution: *When removing the driveshaft, use care not to damage the bushing inside of the ends of the driveshaft.*

Installation

13 Installation is reverse of removal, noting the following items:
  a) *Have a second person help with installation of the driveshaft.*
  b) *Install the rubber discs in the locations marked during removal.*
14 Install the bolts for the rubber discs to the flanges, then install the center support nuts loosely to support the driveshaft.
15 Tighten the bolts attaching the rubber discs to the flanges at the differential and the transfer case to torque listed in this Chapter's Specifications.
16 With the center support in a neutral position, tighten the center support nuts to the torque listed in this Chapter's Specifications.
17 The remainder of installation is reverse of removal.

17 Rear differential - removal and installation

Note: *The following procedures apply to Golf AWD models only.*

Transfer case
Removal

1 Disconnect the cable from the negative terminal of the battery (see Chapter 5, Section 3).
2 Raise the vehicle and support it securely on jackstands.
3 On Golf Wagon models, remove the spare tire and drill two 1-inch (26 mm) holes at the locations marked on the rear compartment floor for access to the differential bolts. Remove these bolts.
4 On all models, remove the exhaust system at the rear of the vehicle (see Chapter 4, Section 14).
Note: *Support the front of the exhaust while the rear section is removed.*
5 Remove the rear stabilizer bar (see Chapter 10).
6 Remove the rear coil springs (see Chapter 10).
7 Remove the rear driveaxles from the vehicle (see Section 13). Jack up the rear suspension as necessary to provide clearance to remove the driveaxles.
8 Mark the driveshaft rear rubber disc relationship to the differential flange and remove the bolts attaching the two together.
9 On Golf R models, remove the driveshaft center support bolts. Detach the rear rubber disc and support the driveshaft.
10 Support the differential using a floor jack; preferably one with a transmission jack head adapter.
11 Loosen the rear axle assembly mounting bolts just behind the upper coil spring mounts.
12 On all models except Golf Wagon, lower the rear axle assembly about 2-inches (40 mm) to gain access for removing the differential rear attaching bolt.
13 On all models, with the differential supported, remove the remaining attaching bolt at the front of the differential.
14 Move the differential rearward and disconnect the driveshaft. Support the driveshaft to prevent damage.
15 Slowly lower the differential while tilting to different angles to gain access to remove the vent lines and disconnect the electrical connectors.
16 Slowly lower and move forward to remove from the vehicle.

Installation

17 Installation is reverse of removal noting the following points.
It is recommended to use new bolts for attaching the differential to the rear axle.
  a) *Align the marks on the rubber disc and the differential flange when installing the driveshaft.*
  b) *Tighten all fasteners to the torque listed in this Chapter's Specifications.*
  c) *Refill or top off the differential fluid as necessary (see Chapter 1, Section 27).*

Driveaxle seals

18 Remove the rear driveaxles (see Section 13).
19 Use a slide hammer to remove the flange and shaft assembly.
20 Pry the driveaxle seal from the differential housing.
21 Coat the new seal contact surfaces with grease.
22 Drive the new seal into the differential, with the open side of the seal facing the differential housing.
23 Pry the old spring clip from the inner end of the flange and shaft assembly and install a new one **(see illustration 12.17).**
24 Install the flange and shaft squarely into the differential housing as far as it will go, and use a dead-blow hammer to secure the locking ring inside of the differential.
25 The remainder of the installation is reverse of removal.
26 Top off the differential fluid as necessary (see Chapter 1, Section 27).

Pinion seal

27 Disconnect the driveshaft from the differential flange and position to the side (see Section 16).
28 Remove the pinion flange nut and use a puller to remove the flange.
29 Pry the pinion seal from the differential housing.
30 Coat the new seal contact surfaces with grease.
31 Drive the new seal into the differential, with the open side of the seal facing the differential housing.
32 Tighten the pinion flange nut to the specification in this Chapter's Specifications.
33 The remainder of the installation is reverse of removal.
34 Top off the differential fluid as necessary (see Chapter 1, Section 27).

18 All-wheel drive control module - replacement

Note: *The following procedures apply to Golf AWD models only.*
Caution: *If the same module is re-installed, no programming is required. If a NEW module is installed, programming is required with a specialized diagnostic tool by a qualified repair facility after installation.*

1 Raise the vehicle and support it securely on jackstands.
2 Disconnect the cable from the negative terminal of the battery (see Chapter 5, Section 3).
3 Locate the module near the front of the differential. Disconnect the two electrical connectors from the module.

Generation VI

4 Place a drain pan under the module to catch any fluid during removal.
5 Remove the bolts attaching the module to the clutch housing and pull the module off.
6 Remove and replace the cover/seal. It may be stuck to the clutch housing or to the module itself.
7 Using a pair of pliers, pull the final drive clutch control valve from the clutch housing.
8 Inspect the control valve O-rings for damage and replace as necessary.
9 Before installing the control module, place the new cover/seal on the module, then install the control valve into the control module.
10 Install the control module with the solenoid valve into the clutch housing.
11 Top off the differential fluid as necessary (see Chapter 1, Section 27).

Generation V

12 Remove the bolts attaching the module to the bracket and remove the module from the vehicle.
13 Installation is reverse of removal.

19 All-wheel drive clutch - removal and installation

Note: *The following procedure applies to Golf models only with both Generation VI and V AWD clutches.*

Removal

1 Raise the vehicle and support it securely on jackstands.
2 Disconnect the cable from the negative terminal of the battery (see Chapter 5, Section 3).
3 Drain the AWD clutch fluid (see Chapter 1, Section 27).
4 Disconnect the vehicle harness connector from the AWD control module.
5 Disconnect the driveshaft from the differential flange and position to the side (see Section 16).
6 Place a drain pan under the clutch housing where it meets the differential, to catch any fluid during removal.
7 Remove the bolts attaching the AWD clutch housing to the differential.
8 Pull the clutch housing (with the module and pump attached) from the vehicle.

Installation

9 Installation is reverse of removal noting the following points.
  a) *Align the marks on the rubber disc and the differential flange when installing the driveshaft.*
  b) *Tighten all fasteners to the torque listed in this Chapter's Specifications.*
  c) *Top off the clutch and differential fluid as necessary (see Chapter 1, Section 27).*

20 All-wheel drive clutch pump - removal and installation

Note: *The following procedure applies to Golf models only with both Generation VI and V AWD clutches.*
Caution: *If the same clutch pump is re-installed, no programming is required. If a NEW clutch pump is installed, programming is required with a specialized diagnostic tool by a qualified repair facility after installation.*

Removal

1 Raise and support the rear of the vehicle on jackstands.
2 Disconnect the cable from the negative terminal of the battery (see Chapter 5, Section 3).
3 Drain the AWD clutch fluid (see Chapter 1, Section 27).
4 Disconnect the clutch pump connector from the AWD control module.
5 Detach the clutch pump harness from the clip.
6 Place a drain pan under the clutch housing to catch any fluid during removal.
7 Remove the bolt securing the clutch pump to the clutch housing.
8 Pull the clutch pump from the housing. Inspect the O-rings and replace if damaged.

Installation

9 Prior to installation, coat the clutch pump O-rings with oil.
10 Installation is reverse of removal.
11 Top off the clutch fluid as necessary (see Chapter 1, Section 27).

Notes

Chapter 9
Brakes

Contents

	Section		Section
Anti-lock Brake System (ABS) and Electronic Stability Program (ESP) - general information	3	Drum brake shoes - replacement	8
		General Information	1
Brake disc - inspection, removal and installation	7	Master cylinder - removal and installation	10
Brake hoses and lines - inspection and replacement	11	Parking brake - check and adjustment	14
Brake hydraulic system - bleeding	12	Power brake booster - check, removal and installation	13
Brake light switch - removal and installation	16	Troubleshooting	2
Brake pedal - removal and installation	15	Vacuum pump (electric) - removal and installation	17
Disc brake caliper - removal and installation	6	Vacuum pump (mechanical) - removal and installation	18
Disc brake pads (front) - replacement	4	Wheel cylinder - removal and installation	9
Disc brake pads (rear) - replacement	5		

Specifications

General
Brake fluid type	See Chapter 1
Parking brake adjustment gap (disc brake models)	0.060 inch (1.5 mm) maximum total, left and right sides combined

Disc brakes
Minimum brake pad thickness	See Chapter 1
Disc minimum thickness	Cast into disc
Disc runout limit	Not specified (see Section 7 of this Chapter)

Rear drum brakes
Shoe friction material minimum thickness	See Chapter 1
Maximum inside diameter	Cast into drum
Maximum out-of-round	Not specified (see Section 8, Step 26 of this Chapter)

Torque specifications
Ft-lbs (unless otherwise indicated)   Nm

Note: *One foot-pound (ft-lb) of torque is equivalent to 12 inch-pounds (in-lbs) of torque. Torque values below approximately 15 ft-lbs are expressed in inch-pounds, because most foot-pound torque wrenches are not accurate at these smaller values.*

	Ft-lbs	Nm
Brake disc retaining screw	35 in-lbs	4
Caliper guide pins/mounting bolts*		
Front		
Jetta/GLI	22	30
Golf/GTI/Type R	26	35
Rear	26	35
Caliper mounting bracket bolts		
Front	147.5	200
Rear		
Step 1	66	90
Step 2	Tighten an additional 1/4-turn (90-degrees)	
Backing plate mounting bolts	8	12
Brake hose-to-caliper banjo bolt	26	35
Brake line-to-rear caliper fitting (CI 38 caliper)	120 in-lbs	14
Master cylinder-to-brake booster nuts	18	25
Brake booster nuts	18	25
Vacuum pump-to-bracket bolts	72 in-lbs	8
Vacuum pump bracket-to-automatic transmission bolts	18	25
Wheel cylinder bolt	72 in-lbs	8
Wheel bolts	See the Chapter 1 Specifications	

* Replace with new fasteners

Chapter 9 Brakes

1 General Information

1 The vehicles covered by this manual are equipped with hydraulically operated front and rear brake systems. The front brakes are disc type and the rear brakes are disc or drum type. Both the front and rear brakes are self adjusting. The disc brakes automatically compensate for pad wear, while the drum brakes incorporate an adjustment mechanism that is activated as the parking brake is applied.

Hydraulic system

2 The hydraulic system consists of two separate circuits. The master cylinder has separate reservoirs for the two circuits, and in the event of a leak or failure in one hydraulic circuit, the other circuit will remain operative. A dual proportioning valve on the firewall provides brake balance between the front and rear brakes.

Power brake booster

3 The power brake booster, utilizing engine manifold vacuum and atmospheric pressure to provide assistance to the hydraulically operated brakes, is mounted on the firewall in the engine compartment.

Parking brake

4 The parking brake operates the rear brakes only, through cable actuation. It's activated by a lever mounted in the center console.

Tire Pressure Monitoring System - TPMS

5 A tire monitoring system is incorporated into the software of the ABS control module. The system will recognize a slow and gradual decrease in tire pressure on a wheel by checking the circumference values of each individual tire by way of the ABS sensors. The TPMS compares the speed and rolling circumference of the individual wheels and converts that information into a tire pressure that is then displayed on the instrument panel. If the tire pressure drops by more than 10 percent, the tire pressure warning light will be illuminated in the instrument panel.

6 After a tire rotation or changing a tire, the 'SET' button (located in the glove box) must be depressed after the ignition is switched on and while the vehicle is stationary. Hold the button in until an audible signal tone is heard. This signal tone confirms that the system will recognize the changes in the tires (rotated or otherwise) the next time you move the vehicle. **Note:** *Some driving distance (approx. 100 yards) may be necessary for the ABS sensors to properly read and turn the low tire pressure light off. If after adjusting the tire pressure (in all wheels) and the light fails to turn off after the reset procedure has been performed, seek professional help.*

Service

7 After completing any operation involving disassembly of any part of the brake system, always test drive the vehicle to check for proper braking performance before resuming normal driving. When testing the brakes, perform the tests on a clean, dry, flat surface. Conditions other than these can lead to inaccurate test results.

8 Test the brakes at various speeds with both light and heavy pedal pressure. The vehicle should stop evenly without pulling to one side or the other.

9 Tires, vehicle load and wheel alignment are factors which also affect braking performance.

Precautions

10 There are some general cautions and warnings involving the brake system on this vehicle:

  a) *Use only brake fluid conforming to DOT 3 specifications.*
  b) *The brake pads and linings contain fibers that are hazardous to your health if inhaled. Whenever you work on brake system components, clean all parts with brake system cleaner. Do not allow the fine dust to become airborne. Also, wear an approved filtering mask.*
  c) *Safety should be paramount whenever any servicing of the brake components is performed. Do not use parts or fasteners that are not in perfect condition, and be sure that all clearances and torque specifications are adhered to. If you are at all unsure about a certain procedure, seek professional advice. Upon completion of any brake system work, test the brakes carefully in a controlled area before putting the vehicle into normal service. If a problem is suspected in the brake system, don't drive the vehicle until it's fixed*

Warning: *Vehicles equipped with the Start/Stop System can restart at any time if the key has been left in the on position during any maintenance or repairs. Be sure to deactivate the Start/Stop system before performing any repairs. A message will be displayed in the instrument cluster indicating the Start/Stop has been disabled.*

2 Troubleshooting

PROBABLE CAUSE	CORRECTIVE ACTION

No brakes - pedal travels to floor

1 Low fluid level 2 Air in system	1 and 2 Low fluid level and air in the system are symptoms of another problem a leak somewhere in the hydraulic system. Locate and repair the leak
3 Defective seals in master cylinder	3 Replace master cylinder
4 Fluid overheated and vaporized due to heavy braking	4 Bleed hydraulic system (temporary fix). Replace brake fluid (proper fix)

Brake pedal slowly travels to floor under braking or at a stop

1 Defective seals in master cylinder	1 Replace master cylinder
2 Leak in a hose, line, caliper or wheel cylinder	2 Locate and repair leak
3 Air in hydraulic system	3 Bleed the system, inspect system for a leak

Chapter 9 Brakes

PROBABLE CAUSE | **CORRECTIVE ACTION**

Brake pedal feels spongy when depressed

Probable Cause	Corrective Action
1 Air in hydraulic system	1 Bleed the system, inspect system for a leak
2 Master cylinder or power booster loose	2 Tighten fasteners
3 Brake fluid overheated (beginning to boil)	3 Bleed the system (temporary fix). Replace the brake fluid (proper fix)
4 Deteriorated brake hoses (ballooning under pressure)	4 Inspect hoses, replace as necessary (it's a good idea to replace all of them if one hose shows signs of deterioration)

Brake pedal feels hard when depressed and/or excessive effort required to stop vehicle

Probable Cause	Corrective Action
1 Power booster faulty	1 Replace booster
2 Engine not producing sufficient vacuum, or hose to booster clogged, collapsed or cracked	2 Check vacuum to booster with a vacuum gauge. Replace hose if cracked or clogged, repair engine if vacuum is extremely low
3 Brake linings contaminated by grease or brake fluid	3 Locate and repair source of contamination, replace brake pads or shoes
4 Brake linings glazed	4 Replace brake pads or shoes, check discs and drums for glazing, service as necessary
5 Caliper piston(s) or wheel cylinder(s) binding or frozen	5 Replace calipers or wheel cylinders
6 Brakes wet	6 Apply pedal to boil-off water (this should only be a momentary problem)
7 Kinked, clogged or internally split brake hose or line	7 Inspect lines and hoses, replace as necessary

Excessive brake pedal travel (but will pump up)

Probable Cause	Corrective Action
1 Drum brakes out of adjustment	1 Adjust brakes
2 Air in hydraulic system	2 Bleed system, inspect system for a leak

Excessive brake pedal travel (but will not pump up)

Probable Cause	Corrective Action
1 Master cylinder pushrod misadjusted	1 Adjust pushrod
2 Master cylinder seals defective	2 Replace master cylinder
3 Brake linings worn out	3 Inspect brakes, replace pads and/or shoes
4 Hydraulic system leak	4 Locate and repair leak

Brake pedal doesn't return

Probable Cause	Corrective Action
1 Brake pedal binding	1 Inspect pivot bushing and pushrod, repair or lubricate
2 Defective master cylinder	2 Replace master cylinder

Brake pedal pulsates during brake application

Probable Cause	Corrective Action
1 Brake drums out-of-round	1 Have drums machined by an automotive machine shop
2 Excessive brake disc runout or disc surfaces out-of-parallel	2 Have discs machined by an automotive machine shop
3 Loose or worn wheel bearings	3 Adjust or replace wheel bearings
4 Loose lug nuts	4 Tighten lug nuts

Brakes slow to release

Probable Cause	Corrective Action
1 Malfunctioning power booster	1 Replace booster
2 Pedal linkage binding	2 Inspect pedal pivot bushing and pushrod, repair/lubricate
3 Malfunctioning proportioning valve	3 Replace proportioning valve
4 Sticking caliper or wheel cylinder	4 Repair or replace calipers or wheel cylinders
5 Kinked or internally split brake hose	5 Locate and replace faulty brake hose

Chapter 9 Brakes

Troubleshooting (continued)

PROBABLE CAUSE	CORRECTIVE ACTION
Brakes grab (one or more wheels)	
1 Grease or brake fluid on brake lining	1 Locate and repair cause of contamination, replace lining
2 Brake lining glazed	2 Replace lining, deglaze disc or drum
Vehicle pulls to one side during braking	
1 Grease or brake fluid on brake lining	1 Locate and repair cause of contamination, replace lining
2 Brake lining glazed	2 Deglaze or replace lining, deglaze disc or drum
3 Restricted brake line or hose	3 Repair line or replace hose
4 Tire pressures incorrect	4 Adjust tire pressures
5 Caliper or wheel cylinder sticking	5 Repair or replace calipers or wheel cylinders
6 Wheels out of alignment	6 Have wheels aligned
7 Weak suspension spring	7 Replace springs
8 Weak or broken shock absorber	8 Replace shock absorbers
Brakes drag (indicated by sluggish engine performance or wheels being very hot after driving)	
1 Brake pedal pushrod incorrectly adjusted	1 Adjust pushrod
2 Master cylinder pushrod (between booster and master cylinder)	2 Adjust pushrod incorrectly adjusted
3 Obstructed compensating port in master cylinder	3 Replace master cylinder
4 Master cylinder piston seized in bore	4 Replace master cylinder
5 Contaminated fluid causing swollen seals throughout system	5 Flush system, replace all hydraulic components
6 Clogged brake lines or internally split brake hose(s)	6 Flush hydraulic system, replace defective hose(s)
7 Sticking caliper(s) or wheel cylinder(s)	7 Replace calipers or wheel cylinders
8 Parking brake not releasing	8 Inspect parking brake linkage and parking brake mechanism, repair as required
9 Improper shoe-to-drum clearance	9 Adjust brake shoes
10 Faulty proportioning valve	10 Replace proportioning valve
Brakes fade (due to excessive heat)	
1 Brake linings excessively worn or glazed	1 Deglaze or replace brake pads and/or shoes
2 Excessive use of brakes	2 Downshift into a lower gear, maintain a constant slower speed (going down hills)
3 Vehicle overloaded	3 Reduce load
4 Brake drums or discs worn too thin	4 Measure drum diameter and disc thickness, replace drums or discs as required
5 Contaminated brake fluid	5 Flush system, replace fluid
6 Brakes drag	6 Repair cause of dragging brakes
7 Driver resting left foot on brake pedal	7 Don't ride the brakes
Brakes noisy (high-pitched squeal)	
1 Glazed lining	1 Deglaze or replace lining
2 Contaminated lining (brake fluid, grease, etc.)	2 Repair source of contamination, replace linings
3 Weak or broken brake shoe hold-down or return spring	3 Replace springs
4 Rivets securing lining to shoe or backing plate loose	4 Replace shoes or pads
5 Excessive dust buildup on brake linings	5 Wash brakes off with brake system cleaner
6 Brake drums worn too thin	6 Measure diameter of drums, replace if necessary
7 Wear indicator on disc brake pads contacting disc	7 Replace brake pads
8 Anti-squeal shims missing or installed improperly	8 Install shims correctly

Chapter 9 Brakes

PROBABLE CAUSE	CORRECTIVE ACTION
Brakes noisy (scraping sound)	
1 Brake pads or shoes worn out; rivets, backing plate or brake	1 Replace linings, have discs and/or drums machined (or replace) shoe metal contacting disc or drum

Brakes chatter

PROBABLE CAUSE	CORRECTIVE ACTION
1 Worn brake lining	1 Inspect brakes, replace shoes or pads as necessary
2 Glazed or scored discs or drums	2 Deglaze discs or drums with sandpaper (if glazing is severe, machining will be required)
3 Drums or discs heat checked	3 Check discs and/or drums for hard spots, heat checking, etc. Have discs/drums machined or replace them
4 Disc runout or drum out-of-round excessive	4 Measure disc runout and/or drum out-of-round, have discs or drums machined or replace them
5 Loose or worn wheel bearings	5 Adjust or replace wheel bearings
6 Loose or bent brake backing plate (drum brakes)	6 Tighten or replace backing plate
7 Grooves worn in discs or drums	7 Have discs or drums machined, if within limits (if not, replace them)
8 Brake linings contaminated (brake fluid, grease, etc.)	8 Locate and repair source of contamination, replace pads or shoes
9 Excessive dust buildup on linings	9 Wash brakes with brake system cleaner
10 Surface finish on discs or drums too rough after machining	10 Have discs or drums properly machined (especially on vehicles with sliding calipers)
11 Brake pads or shoes glazed	11 Deglaze or replace brake pads or shoes

Brake pads or shoes click

PROBABLE CAUSE	CORRECTIVE ACTION
1 Shoe support pads on brake backing plate grooved or	1 Replace brake backing plate excessively worn
2 Brake pads loose in caliper	2 Loose pad retainers or anti-rattle clips
3 Also see items listed under Brakes chatter	

Brakes make groaning noise at end of stop

PROBABLE CAUSE	CORRECTIVE ACTION
1 Brake pads and/or shoes worn out	1 Replace pads and/or shoes
2 Brake linings contaminated (brake fluid, grease, etc.)	2 Locate and repair cause of contamination, replace brake pads or shoes
3 Brake linings glazed	3 Deglaze or replace brake pads or shoes
4 Excessive dust buildup on linings	4 Wash brakes with brake system cleaner
5 Scored or heat-checked discs or drums	5 Inspect discs/drums, have machined if within limits (if not, replace discs or drums)
6 Broken or missing brake shoe attaching hardware	6 Inspect drum brakes, replace missing hardware

Rear brakes lock up under light brake application

PROBABLE CAUSE	CORRECTIVE ACTION
1 Tire pressures too high	1 Adjust tire pressures
2 Tires excessively worn	2 Replace tires
3 Defective proportioning valve	3 Replace proportioning valve

Brake warning light on instrument panel comes on (or stays on)

PROBABLE CAUSE	CORRECTIVE ACTION
1 Low fluid level in master cylinder reservoir (reservoirs with fluid level sensor)	1 Add fluid, inspect system for leak, check the thickness of the brake pads and shoes
2 Failure in one half of the hydraulic system	2 Inspect hydraulic system for a leak
3 Piston in pressure differential warning valve not centered	3 Center piston by bleeding one circuit or the other (close bleeder valve as soon as the light goes out)
4 Defective pressure differential valve or warning switch	4 Replace valve or switch
5 Air in the hydraulic system	5 Bleed the system, check for leaks
6 Brake pads worn out (vehicles with electric wear sensors - small	6 Replace brake pads (and sensors) probes that fit into the brake pads and ground out on the disc when the pads get thin)

9-6 Chapter 9 Brakes

Troubleshooting (continued)

PROBABLE CAUSE | **CORRECTIVE ACTION**

Brakes do not self adjust

Disc brakes

Probable cause	Corrective action
1 Defective caliper piston seals	1 Replace calipers. Also, possible contaminated fluid causing soft or swollen seals (flush system and fill with new fluid if in doubt)
2 Corroded caliper piston(s)	2 Same as above

Drum brakes

Probable cause	Corrective action
1 Adjuster screw frozen	1 Remove adjuster, disassemble, clean and lubricate with high-temperature grease
2 Adjuster lever does not contact star wheel or is binding	2 Inspect drum brakes, assemble correctly or clean or replace parts as required
3 Adjusters mixed up (installed on wrong wheels after brake job)	3 Reassemble correctly
4 Adjuster cable broken or installed incorrectly (cable-type adjusters)	4 Install new cable or assemble correctly

Rapid brake lining wear

Probable cause	Corrective action
1 Driver resting left foot on brake pedal	1 Don't ride the brakes
2 Surface finish on discs or drums too rough	2 Have discs or drums properly machined
3 Also see Brakes drag	

3 Anti-lock Brake System (ABS) and Electronic Stability Program (ESP) - general information

1 The Anti-lock Brake System (ABS) and Electronic Stabilization Program (ESP) are designed to help maintain vehicle steerability, directional stability and optimum deceleration under severe braking or maneuvering conditions and on most road surfaces. The ABS system is primarily designed to prevent wheel lockup during heavy or panic braking situations. It works by monitoring the rotational speed of each wheel and controlling the brake line pressure to each wheel when engaged. Data provided by the ABS wheel speed sensors is shared with the Electronic Stabilization Program. This very sophisticated system helps with traction control, over/under-steering and acceleration control under all driving conditions. Overall, these systems aid in vehicle control and handling. Other systems added to vehicles with ESP are Electronic Differential Lock (EDL) and Acceleration Slip Regulation (ASR).

Tire pressure monitoring system - TPMS

2 A tire pressure monitoring system is incorporated into the software of the ABS control module. The system will recognize a slow and gradual decrease in tire pressure on a wheel by checking the circumference values of each individual tire by way of the ABS sensors. The TPMS software continuously compares the speed and rolling circumference of each individual wheel from the original value stored when the TPMS was set, and converts that information into a tire pressure that is then displayed on the instrument panel. If the tire pressure drops by more than 10 percent, the tire pressure warning light will be illuminated in the instrument panel.

3 See Section 1 for more details on the TPMS operation.

Components

Actuator assembly

4 The actuator assembly is mounted in the engine compartment and consists of an electric hydraulic pump and solenoid valves (see illustration).

a) The electric pump provides hydraulic pressure to charge the reservoirs in the actuator, which supplies pressure to the braking system. The pump and reservoirs are housed in the actuator assembly.

b) The solenoid valves modulate brake line pressure during ABS, ESP or ASR operation.

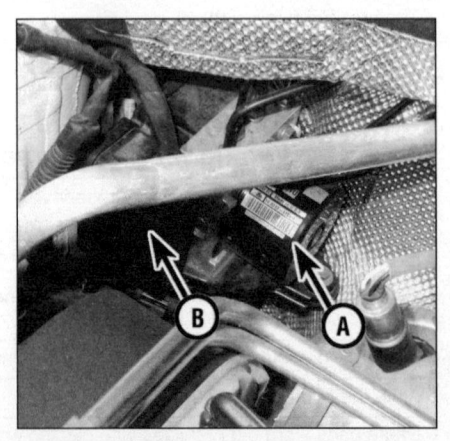

3.4 Typical ABS/ESP actuator (A) and control module (B) assembly

Chapter 9 Brakes

Wheel speed sensors

5   There is a wheel speed sensor for each wheel. Each sensor generates a signal in the form of a low-voltage electrical current or a frequency when the wheel is turning. A variable signal is generated as a result of a square-toothed ring (tone-ring, exciter-ring, reluctor, etc.) that rotates very close to the sensor. The signal is directly proportional to the wheel speed and is interpreted by an electronic module (computer).

6   The front sensors are mounted on the steering knuckles.

7   The rear sensors are mounted in the rear suspension knuckles.

ABS/ESP computer

8   The ABS/ESP control module or computer is mounted with the actuator and is the brain of these systems. The function of the computer is to accept and process information received from the wheel speed sensors to control the hydraulic line pressure, avoiding wheel lock up or wheel spin. The computer also constantly monitors the system, even under normal driving conditions, to find faults within the system.

Diagnosis and repair

9   If a dashboard warning light comes on and stays on while the vehicle is in operation, the ABS or ESP system requires attention. Although special electronic diagnostic testing tools are necessary to properly diagnose the system, you can perform a few preliminary checks before taking the vehicle to a dealer service department.

a) Check the brake fluid level in the reservoir.
b) Verify that the computer electrical connectors are securely connected.
c) Check the electrical connectors at the hydraulic control unit.
d) Check the fuses.
e) Follow the wiring harness to each wheel and verify that all connections are secure and that the wiring is undamaged.

10   If the above preliminary checks do not rectify the problem, the vehicle should be diagnosed by a dealer service department or other qualified repair shop. Due to the complexity of this system, all actual repair work must be done by a qualified automotive technician.

Warning: *Do NOT try to repair an ABS/ESP wiring harness. These systems are sensitive to even the smallest changes in resistance. Repairing the harness could alter resistance values and cause the system to malfunction. If the wiring harness is damaged in any way, it must be replaced.*

Caution: *Make sure the ignition is turned off before unplugging or reattaching any electrical connections.*

Wheel speed sensor - removal and installation

11   Loosen the wheel bolts, raise the vehicle and support it securely on jackstands. Remove the wheel.

12   Make sure the ignition key is turned to the Off position.

13   Trace the wiring back from the sensor, detaching all brackets and clips while noting its correct routing, then disconnect the electrical connector.

14   Remove the mounting bolt and carefully pull the sensor out from the knuckle **(see illustrations)**.

15   Installation is the reverse of the removal procedure. Tighten the mounting fastener securely.

16   Install the wheel and bolts, tightening them securely. Lower the vehicle and tighten the bolts to the torque listed in the Chapter 1 Specifications.

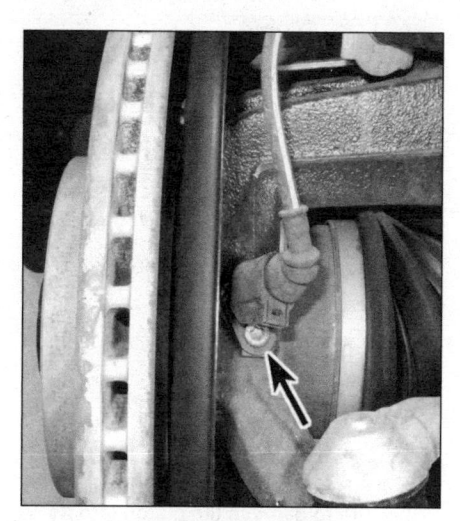

3.14a Front wheel speed sensor location

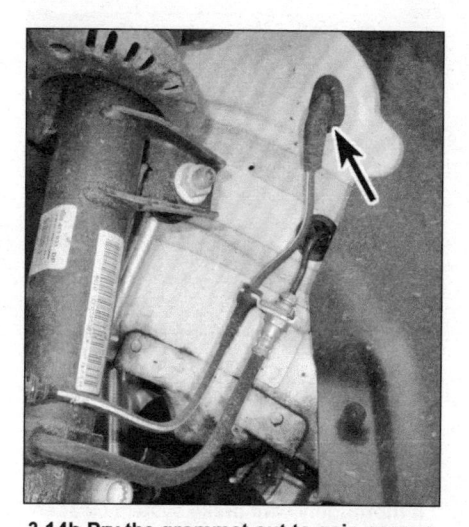

3.14b Pry the grommet out to gain access to the connector for the front wheel speed sensor wire harness

3.14c Rear wheel speed sensor location

3.14d Pry the harness fasteners out in these locations

3.14e Remove the splash shield to gain access to this grommet. Remove the grommet, then pull the harness out to disconnect the rear speed sensor wire connection

4 Disc brake pads (front) - replacement

Warning: *Disc brake pads must be replaced on both front wheels at the same time - never replace the pads on only one wheel. Also, the dust created by the brake system is harmful to your health. Never blow it out with compressed air and don't inhale any of it. An approved filtering mask should be worn when working on the brakes. Do not, under any circumstances, use petroleum-based solvents to clean brake parts. Use brake system cleaner only!*

Warning: *On models with PC57/C60/PC60 calipers, the caliper mounting bolts are self-locking and designed to be used only once. The manufacturer states to replace the caliper mounting bolts any time they are removed.*

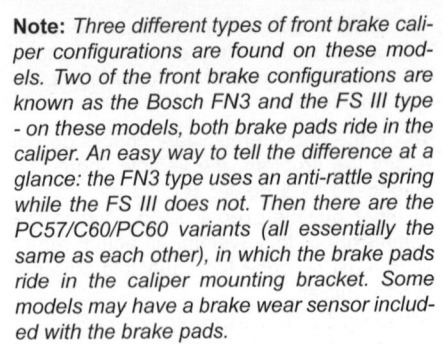

Note: *Three different types of front brake caliper configurations are found on these models. Two of the front brake configurations are known as the Bosch FN3 and the FS III type - on these models, both brake pads ride in the caliper. An easy way to tell the difference at a glance: the FN3 type uses an anti-rattle spring while the FS III does not. Then there are the PC57/C60/PC60 variants (all essentially the same as each other), in which the brake pads ride in the caliper mounting bracket. Some models may have a brake wear sensor included with the brake pads.*

1 Remove the cap from the brake fluid reservoir. Remove about two-thirds of the fluid from the reservoir; reinstall the cap.

Warning: *Brake fluid is poisonous - never siphon it by mouth. Use a suction gun or old poultry baster. If a baster is used, never again use it for the preparation of food.*

Caution: *Brake fluid will damage paint. If any fluid is spilled, wash it off immediately with plenty of clean, cold water.*

2 Loosen the front wheel bolts, raise the front of the vehicle and support it securely on jackstands. Block the wheels at the opposite end.

3 Remove the wheels. Work on one brake assembly at a time, using the assembled brake for reference if necessary.

4 Before removing anything, thoroughly clean the caliper and disc with brake system cleaner (see illustration).

Removal

5 Push the piston back into the bore to allow the caliper to be removed easily, and to make room for the new, thicker pads (see illustration).

6 Disconnect the brake pad wear indicator electrical connector (see illustration) and remove it from the bracket. Carefully unclip the pad retaining spring and remove it from the brake caliper (see illustration).

Note: *If your vehicle is equipped with brake pad wear sensors and you install a set of brake pads that are not equipped with the sensor, the brake wear warning light will remain on in the instrument cluster.*

7 On FN3 and FS III calipers, remove the end caps from the guide bushing to gain access to the caliper guide pins (see illustration).

8 Unscrew the caliper guide pins (FN3 and FS III calipers) or caliper mounting bolts (PC57/C60/PC60 calipers), then lift the caliper away from the mounting bracket. Tie the caliper to the suspension strut using a length of wire; do not allow it to hang unsupported from the flexible brake hose (see illustrations).

9 On FN3 calipers, unclip the inner pad from the caliper piston and remove the outer pad from the mounting bracket (see illustrations). On FS III calipers, unclip both pads from the caliper (see illustrations). On PC57/C60/PC60 calipers, remove both pads from the caliper mounting bracket.

4.4 Before disassembling the brake, wash it thoroughly with brake system cleaner and allow it to dry - position a drain pan under the brake to catch the residue - DO NOT use compressed air to blow off the brake dust!

4.5 Before removing the caliper, be sure to depress the piston into its bore in the caliper with a large C-clamp to make room for the new pads

4.6a Disconnect the brake pad wear indicator sensor connector and remove the sensor connector from the bracket

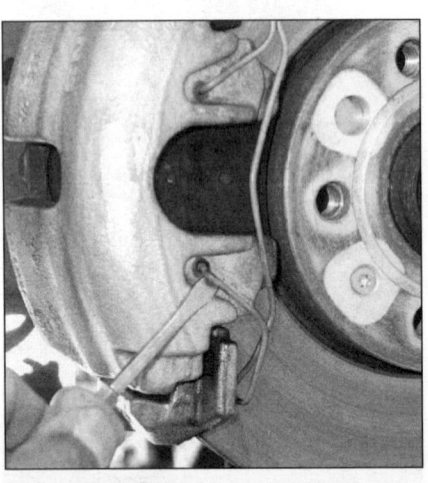

4.6b Unclip the anti-rattle spring (FN3 caliper) and remove it from the caliper

4.7 Remove the protective caps from the guide pins and remove the pins (FN3/FSIII calipers)

Chapter 9 Brakes

Inspection

10 Inspect the brake disc carefully as outlined in Section 7. If machining is necessary, follow the information in that Section to remove the disc, at which time the pads can be removed as well.

11 Prior to fitting the pads, check that the guide pins are free to slide easily in the caliper body bushings (FN3/FS III calipers) or in the caliper mounting bracket (PC57/C60/PC60 cailpers) **(see illustration)**, and are a reasonably tight fit. If they're OK, lubricate them with high-temperature brake grease and reinstall them. Use brake system cleaner to clean the caliper and piston. Inspect the dust seal around the piston for damage, and the piston for evidence of fluid leaks, corrosion or damage. If any of these components require attention, the caliper should be replaced.

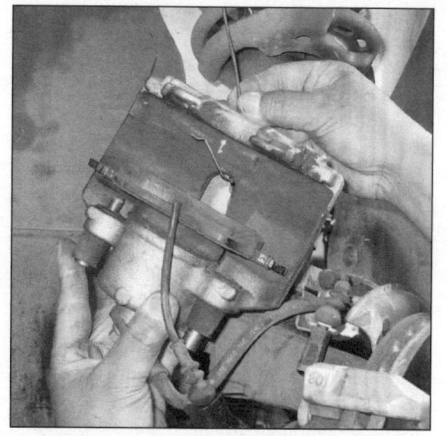

4.8a Remove the caliper from the mounting bracket and support it with a length of wire

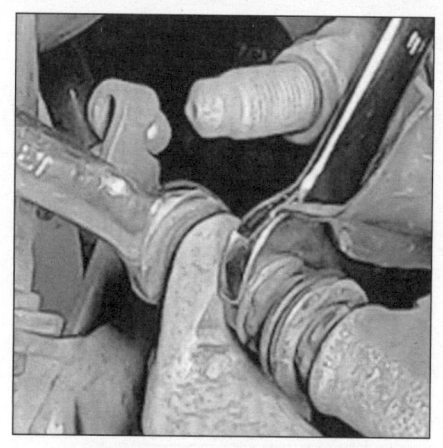

4.8b On PC57/C60/PC60 calipers, counter-hold the guide pins with an open-end wrench while loosening the caliper mounting bolts

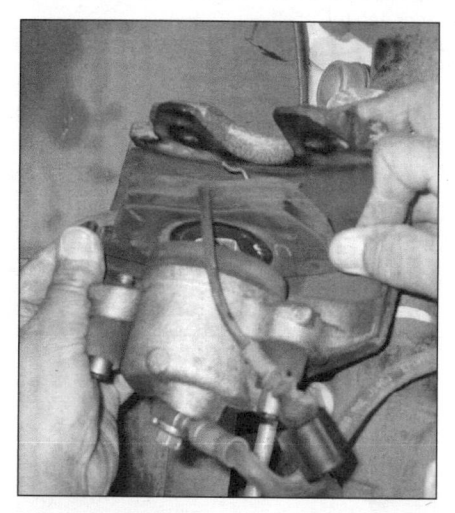

4.9a On FN3 calipers, unclip the inner pad from the caliper piston . . .

4.9b . . . and remove the outer pad from the mounting bracket

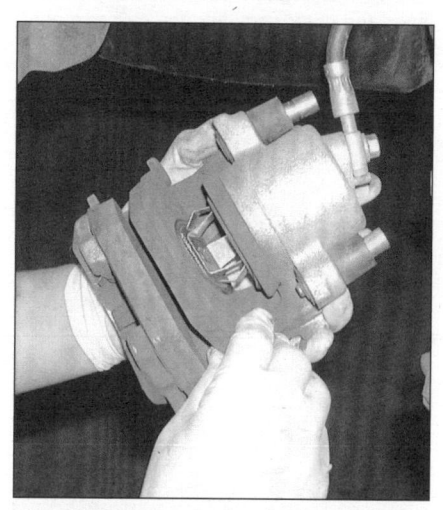

4.9c On FS III calipers, unclip the inner pad from the caliper piston . . .

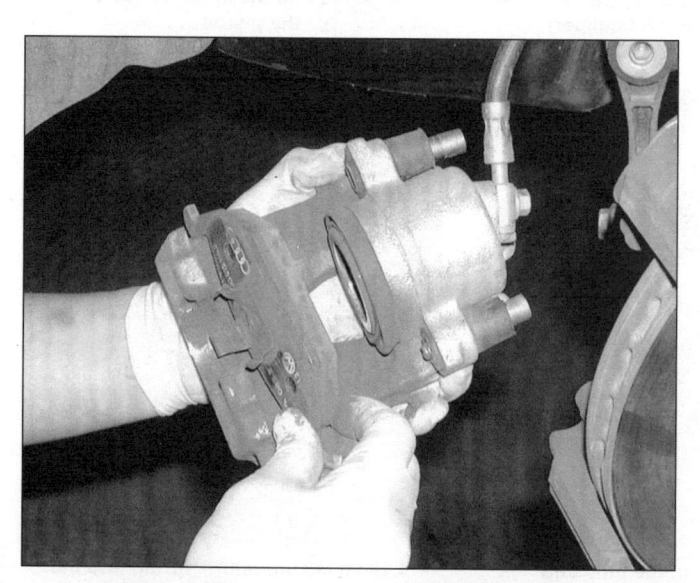

4.9d . . . and the outer pad from the caliper frame

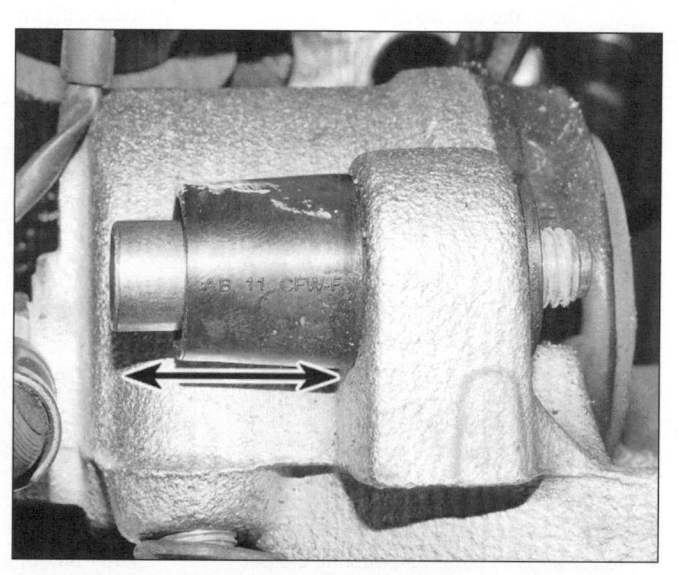

4.11 The guide pins should slide back and forth with no binding or restrictions

9-10   Chapter 9   Brakes

Installation

12 To make room for the new pads, the caliper piston(s) must be pushed back into the cylinder. Either use a piston retraction tool or a C-clamp. As the piston is pushed into its bore, make sure the fluid in the master cylinder does not overflow. If necessary, remove some of the fluid.

13 On FN3 calipers, clip the inner pad into the caliper piston and fit the outer pad to the mounting bracket **(see illustrations)**, ensuring its friction material is against the brake disc. Note that the outer pad has an arrow stamped onto its outer lower edge; this should point in the normal direction of rotation of the brake disc. Remove the adhesive foil backing from the outer pad, if equipped.

14 On FS III calipers, clip the inner pad into the caliper piston and the outer pad into the caliper frame **(see illustrations)**.

15 On PC57/C60/PC60 calipers, place the inner and outer pads into the caliper mounting bracket, making sure the friction material is against the brake disc.

16 Lubricate the caliper guide pins with a light coat of high-temperature brake grease **(see illustration)**. Maneuver the caliper into position, then, on FN3 and FS III calipers install the caliper guide pins and tighten them to the torque listed in this Chapter's Specifications. On PC57/C60/PC60 calipers, lubricate the guide pins and reinstall them, then place the caliper over the brake pads and install the mounting bolts, tightening them to the torque listed in this Chapter's Specifications.

17 On FN3 and FS III calipers, install the end caps to the caliper guide pins.

18 On FS3 calipers, install the anti-rattle spring, ensuring its ends are correctly located in the caliper body holes. Press the inner edge of the spring into position so that its ends are firmly in contact with the surface of the caliper bracket **(see illustration)**.

19 Lock the brake pad wear indicator electrical connector into the bracket and reconnect the electrical connector.

20 Depress the brake pedal repeatedly, until the pads are pressed into contact with the brake disc, and a firm brake pedal feel is obtained.

21 Repeat the above procedure on the remaining front brake caliper.

22 Install the wheels and wheel bolts, then lower the vehicle to the ground and tighten the bolts to the torque listed in the Chapter 1 Specifications.

23 Once again, firmly depress the brake pedal a few times to bring the pads into contact with the disc.

24 Check and, if necessary, top-up the brake fluid level as described in Chapter 1.

25 Test the operation of the brakes carefully before placing the vehicle into normal service.

Warning: *New pads will not give full braking efficiency until they have bedded in. Be prepared for this, and avoid hard braking as much as possible for the first hundred miles or so after pad replacement.*

4.13a Clip the inner pad into the caliper . . .

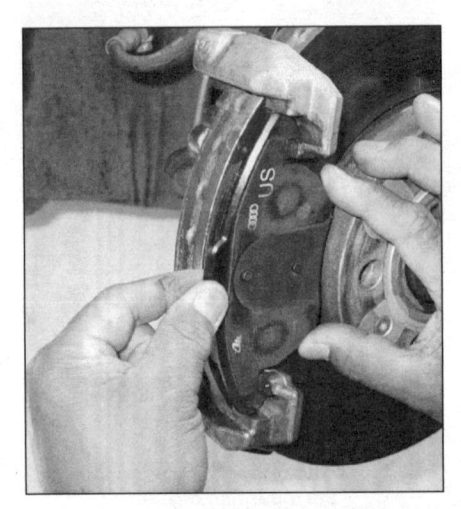

4.13b . . . and place the outer pad into the mounting bracket (FN3 caliper)

4.14a Clip the inner pad into the piston . . .

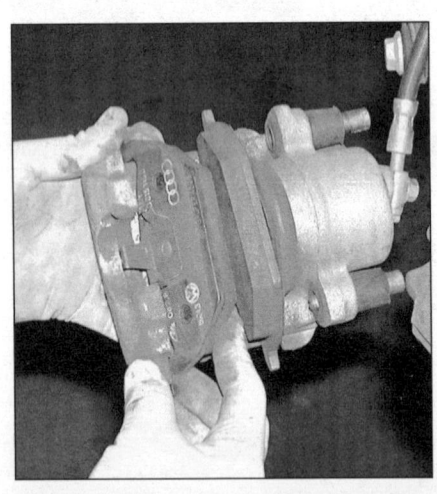

4.14b . . . and the outer pad into the caliper frame (FS III caliper)

4.16 Pull out the guide pins and clean them, then apply a coat of high temperature grease to the pins and reinstall the pins in the caliper

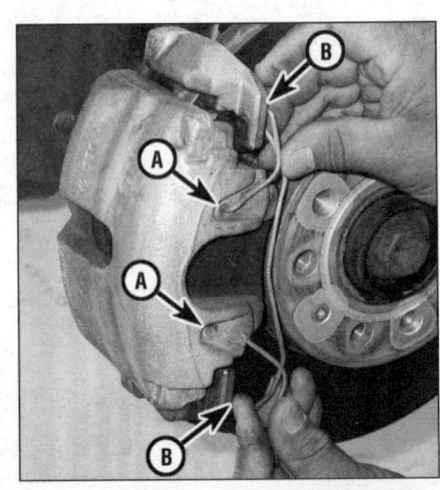

4.18 Install the return spring into the caliper body holes (A), and set the ends of the spring under the caliper mounting bracket (B) (FN3 caliper)

Chapter 9 Brakes

5 Disc brake pads (rear) - replacement

Warning: *Disc brake pads must be replaced on both rear wheels at the same time - never replace the pads on only one wheel. Also, the dust created by the brake system is harmful to your health. Never blow it out with compressed air and don't inhale any of it. An approved filtering mask should be worn when working on the brakes. Do not, under any circumstances, use petroleum-based solvents to clean brake parts. Use brake system cleaner only!*

Warning: *The caliper mounting bolts are self-locking and designed to be used only once. The manufacturer states to replace the caliper mounting bolts any time they are removed.*

Note: *Two different types of rear brake caliper configurations are found on these models. The removal and installation procedures are basically the same. The different rear brake configurations are known as the Bosch ZOH 38 type and the CI 38 type. To tell the difference at a glance, The CI 38 type uses butterfly type anti-rattle springs on the outer edge of the brake pads while the ZOH BIR 38 type uses pad support plates on the caliper mounting bracket.*

1   Loosen the rear wheel bolts. Block the front wheels, then raise the rear of the vehicle and support it securely on jackstands. Remove the rear wheels. Clean the caliper and surrounding area with brake system cleaner **(see illustration)**.

2   Unscrew the caliper mounting bolts while holding the guide pins with an open-end wrench to prevent them from rotating **(see illustration)**. Discard the mounting bolts - new bolts must be used on installation.

3   Lift the caliper away from the brake pads **(see illustration)** and support it with a length of wire. Do not allow the caliper to hang unsupported on the flexible brake hose.

4   Remove the brake pads and, on CI 38 calipers, the anti-rattle springs from the caliper mounting bracket **(see illustrations)**.

5.1 Wash the brake thoroughly with brake system cleaner and allow it to dry

5.2 When removing the mounting bolts, use an open-end wrench to hold the guide pins and prevent them from rotating

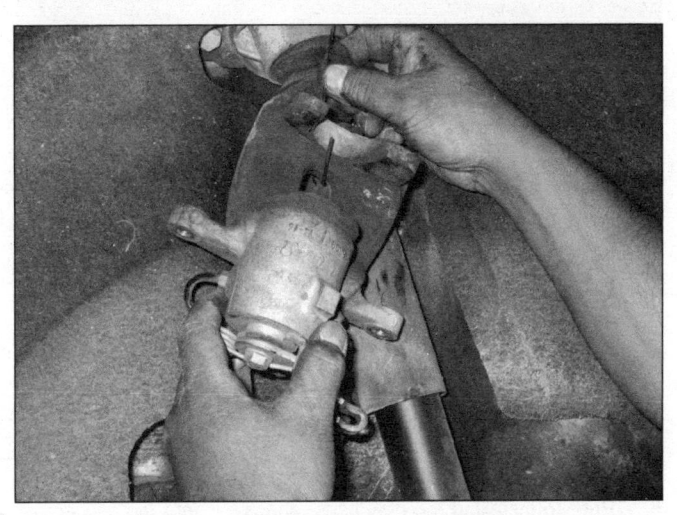

5.3 Remove the caliper from the mounting bracket and support it with a length of wire

5.4a Remove the outer pad . . .

5.4b . . . and the inner pad . . .

5.4c . . . then, on CI 38-type calipers, the anti-rattle springs from the caliper mounting bracket

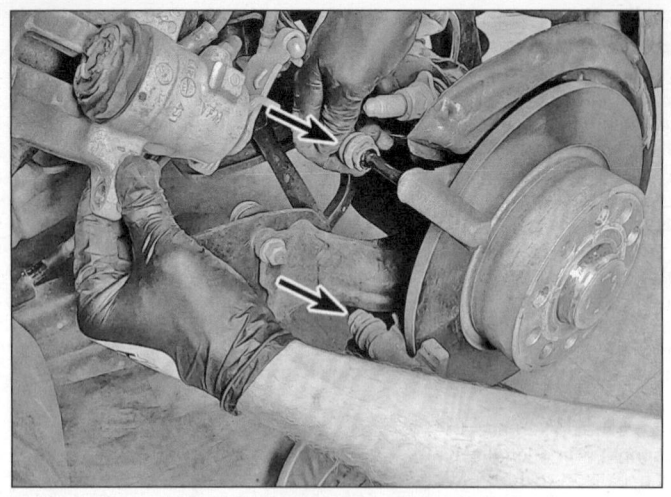

5.5a Remove the caliper guide pins . . .

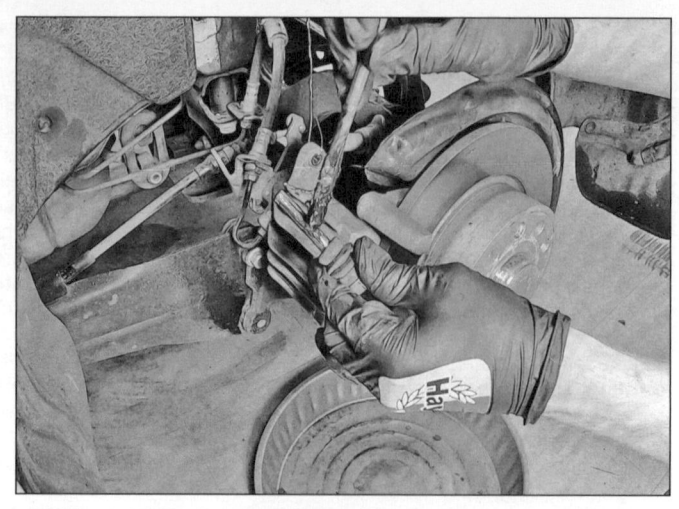

5.5b . . . lubricate them with high-temperature brake grease, then reinstall them

5.6 Using a retraction tool to push and rotate the piston back into the caliper

5   Prior to installing the new pads, check that the anti-rattle springs are not damaged and fit tightly to the caliper mounting bracket (CI 38 calipers only), check that the guide pins are free to slide easily in the caliper bracket, and check that the rubber guide pin boots are undamaged. Pull the guide pins out and inspect them for signs of wear. If they're OK, lubricate them with high-temperature brake grease and reinstall them **(see illustrations)**. Clean the caliper and piston with brake system cleaner. Inspect the dust seal around the piston for damage, and the piston for evidence of fluid leaks, corrosion or damage. If necessary, replace the caliper.

6   To make room for the new brake pads, it will be necessary to retract the piston fully into the caliper bore by rotating it in a clockwise direction using a caliper piston retraction tool (available at most auto parts stores) **(see illustration)**. If the master cylinder reservoir is full, remove some of the brake fluid from the reservoir to avoid spilling the fluid as you compress the caliper piston into the caliper. **Caution:** *Do not attempt to push the piston into the caliper with a C-clamp - it must be rotated into its bore as it is being depressed.*

7   Peel the protective sheet from the pad backing plates (if applicable), then install the pads in the mounting bracket, ensuring that each pad's friction material is facing the brake disc.

8   Slide the caliper back into position over the pads, making sure, on Bosch ZOH BIR 38-type caliper, that the pad anti-rattle springs are correctly positioned against the inner surface of the caliper body and are not jammed in the inspection aperture. **Warning:** *New caliper mounting bolts must be used when the caliper is installed.*

9   Press the caliper into position, then install the new mounting bolts, tightening them to the torque listed in this Chapter's Specifications while preventing the guide pins from turning with an open-end wrench.

10   Repeat the above procedure on the remaining rear brake caliper.

11   Depress the brake pedal repeatedly to force the pads into firm contact with the discs. Once normal pedal feel has returned, check that the discs rotate freely.

12   Install the wheels and wheel bolts, then lower the vehicle to the ground and tighten the wheel bolts to the torque listed in the Chapter 1 Specifications.

13   Check and, if necessary, top-up the brake fluid level as described in Chapter 1.

14   Test the operation of the brakes carefully before placing the vehicle into normal service. **Warning:** *New pads will not give full braking efficiency until they have bedded in. Be prepared for this, and avoid hard braking as much as possible for the first hundred miles or so after pad replacement.*

6   Disc brake caliper - removal and installation

Warning: *Dust created by the brake system is harmful to your health. Never blow it out with compressed air and don't inhale any of it. An approved filtering mask should be worn when working on the brakes. Do not, under any circumstances, use petroleum-based solvents to clean brake parts. Use brake system cleaner only!*

Warning: *The caliper mounting bolts (except on FN3 and FS III calipers) and caliper mounting bracket bolts are self-locking and designed to be used only once. The manufacturer states to replace these bolts any time they are removed.*

Warning: *Always replace the calipers in pairs - never replace just one of them.*

Note: *If you're removing the caliper for access to other components, don't disconnect the hose. Suspend the caliper with a piece of wire to prevent damaging the brake hose.*

Removal

1   Loosen the front or rear wheel bolts, raise the front or rear of the vehicle and support it securely on jackstands. Block the wheels at the opposite end. Remove the front or rear wheel.

Chapter 9 Brakes   9-13

6.2a Work the ball end of the cable off of the lever by pushing the lever in by hand, then slip the ball off of the hooked end of the lever

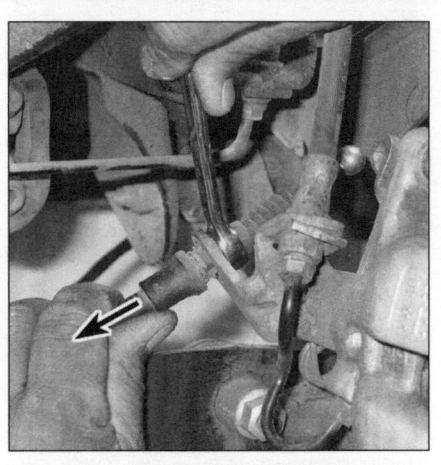

6.2b Use a box-end wrench to compress the retaining tabs (pliers will work too), then detach the cable from the bracket on the caliper

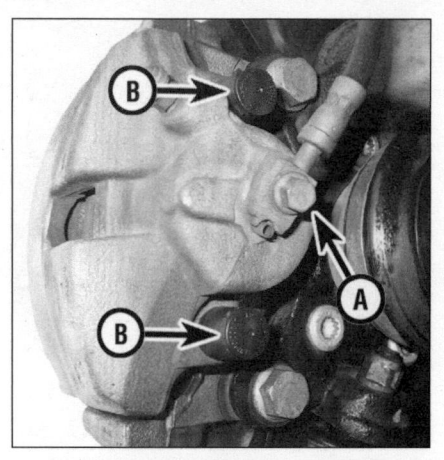

6.3a On all except Cl 38 rear calipers, remove the brake line banjo bolt (A), cap the line and hole, then remove the caps and caliper guide pins (B)

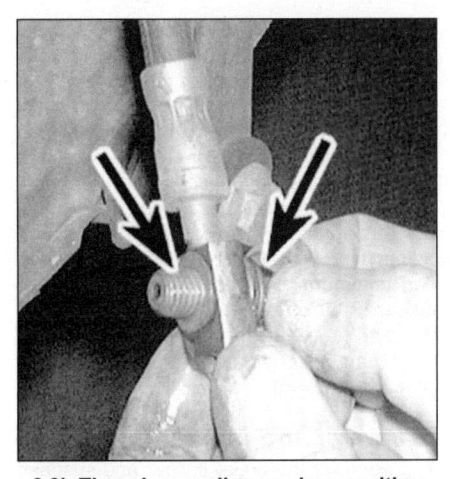

6.3b There is a sealing washer on either side of the front brake hose inlet fitting; be sure to replace these with new ones when reconnecting the hose

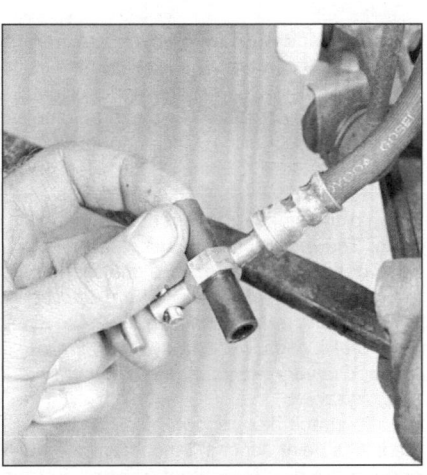

6.3c The brake hose can be plugged using a snug-fitting piece of tubing

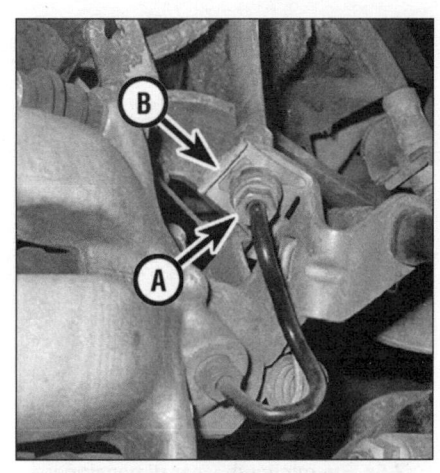

6.4 On Cl 38 rear calipers, unscrew the brake line fitting from the brake hose (A), then remove the clip and detach the hose from the bracket and brake line

2   If you are removing a rear caliper, disconnect the cable from the parking brake lever on the caliper and the bracket (**see illustrations**).

3   If you're removing a front caliper, or a ZOH BIR 38 rear caliper, remove the banjo fitting bolt and disconnect the brake hose from the caliper (**see illustration**). Discard the sealing washers from each side of the hose fitting (**see illustration**). Plug all open fittings to prevent fluid loss and the entry of contaminants (**see illustration**).

4   If you're removing a Cl 38 rear caliper, unscrew the brake line fitting nut from the brake hose fitting using a flare nut wrench, then remove the clip securing the brake hose to its bracket and detach the hose (**see illustration**). Plug the fitting to prevent fluid loss and the entry of contaminants.

5   Remove the caliper guide pins or the mounting bolts and detach the caliper (see Section 4 [front] or Section 5 [rear]).

Installation

6   Installation is the reverse of removal. If you're installing a front caliper, use new sealing washers on each side of the brake hose inlet fitting, and tighten the fasteners to the torque listed in this Chapter's Specifications.

7   Bleed the brake system (see Section 12). Make sure there are no leaks from the hose connections. Pump the brake pedal several times before driving the vehicle, and test the brakes carefully before returning the vehicle to normal service.

7   Brake disc - inspection, removal and installation

Warning: *The caliper mounting bolts (except on FN3 and FS III calipers) and caliper mounting bracket bolts are self-locking and designed to be used only once. The manufacturer states to replace these bolts any time they are removed.*

Inspection

1   Loosen the wheel bolts, raise the vehicle and support it securely on jackstands. Remove the wheel.

2   Remove the brake caliper as outlined in Section 6. It's not necessary to disconnect the brake hose for this procedure. After removing the caliper bolts, suspend the caliper out of the way with a piece of wire. Don't let the caliper hang by the hose and don't stretch or twist the hose.

3   Reinstall the wheel bolts to hold the disc securely against the hub, if necessary. It may be necessary to install washers between the disc and the wheel bolts to take up space.

4   Visually check the disc surface for score marks, cracks and other damage. Light

9-14   Chapter 9   Brakes

7.4 The brake pads on this vehicle were obviously neglected, as they wore down completely and cut deep grooves into the disc - wear this severe means the disc must be replaced

7.5a Use a dial indicator to check disc runout; if the reading exceeds the maximum allowable runout limit, the disc will have to be machined or replaced

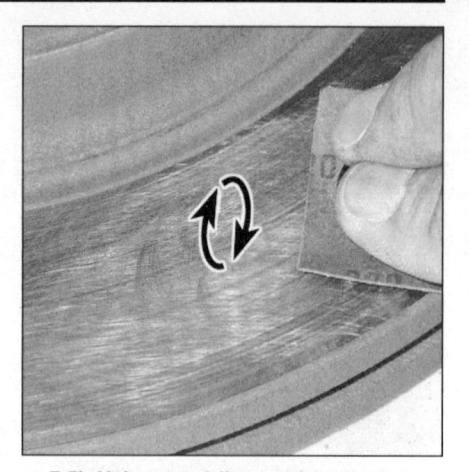

7.5b Using a swirling motion, remove the glaze from the disc surface with sandpaper or emery cloth

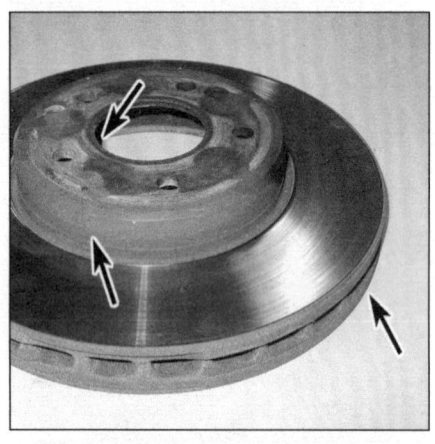

scratches and shallow grooves are normal after use and may not always be detrimental to brake operation. Deep score marks or cracks may require disc refinishing by an automotive machine shop or disc replacement **(see illustration)**. Be sure to check both sides of the disc. If pulsating has been noticed during application of the brakes, suspect disc runout.

Note: *The most common symptoms of damaged or worn brake discs are pulsation in the brake pedal when the brakes are applied or loud grinding noises caused from severely worn brake pads. If these symptoms are extreme, it is very likely that the discs will need to be replaced.*

5   To check disc runout, place a dial indicator at a point about 1/2-inch from the outer edge of the disc **(see illustrations)**. Set the indicator to zero and turn the disc. Although the manufacturer doesn't give a runout specification, an indicator reading that exceeds 0.003

inch could cause pulsation upon brake application and will require disc refinishing by an automotive machine shop or disc replacement.

Note: *If disc refinishing or replacement is not necessary, you can de-glaze the brake pad surface on the disc with emery cloth or sandpaper (use a swirling motion to ensure a non-directional finish)(see illustration).*

6   It's absolutely critical that the disc not be machined to a thickness under the specified minimum thickness. The minimum (or discard) thickness is cast or stamped into the disc **(see illustration)**. The disc thickness can be checked with a micrometer **(see illustration)**.

Removal and installation

7   Remove the brake caliper and suspend it out of the way with a piece of wire (don't disconnect the line/hose). Remove the caliper mounting bracket **(see illustrations)**.

7.6a The minimum wear dimension is typically cast or etched into the disc. Inspect all areas (front, back, edges, etc.) of the disc closely to find this information

7.6b Use a micrometer to measure disc thickness

7.7a Remove the caliper mounting bracket fasteners and detach the mounting bracket - this is a typical front caliper mounting bracket . . .

7.7b . . . and this is a typical rear caliper mounting bracket

Chapter 9 Brakes   9-15

7.8 A disc retaining screw holds the disc to the hub flange

7.9a Clean any rust and corrosion from the areas of the hub flange that contact the disc. A wire brush or sanding tool, designed to be used with a power drill, can make the job a lot easier

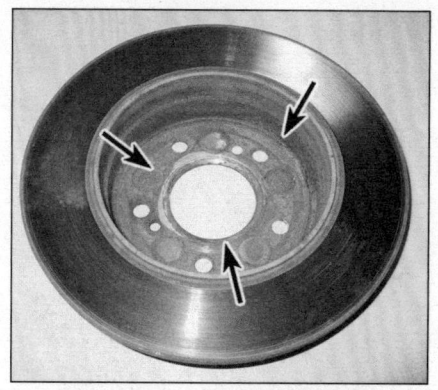

7.9b Clean any rust or corrosion from the area inside the disc that contacts the hub flange. Again, power tools are very useful for this job

8 Remove the disc retaining screw and any wheel bolts installed during inspection, and remove the disc **(see illustration)**. If it's stuck, use a mallet to loosen it from the hub.

9 Clean the hub flange and the inside of the brake disc thoroughly; removing any rust or corrosion **(see illustrations)**. Apply a thin film of anti-seize compound between the hub flange and inside of the disc to prevent rust and corrosion prior to the next brake service.

10 Install the disc onto the hub and tighten the retaining screw securely.

11 Install the brake caliper mounting bracket and tighten the bolts to the torque listed in this Chapter's Specifications.

12 Install the brake pads and caliper, tightening the bolts to the torque listed in this Chapter's Specifications.

13 Install the wheel, then lower the vehicle to the ground. Tighten the wheel bolts to the torque listed in the Chapter 1 Specifications. Depress the brake pedal a few times to bring the brake pads into contact with the disc. Bleeding of the system will not be necessary unless the brake hose was disconnected from the caliper. Check the operation of the brakes carefully before placing the vehicle into normal service. Also, check the parking brake operation and adjust if necessary.

8 Drum brake shoes - replacement

Warning: *Brake shoes must be replaced on both rear wheels at the same time - never replace the shoes on only one wheel, as uneven braking may result. Also, dust created by the brake system is harmful to your health. Never blow it out with compressed air and don't inhale any of it. An approved filtering mask should be worn when working on the brakes. Do not, under any circumstances, use petroleum-based solvents to clean brake parts. Use brake system cleaner only!*

1 Raise the rear of the vehicle and support it securely on jackstands, then chock the front wheels to prevent the vehicle from rolling. Release the parking brake, remove the brake drum retaining screw (similar to the disc retaining screw shown in **illustration 7.8**), then slide the drum off the hub. If the drum won't come off, insert a screwdriver through one of the wheel bolt holes and pry up on the adjuster wedge; this will retract the shoes away from the drum **(see illustrations)**.

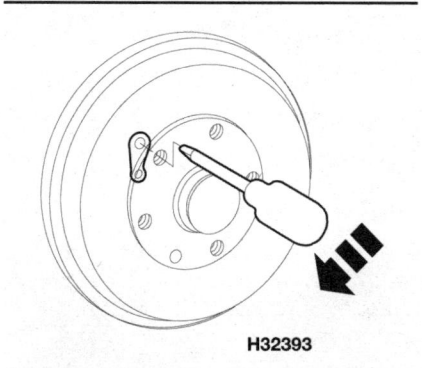

8.1a If the brake drum is tight, release the brake shoes by inserting a flat-bladed screwdriver through a hole in the drum and pry up the adjuster wedge

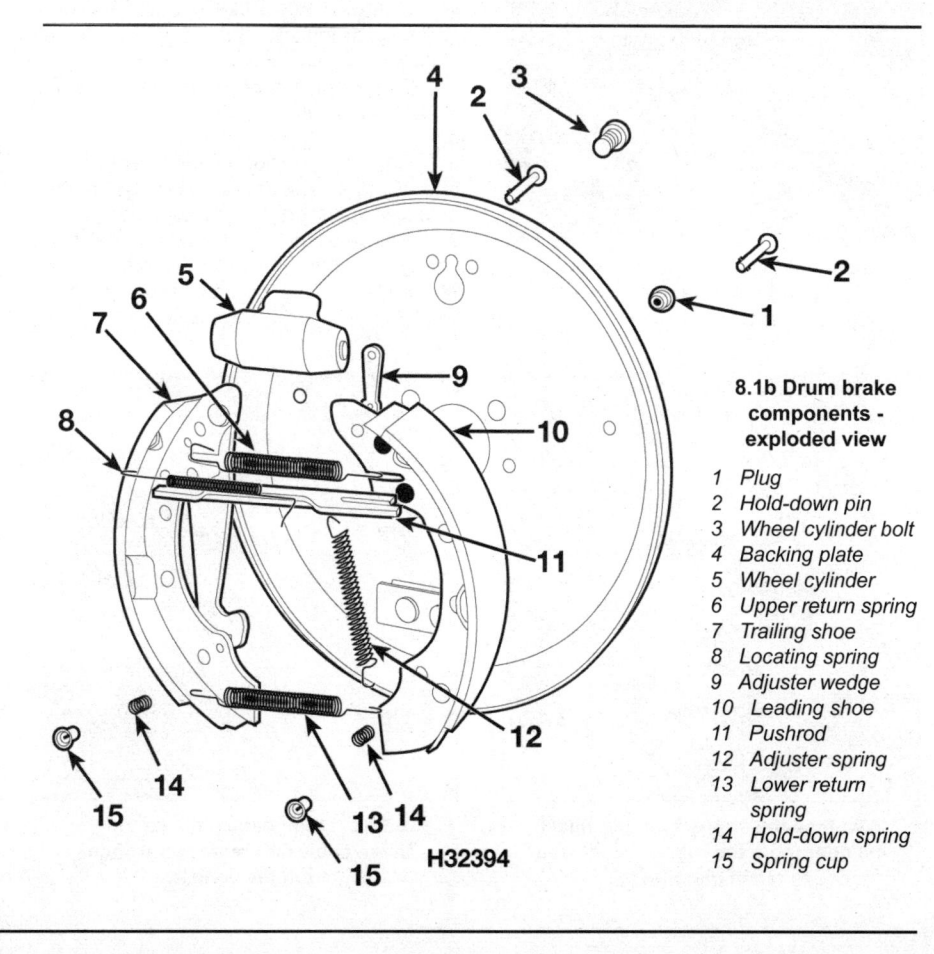

8.1b Drum brake components - exploded view

1. Plug
2. Hold-down pin
3. Wheel cylinder bolt
4. Backing plate
5. Wheel cylinder
6. Upper return spring
7. Trailing shoe
8. Locating spring
9. Adjuster wedge
10. Leading shoe
11. Pushrod
12. Adjuster spring
13. Lower return spring
14. Hold-down spring
15. Spring cup

9-16   Chapter 9 Brakes

8.7a Remove the spring cup . . .

8.7b . . . then remove the hold-down spring . . .

8.7c . . . and withdraw the hold-down pin from the rear of the backing plate

8.8 Unhook the shoes from the lower pivot point and remove the lower return spring

2   Before beginning work, wash off the brake assembly with brake system cleaner and allow the residue to drain into a drip pan.
3   Measure the thickness of the friction material of each brake shoe at several points; if either shoe is worn at any point to the specified minimum thickness or less, all four shoes must be replaced as a set. The shoes should also be replaced if any are fouled with oil or grease; there is no satisfactory way of degreasing friction material once it has been contaminated.
4   If any of the brake shoes are worn unevenly, or fouled with oil or grease, trace and repair the cause before reassembly.
5   To replace the brake shoes, continue as follows. If all is well, install the brake drum.
6   Note the position of the brake shoes and springs, and mark the webs of the shoes, if necessary, to aid installation.
7   Remove the hold-down cups and springs by depressing and turning them 90-degrees. This can be accomplished with a special hold-down spring tool or, if

you're careful, a pair of pliers. With the cups removed, lift off the springs and withdraw the hold-down pins **(see illustrations)**.
8   Ease the shoes out one at a time from the lower pivot point to release the tension of the return spring, then disconnect the lower return spring from both shoes **(see illustration)**.
9   Ease the upper end of both shoes out from their wheel cylinder locations, taking care not to damage the wheel cylinder seals, and disconnect the parking brake cable from the lever on the trailing shoe. The brake shoe assembly can then be maneuvered out of position and away from the backing plate. Do not depress the brake pedal until the brakes are reassembled; wrap a strong rubber band around the wheel cylinder pistons to retain them **(see illustrations)**.
10   Make a note of the correct installed positions of all components **(see illustration)**, then unhook the upper return spring, and disengage the adjuster wedge spring.

8.9a Free the shoes from the wheel cylinder (note the rubber band used to retain pistons) . . .

8.9b . . . then detach the parking brake cable and remove the shoe from the vehicle

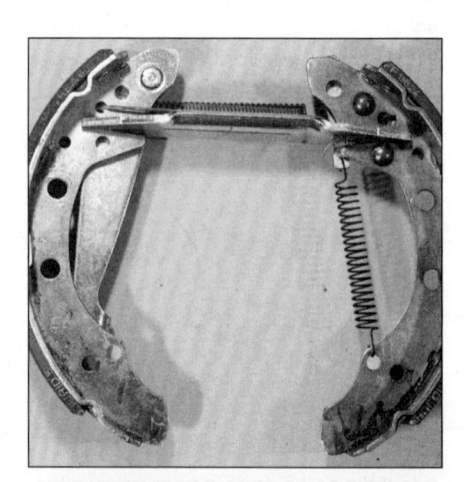

8.10 Prior to disassembly, note the correct installed location of the shoe assembly components

Chapter 9 Brakes

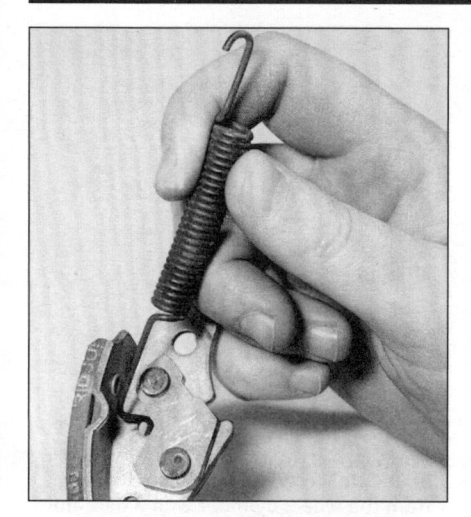

9-17

8.15a Hook the locating spring into the leading shoe . . .

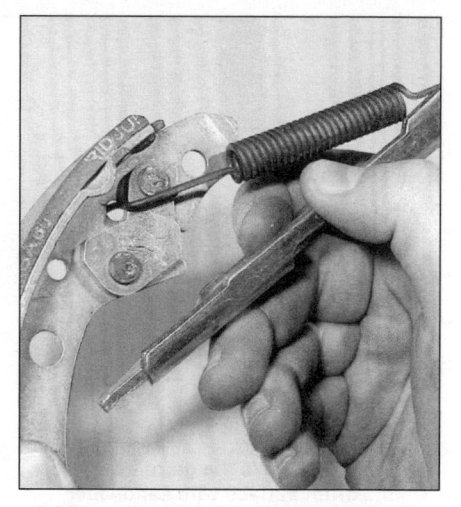

8.15b . . . then engage the pushrod with the opposite end of the spring. . .

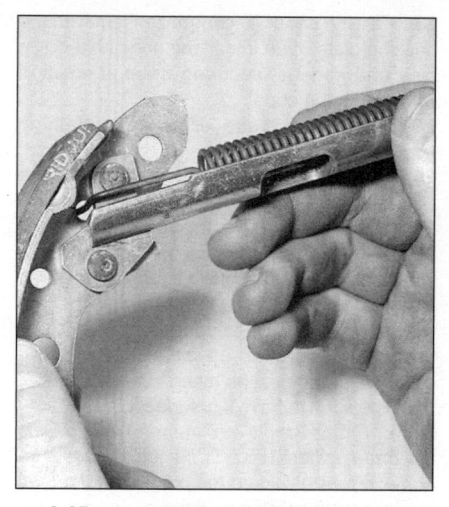

8.15c . . . and pivot the pushrod into position on the shoe

11 Unhook the locating spring, and remove the pushrod from the trailing shoe, together with the wedge key.

12 Examine all components for signs of wear or damage and replace as necessary.

Note: *All return springs should be replaced, regardless of their apparent condition.*

13 Peel back the protective caps, and check the wheel cylinder for fluid leaks or other damage; check that both cylinder pistons are free to move easily. Refer to Section 9, if necessary, for the wheel cylinder replacement procedure.

14 Apply a little high-temperature brake grease to the contact areas of the pushrod and parking brake lever.

15 Hook the locating spring into the leading shoe. Engage the pushrod with the opposite end of the spring, and pivot the pushrod into position on the shoe **(see illustrations)**.

16 Install the adjuster wedge between the leading shoe and pushrod, making sure it is installed correctly **(see illustration)**.

17 Locate the parking brake lever on the trailing shoe in the pushrod, and install the upper return spring using a pair of pliers **(see illustrations)**.

18 Install the spring on the adjuster wedge and hook it onto the leading shoe **(see illustration)**.

19 Clean the backing plate and apply a thin film of high-temperature brake grease or anti-seize compound to the shoe contact areas on the backing plate and to the wheel cylinder pistons and lower pivot point. Do not allow the lubricant to contact the friction material.

20 Remove the rubber band installed on the wheel cylinder, then install the shoe assembly.

21 Connect the parking brake cable to the parking brake lever, then locate the top of the shoes in the wheel cylinder piston slots.

22 Install the lower return spring between the shoes, then lever the bottom of the shoes onto the bottom anchor.

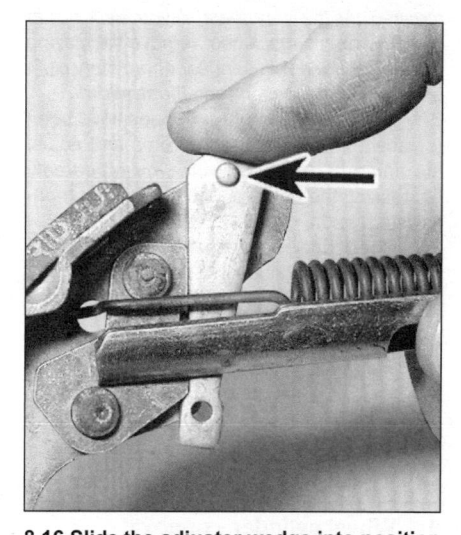

8.16 Slide the adjuster wedge into position in its slot, making sure its raised dot is facing away from the shoe

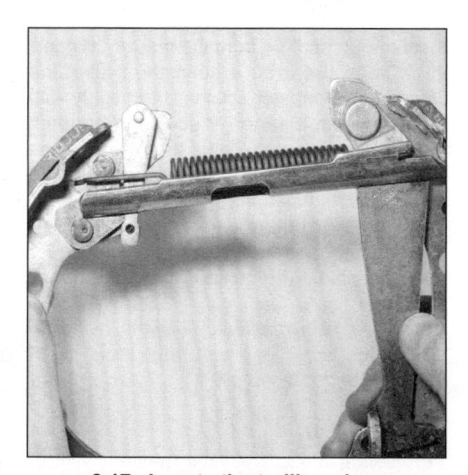

8.17a Locate the trailing shoe in the pushrod . . .

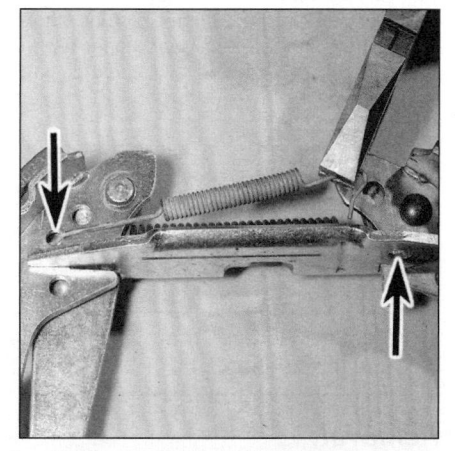

8.17b . . . and hook the upper return spring into position in the trailing shoe and pushrod

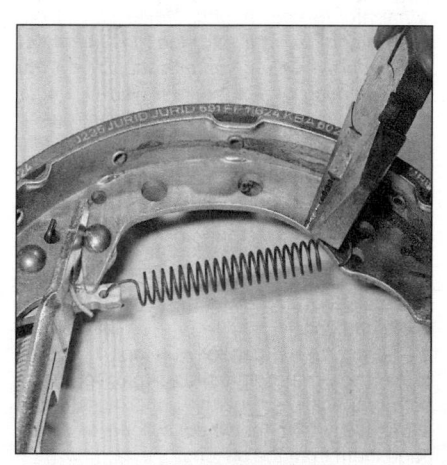

8.18 Fit the spring to the wedge key, and hook it onto the leading shoe

23 Tap the shoes to centralize them with the backing plate, then install the shoe hold-down pins and springs, securing them in position with the spring cups.
24 Install the brake drum and retaining screw.
25 Repeat the above procedure on the other rear brake.
26 Before installing the drum it should be checked for cracks, score marks, deep scratches and hard spots, which will appear as small discolored areas. If the hard spots cannot be removed with sandpaper or emery cloth, or if any of the other conditions listed above exist, the drum must be taken to an automotive machine shop to have it resurfaced.
Note: *Professionals recommend resurfacing the drums whenever a brake job is done. Resurfacing will eliminate the possibility of out-of-round or tapered drums. If the drums are worn so much that they can't be resurfaced without exceeding the maximum allowable diameter (stamped into the drum), then new ones will be required. At the very least, if you elect not to have the drums machined, remove the glazing from the surface with sandpaper or emery cloth using a swirling motion (see illustration).*
27 Once both sets of rear shoes have been replaced, install the brake drums and adjust the lining-to-drum clearance by repeatedly depressing the brake pedal until normal (non-assisted) pedal pressure returns.
28 Check and, if necessary, adjust the parking brake as described in Section 14.
29 Check the brake fluid level as described in Chapter 1.
Note: *New shoes will not give full braking efficiency until they have bedded-in. Be prepared for this, and avoid hard braking as far as possible for the first hundred miles or so after shoe replacement.*

9 Wheel cylinder - removal and installation

Warning: *Dust created by the brake system is harmful to your health. Never blow it out with compressed air and don't inhale any of it. An approved filtering mask should be worn when working on the brakes. Do not, under any circumstances, use petroleum-based solvents to clean brake parts. Use brake system cleaner only!*

Removal

1 Remove the brake drum as described in Section 8. Clean the brake assembly with brake cleaner and allow the residue to drain into a drip pan.
2 Using pliers, carefully unhook the upper brake shoe return spring and remove it from both brake shoes. Pull the upper ends of the shoes away from the wheel cylinder to disengage them from the pistons.
3 Minimize fluid loss by first removing the master cylinder reservoir cap, then tightening it down with a piece of cellophane to obtain an airtight seal.

8.26 Remove the glaze from the drum surface with sandpaper or emery cloth

4 Using a flare nut wrench, if available, unscrew the brake line fitting nut at the wheel cylinder **(see illustration)**. Carefully ease the line out of the wheel cylinder and plug its end to prevent dirt entry. Wipe off any spilled fluid immediately.
5 Unscrew the wheel cylinder retaining bolt from the rear of the backing plate and remove the cylinder; take great care not to allow brake fluid to contaminate the brake shoe linings.

Installation

6 Ensure that the backing plate and wheel cylinder mating surfaces are clean, then spread the brake shoes and maneuver the wheel cylinder into position.
7 Engage the brake line and screw in the fitting nut two or three turns to ensure that the thread has started.
8 Insert the wheel cylinder retaining bolt and tighten it to the torque listed in this Chapter's Specifications. Tighten the brake line fitting nut securely.
9 Remove the cellophane from the master cylinder reservoir cap.
10 Ensure that the brake shoes are correctly located in the cylinder pistons, then carefully install the brake shoe upper return spring, using a screwdriver to stretch the spring into position.
11 Install the drum.
12 Bleed the brake hydraulic system as described in Section 12. Providing precautions were taken to minimize loss of fluid, it should only be necessary to bleed the relevant rear brake.

10 Master cylinder - removal and installation

Removal

Caution: *Brake fluid will damage paint or finished surfaces. Refer to the Precautions in Section 1.*
Caution: *If the battery is disconnected, several*

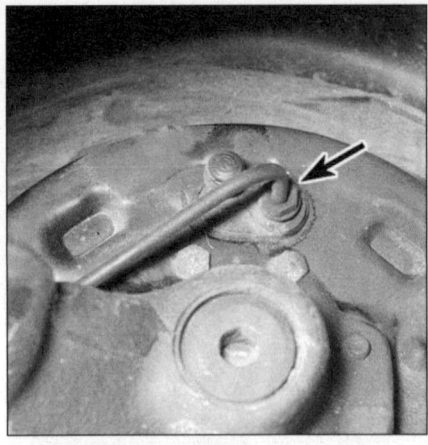

9.4 Unscrew the brake line fitting from the wheel cylinder with a flare-nut wrench to prevent rounding off the corners of the fitting

systems must be re-learned before they will work properly (see Chapter 5, Section 3).
Note: *The master cylinder is not serviceable; replace it with a new or rebuilt unit if it's defective.*

1 Remove the engine cover (see Chapter 1, Section 7).
2 Disconnect the cable from the negative battery terminal (see Chapter 5).
3 Remove the battery and battery tray (see Chapter 5).
4 Unplug the electrical connectors for the brake fluid level warning switch and the brake light switch **(see illustration)**.
5 Remove the reservoir cap and siphon as much fluid as possible from the reservoir with a syringe or equivalent.
6 On manual transaxle models, disconnect and plug the clutch master cylinder hose from the back of the brake master cylinder reservoir.
7 Remove the brake light mounting fastener and the switch (see Section 16).
8 Carefully pull the locking tabs on the reservoir outwards **(see illustration)** and lift the reservoir up and out from the master cylinder.
9 Clean the area around the brake line fittings at the master cylinder thoroughly with brake system cleaner. Place rags beneath the fittings to catch fluid, then detach the brake lines. Cap or plug all openings to prevent contamination.
10 Remove the master cylinder mounting nuts, then detach the master cylinder and heat shield (if equipped) from the power booster.
Note: *If a new master cylinder is being installed, be sure to install new seals when transferring the reservoir.*

Installation

11 Bench bleed the new master cylinder before installing it. Mount the master cylinder in a vise, with the jaws of the vise clamping on the mounting flange.
12 Attach a pair of master cylinder bleeder

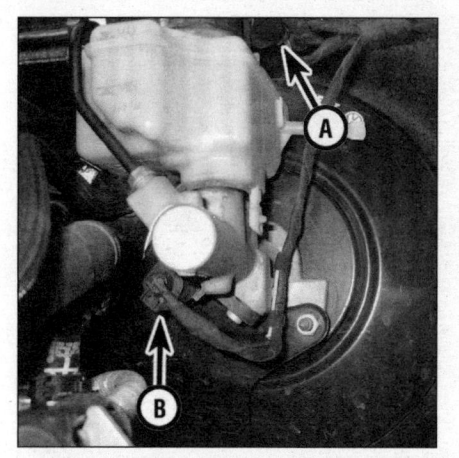

Chapter 9 Brakes

10.4 Disconnect the brake fluid level warning switch (A) and the brake light switch (B)

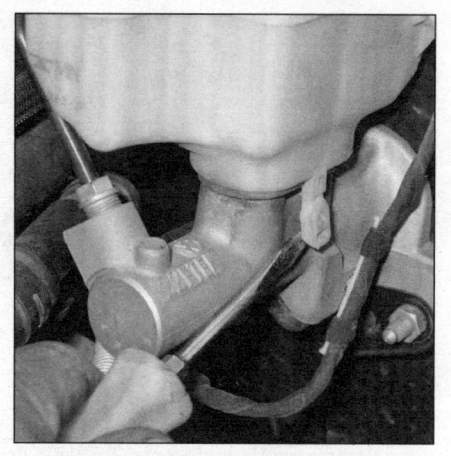

10.8 Carefully pull the locking tabs outwards on each side of the reservoir

10.12 The best way to bleed air from the master cylinder before installing it on the vehicle is with a pair of bleeder tubes that direct brake fluid into the reservoir during bleeding

tubes to the outlet ports of the master cylinder **(see illustration)**.

13 Fill the reservoir with brake fluid of the recommended type (see Chapter 1).

14 Slowly push the pistons into the master cylinder (a large Phillips screwdriver can be used for this) - air will be expelled from the pressure chambers and into the reservoir. Because the tubes are submerged in fluid, air can't be drawn back into the master cylinder when you release the pistons.

15 Repeat the procedure until no more air bubbles are present.

16 Remove the bleed tubes, one at a time, and install plugs in the open ports to prevent fluid leakage and air from entering. Install the reservoir cap.

17 Install the master cylinder over the studs on the power brake booster and tighten the attaching nuts only finger-tight at this time.

18 Thread the brake line fittings into the master cylinder. Since the master cylinder is still a bit loose, it can be moved slightly in order for the fittings to thread in easily. Do not strip the threads as the fittings are tightened.

19 Tighten the mounting nuts to the torque listed in this Chapter's Specifications, then tighten the brake line fittings securely.

20 On manual transaxle models, connect the clutch master cylinder hose from the back of the brake master cylinder reservoir.

Note: *The clutch system will also need bleeding after the brake system has been bled.*

21 Fill the master cylinder reservoir with fluid, then bleed the master cylinder and the brake system as described in Section 12. To bleed the cylinder on the vehicle, have an assistant depress the brake pedal and hold the pedal to the floor. Loosen the fitting to allow air and fluid to escape. Repeat this procedure on both fittings until the fluid is clear of air bubbles.

Caution: *Have plenty of rags on hand to catch the fluid - brake fluid will ruin painted surfaces. After the bleeding procedure is completed, rinse the area under the master cylinder with clean water.*

Warning: *Do not operate the vehicle if you are in doubt about the effectiveness of the brake system. It is possible for air to become trapped in the anti-lock brake system hydraulic control unit, so, if the pedal continues to feel spongy after repeated bleedings or the BRAKE or ANTI-LOCK light stays on, have the vehicle towed to a dealer service department or other qualified shop to be bled with the aid of a scan tool.*

22 Test the operation of the brake system carefully before placing the vehicle into normal service.

11 Brake hoses and lines - inspection and replacement

Inspection

1 Once a year, with the vehicle raised and supported securely on jackstands, the rubber hoses which connect the steel brake lines with the front and rear brake assemblies should be inspected for cracks, chafing of the outer cover, leaks, blisters and other damage. These are important and vulnerable parts of the brake system and inspection should be complete. A light and mirror will be helpful for a thorough check. If a hose exhibits any of the above conditions, replace it with a new one.

Replacement

Front brake hose

2 Loosen the wheel bolts, raise the vehicle and support it securely on jackstands. Remove the wheel.

3 At the frame bracket **(see illustration)**, note how the small tabs of the hose fitting sit in the bracket and keep it from rotating.

4 Support the hose fitting with an open-end wrench, and unscrew the brake line fitting from the hose. Use a flare-nut wrench to prevent rounding off the corners of the nut and be careful not to lose the spring clip on the end of the hose fitting.

5 At the caliper end of the hose, remove the inlet fitting bolt and discard the old sealing washers **(see illustration 6.3a)**. Disconnect the brake hose from the caliper.

6 Remove the spring clip from the bracket at the lower end of the strut, then lift the hose from the bracket.

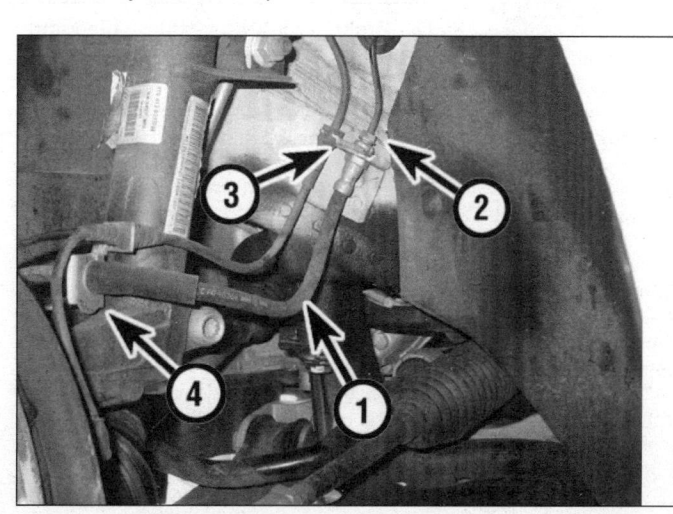

11.3 Typical front brake hose details

1 Brake hose
2 Hose fitting
3 Frame bracket/ spring clip
4 Strut bracket/ spring clip

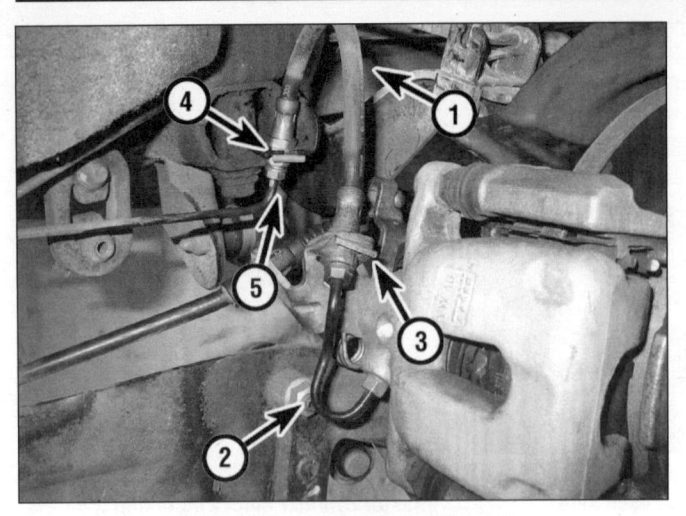

11.12 Typical rear brake hose details

1. Brake hose
2. Caliper brake line
3. Caliper bracket/spring clip
4. Brake hose bracket/spring clip
5. Brake line

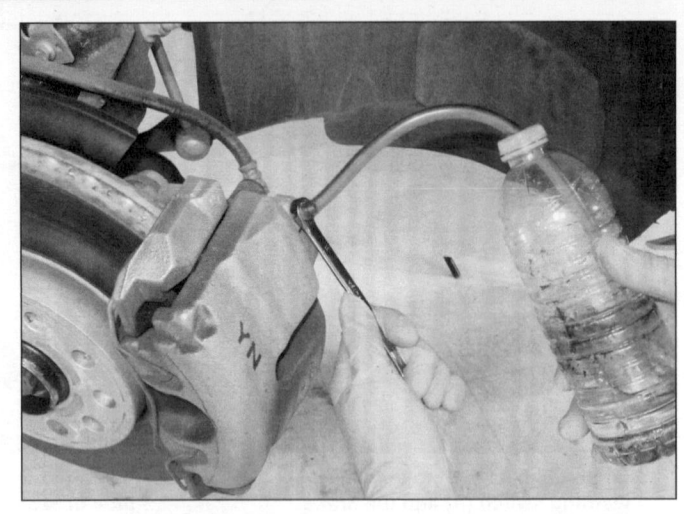

12.8 When bleeding the brakes, a hose is connected to the bleeder valve at the caliper and the other end is submerged in brake fluid. Air will be seen as bubbles in the tube and container. All air must be expelled before moving to the next wheel

7   To install the hose, place the hose into the bracket on the strut and install the spring clip, using new sealing washers and making sure the hose isn't twisted. Tighten the inlet fitting bolt at the caliper to the torque listed in this Chapter's Specifications.

8   Place the brake hose fitting into the frame bracket while making sure the hose isn't twisted between the caliper and the frame bracket.

9   Connect the brake line fitting to the hose, starting the threads by hand. Install the spring clip then tighten the fitting securely.

10   Bleed the caliper (see Section 12).

11   Install the wheel and bolts, lower the vehicle and tighten the bolts to the torque listed in the Chapter 1 Specifications.

Rear brake hose

12   The rear brake hose has a fitting at each end, with a mounting bracket at the body and the caliper, secured by spring clips **(see illustration)**.

13   Support the hose fitting with an open-end wrench, and unscrew the brake line fitting from the hose. Using pliers, remove the spring clip that locks the hose to the bracket.

Note: *Use a flare-nut wrench to prevent rounding off the corners of the nut.*

14   At the caliper end of the hose, use a flare nut wrench to separate the hose fitting from the caliper.

15   Connect the hose fitting to the caliper and tighten it with a flare-nut wrench.

16   On models with rear drum brakes, unscrew the brake line fitting from the wheel cylinder **(see illustration 9.4)**.

17   Detach the line from the plastic clips along the rear axle and remove the line from the vehicle.

18   Installation is the reverse of removal. Tighten the line securely and bleed the caliper or wheel cylinder served by the line that was replaced (see Section 12).

19   Install the wheel and bolts, then lower the vehicle and tighten the bolts to the torque listed in the Chapter 1 Specifications.

Metal brake lines

20   When replacing brake lines, be sure to use the correct parts. Don't use copper tubing for any brake system components. Purchase genuine steel brake lines from a dealer or auto parts store.

21   Prefabricated brake line, with the tube ends already flared and fittings installed, is available at auto parts stores and dealer parts departments.

22   When installing the new line, make sure it's securely supported in the brackets and has plenty of clearance between moving or hot components.

23   After installation, check the master cylinder fluid level and add fluid as necessary. Bleed the brake system (see Section 12) and test the brakes carefully before driving the vehicle in traffic.

12   Brake hydraulic system - bleeding

Warning: *Wear eye protection when bleeding the brake system. If the fluid comes in contact with your eyes, immediately rinse them with water and seek medical attention.*

Note: *Bleeding the hydraulic system is necessary to remove any air that manages to find its way into the system when it's been opened during removal and installation of a hose, line, caliper or master cylinder.*

1   You'll probably have to bleed the system at all four brakes if air has entered it due to low fluid level, or if the brake lines have been disconnected at the master cylinder.

2   If a brake line was disconnected only at a wheel, then only that caliper must be bled. If a brake line is disconnected at a fitting located between the master cylinder and any of the brakes, that part of the system served by the disconnected line must be bled.

3   Raise the vehicle about one foot and support it securely on jackstands.

4   Remove any residual vacuum from the brake power booster by pressing the brake pedal several times with the engine off.

5   Remove the master cylinder reservoir cap and fill the reservoir with brake fluid. Reinstall the cover.

6   Have an assistant on hand, as well as a supply of new brake fluid, a clear container partially filled with clean brake fluid, a length of clear tubing to fit over the bleeder valve and a wrench to open and close the bleeder valve.

Note: *Continue to add fluid while bleeding the system to prevent the fluid level from dropping too low; if this happens, air will enter the master cylinder.*

7   Beginning at the left front wheel, loosen the bleeder valve slightly, then tighten it to a point where it's snug but can still be loosened quickly and easily.

Note: *Use a six-point box-end wrench or socket to loosen the bleeder valve. For bleeder valves that appear to be stuck, clean the area where the valve screws into the caliper with a small wire brush, then apply penetrating oil to the threads and allow it to soak in for awhile.*

8   Place one end of the tubing over the bleeder valve and submerge the other end in brake fluid in the container **(see illustration)**.

9   Have the assistant depress the brake pedal slowly, then hold the pedal down firmly.

10   While the pedal is held down, open the bleeder valve just enough to allow a flow of fluid to leave the valve. Watch for air bubbles to exit the submerged end of the tube. When the fluid flow slows after a couple of seconds, close the valve and have your assistant release the pedal.

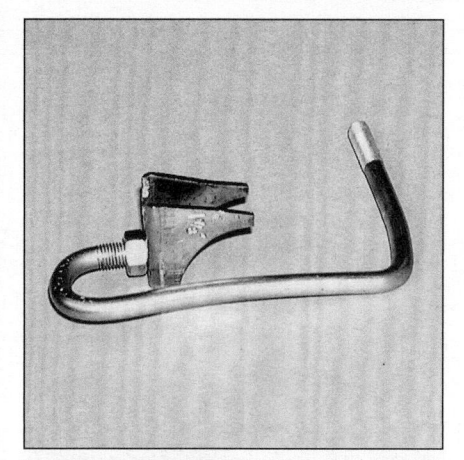

Chapter 9 Brakes

13.11a A tool, fabricated from a modified exhaust clamp, can be used to detach the booster pushrod from the brake pedal

13.11b The tool is used to push on the plastic retaining lugs to release the brake pedal

13.11c The retaining lugs and pedal assembly viewed from the rear

11 Repeat Steps 9 and 10 until no more air is seen leaving the tube, then carefully tighten the bleeder valve and proceed to the right front wheel, the left rear wheel and the right rear wheel, in that order, and perform the same procedure. Be sure to check the fluid in the master cylinder reservoir frequently.
12 Never use old brake fluid. It contains moisture that can boil, rendering the brake system inoperative.
13 Refill the master cylinder with fluid at the end of the operation.
14 Check the operation of the brakes. The pedal should feel solid when depressed, with no sponginess. If necessary, repeat the entire process.
Warning: *Do not operate the vehicle if you are in doubt about the effectiveness of the brake system. It's possible for air to become trapped in the ABS hydraulic control unit, so, if the pedal continues to feel spongy after repeated bleedings or the BRAKE or ABS light stays on, have the vehicle towed to a dealer service department or other qualified repair shop to be bled.*

13 Power brake booster - check, removal and installation

Caution: *If the battery is disconnected, several systems must be re-learned before they will work properly.*
Note: *On some models equipped with an automatic transaxle, a vacuum pump supplies vacuum to the power brake booster in addition to manifold vacuum. The pump is mounted near the ABS/ESP hydraulic unit and has a vacuum hose routed to the brake booster. Vehicles equipped with a brake booster vacuum pump require diagnosis by a dealer service department or other qualified repair shop if the system appears to be defective.*
Note: *The power brake booster is not serviceable; replace it with a new or rebuilt unit if it's defective.*

Note: *The master cylinder is removed along with the power brake booster as an assembly. They are separated after they are removed.*

Operating check
1 Depress the brake pedal several times with the engine off and make sure there's no change in the pedal reserve distance.
2 Depress the pedal and start the engine. If the pedal goes down slightly, operation is normal.

Airtightness check
3 Start the engine and turn it off after one or two minutes. Depress the brake pedal slowly several times. If the pedal depresses less each time, the booster is airtight.
4 Depress the brake pedal while the engine is running, then stop the engine with the pedal depressed. If there's no change in the pedal reserve travel after holding the pedal for 30 seconds, the booster is airtight.

Removal
Caution: *Brake fluid will damage paint or finished surfaces. Refer to the Precautions in Section 1.*

5 With the engine off, press the brake pedal several times to remove any stored vacuum in the power brake booster.
6 Remove the engine cover.
7 Disconnect the cable from the negative battery terminal (see Chapter 5).
8 Remove the battery and battery tray (see Chapter 5).
9 Remove the driver's side knee bolster (see Chapter 11).
10 Remove the brake light switch (see Section 16).
11 Detach the booster pushrod from the brake pedal **(see illustrations)**.
12 Working inside the engine compartment, remove the master cylinder (see Section 10).
13 Remove the vacuum hose from the brake booster **(see illustration)**. Also disconnect the brake booster vacuum sensor electrical connector, if equipped.
14 Remove the clips and the sound insulation panel from around the booster, if equipped.
15 On models equipped with manual transaxles, remove the shift cables and bracket (see Chapter 7A).
Note: *If the booster is being replaced, pull the vacuum sensor from the booster and install it on the new booster.*

13.13 Brake booster check valve

9-22 Chapter 9 Brakes

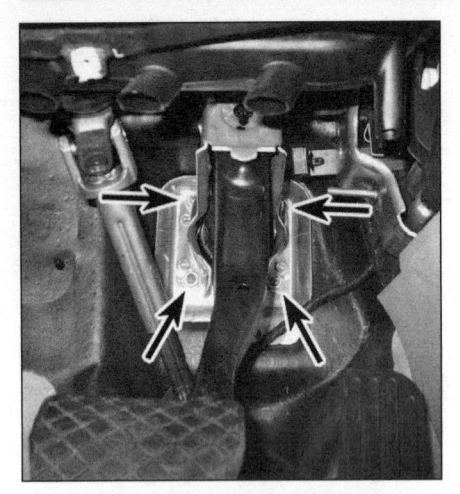

13.17 To detach the brake booster from the firewall, remove the mounting nuts

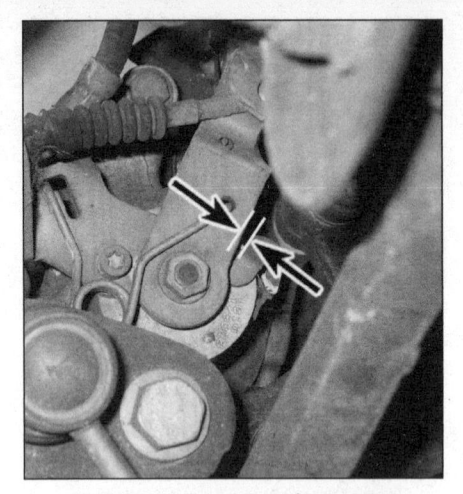

14.9 Measure the gap between the parking brake lever and the stop on the rear caliper

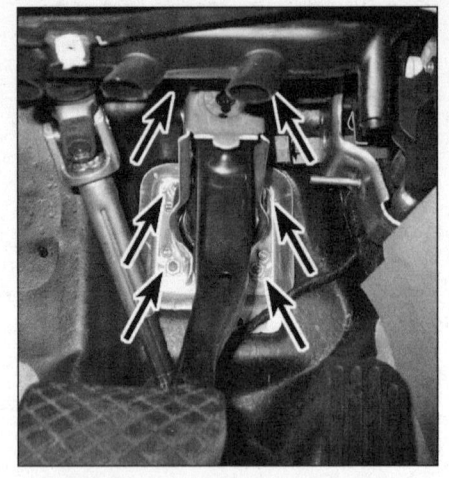

15.5 Brake pedal bracket mounting nut locations (top nuts only visible from under the instrument panel)

16 On models equipped with automatic transaxles, clamp off and disconnect the transaxle cooler line (see Chapter 7B).

17 Working inside of the vehicle, remove the lower trim panel and the booster mounting nuts **(see illustration)**. Carefully remove the booster out from the cowl compartment. Separate the master cylinder from the booster.

Installation

18 Assemble the master cylinder to the power brake booster. Be sure to match the pushrod to the master cylinder correctly during assembly.

19 The remainder of the installation procedures are essentially the reverse of removal.

  a) Tighten all mounting fasteners to the torque values listed in this Chapter's Specifications.
  b) To bleed the master cylinder on the vehicle, have an assistant pump the brake pedal several times slowly, then hold the pedal to the floor. Loosen the line fittings one at a time to allow air and fluid to escape. Repeat this procedure on both fittings until the fluid is clear of air bubbles. Have plenty of rags on hand to catch the fluid - brake fluid will ruin painted surfaces.
  c) Bleed the brake system (see Section 12) and test the operation of the brakes before putting the vehicle into normal service.

Warning: *Do not operate the vehicle if you are in doubt about the effectiveness of the brake system. It's possible for air to become trapped in the ABS hydraulic control unit, so, if the pedal continues to feel spongy after repeated bleedings or the BRAKE or ABS light stays on, have the vehicle towed to a dealer service department or other qualified repair shop to be bled.*

Note: *On models with a manual transaxle, the clutch system will also need bleeding after the brake system has been bled (see Chapter 8).*

14 Parking brake - check and adjustment

Check

1 Pulling up the parking brake lever five clicks should engage the parking brake fully. If the number of clicks is much less, there's a chance the parking brake might not be releasing completely resulting in brake drag. If the number of clicks is much more, the parking brake may not hold the vehicle on an incline.

2 One method of checking the parking brake is to park the vehicle on a steep hill with the parking brake set and the transmission in Neutral (be sure to stay in the vehicle for this check!). If the parking brake cannot prevent the vehicle from rolling, it's in need of adjustment.

Adjustment

Note: *The rear brakes are self-adjusting, and do not require normal maintenance adjustments. Adjustment is only required after replacing brake drums, brake shoes, wheel cylinders or parking brake cables.*

3 Block the front wheels, raise the rear of the vehicle and support it securely on jackstands. Remove the rear wheels.

4 Remove the center console (see Chapter 11, Section 27).

5 Apply the brake pedal several times, then release the pedal. Then apply the parking brake several times and release the parking brake to center the cables.

Rear drum brake models

6 With the parking brake set on the fourth notch of the ratchet mechanism, tighten the adjusting nut until it is difficult to turn both rear wheels. Once this is so, fully release the parking brake lever and check that the wheels/hub rotate freely. Check the adjustment by applying the brake fully and counting the clicks emitted from the parking brake ratchet. If necessary, re-adjust.

7 Once adjustment is correct, hold the adjusting nuts and securely tighten the locknuts.

8 Installation is the reverse of removal.

Rear disc brake models

9 Insert a 1/16-inch (1.5 mm) feeler gauge between each rear caliper lever and the stop **(see illustration)**.

10 Tighten the adjusting nut at the parking brake lever until there is a slight drag on the feeler gauges.

11 Remove the feeler gauges and set the parking brake fully. Release the parking brake and confirm the gap between the caliper lever and the stop is between 3/64 and 1/8-inch (1 to 3 mm).

12 Apply the parking brake and confirm the brake holds.

13 Release the parking brake and confirm that the brakes don't drag when the rear wheels are turned.

14 Installation is the reverse of removal.

15 Brake pedal - removal and installation

1 Remove the driver's side knee bolster for access (see Chapter 11).

2 Disconnect the brake light switch electrical connector (see Section 16), if equipped.

3 Disconnect the brake booster pushrod from the brake pedal (see Section 13).

4 Remove the pedal bracket support retaining nut, if equipped.

5 Loosen the nuts securing the pedal support bracket to the firewall/booster **(see illustration)**, enough to allow the bracket some movement. Do not remove the nuts completely.

6 Remove the pivot shaft nut **(see illustration)** and slide the pivot shaft to the left, until the pedal is free. Remove the pedal and pivot bushings.

7 Installation is the reverse of removal.

Chapter 9 Brakes   9-23

15.6 Location of the pedal pivot shaft nut

16.3 Typical brake light switch location

16 Brake light switch - removal and installation

1 Remove the engine cover (see Chapter 1, Section 7).
2 Remove the air inlet hoses (*see Air filter - removal and installation* in Chapter 4).
3 Disconnect the brake light switch electrical connector **(see illustration)**.
4 Remove the brake light switch mounting fastener.
5 Slide the brake light switch down and out from the retaining tab.
6 Install the switch and tighten the retaining fastener securely. The switch will self-adjust by design when installed correctly.
7 Connect the electrical connector. Confirm that the brake lights are operating properly.
8 Install the air inlet hoses and engine cover.

17 Vacuum pump (electric) - removal and installation

Note: *If the battery is disconnected, several systems must be re-learned before they will work properly (see Chapter 5, Section 3).*
Note: *The vacuum pump is located on the front side of the transaxle on some models, and on the rear side of the transaxle on other models.*
1 Disconnect the cable from the negative battery terminal (see Chapter 5).
2 On models with a front-mounted vacuum pump, remove the engine cover. On models with a rear-mounted vacuum pump, raise the front of the vehicle and support it securely on jackstands, then remove the shield from above the left driveaxle.
3 Disconnect the vacuum line from the pump **(see illustration)**.
4 Disconnect the vacuum pump electrical connector and the harness connectors, and set the harness to the side.

5 Remove the vacuum pump mounting bracket retaining bolts and remove the pump and bracket as an assembly.
6 Remove the vacuum pump-to-bracket retaining fasteners and separate the pump from the bracket.
7 Installation is the reverse of removal.

18 Vacuum pump (mechanical) - removal and installation

Note: *If the battery is disconnected, several systems must be re-learned before they will work properly (see Chapter 5, Section 3).*
1 Relieve the fuel system pressure (see Chapter 4), then disconnect the cable from the negative terminal of the battery (see Chapter 5, Section 3).
2 Remove the engine cover/air filter housing (see Chapter 1, Section 7).
3 Detach the vacuum line from its brackets, then detach the line from the pump.
4 Remove the coolant line bracket bolts, then reposition the coolant line out of the way. **Caution:** *Be careful not to bend the coolant line.*
5 Remove the high-pressure fuel pump (four-cylinder models; see Chapter 4).
6 Remove the three bolts and detach the vacuum pump from the cylinder head.
7 Clean the gasket mating surface on the engine (and the pump, if the same one is to be reinstalled).
8 Install a new gasket to the vacuum pump (RTV sealant can be used to hold it in place).
9 Install the pump onto the engine, making sure the drive dogs on the pump shaft engage with the slots in the sprocket.
10 Install the bolts and tighten them to the torque listed in this Chapter's Specifications.
11 The remainder of installation is the reverse of the removal procedure.

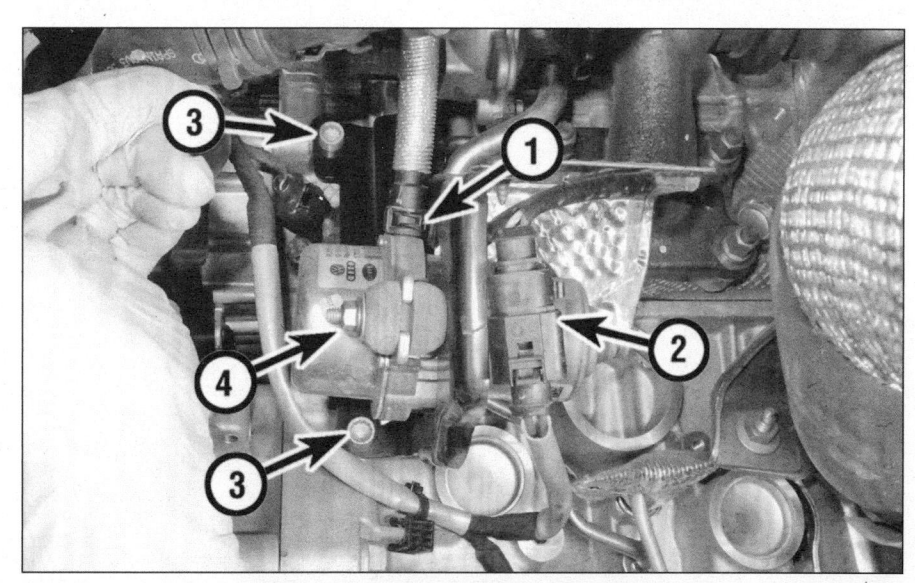

17.3 Vacuum pump details (rear-mounted pump)

1 Vacuum hose connector
2 Electrical connector
3 Mounting bracket bolts
4 Vacuum pump-to-bracket nut

Notes

Chapter 10
Suspension and steering systems

Contents

	Section		Section
Balljoints - check and replacement	8	Stabilizer bar and bushings (rear, multi-link suspension) -	
Coil spring (rear) - removal and installation	10	removal and installation	14
Control arm (front) - removal, bushing replacement		Steering column - removal and installation	19
and installation	7	Steering gear - removal and installation	22
General information and precautions	1	Steering gear boots - replacement	21
Hub and wheel bearing (front) - removal and installation	5	Steering knuckle - removal and installation	4
Hub and wheel bearing (rear) - removal and installation	11	Steering wheel - removal and installation	18
Knuckle (rear, multi-link suspension) - removal and installation	12	Strut assembly - removal, inspection and installation	2
Power steering pump - removal and installation	23	Strut/coil spring assembly - replacement	3
Power steering system - bleeding	24	Stub axle (rear, torsion beam suspension) -	
Rear axle beam (torsion bear suspension) - removal		removal and installation	13
and installation	16	Subframe - removal and installation	17
Rear suspension arms - removal and installation	15	Tie-rod ends - removal and installation	20
Shock absorber (rear) - removal and installation	9	Wheel alignment - general information	26
Stabilizer bar and bushings (front) - removal and installation	6	Wheels and tires - general information	25

Specifications

General

Power steering fluid type .... See Chapter 1

Caution: *All suspension fasteners going through rubber bushings should be tightened to the specified torque at ride height (or simulated ride height) and NOT while supported on jackstands with the suspension hanging free, unless noted in the procedure for that particular component.*

Torque specifications   Ft-lbs (unless otherwise indicated)   Nm

Note: *One foot-pound (ft-lb) of torque is equivalent to 12 inch-pounds (in-lbs) of torque. Torque values below approximately 15 ft-lbs are expressed in inch-pounds, because most foot-pound torque wrenches are not accurate at these smaller values.*

Front suspension

Driveaxle/hub bolt* .... See Chapter 8
Stabilizer bar
   Jetta
      Link nuts* 48   65
      Stabilizer bar clamp-to-subframe bolts*
         Step 1 15   20
         Step 2 .... Tighten an additional 90-degrees (1/4-turn)
   Golf
      Link nuts* 15   20
      Stabilizer bar clamp-to-subframe bolts
         Step 1 15   20
         Step 2 .... Tighten an additional 180-degrees (1/2-turn)

Warning: * *Replace fasteners after removing - do not reuse the old fasteners.*

10-2   Chapter 10 Suspension and steering systems

Torque specifications (continued)   **Ft-lbs (unless otherwise indicated)**   **Nm**

Note: One foot-pound (ft-lb) of torque is equivalent to 12 inch-pounds (in-lbs) of torque. Torque values below approximately 15 ft-lbs are expressed in inch-pounds, because most foot-pound torque wrenches are not accurate at these smaller values.

Front suspension (continued)

Control arm-to-subframe fasteners*
   Jetta
      2014 and earlier models
         Step 1 .... 52 70
         Step 2 .... Tighten an additional 90-degrees (1/4-turn)
      2015 and later models
         Step 1 .... 52 70
         Step 2 .... Tighten an additional 180-degrees (1/2-turn)
   Golf
      Step 1 .... 52 70
      Step 2 .... Tighten an additional 180-degrees (1/2-turn)
Balljoint
   Jetta
      Balljoint-to-control arm nuts
         2015 and earlier models 74 100
         2016 and later models
            Step 1 .... 29.5 40
            Step 2 .... Tighten an additional 45-degrees (1/8-turn)
      Balljoint-to-steering knuckle nut 44 60
   Golf
      Balljoint-to-control arm nuts
         Step 1 .... 29.5 40
         Step 2 .... Tighten an additional 45-degrees (1/8-turn)
      Balljoint-to-steering knuckle nut 44 60
Strut
   Strut-to-steering knuckle nut and bolt*
      Jetta
         Step 1 .... 52 70
         Step 2 .... Tighten an additional 90-degrees (1/4-turn)
      Golf
         Step 1 .... 52 70
         Step 2 .... Tighten an additional 180-degrees (1/2-turn)
   Strut upper mounting bolts*
      Step 1 .... 132 in-lbs 15
      Step 2 .... Tighten an additional 90-degrees (1/4-turn)
   Strut damper shaft nut* 44 60
Subframe
   Subframe-to-body bolts*
      Jetta
         Step 1 .... 37 50
         Step 2 .... Tighten an additional 90-degrees (1/4-turn)
      Golf
         Step 1 .... 52 70
         Step 2 .... Tighten an additional 180-degrees (1/2-turn)
Pendulum support-to-subframe bolt*
   Jetta
      2015 and earlier models
         Step 1 .... 74 100
         Step 2 .... Tighten an additional 90-degrees (1/4-turn)
      2016 and later models
         Step 1 .... 96 130
         Step 2 .... Tighten an additional 90-degrees (1/4-turn)
   Golf
      Step 1 .... 96 130
      Step 2 .... Tighten an additional 90-degrees (1/4-turn)
Pendulum support-to-transmission bolts*
   Step 1 .... 37 50
   Step 2 .... Tighten an additional 90-degrees (1/4-turn)
Wheel bearing/hub assembly-to-steering knuckle bolts*
   Step 1 .... 52 70
   Step 2 .... Tighten an additional 90-degrees (1/4-turn)

Warning: * *Replace fasteners after removing - do not reuse the old fasteners.*

Chapter 10 Suspension and steering systems 10-3

Torque specifications (continued)	Ft-lbs (unless otherwise indicated)	Nm

Rear suspension

Multi-link rear suspension
  Lower control arm-to-subframe bolt/nut* .... 70 / 95
  Lower control arm-to-knuckle bolt/nut*
    2011 (A6) and 2012 models
      Step 1 .... 66 / 90
      Step 2 .... Tighten an additional 90-degrees (1/4-turn)
    2013 and later models
      Step 1 .... 52 / 70
      Step 2 .... Tighten an additional 180-degrees (1/2-turn)
  Upper control arm-to-subframe bolt/nut* .... 70 / 95
  Upper control arm-to-knuckle bolt/nut*
    2014 and earlier models
      Step 1 .... 96 / 130
      Step 2 .... Tighten an additional 90-degrees (1/4-turn)
    2015 and later models
      Step 1 .... 96 / 130
      Step 2 .... Tighten an additional 180-degrees (1/2-turn)
  Tie-rod-to-subframe bolt/nut*
    Step 1 .... 52 / 70
    Step 2 .... Tighten an additional 180-degrees (1/2-turn)
  Tie-rod-to-knuckle bolt/nut*
    Jetta
      Step 1 .... 96 / 130
      Step 2
        2014 and earlier models .... Tighten an additional 90-degrees (1/4-turn)
        2015 and later models .... Tighten an additional 180-degrees (1/2-turn)
    Golf
      Step 1 .... 52 / 70
      Step 2 .... Tighten an additional 180-degrees (1/2-turn)
  Trailing arm bracket-to-body bolts*
    Step 1 .... 37 / 50
    Step 2 .... Tighten an additional 45-degrees (1/8-turn)
  Trailing arm-to-bracket bolt*
    Step 1 .... 66 / 90
    Step 2 .... Tighten an additional 90-degrees (1/4-turn)
  Trailing arm-to-knuckle bolts*
    Jetta
      Step 1 .... 66 / 90
      Step 2 .... Tighten an additional 45-degrees (1/8-turn)
    Golf
      Step 1 .... 52 / 70
      Step 2 .... Tighten an additional 45-degrees (1/8-turn)
  Stabilizer bar bracket bolts*
    2014 and earlier models
      Step 1 .... 18 / 25
      Step 2 .... Tighten an additional 45-degrees (1/8-turn)
    2015 and later models
      Step 1 .... 15 / 20
      Step 2 .... Tighten an additional 90-degrees (1/4-turn)
  Stabilizer bar link nuts* .... 33 / 45
  Stabilizer bar link-to-lower control arm bolt/nut (Golf)*
    Step 1 .... 15 / 20
    Step 2 .... Tighten an additional 180-degrees (1/2-turn)
  Shock absorber upper mounting bolts*
    Step 1 .... 37 / 50
    Step 2 .... Tighten an additional 45-degrees (1/8-turn)
  Shock absorber lower mounting fastener(s)*
    Jetta (bolt)
      2014 and earlier models .... 132 / 180
      2015 and later models
        Step 1 .... 96 / 130
        Step 2 .... Tighten an additional 90-degrees (1/4-turn)
    Golf (bolt/nut)
      Step 1 .... 52 / 70
      Step 2 .... Tighten an additional 180-degrees (1/2-turn)
  Shock absorber damper shaft nut* .... 18 / 25

Warning: * *Replace fasteners after removing - do not reuse the old fasteners.*

10-4   Chapter 10   Suspension and steering systems

Torque specifications (continued)   Ft-lbs (unless otherwise indicated)   Nm

Note: *One foot-pound (ft-lb) of torque is equivalent to 12 inch-pounds (in-lbs) of torque. Torque values below approximately 15 ft-lbs are expressed in inch-pounds, because most foot-pound torque wrenches are not accurate at these smaller values.*

Rear suspension (continued)

Subframe-to-body bolts*
- Jetta
  - 2014 and earlier models
    - Step 1 66   90
    - Step 2 .... Tighten an additional 90-degrees (1/4-turn)
    - Step 3 .... Loosen 360-degrees (1-turn)
    - Step 4 66   90
    - Step 5 .... Tighten an additional 180-degrees (1/2-turn)
  - 2015 and later models
    - Step 1 52   70
    - Step 2 .... Tighten an additional 180-degrees (1/2-turn)
- Golf
  - Long bolts
    - Step 1 52   70
    - Step 2 .... Tighten an additional 180-degrees (1/2-turn)
  - Short bolts
    - Step 1 37   50
    - Step 2 .... Tighten an additional 45-degrees (1/8-turn)

Torsion beam rear suspension
- Rear axle beam-to-bracket pivot bolts/nuts*
  - Jetta
    - Step 1 52   70
    - Step 2 .... Tighten an additional 90-degrees (1/4-turn)
  - Golf
    - Step 1 52   70
    - Step 2 .... Tighten an additional 360-degrees (1-turn)
- Mounting bracket-to-body fasteners*
  - Step 1 37   50
  - Step 2 .... Tighten an additional 45-degrees (1/8-turn)
- Crossbrace fasteners (Jetta only)*
  - Step 1 52   70
  - Step 2 .... Tighten an additional 90-degrees (1/4-turn)

Rear wheel bearing hub-to-stub axle bolt*
- Jetta
  - Step 1 132   180
  - Step 2 .... Tighten an additional 180-degrees (1/2-turn)
- Golf
  - Step 1 147   200
  - Step 2 .... Tighten an additional 90-degrees (1/4-turn)

Stub axle-to-axle beam bolts*
- Step 1 22   30
- Step 2 .... Tighten an additional 90-degrees (1/4-turn)

Shock absorber lower mounting bolt/nut*
- Jetta
  - Step 1 30   40
  - Step 2 .... Tighten an additional 90-degrees (1/4-turn)
- Golf
  - Step 1 52   70
  - Step 2 .... Tighten an additional 180-degrees (1/2-turn)

Shock absorber upper mounting bolts*
- Step 1 37   50
- Step 2 .... Tighten an additional 45-degrees (1/8-turn)

Subframe bolts*
- Step 1 66   90
- Step 2 .... Tighten an additional 90-degrees (1/4-turn)
- Step 3 .... Loosen one full turn (360-degrees)
- Step 4 66   90
- Step 5 .... Tighten an additional 90-degrees (1/4-turn)

Warning: * *Replace fasteners after removing - do not reuse the old fasteners.*

Torque specifications (continued)

Steering

	Ft-lbs (unless otherwise indicated)	Nm
Steering column		
Mounting bolts	15	20
Universal joint pinch bolt/nut		
Jetta	22	30
Golf		
Step 1	15	20
Step 2	Tighten an additional 90-degrees (1/4-turn)	
Steering gear mounting bolts*		
Jetta		
Step 1	37	50
Step 2	Tighten an additional 90-degrees (1/4-turn)	
Golf		
Step 1	70	
Step 2	Tighten an additional 90-degrees (1/4-turn)	
Tie-rod end-to-knuckle nut*		
Step 1	15	20
Step 2	Tighten an additional 90-degrees (1/4-turn)	
Steering wheel bolt*		
Step 1	22	30
Step 2	Tighten an additional 90-degrees (1/4-turn)	
Power steering pump		
Pulley mounting bolts	16	22
Pump-to-bracket bolts	16	22
Pump bracket mounting bolts	16	22
Pressure line fitting	23.5	32

Wheels

Wheel bolts	See Chapter 1	

Warning: *Replace fasteners after removing - do not reuse the old fasteners.*

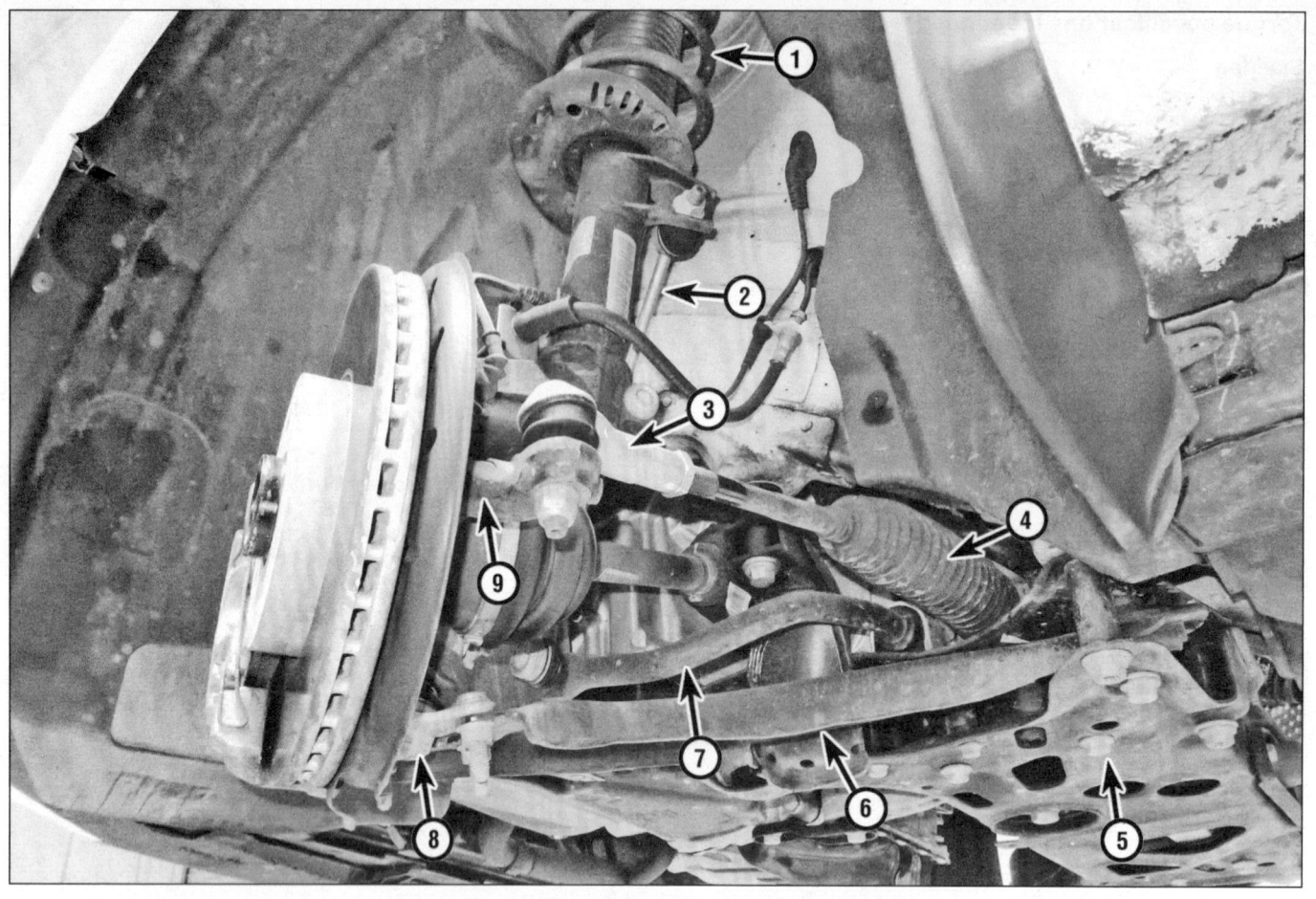

10-6   Chapter 10   Suspension and steering systems

1.1 Front suspension and steering components

1. Strut and spring assembly
2. Stabilizer bar link
3. Tie-rod end
4. Steering gear boot
5. Subframe
6. Control arm
7. Stabilizer bar
8. Balljoint
9. Steering knuckle

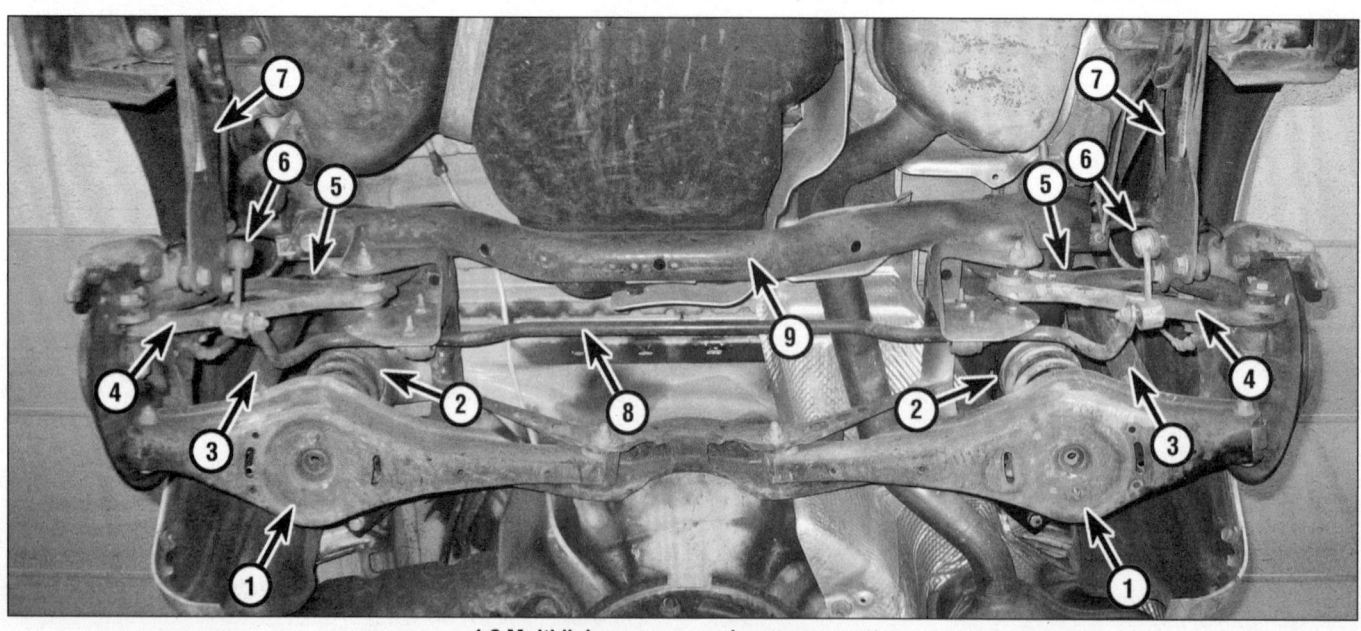

1.2 Multi-link rear suspension components

1. Lower control arm
2. Coil spring
3. Shock absorber
4. Tie-rod
5. Upper control arm
6. Stabilizer bar link
7. Trailing arm
8. Stabilizer bar
9. Subframe

Chapter 10 Suspension and steering systems

1 General information and precautions

Front suspension

1   The front suspension is a MacPherson strut design. The upper end of each strut is attached to the vehicle's body strut support. The lower end of the strut is connected to the upper end of the steering knuckle. The steering knuckle is attached to a balljoint mounted on the outer end of the suspension control arm **(see illustration)**.

Rear suspension

2   The rear suspension on all except some 2011 and later Jetta models consists of a multi-link suspension which incorporates upper and lower control arms (or transverse links), trailing arms, knuckle and hub assemblies, rear tie-rods connected to the knuckles, shock absorbers, coil springs and a stabilizer bar **(see illustration)**. On some 2014 and earlier Jetta models, a solid beam rear axle, shock absorbers and coil springs are used.

Steering

3   The steering column is connected to the steering gear by a universal joint.
4   The steering gear is mounted on the front subframe and is connected by two tie-rods, with balljoints at their outer ends, to the steering arms projecting rearwards from the steering knuckles.
5   On some models, power steering is provided by a hydraulic power steering pump mounted on the engine. Other models use an eletromechanical power steering system, in which an electric motor mounted on the steering gear provides power assistance.

Ride height and ADAS

6   Advanced Driver Assistance Systems (ADAS) is the latest technology developed to automate, adapt or enhance vehicle systems for safety and to increase driver awareness and control. Today's vehicles utilize various satellite, radar, cameras and sensors to determine the conditions and area in which the vehicle is. Given the situation, the ADAS system can control the braking force (either in reverse or forward) to avoid contact with an object that is in its path.
7   With the electronic steering systems now in place on some vehicles, the ADAS system is also capable of slight steering corrections while in cruise control mode to maintain the vehicle in its lane, to parallel parking assistance.
8   Ride height is one of those critical factors for camera and sensor adjustment. Anything from under-inflated tires, the wrong size (diameter) tire, faulty suspension components, weak springs, shocks, heavily loaded trunk, to a cracked windshield can all contribute to a malfunctioning or improperly aligned camera system.
Warning: *It is essential to maintain the factory ride height and stance to insure the safety features built into the vehicle will perform correctly.*

Precautions

9   Frequently, when working on the suspension or steering system components, you may come across fasteners which seem impossible to loosen. These fasteners on the underside of the vehicle are continually subjected to water, road grime, mud, etc., and can become rusted or frozen, making them extremely difficult to remove. In order to unscrew these stubborn fasteners without damaging them (or other components), be sure to use lots of penetrating oil and allow it to soak in for a while. Using a wire brush to clean exposed threads will also ease removal of the nut or bolt and prevent damage to the threads.
10   Sometimes a sharp blow with a hammer and punch will break the bond between a nut and bolt threads, but care must be taken to prevent the punch from slipping off the fastener and ruining the threads. Heating the stuck fastener and surrounding area with a torch sometimes helps too, but isn't recommended because of the obvious dangers associated with fire. Long breaker bars and extension, or cheater, pipes will increase leverage, but never use an extension pipe on a ratchet - the ratcheting mechanism could be damaged. Sometimes tightening the nut or bolt first will help to break it loose.
Warning: *Fasteners that require drastic measures to remove should always be replaced with new ones. Many of the fasteners (nuts and bolts) that are used to mount the suspension components are self-locking (prevailing torque) and designed to be used only once. The manufacturer requires replacement of self-locking fasteners whenever they are loosened or removed.*
11   Since most of the procedures dealt with in this Chapter involve jacking up the vehicle and working underneath it, a good pair of jackstands will be needed. A hydraulic floor jack is the preferred type of jack to lift the vehicle, and it can also be used to support certain components during various operations, but not as the only type of support. Always install jackstands before crawling under a vehicle.
Warning: *Never, under any circumstances, rely on a jack to support the vehicle while working on it. Whenever any of the suspension or steering fasteners are loosened or removed they must be inspected and, if necessary, replaced with new ones of the same part number or of original equipment quality and design. Torque specifications must be followed for proper reassembly and component retention. Never attempt to heat or straighten any suspension or steering components. Instead, replace any bent or damaged part with a new one.*
Warning: *Vehicles equipped with the Start/Stop System can restart at any time if the key has been left in the on position during any maintenance or repairs. Be sure to deactivate the Start/Stop system before performing any repairs. A message will be displayed in the instrument cluster indicating the Start/Stop has been disabled.*

2 Strut assembly - removal, inspection and installation

Warning: *Always replace the struts in pairs - never replace just one of them, as handling peculiarities may result.*

Removal

1   Remove the wheel trim/hub cap and loosen the driveaxle/hub bolt (no more than a 90-degree turn) and the wheel bolts with the vehicle resting on its wheels.
2   Chock the rear wheels of the car, firmly apply the parking brake, then raise the front of the vehicle and support it securely on jackstands. Remove the wheel and driveaxle/hub bolt.
3   Disconnect the stabilizer bar link from the strut **(see illustration)**.

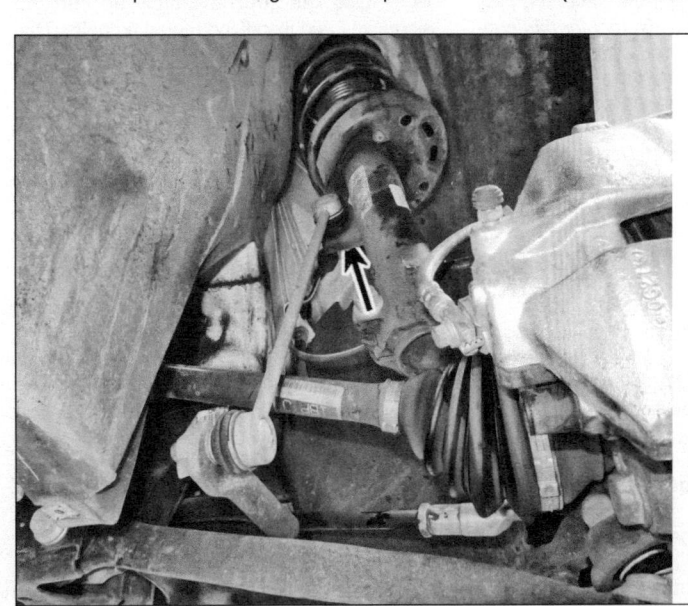

2.3 Location of the stabilizer bar link-to-strut nut

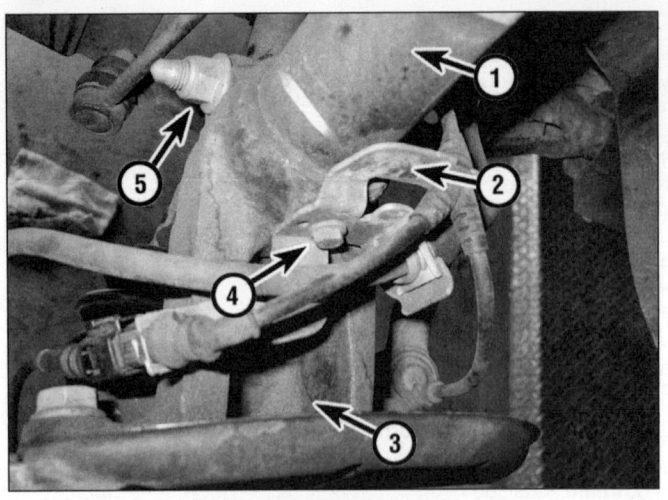

10-8 Chapter 10 Suspension and steering systems

2.4 Strut mounting details

1 Strut	3 Steering knuckle	5 Strut-to-knuckle
2 Bracket	4 Bracket bolt	pinch bolt/nut

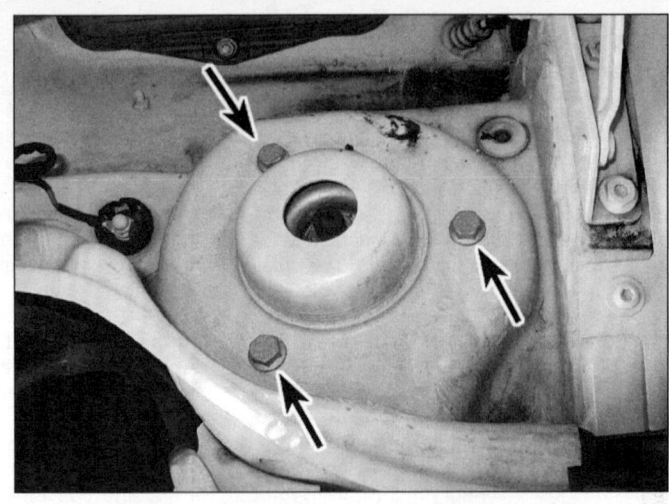

2.10 Remove the strut upper mounting fasteners

4 Remove the fastener and bracket that holds the brake hose and ABS sensor harness connector to the strut **(see illustration)**.
5 Remove the strut-to-knuckle nut and knock the bolt out with a hammer and punch.
6 Remove the balljoint-to-control arm nuts and separate the control arm from the balljoint (see Section 8).
Note: Disconnect the front level control sensor arm-to-control arm nut and separate the sensor arm from the control arm, if equipped.
7 Separate the strut from the steering knuckle (see Chapter 8) and support the driveaxle.
Note: The shock tube can become frozen in the steering knuckle from rust and debris. There is a slit on the back side of the steering knuckle where the strut tube is fitted into the steering knuckle. Using some penetrating oil (or even just plain water) and by wedging a large screwdriver along the slit you can slightly widen the opening so that the strut tube can slide out. There is a specific tool for this purpose, but a screwdriver works just as well. Do not jab the screwdriver into the strut tube.
Note: The shock tube can become lodged in the steering knuckle or fail to completely be seated into place when installing because the angle of the steering knuckle and the angle of the tube are skewed (canted) causing the strut tube to bind in the knuckle. Make sure to note the depth of the strut tube in the steering knuckle before removing it and reinstall it at the same depth.
Caution: The driveaxle must be supported once it's removed from the knuckle or damage may occur to the inner CV joint. Use a length of wire to support the joint.
8 Place a floor jack under the steering knuckle to support it. Raise the jack just enough to carry the load without lifting the vehicle.
9 Remove the cowl panel (see Chapter 11).
10 Support the strut and spring assembly with one hand and remove the three strut-to-shock tower fasteners **(see illustration)**. Guide the strut assembly from the fenderwell.

Inspection
11 Check the strut body for leaking fluid, dents, cracks and other obvious damage that would warrant repair or replacement.
12 Check the coil spring for chips or cracks in the spring coating (this can cause premature spring failure due to corrosion). Inspect the spring seat for cuts, hardness and general deterioration.
13 If any undesirable conditions exist, proceed to the strut disassembly procedure (see Section 4).

Installation
Note: You may find it easier to install the strut assembly with the help of an assistant, as it can be quite heavy and awkward to hold in place by yourself.
14 Guide the strut assembly up into the fenderwell and install the mounting bolts. This is most easily accomplished with the help of an assistant, as the strut is quite heavy and awkward.
Note: There are two arrows on the top of the strut bearing retainer; be sure to align one of the two arrows pointing towards the front of the car.
15 Slide the strut tube into the steering knuckle until it is fully seated (there is a small alignment notch or installation line on the strut tube to indicate when it is fully inserted into the steering knuckle), then insert the bolt with the pointed (threaded) end facing forward. Install the nut and tighten it to the torque listed in this Chapter's Specifications.
16 Reattach the brake hose and ABS speed sensor wiring harness bracket to the strut.
17 Install the driveaxle into the hub splines and install a new driveaxle/hub bolt, tightening it securely. Connect the control arm to the balljoint. Tighten the balljoint-to-control arm fasteners to the torque listed in this Chapter's Specifications.
18 Tighten the three upper mounting bolts to the torque listed in this Chapter's Specifications.
19 Install the wheel and lower the vehicle to the ground.
20 Tighten the driveaxle/hub bolt to the torque listed in the Chapter 8 Specifications, then tighten the wheel bolts to the torque listed in the Chapter 1 Specifications.
21 Reinstall the cowl panel.

3 Strut/coil spring assembly - replacement

Warning: *Always replace the struts or coil springs in pairs - never replace just one of them, as handling peculiarities may result.*
Warning: *Before attempting to disassemble the strut/coil spring assembly, obtain a tool to hold the coil spring in compression. Do not attempt to use makeshift methods. Uncontrolled release of the spring could cause damage and personal injury or even death. Use a high-quality spring compressor, and carefully follow the tool manufacturer's instructions provided with it. After removing the coil spring with the compressor still installed, place it in a safe, isolated area.*

1 If the front suspension strut/coil springs exhibit signs of wear (leaking fluid, loss of damping capability, sagging or cracked coil springs), they should be disassembled and overhauled as necessary. The struts themselves cannot be serviced, and should be replaced if faulty; the springs and related components can be replaced individually. To maintain balanced characteristics on both sides of the vehicle, the components on both sides should be replaced at the same time.
2 With the assembly removed from the vehicle (see Section 2), clean away all external dirt, then mount the assembly in a vise.

Chapter 10 Suspension and steering systems

3.3 Make sure the spring compressor tool is on securely

3.4 Loosen and remove the piston rod nut

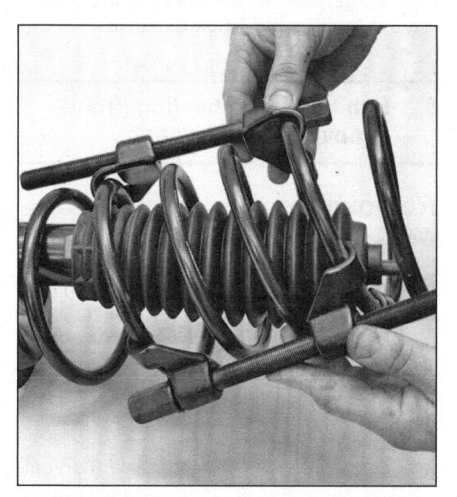

3.5a Remove the upper mount and spring seat, then carefully remove the spring . . .

3.5b . . . followed by the boot . . .

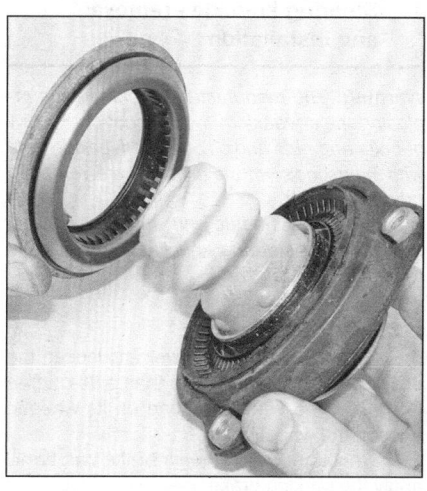

3.5c . . . and the spring seat bearing

3 Install the coil spring compressor tools (ensuring that they are fully engaged), and compress the spring until all tension is relieved from the upper mount **(see illustration)**.

4 Hold the strut piston rod with an Allen key and unscrew the thrust bearing retaining nut with a box-end wrench **(see illustration)**.

5 Withdraw the top mount, upper spring seat and spring, followed by the boot, bearing and bump stop **(see illustrations)**.

6 If a new spring is to be installed, the original spring must now be carefully released from the compressor. If it is to be re-used, the spring can be left in compression.

7 With the strut assembly now completely disassembled, examine all the components for wear and damage, and check the bearing for smoothness of operation. Replace components as necessary.

8 Examine the strut for signs of fluid leakage. Check the piston rod for signs of pitting along its entire length, and check the strut body for signs of damage. Test the operation of the strut, while holding it in an upright position, by moving the piston through a full stroke, then through short strokes of 2 to 4 inches. In both cases, the resistance felt should be smooth and continuous. If the resistance is jerky or uneven, or if there is any visible sign of wear or damage, replacement is necessary.

9 Reassembly is the reverse of disassembly, noting the following points:

a) The coil springs must be installed with the paint mark at the bottom.
b) Make sure that the coil spring ends are correctly located in the upper and lower seats before releasing the compressor **(see illustration)**.
c) Tighten the piston rod nut to the specified torque.

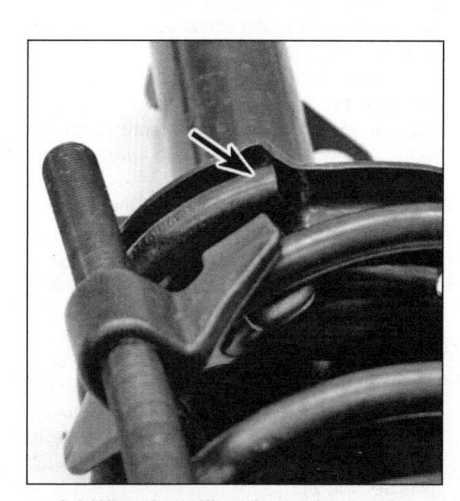

3.9 When installing the spring, make sure the ends fit into the recessed portion of the seats

10-10 Chapter 10 Suspension and steering systems

5.6 Remove the disc shield mounting bolts

5.7 Front hub and bearing mounting bolts

4 Steering knuckle - removal and installation

Warning: *The manufacturer recommends replacing the driveaxle/hub bolt, control arm pivot bolts and nuts and all other self-locking nuts with new ones whenever they are removed (see Section 1).*

Note: *Remove the hub and wheel bearing assembly if the steering knuckle is going to be replaced (see Section 5).*

Removal

1 Remove the wheel cover and loosen the driveaxle/hub bolt (no more than a 90-degree turn) with the vehicle resting on its wheels. Loosen the wheel bolts.
2 Chock the rear wheels of the car, firmly apply the parking brake, then raise the front of the car and support it securely on jackstands. Remove the front wheel.
3 Remove the driveaxle/hub bolt.
4 Remove the ABS wheel speed sensor as described in Chapter 9.
5 Remove the brake caliper (don't disconnect the hose) and brake disc (see Chapter 9). Using a piece of wire or string, tie the caliper to the coil spring - don't let the caliper hang by the hose.
6 Detach the tie-rod end from the steering knuckle (see Section 20).
7 Detach the balljoint from the control arm (see Section 8). Loosen the balljoint-to-steering knuckle nut a few turns, then separate the balljoint from the knuckle (see Section 8).
Note: *Disconnect the front level control sensor arm-to-control arm nut and separate the sensor arm from the control arm, if equipped.*
8 Remove the strut-to-knuckle nut and knock the bolt out with a hammer and punch (see Section 2).
Note: *It may be necessary to spread open the gap where the strut slides into the knuckle.*
9 Carefully pull the knuckle assembly outwards while pushing the driveaxle from the hub. If necessary, tap the CV joint out of the hub using a soft-faced hammer. If this fails to free it from the hub, the joint will have to be pressed out using a puller. Support the driveaxle with wire or equivalent and never let it hang.
Warning: *Be careful not to overextend the inner CV joint.*

Installation

10 Install the balljoint to the knuckle and tighten the new nut to the torque listed in this Chapter's Specifications.
11 Lubricate the splines of the driveaxle with multi-purpose grease.
12 Maneuver the knuckle/hub assembly into position and engage it with the driveaxle stub shaft. Install a new driveaxle/hub bolt, but don't attempt to tighten it yet.
13 Connect the strut to the knuckle, pushing the strut into the bore as far as possible. Install the new nut and bolt, and tighten the bolt and nut to the torque listed in this Chapter's Specifications.
14 Connect the control arm to the balljoint. Install new nuts and tighten them to the torque listed in this Chapter's Specifications.
15 Engage the tie-rod end with the steering knuckle, then install the new retaining nut. Tighten the nut to the torque listed in this Chapter's Specifications.
16 Install the brake disc and caliper, tightening the caliper mounting bracket bolts (if equipped) and caliper guide pins/mounting bolts to the torque listed in the Chapter 9 Specifications.
17 Tighten the driveaxle/hub bolt as securely as possible.
18 Install the ABS wheel speed sensor as described in Chapter 9.
19 Install the wheel and lower the vehicle to the ground.
20 Tighten the driveaxle/hub bolt to the torque listed in Chapter 8 Specifications, then tighten the wheel bolts to the torque listed in the Chapter 1 Specifications.

5 Hub and wheel bearing (front) - removal and installation

Removal

Warning: *The manufacturer recommends replacing the driveaxle/hub bolt, control arm pivot bolts and nuts and all other self-locking nuts with new ones whenever they are removed (see Section 1).*
1 Remove the wheel trim/hub cap and loosen the driveaxle bolt and wheel bolts no more than a 90-degree turn with the vehicle resting on its wheels.
2 Loosen the wheel bolts, raise the vehicle and support it securely on jackstands. Remove the wheel.
3 Remove the brake caliper, the caliper mounting bracket and the brake disc from the hub (see Chapter 9).
Note: *Be sure to support the brake caliper with a length of wire or rope.*
4 Remove the wheel speed sensor (see Chapter 9).
5 Remove the driveaxle/hub bolt and tap the CV joint out of the hub as far as possible using a soft-faced hammer. If this fails to free it from the hub, the joint will have to be pressed out using a puller.
6 Remove the shield from the steering knuckle **(see illustration)**.
7 Remove the hub/bearing assembly mounting bolts from the rear of the steering knuckle **(see illustration)**.
Note: *Clean any debris out of the bolt head recesses before using your spline-drive tool.*
Note: *If the tool you are using cannot engage the fasteners with the driveaxle in place, remove the driveaxle.*
8 Remove the hub/bearing assembly from the steering knuckle and driveaxle. Be careful

Chapter 10 Suspension and steering systems

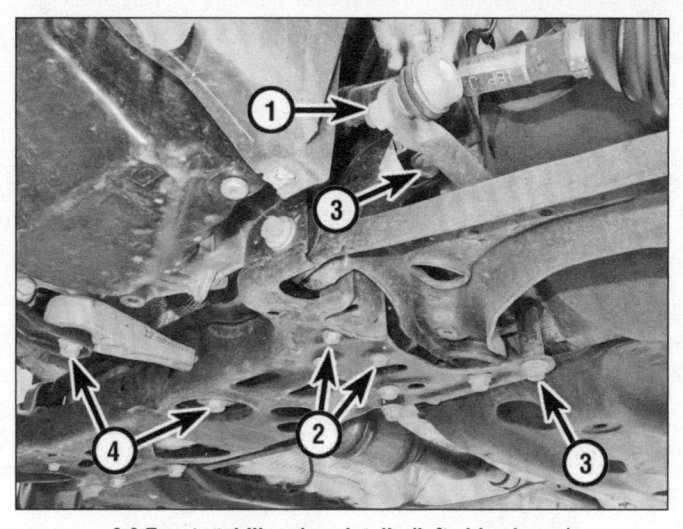

6.6 Front stabilizer bar details (left side shown)

1. Stabilizer bar link-to-bar nut
2. Stabilizer bar clamp bolts
3. Subframe bolts
4. Pendulum mount bolts (two of three shown)

7.3 Balljoint-to-control arm nuts

not to allow the inner CV joint of the driveaxle to overextend.

Installation

9 Make sure that the mounting surfaces inside the steering knuckle and on the driveaxle splines are smooth and free of burrs and nicks prior to installing the hub/bearing assembly.
10 Lubricate the driveaxle splines with multi-purpose grease, then guide the axle into the hub/bearing assembly as the hub/bearing is mated to the steering knuckle. Install a new driveaxle/hub bolt, but don't attempt to tighten it yet.
11 Tighten the hub/bearing bolts to the torque listed in this Chapter's Specifications.
12 Install the brake disc, the caliper mounting bracket and the caliper; tighten the fasteners to the torque values listed in the Chapter 9 Specifications.
13 Install the wheel, remove the jackstands and lower the vehicle.
14 Tighten the driveaxle/hub bolt to the torque listed in the Chapter 8 Specifications.
15 Tighten the wheel bolts to the torque listed in the Chapter 1 Specifications.

6 Stabilizer bar and bushings (front) - removal and installation

Note: *The manufacturer recommends replacing the stabilizer bar clamp bolts with new ones whenever they are removed (see Section 1).*

1 Raise the front of the vehicle and support it securely on jackstands.
2 Remove the lower splash shield below the engine (see Chapter 1, Section 6).
3 Remove the fasteners and detach the stabilizer bar links from the stabilizer bar (see Section 2).
4 Remove the pendulum mount (see Chapter 2A or 2B) and disconnect the steering column U-joint coupler from the steering gear (see Section 22).
5 Remove the exhaust pipe support bracket fasteners from the subframe.
6 Remove the stabilizer bar bushing clamp fasteners **(see illustration)**.
Note: *Disconnect the front level control sensor arm-to-control arm nut and separate the sensor arm from the control arm, if equipped.*
7 Place a floor jack under the subframe, remove the rear subframe fasteners and loosen the front subframe fasteners. Lower the subframe no more than three inches.
8 Detach the stabilizer bar bushing clamps and remove the stabilizer bar from the subframe.
9 Inspect the clamp bushings and the link bushings. If they're cracked, hardened or deteriorated in any way, replace them.
10 Installation is the reverse of the removal procedure. Tighten the fasteners to the torque values listed in this Chapter's Specifications.

7 Control arm (front) - removal, bushing replacement and installation

Warning: *The manufacturer recommends replacing the control arm pivot bolts and nuts, the balljoint nuts and the strut through-bolt/nut with new ones whenever they are removed.*

Removal

1 Remove the wheel trim/hub cap (as applicable) and loosen the driveaxle/hub bolt 1/4-turn with the vehicle resting on its wheels. Loosen the wheel bolts.
2 Raise the front of the vehicle and support it securely on jackstands. Block the rear wheels to prevent the vehicle from rolling off the stands. Remove the wheel and driveaxle/hub bolt.
3 Remove the balljoint-to-control arm nuts **(see illustration)**.
4 Pull the control arm down and separate the control arm from the balljoint.
Note: *Disconnect the front level control sensor arm-to-control arm nut and separate the sensor arm from the control arm, if equipped.*

Right side control arm

Note: *This procedure applies to both left and right control arms on vehicles equipped with a manual transaxle.*

5 Remove the control arm pivot bolt (horizontal) from the front of the arm **(see illustration 7.10)**.
6 Remove the control arm mounting bolt and nut (vertical) from the rear of the arm.
7 Maneuver the arm towards the rear of the vehicle, then pull the front of the control arm out and remove it from the vehicle.

Left side control arm

Note: *This procedure applies to the left control arm on vehicles equipped with an automatic or DSG transaxle.*

8 Remove the pendulum mount (see Chapter 2A or 2B) and disconnect the steering column U-joint coupler from the steering gear.
9 Remove the exhaust pipe support bracket fasteners from the subframe. Also detach the power steering pressure line from the transaxle.

7.10 Control arm mounting bolts

1 Front mounting bolt
2 Rear mounting bolt/nut

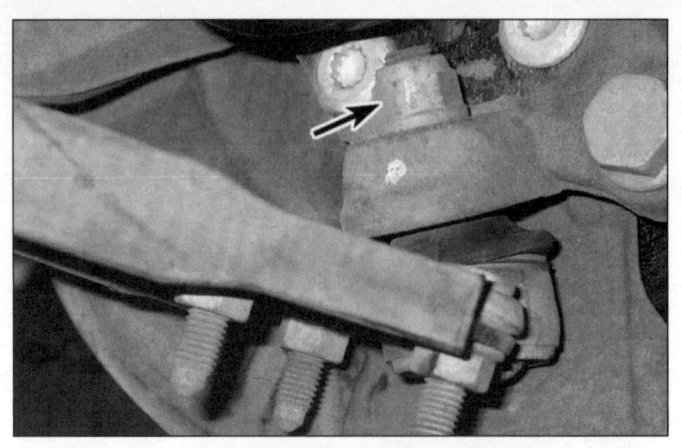

8.8 Loosen the balljoint nut a couple of turns but do not completely remove it at this time

10 Remove the control arm bolt (horizontal) at the front of the control arm and the nut and bolt (vertical) from the rear of the control arm **(see illustration)**.
11 To allow the pivot bolt to be withdrawn, the subframe must be lowered. To do this, support the subframe with a floor jack, unscrew and remove the two outermost subframe-to-body bolts, loosen the front subframe bolts and lower the rear of the subframe, but no more than 3-15/16 inches (10 cm).
12 Remove the control arm from the subframe.

Bushing replacement

13 Thoroughly clean the arm, removing all traces of dirt, thread locking compound and undercoating if necessary, then check carefully for cracks, distortion or any other signs of wear or damage, paying particular attention to the inner pivot bushings.
14 Replacement of the bushings will require the use of a hydraulic press and several spacers and is therefore best entrusted to an automotive machine shop. If such equipment is available, press out the old bushing and install the new one using a spacer which bears only on the bushing outer edge. Ensure the bushing is correctly positioned so that the cavities are aligned with the center axis of the arm.

Installation

15 Installation is the reverse of removal, noting the following points:
 a) Use new control arm and subframe fasteners.
 b) Raise the outer end of the control arm with a floor jack to simulate normal ride height before tightening the control arm-to-subframe fasteners.
 c) Tighten all suspension fasteners to the torque values listed in this Chapter's Specifications.
 d) Lower the vehicle and tighten the wheel bolts to the torque listed in the Chapter 1 Specifications.

 e) Have the front end alignment checked and, if necessary, adjusted.

8 Balljoints - check and replacement

Check

1 Inspect the control arm balljoints for looseness whenever either of them is separated from the steering knuckle. See if you can turn the ballstud in its socket with your fingers. If the balljoint is loose, or if the ballstud can be turned, replace the balljoint. You can also check the balljoints with the suspension assembled as follows.
2 Raise the front of the vehicle and support it securely on jackstands placed under the frame rails.
3 Place a floor jack under the front lower control arm and raise it slightly. Attempt to move the steering knuckle up and down; a large prybar underneath the tire, or a prybar placed between the end of the control arm and the steering knuckle will be helpful. If any play is felt, replace the balljoint.
4 Also, try to move the steering knuckle in-and-out by jerking the tire inwards and outwards while observing the balljoint attachment area of the control arm and steering knuckle. If any play is felt or seen, replace the balljoint.
5 Check the balljoint boot for cracks and tears. If any are present, replace the balljoint.

Replacement

6 Remove the wheel trim/hub cap (as applicable) and loosen the driveaxle/hub bolt 1/4-turn with the vehicle resting on its wheels. Also loosen the wheel bolts.
7 Raise the front of the vehicle and support it securely on jackstands. Block the rear wheels to prevent the vehicle from rolling off the stands. Remove the wheel and driveaxle/hub bolt.
Note: *There are two types of hub bolts used; always replace the driveaxle/hub bolt with the same type.*
8 Loosen the balljoint-to-steering knuckle nut **(see illustration)**, but don't remove it yet.
9 Remove the balljoint-to-control arm nuts (see Section 7).
10 Separate the control arm from the balljoint.
11 Push the driveaxle out of the hub far enough to install a small balljoint separator, then pop the balljoint stud from the steering knuckle **(see illustration)**.

8.11 Separate the balljoint from the steering knuckle (the nut has been loosened but not completely removed - this will prevent the components from separating violently)

Chapter 10 Suspension and steering systems

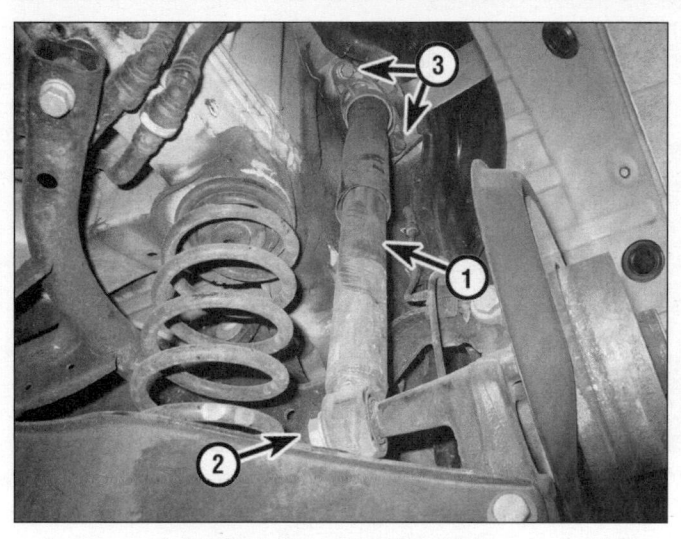

9.3 Rear shock absorber details (Jetta multi-link suspension shown)

1. Rear shock absorber
2. Lower mounting fastener
3. Upper mounting fasteners

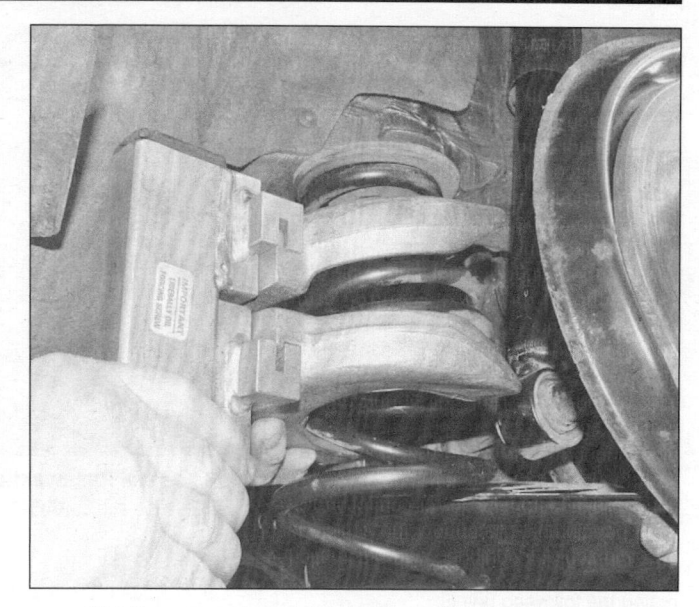

10.3 Compress the rear coil spring with the special tool

12 Unscrew the nut and remove the balljoint from the steering knuckle.
13 Install the new balljoint into the steering knuckle and install the nut, but don't tighten it completely yet.
14 Connect the balljoint to the lower arm and tighten the fasteners to the torque listed in this Chapter's Specifications. Also tighten the balljoint-to-steering knuckle nut to the torque listed in this Chapter's Specifications.
15 Install the wheel and bolts, lower the vehicle and tighten the wheel bolts to the torque listed in the Chapter 1 Specifications.
16 Install a new driveaxle/hub bolt and tighten it to the torque listed in the Chapter 1 Specifications.

9 Shock absorber (rear) - removal and installation

Warning: *The manufacturer recommends replacing the shock absorber lower mounting bolt and nut with new ones whenever they are removed.*
Warning: *Always replace the shock absorbers in pairs - never replace just one of them, as handling peculiarities may result.*

1 Loosen the rear wheel bolts. Chock the front wheels to keep the vehicle from rolling, then raise the rear of the vehicle and support it securely on jackstands. Remove the rear wheel.
2 On models with multi-link rear suspension, support the lower control arm with a floor jack positioned near the outer end of the arm. On models equipped with a torsion beam rear axle, place the floor jack below the coil spring pocket.
Note: *The jack must remain in position throughout the entire procedure.*

3 Remove the shock absorber lower mounting fastener **(see illustration)**.
4 Remove the shock absorber upper mounting fasteners and remove the shock absorber.
5 Guide the shock absorber into position and install the upper mounting bolts and a new lower mounting bolt (and nut, where applicable). Don't tighten the lower mounting fastener(s) yet.
6 Tighten the upper mounting bolts to the torque listed in this Chapter's Specifications.
7 Raise the rear suspension to simulate normal ride height, then tighten the lower mounting bolt/nut to the torque listed in this Chapter's Specifications.
8 Repeat the procedure to replace the other rear shock absorber.
9 Install the wheels and lower the vehicle. Tighten wheel bolts to the torque listed in the Chapter 1 Specifications.

10 Coil spring (rear) - removal and installation

Warning: *Always replace the coil springs in pairs - never replace just one of them.*

1 Loosen the wheel bolts. Chock the front wheels to prevent the vehicle from rolling, then raise the rear of the vehicle and support it securely on jackstands placed under the rocker panel flanges. Remove the wheels.
2 On models with multi-link rear suspension, disconnect the ride height sensor, if applicable.
3 Following the tool manufacturer's instructions, install a spring compressor (which can be obtained at most auto parts stores or equipment yards on a daily rental basis) on the spring and compress it enough to relieve the tension on the other suspension components **(see illustration)**. This can be verified by wiggling the spring.
Warning: *Removing a coil spring is potentially dangerous. Use only a high-quality spring compressor and carefully follow the manufacturer's instructions furnished with the tool. After removing the coil spring, set it aside in a safe, isolated area.*

4 If you're working on a model with multi-link rear suspension, support the lower control arm with a floor jack, then remove the shock absorber lower mounting bolt, detach the stabilizer bar link from the lower control arm, then disconnect the lower control arm from the rear knuckle (see Section 15). Lower the jack to make room for spring removal.
5 Remove the spring and the upper and lower spring seats. Check the spring for cracks and chips, replacing the springs as a set if any defects are found. Also check the upper and lower seats for damage and deterioration, replacing them as necessary.
Note: *If the lower spring seat is removed, it has a location pin that must be inserted back into the lower control arm or axle beam to center the seat.*

6 Installation is the reverse of the removal procedure, but make sure the lower spring seat is positioned with the pin on the seat inserted into the hole in the control arm or axle beam, and the coil springs positioned with the end of the spring up against the spring seat stop.
7 On models with multi-link suspension, raise the lower control arm with the floor jack to simulate normal ride height before tightening the suspension fasteners.
8 Lower the vehicle and tighten the wheel bolts to the torque listed in the Chapter 1 Specifications.

10-14   Chapter 10 Suspension and steering systems

11 Hub and wheel bearing (rear) - removal and installation

1   Loosen the rear wheel bolts, raise the rear of the vehicle and support it securely on jackstands. Remove the wheel.
2   On rear drum brake models, remove the brake drum (see Chapter 9).
3   On rear disc brake models, remove the brake caliper (don't detach the hose), mounting bracket and disc (see Chapter 9). Hang the caliper with a piece of wire - don't let it hang by the brake hose.
4   Remove the dust cap from the hub (see illustration).
Note: *Always replace the dust cap with a new one whenever it's removed.*
5   Unscrew the hub bearing bolt (see illustration) and remove the hub and bearing assembly.
6   Installation is the reverse of removal, noting the following points:
  a) *Make sure the mating surfaces on the knuckle or stub axle and the hub and bearing assembly are clean before installation.*
  b) *Tighten the mounting bolt to the torque listed in this Chapter's Specifications.*
  c) *Disc brake models: Install the brake caliper (see Chapter 9), tightening the mounting bolts to the torque listed in the Chapter 9 Specifications.*
  d) *Drum brake models: Install the brake drum and adjust the parking brake (see Chapter 9).*
  e) *Install the wheel and wheel bolts. Lower the vehicle and tighten the bolts to the torque listed in the Chapter 1 Specifications.*

12 Knuckle (rear, multi-link suspension) - removal and installation

Warning: *The manufacturer recommends replacing all fasteners whenever they are removed.*
1   Loosen the wheel bolts, raise the rear of the vehicle and support it securely on jackstands, then remove the wheel.
2   Remove the rear brake caliper, caliper mounting bracket and brake disc (see Chapter 9). Support the caliper with a length of wire - don't let it hang by the hose.
3   Remove the ABS rear wheel speed sensor (see Chapter 9).
4   Remove the rear coil spring (see Section 10).
5   On Jetta models, detach the shock absorber from the knuckle (see Section 9).
6   Detach the lower control arm from the knuckle (see Section 15).
7   Detach the tie-rod from the knuckle (see Section 15).
8   Detach the upper control arm from the knuckle (see Section 15).

11.4 Remove the dust cap from the rear hub

9   Detach the trailing arm from the knuckle (see Section 15), then remove the knuckle.
Note: *Make reference marks on the adjusting fasteners before removing the arm.*
10   Installation is the reverse of removal, noting the following points:
Caution: *All suspension fasteners going through rubber bushings should be tightened to the specified torque at ride height (or simulated ride height) and NOT while supported on jackstands with the suspension hanging free.*
  a) *Use a new upper control arm-to-rear knuckle bolt, washer and nut, lower control arm-to-rear knuckle nut, trailing arm-to-rear knuckle nuts and bolts, and tie-rod bar-to-rear knuckle nut.*
  b) *Don't tighten any of the suspension fasteners until the rear suspension has been raised to simulate normal ride height (this will prevent bushing distortion). Tighten all suspension fasteners to the torque values listed in this Chapter's Specifications.*
  c) *Install the brake disc, caliper mounting bracket and caliper (see Chapter 9). Tighten the brake fasteners to the torque values listed in the Chapter 9 Specifications.*
  d) *Install the wheel and tighten the bolts securely, then lower the vehicle and tighten the bolts to the torque listed in the Chapter 1 Specifications.*

13 Stub axle (rear, torsion beam suspension) - removal and installation

Warning: *The manufacturer recommends replacing all nuts and bolts whenever they are removed.*
1   Loosen the rear wheel bolts, raise the rear of the vehicle and support it securely on jackstands. Remove the wheel.
2   Loosen the tension from the parking

11.5 Rear hub and bearing mounting bolt

brake, then remove the drum retaining screw and drum (see Chapter 9).
3   Remove the brake shoes and wheel cylinder (see Chapter 9).
4   Remove the dust cap from the hub (see illustration 11.4).
Note: *Always replace the dust cap with a new one whenever it's removed.*
5   Remove the hub bearing assembly from the stub axle (see Section 11).
6   Remove the backing plate-to-axle beam bolts and remove the backing plate and stub axle together. Separate the stub axle from the backing plate.
7   Installation is the reverse of removal, noting the following points:
  a) *Make sure the mating surfaces on the knuckle or stub axle and the hub and bearing assembly are clean before installation.*
  b) *Tighten the mounting bolt to the torque listed in this Chapter's Specifications.*
  c) *Install the wheel cylinder, brake shoes and brake drum, then bleed the brake system (see Chapter 9).*
  d) *Adjust the parking brake (see Chapter 9).*
  e) *Install the wheel and wheel bolts. Lower the vehicle and tighten the bolts to the torque listed in the Chapter 1 Specifications.*

14 Stabilizer bar and bushings (rear, multi-link suspension) - removal and installation

Warning: *The manufacturer recommends replacing all self-locking nuts whenever they are removed.*
1   Loosen the rear wheel bolts, raise the rear of the vehicle and support it securely on jackstands. Block the front wheels to keep the vehicle from rolling off the stands. Remove the rear wheels and lower control arm stone trim protector (if equipped).

Chapter 10 Suspension and steering systems

14.2 Remove the stabilizer link fastener and separate the link from the bar - right side shown, left side identical

14.3 Rear stabilizer bar clamp bolts - right side shown, left side identical

2   Remove the mounting fasteners from the stabilizer bar links, then separate the links from the bar **(see illustration)**.
3   Remove the stabilizer bar clamp mounting fasteners **(see illustration)** and remove the stabilizer bar.
4   Inspect the stabilizer bar bushings and link bushings for cracks, tears and other signs of deterioration. Replace as necessary. Also check the ballstuds at the ends of the links for looseness, replacing the links if necessary.
5   Installation is the reverse of removal. Tighten all fasteners to the torque values listed in this Chapter's Specifications.
6   Tighten the wheel bolts to the torque listed in the Chapter 1 Specifications.

15 Rear suspension arms - removal and installation

Warning: *The manufacturer recommends replacing all fasteners that are removed or loosened.*

1   Loosen the rear wheel bolts, raise the rear of the vehicle and support it securely on jackstands. Block the front wheels to keep the vehicle from rolling off the stands. Remove the rear wheel.

Upper control arm (transverse link)

2   Remove the coil spring (see Section 10).
3   Using a floor jack, support the subframe on the same side from which the upper arm will be removed.
4   Unhook the speed sensor wire from the upper arm.
5   Mark the relationship of the adjuster cam on the upper control arm-to-subframe fastener to the subframe **(see illustration)**.
6   Remove the upper control arm-to-subframe fasteners.
7   Remove the upper arm-to-knuckle fastener, then remove the arm.
Note: *Mark the location of the washer on the arm-to-knuckle fastener; it must reinstalled in the same location, with a gap between the washer and the backing plate.*
8   Inspect the bushings for damage and wear. The bushings can be replaced, but a press and special adapters are required. If the bushings need to be replaced, take the arm to an automotive machine shop.
9   Installation is the reverse of removal, noting the following points:
Caution: *All suspension fasteners going through rubber bushings should be tightened to the specified torque at ride height (or simulated ride height) and NOT while supported on jackstands with the suspension hanging free.*

a) Use new fasteners (see the Warning at the beginning of this Section).
b) Don't tighten any of the suspension fasteners until the rear suspension has been raised to simulate normal ride height (this will prevent bushing distortion). Be sure to align the adjuster cam with its mark made during removal. Tighten all suspension fasteners to the torque values listed in this Chapter's Specifications.
c) Install the wheel and wheel bolts, then lower the vehicle and tighten the wheel bolts to the torque listed in the Chapter 1 Specifications.
d) Have the rear wheel alignment checked and, if necessary, adjusted.

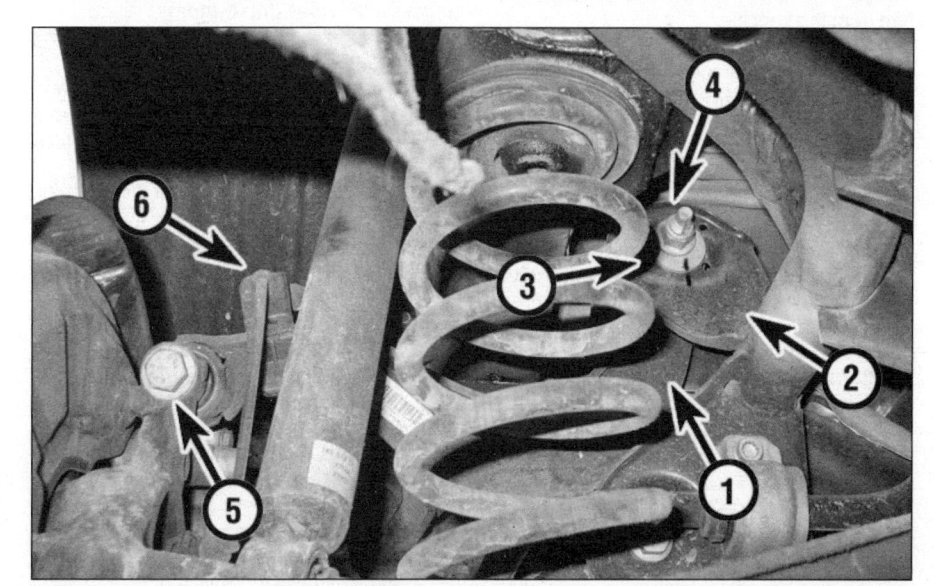

15.5 Upper control arm (transverse link) details

1   Upper control arm (transverse link)
2   Subframe
3   Adjuster cam
4   Upper control arm-to-subframe fastener
5   Upper control arm-to-knuckle fastener
6   Speed sensor wire clip

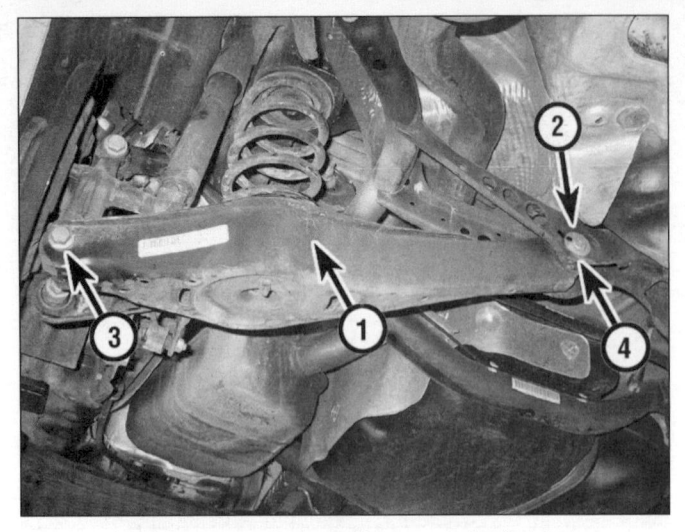

10-16   Chapter 10 Suspension and steering systems

15.14 Lower control arm (transverse link) details
1 Lower control arm (transverse link)
2 Adjuster cam (one on each side)
3 Control arm-to-knuckle fastener
4 Control arm-to-subframe fastener

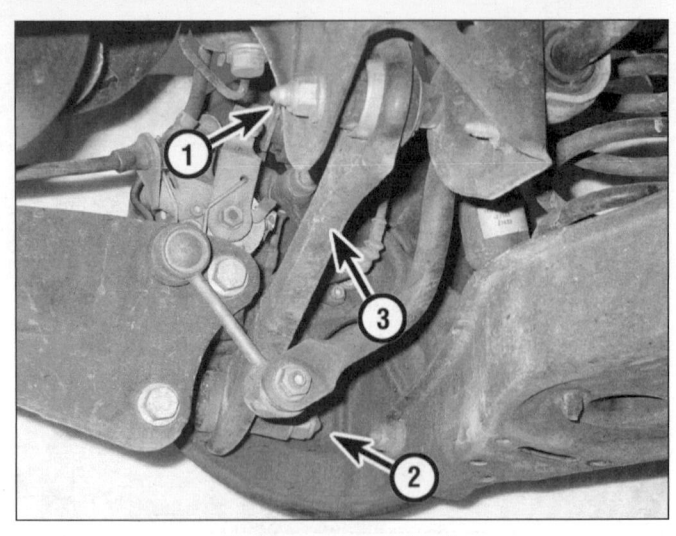

15.20 Rear tie-rod details
1 Tie-rod-to-subframe fastener
2 Tie-rod-to-knuckle fastener
3 Tie-rod

Lower control arm (transverse link)

10 On vehicles equipped with automatic leveling headlamps, disconnect the vehicle level sensor linkage from the lower control arm.
Note: *On some models it will be necessary to disconnect and lower the rear part of the exhaust system.*
11 Remove the rear coil spring (see Section 10).
12 Mark the relationship of the adjuster cam to the subframe.
13 Remove the lower control arm-to-knuckle mounting fastener and separate the arm from the knuckle.
14 Remove the lower control arm-to-subframe mounting fastener, the front mounting nut, the eccentric washer and the eccentric bolt (adjuster cam) **(see illustration)**.
15 Detach the arm from the subframe and remove it from the vehicle.
16 Inspect the bushings for damage and wear. The bushings can be replaced, but special tools are required. If the bushings need to be replaced, take the arm to an automotive machine shop.
17 Installation is the reverse of removal, noting the following points:
Caution: *All suspension fasteners going through rubber bushings should be tightened to the specified torque at ride height (or simulated ride height) and NOT while supported on jackstands with the suspension hanging free.*

a) *Use new fasteners (see the Warning at the beginning of this Section).*
b) *Don't tighten any of the suspension fasteners until the rear suspension has been raised to simulate normal ride height (this will prevent bushing distortion). Be sure to align the adjuster cam with its mark made during removal.* *Tighten all suspension fasteners to the torque values listed in this Chapter's Specifications.*
c) *Install the wheel and wheel bolts, then lower the vehicle and tighten the wheel bolts to the torque listed in the Chapter 1 Specifications.*
d) *Have the rear wheel alignment checked and, if necessary, adjusted.*

Tie-rod

18 Remove the coil spring (see Section 10).
19 Disconnect the stabilizer links and stabilizer bar brackets (see Section 14).
20 Remove the tie-rod-to-subframe mounting fasteners **(see illustration)**.
21 Remove the tie-rod-to-knuckle mounting fasteners, then remove the tie-rod.
22 Inspect the bushings for damage and wear. The bushings are not replaceable separately; if they are damaged, the tie-rod must be replaced.
23 Installation is the reverse of removal, noting the following points:
Caution: *All suspension fasteners going through rubber bushings should be tightened to the specified torque at ride height (or simulated ride height) and NOT while supported on jackstands with the suspension hanging free.*

a) *Use new fasteners (see the Warning at the beginning of this Section).*
b) *Don't tighten the tie-rod mounting fasteners until the suspension has been raised to simulate normal ride height.*
c) *Tighten the wheel bolts to the torque listed in the Chapter 1 Specifications.*
d) *Have the rear wheel alignment checked and, if necessary, adjusted.*

Trailing arm

24 Remove the rear coil spring (see Section 10).
25 Remove the parking brake cable bracket fastener and separate the bracket from the trailing arm.
Note: *Some models are equipped with a parking brake cable bracket mounted to the trailing arm with a rivet. To remove the rivet, push the center of the rivet in and remove the bracket from the arm.*
26 Disconnect the stabilizer bar link from the trailing arm (see Section 14).
27 Remove the trailing arm-to-knuckle mounting fasteners, then remove the trailing arm bracket-to-body fasteners **(see illustration)**.
28 Detach the arm from the knuckle, then from the body and remove it from the vehicle.
29 Once the trailing arm is out of the vehicle, remove the trailing arm-to-bracket through-bolt/nut and separate the arm from the bracket.
30 Inspect the bushings for damage and wear. The bushings can be replaced, but special tools are required. If the bushings need to be replaced, take the arm to an automotive machine shop.
31 Installation is the reverse of removal, noting the following points:
Caution: *All suspension fasteners going through rubber bushings should be tightened to the specified torque at ride height (or simulated ride height) and NOT while supported on jackstands with the suspension hanging free.*

a) *Use new fasteners (see the Warning at the beginning of this Section).*
b) *Don't tighten any of the suspension fasteners until the rear suspension has been raised to simulate normal ride height (this will prevent bushing distortion). Tighten all suspension fasteners to the torque values listed in this Chapter's Specifications.*

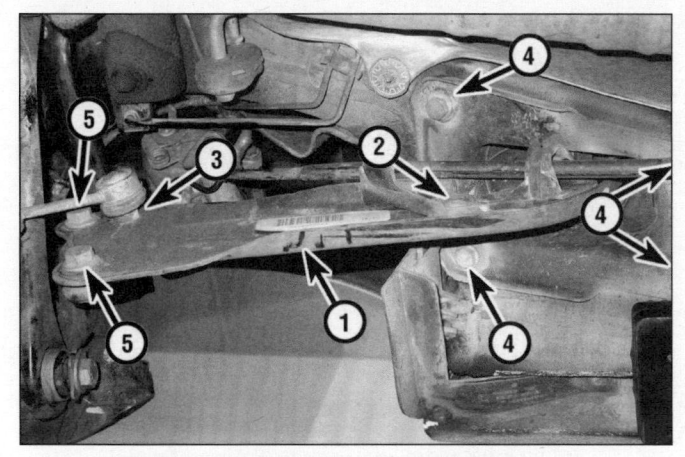

Chapter 10 Suspension and steering systems   10-17

15.27 Trailing arm details

1. Trailing arm
2. Parking brake cable bracket (with push-pin fastener)
3. Stabilizer bar link-to-trailing arm (nut on opposite side)
4. Trailing arm bracket fasteners (two of four fasteners shown)
5. Trailing arm-to-knuckle fasteners

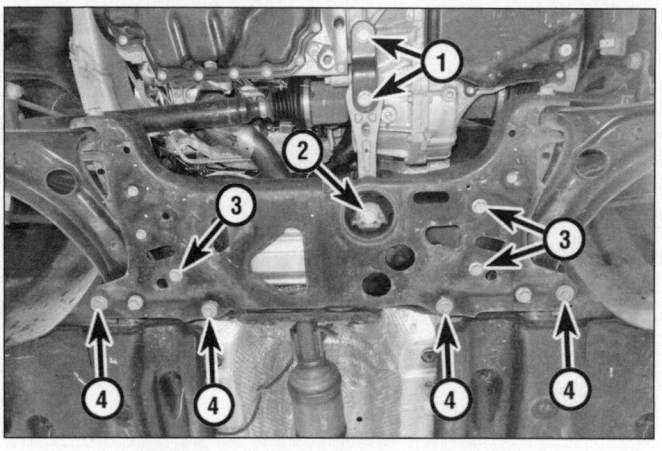

17.9a Front subframe details

1. Pendulum mount-to-transaxle bolts
2. Pendulum mount-to-subframe bolt
3. Steering gear mounting bolts (electric power steering gear)
4. Subframe rear mounting bolts

c) Install the wheel and wheel bolts, then lower the vehicle and tighten the wheel bolts to the torque listed in the Chapter 1 Specifications.
d) Have the rear wheel alignment checked and, if necessary, adjusted.

16 Rear axle beam (torsion bear suspension) - removal and installation

Warning: *The manufacturer recommends replacing all fasteners that are removed or loosened.*

Warning: *If the vehicle is raised using a hoist or lift, the rear of the vehicle must be tied to the lifting points to prevent it from lifting off of the rear lifting points and falling off once the rear axle beam is removed.*

1 Loosen the rear wheel bolts, raise the rear of the vehicle and support it securely on jackstands. Block the front wheels to keep the vehicle from rolling off the stands. Remove the rear wheels.
2 Release the parking brake and loosen the parking brake adjustment nut until the cables can be disconnected from the brake cable equalizer (see Chapter 9).
3 Disconnect the rear wheel speed sensors (see Chapter 9).
4 Remove the coil springs (see Section 10).
5 Using a floor jack, raise the axle beam until the crossbrace is in a horizontal position. Remove the crossbrace-to-rear axle fasteners and separate the crossbrace from the rear axle.
6 Disconnect the brake lines from each side of the rear axle, then remove the clips attaching the lines to the rear axle (see Chapter 9).
7 Remove the rear axle bracket fasteners from the right side, then remove just the through-bolt/nut from the left side.

Caution: *The rear axle mounting bracket fasteners are different lengths - be sure to mark them as you remove them for installation.*

8 Remove the rear shock absorber-to-rear axle fasteners and lower the rear axle from the vehicle.
9 Installation is the reverse of removal, noting the following points:

Caution: *All suspension fasteners going through rubber bushings should be tightened to the specified torque at ride height (or simulated ride height) and NOT while supported on jackstands with the suspension hanging free.*

a) Use new fasteners (see the Warning at the beginning of this Section).
b) Don't tighten the rear axle mounting fasteners until the suspension has been raised to simulate normal ride height.
c) Bleed the brake system (see Chapter 9).
d) Tighten the wheel bolts to the torque listed in the Chapter 1 Specifications.
e) Have the rear wheel alignment checked and, if necessary, adjusted.

17 Subframe - removal and installation

Warning: *The manufacturer recommends replacing the subframe mounting bolts and all steering and suspension fasteners with new ones whenever they are removed.*

Front

1 Loosen the front wheel bolts, then raise the front of the vehicle and support it securely on jackstands. Remove the wheels.
2 Remove the lower splash shield below the engine.
3 Remove the heat shield fasteners and heat shield from the subframe.
4 Remove the exhaust system bracket fasteners from the subframe.

17.9b Subframe front mounting bolt (left side shown)

5 Remove the pendulum mount (see Chapter 2B or 2A).
6 Disconnect the control arms from the balljoints (see Section 7).
7 Disconnect the tie-rod ends from the steering knuckles (see Section 20).
8 Disconnect the stabilizer bar links and remove the stabilizer bar clamp fasteners (see Section 6).
9 Remove the steering gear-to-subframe fasteners **(see illustrations)**. Support the steering gear from above with a length of wire or rope.

Note: *The steering gear can be left in-vehicle or removed with the subframe. This procedure explains how to remove the subframe while leaving the steering gear in the vehicle. To remove the steering gear with the subframe, see Section 22.*

Note: *Models with hydraulic power steering have two right-side mounting bolts.*

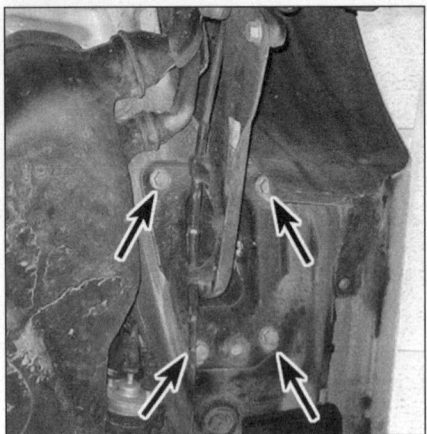

10-18 Chapter 10 Suspension and steering systems

17.21 Trailing arm bracket-to-body fastener locations - multi-link rear suspension

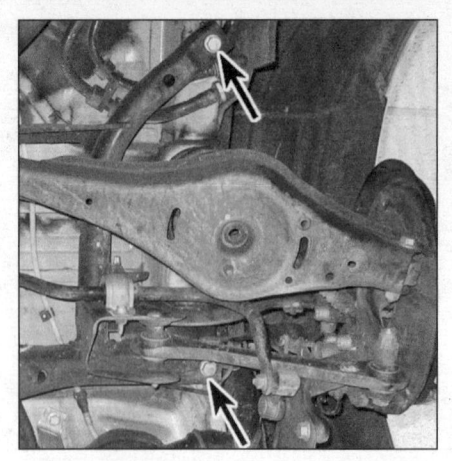

17.22 Rear subframe fastener locations (right side shown, left side similar) - multi-link rear suspension

18.5 Rotate the steering wheel 90-degrees from center (counterclockwise), insert a flat-blade screwdriver into the right-side opening on the steering wheel and twist it toward the left to release the retainer

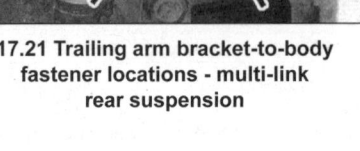

10 Support the subframe with a floor jack, then loosen and withdraw the subframe securing bolts and lower the subframe from the vehicle.

11 Installation is the reverse of removal, noting the following points:

a) Use new fasteners (see the Warning at the beginning of this Section), tightening them to the torque values listed in this Chapter's Specifications.
b) Don't tighten the control arm-to-subframe fasteners (if removed) until the suspension has been raised to simulate normal ride height.
c) Tighten the wheel bolts to the torque listed in the Chapter 1 Specifications.
d) Have the front wheel alignment checked and, if necessary, adjusted.

Rear

Multi-link suspension

Warning: *If the vehicle is raised using a hoist or lift, the rear of the vehicle must be tied to the lifting points to prevent it from lifting off of the rear lifting points and falling off once the rear subframe is removed.*

12 Loosen the rear wheel bolts, raise the rear of the vehicle and support it securely on jackstands. Block the front wheels to keep the vehicle from rolling off the stands. Remove the rear wheels.

13 Remove the exhaust system (see Chapter 4).

14 On vehicles equipped with automatic leveling headlamps, disconnect the vehicle level sensor linkage from the lower control arm.

15 Remove the ABS wheel speed sensors (see Chapter 9).

16 Remove the coil springs (see Section 10). Also detach the upper control arms and tie rods from the rear knuckles.

17 Remove the shock absorber-to-knuckle fastener (Jetta) or shock absorber-to-lower control arm nut and bolt (Golf) (see Section 9).

18 Remove the rear brake calipers and parking brake cable housing fasteners (see Chapter 9). Support the calipers with lengths of wire - don't let them hang by the hose.

19 Remove the ABS speed sensor wire harnesses from the brackets on the upper control arms.

20 Remove the brake hose retaining clips from both sides of the vehicle (see Chapter 9).

21 Support the subframe and suspension assembly with a floor jack and slowly remove the trailing arm-to-body bracket fasteners **(see illustration)**.

22 Remove the subframe fasteners and slowly lower the subframe assembly from the vehicle **(see illustration)**.

23 Installation is the reverse of removal, noting the following points:

Caution: *All suspension fasteners going through rubber bushings should be tightened to the specified torque at ride height (or simulated ride height) and NOT while supported on jackstands with the suspension hanging free.*

a) Use new fasteners (see the Warning at the beginning of this Section), tightening them to the torque values listed in this Chapter's Specifications.
b) Don't tighten the control arms or tie-rod fasteners until the suspension has been raised to simulate normal ride height.
c) Tighten the wheel bolts to the torque listed in the Chapter 1 Specifications.
d) Have the rear wheel alignment checked and, if necessary, adjusted.

Torsion beam suspension

24 Remove the rear axle beam (see Section 16).

25 Remove the subframe fastener and lower the subframe.

26 Installation is the reverse of removal, noting the following points:

Caution: *All suspension fasteners going through rubber bushings should be tightened to the specified torque at ride height (or simulated ride height) and NOT while supported on jackstands with the suspension hanging free.*

a) Use new fasteners (see the Warning at the beginning of this Section), tightening them to the torque values listed in this Chapter's Specifications.
b) Tighten the wheel bolts to the torque listed in the Chapter 1 Specifications.
c) Have the rear wheel alignment checked and, if necessary, adjusted.

18 Steering wheel - removal and installation

Warning: *These models are equipped with airbags. Always disable the airbag system before working in the vicinity of any airbag system component to avoid the possibility of accidental deployment of the airbag(s), which could cause personal injury (see Chapter 12).*

Warning: *Do not use a memory saving device to preserve the PCM's memory when working on or near airbag system components.*

Caution: *If the battery is disconnected, several systems must be re-learned before they will work properly (see Chapter 5, Section 3).*

Removal

1 Park the vehicle with the wheels pointing straight ahead. Disconnect the cable from the negative battery terminal (see Chapter 5).

2014 and earlier models

2 Place the steering wheel in its lowest position and extend it out to the maximum distance.

3 Turn the wheel counterclockwise about 90 degrees.

4 Locate the access hole and insert a small flat-bladed screwdriver into the access hole.

5 Rotate the screwdriver to the left (towards the driver's door) to release the retainer **(see illustration)**.

Chapter 10 Suspension and steering systems

18.6 Rotate the steering wheel 180-degrees to the right, insert the screwdriver into the left side opening and twist it towards the right to unlock the left side retainer, then carefully lift the airbag module off of the steering wheel

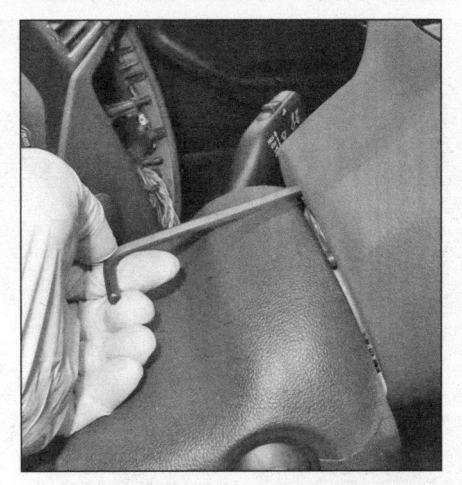

18.10 Insert an Allen wrench into the hole in the back of the steering wheel to disengage the airbag spring retainer

18.17 Remove the steering wheel mounting fastener; a 12 mm, 12 point spline-drive bit is required

6 Rotate the steering wheel 180-degrees clockwise and perform the same procedure again **(see illustration)**.
7 Lift the airbag from the steering wheel far enough to disconnect the electrical connectors.
8 Remove the airbag from the steering wheel.
Warning: *When carrying the airbag module, keep the driver's (trim) side of it away from your body and, when you set it down, make sure the driver's side is facing up.*

2015 and later models

9 Rotate the steering wheel 90-degrees from center counterclockwise.
10 Insert a 5 or 6 mm Allen wrench or punch into the access hole **(see illustration)**.
11 Hold the tool parallel to the steering column as you slide it into the access hole.
12 Slightly lift the tool upwards (away from the steering column) causing the internal spring to flex and release the end of the spring from the catch.
13 Rotate the steering wheel 180-degrees and perform the same procedure again.
14 Carefully lift the airbag from the steering wheel far enough to disconnect the electrical connectors.
15 Remove the airbag from the steering wheel.
Note: *If the steering wheel has never been removed, there will be a thin layer of plastic covering the access hole for the tool. Push the tool through the thin layer.*
16 Set the module aside in a safe, isolated area, with the airbag side of the module facing UP.
Warning: *When carrying the airbag module, keep the driver's (trim) side of it away from your body and, when you set it down, make sure the driver's side is facing up.*

All models

17 Remove the steering wheel bolt using a 12 mm, 12-point spline-drive bit **(see illustration)**.
18 Mark the position of the steering wheel to the steering shaft if marks don't already exist or don't line up **(see illustration)**.
19 Lift the steering wheel from the shaft, noting how any electrical wire harnesses are routed. If the steering wheel is tight, tap it up near the center using the palm of your hand, or twist it from side-to-side while pulling upwards.
Warning: *Don't hammer on the shaft to remove the wheel. Don't allow the steering shaft to turn after the steering wheel is removed or damage to the airbag clockspring could occur. Also, do not allow the hub of the clockspring to turn.*
20 Place a piece of tape across the clockspring to prevent it from rotating.

Installation

21 Install the wheel onto the steering shaft, aligning the index marks and routing any wiring harnesses.
22 Install a *new* steering wheel bolt and tighten it to the torque listed in this Chapter's Specifications.
Warning: *Do not reuse the steering wheel bolt. The manufacturer recommends replacing the bolt with a new one.*
23 Connect the electrical connectors for the horn and the airbag module.
24 Position the airbag module onto the steering wheel and lock the module in place.

19 Steering column - removal and installation

Warning: *These models are equipped with airbags. Always disable the airbag system before working in the vicinity of any airbag system component to avoid the possibility of accidental deployment of the airbag(s), which*

18.18 After removing the bolt, check for alignment marks on the steering wheel and steering shaft - if there aren't any, make your own

could cause personal injury (see Chapter 12).
Warning: *Do not use a memory saving device to preserve the ECM's memory when working on or near airbag system components.*
Caution: *If the battery is disconnected, several systems must be re-learned before they will work properly, (see Chapter 5, Section 3).*

Removal

1 Park the vehicle with the wheels pointing straight ahead. For models with an automatic transmission, place the shift lever in PARK. Disconnect the cable from the negative battery terminal (see Chapter 5). Disable the airbag system (see Chapter 12).
2 Remove the steering column covers (see Chapter 11).
3 Remove the steering wheel (see Section 18). Prevent the steering shaft from turning.
Warning: *If the steering shaft is not held stationary, the airbag clockspring could be damaged.*
Warning: *Place a piece of tape across the airbag clockspring to prevent it from rotating.*

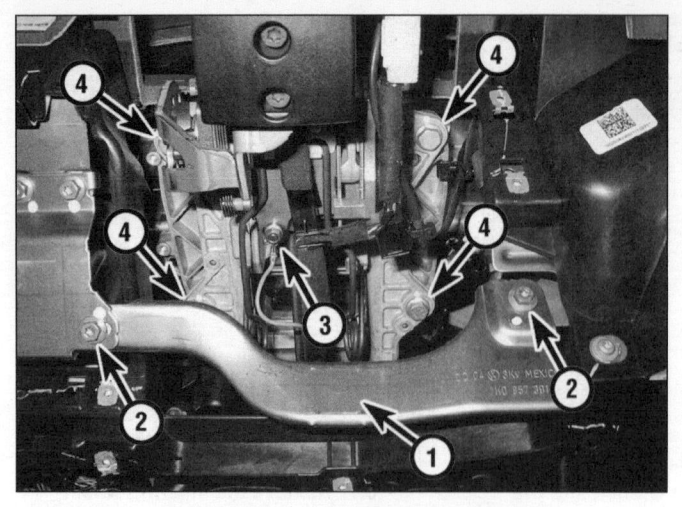

10-20 Chapter 10 Suspension and steering systems

19.11 Steering column mounting details

1 Instrument panel brace
2 Brace fasteners
3 Ground wire
4 Steering column fasteners

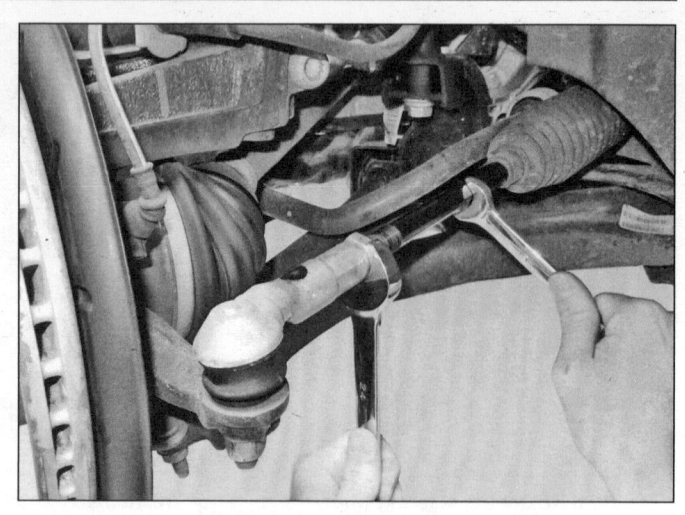

20.2a Loosen the jam nut . . .

20.2b . . . then mark the position of the tie-rod end in relation to the threads

20.4 Disconnect the tie-rod end from the steering knuckle arm with a puller

4   Remove the lower instrument panel trim (under the steering column) (see Chapter 11).
5   Remove the multi-function switch (see Chapter 12).
6   Disconnect the wiring harness clips, harness connectors and ground wire from the steering column.
7   Press the tabs on both sides of the cable guide cover towards the inside and remove the cover from the guide on the steering column.
8   On manual transmission models, remove the clutch pedal crash brace fastener and brace from the column, if equipped.
9   Secure the lower section of the steering column to the upper section by passing a length of wire through the hole in the lower section of the column and through the spring on the upper section.
10   Remove the footwell trim panel and mark the relationship of the steering shaft lower universal joint to the steering gear input shaft, then remove the pinch bolt securing the universal joint (see Section 22).

11   Remove the steering column mounting bolts **(see illustration)**, then separate the universal joint from the steering gear input shaft and guide the column out from the instrument panel.

Installation

12   Guide the column into position, connecting the U-joint with the steering gear input shaft. Install the pinch bolt, but don't tighten it yet.
13   Install the steering column mounting bolts, but don't tighten them yet.
14   Remove the wire installed in Step 10. If a new steering column was installed, remove the protective materials used for transport from the column.
15   The remainder of installation is the reverse of the removal procedure, noting the following points:

a)   When installing the steering wheel, be sure the airbag clockspring is centered, and tighten the steering wheel bolt to the torque listed in this Chapter's Specifications.

b)   With the steering column in position, tighten the column mounting bolts to the torque listed in this Chapter's Specifications.
c)   Tighten the U-joint pinch bolt to the torque listed in this Chapter's Specifications after the steering column mounting fasteners have been tightened.

20 Tie-rod ends - removal and installation

Removal

1   Loosen the front wheel bolts. Apply the parking brake, raise the front of the vehicle and support it securely on jackstands. Remove the front wheel.
2   Hold the tie-rod with a pair of locking pliers or wrench and loosen the jam nut enough to mark the position of the tie-rod end in relation to the threads **(see illustrations)**.
3   Loosen (but don't remove) the nut on the tie-rod end stud.
4   Disconnect the tie-rod end from the steering knuckle arm with a puller **(see illustration)**. Remove the nut and detach the tie-rod end.
5   Unscrew the tie-rod end from the tie-rod.

Installation

6   Thread the tie-rod end on to the marked position and insert the stud into the steering knuckle arm. Tighten the jam nut securely.
7   Install the nut on the stud and tighten it to the torque listed in this Chapter's Specifications.
8   Install the wheel and wheel bolts. Lower the vehicle and tighten the wheel bolts to the torque listed in the Chapter 1 Specifications.
9   Have the alignment checked and, if necessary, adjusted.

Chapter 10 Suspension and steering systems

10-21

21.3 The outer ends of the steering gear boots are secured by spring-type clamps (A); they're easily released with a pair of pliers. The inner ends of the steering gear boots are retained by boot clamps which must be cut off and discarded (B)

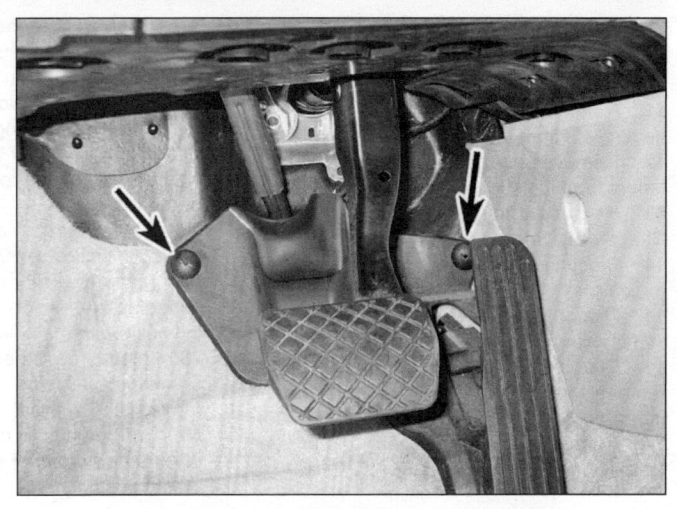

22.6a Remove the footwell cover fasteners, then remove cover . . .

22.6b . . . and remove the pinch bolt at the base of the steering column

21 Steering gear boots - replacement

1   Loosen the lug nuts, raise the vehicle and support it securely on jackstands. Remove the wheel.
2   Remove the tie-rod end and jam nut (see Section 20).
3   Remove the outer steering gear boot clamp with a pair of pliers **(see illustration)**. Cut off the inner boot clamp with a pair of diagonal cutters. Slide off the boot.
4   Before installing the new boot, wrap the threads and serrations on the end of the steering rod with a layer of tape so the small end of the new boot isn't damaged.
5   Slide the new boot into position on the steering gear until it seats in the groove in the steering rod and install new clamps.
6   Remove the tape and install the tie-rod end (see Section 20).
7   Install the wheel and lug nuts. Lower the vehicle and tighten the lug nuts to the torque listed in the Chapter 1 Specifications.

22 Steering gear - removal and installation

Warning: *These models are equipped with airbags. Always disable the airbag system before working in the vicinity of any airbag system component to avoid the possibility of accidental deployment of the airbag, which could cause personal injury (see Chapter 12). Also, don't allow the steering wheel to turn after the steering gear has been removed. To prevent this, pass the seat belt through the steering wheel and plug it into its latch.*
Warning: *Do not use a memory saving device to preserve the ECM's memory when working on or near airbag system components.*
Caution: *If the battery is disconnected, several systems must be re-learned before they will work properly, see Chapter 5, Section 3.*
Note: *The steering gear cannot be repaired. If the steering gear is worn or damaged it must be replaced.*

1   Park the vehicle with the front wheels pointing straight ahead. Apply the parking brake and chock the rear wheels, then loosen the front wheel bolts. Raise the front of the vehicle and support it securely on jackstands. Remove both front wheels.
2   Remove the battery from the engine compartment (see Chapter 5).
3   Turn the steering wheel to the center position, then remove the ignition key to engage the steering lock.
Warning: *Ensure that the steering column remains in the straight-ahead position throughout the remainder of this procedure, or the airbag contact unit may become misaligned, leading to the failure of the airbag system.*
4   Remove the driver's side under-dash panel (see Chapter 11).
5   Secure the lower section of the steering column to the upper section as described in Section 19.
6   Remove footwell cover fasteners and cover **(see illustration)**, then the pinch bolt securing the universal joint to the input shaft of the steering gear **(see illustration)**.
Warning: *Do not allow the upper and lower sections of the steering column to become separated while the steering column is detached from the steering gear, as this can cause the internal components to become detached and misaligned.*
7   Pull the steering column universal joint off the steering gear input shaft and move it to one side. Pull the input shaft cover from the firewall and into the vehicle, if equipped.

Models equipped with hydraulic power steering

8   Siphon the fluid from the power steering fluid reservoir. If a suitable implement is not readily available to siphon the fluid from the system, it can be drained into a container when the hydraulic lines are detached from the steering gear.
9   Apply hose clamps to the fluid lines leading to and from the steering gear. Take care to avoid causing damage to the hoses by pinching.

All models

10 Remove the lower splash shield below the engine (see Chapter 1, Section 6).
11 Remove the heat shield fasteners and heat shield from the subframe, if equipped.
12 Remove the exhaust system bracket fasteners from the subframe.
13 Remove the pendulum mount (see Chapter 2B, Section 15).
14 Disconnect the lower control arms from the balljoints (see Section 8).
15 Disconnect the tie-rod ends from the steering knuckles (see Section 20).
16 Disconnect the stabilizer bar links and remove the stabilizer bar clamps fasteners (see Section 6).
17 Remove the steering gear-to-subframe fasteners **(see illustration 17.9)**.

Models equipped with electro-mechanical steering gear

18 Support the subframe with a floor jack. Remove the subframe rear bolts and loosen the subframe front mounting bolts **(see illustrations 17.9a and 17.9b)**.
Note: *Make sure no electrical wiring is being stretched while lowering the subframe.*
19 Remove the steering gear heat shield fasteners and heat shield.
20 Disconnect the cable guide clips from the subframe.
21 Disconnect the electrical connectors to the steering gear. Remove the steering gear mounting bolts **(see illustration 17.9a)**.
22 Carefully lower the rear of the subframe using the floor jack, enough to lift the steering gear off the subframe and out of the vehicle.
23 Install the steering gear on to the subframe and connect the electrical connectors.
Note: *To avoid damaging the steering gear control module, do not set the steering gear down on top of the control module.*
24 Align the steering gear mounting holes with the holes in the subframe, then install the mounting bolts and tighten the bolts to the torque listed in this Chapter's Specifications.
25 The remainder of the installation procedure is a reversal of removal, noting the following points:
a) Refer to Section 17 to reconnect the subframe.
b) Refer to Section 20 to reconnect the tie-rod ends.
c) Tighten the wheel bolts to the torque listed in the Chapter 1 Specifications.
d) Finally, have the wheel alignment checked and, if necessary, adjusted.

Models equipped with hydraulic steering gear

26 Unscrew the fittings and disconnect the fluid supply and return lines from the steering gear. Clean the connections before they are detached. Drain any fluid remaining in the system into a container for disposal. Tie the lines back away from the work area, and seal off their ends to prevent further leakage and keep dirt from entering the system.
27 Support the subframe with a floor jack. Remove the subframe rear mounting bolts and loosen subframe front mounting bolts **(see illustrations 17.9a and 17.9b)**. Remove the steering gear mounting bolts and lower the rear of the subframe enough to lift the steering gear off the subframe and out of the vehicle.
Note: *Make sure no electrical wiring is being stretched while lowering the subframe.*
28 Maneuver the steering gear onto the subframe and align the steering gear mounting holes with the holes in the subframe.
29 Install the steering gear-to-subframe mounting bolts and tighten the bolts to the torque listed in this Chapter's Specifications.
30 The remainder of the installation procedure is a reversal of removal, noting the following points:
a) *Connect the pressure and return line fittings to the steering gear. Tighten the pressure line securely.*
b) *Refer to Section 20 to reconnect the tie-rod ends.*
c) *Top up the fluid level as described in Chapter 1 and bleed the system as described in Section 1.*
d) *Tighten the wheel bolts to the torque listed in the Chapter 1 Specifications.*
e) *Finally, have the wheel alignment checked and, if necessary, adjusted.*

23 Power steering pump - removal and installation

1 Raise the front of the vehicle and support it securely on jackstands.
2 Remove the lower splash shield below the engine (see Chapter 1, Section 6).
3 Loosen the power steering pump pulley bolts, then remove the drivebelt (see Chapter 1).
4 Using brake hose clamps, clamp both the supply and return hoses near the power steering fluid reservoir. This will minimize fluid loss during subsequent operations.
5 Wipe clean the area around the power steering pressure and return line fittings.
6 Unscrew and disconnect the pressure line from the pump; be prepared for fluid spillage, and position a container beneath the pipe while unscrewing the line. Plug the end of the line and the steering pump orifice, to minimize fluid leakage and to keep dirt out of the hydraulic system.
7 Loosen the clamp and disconnect the fluid supply hose from the rear of the power steering pump. Plug the end of the hose and cover the pump fluid port to prevent contamination.
8 Remove the pump pulley bolt and pulley, then remove the pump mounting fasteners and detach the pump from its bracket.
9 If the power steering pump is faulty, it must be replaced. The pump is a sealed unit and cannot be overhauled.
10 If a new pump is to be installed, it must be primed with fluid first, to ensure adequate lubrication during its initial stages of operation. Failure to do this could cause noisy operation and may lead to early pump failure. To prime the pump, pour the specified grade of hydraulic fluid into the fluid supply port on the pump, and simultaneously rotate the pump pulley. When the fluid exits from the pressure port, it is primed and ready for use.
11 Maneuver the pump into position, then install its mounting bolts and tighten them to the torque listed in this Chapter's Specifications.
12 Install the pulley to the pump, then install the pulley mounting bolts and tighten them securely at this time.
13 Connect the pressure line to the pump and carefully screw the line in to the pump. Ensure the line is correctly routed, then tighten the line to the torque listed in this Chapter's Specifications.
14 Reconnect the supply hose to the pump and secure it in position with the retaining clip. Remove the hose clamps used to minimize fluid loss.
15 Install the power steering pump drivebelt (see Chapter 1) and tighten the pulley pump fasteners to the torque listed in this Chapter's Specifications.
16 Install the under-vehicle splash shield. Reconnect the negative battery cable.
17 Top up the hydraulic system (see Chapter 1), then bleed the system as described in Section 24.

24 Power steering system - bleeding

1 Following any operation in which the power steering fluid lines have been disconnected, the power steering system must be bled to remove all air and obtain proper steering performance.
2 With the front wheels in the straight ahead position, check the power steering fluid level and, if low, add fluid until it is up to the MIN mark on the dipstick.
3 Raise the front of the vehicle and support it securely on jackstands.
4 Turn the steering wheel back-and-forth repeatedly, lightly hitting the stops.
Note: *Don't hold the steering wheel in the full-right or full-left position, as this could damage the pump.*
5 Start the engine and allow it to run at fast idle. Recheck the fluid level and add more if necessary until it is up to the MIN mark.
6 Bleed the system by turning the wheels from side to side, just barely contacting the stops. This will work the air out of the system. Keep the reservoir full of fluid as this is done.
7 When the air is worked out of the system and the fluid level stabilizes, return the wheels to the straight ahead position and leave the vehicle running for several more minutes before shutting it off. Lower the vehicle.
8 Road test the vehicle to be sure the steering system is functioning normally and noise-free.
9 Recheck the fluid level to be sure it is up to the HOT mark on the dipstick while the engine is at normal operating temperature. Add fluid if necessary (see Chapter 1).

Chapter 10 Suspension and steering systems

25 Wheels and tires - general information

1   All vehicles covered by this manual are equipped with metric-sized steel belted radial tires **(see illustration)**. Use of other size or type of tires may affect the ride and handling of the vehicle. Don't mix different types of tires, such as radials and bias belted, on the same vehicle as handling may be seriously affected. It's recommended that tires be replaced in pairs on the same axle, but if only one tire is being replaced, be sure it's the same size, structure and tread design as the other.

2   Because tire pressure has a substantial effect on handling and wear, the pressure on all tires should be checked at least once a month or before any extended trips (see Chapter 1).

3   Wheels must be replaced if they are bent, dented, leak air, have elongated bolt holes, are heavily rusted, out of vertical symmetry or if the wheel bolts won't stay tight. Wheel repairs that use welding or peening are not recommended.

4   Tire and wheel balance is important in the overall handling, braking and performance of the vehicle. Unbalanced wheels can adversely affect handling and ride characteristics as well as tire life. Whenever a tire is installed on a wheel, the tire and wheel should be balanced by a shop with the proper equipment.

26 Wheel alignment - general information

1   A wheel alignment refers to the adjustments made to the wheels so they are in proper angular relationship to the suspension and the ground. Wheels that are out of proper alignment not only affect vehicle control, but also increase tire wear.

2   The angles normally measured are camber, caster and toe-in **(see illustration)**. Camber and caster are not always adjustable but are usually checked to see if any suspension components are worn or damaged. The toe-in angle is commonly adjusted on all vehicles in the front and on the rear of vehicles with independent rear suspension.

3   Getting the proper wheel alignment is a very exacting process, one in which complicated and expensive machines are necessary to perform the job properly. Because of this, you should have a technician with the proper equipment perform these tasks. We will, however, use this space to give you a basic idea of what is involved with a wheel alignment so you can better understand the process and deal intelligently with the shop that does the work.

4   Toe-in is the turning in of the wheels. The purpose of a toe specification is to ensure parallel rolling of the wheels. In a vehicle with zero toe-in, the distance between the front edges of the wheels will be the same as the distance between the rear edges of the wheels. The actual amount of toe-in is normally only a fraction of an inch. On the front end, toe-in is controlled by the tie-rod end position on the tie-rod. Incorrect toe-in will cause the tires to wear improperly by making them scrub against the road surface.

5   Camber is the tilting of the wheels from vertical when viewed from one end of the vehicle. When the wheels tilt out at the top, the camber is said to be positive (+). When the wheels tilt in at the top the camber is negative (-). The amount of tilt is measured in degrees from vertical and this measurement is called the camber angle. This angle affects the amount of tire tread that contacts the road and compensates for changes in the suspension geometry when the vehicle is cornering or traveling over an undulating surface.

6   Caster is the tilting of the front steering axis from the vertical. A tilt toward the rear is positive caster and a tilt toward the front is negative caster.

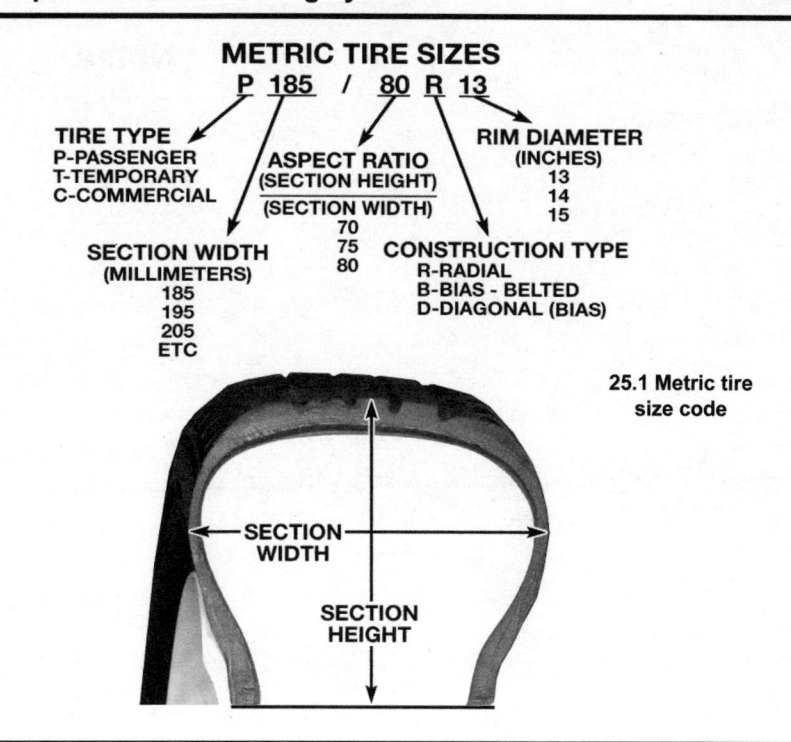

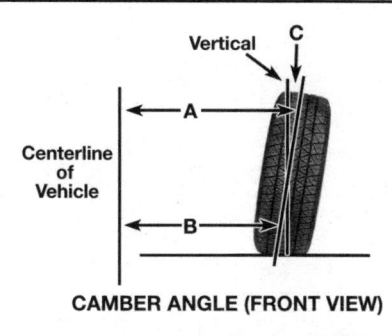

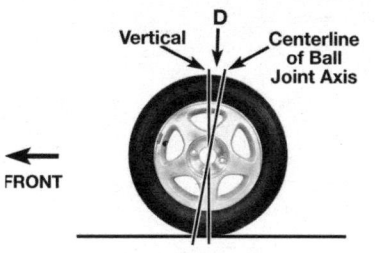

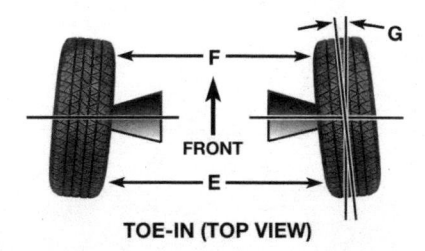

25.1 Metric tire size code

METRIC TIRE SIZES
P 185 / 80 R 13

TIRE TYPE
P-PASSENGER
T-TEMPORARY
C-COMMERCIAL

ASPECT RATIO
(SECTION HEIGHT)
(SECTION WIDTH)
70
75
80

RIM DIAMETER (INCHES)
13
14
15

SECTION WIDTH (MILLIMETERS)
185
195
205
ETC

CONSTRUCTION TYPE
R-RADIAL
B-BIAS - BELTED
D-DIAGONAL (BIAS)

A minus B = C (degrees camber)
D = degrees caster
E minus F = toe-in (measured in inches)
G = toe-in (expressed in degrees)

26.2 Wheel alignment details

Notes

Chapter 11
Body

Contents

	Section
Body repair - major damage	4
Body repair - minor damage	3
Bumpers and bumper covers - removal and installation	12
Center console - removal and installation	27
Cowl cover - removal and installation	16
Door lock cylinder, outside handle and latch - removal and installation	23
Door - removal, installation and adjustment	22
Door trim panels - removal and installation	21
Door window glass - removal and installation	24
Door window regulator and motor - removal and installation	25
Fastener and trim removal	5
Fenderwell splash shields - removal and installation	14
Front fender - removal and installation	15
General information	1
Hinges and locks - maintenance	7
Hood latch and support strut - removal and installation	11
Hood latch release cable, release lever and safety lever - removal and installation	9
Hood - removal, installation and adjustment	10
Instrument panel and crossbeam - removal and installation	30
Interior trim panels - removal and installation	28
Liftgate (coupe and wagon models) - removal, installation and adjustment	19
Liftgate latch, actuator and support struts (coupe and wagon models) - removal and installation	20
Mirrors - removal and installation	26
Radiator grille and radiator shutter assembly - removal and installation	13
Rear parcel shelf - removal and installation	34
Repairing minor paint scratches	2
Seat belts - removal and installation	33
Seats - removal and installation	31
Steering column covers - removal and installation	29
Sunroof - check, adjustments, removal and installation	32
Trunk lid latch, support struts and release switch/backup camera assembly (sedan models) - removal and installation	18
Trunk lid (sedan models) - removal, installation and adjustment	17
Upholstery, carpets and vinyl trim - maintenance	6
Windshield and fixed glass - replacement	8

Specifications

Torque specifications

Note: *One foot-pound (ft-lb) of torque is equivalent to 12 inch-pounds (in-lbs) of torque. Torque values below approximately 15 ft-lbs are expressed in inch-pounds, because most foot-pound torque wrenches are not accurate at these smaller values.*

Torque specifications	Ft-lbs (unless otherwise indicated)	Nm
Side impact protection bar bolts	15	20
Doors (front and rear)		
Door retaining pinch bolts	17	23
Door bracket bolts*		
Bracket-to-door bolts	37	50
Bracket-to-body bolts		
Original A-pillar	37	50
If A-pillar is being replaced		
First pass	15	20
Second pass	Tighten an additional 90-degrees	
Door strap fasteners		
Strap-to-body bolt	22	30
Strap-to-door bolt	84 inch-lbs	9
Window regulator mounting nuts	72 inch-lbs	8
Window motor mounting screws	48 inch-lbs	4
Window glass retaining screws (front)	72 inch-lbs	8

* Replace bolt if fully removed.

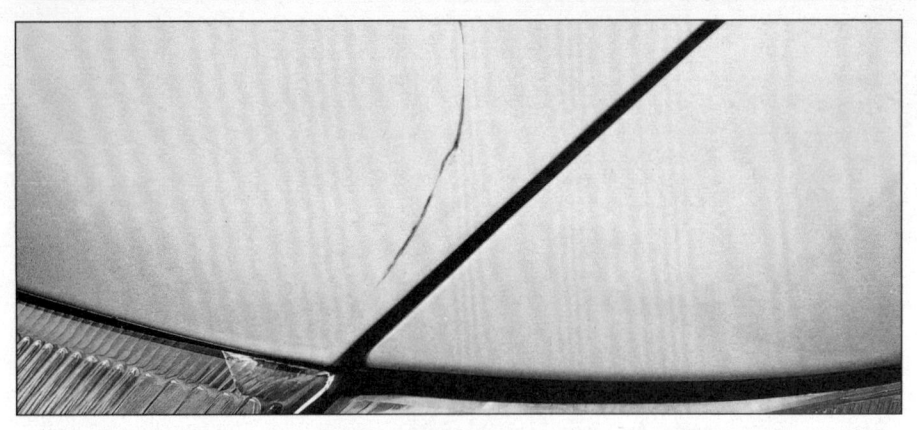

2.1a Make sure the damaged area is perfectly clean and rust free. If the touch-up kit has a wire brush, use it to clean the scratch or chip. Or use fine steel wool wrapped around the end of a pencil. Clean the scratched or chipped surface only, not the good paint surrounding it. Rinse the area with water and allow it to dry thoroughly

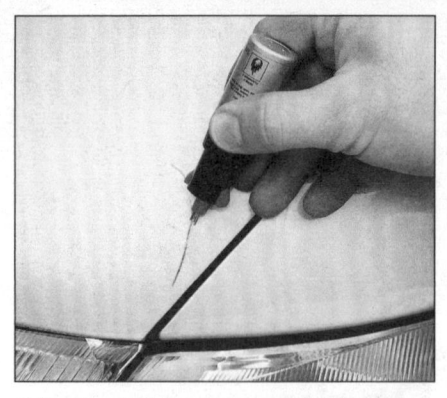

2.1b Thoroughly mix the paint, then apply a small amount with the touch-up kit brush or a very fine artist's brush. Brush in one direction as you fill the scratch area. Do not build up the paint higher than the surrounding paint

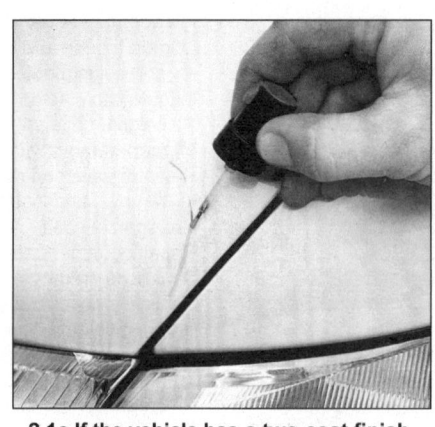

2.1c If the vehicle has a two-coat finish, apply the clear coat after the color coat has dried

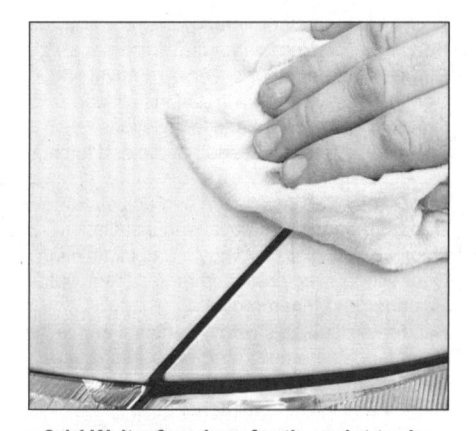

2.1d Wait a few days for the paint to dry thoroughly, then rub out the repainted area with a polishing compound to blend the new paint with the surrounding area. When you're happy with your work, wash and polish the area

1 General Information

Warning: *The models covered by this manual are equipped with a Supplemental Restraint System (SRS), more commonly known as airbags. Always disable the airbag system before working in the vicinity of any airbag system components to avoid the possibility of accidental deployment of the airbags, which could cause personal injury (see Chapter 12).*

1   Certain body components are particularly vulnerable to accident damage and can be unbolted and repaired or replaced. Among these parts are the hood, doors, tailgate, liftgate, bumpers and front fenders.
2   Only general body maintenance practices and body panel repair procedures within the scope of the do-it-yourselfer are included in this Chapter.

2 Repairing minor paint scratches

1   No matter how hard you try to keep your vehicle looking like new, it will inevitably be scratched, chipped or dented at some point. If the metal is actually dented, seek the advice of a professional. But you can fix minor scratches and chips yourself **(see illustrations)**. Buy a touch-up paint kit from a dealer service department or an auto parts store. To ensure that you get the right color, you'll need to have the specific make, model and year of your vehicle and, ideally, the paint code, which is located on a special metal plate under the hood or in the door jamb.

3 Body repair - minor damage

Plastic body panels

1   The following repair procedures are for minor scratches and gouges. Repair of more serious damage should be left to a dealer service department or qualified auto body shop. Below is a list of the equipment and materials necessary to perform the following repair procedures on plastic body panels.

Wax, grease and silicone removing solvent
Cloth-backed body tape
Sanding discs
Drill motor with three-inch disc holder
Hand sanding block
Rubber squeegees
Sandpaper
Non-porous mixing palette
Wood paddle or putty knife
Wood paddle or putty knife
Curved-tooth body file
Flexible parts repair material

Flexible panels (bumper trim)

2   Remove the damaged panel, if necessary or desirable. In most cases, repairs can be carried out with the panel installed.
3   Clean the area(s) to be repaired with a wax, grease and silicone removing solvent applied with a water-dampened cloth.
4   If the damage is structural, that is, if it extends through the panel, clean the backside of the panel area to be repaired as well. Wipe dry.
5   Sand the rear surface about 1-1/2 inches beyond the break.
6   Cut two pieces of fiberglass cloth large enough to overlap the break by about 1-1/2 inches. Cut only to the required length.
7   Mix the adhesive from the repair kit according to the instructions included with the kit, and apply a layer of the mixture approximately 1/8-inch thick on the backside of the panel. Overlap the break by at least 1-1/2 inches.
8   Apply one piece of fiberglass cloth to the adhesive and cover the cloth with additional adhesive. Apply a second piece of fiberglass cloth to the adhesive and immediately cover the cloth with additional adhesive in sufficient quantity to fill the weave.
9   Allow the repair to cure for 20 to 30 minutes at 60-degrees to 80-degrees F.
10   If necessary, trim the excess repair material at the edge.
11   Remove all of the paint film over and around the area(s) to be repaired. The repair material should not overlap the painted surface.
12   With a drill motor and a sanding disc (or a rotary file), cut a "V" along the break line

approximately 1/2-inch wide. Remove all dust and loose particles from the repair area.

13 Mix and apply the repair material. Apply a light coat first over the damaged area; then continue applying material until it reaches a level slightly higher than the surrounding finish.

14 Cure the mixture for 20 to 30 minute at 60-degrees to 80-degrees F.

15 Roughly establish the contour of the area being repaired with a body file. If low areas or pits remain, mix and apply additional adhesive.

16 Block sand the damaged area with sandpaper to establish the actual contour of the surrounding surface.

17 If desired, the repaired area can be temporarily protected with several light coats of primer. Because of the special paints and techniques required for flexible body panels, it is recommended that the vehicle be taken to a paint shop for completion of the body repair.

Steel body panels

Repairing simple dents

18 When repairing dents, the first job is to pull the dent out until the affected area is as close as possible to its original shape. There is no point in trying to restore the original shape completely as the metal in the damaged area will have stretched on impact and cannot be restored to its original contours. It is better to bring the level of the dent up to a point that is about 1/8-inch below the level of the surrounding metal. In cases where the dent is very shallow, it is not worth trying to pull it out at all.

19 If the backside of the dent is accessible, it can be hammered out gently from behind using a soft-face hammer. While doing this, hold a block of wood firmly against the opposite side of the metal to absorb the hammer blows and prevent the metal from being stretched.

20 If the dent is in a section of the body which has double layers, or some other factor makes it inaccessible from behind, a different technique is required. Drill several small holes through the metal inside the damaged area, particularly in the deeper sections. Screw long, self-tapping screws into the holes just enough for them to get a good grip in the metal. Now pulling on the protruding heads of the screws with locking pliers can pull out the dent.

21 The next stage of repair is the removal of paint from the damaged area and from an inch or so of the surrounding metal. This is easily done with a wire brush or sanding disk in a drill motor, although it can be done just as effectively by hand with sandpaper. To complete the preparation for filling, score the surface of the bare metal with a screwdriver or the tang of a file or drill small holes in the affected area. This will provide a good grip for the filler material. To complete the repair, see the Section on filling and painting.

Repair of rust holes or gashes

22 Remove all paint from the affected area and from an inch or so of the surrounding metal using a sanding disk or wire brush mounted in a drill motor. If these are not available, a few sheets of sandpaper will do the job just as effectively.

23 With the paint removed, you will be able to determine the severity of the corrosion and decide whether to replace the whole panel, if possible, or repair the affected area. New body panels are not as expensive as most people think and it is often quicker to install a new panel than to repair large areas of rust.

24 Remove all trim pieces from the affected area except those which will act as a guide to the original shape of the damaged body, such as headlight shells, etc. Using metal snips or a hacksaw blade, remove all loose metal and any other metal that is badly affected by rust. Hammer the edges of the hole in to create a slight depression for the filler material.

25 Wire-brush the affected area to remove the powdery rust from the surface of the metal. If the back of the rusted area is accessible, treat it with rust inhibiting paint.

26 Before filling is done, block the hole in some way. This can be done with sheet metal riveted or screwed into place, or by stuffing the hole with wire mesh.

27 Once the hole is blocked off, the affected area can be filled and painted. See the following subsection on filling and painting.

Filling and painting

28 Many types of body fillers are available, but generally speaking, body repair kits which contain filler paste and a tube of resin hardener are best for this type of repair work. A wide, flexible plastic or nylon applicator will be necessary for imparting a smooth and contoured finish to the surface of the filler material. Mix up a small amount of filler on a clean piece of wood or cardboard (use the hardener sparingly). Follow the manufacturer's instructions on the package, otherwise the filler will set incorrectly.

29 Using the applicator, apply the filler paste to the prepared area. Draw the applicator across the surface of the filler to achieve the desired contour and to level the filler surface. As soon as a contour that approximates the original one is achieved, stop working the paste. If you continue, the paste will begin to stick to the applicator. Continue to add thin layers of paste at 20-minute intervals until the level of the filler is just above the surrounding metal.

30 Once the filler has hardened, the excess can be removed with a body file. From then on, progressively finer grades of sandpaper should be used, starting with a 180-grit paper and finishing with 600-grit wet-or-dry paper. Always wrap the sandpaper around a flat rubber or wooden block, otherwise the surface of the filler will not be completely flat. During the sanding of the filler surface, the wet-or-dry paper should be periodically rinsed in water. This will ensure that a very smooth finish is produced in the final stage.

31 At this point, the repair area should be surrounded by a ring of bare metal, which in turn should be encircled by the finely feathered edge of good paint. Rinse the repair area with clean water until all of the dust produced by the sanding operation is gone.

32 Spray the entire area with a light coat of primer. This will reveal any imperfections in the surface of the filler. Repair the imperfections with fresh filler paste or glaze filler and once more smooth the surface with sandpaper. Repeat this spray-and-repair procedure until you are satisfied that the surface of the filler and the feathered edge of the paint are perfect. Rinse the area with clean water and allow it to dry completely.

33 The repair area is now ready for painting. Spray painting must be carried out in a warm, dry, windless and dust free atmosphere. These conditions can be created if you have access to a large indoor work area, but if you are forced to work in the open, you will have to pick the day very carefully. If you are working indoors, dousing the floor in the work area with water will help settle the dust that would otherwise be in the air. If the repair area is confined to one body panel, mask off the surrounding panels. This will help minimize the effects of a slight mismatch in paint color. Trim pieces such as chrome strips, door handles, etc., will also need to be masked off or removed. Use masking tape and several thickness of newspaper for the masking operations.

34 Before spraying, shake the paint can thoroughly, then spray a test area until the spray painting technique is mastered. Cover the repair area with a thick coat of primer. The thickness should be built up using several thin layers of primer rather than one thick one. Using 600-grit wet-or-dry sandpaper, rub down the surface of the primer until it is very smooth. While doing this, the work area should be thoroughly rinsed with water and the wet-or-dry sandpaper periodically rinsed as well. Allow the primer to dry before spraying additional coats.

35 Spray on the top coat, again building up the thickness by using several thin layers of paint. Begin spraying in the center of the repair area and then, using a circular motion, work out until the whole repair area and about two inches of the surrounding original paint is covered. Remove all masking material 10 to 15 minutes after spraying on the final coat of paint. Allow the new paint at least two weeks to harden, then use a very fine rubbing compound to blend the edges of the new paint into the existing paint. Finally, apply a coat of wax

4 Body repair - major damage

1 Major damage must be repaired by an auto body shop specifically equipped to perform body and frame repairs. These shops have the specialized equipment required to do the job properly.

2 If the damage is extensive, the frame must be checked for proper alignment or the vehicle's handling characteristics may be adversely affected and other components may wear at an accelerated rate.

3 Due to the fact that all of the major body components (hood, fenders, etc.) are separate and replaceable units, any seriously damaged components should be replaced rather than repaired. Sometimes the components can be found in a wrecking yard that specializes in used vehicle components, often at considerable savings over the cost of new parts.

11-4   Chapter 11   Body

These photos illustrate a method of repairing simple dents. They are intended to supplement *Body repair - minor damage* in this Chapter and should not be used as the sole instructions for body repair on these vehicles.

1 If you can't access the backside of the body panel to hammer out the dent, pull it out with a slide-hammer-type dent puller. In the deepest portion of the dent or along the crease line, drill or punch hole(s) at least one inch apart . . .

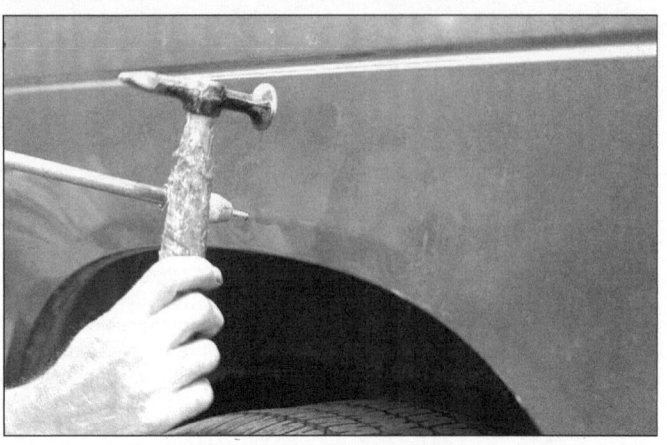

2 . . . then screw the slide-hammer into the hole and operate it. Tap with a hammer near the edge of the dent to help 'pop' the metal back to its original shape. When you're finished, the dent area should be close to its original contour and about 1/8-inch below the surface of the surrounding metal

3 Using coarse-grit sandpaper, remove the paint down to the bare metal. Hand sanding works fine, but the disc sander shown here makes the job faster. Use finer (about 320-grit) sandpaper to feather-edge the paint at least one inch around the dent area

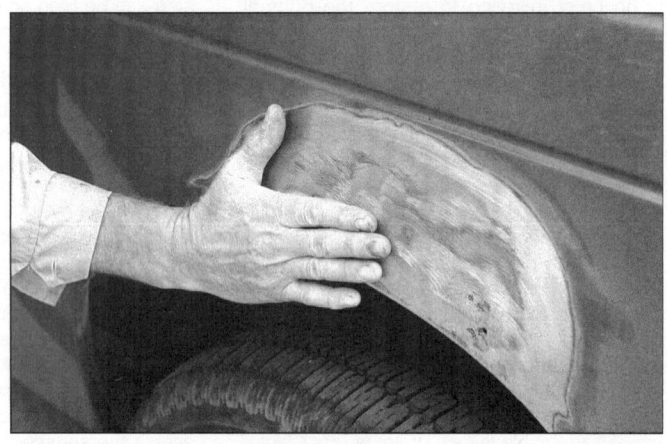

4 When the paint is removed, touch will probably be more helpful than sight for telling if the metal is straight. Hammer down the high spots or raise the low spots as necessary. Clean the repair area with wax/silicone remover

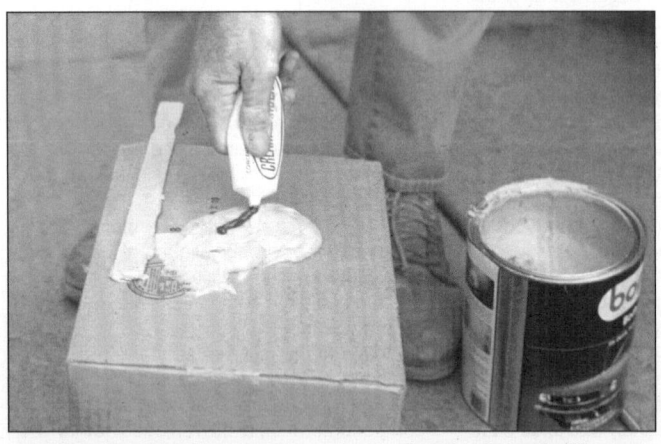

5 Following label instructions, mix up a batch of plastic filler and hardener. The ratio of filler to hardener is critical, and, if you mix it incorrectly, it will either not cure properly or cure too quickly (you won't have time to file and sand it into shape)

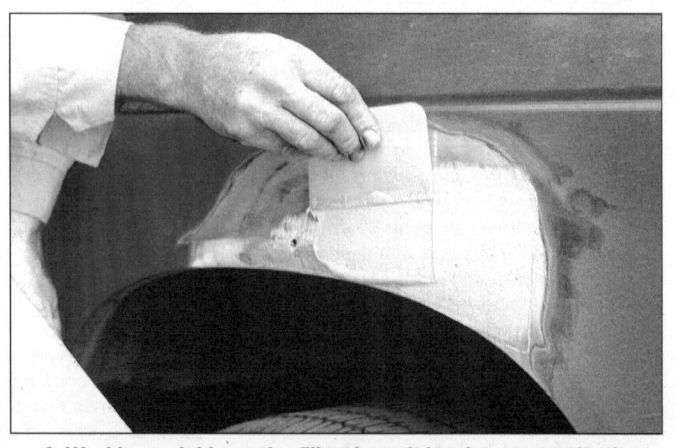

6 Working quickly so the filler doesn't harden, use a plastic applicator to press the body filler firmly into the metal, assuring it bonds completely. Work the filler until it matches the original contour and is slightly above the surrounding metal

Chapter 11 Body

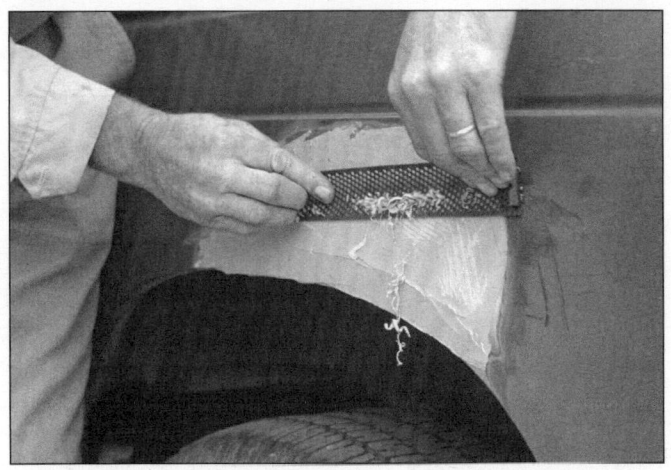

11-5

7 Let the filler harden until you can just dent it with your fingernail. Use a body file or Surform tool (shown here) to rough-shape the filler

8 Use coarse-grit sandpaper and a sanding board or block to work the filler down until it's smooth and even. Work down to finer grits of sandpaper - always using a board or block - ending up with 360 or 400 grit

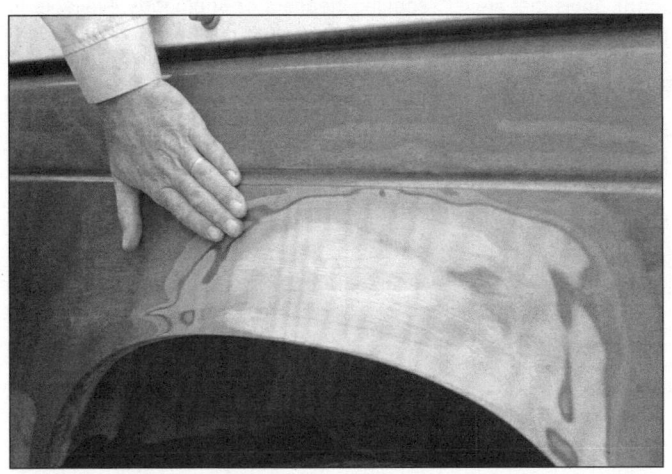

9 You shouldn't be able to feel any ridge at the transition from the filler to the bare metal or from the bare metal to the old paint. As soon as the repair is flat and uniform, remove the dust and mask off the adjacent panels or trim pieces

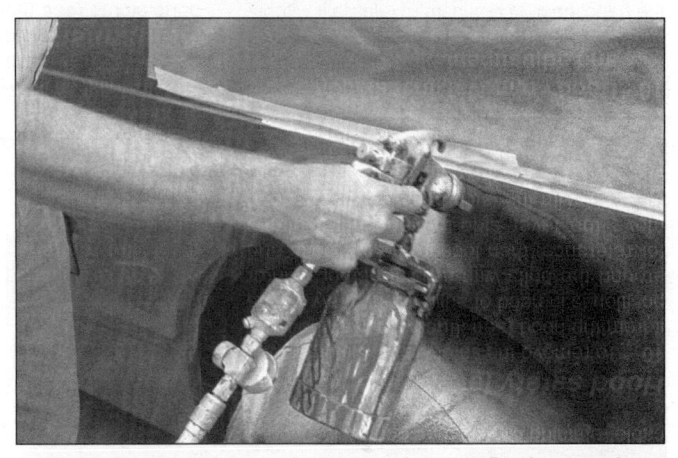

10 Apply several layers of primer to the area. Don't spray the primer on too heavy, so it sags or runs, and make sure each coat is dry before you spray on the next one. A professional-type spray gun is being used here, but aerosol spray primer is available inexpensively from auto parts stores

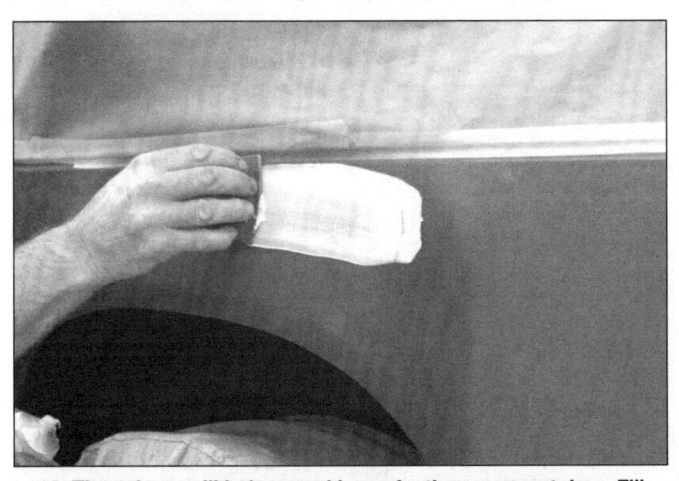

11 The primer will help reveal imperfections or scratches. Fill these with glazing compound. Follow the label instructions and sand it with 360 or 400-grit sandpaper until it's smooth. Repeat the glazing, sanding and respraying until the primer reveals a perfectly smooth surface

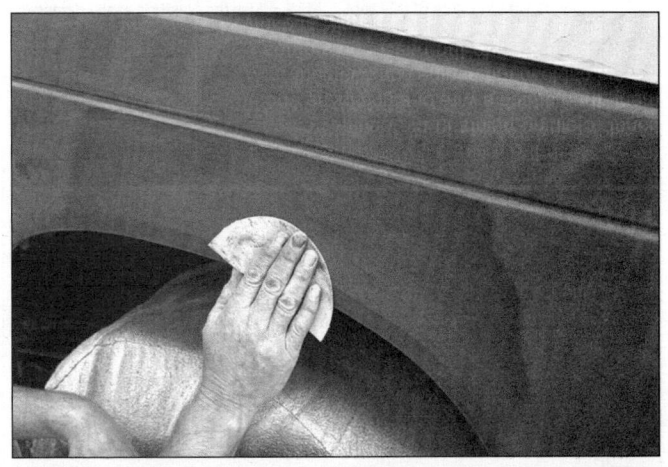

12 Finish sand the primer with very fine sandpaper (400 or 600-grit) to remove the primer overspray. Clean the area with water and allow it to dry. Use a tack rag to remove any dust, then apply the finish coat. Don't attempt to rub out or wax the repair area until the paint has dried completely (at least two weeks)

5 Fastener and trim removal

1   There is a variety of plastic fasteners used to hold trim panels, splash shields and other parts in place in addition to typical screws, nuts and bolts. Once you are familiar with them, they can usually be removed without too much difficulty.

2   The proper tools and approach can prevent added time and expense to a project by minimizing the number of broken fasteners and/or parts.

3   The following illustration shows various types of fasteners that are typically used on most vehicles and how to remove and install them (see illustration). Replacement fasteners are commonly found at most auto parts stores, if necessary.

4   Trim panels are typically made of plastic and their flexibility can help during removal. The key to their removal is to use a tool to pry the panel near its retainers to release it without damaging surrounding areas or breaking-off any retainers. The retainers will usually snap out of their designated slot or hole after force is applied to them. Stiff plastic tools designed for prying on trim panels are available at most auto parts stores (see illustration). Tools that are tapered and wrapped in protective tape, such as a screwdriver or small pry tool, are also very effective when used with care.

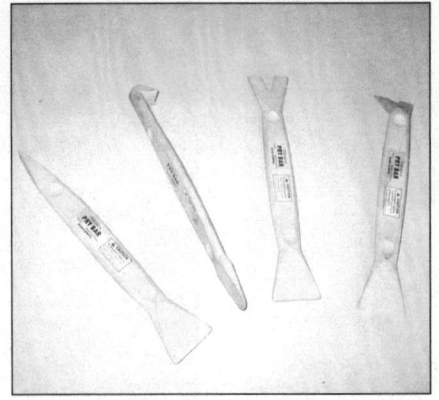

5.4 These small plastic pry tools are ideal for prying off trim panels

6 Upholstery, carpets and vinyl trim - maintenance

Upholstery and carpets

1   Every three months remove the floormats and clean the interior of the vehicle (more frequently if necessary). Use a stiff whiskbroom to brush the carpeting and loosen dirt and dust, then vacuum the upholstery and carpets thoroughly, especially along seams and crevices.

2   Dirt and stains can be removed from carpeting with basic household or automotive carpet shampoos available in spray cans. Follow the directions and vacuum again, then use a stiff brush to bring back the "nap" of the carpet.

3   Most interiors have cloth or vinyl upholstery, either of which can be cleaned and maintained with a number of material-specific cleaners or shampoos available in auto supply stores. Follow the directions on the product for usage, and always spot-test any upholstery cleaner on an inconspicuous

Fasteners

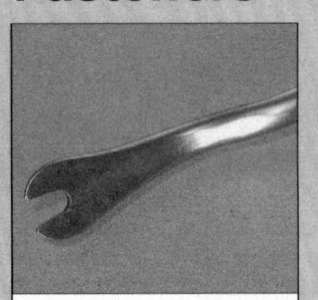

This tool is designed to remove special fasteners. A small pry tool used for removing nails will also work well in place of this tool

A Phillips head screwdriver can be used to release the center portion, but light pressure must be used because the plastic is easily damaged. Once the center is up, the fastener can easily be pried from its hole

Here is a view with the center portion fully released. Install the fastener as shown, then press the center in to set it

This fastener is used for exterior panels and shields. The center portion must be pried up to release the fastener. Install the fastener with the center up, then press the center in to set it

This type of fastener is used commonly for interior panels. Use a small blunt tool to press the small pin at the center in to release it . . .

. . . the pin will stay with the fastener in the released position

Reset the fastener for installation by moving the pin out. Install the fastener, then press the pin flush with the fastener to set it

This fastener is used for exterior and interior panels. It has no moving parts. Simply pry the fastener from its hole like the claw of a hammer removes a nail. Without a tool that can get under the top of the fastener, it can be very difficult to remove

area (bottom edge of a backseat cushion) to ensure that it doesn't cause a color shift in the material.

4 After cleaning, vinyl upholstery should be treated with a protectant.
Note: *Make sure the protectant container indicates the product can be used on seats - some products may make a seat too slippery.*
Caution: *Do not use protectant on vinyl-covered steering wheels.*

5 Leather upholstery requires special care. It should be cleaned regularly with saddle-soap or leather cleaner. Never use alcohol, gasoline, nail polish remover or thinner to clean leather upholstery.

6 After cleaning, regularly treat leather upholstery with a leather conditioner, rubbed in with a soft cotton cloth. Never use car wax on leather upholstery.

7 In areas where the interior of the vehicle is subject to bright sunlight, cover leather seating areas of the seats with a sheet if the vehicle is to be left out for any length of time.

Vinyl trim

8 Don't clean vinyl trim with detergents, caustic soap or petroleum-based cleaners. Plain soap and water works just fine, with a soft brush to clean dirt that may be ingrained. Wash the vinyl as frequently as the rest of the vehicle.

9 After cleaning, application of a high-quality rubber and vinyl protectant will help prevent oxidation and cracks. The protectant can also be applied to weather-stripping, vacuum lines and rubber hoses, which often fail as a result of chemical degradation, and to the tires.

7 Hinges and locks - maintenance

1 Once every 3000 miles, or every three months, the hinges and latch assemblies on the doors, hood and trunk should be given a few drops of light oil or lock lubricant. The door latch strikers should also be lubricated with a thin coat of grease to reduce

9.2a Pry the hood release cable from above the LH-side headlight housing . . .

wear and ensure free movement. Lubricate the door and trunk locks with spray-on graphite lubricant.

8 Windshield and fixed glass - replacement

1 Replacement of the windshield and fixed glass requires the use of special fast-setting adhesive/caulk materials and some specialized tools and techniques. These operations should be left to a dealer service department or a shop specializing in glass work.

9 Hood latch release cable, release lever and safety lever - removal and installation

Note: *The hood latch release cable is divided into two sections on this model - one section is attached to the hood latch, and the other section is connected to the*

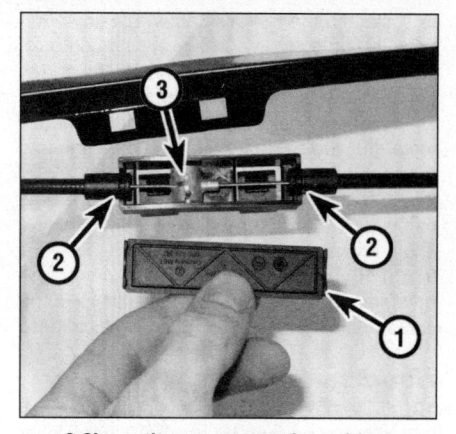

9.2b . . . then separate the cables.

1. Pry off the cover with a screwdriver
2. Release the cable housing from either side of the coupling
3. Swivel and detach the cable ends

release lever inside the vehicle.
Caution: *Do not close the hood with the release cable detached from the release lever or hood latch.*

Latch release cable and interior release lever

1 Open the hood.
2 Pry the release cable coupling off the metal support above the LH-side headlamp housing, then open the plastic cover and separate the cables **(see illustrations)**.
3 From inside the vehicle, insert a small screwdriver into the gap between operating lever and clip, then pry the clip back and remove the lever **(see illustration)**.
4 Pry up the front section of the sill trim and driver's side kick panel (see Section 28).
5 Pull the cable housing from the release lever and disconnect the cable end from the release arm **(see illustration)**.

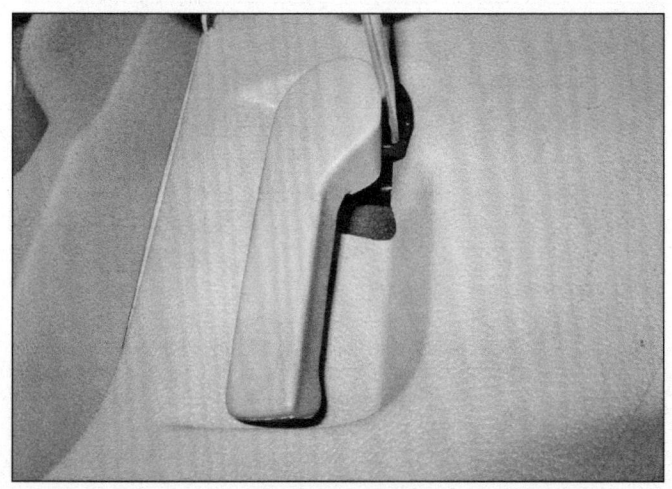

9.3 Insert a screwdriver into the gap and pry back the retaining clip

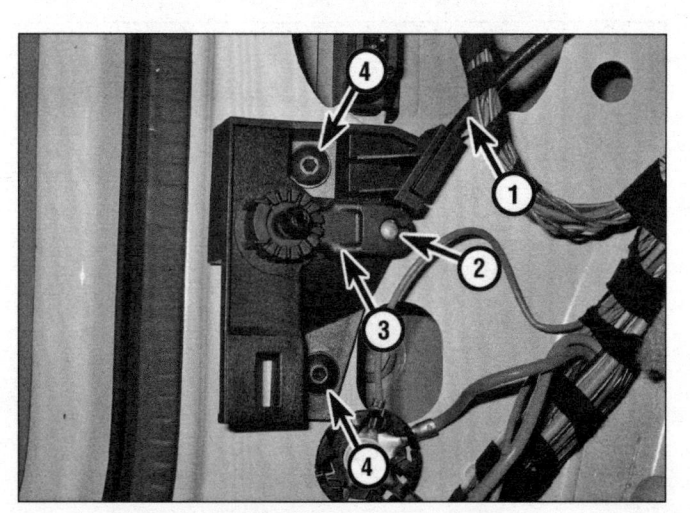

9.5 Hood release handle details

1. Hood release cable housing
2. Hood release cable end
3. Release lever arm
4. Release lever mounting fasteners

6 Detach the release cable from any additional retaining points in the engine compartment, then attach a wire or string to the end of the old cable in the engine compartment.
7 Working inside the passenger compartment, pull the cable through the cowl and into the passenger compartment.
8 Connect the string or wire to the new cable and pull it through the cowl into the engine compartment.
9 The remainder of installation is the reverse of removal. Be sure to fasten all of the cable retaining clips in their original locations.

Hood safety lever

10 To remove the safety lever, simply unbolt it from the hood **(see illustration)**. If the lever position is in need of adjustment, draw circles around the bolts with a paint pen or marker for reference, then loosen the two bolts and move the lever side-to-side until it aligns squarely with the latch below when the hood is closed. Tighten the bolts securely.

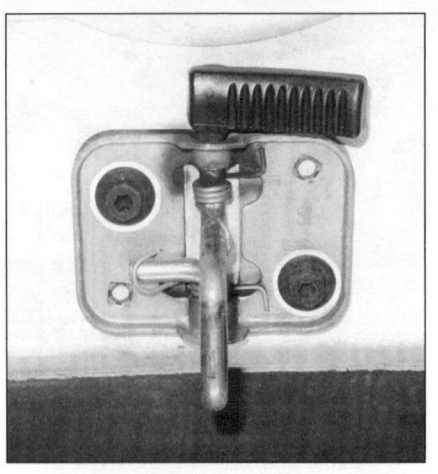

10 Hood - removal, installation and adjustment

Note: *The hood is awkward to remove and install - at least two people should perform this procedure.*

Removal and installation

1 Use blankets or pads to cover the cowl area of the body and the fenders.
2 Pry out and disconnect the harness / hoses from the washer jets at the bottom of the hood, then pry out the grommet and pull the harness / hoses out of the hood.
3 Make alignment marks around the hinge plates using a permanent marker to insure the same installation **(see illustration)**.
4 Have an assistant support the hood, then detach the upper end of the hood support strut if equipped (see Section 11).
5 Remove the hinge-to-hood nuts and lift off the hood.
6 Installation is the reverse of removal.

Adjustment

7 Fore-and-aft and side-to-side adjustment of the hood is done by moving the hood in relation to the hinge plates after loosening the nuts.
8 Loosen the nuts and move the hood into correct alignment. Move it only a little at a time. Tighten the hinge nuts and carefully lower the hood to check the alignment.
9 Adjust the hood bumpers on the hood by turning them in or out to make the hood flush with the fenders when closed **(see illustrations)**.
10 The hood latch assembly or safety lever can also be adjusted up-and-down and side-to-side after loosening the nuts (see Section 9 and Section 11).

9.10 Draw circles around the bolts for reference when adjusting the hood safety lever

10.3 Hood hinge details

11 Hood latch and support strut - removal and installation

Latch

1 Open the hood and remove the front bumper cover (see Section 12).
Note: *Some models may not require the front bumper cover to be removed to access the latch mounting bolts.*
2 Open the release coupling and disconnect the hood latch cable from the mid-way disconnection point (see Section 9).
3 Remove the latch fasteners, disconnect the electrical connector from the latch assembly, then lift the latch assembly out of the radiator support **(see illustrations)**.
4 With the latch removed, disconnect the release cable and hood latch switch from the latch assembly (see illustrations). Transfer the components to the new latch as necessary.
5 Installation is the reverse of removal.

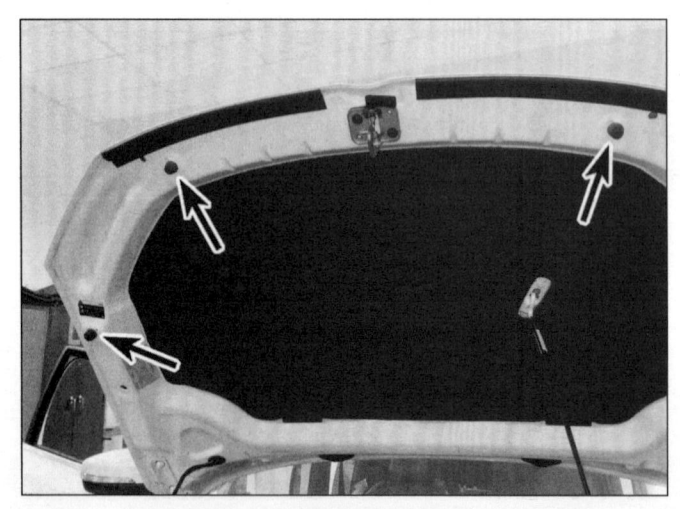

10.9a Locations of the hood bumpers on hood (one not shown)

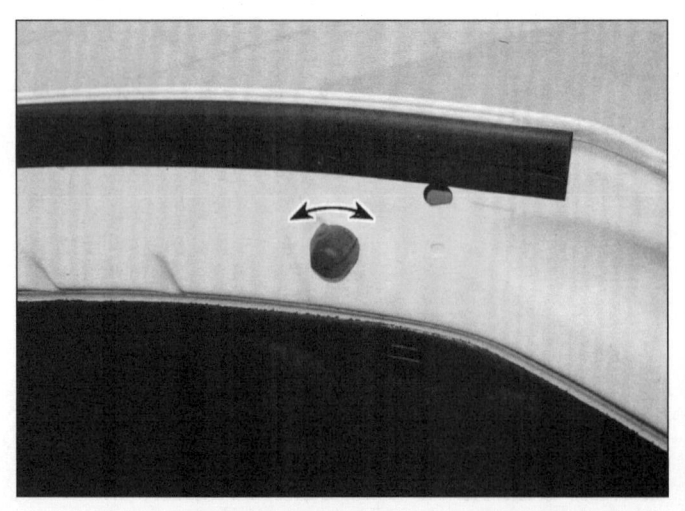

10.9b Adjust the hood height by screwing the hood bumpers in or out

Chapter 11   Body

11-9

11.3a Hood latch mounting fasteners

11.3b Hood latch electrical connector - disconnect and release from the bracket (headlight housing removed for clarity)

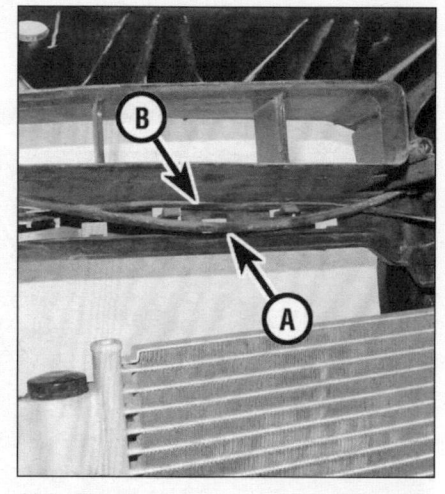

11.3c Release the hood latch release cable (A) and the latch wiring harness (B) from the clips along the radiator support

Support strut (if equipped)

6   Open the hood and support it securely.

7   Use a small screwdriver to carefully lift the retaining springs **(see illustration)** at both ends of the support strut. Then pry or pull the strut from the ballstud to detach it from the vehicle.

8   Installation is the reverse of removal.

12   Bumpers and bumper covers - removal and installation

Note: *Front and rear bumpers are composed of a fascia, or bumper (exterior) cover, and a metal structural beam. This procedure is performed based on a later model VW Jetta, in which there may be slight variations for the removal sequence on other models.*

11.3d Lift the latch assembly out of the radiator support

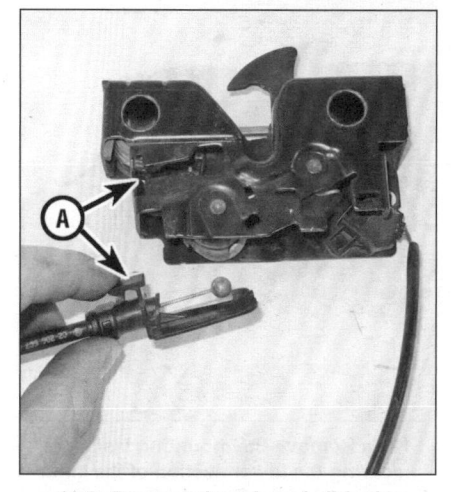

11.4a Depress the tab and slide the cable end out of the bracket (A)

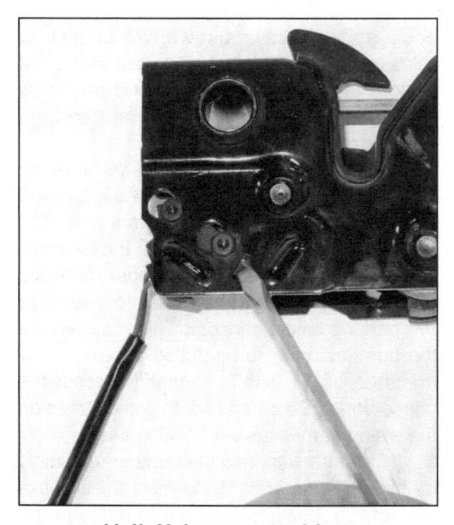

11.4b Using a screwdriver, depress the tab...

11.4c...then slide the hood latch switch out of the latch

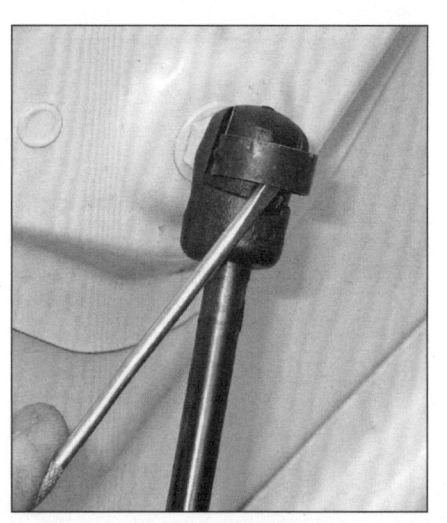

11.7 Pry the clip off to release the hood support strut

11-10 Chapter 11 Body

12.2 Bumper cover lower mounting fasteners

12.3 Bumper cover upper mounting fasteners

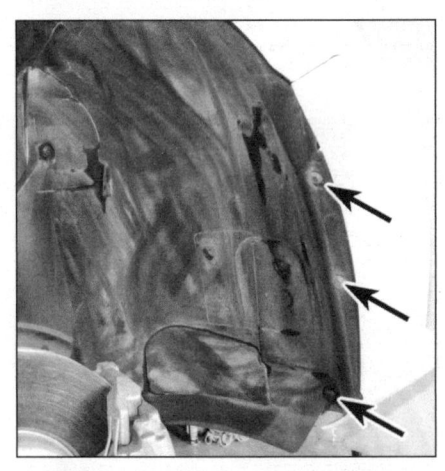

12.4a Remove the mounting fasteners securing the lower section of the inner fender splash shield to the bumper cover (right side shown, left side identical) . . .

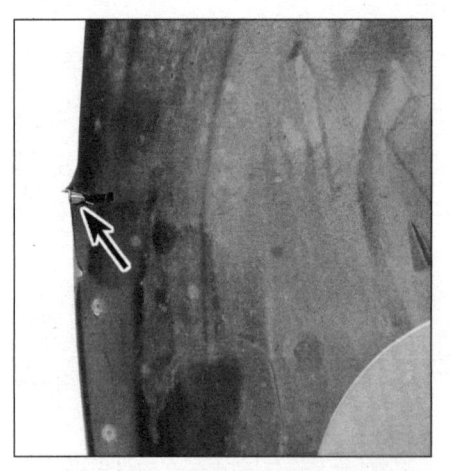

12.4b . . . then peel the splash shield back enough to access the fenderwell bumper fastener (left side shown, right side identical)

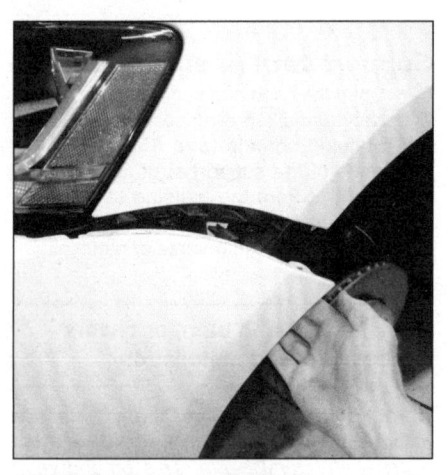

12.6 Pull out and unclip the bumper cover from the fender at each side

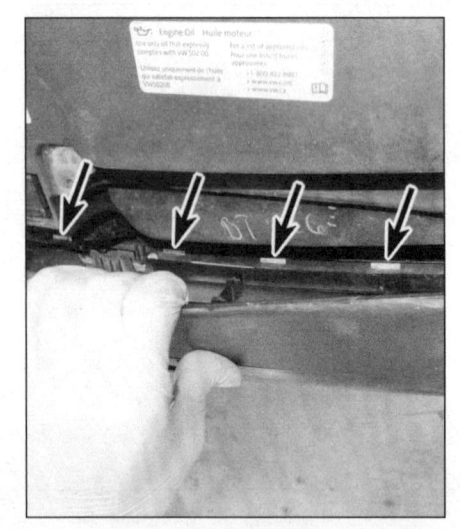

12.7 Pull the grille forward and release the bumper cover retaining tabs at the bottom of the grille

Front bumper cover

1   Raise the front of the vehicle and support it securely on jackstands. Disconnect the cable from the negative battery terminal (see Chapter 5).
2   Remove the lower bumper cover fasteners **(see illustration)**.
3   Open the hood and remove the upper mounting fasteners **(see illustration)**.
4   Remove the mounting fasteners and peel back the lower section of the fenderwell splash shield enough to access the fenderwell bumper fastener at each side **(see illustrations)**.
5   If equipped with a front spoiler (diffuser), remove the spoiler mounting fasteners from below and remove the spoiler from the bumper cover.
6   Carefully pull out and unclip the rear edges of the bumper cover at each side, separating the bumper cover from the fenderwell **(see illustration)**.
7   Pull the grille slightly forward for access, then use a screwdriver or pick to release the retaining tabs located along the bottom of the grille **(see illustration)**. This will detach the bumper cover from the front bumper reinforcement.
Note: *This Step of detaching the bumper cover from the vehicle is more easily accomplished with the help of an assistant.*
8   If equipped, disconnect the electrical connector(s) from any bumper-mounted electrical components, such as fog lights or the parking assist sensors. This may require the bumper cover to be pulled out and away from the vehicle slightly, enough to access the connector(s). Be careful not to place stress on the wiring harnesses when doing this.
9   Carefully withdraw the bumper assembly by sliding it squarely away from the front of the vehicle, being careful not to damage the paint. If equipped with headlight wipers, pull the bumper out to the point where the washer hose can be disconnected before fully withdrawing the bumper.

Chapter 11   Body   11-11

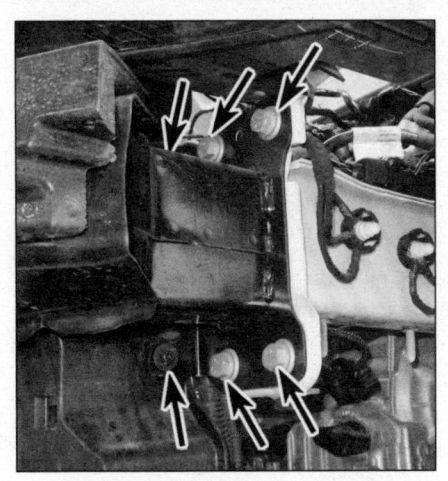

12.11 Remove the bumper support beam bracket bolts from both sides of the vehicle

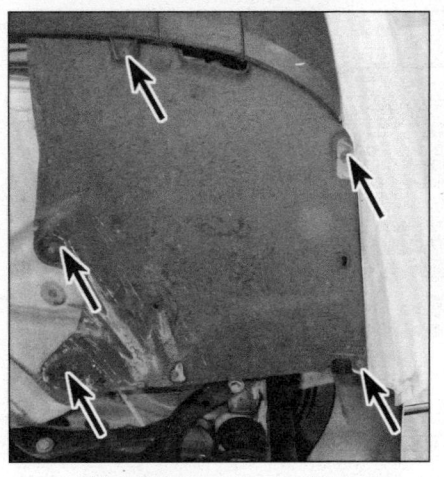

12.14a Evaporator canister shield fasteners

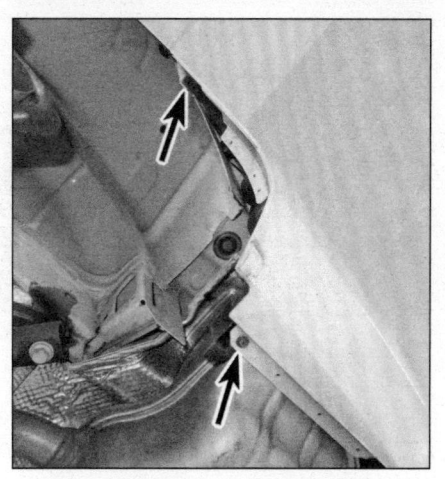

12.14b Rear bumper cover wheelwell fasteners (wheel and wheelwell liner removed for clarity)

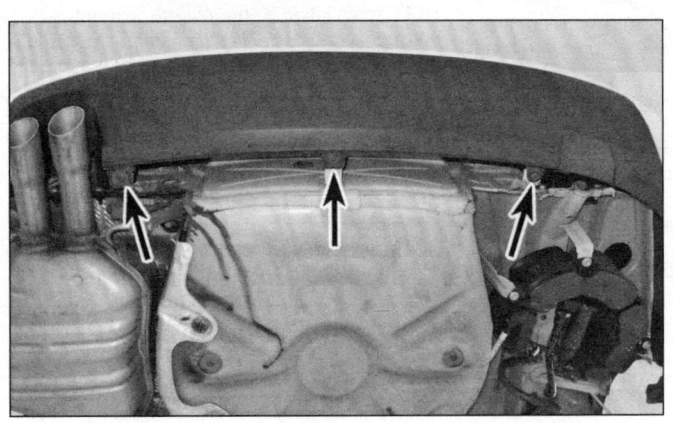

12.15 Rear bumper cover lower mounting fasteners

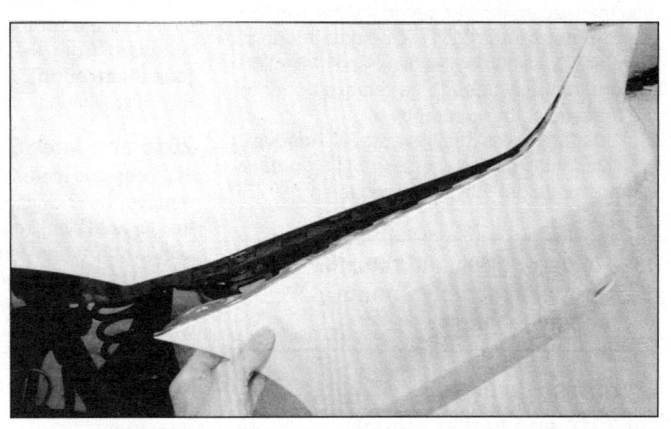

12.16 Unclip the trailing edge of the bumper cover from its retaining bracket, then pull the rear bumper away from the vehicle

10   Installation is the reverse of removal. Loosely install all threaded fasteners, to obtain proper alignment, before fully tightening them.
Note: *If equipped with parking assist, the parking assist system may need to be calibrated after the bumper cover is installed. Visit your local independent repair shop or dealership that has the appropriate equipment and is capable of calibrating the system.*

Front bumper support beam

11   To remove the bumper support beam from the vehicle, remove the bolts from each side that secure the bumper to the frame **(see illustration)**.
12   Installation is the reverse of removal. Loosely install all fasteners, to obtain proper alignment, before fully tightening them.

Rear bumper cover

13   Remove the body-mounted taillight housing (see Chapter 12).
14   Working at the rear wheelwells, remove the fasteners from the liner at the lower rear corner, and on the right-hand side, remove the evaporator canister shield. Then, fold the wheelwell liner back and remove the fasten-

ers securing the bumper cover to the wheel-well **(see illustrations)**.
15   Remove the rear bumper cover lower fasteners from under the bumper **(see illustration)**.
16   Disengage the sides of the rear bumper cover by grasping the upper edge of the bumper and pulling it away from the rear quarter panel to release it **(see illustration)**.
17   Release the bumper cover upper tabs from the taillight housing recess **(see illustration)**.
Note: *This Step of detaching the bumper cover from the vehicle is more easily accomplished with the help of an assistant.*
18   If equipped, disconnect the electrical connector(s) from any bumper-mounted electrical components, such as the parking assist sensors. This may require the bumper cover to be pulled out and away from the vehicle slightly, enough to access the connector(s). Be careful not to place stress on the wiring harnesses when doing this.
19   Carefully withdraw the bumper assembly by sliding it squarely away from the rear of the vehicle, being careful not to damage the paint.

20   Installation is the reverse of removal. Loosely install all threaded fasteners, to obtain proper alignment, before fully tightening them.
Note: *If equipped with parking assist, the parking assist system may need to be calibrated after the bumper cover is installed. Visit your local independent repair shop or dealership that has the appropriate equipment and is capable of calibrating the system.*

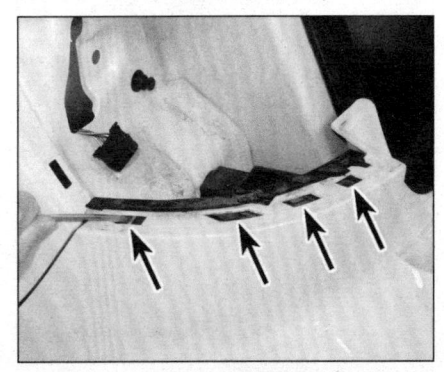

12.17 Using a small flat tip screwdriver, pry up on the tabs to release the bumper cover

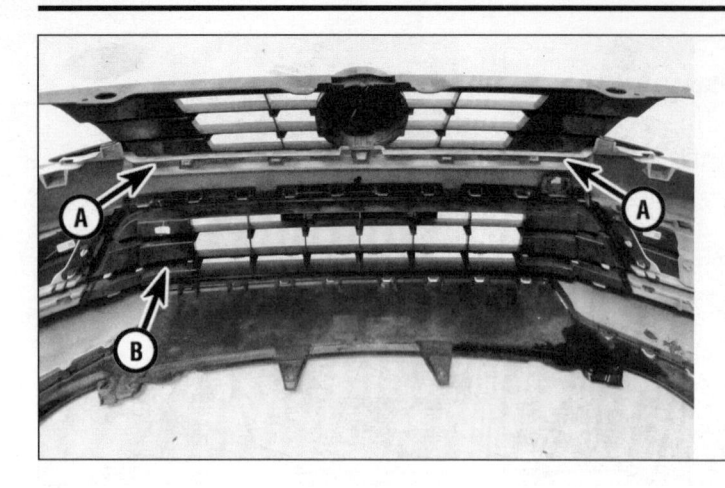

13.5 Upper grille tabs / hooks (between A arrows), and lower grille (B)

Rear bumper support beam

21 To remove the bumper support beam from the vehicle, remove the bolts from each side that secure the bumper to the frame (similar to **illustration 12.11**). On some models, there are brackets for the blind-spot detection system that may need to be removed prior to the removal of the support beam.

22 Installation is the reverse of removal. Loosely install all fasteners, to obtain proper alignment, before fully tightening them.

13 Radiator grille and radiator shutter assembly - removal and installation

Radiator grille

All 2014 and earlier models

1 Open the hood and remove the fasteners along the upper grill.

2 Use a small screwdriver to disengage the locking tabs along the lower edge of the grille, then remove the grille by pulling it forward.

3 Installation is the reverse of removal.

2015 and later Jetta / GLI models

4 Remove the front bumper cover (see Section 12).

5 Pry up on the tabs and hooks securing the upper grille to the bumper cover, then pull the grille forward to separate it from the bumper cover. The lower grille is removed similar to that of the upper grille, but moved rearward from the bumper cover, instead **(see illustration)**.

6 Installation is the reverse of removal.

2015 and later Golf / GTI models

7 For the removal of the upper grille, remove the 2 bolts from each end on top of the grille. Then, release the grille retaining clips along the inside of the grille while moving it forward slightly. Finally, separate the grille from the bumper cover by pulling it forward and upward.

8 For the removal of the lower grille, first remove the bumper cover (see Section 12), then remove it as described in **illustration 13.5**).

Radiator shutter assembly (if equipped)

Note: *The radiator shutter assembly is used to control air flow into the radiator and engine. It is designed to promote better engine cooling functionality when they are open, and greater vehicle aerodynamic properties when they are closed.*

Note: *To maximize the cooling capacity of the vehicle, the radiator shutters need to be in operating condition at all times. Service codes can be stored if the radiator shutter motor is not operating correctly.*

9 Remove front bumper cover (see Section 12).

10 Disconnect the electrical connections to the shutter assembly, located near the left-hand end of the radiator / shutter assembly. Also unclip the electrical connector brackets.

11 Remove the bolts securing the shutter assembly to the radiator at each side.

12 Pull the shutter assembly along with the surrounding seals forward enough to gain access for separating the shutter from the seals. Brace the shutter assembly from falling out as you separate the seals surrounding the shutter assembly - there are several clips around the shutter that require disengagement.

13 Remove the shutter assembly by sliding it out towards the bottom of the vehicle.

14 Installation is the reverse of removal.

Note: *When installing the shutter assembly, pay close attention to the seals around the unit - making sure they are seated against the shutter and bumper carrier area. There should not be any gaps present between the shutter and the radiator.*

14 Fenderwell splash shields - removal and installation

Note: *Some models have slight variations in the actual fastener locations, but are similar to the locations shown in this procedure - which was photographed on a later model VW Jetta*

Front

1 Remove the bolts and expanding nuts securing the fenderwell splash shield and mud guards **(see illustrations)**. Remove the mud guards, then remove the splash shield from the fenderwell, being careful not to tear the splash shield.

2 Installation is the reverse of removal. Avoid overtightening the fasteners.

14.1a Front fenderwell splash shield and mud guard fasteners (rear)

14.1b Front fenderwell mud guard fastener (rear)

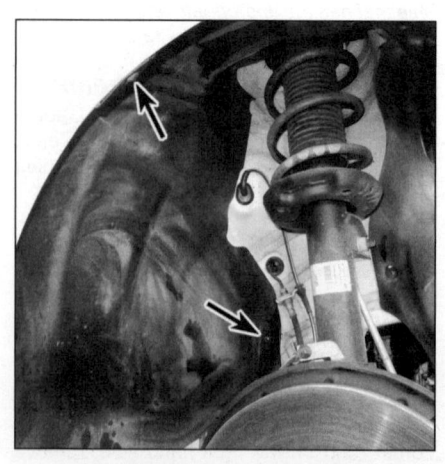

14.1c Front fenderwell splash shield fasteners (rear)

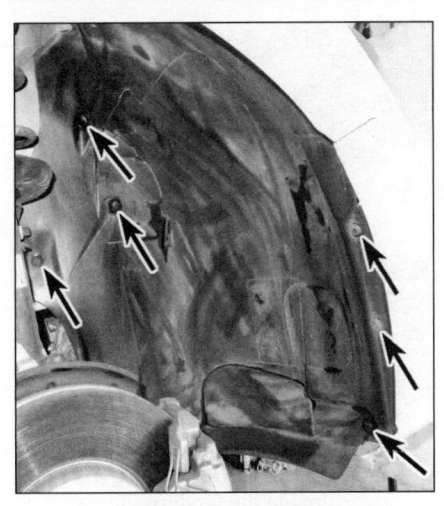

Chapter 11 Body

14.1d Front fenderwell splash shield fasteners (front)

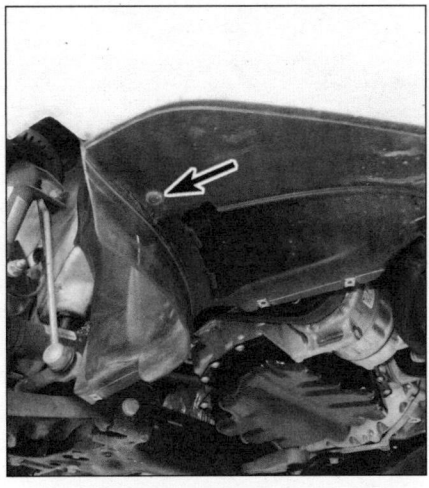

14.1e Front fenderwell splash shield fastener (front)

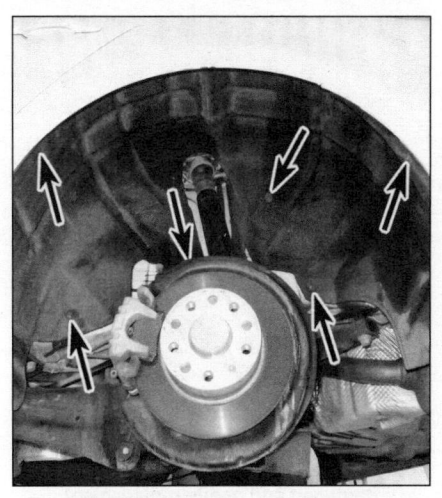

14.3a Rear fenderwell splash shield main fasteners (locations may vary)

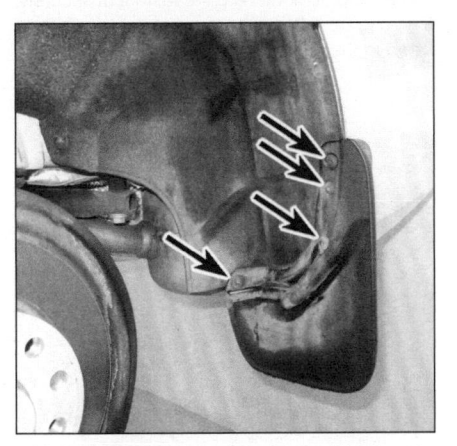

14.3b Rear fenderwell splash shield and mud guard fasteners (rear)

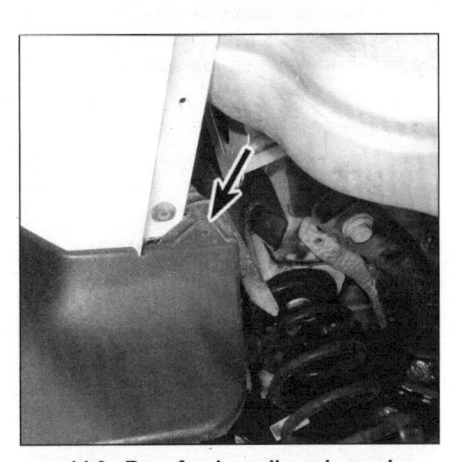

14.3c Rear fenderwell mud guard fastener (rear)

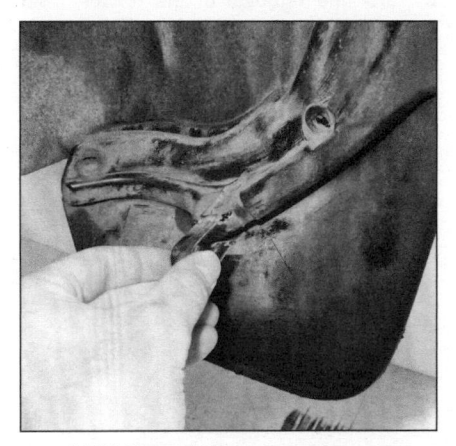

14.3d Pry-out the rear mud guard retaining clip with a screwdriver, then remove the mud guard

Rear

Warning: *Some models may use a gas-tight expanding nut or nuts that retain part of the splash shield, but also seal off the vehicle's interior from harmful exhaust gases. If a gas-tight expanding nut is damaged in any way, replace the nut.*

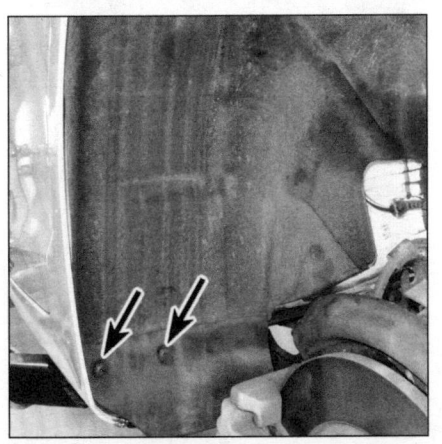

3   Remove the bolts, expanding nuts, and, if used, the expanding gas-tight nut(s) securing the fenderwell splash shield and mud guards **(see illustrations)**. Remove the mud guards, then remove the splash shield from the fenderwell, being careful not to tear the splash shield.

4   Installation is the reverse of removal. Avoid overtightening the fasteners.

15 Front fender - removal and installation

Note: *It's a good idea to have an assistant support the fender while it's being moved away from the vehicle to prevent damage to the surrounding body panels. The fender fastener locations may vary slightly depending on the model and trim level.*

1   Loosen the wheel bolts, then raise the front of the vehicle and support it securely on jackstands. Remove the wheel.
2   Remove the lower splash shield below the engine (see Chapter 1, Section 6).
3   Remove the front bumper cover (see Section 12) and cowl cover (see Section 16).
4   Remove the inner fender splash shield fasteners, then remove the splash shield (see Section 14).
5   Remove the headlamp assembly for the fender being replaced (see Chapter 12).

14.3e Rear fenderwell splash shield and mud guard fasteners (front)

14.3f Rear fenderwell mud guard fastener (front)

11-14   Chapter 11   Body

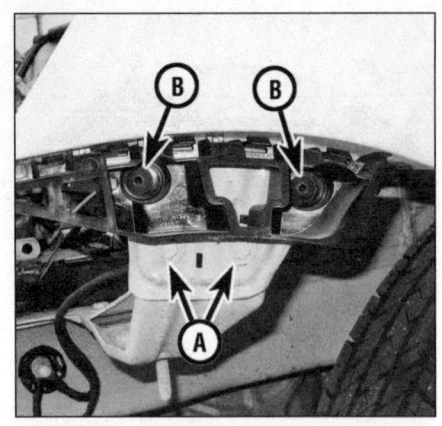

15.6a Remove the fender front fasteners (A) - only remove the plastic moulding (B) if it is damaged and needs replacement

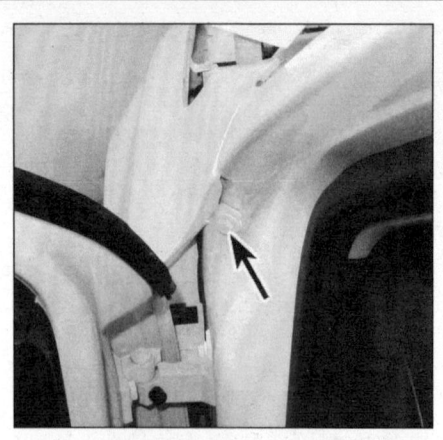

15.6b Open the door to expose the upper-rear fender fastener

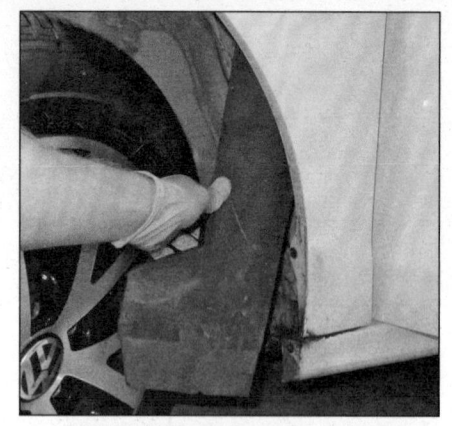

15.6c Turn the wheel as necessary, then remove the foam piece from inside the fenderwell

6   For the removal of the fender, refer to the illustrations below, following them in sequence **(see illustrations)**.

7   Installation is the reverse of removal. Loosely install all fender fasteners, then align the fender evenly with the surrounding body panels prior to tightening the fasteners securely.

8   Tighten the wheel bolts to the torque listed in the Chapter 1 Specifications.

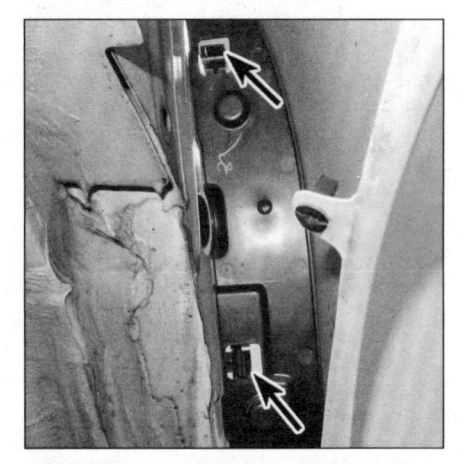

15.6d Depress the tabs and remove the plastic piece . . .

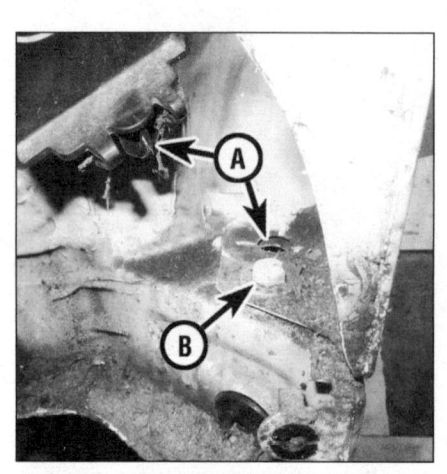

15.6e . . . then lift out the plastic piece, noting the placement of the notch and corresponding hole (A) - Then remove the rocker panel bolt (B)

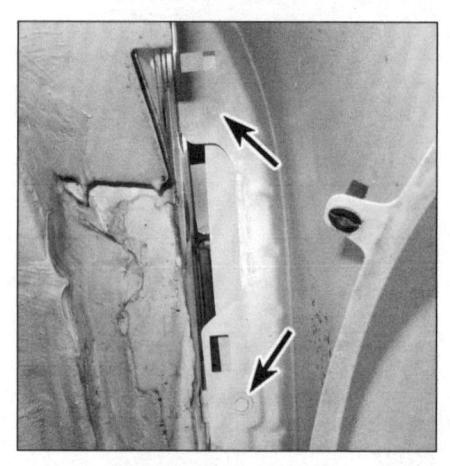

15.6f Remove the inner fenderwell bolts

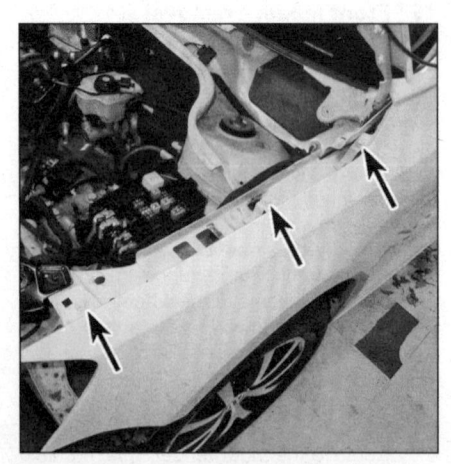

15.6g Remove the bolts along the top of the fender

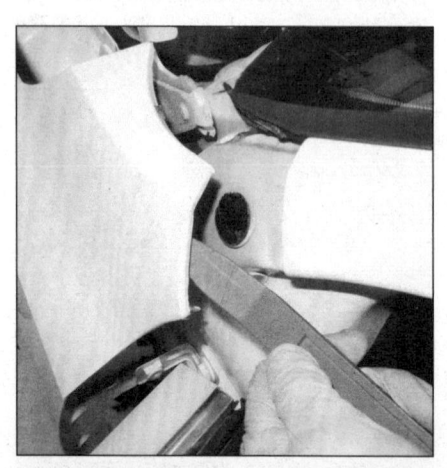

15.6h Use a plastic trim tool to carefully pry the upper-rear edge of the fender free from the adhesive

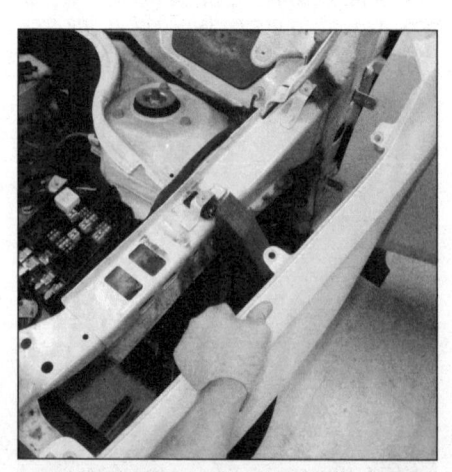

15.6i Carefully remove the fender while avoiding scratching the paint - this step may be easier accomplished with the help of an assistant

Chapter 11 Body   11-15

15.6j Reuse the fender shape retaining / dampening foam for installation - make sure it is correctly oriented, as shown

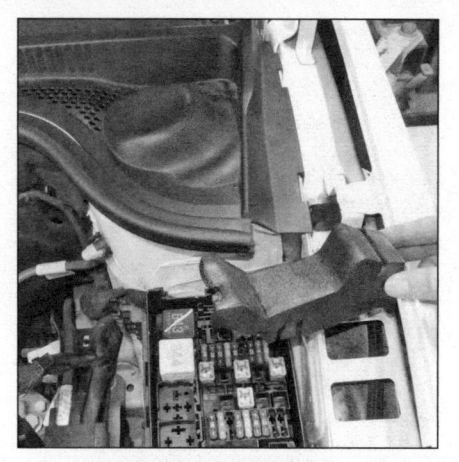

16.2a Remove the left side cowl cover foam seal . . .

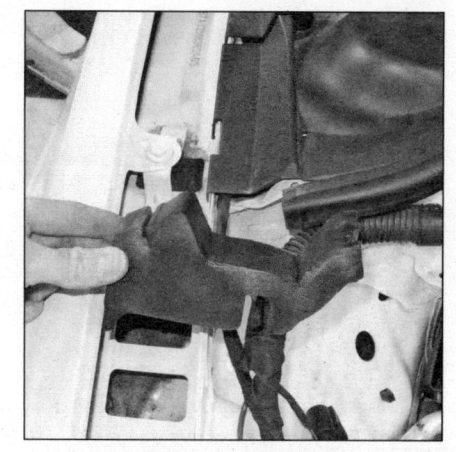

16.2b . . . and the right side seal, noting how they're installed.

16 Cowl cover - removal and installation

1   Open the hood and remove the windshield wiper arms (see Chapter 12).

2   Remove the side foam seals from each end of the cowl covers **(see illustrations)**.

3   Remove the rubber cowl cover retaining seal that runs along the bulkhead **(see illustration)**.

4   Remove the cowl covers, starting with the left side cover **(see illustrations)**.

Caution: *Do not use any sharp or metal tools to pry the cowling cover upward - You can easily damage the windshield in the process*

5   Installation is the reverse of removal. Use only hand pressure to press the cowling covers into the retainers. Avoid using a hammer or any sharp instrument to install the cowling covers - if done correctly, the covers should fit evenly into the groove on the windshield by working them in place by hand only.

16.3 Remove the rubber seal

16.4a Carefully release the left cowl cover from the groove along the bottom of the windshield - this may require the end of the cover to the pulled up sharply at first . . .

16.4b . . . then guide out the left cowl cover from the engine compartment

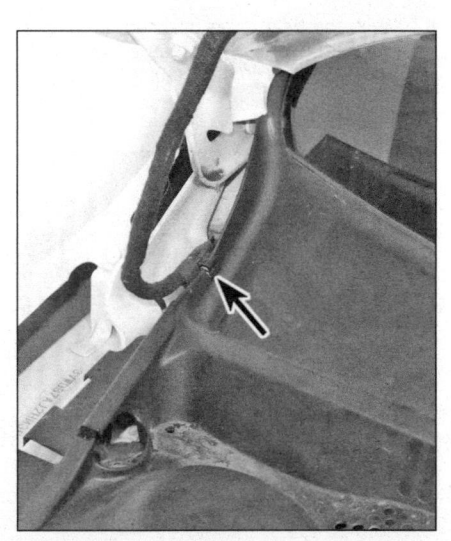

16.4c Release the washer hose harness clip from the right side cowl cover . . .

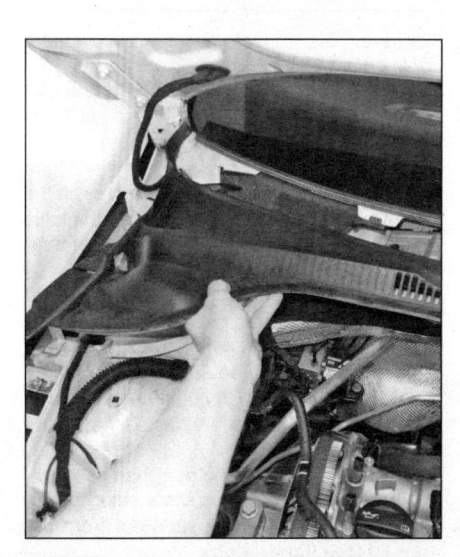

16.4d . . . then guide out the right cowl cover from the engine compartment

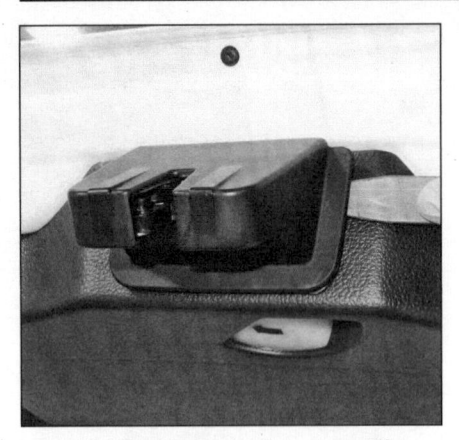

11-16 Chapter 11 Body

17.2 Pry the protective cover off of the trunk latch

17.3 Remove the trunk lid trim panel mounting screws from the grab handle recesses (one on each side)

17.4 Use a plastic trim tool to pry off the trunk lid trim panel

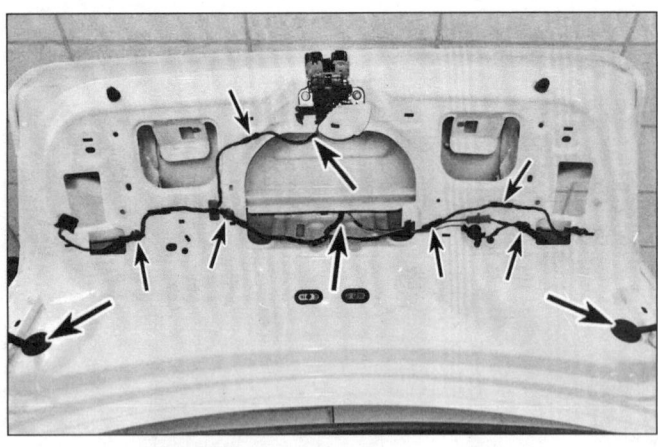

17.5 Trunk lid electrical harness connectors, retaining points, and grommets

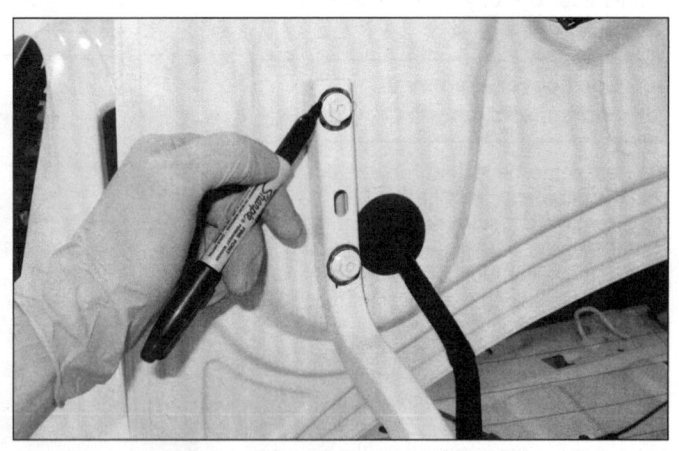

17.6 Mark the relationship of the trunk lid by drawing around the outside of each hinge mounting bolt

17 Trunk lid (sedan models) - removal, installation and adjustment

Warning: *Disconnect the negative battery cable before beginning (see Chapter 5).*

Note: *The trunk lid is heavy and somewhat awkward to remove and install - at least two people should perform this procedure.*

1 Open the trunk lid and cover the edges of the trunk compartment with pads or cloths to protect the painted surfaces when the lid is removed. Disconnect the cable from the negative battery terminal (see Chapter 5).
2 Carefully pry the cover off of the trunk latch with a plastic trim tool **(see illustration)**.
3 Remove the trim panel mounting fasteners from the trunk lid **(see illustration)**.
4 Use a trim tool to pry off the trunk lid trim panel **(see illustration)**, releasing it from the various retaining clips.
5 Disconnect the electrical harness connectors, pry out the harness sealing grommets, and pull the harnesses out from the trunk lid **(see illustration)**.
6 Mark the relationship of the trunk lid by drawing around the outside of each hinge mounting bolt with a marker **(see illustration)**. This will aid in easier alignment when installing.
7 Remove the trunk lid support struts (see Section 18).
8 With the help of an assistant to support the trunk lid, remove the hinge-to-trunk lid retaining nuts, and lift the lid off the hinges.
9 Installation is the reverse of removal. Check the trunk lid for correct alignment. If necessary, loosen the hinge bolts to adjust, then retighten them; there should be an even gap of approximately 1/8 inch (3 mm) between the outside edge of the trunk lid and the surrounding bodywork. Adjust the trunk lid height (if necessary) by rotating the rubber bumper stop on either side, in or out as necessary **(see illustrations)**.

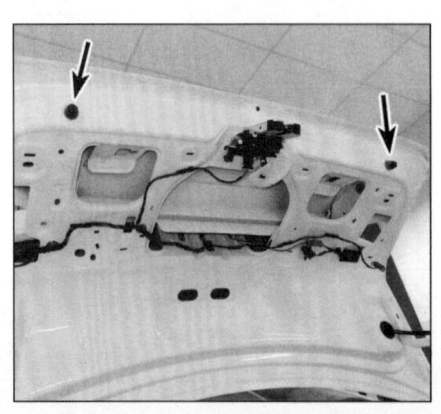

17.9a Rubber bumper height-adjustment stop locations (trunk lid)

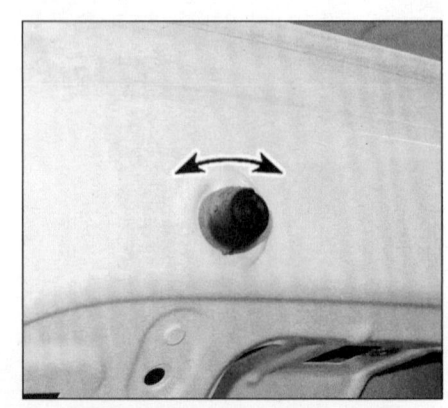

17.9b Rotate the bumper stop as necessary to adjust the height of the trunk lid when closed

Chapter 11 Body

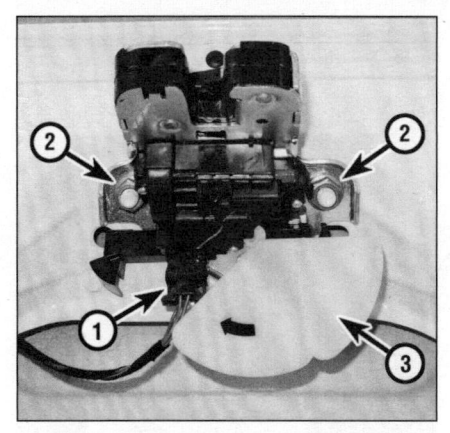

18.2 Trunk latch details

1. Electrical connector
2. Mounting bolts
3. Trunk interior release lever (part of latch assembly)

18 Trunk lid latch, support struts and release switch/backup camera assembly (sedan models) - removal and installation

Warning: *Disconnect the negative battery cable before beginning (see Chapter 5).*
Note: *On models equipped, the trunk lock cylinder is built into the trunk latch/actuator - it is not available for separate replacement, as it must be replaced as a whole unit.*

Latch

1 Open the trunk and disconnect the cable from the negative battery terminal (see Chapter 5). Remove the inner trim panel from the trunk lid (see Section 17).
Note: *Some models may have a plastic trim cover that fits over the latch, that requires removal to access all mounting bolts. This cover can be removed by depressing the retaining tab and sliding it off the latch.*
2 Disconnect the electrical connector from the latch **(see illustration)**.

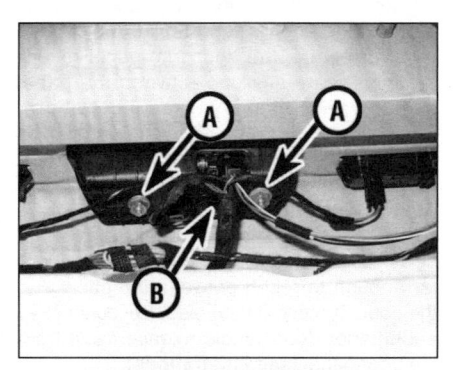

18.9 Trunk release switch / backup camera electrical connector and retaining nut locations

A *Retaining nuts*
B *Electrical connector(s)*

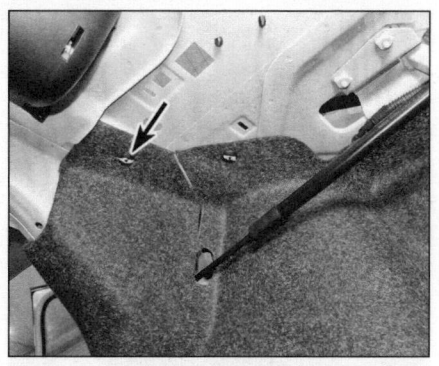

18.5a Remove this carpet retaining clip . . .

3 Outline the installed position of the latch bracket with a marker. Remove the mounting nuts, then remove the latch assembly.
4 Installation is the reverse of removal.
Note: *When reinstalling the trunk lid latch, align the marks made during removal.*

Support struts

Warning: *Have an assistant support the trunk upright while the lift supports are in the process of being replaced, or injury may occur.*
5 Open the trunk, then remove the upper carpet retaining pin. Peel back the carpet to gain access to the inner lift support strut retaining clip **(see illustrations)**. While supporting the trunk, detach the support strut ends and remove it from the trunk.
6 For more information on prying and releasing the support strut retaining clips, see Section 11.
7 Installation is the reverse of removal.

Release switch and backup camera assembly

8 Disconnect the cable from the negative battery terminal (see Chapter 5), then remove trunk lid trim panel (see Section 17).
9 Disconnect the electrical connectors and remove the switch retaining nuts **(see illustration)**.
10 Remove the switch assembly from the outside of the trunk lid.
11 Installation is the reverse of removal.

19.8 Liftgate hinge fasteners

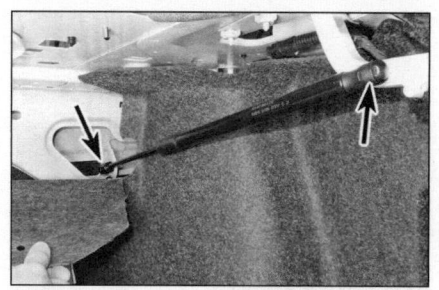

18.5b . . . then peel back the carpet to allow access to both support strut clips.

19 Liftgate (coupe and wagon models) - removal, installation and adjustment

Warning: *Disconnect the negative battery cable before beginning (see Chapter 5).*
Note: *The liftgate is heavy and somewhat awkward to remove and install - at least two people should perform this procedure.*

1 Open the liftgate, then disconnect the cable from the negative battery terminal (see Chapter 5). Remove the upper trim paneling bordering the window by using a plastic trim tool to release the paneling retaining clips.
2 Remove the fasteners located near the grab handles, then carefully pry the lower section of the trim panel free from the liftgate by grasping the grab handles and pulling down sharply. A trim tool can be used to release the remaining clips.
3 Disconnect the wiring from the liftgate components (lock switch, wiper motor, and rear light units) at the connectors. Note the routing and attachment locations of the wires, then pry out the grommets and feed the wiring out through the grommet holes. Remove the center-high mounted brake light (see Chapter 12).
4 Mark the relationship between the liftgate and its hinges using a marker.
5 With an assistant to help support the liftgate, use a small screwdriver to carefully lift the retaining springs at both ends of the support struts. These are the same type of struts mentioned in Section 18.
6 While disengaging the retaining springs, pry or pull the strut from the ballstud to detach it from the vehicle. Repeat this at the opposite end.
7 Carefully pry the hinge covers off and remove the covers (if equipped).
8 Remove the liftgate-to-hinge fasteners **(see illustration)**, and carry the liftgate clear of the vehicle.
9 Installation is the reverse of removal. Check that the liftgate is correctly aligned before fully tightening the liftgate hinge bolts.
10 The latch assembly can also be adjusted up-and-down after loosening the nuts.
11 The closed position of the liftgate can be adjusted by altering the positions of the rubber liftgate stops mounted near the top of the liftgate opening.
12 Unclip the plastic cover from the liftgate stop and adjust the center adjusting screw in or out as necessary.

11-18　　　　　　　　　　　　　　　　Chapter 11　Body

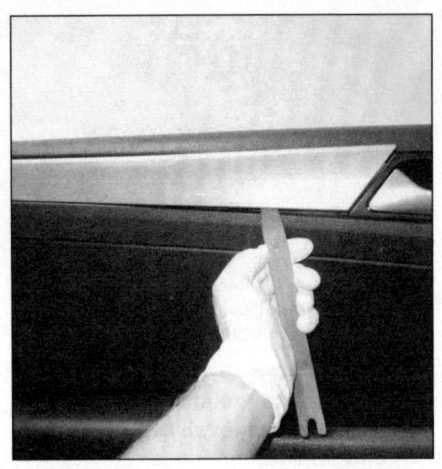

21.1a Pry the long upper decorative trim panel free it from the retaining clips

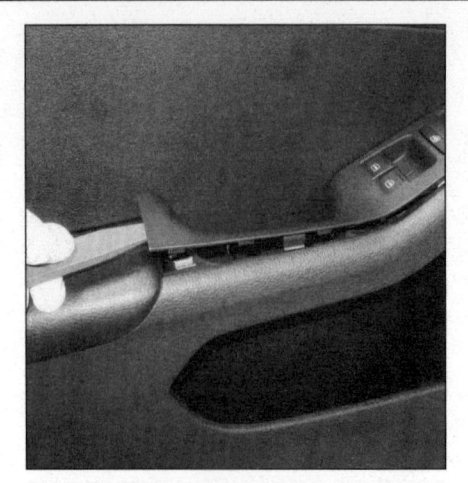

21.1b Pry up the window switch assembly from the rear . . .

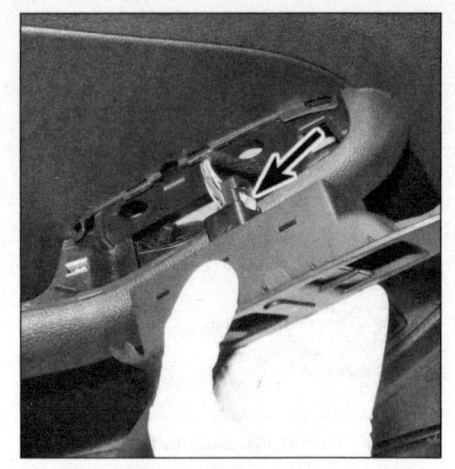

21.1c . . . then disconnect the electrical connector, firmly pushing the colored tab to release the connector

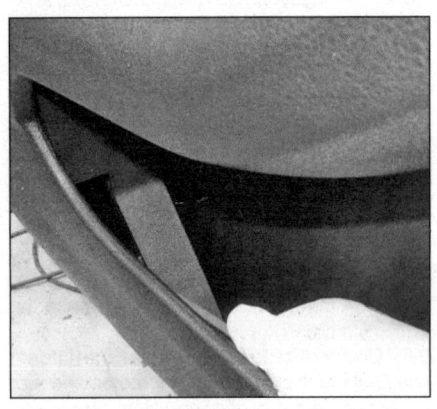

21.1d Pry the lower trim panel out using a hooked trim tool . . .

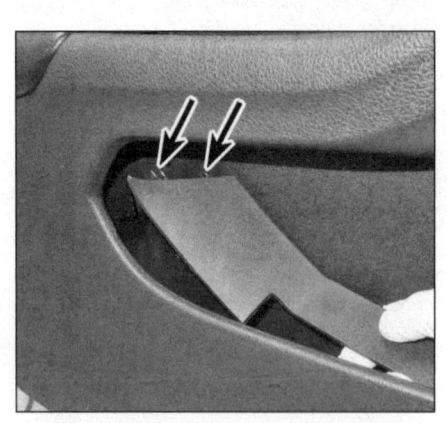

21.1e . . . then remove the lower panel, taking note of the upper tab locations

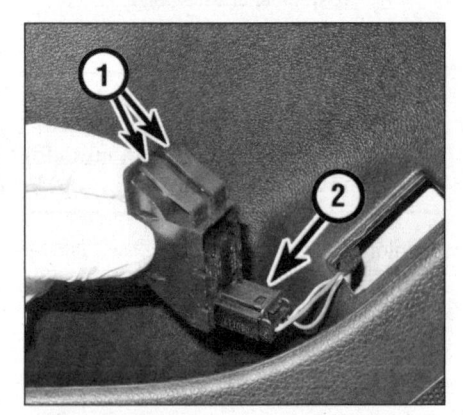

21.1f Pry out the switch located at the lower end of the door panel (driver's side), releasing it from the retaining tabs (1), then disconnect the connector (2)

20   Liftgate latch, actuator and support struts (coupe and wagon models) - removal and installation

Note: *On models equipped, the liftgate lock cylinder is built into the latch release actuator - it is not available for separate replacement, as the whole unit must be replaced.*

1   Open up the liftgate, then disconnect the cable from the negative battery terminal (see Chapter 5), and remove the trim panel as described in Section 19.

Note: *If the lock is inoperative, the liftgate can be opened manually by operating the emergency access lever from inside the vehicle - there may be a trim panel that needs to be removed to access this lever.*

Liftgate latch

2   Disconnect the wiring from the latch.
3   Remove the retaining fasteners and remove the latch from the liftgate.
4   Installation is the reverse of removal.

Liftgate handle/release actuator/rearview camera

5   If equipped, and if it blocks access, remove the tailgate wiper motor as described in Chapter 12.
6   Disconnect the wiring from the handle/release unit.
7   Remove the retaining fasteners holding the unit to the liftgate and locking brackets. Turn the unit clockwise and remove it. If equipped, disconnect the drain hose from the unit.
8   Installation is a reversal of removal - however, before installing the trim panel, check the operation of the lock components.

Support struts (all models)

9   The liftgate support struts are removed in the same manner as the hood support struts. Refer to Section 11 for this procedure.

21   Door trim panels - removal and installation

Warning: *Disconnect the negative battery cable before beginning (see Chapter 5).*

Front door trim panel

1   Remove the necessary decorative trim panels by carefully inserting a screwdriver or plastic trim tool under the trim edges and prying them out, being careful not to damage the trim pieces. Disconnect any electrical connectors as you proceed **(see illustrations)**.
2   Remove the screws securing the door panel to the door **(see illustrations)**.
3   To release the door trim panel retaining clips, carefully pry between the panel and door with a trim tool **(see illustration)**. Work around the outside of the panel, and when all the retaining clips are released, lift the door trim panel upwards and off the window slot.
4   With the door panel detached from the door, pull the door panel away from the door far enough to reach between the panel and the door. Disconnect the electrical connectors and interior door handle release cable from the door panel **(see illustrations)**.
5   Installation is the reverse of removal. Ensure that the wiring and connections are secure and correctly routed, clear of the window regulator and latch/lock components.

Chapter 11 Body   11-19

21.2a Upper end door panel screw

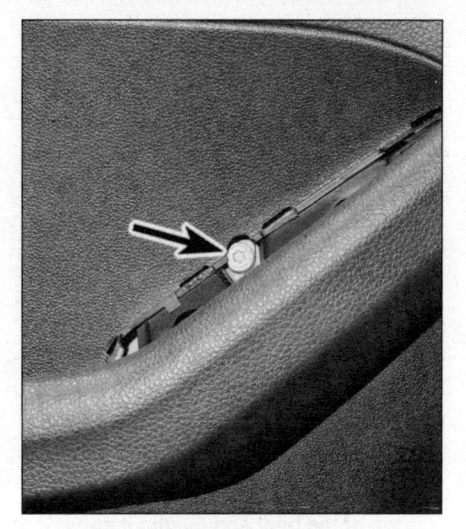

21.2b Window switch recess door panel screw

21.2c Grab handle door panel screw

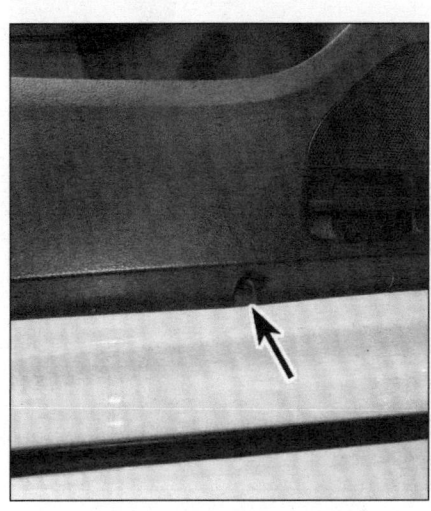

21.2d Door panel lower screw

21.3 Use a trim tool to pry the door trim panel away from the door, working around the perimeter until all the fasteners have been released

21.4a Disconnect the door electrical connectors . . .

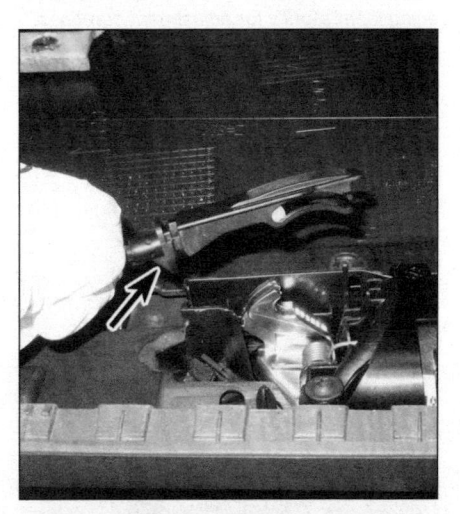

21.4b . . . then slide out the end of the interior handle cable housing and unhook it from the door panel

21.4c If the interior door handle/lock/unlock switch needs replacement, remove these bolts and depress the tabs to remove the assembly

Chapter 11 Body

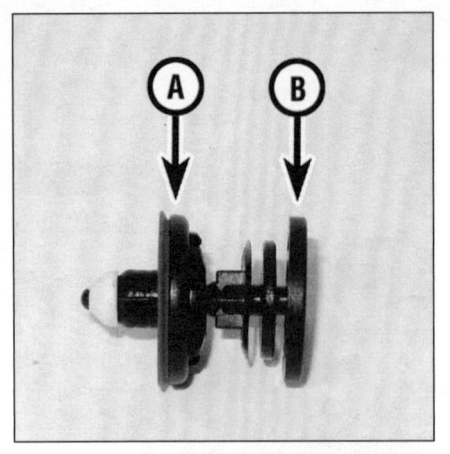

21.6 Before installing the door panel, make sure these clips are in the "install" position with the donut ring (A) pushed away from the main body (B)

22.1a Separate the rubber boot from the vehicle body by pressing the tab on the top of the boot and pulling it away from the door . . .

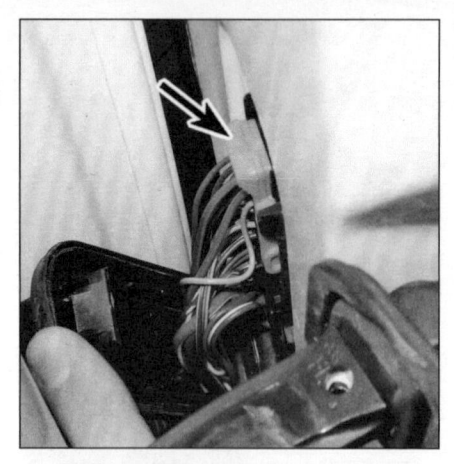

22.1b . . . then rotate the connector locking lever downwards . . .

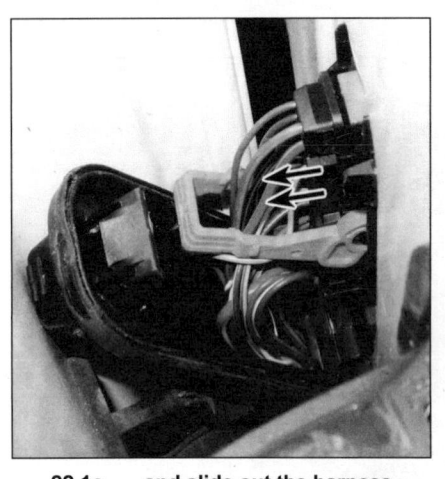

22.1c . . . and slide out the harness connector towards the door

22.2a Door mounting details

1. Door strap-to-vehicle body bolt
2. Door retaining pinch bolts (remove protective caps)
3. Door strap mounting bolts (remove only if replacing door strap)

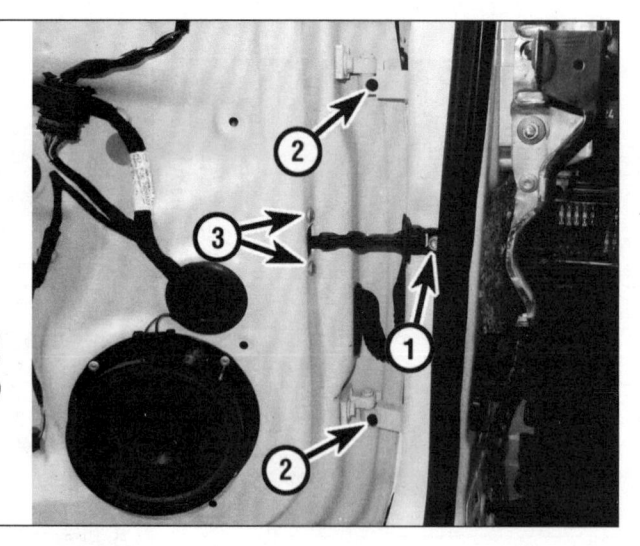

6   On some models - The plastic pressure clips used on these doors are reusable, if not damaged. However, the stud portion of the pressure clip has to be in the open position to allow it to be installed. When the pressure clip is in the closed position, the donut section pushes inward on the main body of the pressure clip allowing it to expand in the retaining hole of the door, which holds the door panel in place **(see illustration)**. Check the door panel for any missing pressure clips that may have fallen out when removing - install or replace the clips onto the panel where necessary.

Rear door trim panel

7   The rear door trim panels are removed/installed in a similar fashion as the front door trim panels - refer to Steps 1 thru 5 for this procedure.

22 Door - removal, installation and adjustment

Note: *The following procedure applies to both front and rear doors, passenger and driver side.*

Note: *The hinge bolts must always be replaced if loosened, with the exception of loosening them slightly for adjustment purposes.*
Warning: *Disconnect the negative battery*

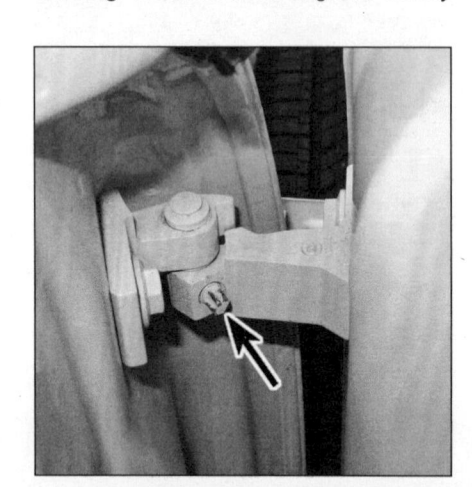

22.2b Pry off the covers, then loosen the upper and lower door retaining pinch bolts

cable before beginning (see Chapter 5).

1   Open the door and disconnect the door harness electrical connector **(see illustrations)**.

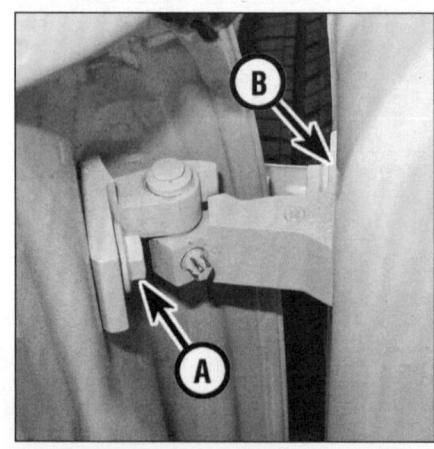

22.4a Hinge-to-door bolt (A) and hinge-to-body bolt (B- one shown, and another accessed from inside the vehicle)

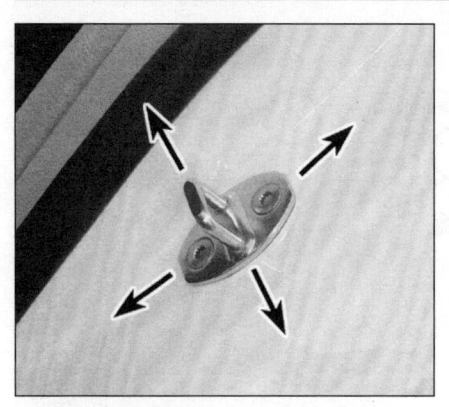

Chapter 11 Body

22.4b The striker can be loosened and moved slightly to achieve secure latch engagement

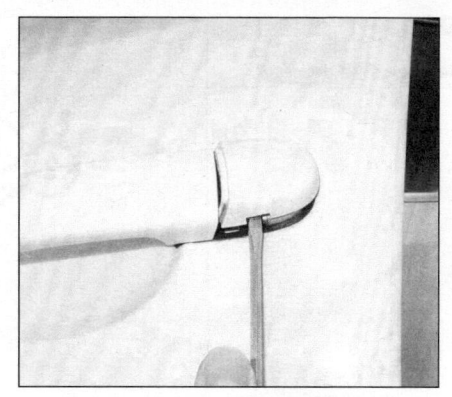

23.1 Pry off the lock cylinder cover

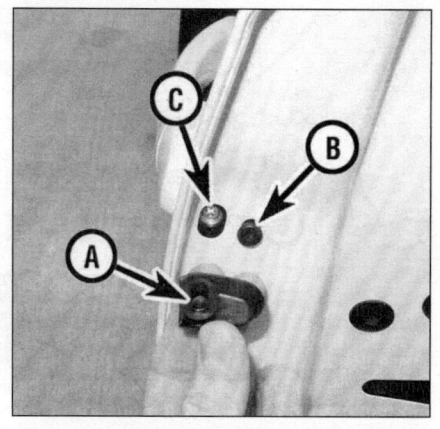

23.2a Door lock cylinder mounting details

   A   Rubber plug
   B   Locking bolt
   C   Lock cylinder retaining bolt

2   Remove the door strap mounting bolt, then push the door strap out of the way and into the door. Loosen the upper and lower door pinch bolts, then with the help of an assistant, carefully lift the door upwards and out of the brackets to remove it **(see illustrations)**. Replace the door strap, if necessary.

3   Installation is the reverse of removal. Tighten the door-related fasteners to the torque settings listed in this Chapter's Specifications.

4   Following installation of the door, check the alignment and adjust it if necessary as follows:

   a) Up-and-down and in-and-out adjustments are made by loosening the hinge-to-door bolts and moving the door as necessary **(see illustration)**.
   b) Forward-and-backward adjustments are made by loosening the hinge-to-body bolts and moving the door as necessary. To access the interior hinge-to-body bolt on the driver's side, remove and reposition the interior fuse box assembly (see Chapter 12), and on the passenger's side, remove the glove box (see Section 28).
   c) The door lock striker can also be adjusted both up-and-down and sideways to provide positive engagement

with the lock mechanism. This is done by loosening the fasteners and moving the striker as necessary **(see illustration)**.

23 Door lock cylinder, outside handle and latch - removal and installation

Lock cylinder

Note: *This Section is shown on the driver's door - The passenger and rear door components are similar.*

Warning: *Disconnect the negative battery cable before proceeding (see Chapter 5).*

1   Pry the lock cylinder cover off of the outside door handle and lock cylinder **(see illustration)**.

2   Open the driver's door and remove the rubber access plug from the end of the door. Remove the locking bolt, then loosen the lock cylinder retaining bolt until the lock cylinder can be released **(see illustration)**.

3   Installation is the reverse of removal.

Outside handle

4   Remove the lock cylinder as described in Steps 1 and 2.

5   Grasp and slide the entire handle

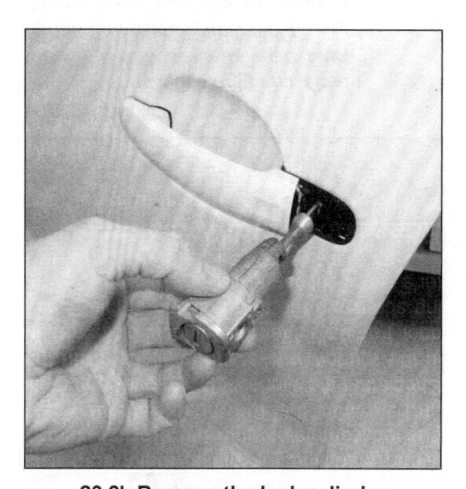

23.2b Remove the lock cylinder from inside the door

towards the rear of the door until the release mechanism is disconnected, then pull out the rest of the handle and disconnect the electrical connector, if equipped **(see illustrations)**.

6   Installation is the reverse of removal.

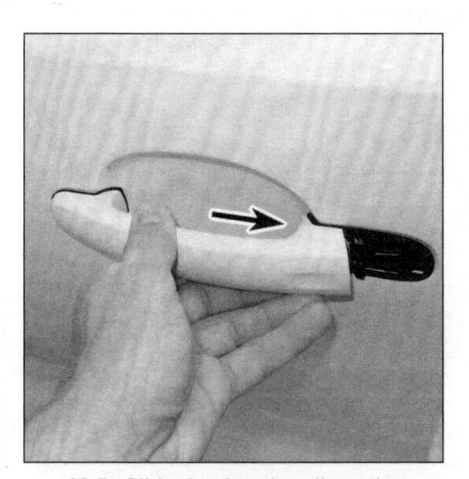

23.5a Slide the door handle to the rear of the door . . .

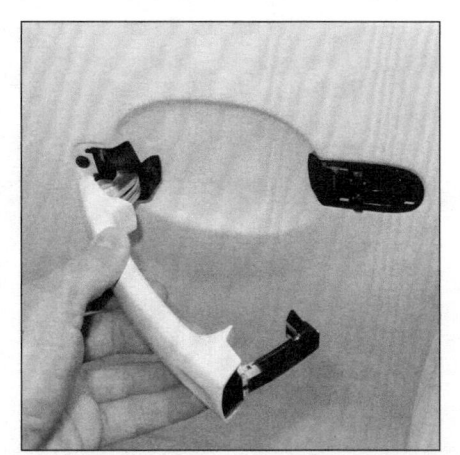

23.5b . . . then rotate the end of the door handle outwards . . .

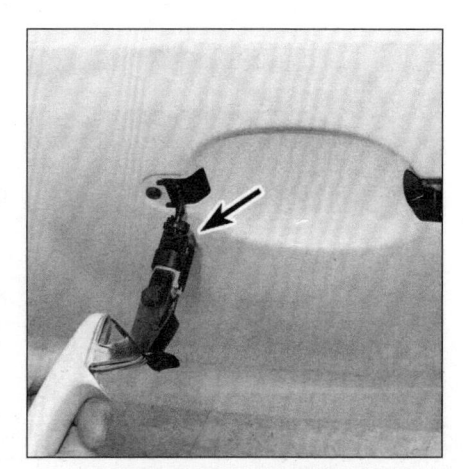

23.5c . . . and disconnect the electrical connector (if equipped)

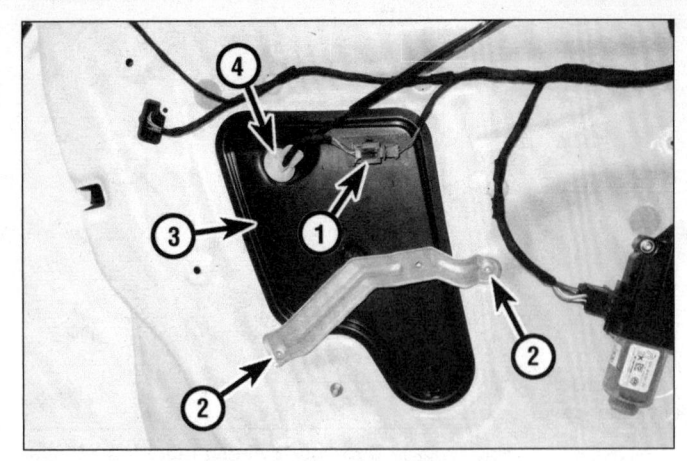

23.9 Door inner access panel details

1. Disconnect the electrical connector and pry off the connector from the panel
2. Remove the mounting bolts and door panel brace
3. Pry out the plastic panel from the door
4. Detach the grommet from the panel, then slide the panel off of the cable

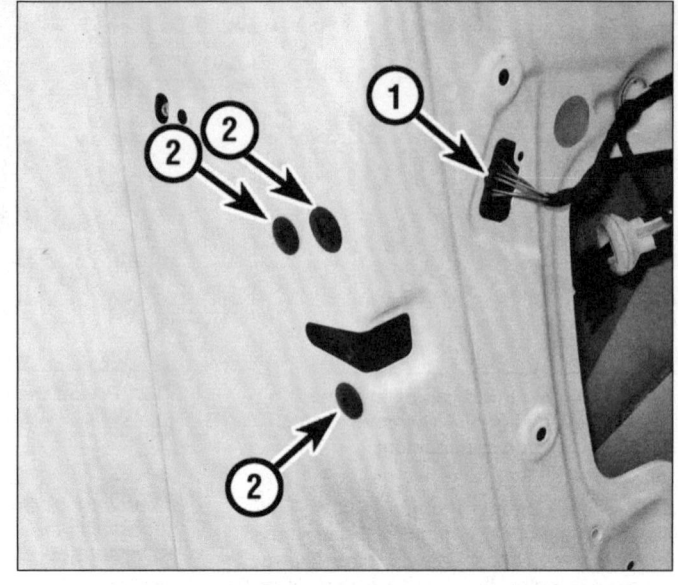

23.11 Disconnect the electrical connector (1) and remove the mounting bolts (2) from the latch

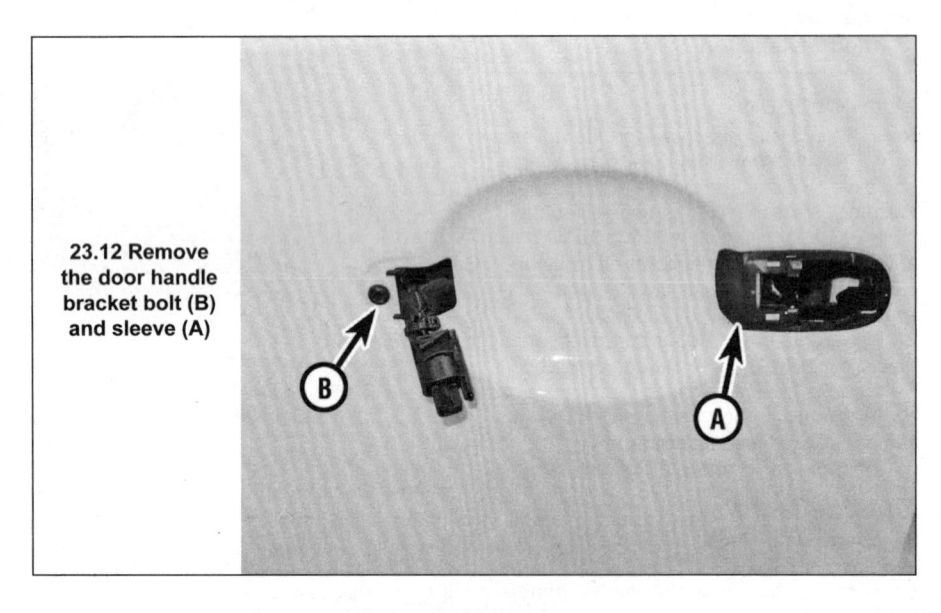

23.12 Remove the door handle bracket bolt (B) and sleeve (A)

Latch and outside handle bracket

Removal

7   Remove the door trim panel (see Section 21).
8   Remove the lock cylinder and outside door handle (see previous Steps).
9   Remove the door inner access panel **(see illustration)**.
10 Remove the door window glass (see Section 24), motor and regulator (see Section 25).
11 Disconnect the latch electrical connector and remove the mounting bolts **(see illustration)**.
12 Remove the outside door handle bracket retaining bolt and dampening sleeve **(see illustration)**.
13 Guide the latch and outside handle bracket assembly out from the opening in the door **(see illustrations)**.
14 Installation is the reverse of removal.

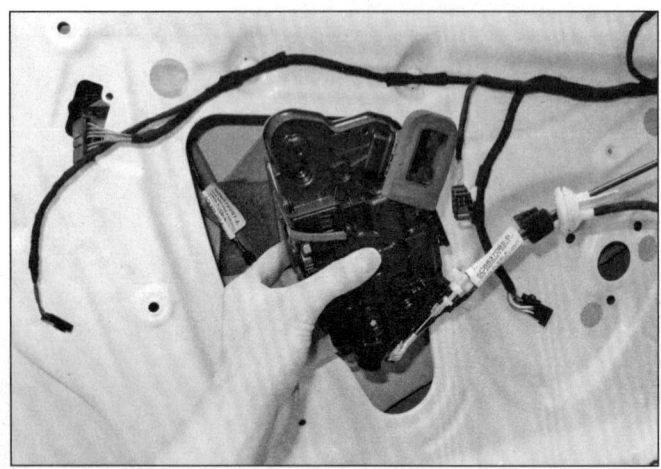

23.13a Remove the latch portion of the assembly . . .

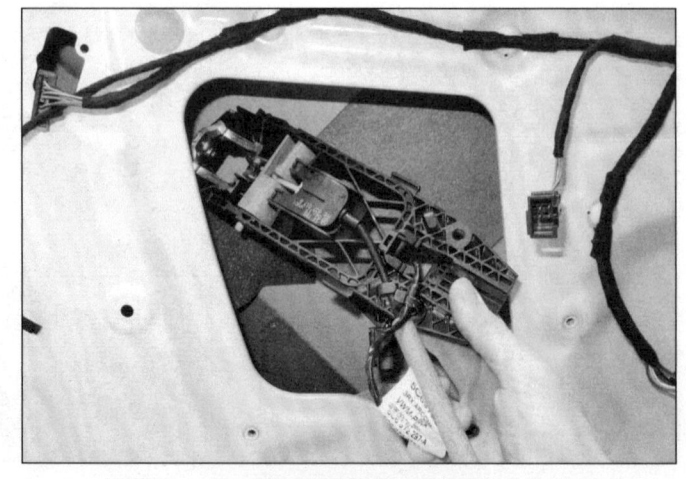

23.13b . . . then the outside handle bracket portion from the door opening.

Chapter 11 Body

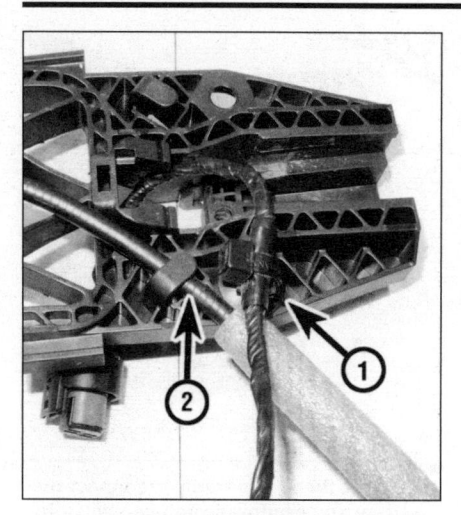

23.15a On the handle bracket, pry off the wiring harness clip (1) and free the cable from the bracket (2) . . .

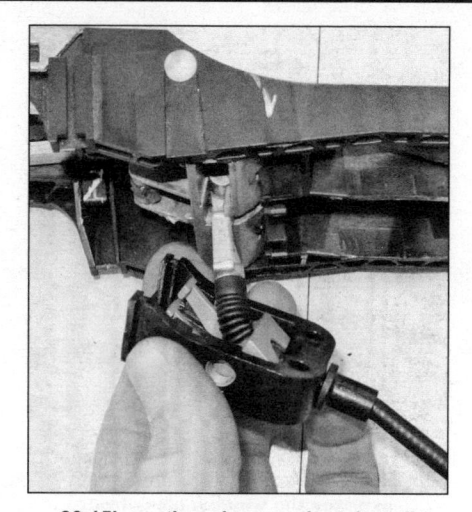

23.15b . . . then depress the tab, pull out the cable housing end piece and disconnect the cable end from the mechanism.

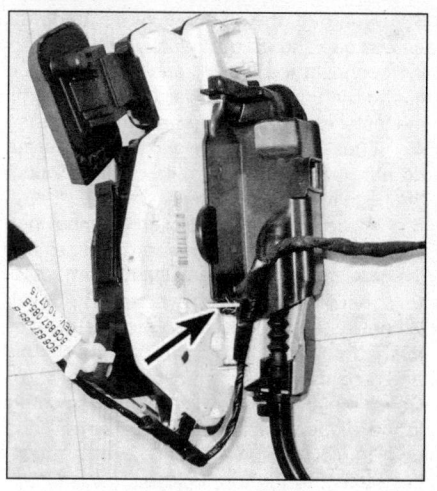

23.15c At the latch assembly, first detach the wiring harness clip . . .

Disassembly

15 To disconnect the cables from the handle bracket and latch assemblies, follow the accompanying photos in sequence (**see illustrations**).

16 Transfer the necessary old components onto the new ones. Reassembly is the reverse of disassembly.

---

24 Door window glass - removal and installation

Warning: *Disconnect the negative battery cable before proceeding (see Chapter 5).*

Front door window glass

1 Remove the door trim panel (see Section 21).

2 Remove the inner door access panel (see Section 23).

23.15d . . . then depress the tabs and slide off the plastic cover . . .

Note: *If the battery cannot be reconnected to raise or lower the glass, a pair of pointed-tip spreading circlip pliers can be used to brace and rotate the cable reel gear - when doing this, be careful not to chip the gear teeth.*

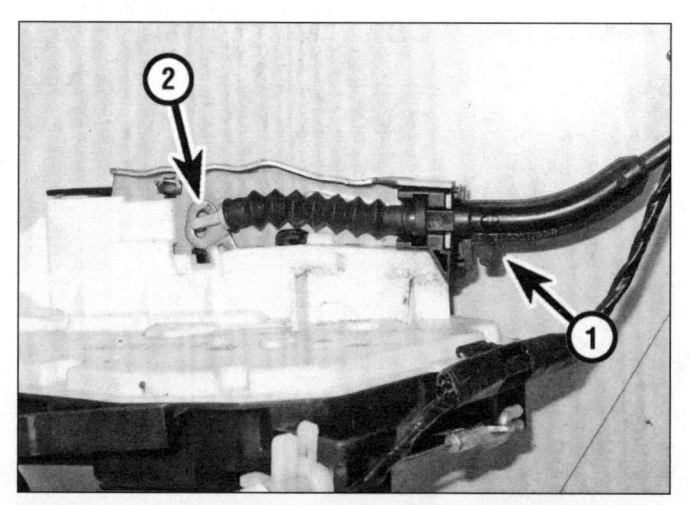

23.15e . . . rotate the tab (1) 90-degrees and slide out the cable housing, disconnect the cable end (2) . . .

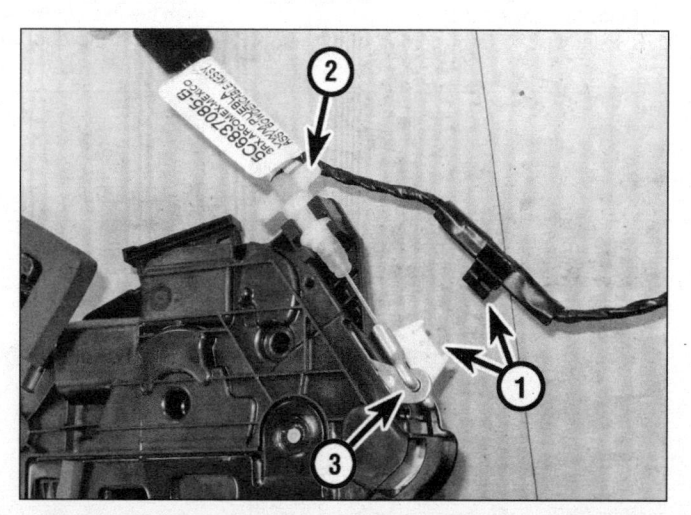

23.15f . . . finally, detach the wiring harness clip (1), rotate the tabs (2) 90-degrees and slide out the cable housing, then disconnect the cable end (3)

11-24　　　　　　　　　　　　　　　　　Chapter 11　Body

3   Remove the window glass fastener access cap and access panel, then temporarily reconnect the battery and window switch control unit. Raise or lower the window until the fasteners are visible (see illustrations).
4   Loosen the window glass fasteners but do not remove them completely. Disconnect the negative battery terminal.
5   Raise the glass up from the clamps, then tilt the front of the glass downwards and remove the glass from the door (see illustration).
6   Installation is the reverse of removal. When installing the glass, make sure it is pushed completely into the bracket clamps and against the rearmost window track. Check that the window glass moves smoothly in the window tracks and is aligned with the door frame. Tighten the window glass retaining screws to the torque listed in this Chapter's specifications.

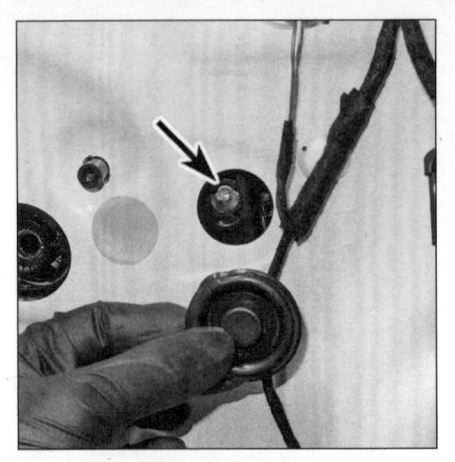

Rear door window glass

7   Remove the rear door trim panel (see Section 21).
8   Remove the inner and outer window sill strips by carefully pulling up at each end - avoid bending the strips, as they cannot be reformed and can become distorted.
Note: *If the battery cannot be reconnected to raise or lower the glass, a pair of pointed-tip spreading circlip pliers can be used to brace and rotate the cable reel gear - when doing this, be careful not to chip the gear teeth.*
9   Remove the window glass fastener access cap (see illustration), then reconnect the battery and window switch and lower the window until the plastic expander plug and pin is visible in access hole - this is what retains the window glass to the regulator mount.
10 Screw in a 5 mm x 70 mm long bolt into the plastic center of the expander pin and pull out the bolt (with pin) (see illustration).
11 Screw in an 8 mm x 80 mm long bolt into the expander plug and pull the plug out of the clamping bracket. When screwing the bolt into the expander plug, do not use excessive force or it may fall inside the door.
12 Remove the front portion of the upper window guide/seal (closest to the B-pillar). There is a pin at the corner of the window frame to be removed, then carefully peel the front portion of the guide off of the window frame, being careful not to distort it.
13 Carefully remove the window glass from the door, lifting it up and tilting it as necessary.
14 Installation is the reverse of removal. Install the expander plug, then the expander pin into the plug with the window glass removed. Press the window glass carefully but firmly into the regulator bracket until is it secured.

Rear door fixed window glass

15 Remove the rear door trim panel (see Section 21).
16 Lower the window, and if neccessary, remove the door window glass (see Steps 7 through 13).
17 Remove the window guide as described in Step 12, carefully peeling the remaining section off of the window frame and away from the fixed glass.
18 With the door trim panel removed, pry out the two grommets located below the fixed glass. Remove the bolts exposed by the grommet holes.
19 Pull and remove the fixed window outward, away from the door until the clips are released. Use a plastic trim tool to gently pry if necessary.
20 Installation is a reversal of removal. Apply pressure on the fixed window when installing until the clips are secure.

25   Door window regulator and motor - removal and installation

Warning: *Disconnect the negative battery cable before beginning (see Chapter 5).*
Note: *This procedure applies to the front and rear window regulator/motor.*

Window regulator

Removal
1   Remove the door trim panel (see Section 21).

24.3a Remove the grommet that exposes the window glass front mounting clamp bolt . . .

24.3b . . . then temporarily reconnect the battery and window switch control unit. Raise or lower the window until both fasteners are visible.

24.5 Carefully lift the window glass out of the door, tilting it as necessary

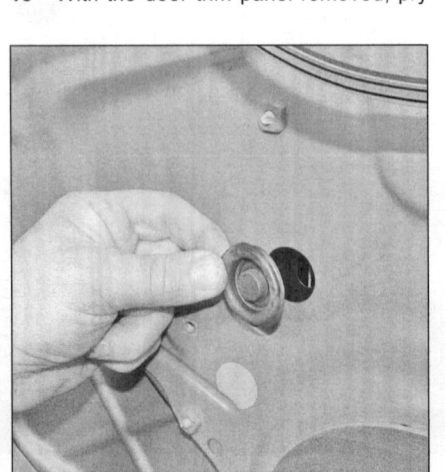

24.9 Remove the access cap

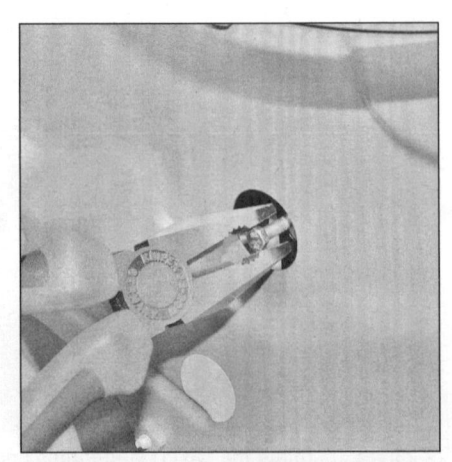

24.10 Use a 5 mm bolt to pull out the plastic center of the expander pin

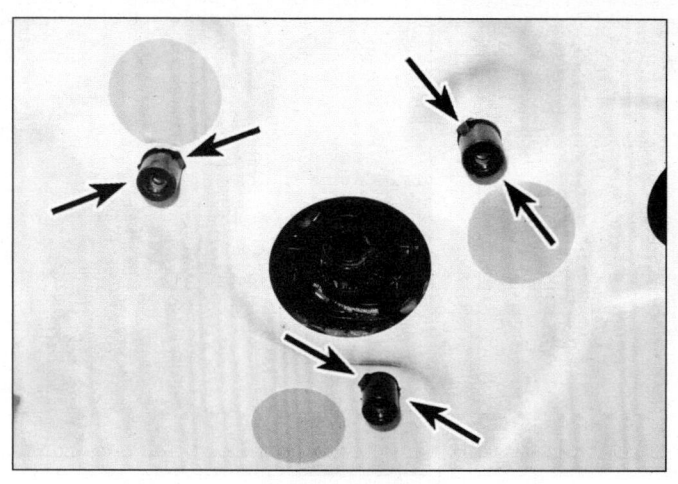

Chapter 11 Body

11-25

25.4 Release the tabs on the cable reel studs and push it into the door

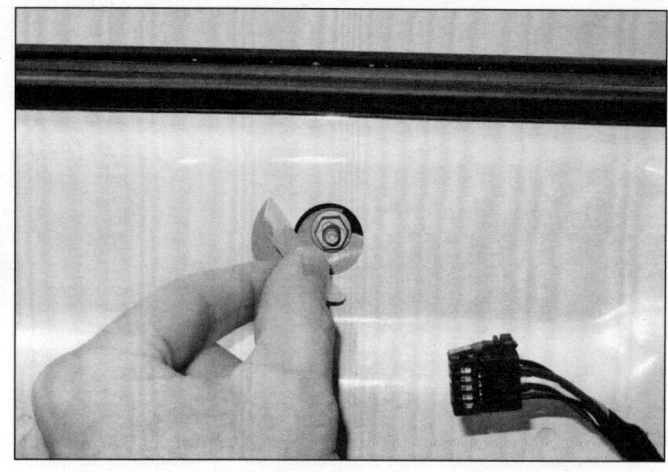

25.5a Peel back the regulator mounting nut access flaps . . .

2   Remove the door window glass (see Section 24).
3   Unplug the electrical connector and remove the power window motor **(see illustration 25.10)**.
4   Release the tabs on the cable reel and push the reel into the door **(see illustration)**.
5   Peel back the access flaps and remove the window regulator mounting nuts, pushing the regulator sections into the door **(see illustrations)**.
6   Guide the window regulator as an assembly out from the door opening **(see illustration)**.
7   Installation is the reverse of removal. Tighten the window regulator mounting nuts to the torque setting listed in this Chapter's Specifications.

Window motor

Removal

8   Remove the door trim panel (see Section 21).
9   With the window fully closed, use masking tape to hold the glass up in position - be careful not to let the glass fall down.
10   Disconnect the electrical connector from the motor, then remove the mounting screws and remove the motor **(see illustration)**.

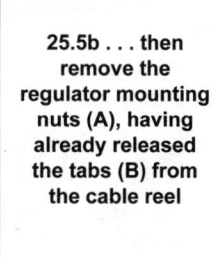

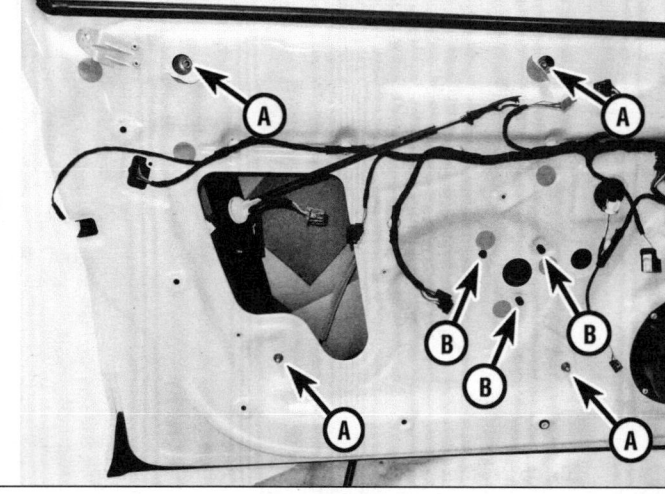

25.5b . . . then remove the regulator mounting nuts (A), having already released the tabs (B) from the cable reel

11   Installation in the reverse of removal. Tighten the window motor mounting screws to the torque setting listed in this Chapter's Specifications.

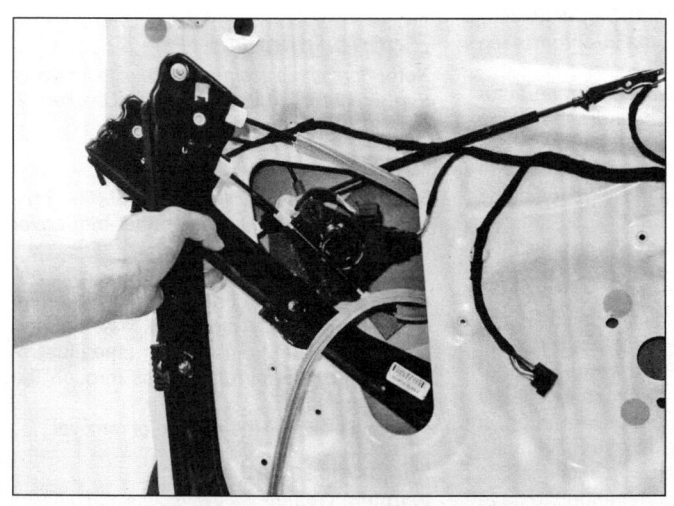

25.6 Remove the window regulator through the opening in the door

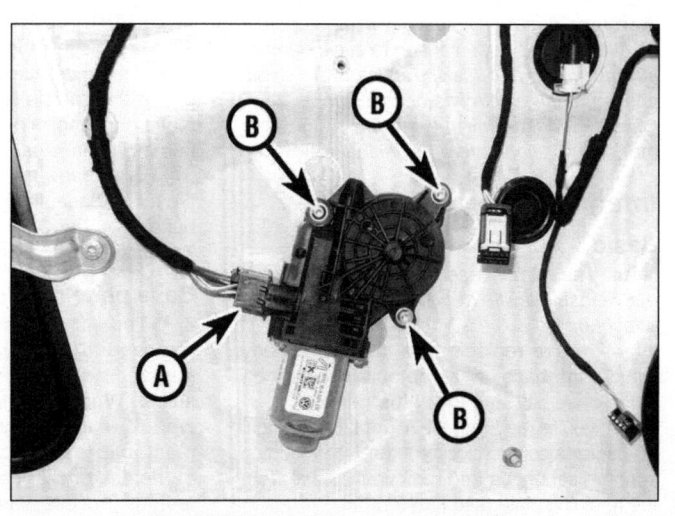

25.10 Window motor mounting details

A   Electrical connector
B   Mounting screws

26.1 Pull the mirror body off of the ball stud

26.2 Rotate the mirror base 90-degrees counterclockwise to remove it

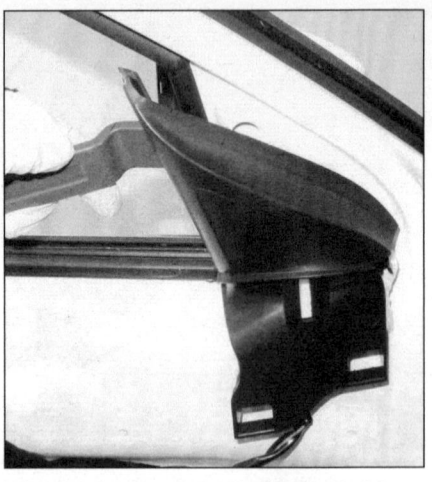

26.10a Pry the sail trim panel off of the door, starting at the top . . .

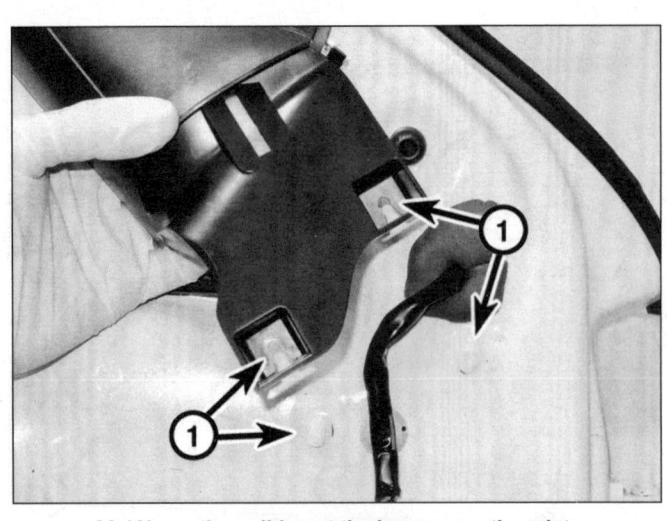

26.10b . . . then slide out the lower mounting slots from the clips (1)

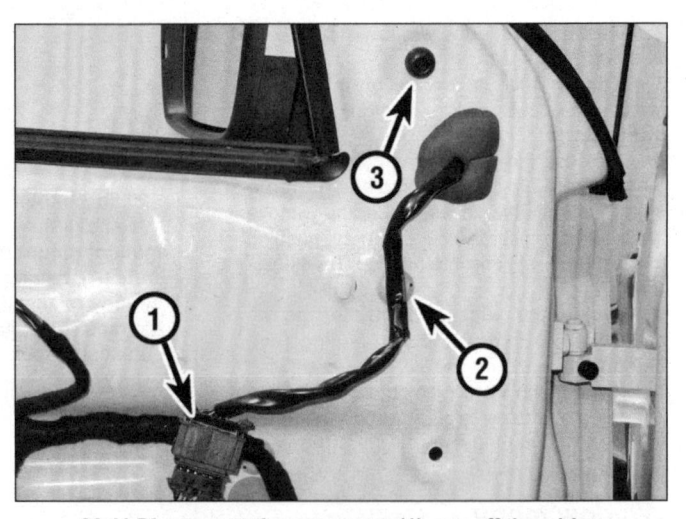

26.11 Disconnect the connector (1), pry off the wiring harness clip (2), then remove the mirror mounting bolt (3)

26 Mirrors - removal and installation

Warning: *Automatic dimming mirrors contain electrolyte fluid - use caution to avoid damaging the mirror and leaking electrolyte - avoid contact of the electrolyte or its fumes*
Caution: *Make sure the ignition switch is in the OFF position before beginning.*

Interior mirror

Basic

Note: *The mirror base can be removed from the windshield with or without the mirror body attached.*

1   If you're replacing just the mirror body without the base, it can be pulled sharply off of the base ball stud **(see illustration)**.
2   If you're replacing the mirror base, rotate it 90-degrees counterclockwise until the retaining spring disengages and remove the base from the retaining plate **(see illustration)**.
3   To install, align and engage the mirror base with the retaining plate, then rotate it clockwise until the base locks in place. Press the mirror body firmly into the ball stud.

With rain sensor and auto dimmer

4   Pry off the trim cap halves from each side of the mirror base.
5   If the vehicle is equipped with an automatic dimming mirror, disconnect the wiring harness connector.
6   Disengage the clips from each side of the mirror base, then slide the mirror off of the base plate.
7   Installation is the reverse of removal.

Base plate replacement

8   The mirror base, in most cases, is removed by applying heat with a heat gun and gently prying it off from the windshield. When installing the base, a special type of rearview mirror adhesive, found at most auto parts stores, is used. Because of the risk of cracking the windshield in this tedious process, it is recommended to have this procedure performed by an automotive glass or a qualified automotive repair shop. Depending on the style of interior mirror, the replacement process may require the windshield to be removed.

Exterior mirror

Note: *For the replacement of the exterior mirror turn signal light (if equipped), see Chapter 12, Section 19.*

Mirror assembly

9   Remove the door panel (see Section 21).
10 Carefully pry the sail panel trim cover from the door **(see illustrations)**.
11 Disconnect the mirror electrical connector, pry the wiring harness clip free, then remove the mirror mounting fastener and detach the mirror from the door **(see illustration)**. Work the mirror harness through the hole in the door.
12 Installation is the reverse of removal.

Mirror glass and motor

Warning: *Wear protective gloves and glasses when replacing the mirror glass - it may crack or shatter and cause injury.*

Chapter 11 Body 11-27

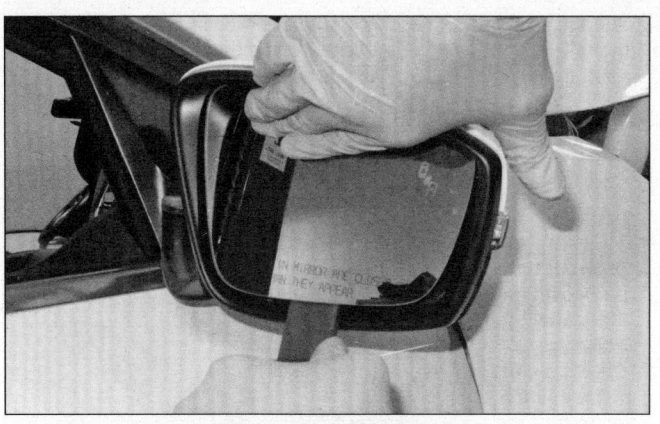

26.13 Remove the mirror glass by prying at the bottom and pressing on the top at the same time

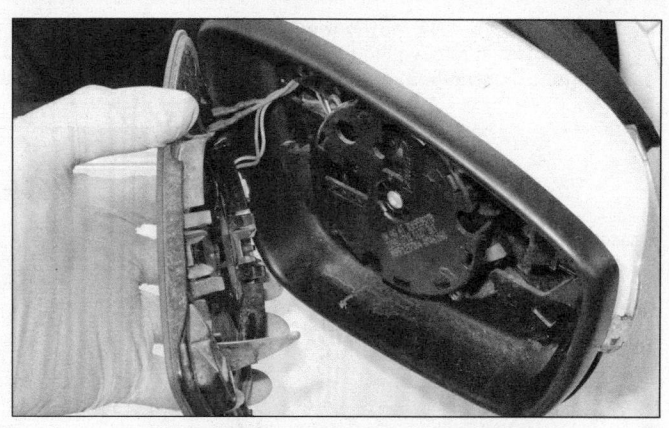

26.14a Pull the glass away from the housing (toward the vehicle) . . .

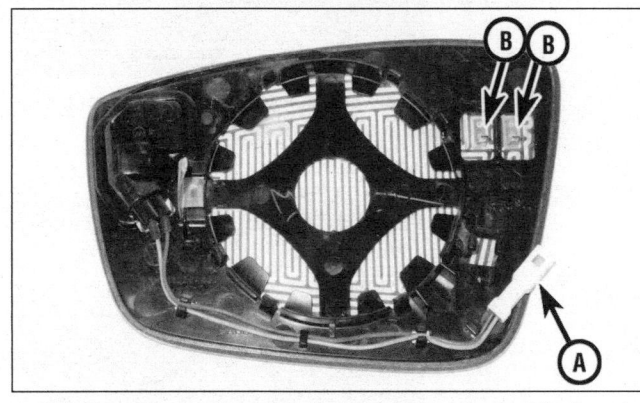

26.14b . . . then disconnect the connectors for the blind-spot light (A) and the heating element prongs (B)

26.15 Mirror motor retaining bolt

13 Using a plastic trim tool, pry the mirror glass from the bottom while at the same time press by hand at the upper-left corner of the glass **(see illustration)**. This will disengage the glass retaining clips on the inside. Use caution to avoid cracking the glass.

14 Swivel the glass away from the mirror housing and disconnect the electrical connectors **(see illustrations)**.

15 To remove the mirror motor- remove the motor retaining bolt in the center **(see illustration)**.

16 Carefully pry out the motor (using caution not to bend the inside connector when prying), then pull the motor out and disconnect the electrical connector.

17 Installation is the reverse of removal.

27 Center console - removal and installation

Warning: *The models covered by this manual are equipped with a Supplemental Restraint System (SRS), more commonly known as airbags. Always disable the airbag system before working in the vicinity of any airbag system components to avoid the possibility of accidental deployment of the airbags, which could cause personal injury (see Chapter 12).*

Note: *The console layout may vary from one model's trim level to another. Some models may have more or less electrical connections or fasteners not covered in this procedure. Keep a lookout for any additional electrical connectors or fasteners as you proceed with the center console removal.*

1 Place the shift lever in the neutral position and pull the parking brake handle to engage it.

2 On manual transaxle models, remove the gear shift knob and boot (see Chapter 7A).

3 On automatic transmission models, remove the shifter trim panel and shift lever (see Chapter 7B).

4 On vehicles equipped with a CD changer that interferes with access to the front of the console, remove the CD changer.

5 Remove the side trim panels from the center console **(see illustrations)**.

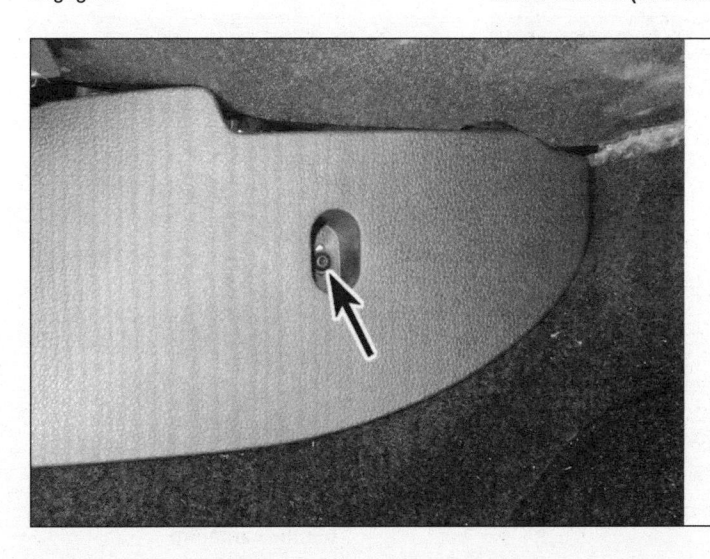

27.5a Remove the fastener . . .

11-28   **Chapter 11   Body**

6   Move the seats all the way forward, then pry off the caps and remove the rear console fasteners **(see illustrations)**.

7   Remove the console storage compartment fasteners **(see illustrations)**.

8   Remove the start/stop and front accessory switch panel **(see illustrations)**.

9   Remove the center console front retaining fasteners **(see illustrations)**.

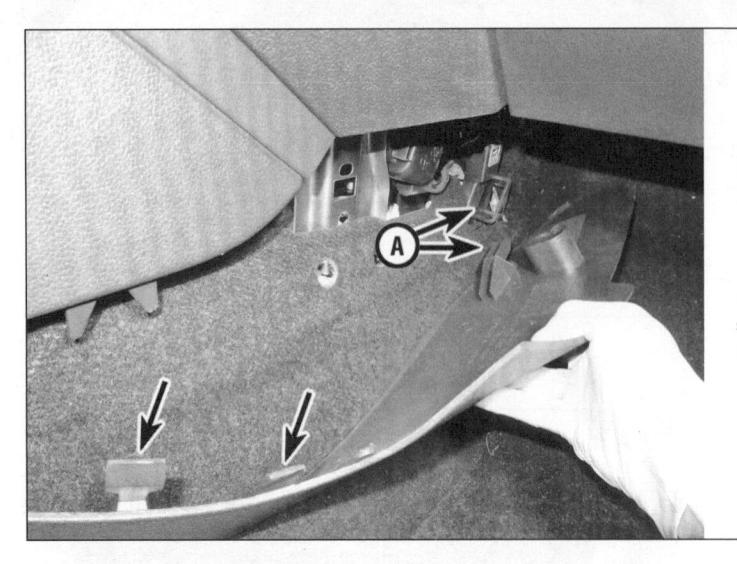

27.5b . . . then pull down on the panel to disengage the clips and pull rearward, separating the front tab from the slot (A)

27.5c Remove the fastener from the opposite end panel . . .

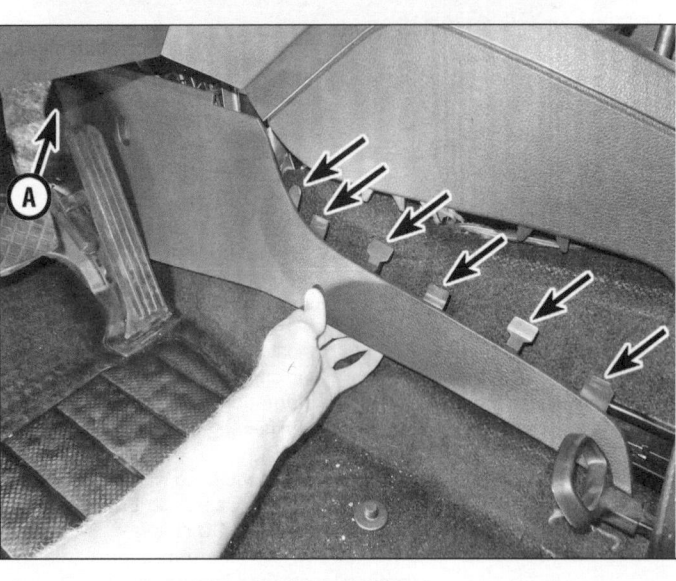

27.5d . . . pulling it down and rearward from the slot (A)

27.6a Remove the left side console fastener . . .

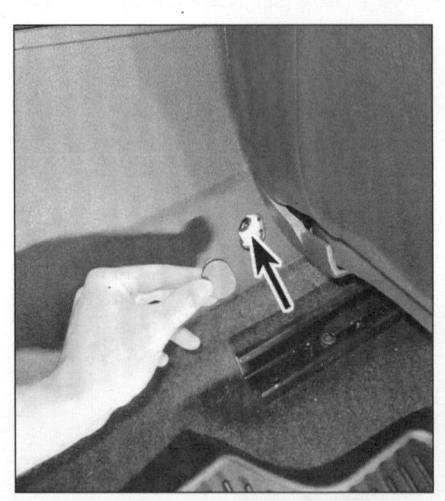

27.6b . . . then the right side fastener

27.7a Open the lid and pull out the carpeting . . .

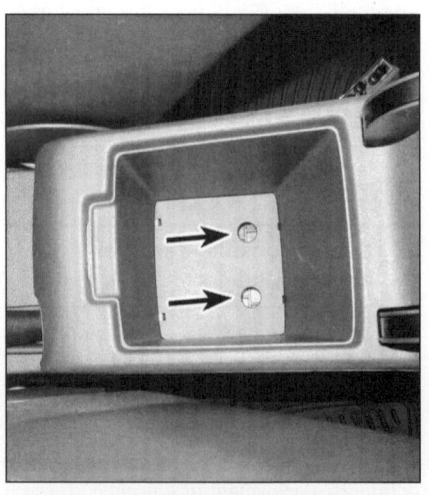

27.7b . . . grasp the holes and lift out the bottom panel . . .

Chapter 11 Body

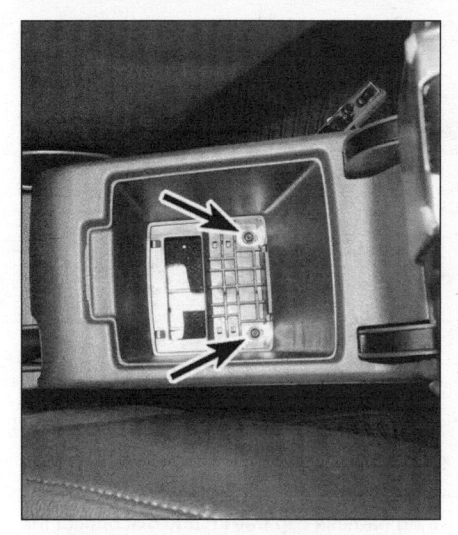

11-29

27.7c . . . then remove the storage compartment fasteners

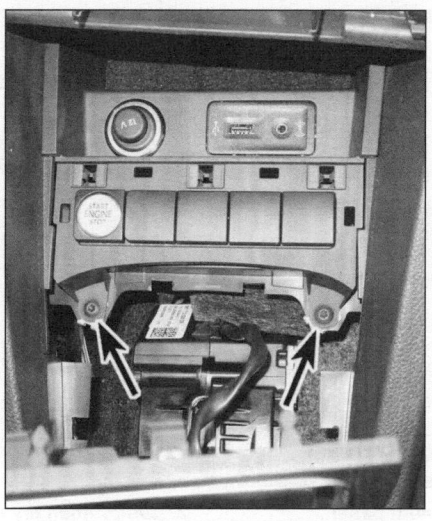

27.8a Remove the start/stop switch panel fasteners . . .

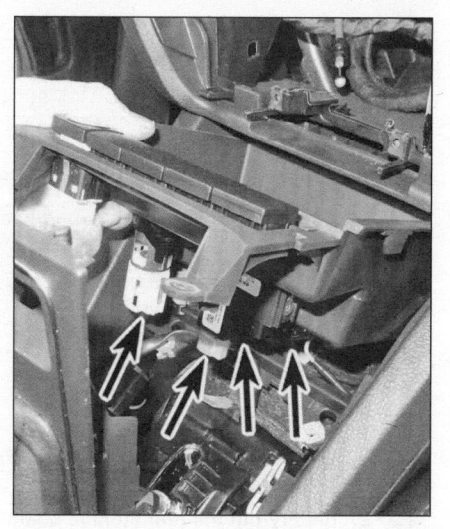

27.8b . . . then pry it up to release the clips and disconnect the electrical connectors

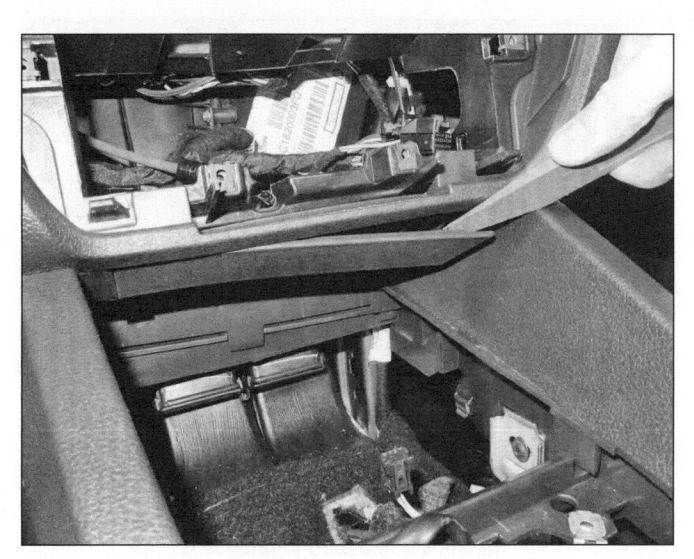

27.9a Pry off the small trim panel at the front . . .

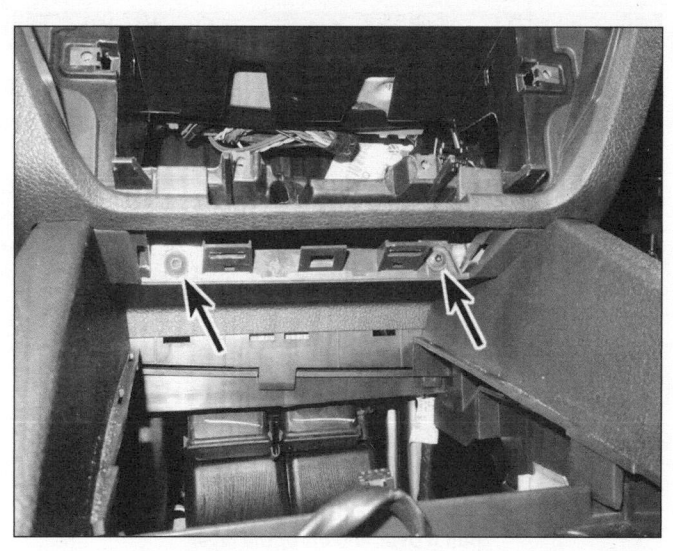

27.9b . . . then remove the console upper-front retaining screws

27.9c Remove the lower-front retaining screw from the right side . . .

27.9d . . . then remove the screw from the left side.

11-30　　　　　　　　　　　　　　　Chapter 11　Body

27.10a Pry-out the rear storage compartment panel . . .

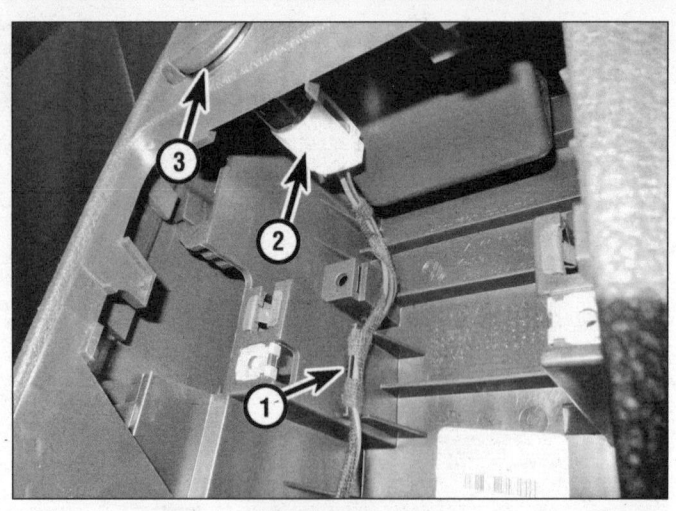

27.10b . . . then pry the wiring harness clip free (1) and disconnect the connector (2). Note that it may be easier to push the cigarette lighter clear through the panel (3) to gain better access to the connector

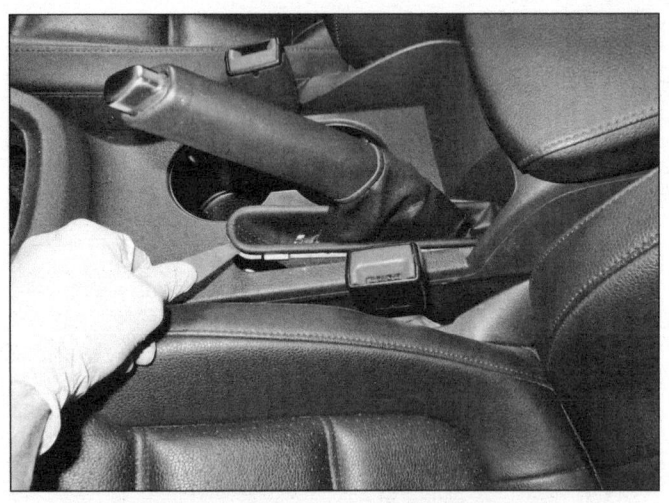

27.11 Pry up on the parking brake trim and pull it over and off of the lever

27.12 Lift up on the center console from the back, then pull it rearward to remove it

28.2 Pry out the end cap to release the retaining clips

10   Pry-out the rear storage compartment panel and disconnect the cigarette lighter connector **(see illustrations)**.

11   Pry up on the parking brake lever trim and pull it over and off the lever to remove it **(see illustration)**.

12   Lift the center console at the back, making sure all electrical connections have been disconnected, then pull it rearward and carry it out of the vehicle **(see illustration)**.

13   Installation is the reverse of removal. Ensure that the clips at the front lower edges of the console engage with the instrument panel support trimwork.

28   Interior trim panels - removal and installation

Warning: *The models covered by this manual are equipped with a Supplemental Restraint System (SRS), more commonly known as airbags. Always disable the airbag system before working in the vicinity of any airbag system components to avoid the possibility of accidental deployment of the airbags, which could cause personal injury (see Chapter 12).*

Caution: *If the battery is disconnected, several systems must be re-learned before they will work properly (see Chapter 5, Section 3).*

Warning: *Disconnect the negative battery cable before beginning (see Chapter 5).*

Note: *These procedures may vary slightly regarding model and trim level.*

Dashboard end caps

1   The end caps are held in place by pressure clips.

2   Pry out the cover with a trim tool to release the mounting clips **(see illustration)**.

3   Installation is the reverse of removal.

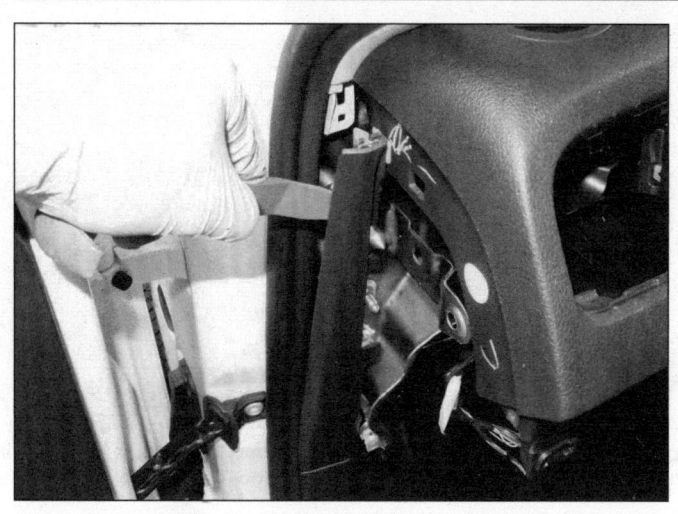

Chapter 11 Body

28.5 Pry out the lower A-pillar trim panel

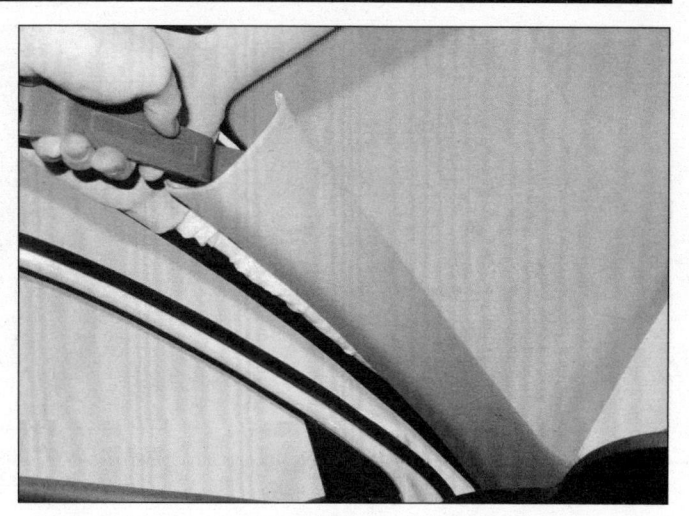

28.8a Pry-out the A-pillar trim panel, starting from the top . . .

Pillar trim panels

Lower A-pillar

4 Remove the dashboard end caps **(see illustration 28.2)**.

5 Pry the lower A-pillar trim out from the instrument panel **(see illustration)**.

6 Installation is the reverse of removal.

Upper A-pillar

Note: *The A-pillar trim starts from the windshield and stops at the top of the door.*

7 Remove the center A-pillar trim (see Steps 4 and 5).

8 Starting from the top, carefully pry the A-pillar trim panel from the windshield A support pillar **(see illustrations)**. Be careful not to distort the trim panel or headliner in the process.

9 Installation is the reverse of removal.

B-pillar

Upper B-pillar trim

10 Pry-out the emblem, then remove the screw behind the air bag emblem **(see illustration)**.

11 Pull the bottom section of the B-pillar trim outward to detach the lower clips **(see illustrations)**.

12 Remove the B-pillar trim from the door moulding.

13 If it is necessary to completely remove the upper B-pillar trim, remove the seat belt lower anchor bolt (see Section 33) and guide the seat belt through the seat belt slot.

14 Installation is the reverse of removal.

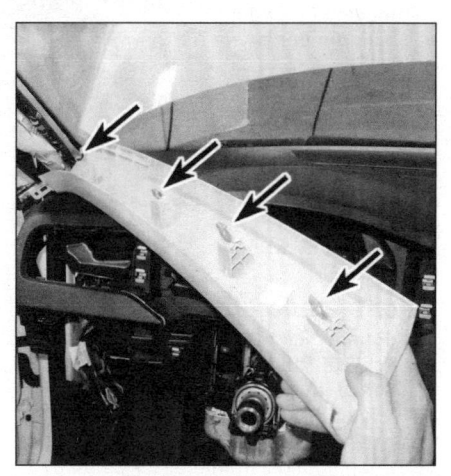

28.8b . . . then work your way down until all the clips have been released.

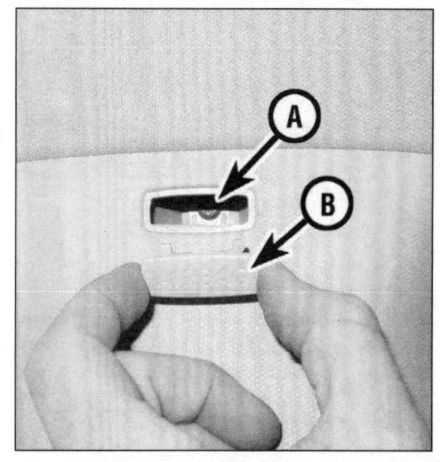

28.10 Fastener locations
A Fastener   B Air bag emblem

28.11a A flat trim tool works best to free the lower pressure clips of the B-pillar trim . . .

28.11b . . . then remove the trim panel

11-32   Chapter 11   Body

28.16a Pry-out the B-pillar trim to release the clips . . .

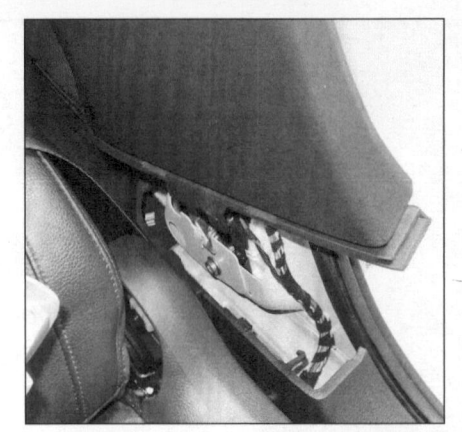

28.16b . . . disengaging it from the door sill below

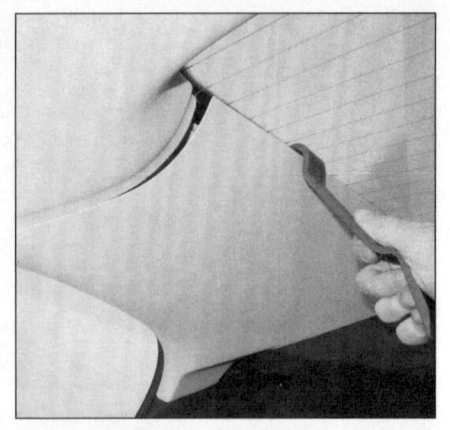

28.18a Pull the C-pillar trim free with a hooked trim tool

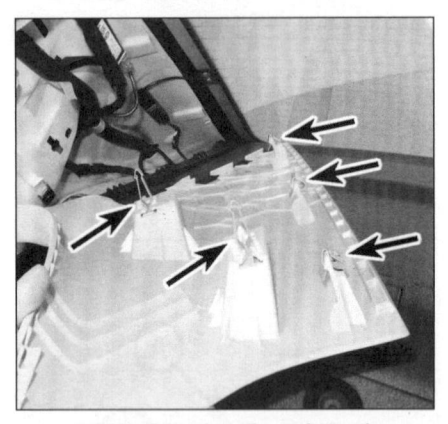

28.18b With the clips released, remove the C-pillar trim

28.20 Pry up the front section of the sill trim panel

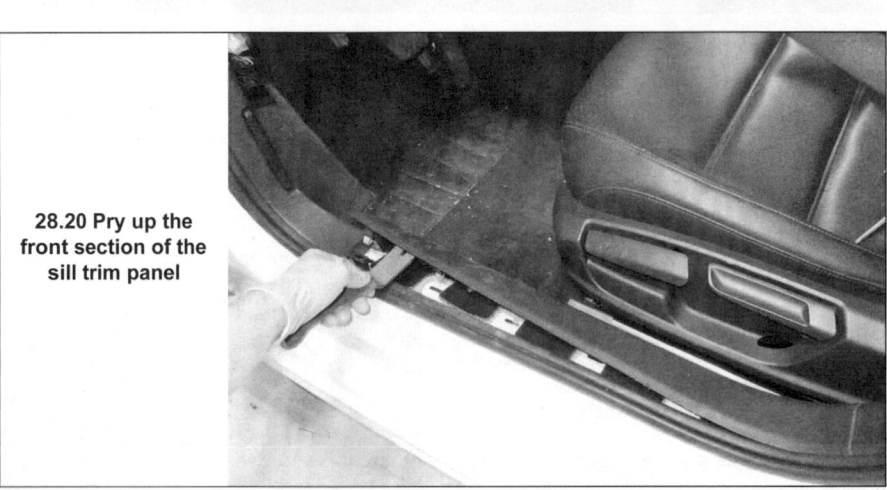

Lower B-pillar trim

15   Remove the upper B-pillar trim panel (see Steps 10 thru 13).
16   Pry the lower B-pillar trim panel free from the retaining clips, disengaging it from the door sill below **(see illustrations)**.
17   Installation is the reverse of removal.

C-pillar

18   Pry the C-pillar free with a trim tool **(see illustrations)**.

Door sill and kick panel

19   Remove the dashboard end caps **(see illustration 28.2)** and lower A-pillar trim panel **(see illustration 28.5)**.
20   Pry up the front section of the door sill trim panel from the floorboard **(see illustration)**, separating it from the kick panel.
21   On the driver's side, remove the hood release handle (see Section 11), then remove the fastener and pry out the kick panel **(see illustrations)**.

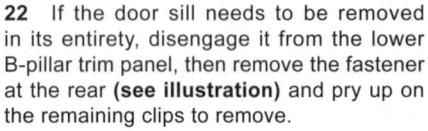

22   If the door sill needs to be removed in its entirety, disengage it from the lower B-pillar trim panel, then remove the fastener at the rear **(see illustration)** and pry up on the remaining clips to remove.
23   Installation is the reverse of removal.

Driver's knee bolster (driver's side storage compartment)

24   Remove the left dashboard end cap to access the side fasteners **(see illustration 28.2)**.

28.21a Remove the kick panel fastener (driver's side shown) . . .

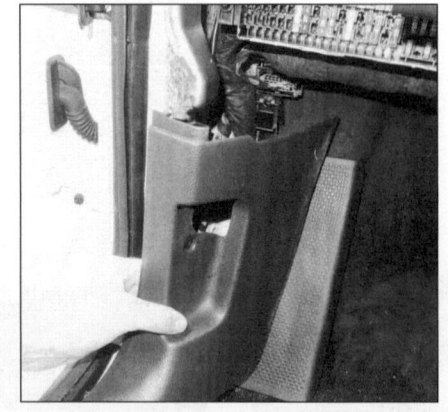

28.21b . . . then pry out on the kick panel to release the retaining clips

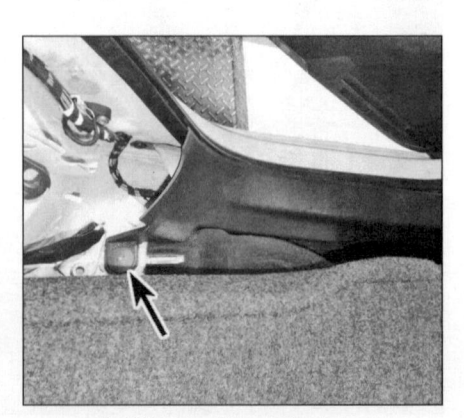

28.22 Door sill rear fastener

Chapter 11 Body

11-33

28.26a Remove the left end fastener . . .

28.26b . . . remove the headlight switch recess fastener . . .

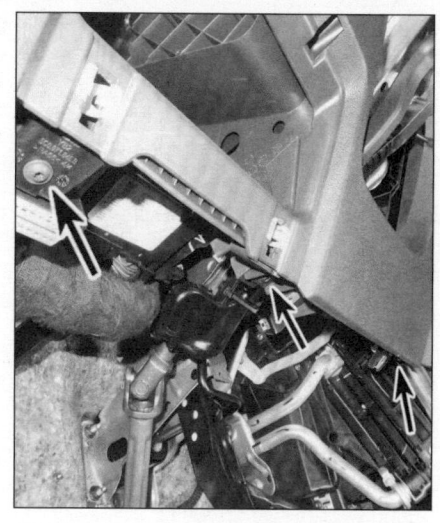

28.26c . . . remove the lower fasteners . . .

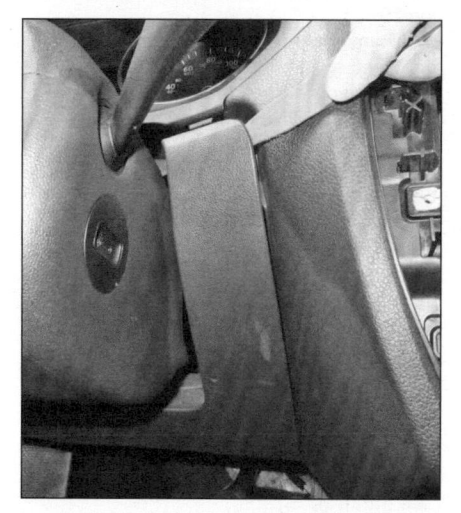

28.26d . . . work around the knee bolster to pry it free from any retaining clips . . .

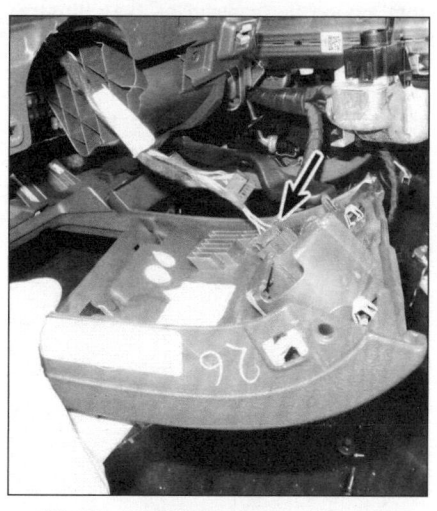

28.26e . . . disconnect any attached electrical connectors, then remove the knee bolster panel

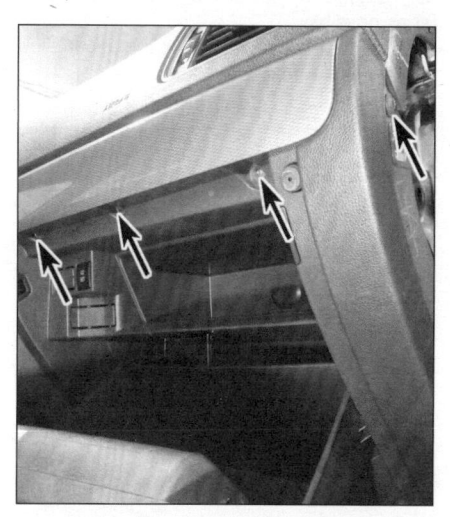

28.29a Glove box upper fasteners

25 Remove the headlight switch (see Chapter 12).
Note: *On models equipped with a storage drawer, remove the drawer to access the lower mounting fasteners.*
26 Remove the driver's knee bolster retaining screws **(see illustrations)**.
27 Installation is the reverse of removal.

Glove box

28 Remove the right dashboard end cap **(see illustration 28.2)**.
29 Remove the glove box fasteners **(see illustrations)**.
30 Slide the glove box out and separate the glove box from the instrument panel, then disconnect the electrical connector **(see illustration)**.
31 Installation is the reverse of removal.

Instrument cluster trim

32 Remove the hazard flasher (see Chapter 12).

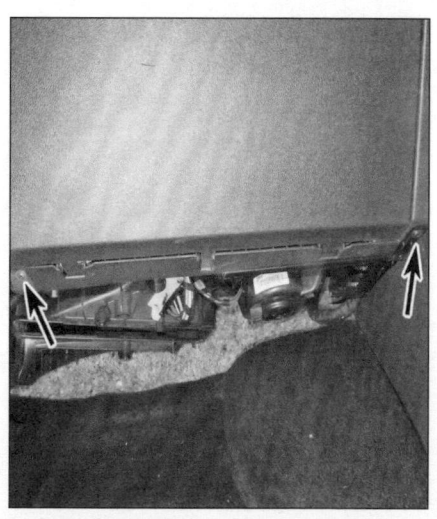

28.29b Glove box lower fasteners

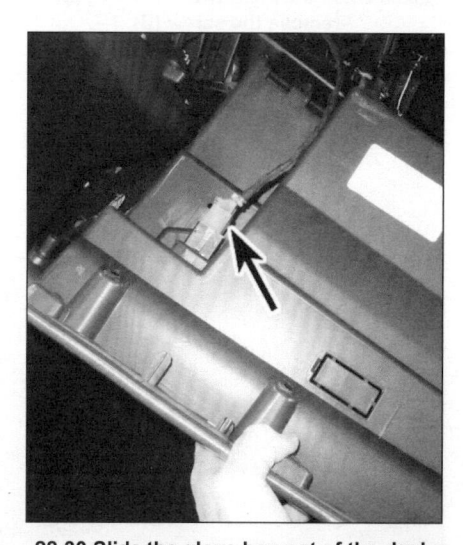

28.30 Slide the glove box out of the dash and disconnect any electrical connections

11-34   Chapter 11   Body

28.35a Use the trim tool around the top and bottom to release all of the pressure clips . . .

28.35b . . . working your way around the entire instrument cluster trim to release all of the clips

28.37 Remove the instrument cluster trim and steering column upper cover together as one piece

33   Extend and lower the steering wheel as far as possible.
34   Using a trim tool, release the upper clips from the knee bolster trim.
35   With the same trim tool, release the clips from around the vent grilles **(see illustrations)**.
36   Separate the upper half of the steering column cover to the lower cover (see Section 29).
37   Remove the entire instrument cluster trim together with the upper half of the steering column cover **(see illustration)**.
38   Installation is the reverse of removal.

Sun Visor

Sun visor center attaching clip
39   Open the protective cover concealing the single screw, and remove the screw **(see illustration)**.
40   Installation is the reverse of removal.

Sun visor and mount
Note: *The protective cap also acts as the locking mechanism to hold the sun visor mounting clip in place.*
41   Pry the protective cap off from the visor **(see illustration)**.
42   Pry down on the visor mount to release the clip, then rotate the visor downward to release the rear mounting hook **(see illustrations)**.

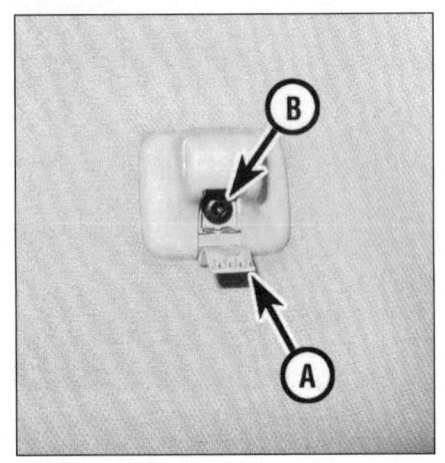

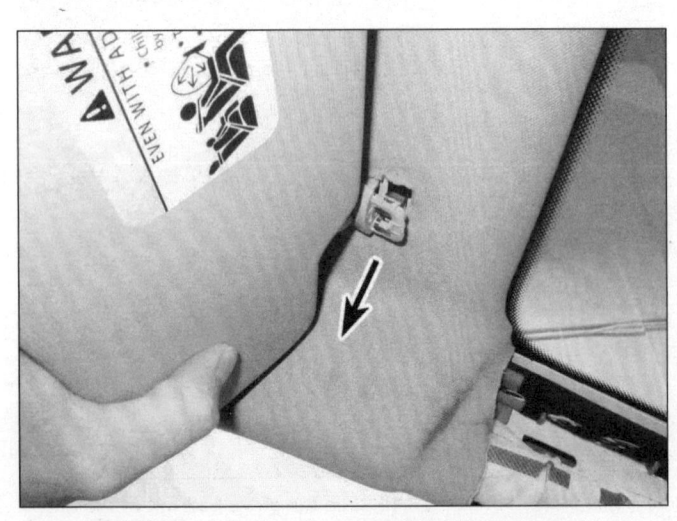

28.39 Open the protective cover (A) and remove the screw (B)

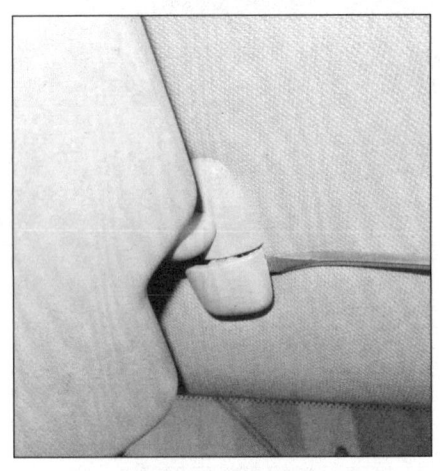

28.41 Pry off the protective cover

28.42a Pry the sun visor mounting clip downward to release the clip . . .

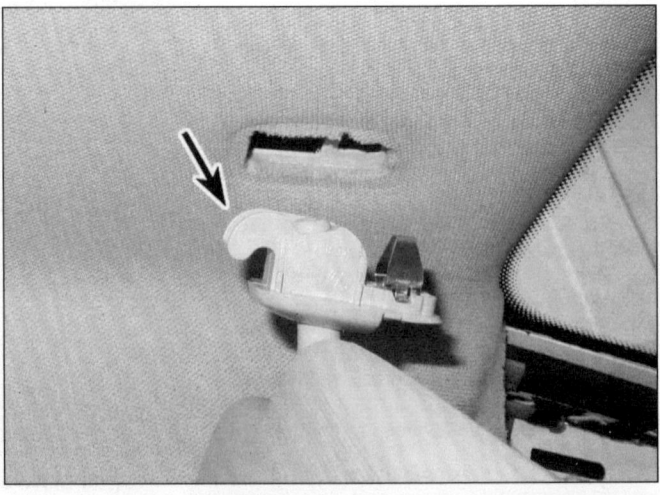

28.42b . . . then rotate the visor down to release it from the hook

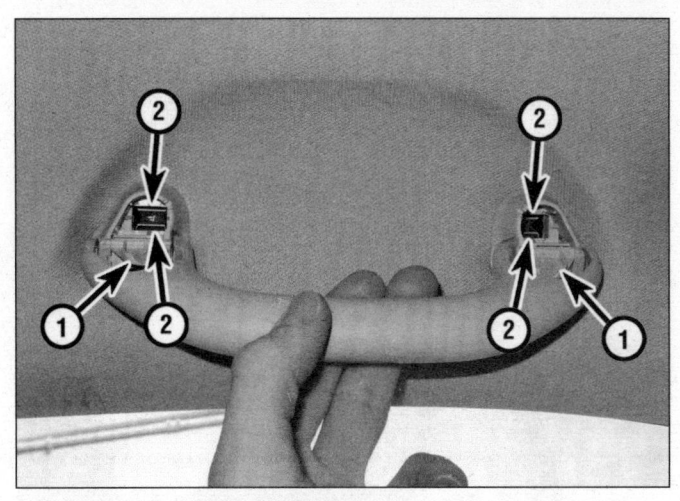

Chapter 11 Body   11-35

28.45 Pry open the grab handle covers (1) and disengage the retaining clips (2) with a screwdriver or needle-nose pliers

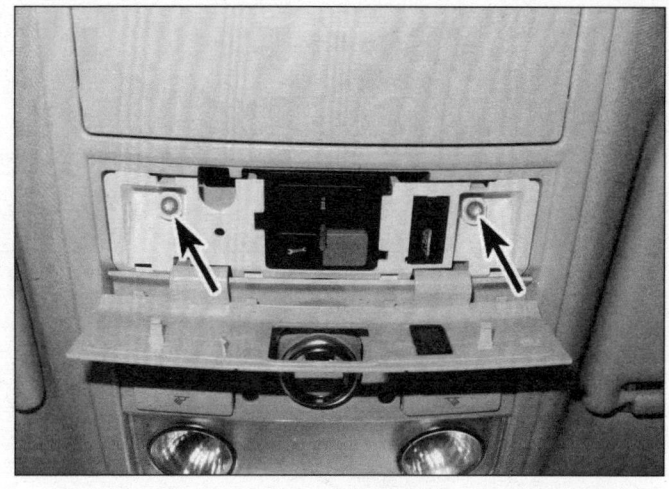

28.47 Overhead console control panel mounting screws

43 To install, place the hooked edge of the sun visor mount into place and snap the clip into the roof bracket. The remainder of the installation is the reverse of removal.

Overhead Grab Handles

44 Fold the grab handle down and open the protective caps hiding the retaining clips.
Note: *Each end of the grab handle (front or rear) has a pressure-type clip inside. The pressure clips are released by squeezing with needle-nose pliers or prying with a screwdriver(s).*
45 Using a small flat tip screwdriver or needle-nose pliers, disengage the grab handle retaining clips **(see illustration)**.
46 To install, align the grab handle to the openings in the roof and press securely into place, along with the protective covers.

Overhead console and dome light assemblies

Control assembly

47 Pry the cover open and remove the fasteners securing the overhead control panel in place **(see illustration)**.

48 Disconnect the electrical connections and remove the overhead control panel **(see illustration)**.

Overhead storage trim panel

49 Remove the fasteners securing the overhead storage trim panel to the roof **(see illustrations)**. Remove the storage trim panel.

Rear dome light housing

50 To remove the rear dome light housing, pry down and release the retaining clips while being careful not to break them, then disconnect the electrical connector by releasing the tabs on each end **(see illustration)**.
51 Installation is the reverse of removal.

28.48 Overhead console control panel electrical connectors

28.49a Remove the rear fasteners . . .

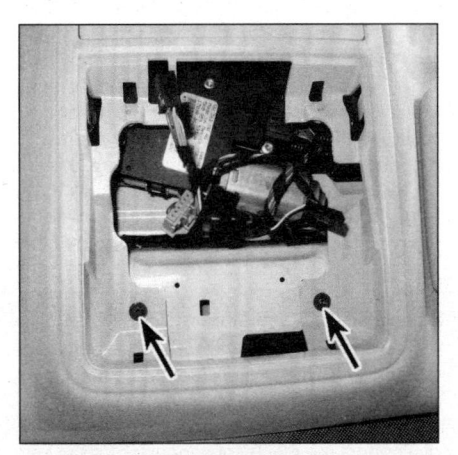

28.49b . . . then the front fasteners

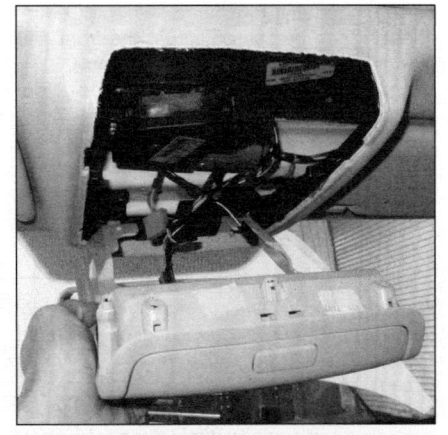

28.49c Pry the panel down from the rear, then remove the overhead storage trim panel

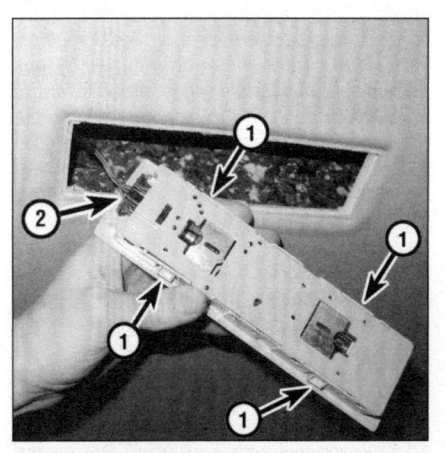

28.50 Release the retaining clips (1) and disconnect the electrical connector (2) from the rear dome light housing

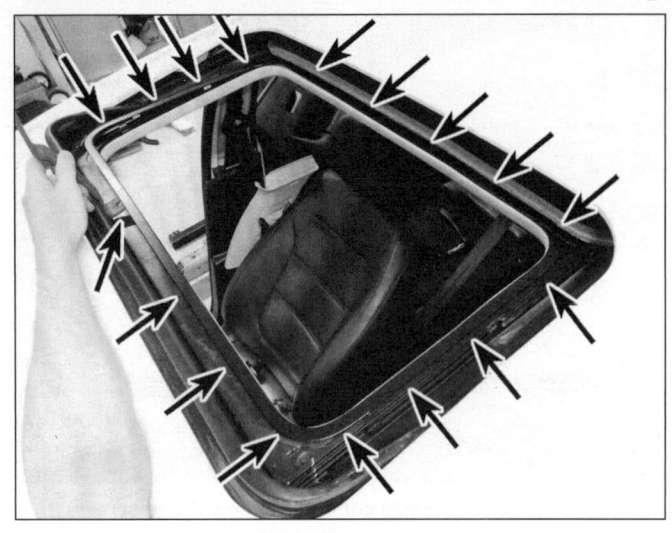

28.56 Use a trim tool to release the sunroof retaining trim from the headliner

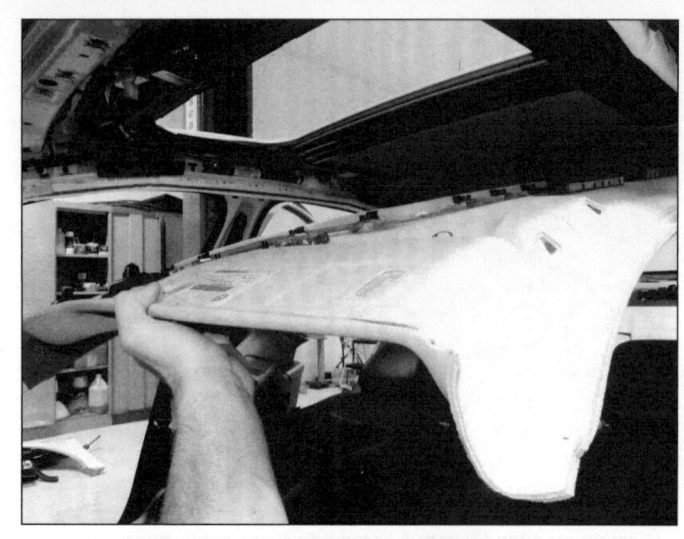

28.58a Carefully release the pressure clips at the front of the headliner . . .

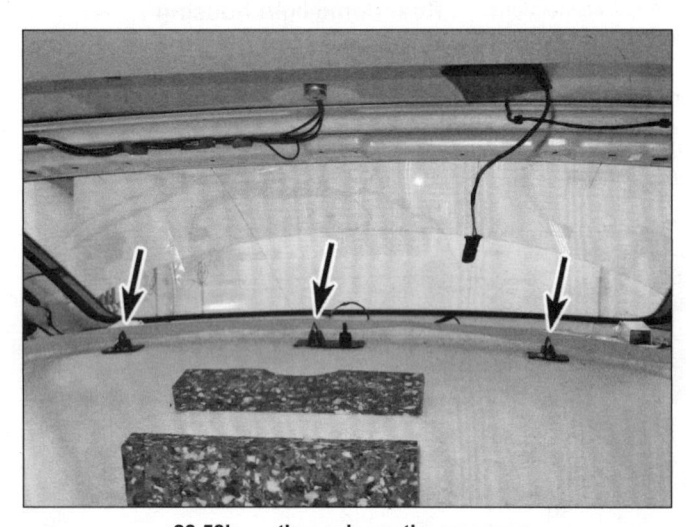

28.58b . . . then release the pressure clips at the rear.

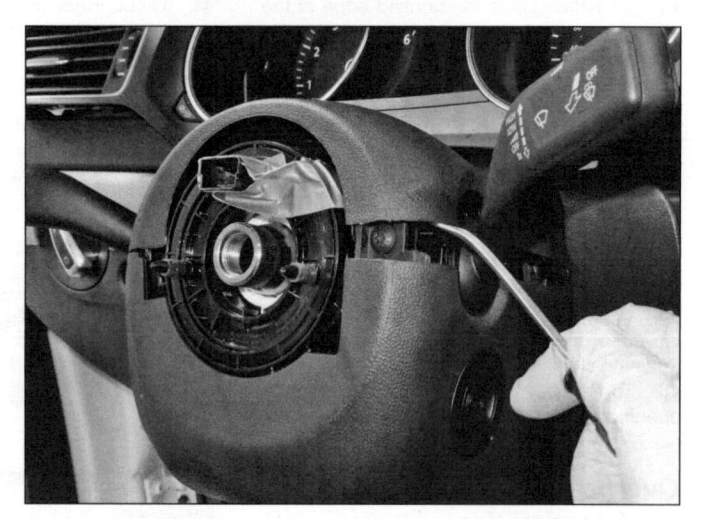

29.2 Pry near the corners to remove the upper trim cover

Headliner

Caution: *DO NOT bend or crease the headliner as it is being removed. Place the headliner in a safe area to prevent damage, if reusing.*
Note: *The headliner can be removed by one person, but if possible, have an assistant to help with supporting the headliner as it is being removed to avoid creasing.*

52 Open the sunroof, if equipped.
53 Turn off all accessories, and disconnect the negative battery cable.
54 Remove the necessary trim panels from the A-pillar, B-pillar, C-pillar, sun visors, overhead grab handles, overhead storage, dome light, and interior rear view mirror (see previous Steps).
55 Move the front seats forward, and the seat backs as far down as possible. If more room is needed, remove the headrests.
56 On models with a sunroof- Pry up the clips and separate the frame retaining piece from the headliner **(see illustration)**.
57 Work the door sealing rubber trim off the edge of the headliner as needed.
58 Using an extended prying tool as necessary, carefully go around the outer edge, releasing the pressure clips at the front and rear of the headliner **(see illustrations)**.
59 Lower the headliner and guide it out through the passenger's side rear door.
60 Installation is the reverse of removal.

29 Steering column covers - removal and installation

Warning: *The models covered by this manual are equipped with a Supplemental Restraint System (SRS), more commonly known as airbags. Always disable the airbag system before working in the vicinity of any airbag system components to avoid the possibility of accidental deployment of the airbags, which could cause personal injury (see Chapter 12).*

Warning: *Disconnect the cable from the negative battery terminal before proceeding (see Chapter 5).*

1 Remove the steering wheel (see Chapter 10).
2 Carefully pry the top steering column cover up with a small flat screwdriver or trim tool **(see illustration)**.
3 Tilt the upper steering column cover up to expose the flexible gap trim clips. Disengage the gap trim clips, then remove the steering column upper cover. Take note of the position of the upper cover studs at the rear - they must fit into the lower column grooves when installing **(see illustration)**.
4 Remove the fasteners attaching the lower trim cover to the steering column **(see illustrations)**.
5 Before removing the lower trim cover, disconnect the electrical connector for the key transponder receiver **(see illustration)**.
6 Installation is the reverse of removal.

Chapter 11 Body

11-37

29.3 Tilt the upper column cover up to expose the flexible trim clips (B) - Also note the position of the cover studs (A) for proper installation

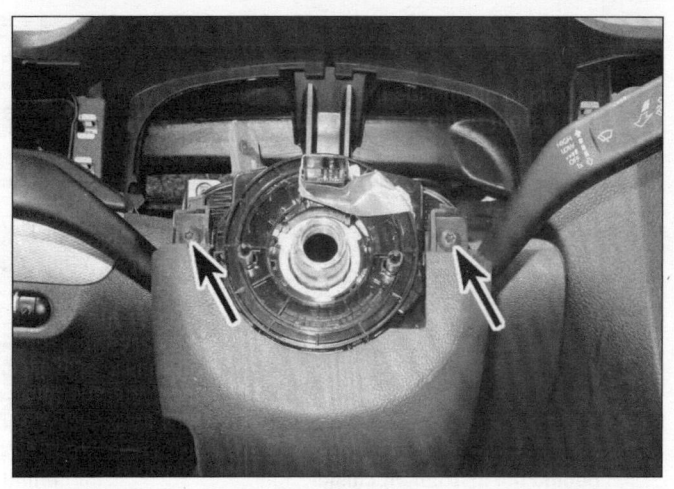

29.4a Two fasteners are hidden by the upper trim cover

29.4b One fastener is on the bottom of the lower trim cover

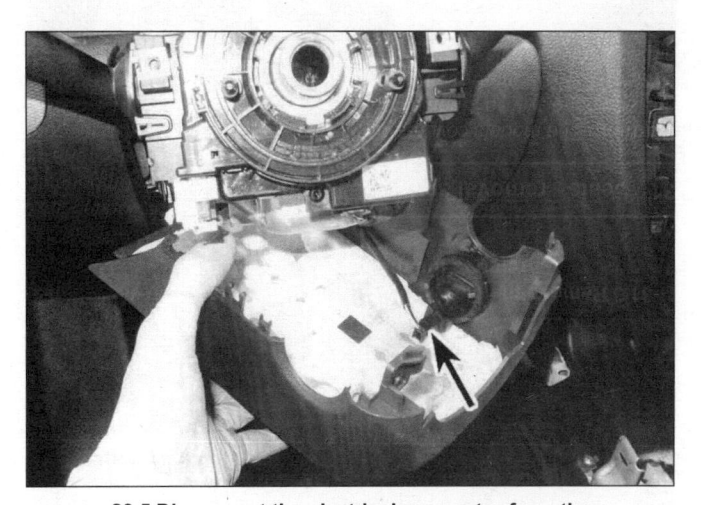

29.5 Disconnect the electrical connector from the key transponder receiver

30 Instrument panel and crossbeam - removal and installation

Warning: *The models covered by this manual are equipped with a Supplemental Restraint System (SRS), more commonly known as airbags. Always disable the airbag system before working in the vicinity of any airbag system components to avoid the possibility of accidental deployment of the airbags, which could cause personal injury (see Chapter 12).*

Caution: *This is a difficult procedure for the home mechanic, involving tedious disassembly and the disconnection/reconnection of numerous electrical connectors. If you do attempt this procedure, make sure you take good notes and mark all matching connectors (and their wiring harness clip points) to aid reassembly.*

Caution: *The entire instrument panel or cross beam assembly can be awkward and heavy. Have an assistant to help with removal.*

Caution: *When battery is disconnected, several systems must be re-learned before they will work properly (see Chapter 5, Section 3).*

1 Disconnect the negative battery cable (see Chapter 5).

Instrument panel

2 Remove the center console (see Section 27).
3 Remove the steering wheel (see Chapter 10).
4 Remove the steering column covers (see Section 29).
5 Remove all of the dashboard interior trim panels (see Section 28).
6 Remove the audio components, the multi-function switch and the instrument cluster (see Chapter 12).
7 If equipped, carefully pry out the optical light sensor at the top of the instrument panel, disconnect the sensor electrical connector and remove the sensor.
8 Remove the remaining screw fasteners securing the instrument panel, disconnecting any electrical connectors still attached to the panel **(see illustrations)**.

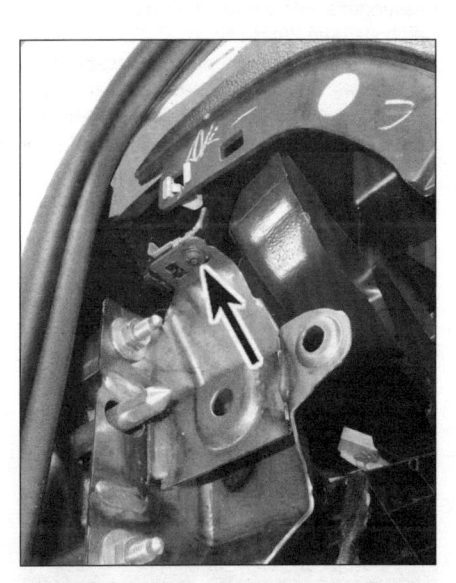

30.8a Remove the bolt at the left end of the instrument panel support and from the area behind the knee bolster

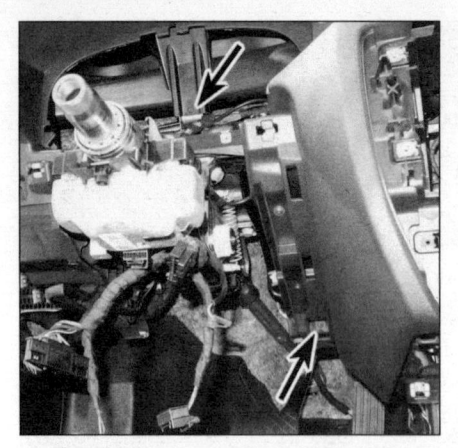

30.8b Remove the bolt from the area behind the instrument cluster

30.8c Remove the bolts from the area behind the radio

30.8d Remove the bolt at the right end of the instrument panel support and from the area behind the glove box

30.9 Airbag electrical connectors (1) and mounting bolts (2)

30.10 Carefully guide out the instrument panel

9   Pry out the safety tab and disconnect the electrical connectors (on both sides) for the passenger's side airbag, then remove the airbag bolts **(see illustration)**.
10 Pull the instrument panel rearward, making sure all electrical connections have been disconnected, and guide it out of the vehicle, preferably with the help of an assistant **(see illustration)**.
11 Installation is the reverse of removal.

Crossbeam

12 Remove the instrument panel (see above Steps).
13 Remove the steering wheel, steering column and steering column module (see Chapter 10).
14 Remove the air ducting from the top of the HVAC module **(see illustrations)**.
15 Remove the left side cowl panel cover (see Section 16).

16 Remove the crossbeam fastener from the outside of the vehicle **(see illustration)**.
17 Remove the crossbeam lower-center fasteners **(see illustrations)**.
Note: *Use a marking pen to draw around the crossbeam side nuts, noting the installed height and left/right position of the crossbeam for installation*
18 After marking the position of the crossbeam, remove the crossbeam side nuts **(see illustrations)**.
Note: *There may be more frontmost fasteners (behind the crossbeam) that are located underneath the arrows is the illustrations below.*
19 Remove the frontmost fasteners from behind the crossbeam **(see illustrations)**.
20 Detach any harness clips attached to the crossbeam (noting their locations), then with the help of an assistant, carefully guide the crossbeam out of the vehicle.
21 Installation is the reverse of removal. Guide the crossbeam onto the locating pins shown in **illustrations 30.18a and 30.18b**, then align the crossbeam with the bolt head marks made before removing. Tighten the fasteners securely.

30.14a Pull off and remove the front air duct . . .

30.14b . . . then remove the pushpin fasteners and pull up the rear duct

Chapter 11 Body

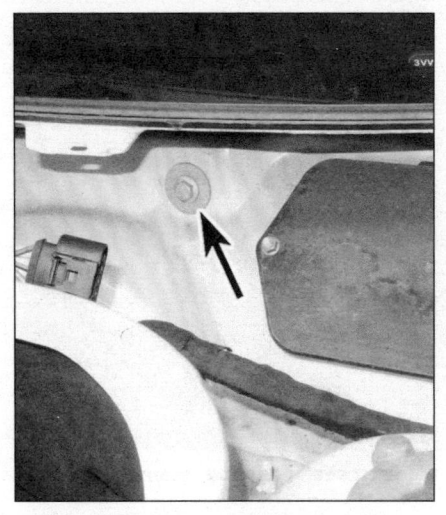

11-39

30.16 Remove the crossbeam fastener from outside of the vehicle (wiper motor removed for clarity)

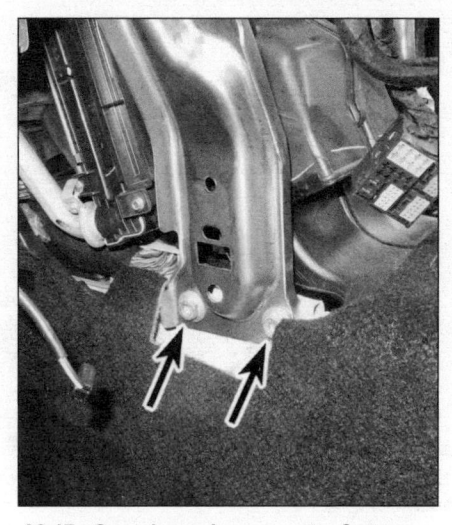

30.17a Crossbeam lower-center fasteners (left side)

30.17b Crossbeam lower-center fasteners (right side)

30.18a Crossbeam side mounting nut (right side)

30.18b Crossbeam side mounting nuts (left side)

30.19a Remaining crossbeam fasteners (far left side)

30.19b Remaining crossbeam fasteners (middle-left side)

30.19c Remaining crossbeam fasteners (middle-right side)

30.19d Remaining crossbeam fasteners (far right side)

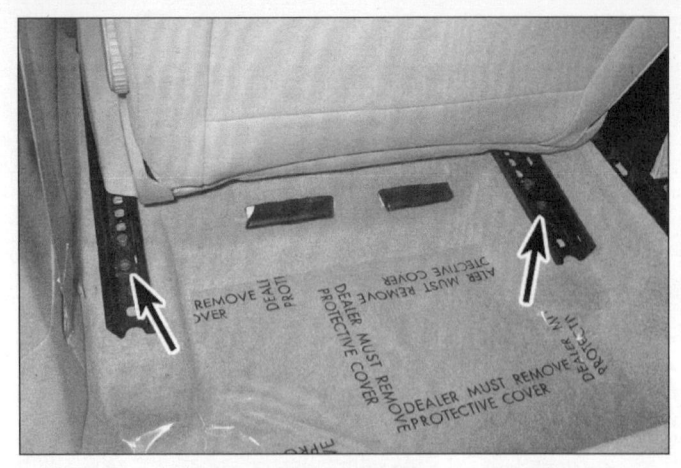

11-40   Chapter 11   Body

31.3 Remove the plastic buttons in the seat track covers, then remove the rear mounting fasteners

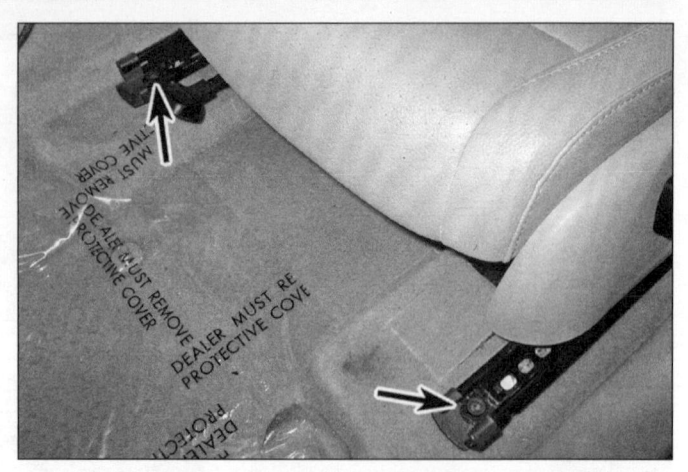

31.4 Remove the front seat mounting fasteners

31 Seats - removal and installation

Warning: *The models covered by this manual are equipped with a Supplemental Restraint System (SRS), more commonly known as airbags. Always disable the airbag system before working in the vicinity of any airbag system components to avoid the possibility of accidental deployment of the airbags, which could cause personal injury (see Chapter 12).*
Warning: *Disconnect the cable from the negative battery cable from the terminal before proceeding (see Chapter 5).*

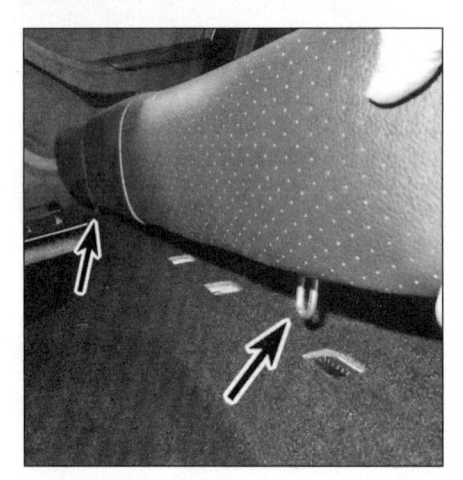

Front

1   Disconnect the negative battery cable (see Chapter 5).
2   Remove the drawer from the seat, if equipped.
3   Slide the seat forward and remove the seat track covers, remove the rear seat track fasteners and unplug any electrical connectors attached to the seat **(see illustration)**.
4   Slide the seat all the way to the back, remove the seat track covers, unplug any electrical connectors and remove the front seat track fasteners **(see illustration)**. Remove the seat from the vehicle.
5   Installation is the reverse of removal.

Rear

Seat cushion

6   At the front of the seat cushion, pull the seat up sharply at each end to disengage the hooks **(see illustration)**.
7   Push rearward from the front of the cushion, and at the same time push the rear part of the cushion downward to release the hook at the rear **(see illustrations)**. Repeat this at the opposite end.
8   Remove the rear seat cushion from the vehicle.
9   Installation is the reverse of removal.

Seat back

10   Remove the seat cushion as described in Steps 6 through 8.
11   Fold the backrests forwards, then pull back the carpet. Remove the cover from the center mount and remove the clamp fastener **(see illustration)**.
12   Remove the right backrest by lifting it from the center mount and sliding it off the outer mounting pin **(see illustrations)**.
13   Remove the left backrest by lifting it from the center mounting and sliding it off the outer mounting pin.
14   Remove the side cushion fasteners and pull the cushion out of the mounts.
15   Installation is the reverse of removal. Tighten the seat back fasteners securely.

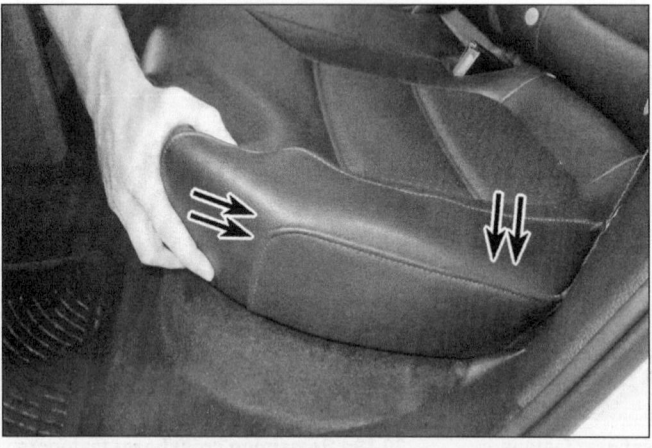

31.6 Pull the front of the seat up sharply at each end to disengage the hooks

31.7a Push the seat cushion rearward and downward at the same time . . .

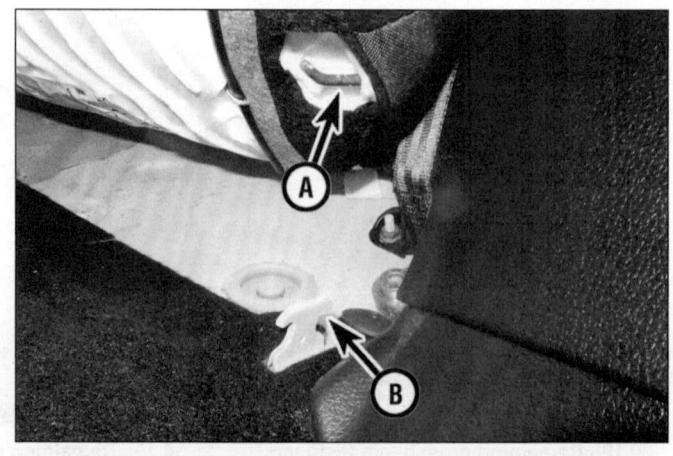

31.7b . . . to disengage the catch (A) from the hook (B)

Chapter 11 Body

11-41

31.11 Pry up the trim cover and remove the seat back fastener

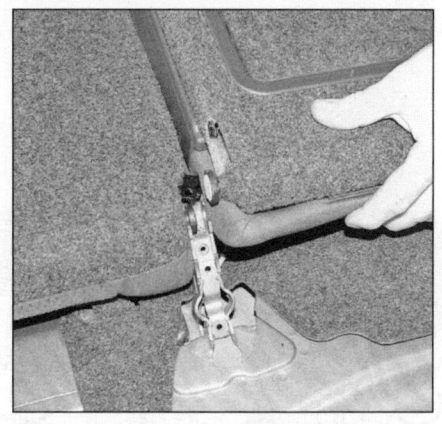

31.12a Lift the seat back off the center mount . . .

31.12b . . . then slide it off the outer mounting pin

32 Sunroof - check, adjustments, removal and installation

Caution: *Only remove the sunroof module (motor) when the sunroof is completely closed.*
Caution: *Failure to initialize the sunroof after replacing sunroof-related components may result in improper operation or breakage.*

Check

1 If the sunroof motor fails to operate, first check the relevant fuse. If the fault cannot be traced and rectified, the sunroof can be opened and closed manually using an Allen key to turn the motor spindle **(see illustration 32.24)** (an Allen key is supplied with the vehicle, and should be in the vehicle tool kit in the trunk). To gain access to the motor, remove the overhead console (see Section 28). Insert the Allen key fully into the motor opening (against spring pressure). Rotate the key to move the sunroof to the required position.

Adjusting and initializing

Initializing

2 Have all doors and windows closed and turn the key to the On position.
3 Pull down on the sunroof control switch and HOLD it there for 10 seconds.
4 The sunroof will tilt and move to the rear in its full open position. Then the sunroof will move forward to the fully closed position.
5 Hold the switch down the entire time. When the sunroof comes to a complete stop the initialization has been completed.

Motor closed position adjust

Note: *If the motor memory positions (e.g., "Open, Closed, Tilt-up") are out of alignment with the track cables, this procedure should correct this problem if all is functioning correctly otherwise.*

6 If necessary, manually adjust the sunroof to the closed position using an Allen key on the motor.
7 Remove the motor mounting bolts **(see illustration 32.24)**, leaving the electrical connector connected while supporting the motor.

32.24 Sunroof motor details

A Electrical connector
B Mounting bolts
C Sunroof manual position override (Allen fitting)

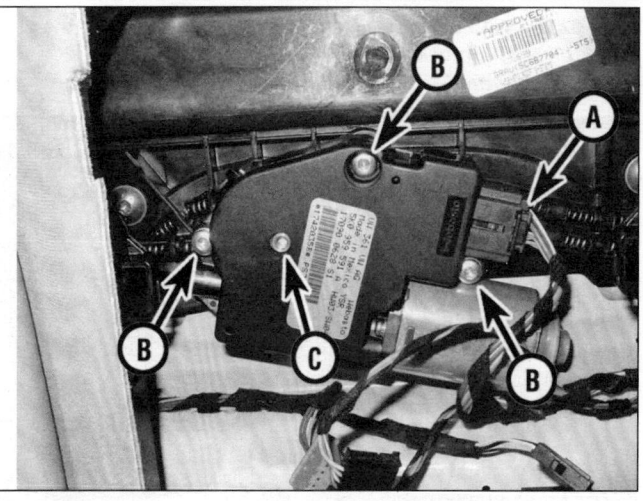

8 Set the sunroof control to the CLOSED position with the motor removed (the motor should operate until it's in the closed position).
9 With the sunroof closed, install the motor. It should now open/close and tilt normally if the assembly is mechanically sound.

Glass adjustment / replacement

Height position adjust (and if necessary, replacement)

10 Tilt the sunroof up.
11 Remove the boot covers at each end-free the boot lower clips, then loosen the front screws a few turns and slide the boot rearward and off the sunroof track.
12 Operate the sunroof glass back to the CLOSED position.
13 Loosen the sunroof glass-to-track retaining bolts - DO NOT remove the bolts, unless you're replacing the glass.
14 If the glass was removed, install the glass and screw-in the mounting screws by hand but don't tighten them.
15 Move the front of the glass by hand until it is flush and even with the top of the roof, then tighten just the front bolts. This Step may be easier with 2 people involved.
16 Move the rear of the glass by hand until it is flush and even with the roof, then tighten the rear and center bolts.
17 Tilt the sunroof back up.
18 Install the boot covers, tightening the screws and clipping them back into place. Operate the sunroof as normal, checking that the glass is even with the vehicle body from above.

Front-to-back position adjust

19 Remove the sunroof glass and motor (see Steps below).
20 Position the track by hand so that the small notch/mark on top (at each side) aligns with the stationary guide stud.
21 Adjust the closed position of the motor (see Steps above), then install the motor (see Steps below).
22 Install the glass (see Steps below).

Removal

Sunroof motor

23 Remove the overhead console / storage assembly (see Section 28).
24 With the sunroof in the closed position, remove the motor mounting bolts and disconnect the electrical connector **(see illustration)**. The sunroof open/closed position can also be adjusted manually with an Allen wrench or socket.
25 Installation is the reverse of removal.

11-42   Chapter 11   Body

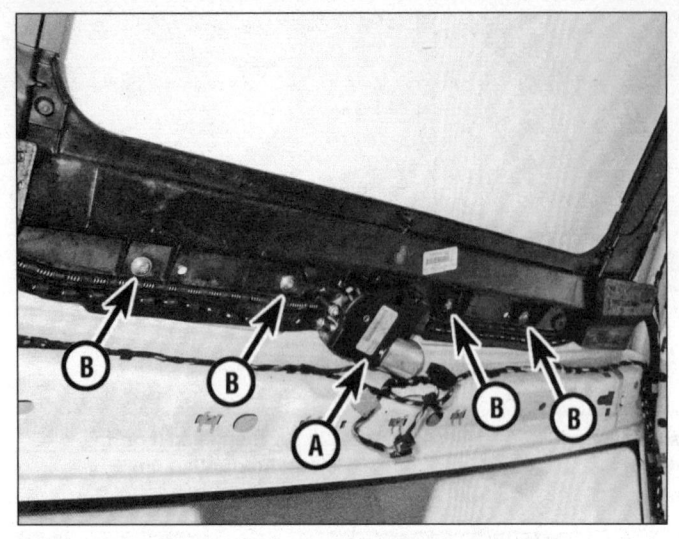

32.28a With the motor (A) already removed, remove the front track mounting fasteners (B)

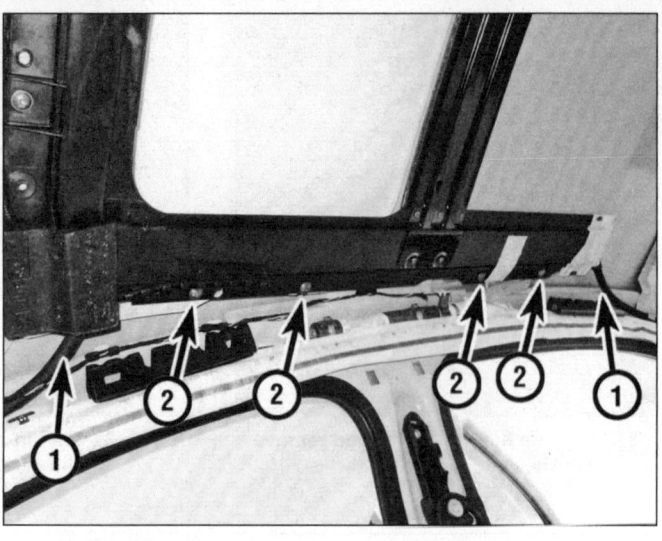

32.28b Disconnect the drain hoses (1), then remove the side mounting bolts (2) (right side shown)

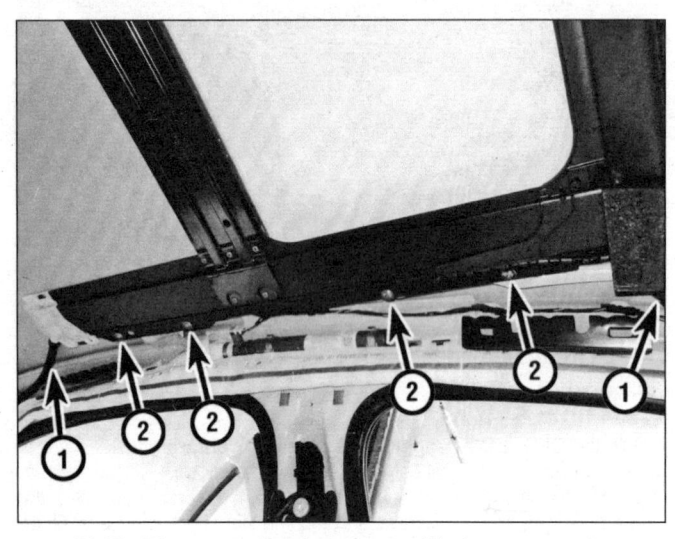

32.28c Disconnect the drain hoses (1), then remove the side mounting bolts (2) (left side shown)

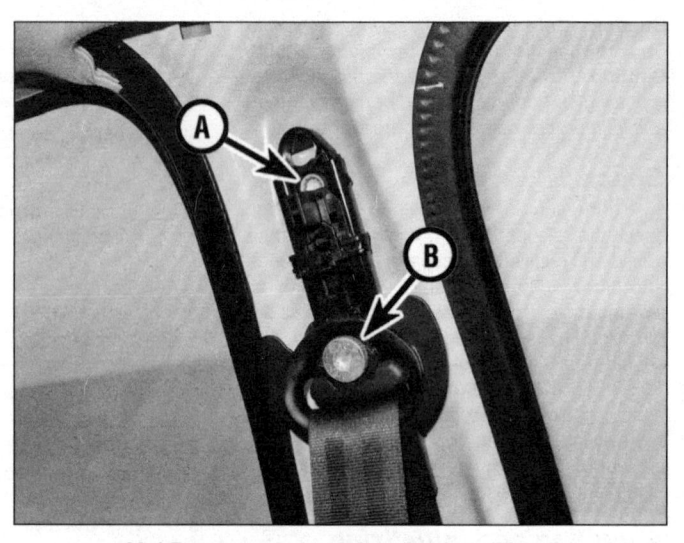

33.4 Front seat belt upper guide bolt (B) and height adjuster bolt (A)

Sunroof track assembly

26   Remove the headliner (see Section 28).
27   Remove the sunroof motor **(see illustration 32.24)**.
28   Remove the drain hoses from the four corners of the unit, then remove the track assembly mounting bolts **(see illustrations)**. Remove the sunroof track through an accessible door opening.
29   Installation is the reverse of removal. Initialize and adjust the sunroof assembly as necessary once installed (see Steps above).

33   Seat belts - removal and installation

Warning: *The models covered by this manual are equipped with a Supplemental Restraint System (SRS), more commonly known as airbags. Always disable the airbag system before working in the vicinity of any airbag system components to avoid the possibility of accidental deployment of the airbags, which could cause personal injury (see Chapter 12).*
Warning: *Do not use electrical test equipment on the belt tensioner system; it could cause the pyrotechnic belt tensioner to discharge.*
Warning: *An auxiliary voltage input device (memory saver) must not be used when working near airbag system components.*
Warning: *Never strike the pillars or floorpan with a hammer or use an impact-driver tool on or around airbag system components.*
Caution: *If the battery is disconnected, several systems must be re-learned before they will work properly (see Chapter 5, Section 3).*
1   Disconnect the negative battery cable (see Chapter 5).

Front seat belts

Note: *This procedure applies to the driver and passenger side front seat belts.*
2   Open the front and rear doors and remove the door sill trim panel (see Section 28).
Note: *The sill trim panel is one piece that extends across the front to the rear door.*
3   Remove the upper and lower B-pillar trim panels (see Section 28).
4   Remove the upper seat belt guide bolt and detach the guide from the belt height adjuster **(see illustration)**.
5   Pull up on the seat belt wire catch and remove the catch.
6   Disconnect the electrical connectors from the seat belt pretensioner, remove the lower anchor bolt and remove the pretensioner reel assembly bolt **(see illustration)**. Remove the seat belt.

Chapter 11 Body   11-43

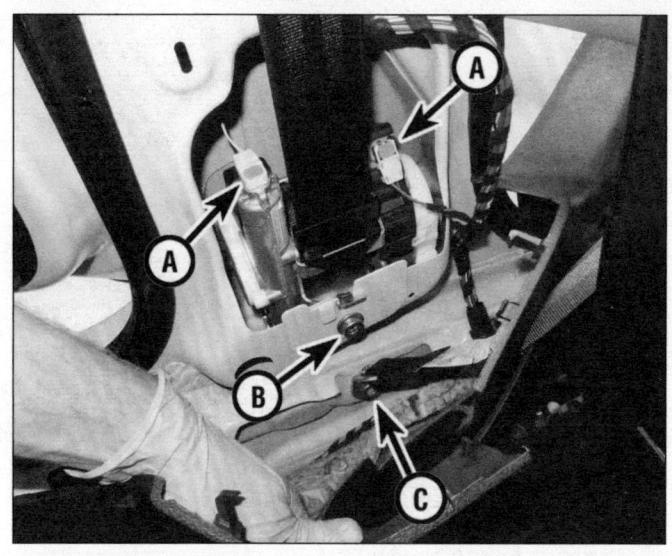

33.6 Front seat belt pretensioner connectors (A), pretensioner reel assembly bolt (B) and lower anchor bolt (C)

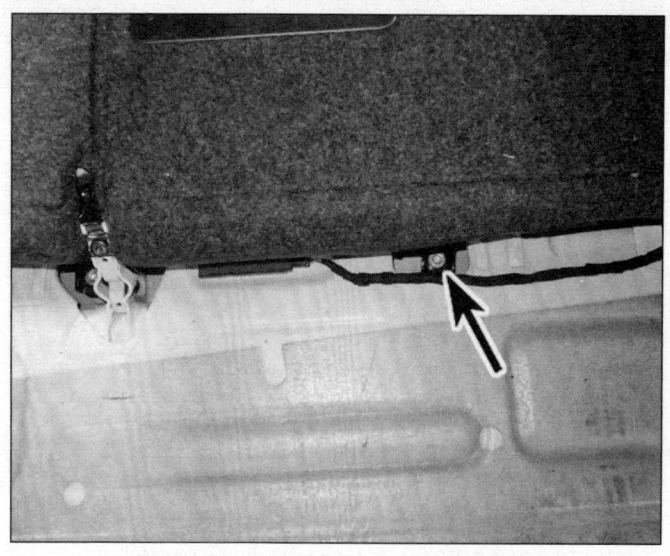

33.14a Lower anchor point fastener (middle belt)

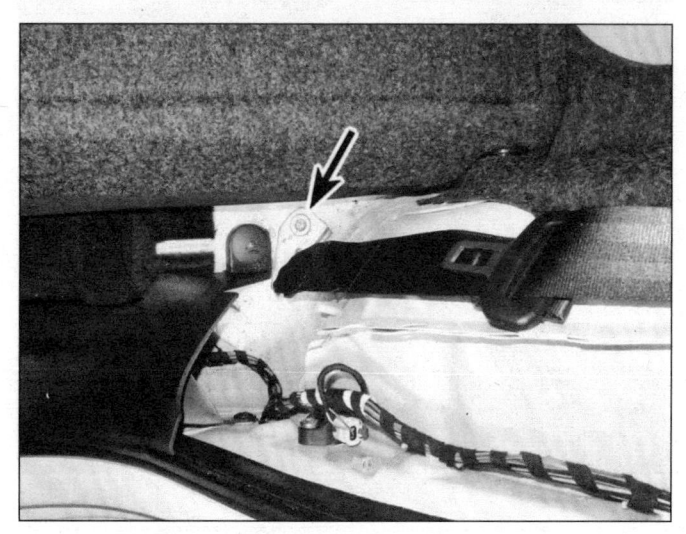

33.14b Lower anchor point fastener (side belt)

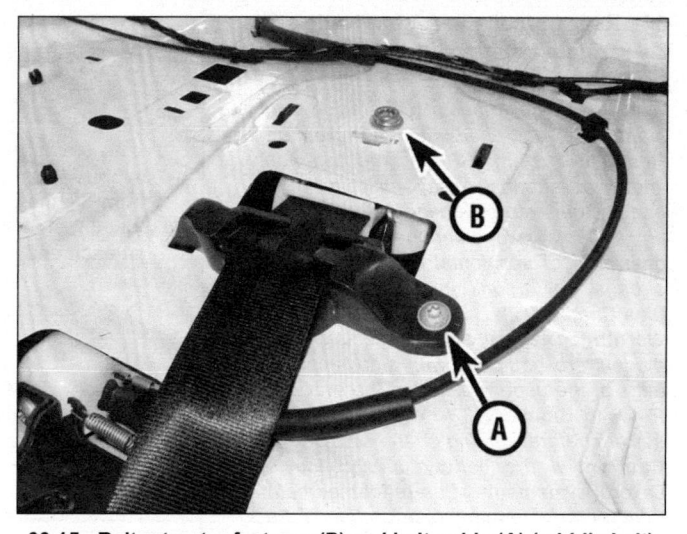

33.15a Belt retractor fastener (B) and belt guide (A) (middle belt)

7 Installation is the reverse of removal. Inspect and clean the bolt threads before the mounting bolts. Make sure the belt is not twisted before installing the lower seat mount side.

Belt buckle replacement

8 Remove the front seat (see Section 31).
9 Remove the buckle fastener and buckle from the base of the seat.
10 Installation is the reverse of removal.

Rear seat belts

11 Remove the rear seat cushion (see Section 31).
12 Remove the rear parcel shelf trim on sedans (see Section 34).
13 Disconnect the electrical connector from the belt retractor (accessed from the trunk).

14 Remove the lower anchor point fastener and belt end **(see illustrations)**.
15 Remove the belt retractor fastener, and if necessary on the middle belt, the belt guide **(see illustrations)**. Remove the belt retractor and seat belt from the trunk.
16 Installation is the reverse of removal. Make sure the belt is not twisted before installing the lower seat mount side.

Belt buckle replacement

17 Remove the seat cushion and seat back (see Section 31).
18 Remove the buckle fastener and buckle from the floorpan.
19 Installation is the reverse of removal. Make sure the belt is not twisted before installing the lower mount side.

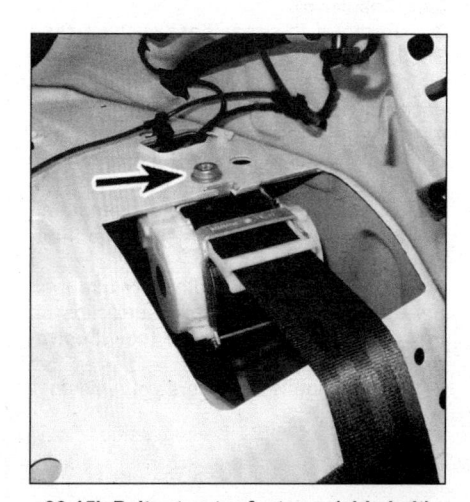

33.15b Belt retractor fastener (side belt)

11-44   Chapter 11   Body

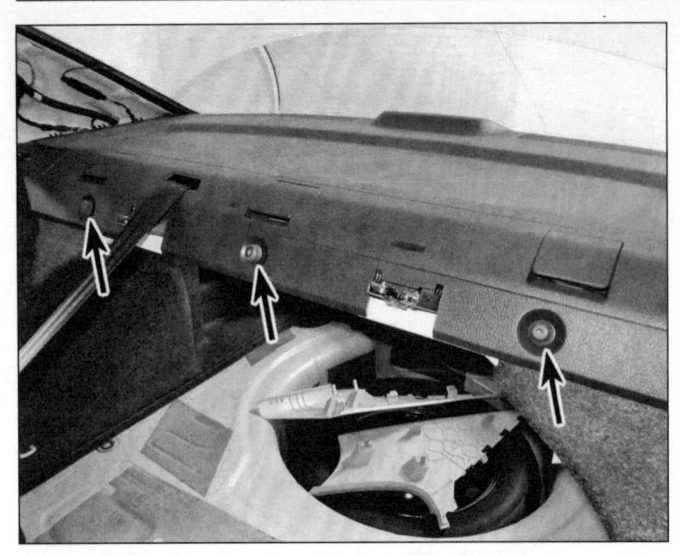

34.4 Parcel shelf front pushpin fasteners

34.5a Pry-out the side trim panels . . .

34 Rear parcel shelf - removal and installation

Warning: *The models covered by this manual are equipped with a Supplemental Restraint System (SRS), more commonly known as airbags. Always disable the airbag system before working in the vicinity of any airbag system components to avoid the possibility of accidental deployment of the airbags, which could cause personal injury (see Chapter 12).*

Warning: *The front and rear seats on some models are equipped with side-impact airbags at the upper outside of the seat back. Refer to Chapter 12 to disable the airbag system before working on the seats.*

Caution: *If the battery is disconnected, several systems must be re-learned before they will work properly (see Chapter 5, Section 3).*

1   Disconnect the negative battery cable (see Chapter 5).
2   Remove the rear seat cushion (see Section 31).
3   Remove the C-pillar trim panels from each side (see Section 28).
4   Pry-out the parcel shelf pushpin fasteners along the front **(see illustration)**.
5   Pry-out and remove the seat side trim panels, then remove the remaining exposed fasteners at each side **(see illustrations)**.
6   Using a long screwdriver or trim tool, pry-up and release the pressure clips on the inside of the parcel shelf **(see illustration)**. Remove the shelf.
7   Installation is the reverse of removal.

34.5b . . . then remove the exposed fasteners at each end

34.6 Pry-up and release the parcel shelf inner pressure clips

Notes

Notes

Chapter 12
Chassis electrical system

Contents

	Section		Section
Airbag - general information and component removal and installation	25	Ignition switch and key lock cylinder - replacement	8
Antenna - removal and installation	13	Instrument cluster - removal and installation	11
Bulb replacement	19	Instrument panel switches - replacement	10
Circuit breakers - general information	4	Key fob - battery replacement and transmitter programming	7
Cruise control system - description and check	22	Power door lock system - general information	24
Electrical connectors - general information	6	Power window system - general information	23
Electrical troubleshooting - general information	2	Radio and speakers - removal and installation	12
Fuse box locations and general fuse information	3	Rear window defogger - check and repair	14
General Information	1	Relays - general information	5
Headlight bulbs - replacement	16	Steering column switches - replacement	9
Headlight housing - removal and installation	15	Taillight housing - removal and installation	18
Headlights - adjustment	17	Wipers	21
Horns - replacement	20	Wiring diagrams - general information	26

1 General Information

1 The electrical system is a 12-volt, negative ground type. Power for the lights and all electrical accessories is supplied by a lead/acid battery, which is charged by the alternator.
2 This Chapter covers repair and service procedures for the various electrical components not associated with the engine. Information on the battery, alternator and starter motor can be found in Chapter 5.
3 It should be noted that when portions of the electrical system are serviced, the negative battery cable should be disconnected from the battery to prevent electrical shorts and/or fires.

2 Electrical troubleshooting - general information

1 A typical electrical circuit consists of an electrical component, any switches, relays, motors, fuses, fusible links or circuit breakers related to that component and the wiring and connectors that link the component to both the battery and the chassis. To help you pinpoint an electrical circuit problem, wiring diagrams are included at the end of this Chapter.
2 Before tackling any troublesome electrical circuit, first study the appropriate wiring diagrams to get a complete understanding of what makes up that individual circuit. Trouble spots, for instance, can often be narrowed down by noting if other components related to the circuit are operating properly. If several components or circuits fail at one time, chances are the problem is in a fuse or ground connection, because several circuits are often routed through the same fuse and ground connections.
3 Electrical problems usually stem from simple causes, such as loose or corroded connections, a blown fuse, a melted fusible link or a failed relay. Visually inspect the condition of all fuses, wires and connections in a problem circuit before troubleshooting the circuit.
4 If test equipment and instruments are going to be utilized, use the diagrams to plan ahead of time where you will make the necessary connections in order to accurately pinpoint the trouble spot.

12-2   Chapter 12   Chassis electrical system

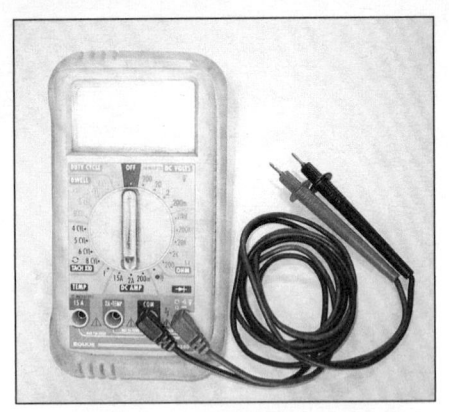

2.5a The most useful tool for electrical troubleshooting is a digital multimeter that can check volts, amps, and test continuity

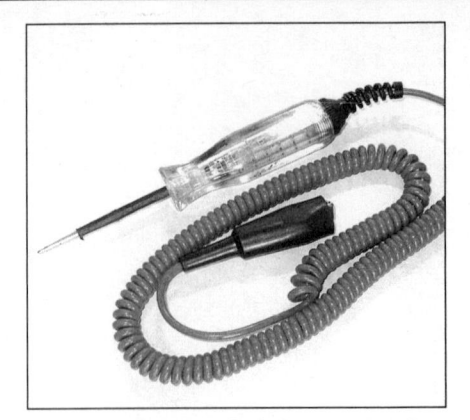

2.5b A simple test light is a very handy tool for testing voltage

Continuity check

9   A continuity check is done to determine if there are any breaks in a circuit - if it is passing electricity properly. With the circuit off (no power in the circuit), a self-powered continuity tester or multimeter can be used to check the circuit. Connect the test leads to both ends of the circuit (or to the power end and a good ground), and if the test light comes on the circuit is passing current properly **(see illustration)**. If the resistance is low (less than 5 ohms), there is continuity; if the reading is 10,000 ohms or higher, there is a break somewhere in the circuit. The same procedure can be used to test a switch, by connecting the continuity tester to the switch terminals. With the switch turned On, the test light should come on (or low resistance should be indicated on a meter).

Finding an open circuit

10   When diagnosing for possible open circuits, it is often difficult to locate them by sight because the connectors hide oxidation or terminal misalignment. Merely wiggling a connector on a sensor or in the wiring harness may correct the open circuit condition. Remember this when an open circuit is indicated when troubleshooting a circuit. Intermittent problems may also be caused by oxidized or loose connections.

11   Electrical troubleshooting is simple if you keep in mind that all electrical circuits are basically electricity running from the battery, through the wires, switches, relays, fuses and fusible links to each electrical component (light bulb, motor, etc.) and to ground, from which it is passed back to the battery. Any electrical problem is an interruption in the flow of electricity to and from the battery.

5   The basic tools needed for electrical troubleshooting include a circuit tester or voltmeter (a 12-volt bulb with a set of test leads can also be used), a continuity tester, which includes a bulb, battery and set of test leads, and a jumper wire, preferably with a circuit breaker incorporated, which can be used to bypass electrical components **(see illustrations)**. Before attempting to locate a problem with test instruments, use the wiring diagram(s) to decide where to make the connections.

Voltage checks

6   Voltage checks should be performed if a circuit is not functioning properly. Connect one lead of a circuit tester to either the negative battery terminal or a known good ground. Connect the other lead to a connector in the circuit being tested, preferably nearest to the battery or fuse **(see illustration)**. If the bulb of the tester lights, voltage is present, which means that the part of the circuit between the connector and the battery is problem free. Continue checking the rest of the circuit in the same fashion. When you reach a point at which no voltage is present, the problem lies between that point and the last test point with voltage. Most of the time the problem can be traced to a loose connection. Note: Keep in mind that some circuits receive voltage only when the ignition key is in the Accessory or Run position.

Finding a short

7   One method of finding shorts in a circuit is to remove the fuse and connect a test light or voltmeter in place of the fuse terminals. There should be no voltage present in the circuit. Move the wiring harness from side-to-side while watching the test light. If the bulb goes on, there is a short to ground somewhere in that area, probably where the insulation has rubbed through. The same test can be performed on each component in the circuit, even a switch.

Ground check

8   Perform a ground test to check whether a component is properly grounded. Disconnect the battery and connect one lead of a continuity tester or multimeter (set to the ohms scale), to a known good ground. Connect the other lead to the wire or ground connection being tested. If the resistance is low (less than 5 ohms), the ground is good. If the bulb on a self-powered test light does not go on, the ground is not good.

3   Fuse box locations and general fuse information

Fuse box locations

1   The electrical circuits of the vehicle are protected by a combination of fuses, circuit

2.6 In use, a basic test light's lead is clipped to a known good ground, then the pointed probe can test connectors, wires or electrical sockets - if the bulb lights, the circuit being tested has battery voltage

2.9 With a multimeter set to the ohm scale, resistance can be checked across two terminals - when checking for continuity, a low reading indicates continuity, a high reading or infinity indicates high resistance or lack of continuity

Chapter 12 Chassis electrical system

breakers and fusible links. The main fuse/relay panel is in the engine compartment **(see illustrations 3.01c and 3.01d)**, while the interior fuse/relay panel is located inside the passenger compartment **(see illustrations 3.01a and 3.01b)** Each of the fuses is designed to protect a specific circuit. The various circuits are identified in the owners manual, on the fuse panel itself, the fuse box lid, or on the legend of the wiring diagram.

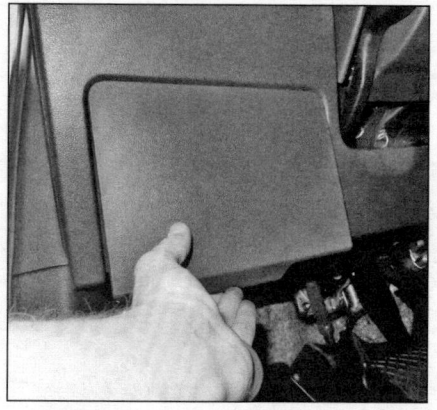

Fuse general information

2   Several sizes of fuses are employed in the fuse blocks. There are small, medium and large sizes of the same design, all with the same blade terminal design. The medium and large fuses can be removed with your fingers, but the small fuses require the use of pliers or the small plastic fuse-puller tool found in most fuse boxes.

3   If an electrical component fails, always check the fuse first. The best way to check the fuses is with a test light. Check for power at the exposed terminal tips of each fuse. If power is present at one side of the fuse but not the other, the fuse is blown. A blown fuse can also be identified by visually inspecting it **(see illustration)**.

4   Be sure to replace blown fuses with the correct type. Fuses (of the same physical size) of different ratings may be physically interchangeable, but only fuses of the proper rating should be used. Replacing a fuse with one of a higher or lower value than specified is not recommended. Each electrical circuit needs a specific amount of protection. The amperage value of each fuse is molded into the top of the fuse body.

5   If the replacement fuse immediately fails, don't replace it again until the cause of the problem is isolated and corrected. In most cases, this will be a short circuit in the wiring caused by a broken or deteriorated wire.

Note: *Many systems have gone from fuses in the fuse box to non-replaceable (internal) fuse systems or computer amperage controlled internal system analysis. In other words, as the amperage for that particular system increases past the preset safe usage level, the computer will drop the voltage output (turn off the controlled signal) for that particular system. At that point, the system is off (off line) and will not start up again until the next key cycle. If (when the key is cycled back to the on position) the system amperage load is still not within specifications the computer will shut that particular system off again. Check the wiring diagrams to determine which systems are utilizing this type of feature as well as the fuse box index.*

High amperage fuses

6   Some circuits are protected by high amperage fuses, known as "SA" fuses, which are fastened to the studs on the front of the underhood fuse/relay box. These are used in circuits which are not ordinarily fused, or which carry high current, such as the circuit between the alternator and the battery. If you have to replace an SA fuse, make sure that you replace it with one of the same specification. If the replacement fuse blows in the same circuit, make sure that you troubleshoot the circuit in which the fuse melted BEFORE installing another one.

3.1a The interior fuse box is located at the left end of the instrument panel, under a cover

3.1b Interior fuse box

4 Circuit breakers - general information

1   Circuit breakers protect certain circuits, such as the power windows or heated seats. Depending on the vehicle's accessories, there may be one or two circuit breakers, located in the fuse/relay box in the engine compartment.

2   Because the circuit breakers reset automatically, an electrical overload in a circuit breaker-protected system will cause the circuit to fail momentarily, then come back on. If the circuit does not come back on, check it immediately.

3   For a basic check, pull the circuit breaker up out of its socket on the fuse panel, but just far enough to probe with a voltmeter. The breaker should still contact the sockets. With the voltmeter negative lead on a good chassis ground, touch each end prong of the circuit breaker with the positive meter probe. There should be battery voltage at each end. If there is battery voltage only at one end, the circuit breaker must be replaced.

4   Some circuit breakers must be reset manually.

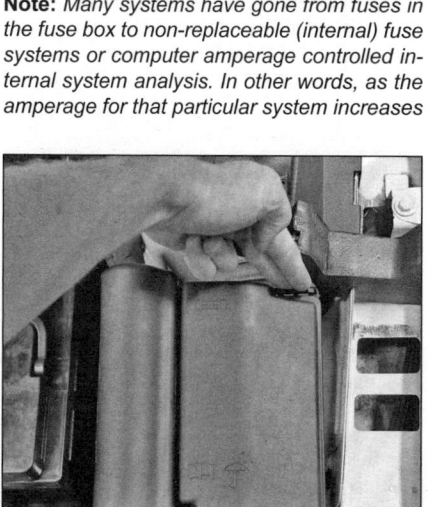

3.1c The engine compartment fuse box is located in the right-front corner of the engine compartment under this cover

3.1d Engine compartment fuse box

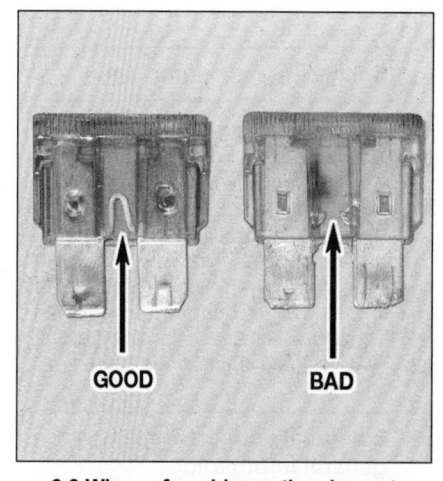

3.3 When a fuse blows, the element between the terminals melts

Electrical connectors

Most electrical connectors have a single release tab that you depress to release the connector

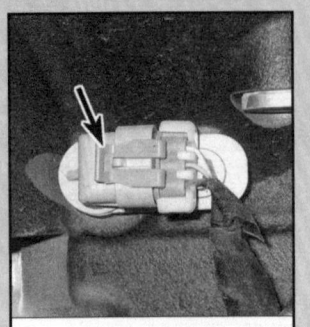

Some electrical connectors have a retaining tab which must be pried up to free the connector

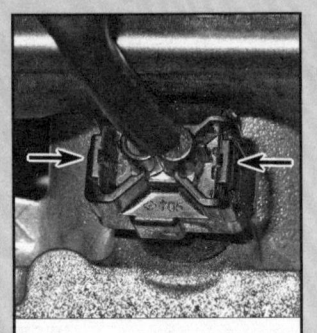

Some connectors have two release tabs that you must squeeze to release the connector

Some connectors use wire retainers that you squeeze to release the connector

Critical connectors often employ a sliding lock (1) that you must pull out before you can depress the release tab (2)

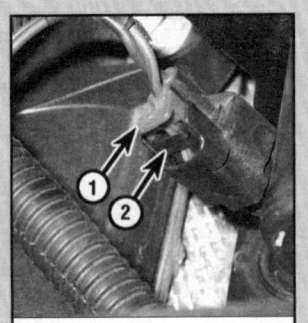

Here's another sliding-lock style connector, with the lock (1) and the release tab (2) on the side of the connector

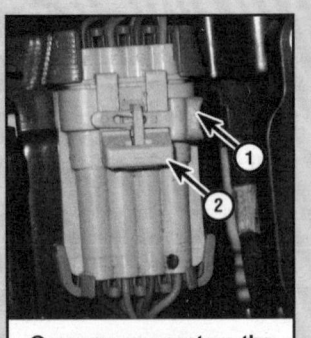

On some connectors the lock (1) must be pulled out to the side and removed before you can lift the release tab (2)

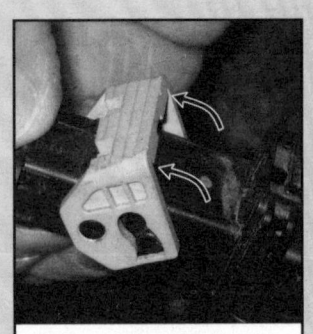

Some critical connectors, like the multi-pin connectors at the Powertrain Control Module employ pivoting locks that must be flipped open

5 Relays - general information

1 Several electrical accessories in the vehicle, such as the fuel injection system, horns, starter, and fog lamps use relays to transmit the electrical signal to the component. Relays use a low-current circuit (the control circuit) to open and close a high-current circuit (the power circuit). If the relay is defective, that component will not operate properly. Most relays are mounted in the engine compartment fuse/relay box **(see illustrations 3.1d)**.

6 Electrical connectors - general information

1 Most electrical connections on these vehicles are made with multiwire plastic connectors. The mating halves of many connectors are secured with locking clips molded into the plastic connector shells. The mating halves of some large connectors, such as some of those under the instrument panel, are held together by a bolt through the center of the connector.

2 To separate a connector with locking clips, use a small screwdriver to pry the clips apart carefully, then separate the connector halves. Pull only on the shell, never pull on the wiring harness as you may damage the individual wires and terminals inside the connectors. Look at the connector closely before trying to separate the halves. Often the locking clips are engaged in a way that is not immediately clear. Additionally, many connectors have more than one set of clips.

3 Each pair of connector terminals has a male half and a female half. When you look at the end view of a connector in a diagram, be sure to understand whether the view shows the harness side or the component side of the connector. Connector halves are mirror images of each other, and a terminal shown on the right side end-view of one half will be on the left side end-view of the other half.

4 It is often necessary to take circuit voltage measurements with a connector connected. Whenever possible, carefully insert a small straight pin (not your meter probe) into the rear of the connector shell to contact the terminal inside, then clip your meter lead to the pin. This kind of connection is called "backprobing." When inserting a test probe into a terminal, be careful not to distort the terminal opening. Doing so can lead to a poor connection and corrosion at that terminal later. Using the small straight pin instead of a meter probe results in less chance of deforming the terminal connector.

Chapter 12 Chassis electrical system

7.1 Separate the key fob battery cover

7.3 Before removing the battery, pay close attention to the polarity and position of the battery

8.5 Disconnect the electrical connector for the electronic immobilizer induction coil

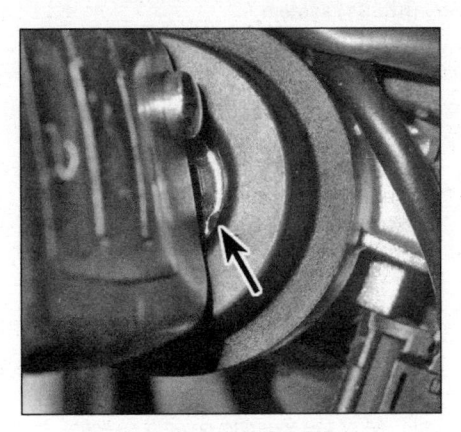

8.6 Access hole location

8.7 Insert a paper clip into the access hole and pull out the assembly

7 Key fob - battery replacement and transmitter programming

Note: *If you have a question about the transmitter strength, take your key fob to any repair shop that has a TPMS tester. The TPMS testers all have a function to test transmitter strength. This will help determine if a new battery is needed or if the transmitter is defective and needs replaced, as well as the actual strength of the transmitter compared to the other transmitters that work with your vehicle. Some auto parts stores also provide this service.*

Battery replacement

1 Use a small flat screwdriver or a coin to separate the upper and lower half of the key fob transmitter **(see illustration)**.
2 Slide the mechanical key and the electronic (upper) half of the key fob out of the lower half of the key fob assembly.
3 Remove the battery out of the key fob. Pay close attention to the polarity marks on the battery and install the new one in the same orientation **(see illustration)**.
4 With the battery installed, slide the upper half back into the lower half of the key fob assembly.
5 Check the operation of the key fob.

Transmitter programming (synchronization)

6 Original equipment key fob programming (synchronization) requires the use of the proper type of scanner and information, as well as a locksmith license. Seek professional help in order to complete this procedure.
7 Aftermarket key fobs are available for some vehicles. The new fob will include instructions on how to synchronize the fob to the vehicle.

8 Ignition switch and key lock cylinder - replacement

Warning: *The models covered by this manual are equipped with a Supplemental Restraint System (SRS), more commonly known as airbags. Always disable the airbag system before working in the vicinity of any airbag system components to avoid the possibility of accidental deployment of the airbags, which could cause personal injury (see Section 25).*
Note: *If the battery is disconnected, several systems must be re-learned before they will work properly (see Chapter 5, Section 3).*

1 The steering lock housing is located under the steering column electronic control module on the steering column (switch on the left, key lock cylinder on the right). It is comprised of a cast-metal housing, the ignition lock cylinder, the ignition switch and the electronic immobilizer induction coil.
2 Disconnect the cable from the negative terminal of the battery (see Chapter 5).
3 Remove the steering column covers (see Chapter 11), steering wheel and clockspring (see Chapter 10), and steering column switch assembly (see Section 9).

Key lock cylinder

4 On models with automatic transaxles, the shift linkage must be in Park before removing/installing the lock cylinder. Insert the key into the lock and turn it to the On position.
5 Disconnect the electrical connector from the electronic immobilizer induction coil **(see illlustration)**.
6 To remove the key lock cylinder, keep the key in the lock and rotated to the On position and align the access hole with the mark on the trim **(see illustration)**.
7 Insert the straightened end of a large paper clip into the hole in the cylinder **(see illustration)** and pull the key, lock cylinder and induction coil from the lock cylinder housing as an assembly.
Note: *The electronic immobilizer induction coil is integrated into the lock cylinder and can't be replaced separately.*
8 When installing the key lock cylinder, align it in its original position (still in the On position with the key in), then push the cylinder in until it snaps in place. Remove the paper clip, turn the key to the Lock position and remove the key.
9 Installation is the reverse of removal.

Ignition switch

Without keyless entry (KESSY) system

10 Loosen the steering column mounting bolts and lower the steering column (see Chapter 10).

12-6   Chapter 12 Chassis electrical system

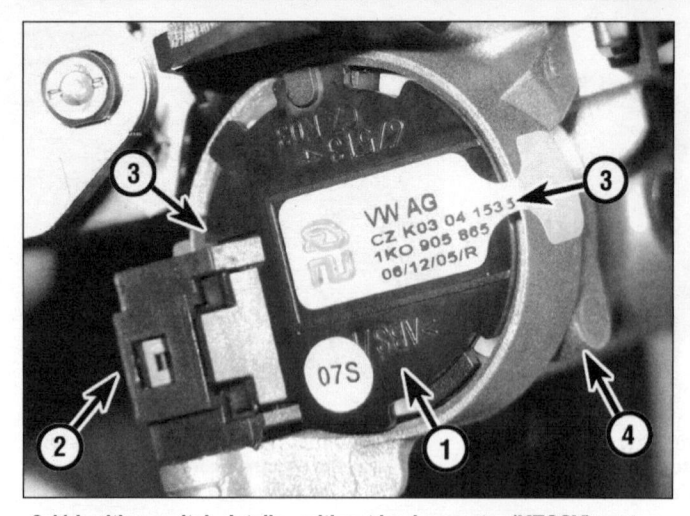

8.11 Ignition switch details - without keyless entry (KESSY) system

1 Ignition switch
2 Electrical connector
3 Ignition switch locking tabs
4 Lock housing

8.21 Lock housing shear bolt locations

11 Disconnect the electrical connector to the ignition switch **(see illustration)**.
12 Unlock the switch by depressing the tabs in the lock housing using a small screwdriver.
13 Pull the ignition switch out from the lock housing.

With keyless entry (KESSY) system

14 Instead of an ignition switch in the steering column, vehicles equipped with the keyless entry system (KESSY) have a start system button in the center console.
15 Remove the center console (see Chapter 11).
16 Disconnect the electrical connector from the switch.
17 Open the tabs on the switch and remove it from the console.
18 Installation is the reverse of removal.

Lock housing

19 Remove the steering column covers, trim and electrical connections, then loosen the steering column mounting bolts and lower the steering column (see Chapter 10).
20 Disconnect the electrical connector to the ignition switch and the electronic immobilizer induction coil.
21 To remove the housing, locate the shear bolt heads **(see illustration)**, then drill the bolt heads out from the top of the steering column bracket.
Caution: *The mounting brackets are easily damaged; do not use a chisel to remove the shear head bolts.*
22 Installation is the reverse of removal. Tighten the new shear bolts until their heads break off.

9 Steering column switches - replacement

Warning: *The models covered by this manual are equipped with a Supplemental Restraint System (SRS), more commonly known as airbags. Always disable the airbag system before working in the vicinity of any airbag system components to avoid the possibility of accidental deployment of the airbags, which could cause personal injury (see Section 25).*
Note: *If the battery is disconnected, several systems must be re-learned before they will work properly (see Chapter 5, Section 3).*
Note: *The steering column switch base carrier is located on the top of the steering column. All of the steering column switches are mounted to the base carrier, including the Steering Column Electronic Control Module (SCECM). Which includes the turn signals, headlight dimmer, windshield wiper/washer, airbag clockspring, steering angle sensor, radio controls, voice recognition, telephone, climate control functions, lane assistance activation, and cruise control.*
Note: *The turn signal, headlight dimmer, windshield wiper/washer and cruise control function switches cannot be disassembled and must be replaced as a complete unit.*
1 Place the front wheels and steering wheel in a straight forward position with the steering wheel telescoped out to its furthest possible position.
2 Disconnect the cable from the negative terminal of the battery (see Chapter 5).
3 Remove the steering wheel (see Chapter 10).
4 Remove the steering column covers (see Chapter 11).
5 Discharge any static electricity that has built up in your body before removing any electrical connection around the steering column by touching any grounded metal attached to the vehicle.

Steering Column Electronic Control Module (SCECM)

6 Detach the electrical connections to the SCECM and switches.
7 Place a strip of tape across the SCECM housing and clockspring hub to prevent the clockspring from turning, then remove the three fasteners securing the SCECM to the steering column **(see illustrations)**.

9.7a Remove the three Torx screws securing the SCECM to the column

9.7b Slide the SCECM off of the steering column

Chapter 12 Chassis electrical system

12-7

9.9 Remove the lower retaining screw, if applicable.

9.10 With the tabs released, slide the combination switch off of the steering column

10.1a If needed, remove the trim from around the headlight switch with a flat trim tool

8 Installation is the reverse of removal.
Warning: *Be sure to align the arrows on the clockspring portion of the SCCM before installing or securing the fasteners.*

Steering column combination switch

Note: *Some models use a screw and tabs to fasten the combination switch to the steering column base. The screw is located on the bottom side of the combination switch.*

9 Remove the SCECM as described in the previous step, then release the fasteners and (if applicable) the retaining screw on the bottom of the combination switch **(see illustration)**.
10 Disconnect any remaining electrical connections.
11 Installation is the reverse of removal.

10 Instrument panel switches - replacement

Warning: *The models covered by this manual are equipped with a Supplemental Restraint System (SRS), more commonly known as airbags. Always disable the airbag system before working in the vicinity of any airbag system components to avoid the possibility of accidental deployment of the airbags, which could cause personal injury (see Section 25).*

Headlight switch

1 Turn the headlight switch knob counterclockwise until it stops at the zero position. Push in on the switch and twist it to the right (clockwise) until it stops, then withdraw it from the instrument panel **(see illustrations)**.

10.1b Push in on the switch button. . .

10.1c. . . then rotate the knob clockwise until it stops. . .

10.1d. . . and pull the switch out far enough to expose the electrical connector

10.1e Disconnect the electrical connector and remove the switch

12-8   Chapter 12   Chassis electrical system

10.6 Pry out and remove the hazard flasher switch using a small screwdriver

11.3 Instrument cluster fasteners

Dash light dimmer switch

3   Remove the driver's side dashboard trim panel (see Chapter 11) and disconnect the electrical connector.
4   Squeeze the switch retaining tabs together and remove the dimmer switch.
5   Installation is the reverse of removal.

Hazard flasher switch

6   Using a small screwdriver, pry out **(see illustration)** and disconnect the electrical connectors to the front passenger's airbag indicator lamp and the hazard flasher switch.
7   Installation is the reverse of removal.

Center console switches

8   Depending on the options of the vehicle, there may be one or more switches on the center console, including seat heaters and rear window defogger.
9   All of the aforementioned switches are located in the switch panel at the top of the center console.
10   Remove the transmission shift lever cover (see Chapter 11).
11   Remove the switch panel fasteners, and lift up the panel.
12   All of the switches are removed the same way. Disconnect the electrical connector and push the switch out from the back side of the panel.
13   Installation is the reverse of removal.

11   Instrument cluster - removal and installation

Warning: *The models covered by this manual are equipped with a Supplemental Restraint System (SRS), more commonly known as airbags. Always disable the airbag system before working in the vicinity of any airbag system components to avoid the possibility of accidental deployment of the airbags, which could cause personal injury (see Section 25).*
Note: *If the battery is disconnected, several systems must be re-learned before they will work properly (see Chapter 5, Section 3).*

Note: *The instrument cluster is not serviceable and must be replaced as a complete unit. The instrument cluster control module data must be saved using a factory scan tool prior to removal. The new cluster must be programmed with a factory scan tool. For this reason it is best to have a VW dealer (or other qualified repair shop) perform this job if replacement of the unit is required.*
1   Disconnect the negative cable from the battery (see Chapter 5).
2   Remove the instrument panel trim and with upper steering column trim cover (see Chapter 11).
3   Remove the screws securing the cluster to the instrument panel **(see illustration)**.
4   Pull the instrument cluster out far enough to disconnect the electrical connection **(see illustrations)**.
Note: *The instrument cluster is plugged in to the dash; pull the cluster straight out to prevent damaging the electrical connector pins.*
5   Installation is the reverse of removal.

12   Radio and speakers - removal and installation

Warning: *The models covered by this manual are equipped with a Supplemental Restraint System (SRS), more commonly known as airbags. Always disable the airbag system before working in the vicinity of any airbag system components to avoid the possibility of accidental deployment of the airbags, which could cause personal injury (see Section 25).*
Note: *If the battery is disconnected, several systems must be re-learned before they will work properly (see Chapter 5, Section 3).*
Note: *The audio system is part of the diagnostic network of the vehicle. Any problems with the radio, antenna or speakers may set a trouble code that can be retrieved with a scan tool.*

11.4a The electrical connection is secured with a lever. This photo shows the lever in the locked position. Lift the lever and rotate it to the opposite end of the connector to remove the electrical connection

11.4b This photo shows the electrical connection in the unlocked position. The lever must be in this position to re-install the electrical connector. Rotate the lever to the locked position only when the connector is fully seated into the cluster connection

Chapter 12 Chassis electrical system

12.4a Remove the radio and A/C control trim panel with a flat trim tool . . .

12.4b . . . then carefully lift the radio and A/C control trim panel out from the instrument panel

Radio

Caution: *On models equipped with the "Premium 8" radio, the transport mode must be activated before removing and deactivated after installation.*

1 Turn the ignition switch to the Off position and remove the key. Also turn the radio to Off.
2 On models with a "Premium 8" radio, turn the unit On, then press and hold the CD player forward, reverse and skip buttons at the same time. After five seconds, "CDC transportation safeguard activated" will appear in the radio display.
3 If needed for maneuvering room, remove the instrument panel center trim and vent panel (see Chapter 11).
Note: *Use a plastic trim tool to pry off the radio and A/C control trim panel.*
4 Remove the radio and A/C control trim panel fasteners and pry the trim panel off of the instrument panel **(see illustrations)**.
5 Remove the radio fasteners, pull the radio out from the instrument panel, release the electrical connector lock, and disconnect the connector from the back of the radio **(see illustrations)**.
6 Installation is the reverse of removal. Deactivate the anti-theft system. On "Premium 8" radio, cancel the "CDC transportation safeguard activated" display; turn the unit On, then press and hold the CD player forward, reverse and skip buttons at the same time until the "deactivation" button appears on the touch screen. Press the "deactivation" button on the screen and transportation safeguard is deactivated.

Satellite radio

7 Turn the ignition switch to the Off position and remove the key.

Sedan models

8 Remove the rear package tray (see Chapter 11) and loosen the satellite radio receiver mounting screws.
9 Open the trunk and slide the receiver back until the screws come through the openings in the sheet metal.
10 Disconnect the electrical connectors and remove the unit.
11 Installation is the reverse of the removal procedure.

All other models

12 Remove the passenger's side front seat (see Chapter 11).
13 Remove the mounting fasteners, then disconnect the electrical connectors and remove the unit.
14 Installation is the reverse of the removal procedure.

CD changer

15 Turn the ignition switch to the Off position and remove the key.

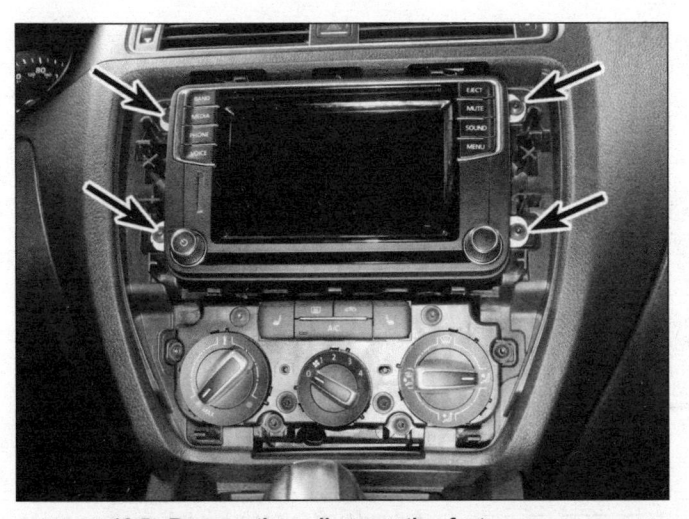

12.5a Remove the radio mounting fasteners . . .

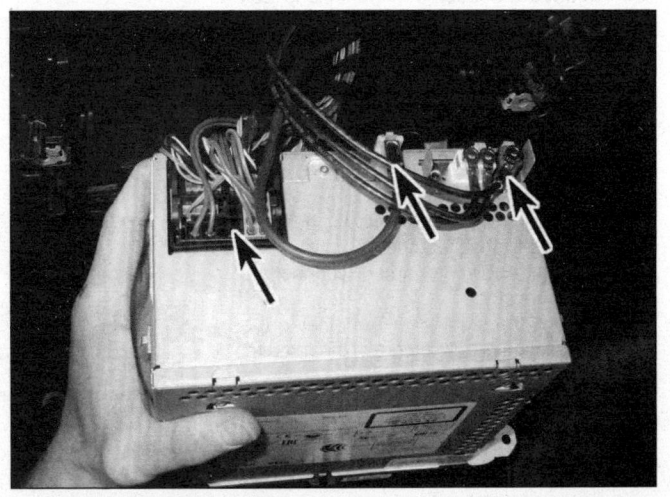

12.5b . . . then disconnect the electrical connectors and the antenna connections from the rear of the radio

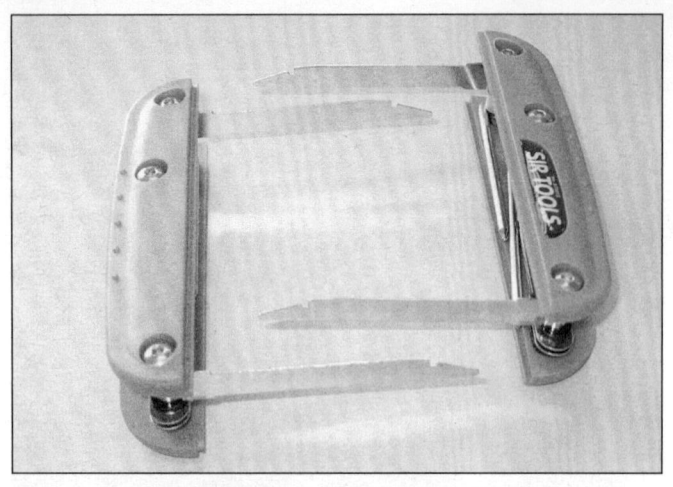

12.17 Typical CD changer removal tools

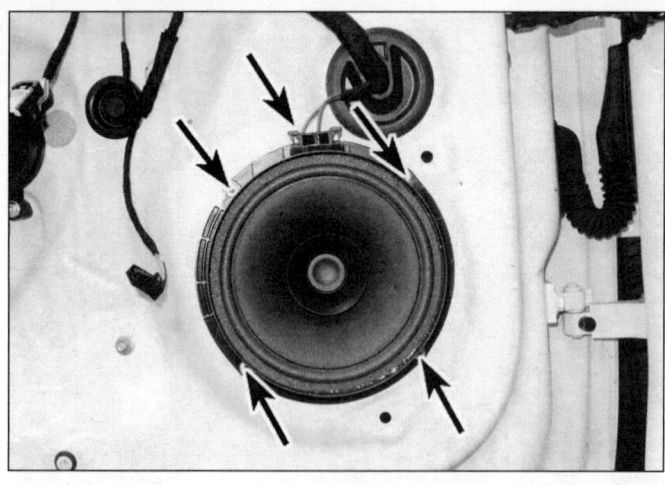

12.21 Disconnect the electrical connector and drill out the rivets

16 Open the armrest and remove any CDs in the changer.
17 Push the two tools into the slots on either side of the CD changer **(see illustration)**.
18 Pull the changer out of the console, disconnect the connectors, and remove the CD changer from the vehicle.
19 Installation is the reverse of the removal procedure.

Speakers

Door bass speakers

20 Remove the door panel (see Chapter 11).
21 Disconnect the electrical connector at the speaker, then drill the rivets out and remove the speaker **(see illustration)**.
22 Installation is the reverse of the removal procedure.
Note: *Do not use sheet metal screws to replace the rivets.*

Door midrange and treble speakers

Note: *On rear door treble speakers, the plastic tab clips must be cut off and melted back in place using a soldering gun.*
23 Remove the door trim panel and sail panel (see Chapter 11).
24 Disconnect the electrical connector at the speaker, remove the mounting fasteners and remove the speaker from the sail panel.
25 Installation is the reverse of the removal procedure.

Subwoofer (sedans)

26 Open the trunk and remove the subwoofer-to-parcel shelf fasteners.
27 Lower the subwoofer and disconnect the electrical connector.
28 Remove the subwoofer from inside the trunk. Installation is the reverse of the removal procedure.

13 Antenna - removal and installation

Note: *There are two types of antennas used on these models: a grid type and roof-mounted type. The grid type is an integral component of the rear window (or windows, depending on model). To replace these antennas you must replace the rear window or side window(s).*
1 Remove the rear headliner trim and D-pillar trim from both sides.
2 Remove the rear reading light or luggage compartment light (wagons) housing (see Section 19).
3 Disconnect the antenna cable(s).
4 Carefully pull down the headliner at the rear, or remove the entire headliner (see Chapter 11).
5 Remove the mounting nut and antenna from the roof **(see illustration)**.
6 Installation is reverse of removal.

14 Rear window defogger - check and repair

1 The rear window defogger consists of a number of horizontal heating elements baked onto the inside surface of the glass.

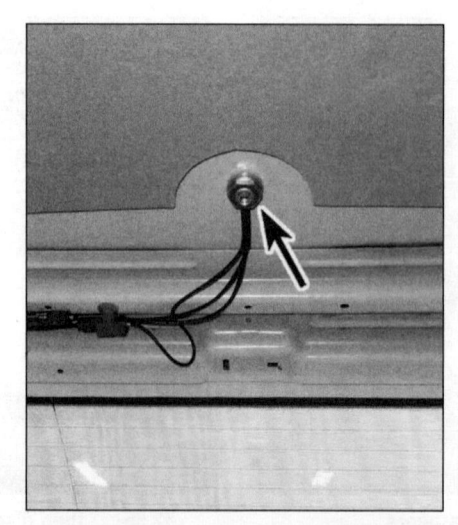

13.5 Disconnect the electrical connections and remove the mounting nut

Power is supplied through a large fuse from the fuse/relay box in the dash area. Refer to the wiring diagrams at the end of Chapter 12. The heater is controlled by the instrument panel switch.
2 Small breaks in the element can be repaired without removing the rear window.

Check

3 Turn the ignition switch and defogger switch to the On position.
4 Using a voltmeter, place the positive probe against the defogger grid positive terminal and the negative probe against the ground terminal. If battery voltage is not indicated, check the fuse, defogger switch, defogger relay and related wiring. If voltage is indicated, but all or part of the defogger doesn't heat, proceed with the following tests.
5 When measuring voltage during the next two tests, wrap a piece of aluminum foil around the tip of the voltmeter positive probe and press the foil against the heating element with your finger **(see illustration)**. Place the negative probe on the defogger grid ground terminal.

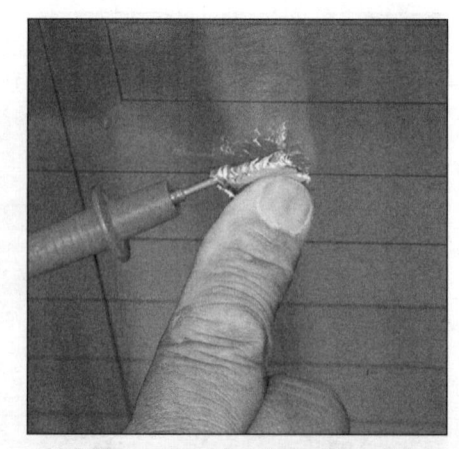

14.5 When measuring the voltage at the rear window defogger grid, wrap a piece of aluminum foil around the positive probe of the voltmeter and press the foil against the wire with your finger

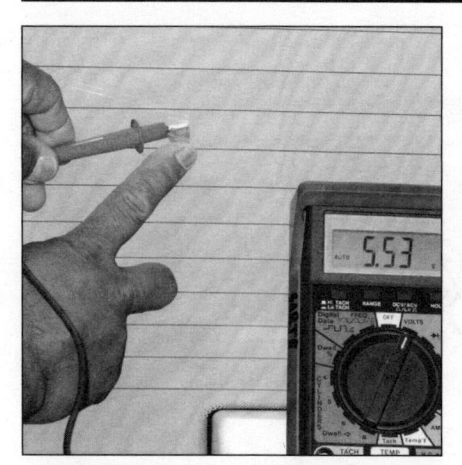

Chapter 12 Chassis electrical system

14.6 To determine if a heating element has broken, check the voltage at the center of each element - if the voltage is 6-volts, the element is unbroken

14.8 To find the break, place the voltmeter negative lead against the defogger ground terminal, place the voltmeter positive lead with the foil strip against the heat wire at the positive terminal end and slide it toward the negative terminal end - the point at which the voltmeter deflects from several volts to zero volts is the point at which the wire is broken

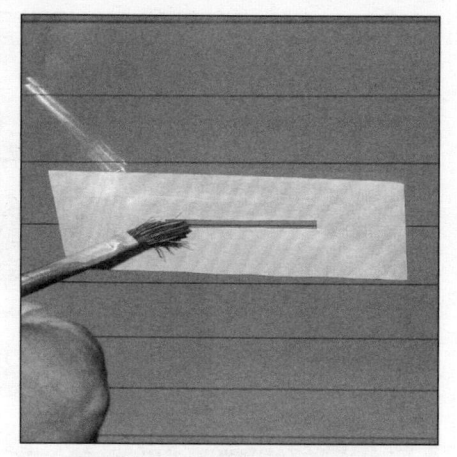

14.14 To use a defogger repair kit, apply masking to the inside of the window at the damaged area, the brush on the special conductive coating

6   Check the voltage at the center of each heating element **(see illustration)**. If the voltage is 5 to 6 volts, the element is okay (there is no break). If the voltage is 0 volts, the element is broken between the center of the element and the positive end. If the voltage is 10 to 12 volts, the element is broken between the center of the element and the ground side. Check each heating element.

7   If none of the elements are broken, connect the negative probe to a good chassis ground. The voltage reading should stay the same - if it doesn't, the ground connection is bad.

8   To find the break, place the voltmeter negative probe against the defogger ground terminal. Place the voltmeter positive probe with the foil strip against the heating element at the positive side and slide it toward the negative side. The point at which the voltmeter deflects from several volts to zero is the point where the heating element is broken **(see illustration)**.

Repair

9   Repair the break in the element using a repair kit specifically for this purpose, such as DuPont paste No. 4817 (or equivalent). The kit includes conductive plastic epoxy.

10   Before repairing a break, turn off the system and allow it to cool for a few minutes.

11   Lightly buff the element area with fine steel wool, then clean it thoroughly with rubbing alcohol.

12   Use masking tape to mask off the area being repaired.

13   Thoroughly mix the epoxy, following the kit instructions.

14   Apply the epoxy material to the slit in the masking tape, overlapping the undamaged area by about 3/4-inch on either end **(see illustration)**.

15   Allow the repair to cure for 24 hours before removing the tape and using the system.

15   Headlight housing - removal and installation

Models with halogen bulbs

Note: *On some later models, the headlight housing electrical connector can only be disconnected after the housing has been moved forward.*

1   Disconnect the electrical harness connector from the headlight housing **(see illustration)**.

2   Remove the front bumper cover (see Chapter 11).

3   Remove the headlight housing fasteners **(see illustration)**.

15.1 Disconnect the electrical connector from the headlight housing

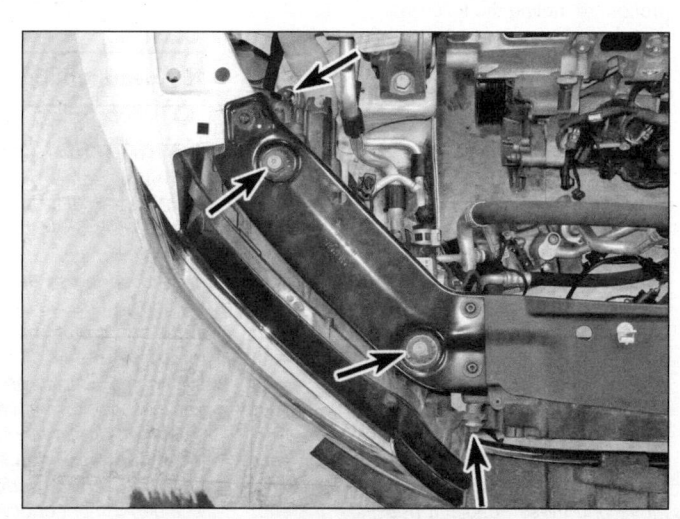

15.3 Headlight housing fasteners

Chapter 12 Chassis electrical system

15.4 Carefully guide the headlamp housing out horizontally, straight forward

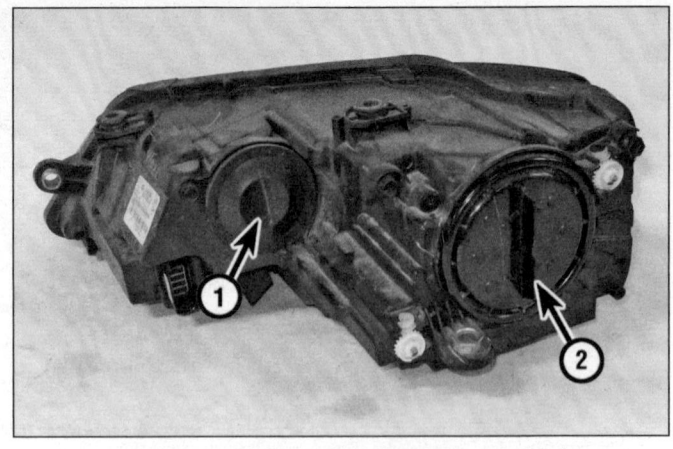

6.3a There are two separate covers protecting the high- and low-beam bulbs

1 High-beam bulb cover   2 Low-beam bulb cover

16.3b Remove the rubber dust cover for access to the high-beam bulb

16.3c Rotate the plastic cover a quarter-turn counterclockwise to remove it for access to the low-beam bulb

16.3d Grasp the back of the bulb fixture, rotate a quarter-turn counterclockwise and remove the bulb from the housing. Unplug the electrical connector from the bulb holder

4   Pull the headlight housing out horizontally **(see illustration)**.
5   Installation is the reverse of the removal procedure, noting the following:
  a) Install the two main upper screws first.
  b) Install the inner, forward screw second.
  c) Install the rear screw third.
  d) Refer to Section 17 for headlight adjusting procedures.

Models with xenon (HID) bulbs

Warning: *Some models use High Intensity Discharge (HID) bulbs instead of halogen bulbs. These can be identified by the high-voltage warning sticker on the headlight housing. According to the manufacturer, the high voltages produced by this system can be fatal in the event of a shock. Also, the voltage can remain in the circuit even after the headlight switch has been turned to Off and the ignition key has been removed. Therefore, for your safety, we don't recommend that you try to remove one these headlight housings. Instead, have this service performed by a dealer service department or other qualified repair shop.*

Warning: *Never attempt to check for voltage at the bulb socket of an HID headlight.*

16   Headlight bulbs - replacement

Xenon (HID) bulbs

Warning: *Some models use High Intensity Discharge (HID) bulbs instead of halogen bulbs. These can be identified by the high-voltage warning sticker on the headlight housing. According to the manufacturer, the high voltages produced by this system can be fatal in the event of a shock. Also, the voltage can remain in the circuit even after the headlight switch has been turned to Off and the ignition key has been removed. Therefore, for your safety, we don't recommend that you try to remove one these headlight housings. Instead, have this service performed by a dealer service department or other qualified repair shop.*

Warning: *Never attempt to check for voltage at the bulb socket of an HID headlight.*

Halogen bulbs

Warning: *Halogen bulbs are gas-filled and under pressure and may shatter if the surface is scratched or the bulb is dropped. Wear eye protection and handle the bulbs carefully, grasping only the base whenever possible. Don't touch the surface of the bulb with your fingers because the oil from your skin could cause it to overheat and fail prematurely. If you do touch the bulb surface, clean it with rubbing alcohol.*

1   The following sequence of photos was taken with the headlamp housing removed. You DO NOT need to remove the housing to replace the bulbs.
2   Open the hood and gain access to the back of the headlamp housing.
3   The following photos will lead you through the procedure of halogen bulb replacement **(see illustrations)**. The high and low beam bulbs are removed in the same manner.
4   Installation is the reverse of removal.

17 Headlights - adjustment

Warning: *The headlights must be aimed correctly. If adjusted incorrectly, they could temporarily blind the driver of an oncoming vehicle and cause an accident or seriously reduce your ability to see the road. The headlights should be checked for proper aim every 12 months and any time a new headlight is installed or front-end bodywork is performed. The following procedure is only an interim step to provide temporary adjustment until the headlights can be adjusted by a properly equipped shop.*

Note: *Some models are equipped with a headlight leveling system. This adjustment procedure will not apply to those models. Have the headlights adjusted by a dealer service department or other qualified repair shop.*

1 There are several methods of adjusting the headlights. The simplest method requires an open area with a blank wall and a level floor **(see illustration)**.

2 Position masking tape vertically on the wall in reference to the vehicle centerline and the centerlines of both headlights.

3 Position a horizontal tape line in reference to the centerline of all the headlights.

Note: *It may be easier to position the tape on the wall with the vehicle parked only a few inches away.*

4 Adjustment should be made with the vehicle parked 25 feet from the wall, sitting level, the gas tank half-full and no unusually heavy load in the vehicle. Make the adjustments with the headlights set on low beam.

5 Position the high intensity zone so it is two inches below the horizontal line and two inches to the side of the vertical headlight line, away from oncoming traffic. Turn the adjustment screws until the desired position has been achieved **(see illustrations)**.

6 With the high beams on, the high intensity zone should be vertically centered with the exact center just below the horizontal line.

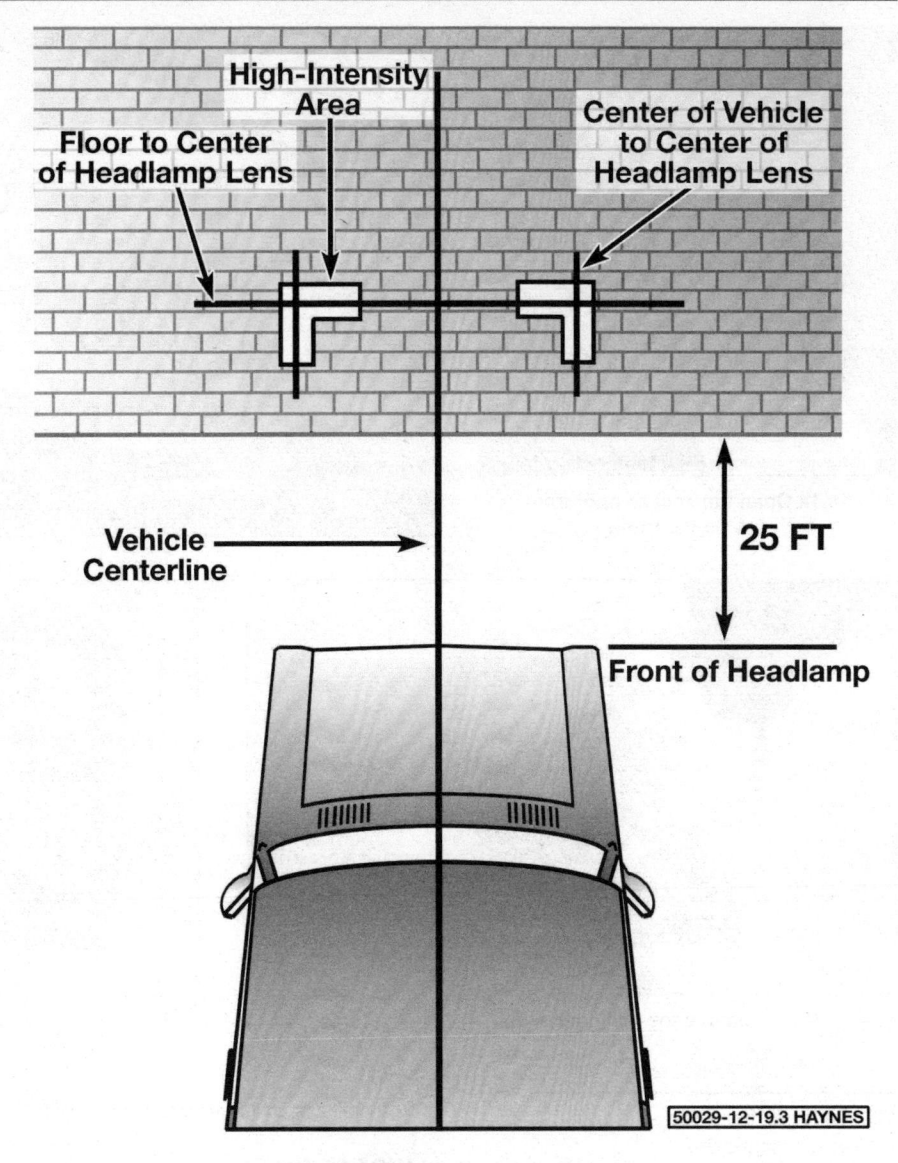

17.1 Headlight adjustment details

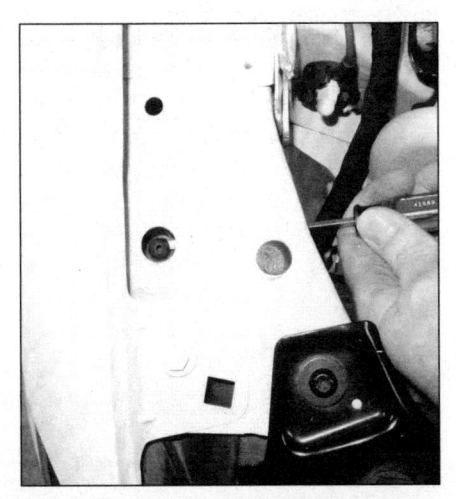

17.5a To make up-or-down down adjustments, remove the plug using a thin screwdriver . . .

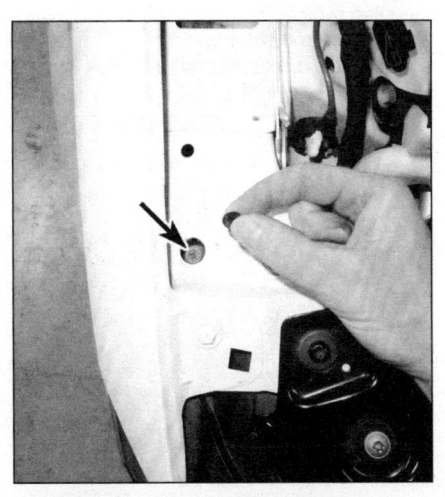

17.5b . . . then turn the screw with a Phillips screwdriver. Reinstall the plug after adjustment

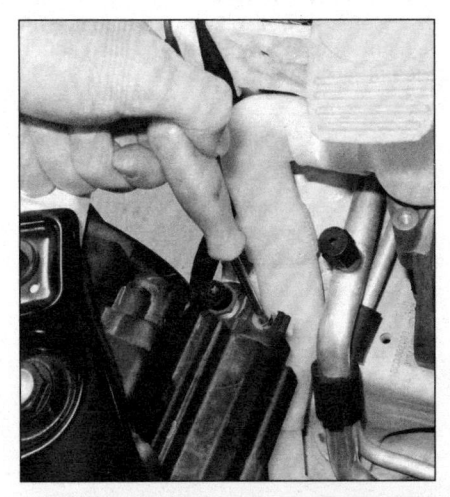

17.5c To make side-to-side adjustments, turn the screw at he rear of the headlight housing

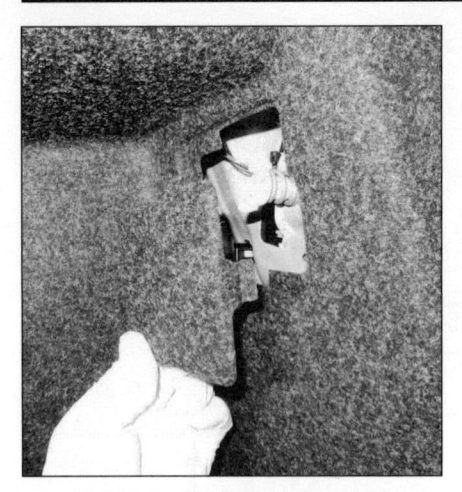

Chapter 12 Chassis electrical system

18.1a Open the access door from inside the trunk

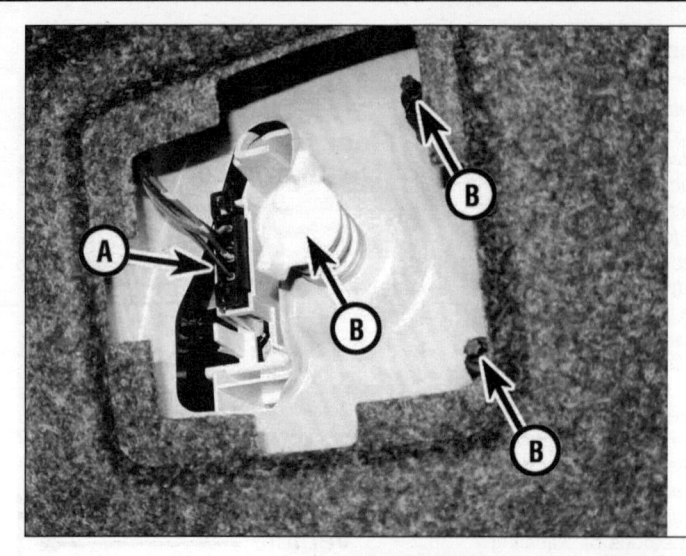

18.1b Disconnect the electrical connection (A) and the fasteners (B) securing the taillight housing to the vehicle

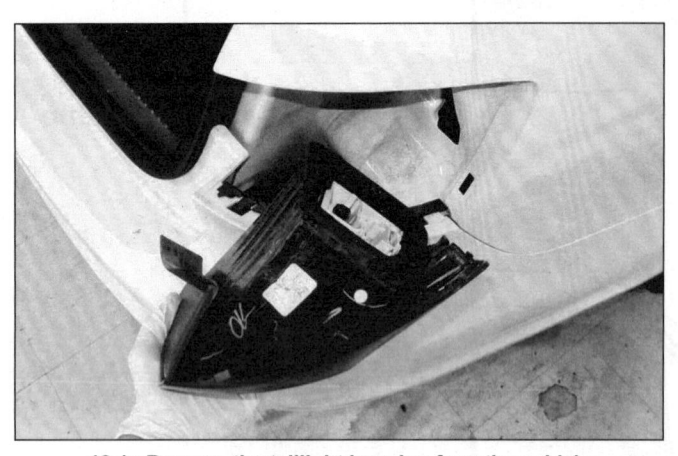

18.1c Remove the taillight housing from the vehicle

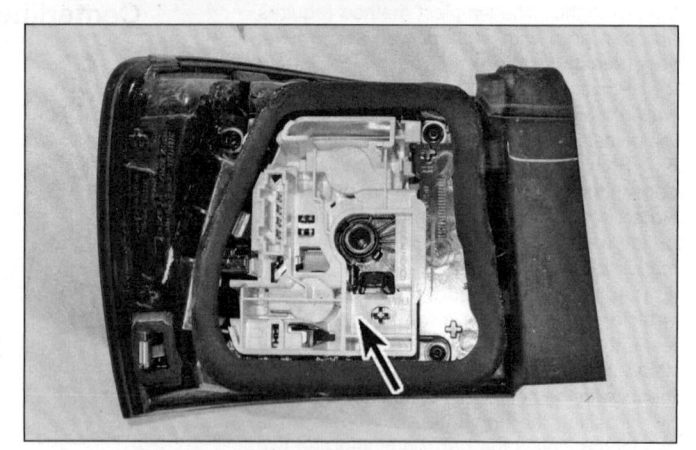

18.1d On vehicles without LED lighting, the inner section of the taillight housing holds two replaceable bulbs

18 Taillight housing - removal and installation

Fender mounted

1   Refer to the following sequence for taillight housing removal and installation **(see illustrations)**.

2   Installation is the reverse of removal.

Trunk mounted

3   Refer to the following sequence for trunk mounted taillight housing removal and installation **(see illustrations)**.
4   Installation is the reverse of removal.

19 Bulb replacement

Warning: *Bulbs can remain hot for up to 20 minutes after they're turned off. Be sure bulbs are off and cool before you touch them.*

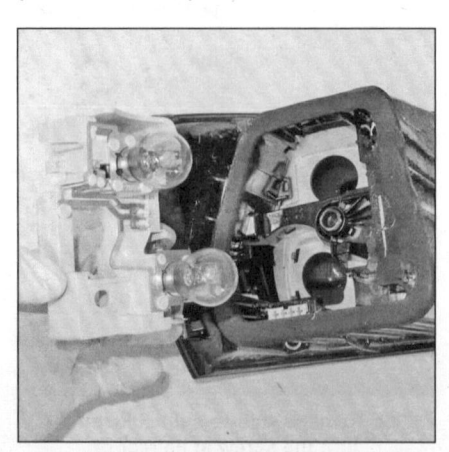

18.1e Push in on the tabs to remove the bulb housing

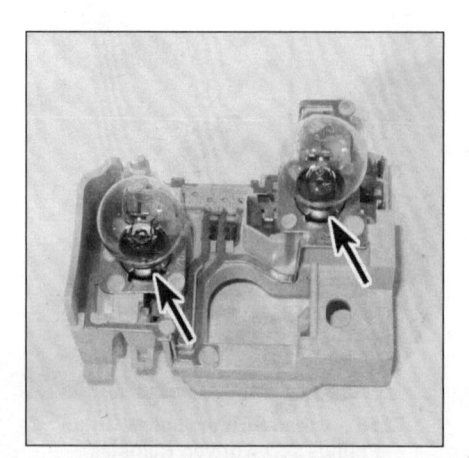

18.1f Remove the bulbs with a counterclockwise twist

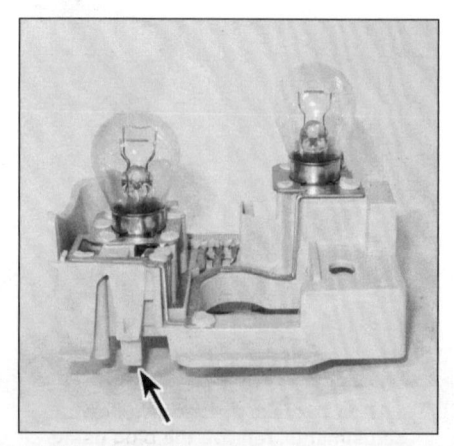

18.1g Detail view of the tab to remove the bulb housing

Chapter 12 Chassis electrical system

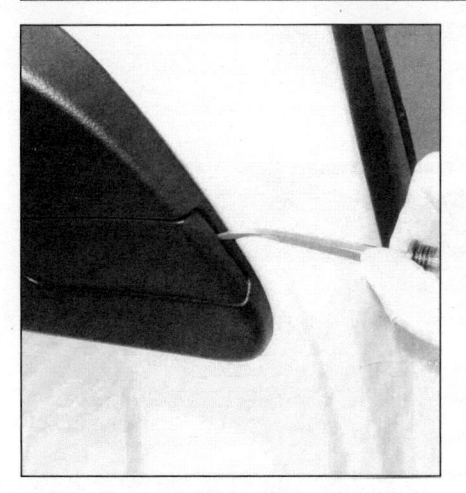

12-15

18.3a Open the trunk and pry the cover off of the trim to expose the back of the taillight housing

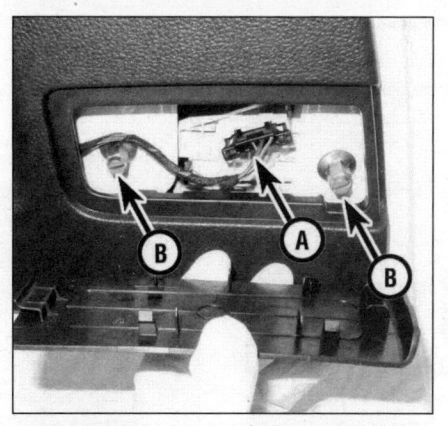

18.3b Remove the trim access panel and disconnect the electrical connection and remove the mounting fasteners
A Electrical connection
B Mounting fasteners

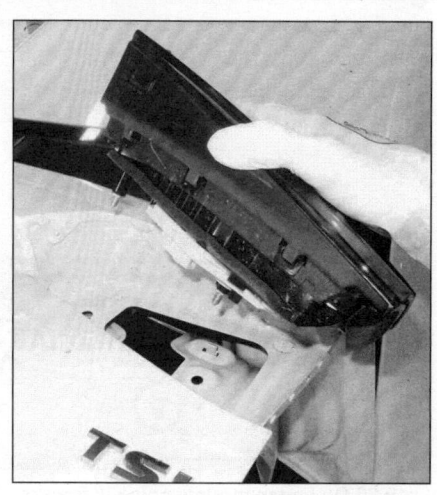

18.3c Remove the taillight housing

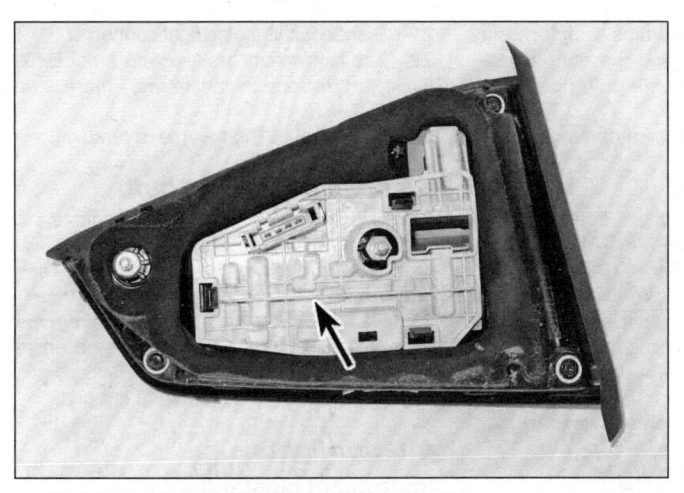

18.3d Vehicles without LED tail lamps will have a removable taillight bulb housing

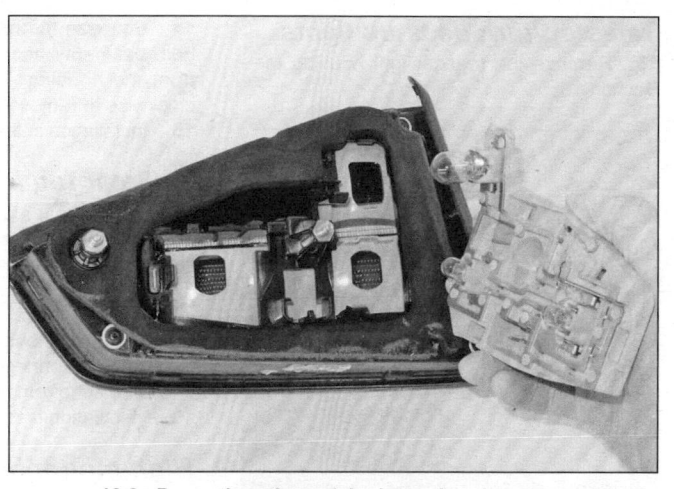

18.3e Press the release tabs inward and remove the taillight bulb housing

Front turn signal light

1 Remove any interfering engine compartment covers/air intake ducts for access to the headlight housing.
2 Rotate the bulb holder counterclockwise and remove the bulb holder and bulb from the contacts.
3 Push the bulb into the holder, rotate the bulb counterclockwise and remove the bulb from the holder.
4 Installation is the reverse of the removal procedure.

Front parking light

5 Remove any interfering engine compartment covers/air intake ducts for access to the headlight housing.
6 Open the plastic cover for the high-beam bulb on the back of the headlight housing.
7 Pull the parking light bulb holder out of the headlight housing, then pull the bulb from the holder.
8 Installation is the reverse of removal.

Side-marker lights

9 Remove the under-vehicle splash shield (see Chapter 1, Section 6).
10 Press the retaining tab forward and remove the marker light from the front bumper cover.
11 Twist the bulb holder counterclockwise to remove it.

Note: *The marker bulb can be removed from the marker light from under the vehicle without removing the marker light.*

12 Installation is the reverse of the removal procedure. Make sure the tab end of the light housing goes in first, then snap the housing in place until the tab is engaged.

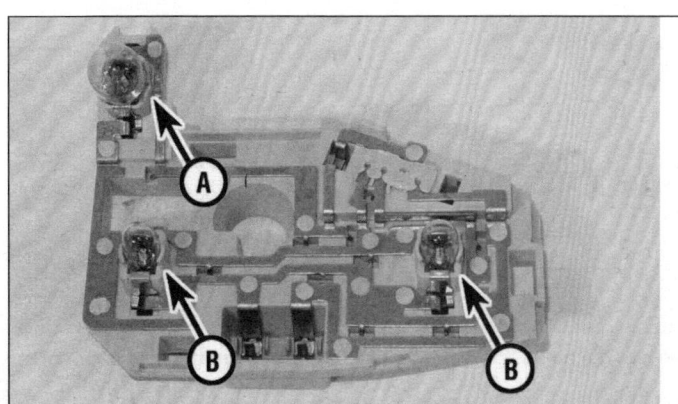

18.3f Taillight bulb housing contains three separate replaceable bulbs
A Reverse lamp
B Marker lamps

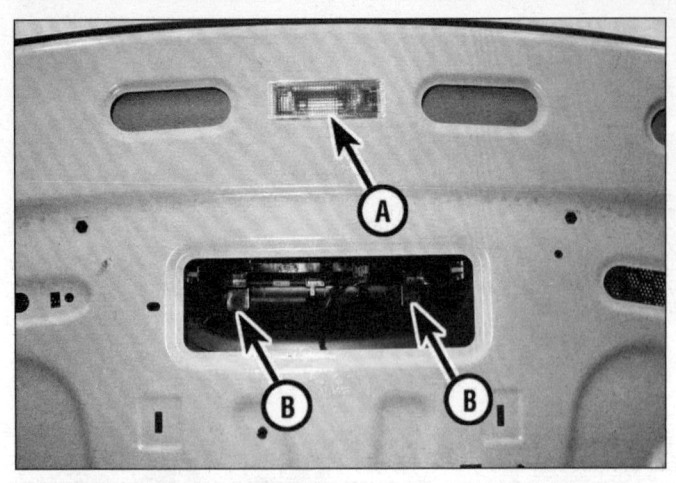

12-16   **Chapter 12   Chassis electrical system**

19.20 On sedan models come in from the trunk to gain access

A   Trunk dome light   B   High-mount brake light mounting screws

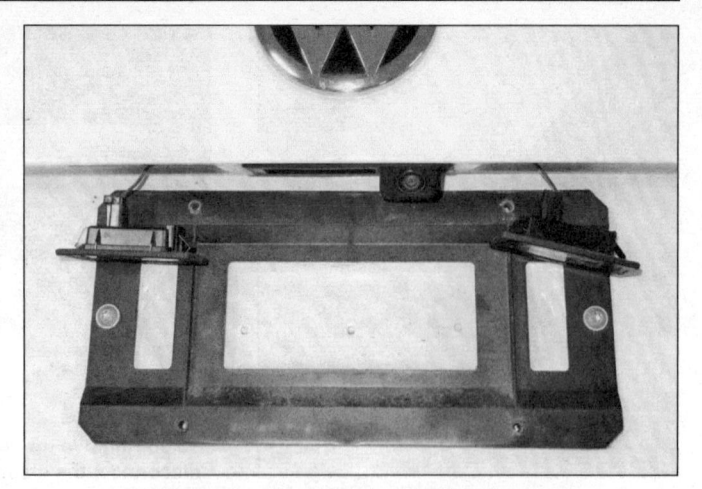

19.24 Remove the license plate light housing. LED type, replace the entire fixture

Tail/stop/turn/back-up lights

13   Remove the rear taillight housing (see Section 18).

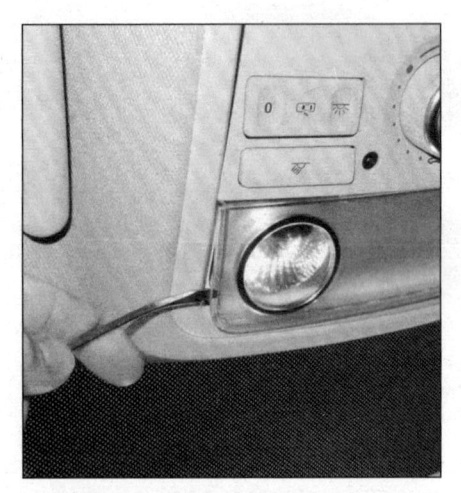

14   Squeeze the release tabs to unlock the bulb assembly and remove the bulb holder from the housing. Twist the bulbs counterclockwise to remove them.

15   Installation is the reverse of removal.

Tail/rear fog light

16   The tail/rear fog lights are located on the trunk lid or hatch (see Section 18). Open the lid or hatch and the trim access panel.

17   On Jetta models, spread the release clips apart to unlock the bulb assembly.

18   Pull out the bulb housing and disconnect the electrical connectors. Twist the bulb counterclockwise to remove it.

19   Installation is the reverse of removal.

High-mount brake light

Sedans

20   Remove the trim (if applicable) from the underside of the parcel tray (inside the trunk) **(see illustration)**, remove the screws securing the brake light housing and pull downwards to remove.

21   Disconnect the electrical connector.

22   The high-mount brake light is an LED fixture type. Replace the entire unit - there is no replaceable bulb.

23   Installation is the reverse of removal.

License plate light bulb

24   Use a small screwdriver to remove the license plate light housing mounting screws, then tilt and pull out the light housing **(see illustration)**.

25   Remove the bulb (if applicable) from the bulb holder. For LED type, replace the entire fixture.

26   Installation is the reverse of removal.

Interior lights

Instrument cluster lights

27   The instrument cluster is illuminated by LEDs that are part of the printed circuit board. There are no user-replaceable bulbs behind the instrument cluster. Consult with a dealer service department or other qualified repair shop; some facilities might be equipped to

19.28a Carefully pry off the lens cover using a small flat-blade screwdriver . . .

19.28b . . . then remove the bulbs

19.30a Carefully pry the lens cover off the light fixture using a small flat-blade screwdriver

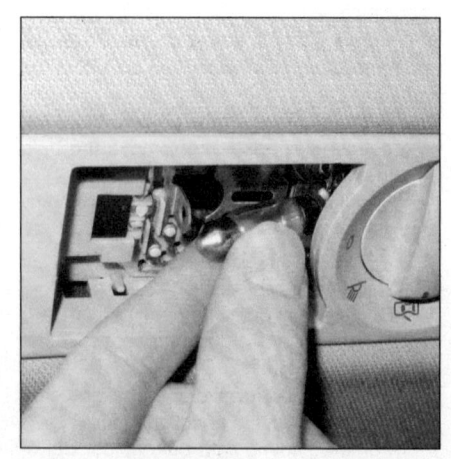

19.30b Remove the bulb and replace with the same type

Chapter 12 Chassis electrical system

19.38 Pry the light housing down . . .

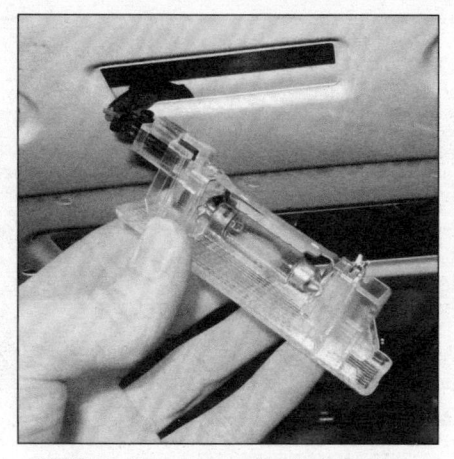

19.39 . . . then remove the festoon-type bulb from between the terminals

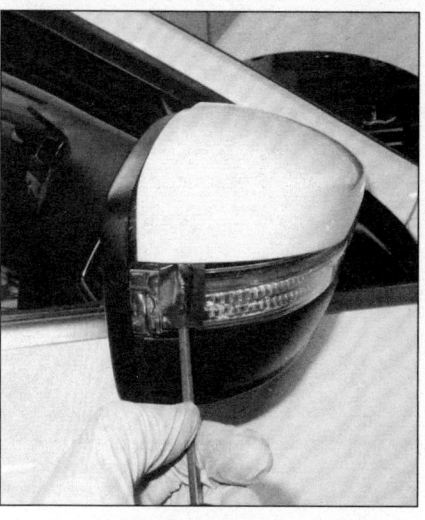

19.41a Using a screwdriver with tip covered in electrical tape or trim tool, carefully pry the mirror shell upward at the end . . .

repair the cluster, or might offer an exchange program. If the instrument cluster is replaced, it will have to be programmed with a proprietary scan tool.

Front dome/reading lights

Note: *Not all models have a reading/map light lens that can be removed. If the lens can't be removed, pry the trim panel off at the rear of the housing, then remove the fasteners and swing the overhead console down. Remove the bulb holders and replace the bulbs.*

28 Remove the dome/reading light lens **(see illustration)**, then remove the bulb(s) by pulling straight out **(see illustration)**.
29 Installation is the reverse of removal.

Center dome lights

30 Remove the lens cover by prying near the edge, then remove the bulb **(see illustrations)**.
Warning: *If necessary, pry only on the bulb terminals. Never pry on the glass of a festoon-type bulb.*
31 Installation is the reverse of removal.

Vanity lights

32 Pry the light out of the headliner carefully with a small screwdriver.

33 Unhook the bulb cover, then remove the bulb by pulling it straight out.
34 Installation is the reverse of removal.

Glovebox light

35 Open the glovebox door and pry the light housing out of the glovebox carefully with a small screwdriver.
36 Remove the bulb by pulling it straight out of the bulb holder.
37 Installation is the reverse of removal.

Luggage compartment lights

38 Pry the light assembly from the trunk **(see illustration)**.
39 Remove the bulb and replace with the same type **(see illustration)**.
Warning: *If necessary, pry only on the bulb terminals. Never pry on the glass of a festoon-type bulb.*
40 Installation is the reverse of removal.

Mirror turn signal lights

Note: *The mirror signal lights consist of an LED assembly - it is replaced as a unit*
41 To remove the mirror signal lights, follow the accompanying illustrations and captions **(see illustrations)**:
42 Installation is the reverse of removal.

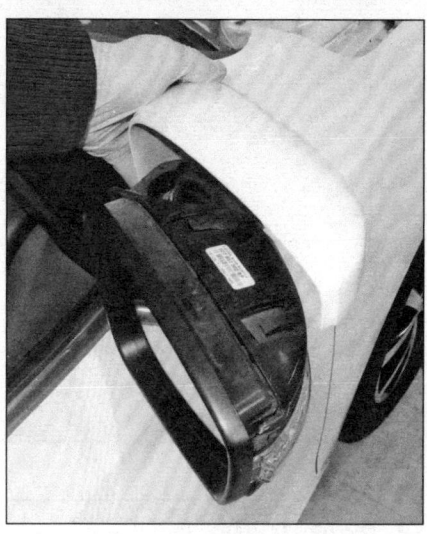

19.41b . . . then remove the shell

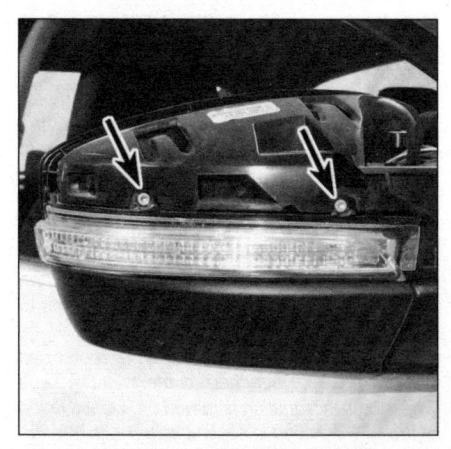

19.41c Remove the mirror signal retaining screws . . .

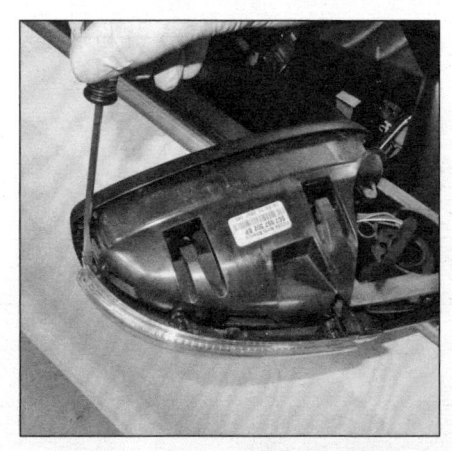

19.41d . . . then carefully pry it out at the end . . .

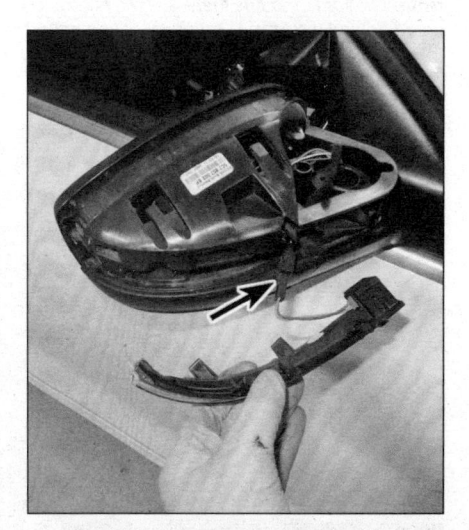

19.41e . . . disconnect the electrical connector and remove the signal unit counterclockwise, then pull it out

Bulb removal

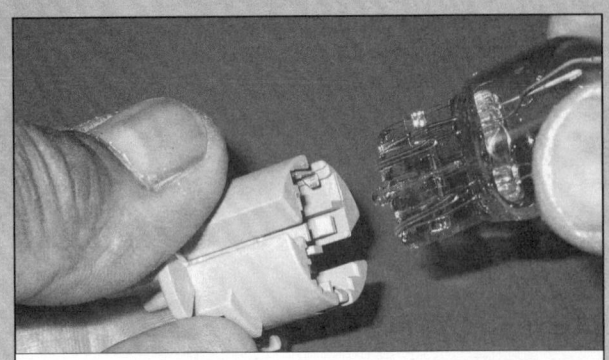

To remove many modern exterior bulbs from their holders, simply pull them out

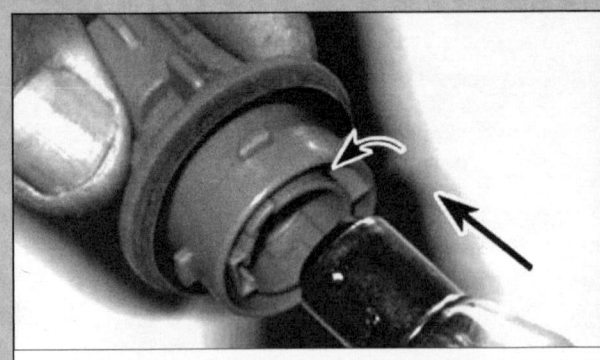

On bulbs with a cylindrical base ("bayonet" bulbs), the socket is spring-loaded; a pair of small posts on the side of the base hold the bulb in place against spring pressure. To remove this type of bulb, push it into the holder, rotate it 1/4-turn counterclockwise, then pull it out

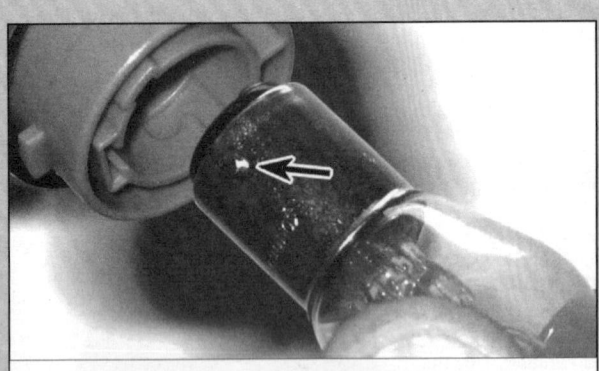

If a bayonet bulb has dual filaments, the posts are staggered, so the bulb can only be installed one way

To remove most overhead interior light bulbs, simply unclip them

20 Horns - replacement

Note: *On some models there are two horns, located behind the front bumper cover and bolted to each end of the front bumper reinforcement bar. The high-tone horn is on the passenger's side, and the low-tone horn is on the driver's side. Other models only have one horn.*

1 Remove the bumper cover (see Chapter 11).

Note: *In some applications, the horn can be accessed from underneath without the bumper cover being removed.*

2 Disconnect the electrical connector, remove the fastener and detach the horn **(see illustration)**.

3 Installation is the reverse of removal.

20.2 Typical horn details - right side shown

1 Low-tone horn
2 Electrical connector
3 Mounting fastener

21 Wipers

Removal and installation

Wiper arm

1 Place the wiper arms into the service position (parked position) before removal.

2 The windshield wiper arms must be in the "winter/service position" before removing the wiper blades. The "service position" is activated when you have operated the windshield wipers in the "one-touch wiping" position and then off for at least 10 seconds to allow the arms to come to their complete resting position.

Caution: *On some models, the wiper motor can only be properly attached to the wiper linkage when the APP function is turned off. Deactivation of the APP function is accomplished with a factory level scanner. See the appropriate repair facility capable of performing the procedure for you.*

Chapter 12 Chassis electrical system

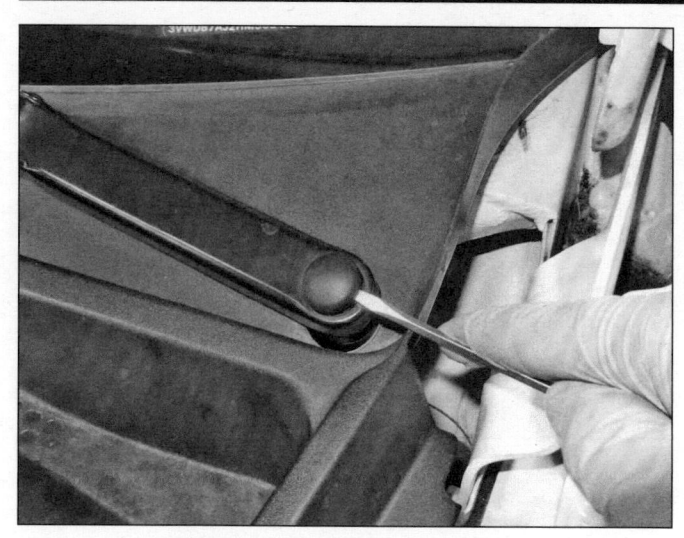

21.3 Pry the caps from the wiper arms

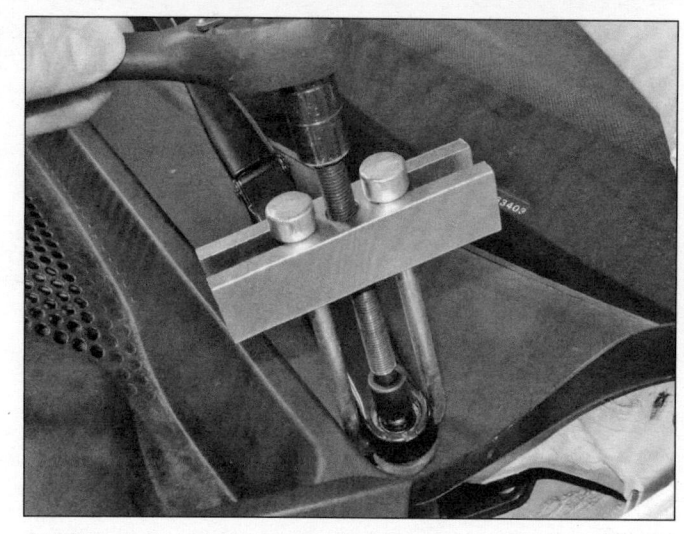

21.6 A puller is used to remove the wiper arm from the wiper shaft

3   Pry the cap from the end of the wiper arms **(see illustration)**.
4   Mark the relationship of the wiper shaft to the wiper arm before continuing.
5   Loosen the locking nut a few turns, but do not completely remove the nut.
6   Use a wiper arm puller or equivalent to release the wiper arm from the wiper shaft **(see illustration)**.
Warning: *The wiper arm shaft can get damaged when removing the windshield wiper arms without using a puller.*
7   With the wiper arm released from the shaft, remove the tool and then remove the locking nut.
8   Lift the wiper arm from the wiper shaft.
9   Installation is the reverse of removal.

Front wiper motor
10   Run the wipers through at least one full sweep, then turn them off. Allow the wiper arms to return to the park position for at least 10 seconds. Disconnect the cable from the negative terminal of the battery (see Chapter 5).
11   Remove the wiper arms.
12   Remove the windshield cowl cover and weatherstrip (see Chapter 11).
13   Disconnect the wiper motor electrical connector and remove the windshield wiper motor/linkage assembly fasteners **(see illustration)**.
14   Lift the windshield wiper motor assembly from the cowl area.
15   Remove the wiper motor-to-linkage nut, then the motor-to-bracket fasteners and separate the motor from the linkage bracket **(see illustration)**.
16   Installation is the reverse of removal.

Rear wiper
17   Turn the ignition switch to the Off position and remove the key.
18   Mark the position of the wiper arm on the glass, then carefully pry the base cap off.
19   Detach the rear spray jet and loosen, but do not remove, the wiper arm nut.
20   Carefully rock the wiper arm back and forth until it is loose and remove nut and wiper arm.
21   Remove the inner trim panel from the hatch/liftgate (see Chapter 11).

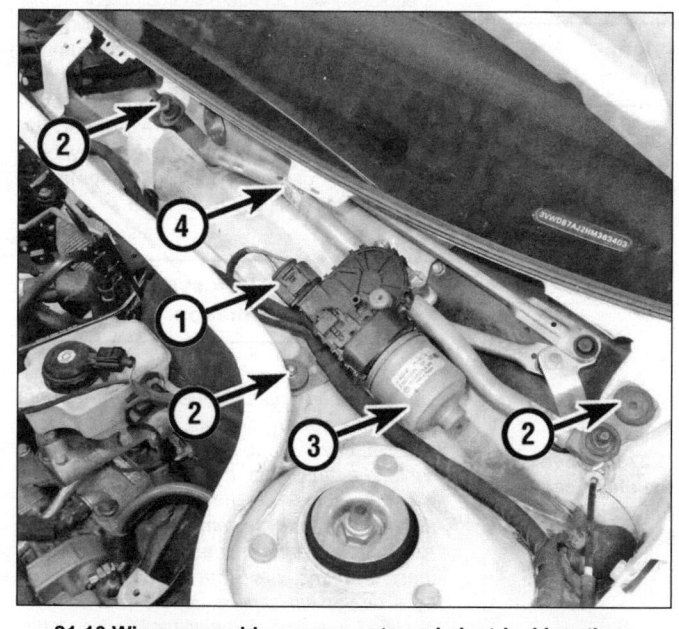

21.13 Wiper assembly components and electrical locations

1   Electrical connector
2   Mounting fasteners
3   Wiper motor
4   Linkage arm assembly

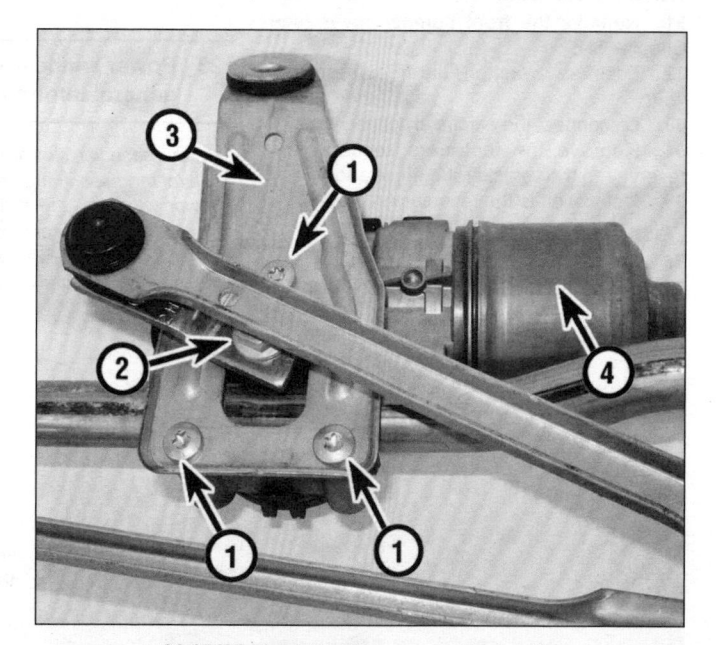

21.15 Windshield wiper assembly details

1   Motor-to-bracket fasteners
2   Motor-to-linkage nut
3   Linkage bracket
4   Wiper motor

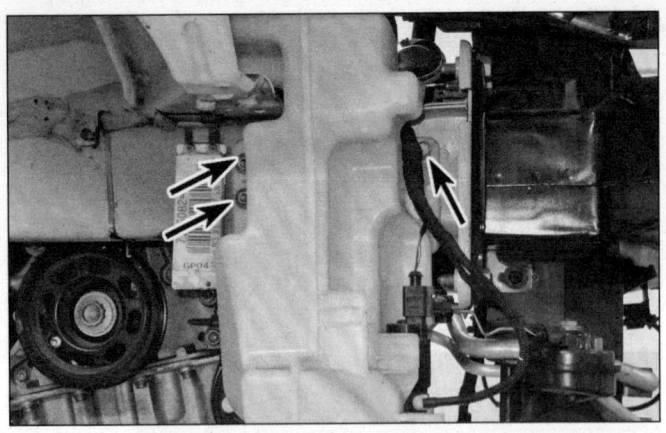

12-20 Chapter 12 Chassis electrical system

21.34a Remove the lower fasteners securing the reservoir to the vehicle

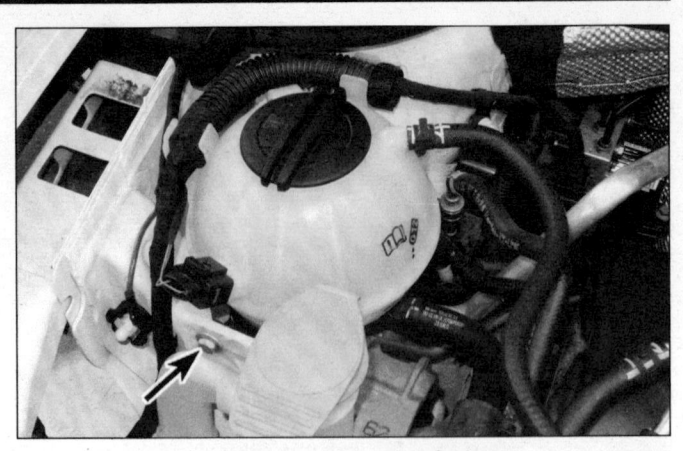

21.34b Remove the upper fastener

22 Disconnect the wiper motor harness connector and remove the wiper motor mounting fasteners.
23 Lift the wiper motor assembly from the hatch/liftgate.
24 Installation is the reverse of removal.

Washer pump

25 Remove the right air grille from the front bumper.
26 Place a drain pan under the washer fluid reservoir to catch the washer fluid.
27 Remove the hose from the washer pump.
28 Disconnect the electrical connection from the washer pump.
29 Grasp the washer pump and pull straight upwards out of the grommet in the washer fluid reservoir.
30 Installation is the reverse of removal.

Washer fluid reservoir

31 Remove the front bumper cover (see Chapter 11).
32 Drain any washer fluid from the reservoir.
33 Disconnect the electrical connection.
34 Remove the fasteners securing the reservoir to the vehicle (see illustrations).
35 Installation is the reverse of removal.

Washer nozzles

36 Push the nozzle upward and remove it.
37 Disconnect the electrical connections (if equipped) and remove the hose (see illustration).
38 Installation is the reverse of removal.

22 Cruise control system - description and check

1 These vehicles have an electrically controlled throttle body. The accelerator pedal communicates with the throttle body through the Powertrain Control Module (PCM). The PCM also controls the cruise control system, which is now an integral function of the electronic throttle control system. If the system malfunctions, begin diagnosis by checking to see if any trouble codes have been set (see Chapter 6). If that doesn't lead to the problem, take it to a dealer service department or other qualified repair shop for further diagnosis.

23 Power window system - general information

Note: *These vehicles are equipped with various control modules that govern the door locks, the power windows, the ignition lock and security system, the interior and exterior lights, the headlights, the horn, the windshield wipers/washers, the heating/air conditioning system, the audio*

21.37 Remove the electrical connection and hose, then remove the washer nozzle

system and the power mirrors. In the event of a malfunction with one of these systems, have the vehicle diagnosed by a dealership service department or other qualified automotive repair facility if no obvious problems are found.

1 The power window system operates electric motors, mounted on the doors, which lower and raise the windows. The system consists of the control switches, the motors, regulators, glass mechanisms and associated wiring.
2 The power windows can be lowered and raised from the master control switch by the driver or by the switch located at the passenger's window. Each window has a separate motor that is reversible. The position of the control switch determines the polarity and therefore the direction of operation.
3 The circuit is protected by fuses and a circuit breaker. Check the fuses in the fuse panel at the left end of the instrument panel. Each motor is equipped with an internal circuit breaker; this prevents one stuck window from disabling the whole system. Refer to the wiring diagrams at the end of this manual. Problems within this system can only be diagnosed with a professional-grade scan tool. If you have eliminated the obvious causes of a problem, have the vehicle checked at a dealership service department or other properly equipped repair shop.

24 Power door lock system - general information

Note: *These vehicles are equipped with various control modules that govern the door locks, the power windows, the ignition lock and security system, the interior and exterior lights, the headlights, the horn, the windshield wipers/washers, the heating/air conditioning system, the audio system and the power mirrors. In the event of a malfunction with one of these systems, have the vehicle diagnosed by a dealership service department or other qualified automotive repair facility if no obvious problems are found.*

1 The central locking system uses an actuator (motor) integrated into each door lock.

Chapter 12 Chassis electrical system

12-21

The actuators are not serviceable separately and if one is bad the entire lock assembly must be replaced. The power door lock systems operate bi-directional motors. The first motor locks the exterior door and the second motor locks the interior door latch assembly, called the safe function. This no longer allows the doors to open from the interior door handles when the doors are locked.

Note: *In the event of an accident in which the airbags have been deployed, the locking system control module will open all locked doors.*

2   The central locking system control module is located behind the glove box. Each door also has a control module located towards the front of each door. Problems within these modules can only be diagnosed with a factory scan tool. If you have eliminated the obvious causes of a problem, have the vehicle checked at a dealership service department or other properly equipped repair shop.

25 Airbag - general information and component removal and installation

Warning: *If your vehicle is ever involved in a flood, or the interior carpeting is soaked for any reason, disconnect the battery and do not start the vehicle until the airbag system can be checked by your dealer. If the SRS system is subjected to flooding, the airbags could go off upon starting the vehicle, even without an accident taking place.*

1   These models are equipped with a Supplemental Restraint System (SRS), more commonly known as airbags, designed to protect the driver and the passenger from serious injury in the event of a head-on or side collision. All models have a diagnostic control unit, located on the floor under the center console.

Airbag modules

Driver's airbag

2   The airbag inflator module contains a housing incorporating the cushion (airbag) and inflator unit, mounted in the center of the steering wheel. The inflator assembly is mounted on the back of the housing over a hole through which gas is expelled, inflating the bag almost instantaneously when an electrical signal is sent from the system. A spiral cable (or clockspring) assembly on the steering column under the steering wheel carries this signal to the module. This clockspring can transmit an electrical signal regardless of steering wheel position.

Passenger's airbag

3   The airbag is mounted inside the right side of the instrument panel, in the area above the glove box. It's similar in design to the driver's airbag, except that it's larger than the steering wheel unit. The trim cover (on the side of the instrument panel that faces toward the passenger) is textured and colored to match the instrument panel and has a molded seam that splits open when the bag inflates.

Side impact airbags

4   Some models are equipped with side-impact airbags located in the outer part of the front seat backs. Additionally some models have side impact airbags integrated into the outside rear seat bolsters.

Side curtain airbags

5   In addition to the side-impact airbags, extra side-impact protection is also provided by side-curtain airbags on some models. These are long airbags that, in the event of a side impact, come out of the headliner at each side of the car and come down between the side windows and the seats. They are designed to protect the heads of both front seat and rear seat passengers.

Sensing and diagnostic module

6   The sensing and diagnostic module supplies the current to the airbag system in the event of a collision, even if battery power is cut off. It checks this system every time the vehicle is started, causing the "AIR BAG" light to go on, then off, if the system is operating properly. If there is a fault in the system, the light will go on and stay on, flash, or the dash will make a beeping sound. If this happens, the vehicle should be taken to your dealer immediately for service. This module is mounted under the center console. There is also a roll-over sensor located directly behind it on later models.

Seat belt pre-tensioners

7   All models are equipped with pyrotechnic (explosive) units in the front seat belt retracting mechanisms. During an impact that would trigger the airbag system, the airbag control unit also triggers the seat belt retractors. When the pyrotechnic charges go off, they accelerate the retractors to instantly take up any slack in the seat belt system to more fully prepare the driver and front seat passenger for impact.

8   The airbag system should be disabled any time work is done to or around the seats.

Warning: *Never strike the pillars or floorpan with a hammer or use an impact-driver tool in these areas unless the system is disabled.*

Disarming the system and other precautions

Warning: *Failure to follow these precautions could result in accidental deployment of the airbag and personal injury.*

Warning: *Any time you are working in the vicinity of airbag wiring or components, DISARM THE SRS SYSTEM.*

Warning: *An auxiliary voltage input device (memory saver) must NOT be used when working near airbag system components.*

9   Whenever working in the vicinity of the steering wheel, steering column or any of the other SRS system components, the system must be disarmed.

To disarm the airbag system:

a) *Turn the steering wheel to the straight-ahead position and turn the ignition switch to the Lock position, then remove the key.*
b) *Disconnect the negative battery cable (see Chapter 5).*
c) *Wait at least two minutes for the back-up power supply to be depleted.*
d) *Before touching any airbag system component, ground yourself to a metal part of the vehicle to discharge any static electricity built up in your body.*

To re-arm the airbag system:

a) *Turn the ignition switch to the On position.*
b) *Make sure there is nobody inside the vehicle and that there are no objects near any of the airbag modules, then reconnect the cable to the negative terminal of the battery.*
c) *Turn the ignition switch to the Off position, wait ten seconds, then turn it to the On position. The AIR BAG light on the instrument panel should come on continuously for about six seconds, then turn off.*

Note: *The light might take up to 30 seconds to come on after the key is turned to the On position (during this time the Restraints Control Module is performing a self-check of the system). If the light fails to come on, or if it flashes, or if a chime sounds in patterns of five sets of five beeps, have the vehicle diagnosed by a dealer service department or other qualified repair shop.*

Whenever handling an airbag module:

Warning: *Always keep the airbag opening (the trim side) pointed away from your body. Never place the airbag module on a bench or other surface with the airbag opening facing the surface. Always place the airbag module in a safe location with the airbag opening facing up.*

Warning: *Never measure the resistance of any SRS component. An ohmmeter has a built-in battery supply that could accidentally deploy the airbag.*

Warning: *Never use electrical welding equipment on a vehicle equipped with an airbag without first disconnecting the electrical connector for each airbag.*

Warning: *Never dispose of a live airbag module. Return it to a dealer service department or other qualified repair shop for safe deployment and disposal.*

Component removal and installation

Driver's side airbag module and spiral cable

10   Refer to Chapter 10, Section 18, for the driver's side airbag module and clockspring removal and installation procedures.

Passenger's airbag module and other airbag modules

11   Even if you ever have to remove the instrument panel, it's not necessary to remove the passenger's airbag module to do so; it can simply remain installed in the instrument panel. We don't recommend removing any of the other airbag modules either. These jobs are best left to a professional.

26 Wiring diagrams - general information

1 Since it isn't possible to include all wiring diagrams for every year and model covered by this manual, the following diagrams are those that are typical and most commonly needed.
2 Prior to troubleshooting any circuits, check the fuses and circuit breakers (if equipped) to make sure they're in good condition. Make sure the battery is properly charged and check the cable connections (see Chapter 1).
3 When checking a circuit, make sure that all connectors are clean, with no broken or loose terminals. When disconnecting a connector, do not pull on the wires. Pull only on the connector housings themselves.

Wiring Diagrams - List

1a. Starting and Charging - GOLF
1b. Starting and Charging - JETTA up to June 2014 (CPKA CPRA) (1.8L)
1c. Starting and Charging - JETTA up to June 2014 (CPLA CPPA) (2.0L TFSI)
1d. Starting and Charging - JETTA up to June 2014 (CBPA) (2.0L 8V)
1e. Starting and Charging - JETTA up to June 2014 (CBTA) (CBUA) (CCTA) (CBFA) (2.5L and 2.0L TSI)
1f. Starting and Charging - JETTA from July 2014 (CPKA CPRA CPLA CPPA) (1.8L and 2.0 TFSI)
1g. Starting and Charging - JETTA from July 2014 (CZTA CBTA CBUA CBPA) (1.4L 2.5L and 2.0 8V)
2a. AC Heating & Cooling - GOLF Climatic
2b. AC Heating & Cooling - GOLF Climatronic
2c. AC Heating & Cooling - GOLF Heated seats with sport seats
2d. AC Heating & Cooling - GOLF Heated seats and cooling fan
2e. AC Heating & Cooling - JETTA Manual AC
2f. AC Heating & Cooling - JETTA Climatic
2g. AC Heating & Cooling - JETTA Climatronic
2h. AC Heating & Cooling - JETTA Heated seats and cooling fan Up to June 2014
2i. AC Heating & Cooling - JETTA Heated seats and cooling fan From July 2014
3a. Power Windows - GOLF
3b. Power windows - JETTA With manual rear windows
3c. Power windows - JETTA Low equipment
3d. Power windows - JETTA High equipment From July 2014
4a. Power door Locks - GOLF
4b. Power door Locks - GOLF Keyless
4c. Power door locks - JETTA Models with rear manual window regulators
4d. Power door locks - JETTA Low equipment Up to June 2014
4e. Power door locks - JETTA High equipment Up to June 2014
4f. Power door locks - JETTA Low equipment From July 2014
4g. Power door locks - JETTA High equipment From July 2014
4h. Power Door Locks - JETTA Keyless
5a. Wiper Washer - GOLF
5b. Wiper Washer - JETTA Low equipment Up to June 2014
5c. Wiper Washer - JETTA High equipment Up to June 2014
5d. Wiper Washer - JETTA Low equipment From July 2014
5e. Wiper Washer - JETTA High equipment From July 2014
6a. Exterior Lights - GOLF
6b. Exterior Lights - GOLF Bixenon
6c. Exterior Lights - GOLF LED tail lights
6d. Exterior lights - JETTA Low Equipment up to June 2014
6e. Exterior lights - JETTA High Equipment up to June 2014
6f. Exterior lights - JETTA Low Equipment from July 2014
6g. Exterior lights - JETTA High Equipment from July 2014
6h. Exterior lights - JETTA Tail cluster with LED and Xenon lights Up to June 2014
6i. Exterior lights - JETTA Tail cluster with LED and Xenon lights From July 2014
7a. Interior Lights - GOLF
7b. Interior lights - JETTA Up to June 2014 - B&W
7c. Interior lights - JETTA From July 2014 - B&W
8a. Sound System - GOLF Radio and Navigation systems - Hatchback
8b. Sound System - GOLF Radio and Navigation systems - Variant
8c. Sound System - GOLF Speakers (without Dynaudio)
8d. Sound System - GOLF Speakers (with Dynaudio) and Mobile telephone systems
8e. Sound System - JETTA RNS and RCD without VW sound system front speakers (Up to June 2014)
8f. Sound System - JETTA VW Sound System front speakers and Subwoofer (Up to June 2014)
8g. Sound System - JETTA RNS and RCD without VW sound system front speakers (From July 2014)
8h. Sound System - JETTA VW Sound System front speakers and Subwoofer (From July 2014)
8i. Sound System - JETTA Mobile telephone systems and Multimedia Media-IN
9a. Fuel Pump - GOLF
9b. Fuel Pump - JETTA Up to June 2014
9c. Fuel Pump - JETTA From July 2014
10a. FUSES & RELAYS - GOLF
10b. FUSES & RELAYS - GOLF
10c. FUSES & RELAYS - GOLF
10d. FUSES & RELAYS - GOLF
10e. FUSES & RELAYS – GOLF
10f. FUSES & RELAYS - JETTA
10g. FUSES & RELAYS - JETTA
10h. FUSES & RELAYS - JETTA
10i. FUSES & RELAYS - JETTA
10j. FUSES & RELAYS - JETTA
10k. FUSES & RELAYS - JETTA
10l. FUSES & RELAYS - JETTA
10m. FUSES & RELAYS - JETTA
10n. FUSES & RELAYS - JETTA
10o. FUSES & RELAYS - JETTA
10p. FUSES & RELAYS - JETTA
10q. FUSES & RELAYS - JETTA

Chapter 12 Chassis electrical system

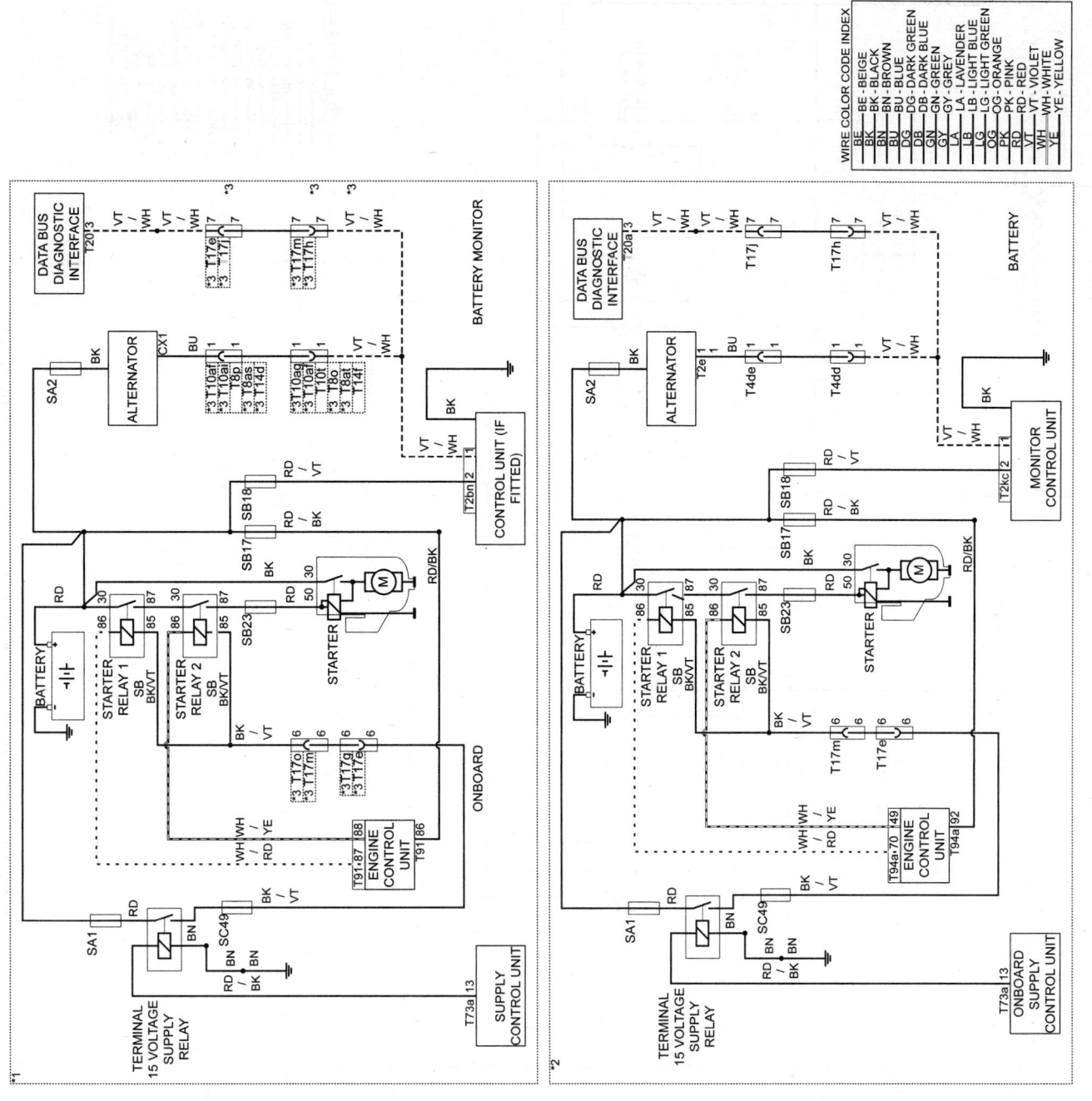

1a. Starting and Charging - GOLF

*1 Except 1.4L engines
*2 For 1.4L engines
*3 According to equipment

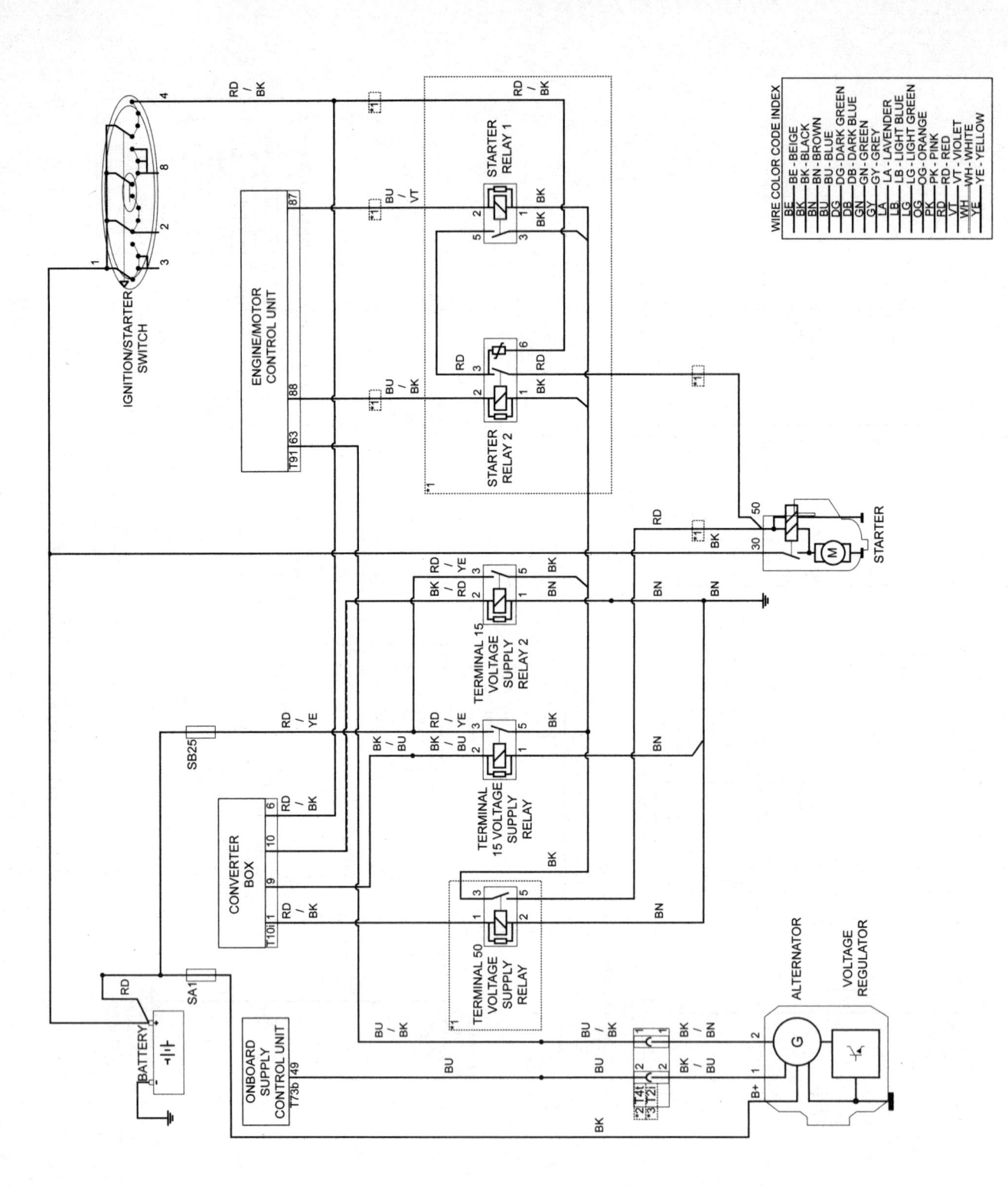

1b. Starting and Charging - JETTA up to June 2014 (CPKA CPRA) (1.8L)

*1 According to equipment
*2 With air conditioning
*3 Without air conditioning

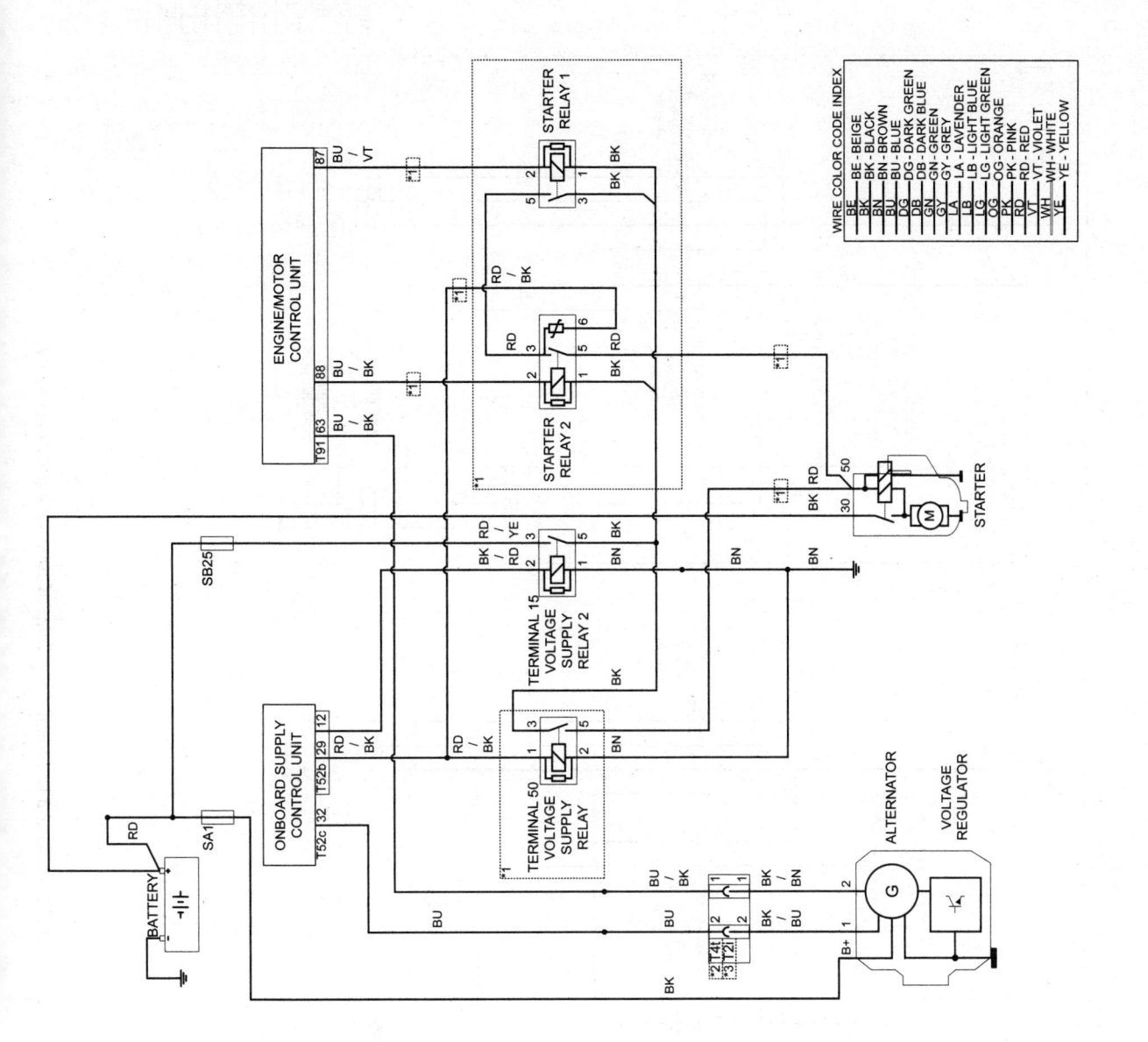

Chapter 12 Chassis electrical system

12-25

1c. Starting and Charging - JETTA up to June 2014 (CPLA CPPA) (2.0L TFSI)

*1 According to equipment
*2 With air conditioning
*3 Without air conditioning

12-26   Chapter 12   Chassis electrical system

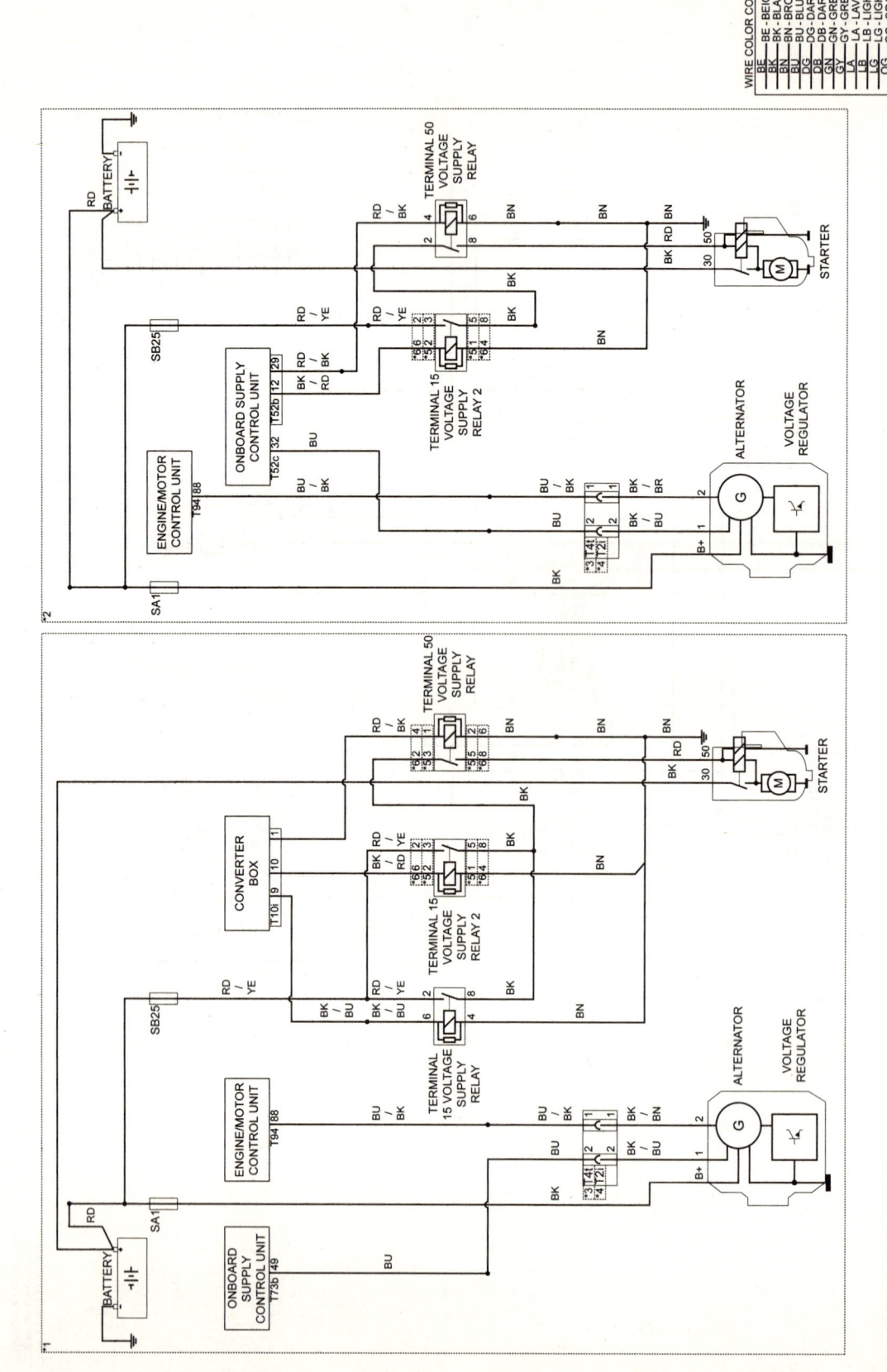

1d. Starting and Charging - JETTA up to June 2014 (CBPA) (2.0L 8V)

Chapter 12 Chassis electrical system

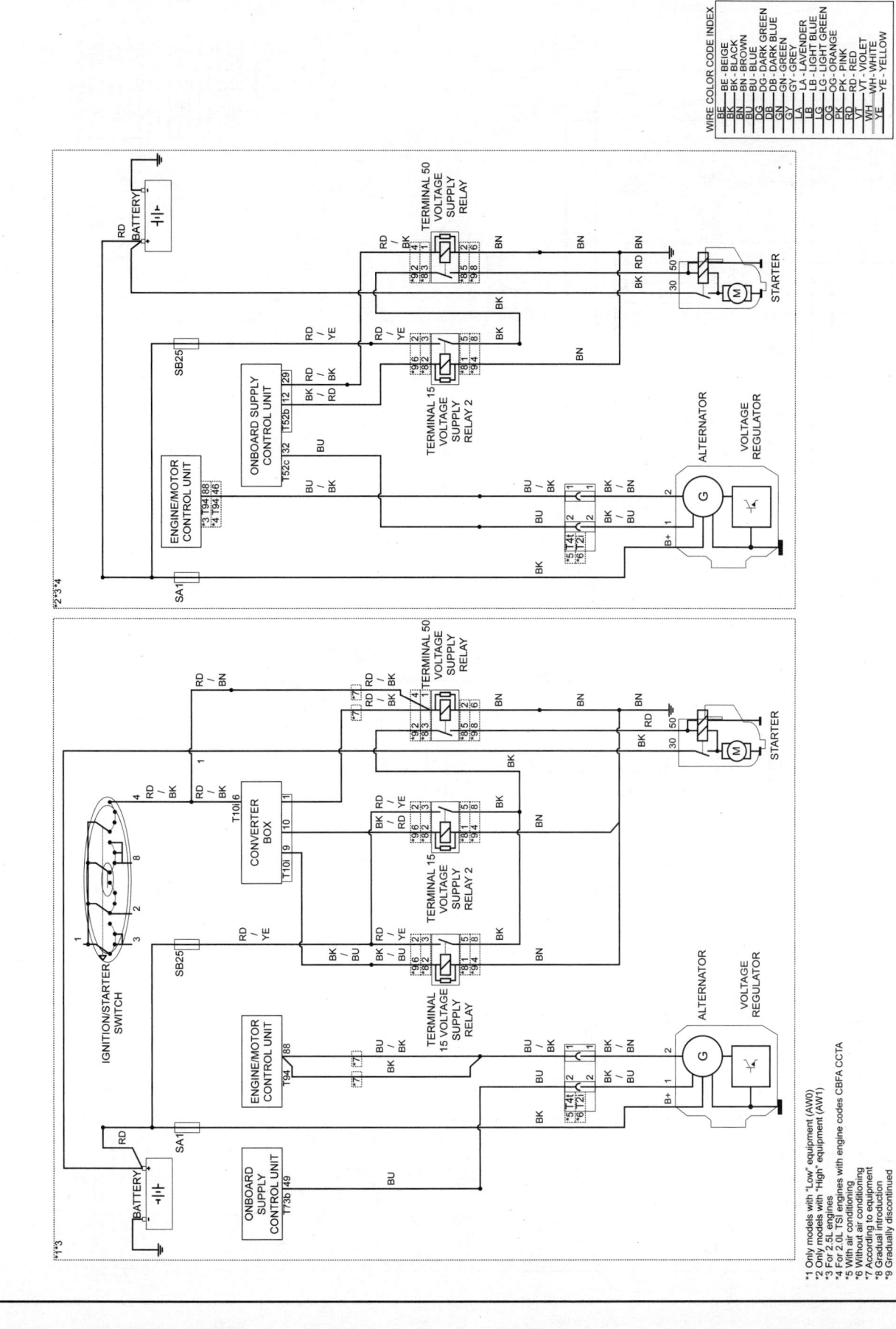

1e. Starting and Charging - JETTA up to June 2014 (CBTA) (CBUA) (CCTA) (CBFA) (2.5L and 2.0L TSI)

*1 Only models with "Low" equipment (AW0)
*2 Only models with "High" equipment (AW1)
*3 For 2.5L engines
*4 For 2.0L TSI engines with engine codes CBFA CCTA
*5 With air conditioning
*6 Without air conditioning
*7 According to equipment
*8 Gradual introduction
*9 Gradually discontinued

Chapter 12 Chassis electrical system

1f. Starting and Charging - JETTA from July 2014 (CPKA CPRA CPLA CPPA) (1.8L and 2.0 TFSI)

*1 For 1.8L engines
*2 For 2.0L TFSI engines
*3 For models with air conditioning
*4 For models without air conditioning
*5 According to equipment
*6 For models without secondary air pump
*7 For models with secondary air pump

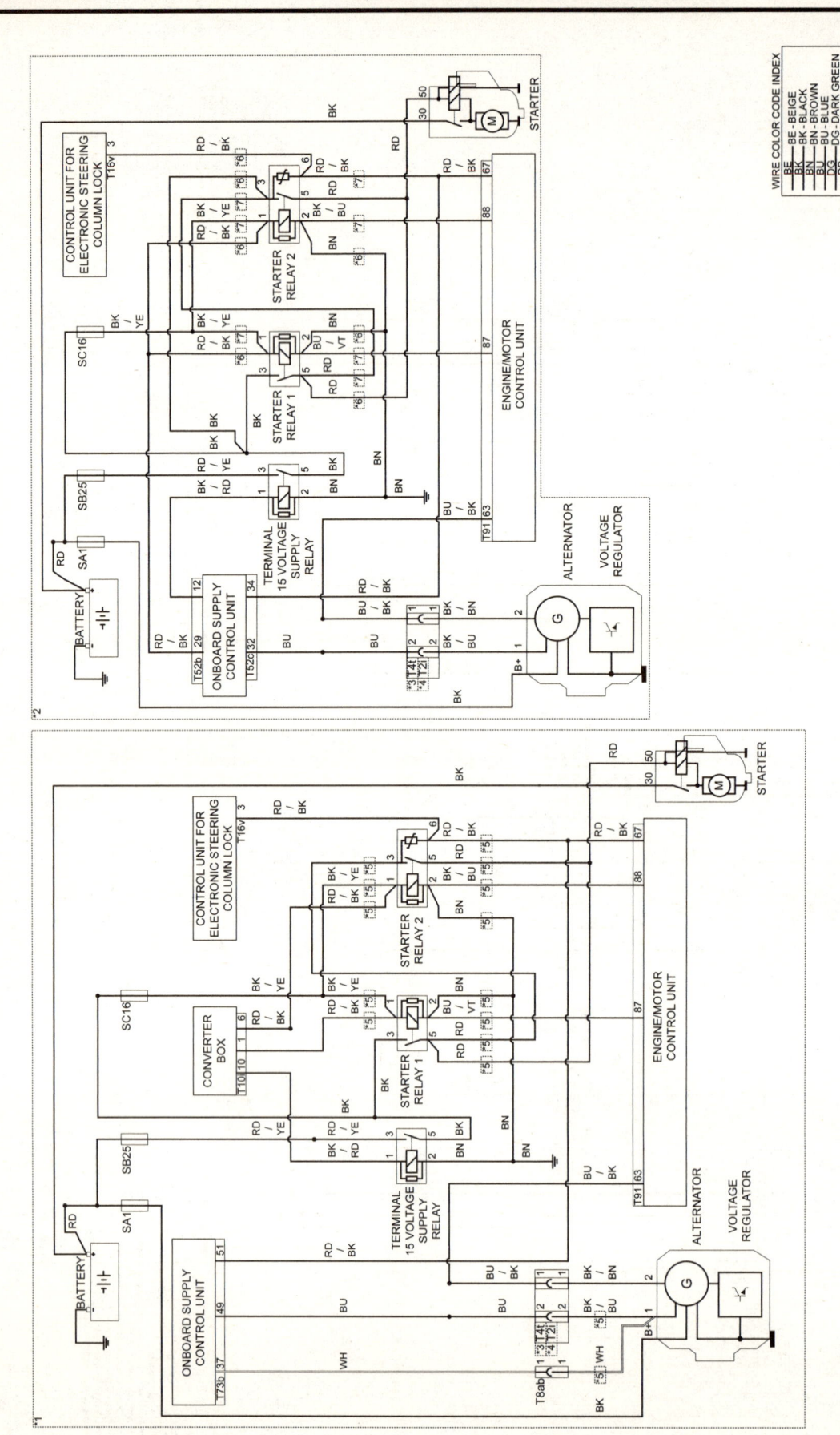

Chapter 12 Chassis electrical system

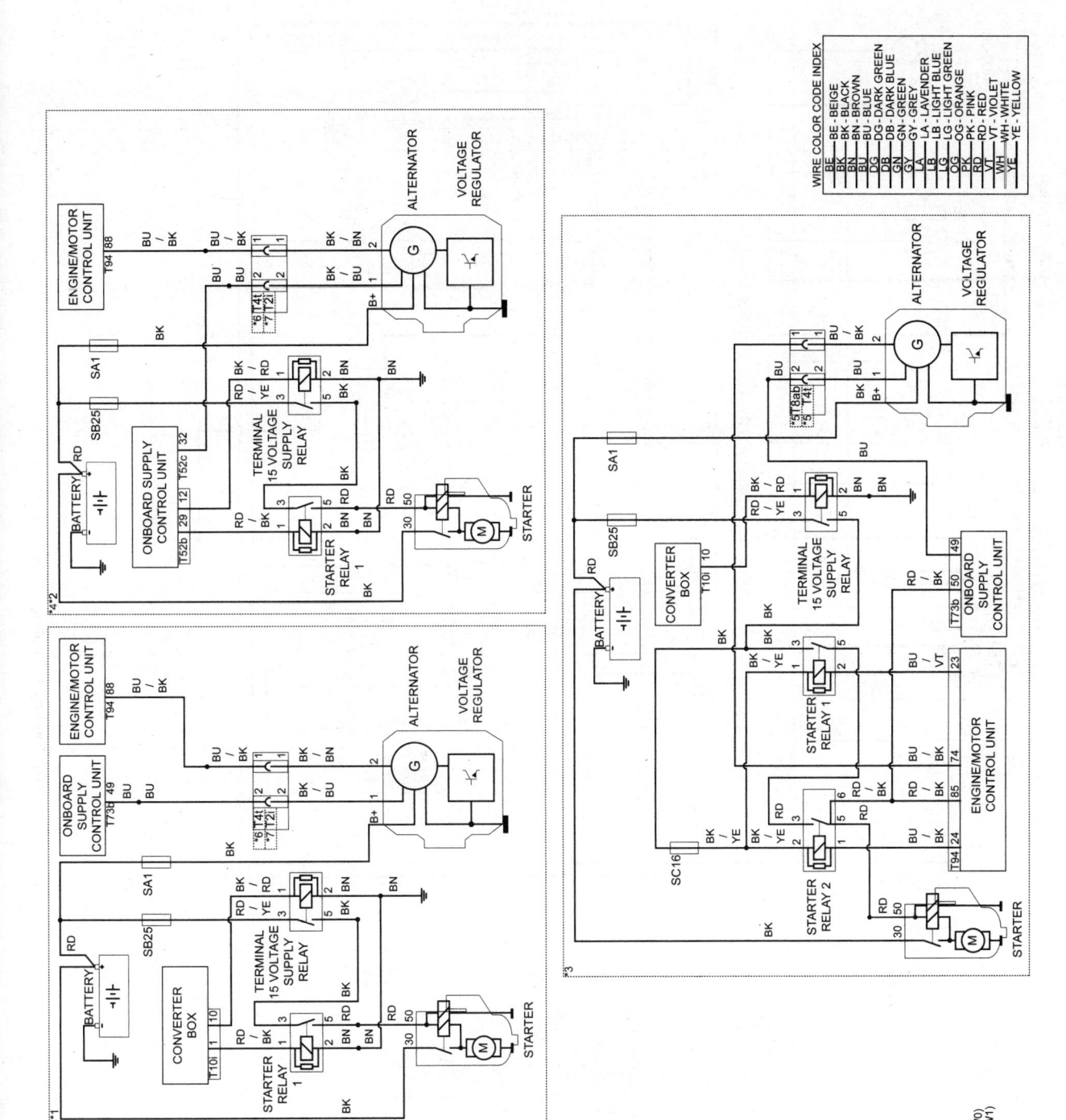

1g. Starting and Charging - JETTA from July 2014 (CZTA CBTA CBUA CBPA) (1.4L 2.5L and 2.0 8V)

*1 Only models with "Low" equipment (AW0)
*2 Only models with "High" equipment (AW1)
*3 For 1.4L engines
*4 Except 1.4L engines
*5 According to equipment
*6 With air conditioning
*7 Without air conditioning

Chapter 12 Chassis electrical system

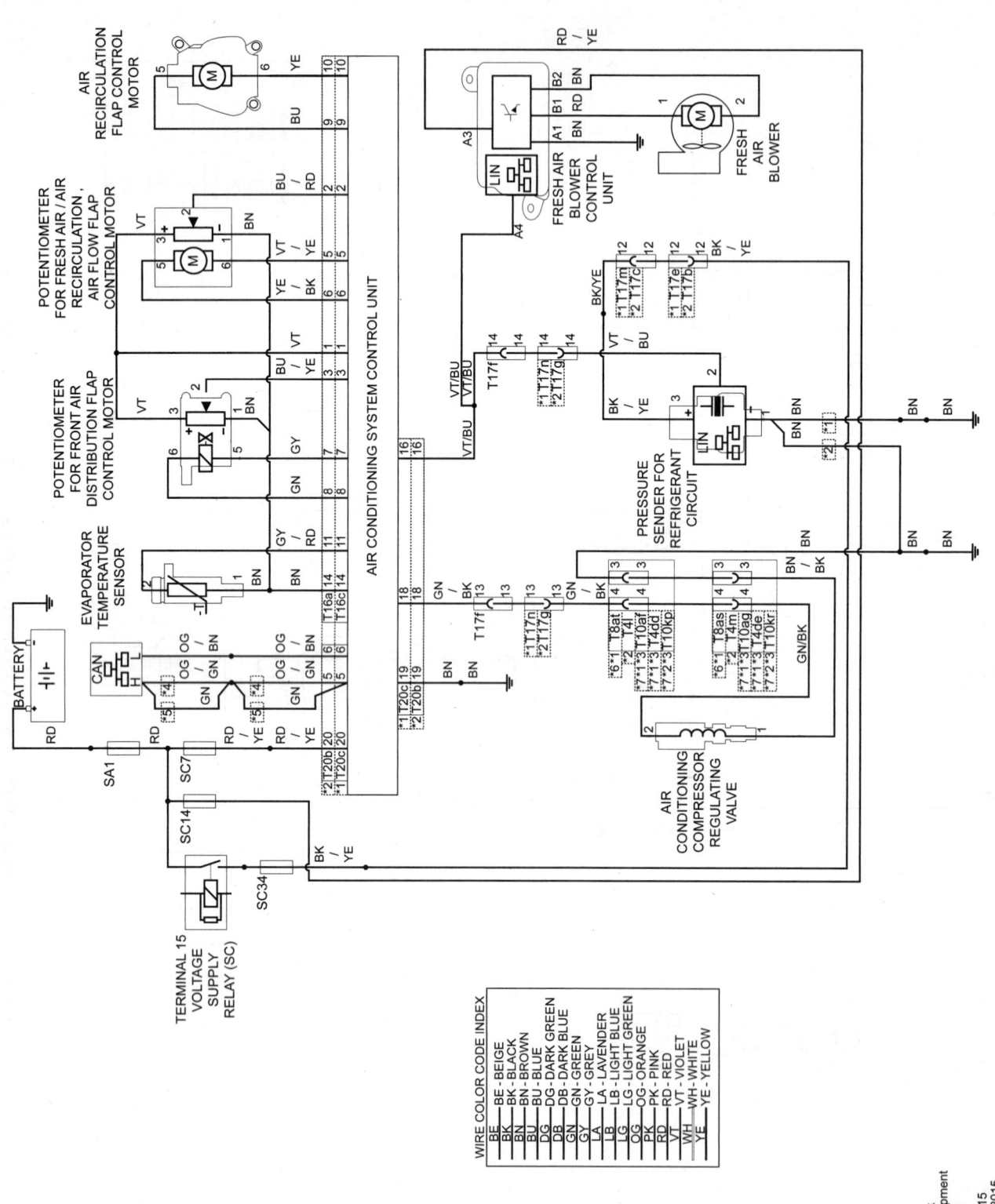

2a. AC Heating & Cooling - GOLF Climatic

Chapter 12 Chassis electrical system

2b. AC Heating & Cooling - GOLF Climatronic

*1 Hatchback
*2 Except hatchback
*3 According to equipment
*4 Up to June 2018
*5 From July 2018
*6 Up to October 2015
*7 From November 2015

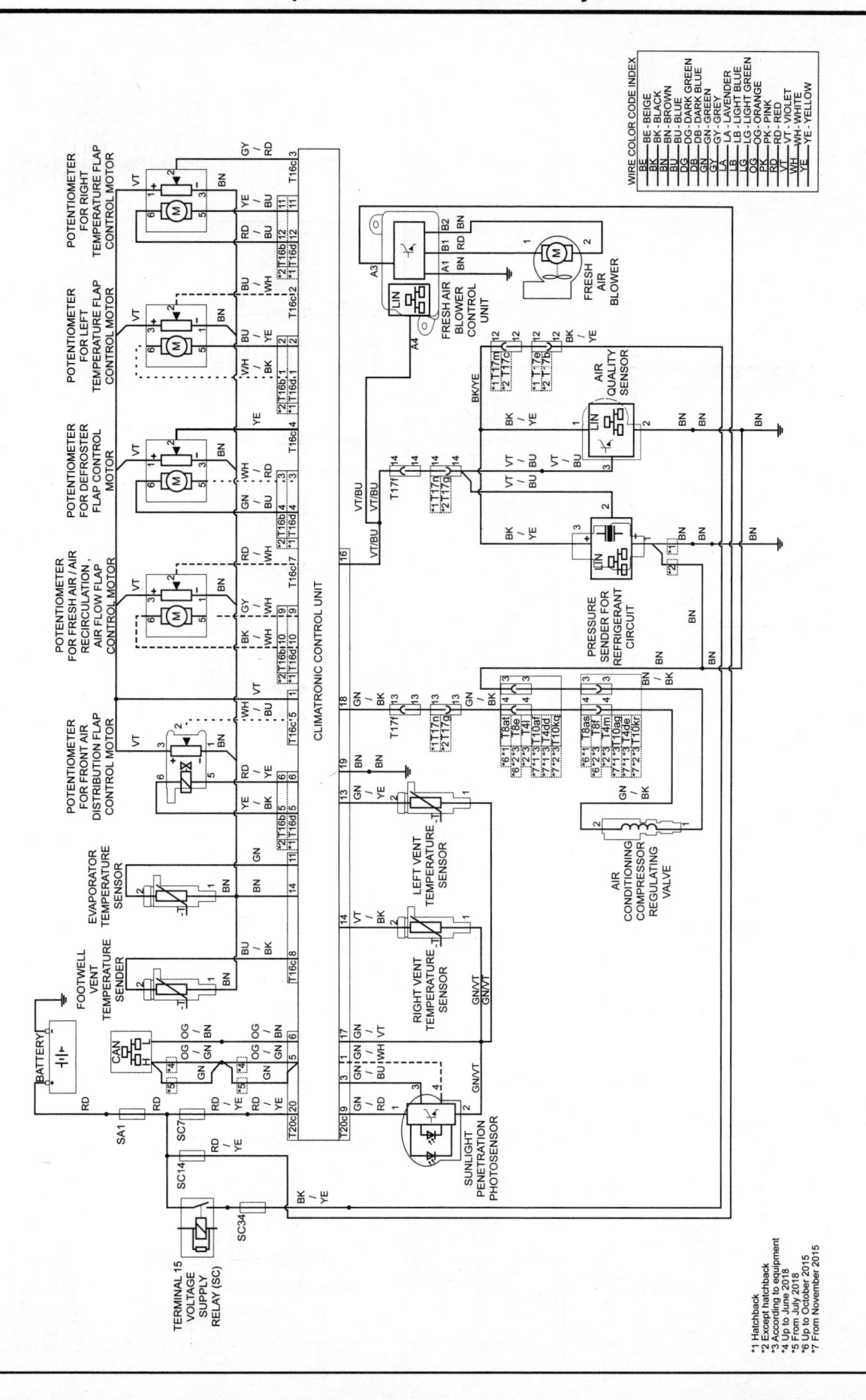

12-32   Chapter 12   Chassis electrical system

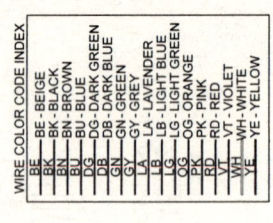

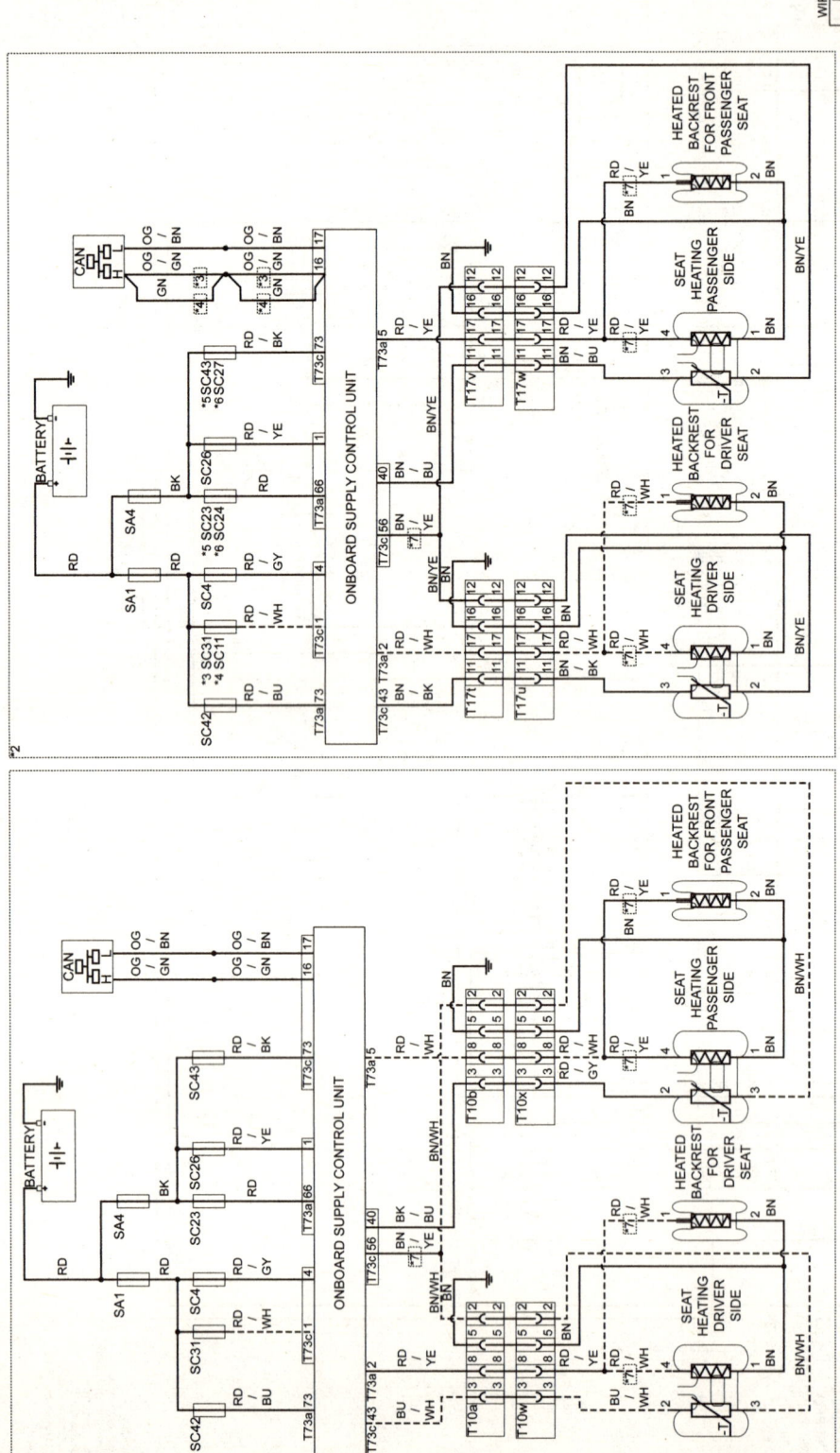

2c. AC Heating & Cooling - GOLF Heated seats with sport seats

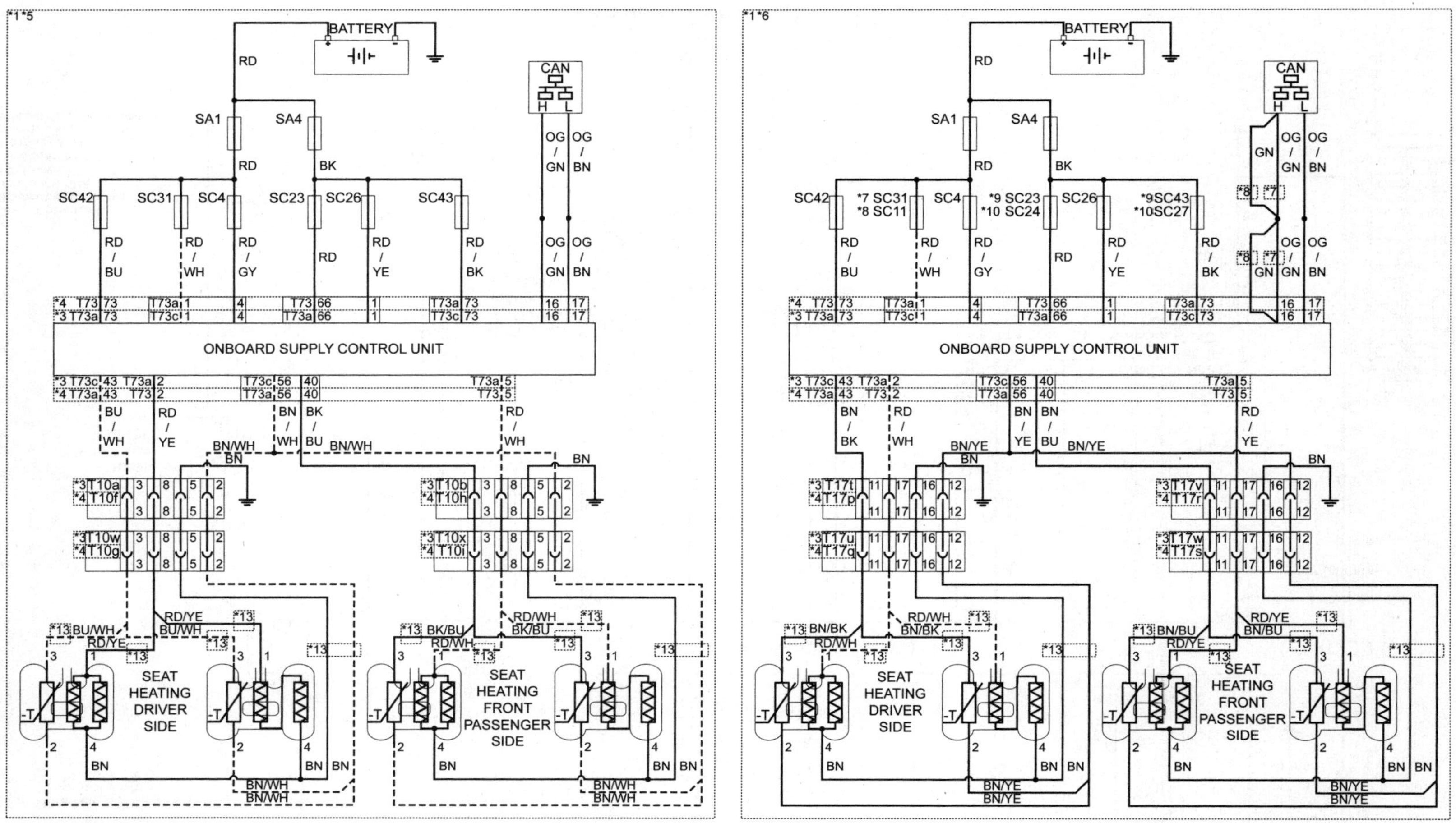

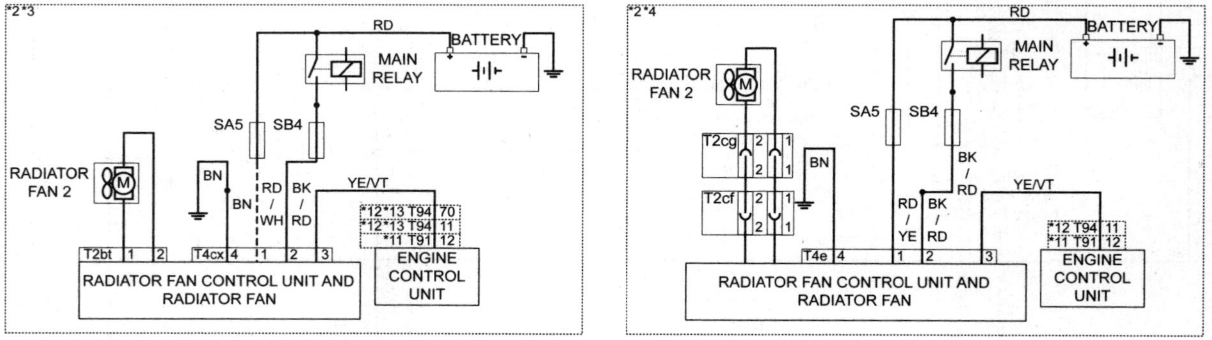

Chapter 12 Chassis electrical system

12-33

2d. AC Heating & Cooling - GOLF Heated seats and cooling fan

*1 Heated seats
*2 Cooling fan
*3 Hatchback
*4 Except hatchback
*5 Up to October 2016
*6 From November 2016
*7 Up to June 2018
*8 From July 2018
*9 Up to April 2017
*10 From May 2017
*11 For 2.0L and 1.8L engines
*12 For 1.4L engines
*13 According to equipment

12-34 Chapter 12 Chassis electrical system

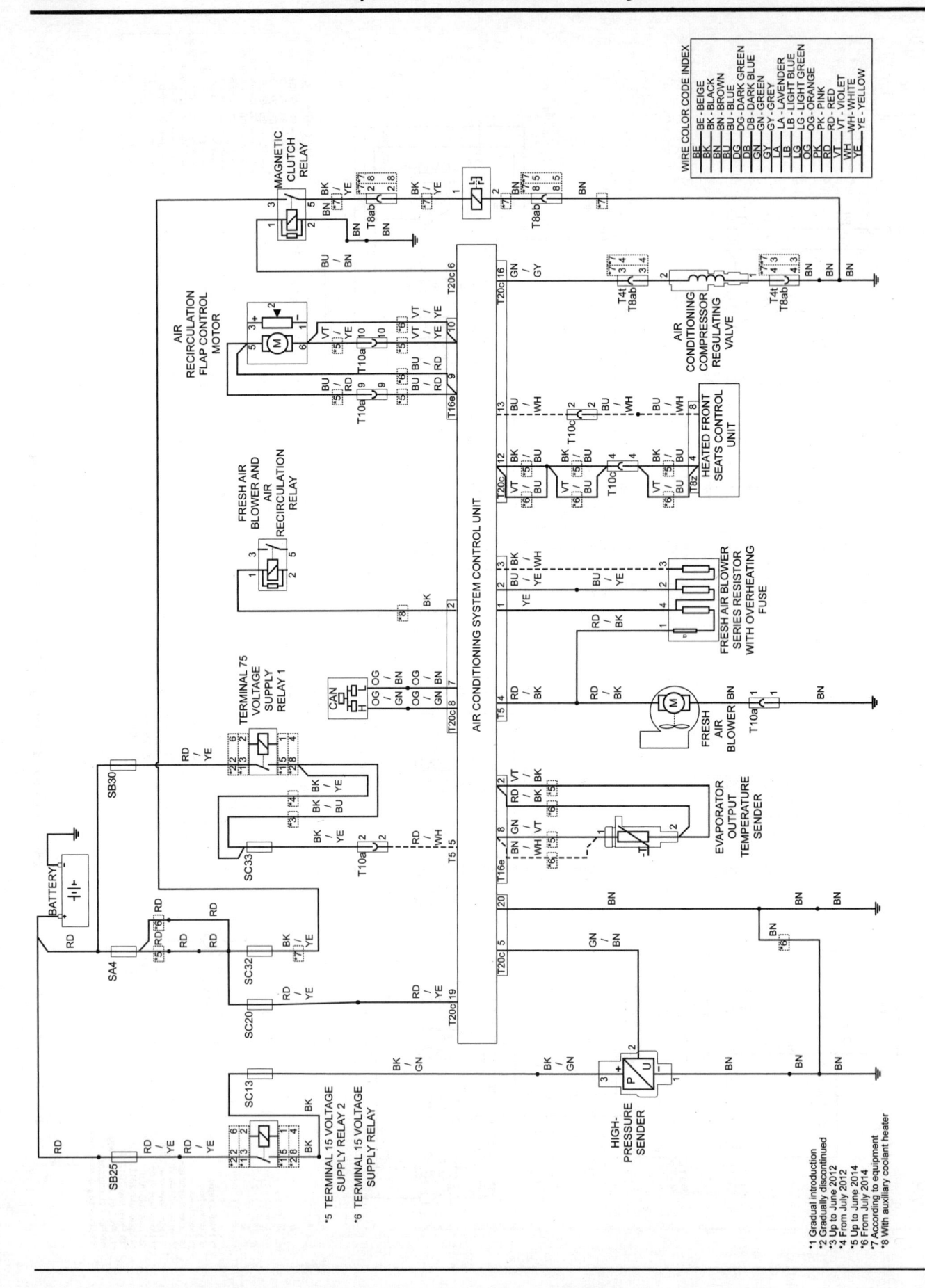

2e. AC Heating & Cooling - JETTA Manual AC

Chapter 12 Chassis electrical system

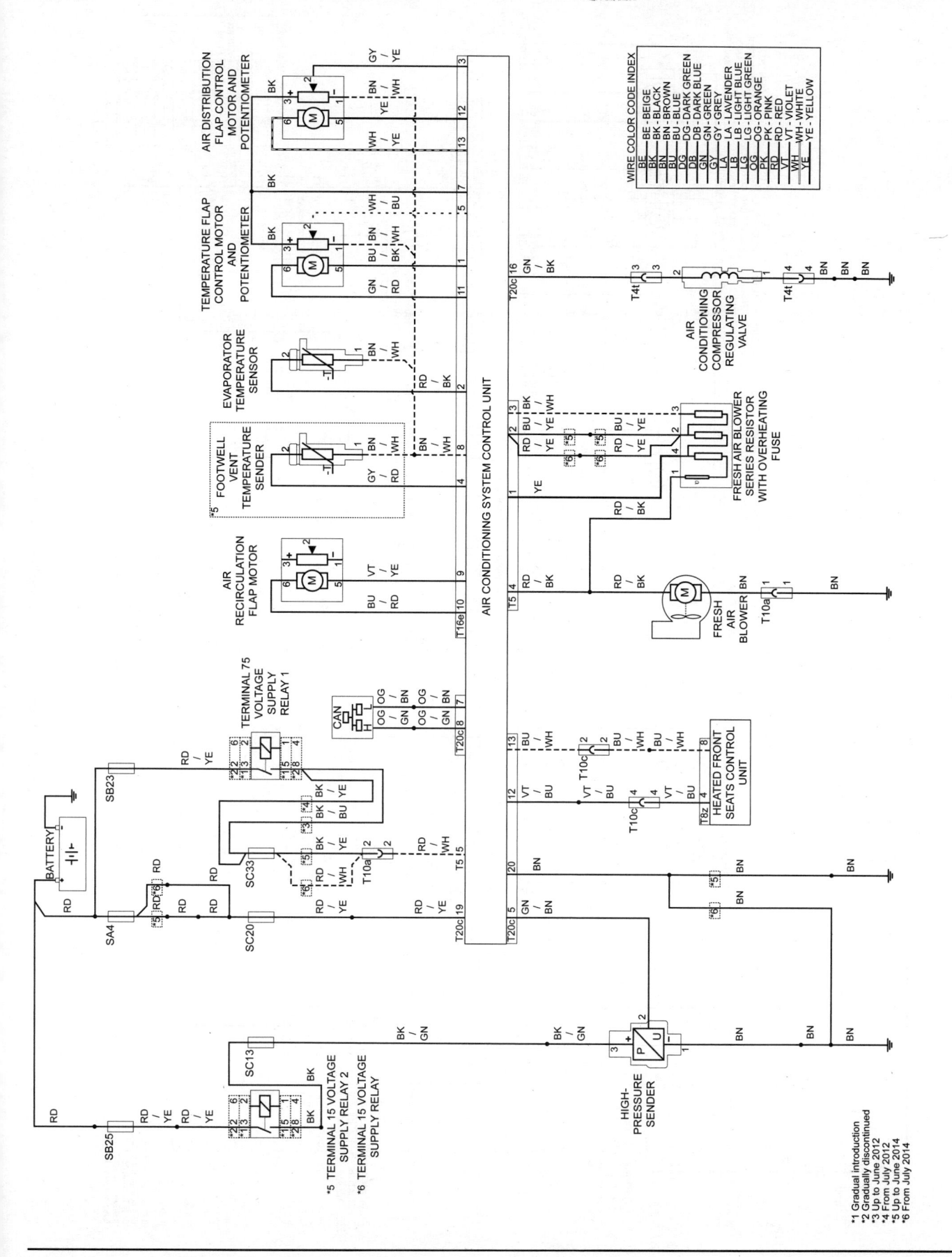

2f. AC Heating & Cooling - JETTA Climatic

*1 Gradual introduction
*2 Gradually discontinued
*3 Up to June 2012
*4 From July 2012
*5 Up to June 2014
*6 From July 2014

12-36 Chapter 12 Chassis electrical system

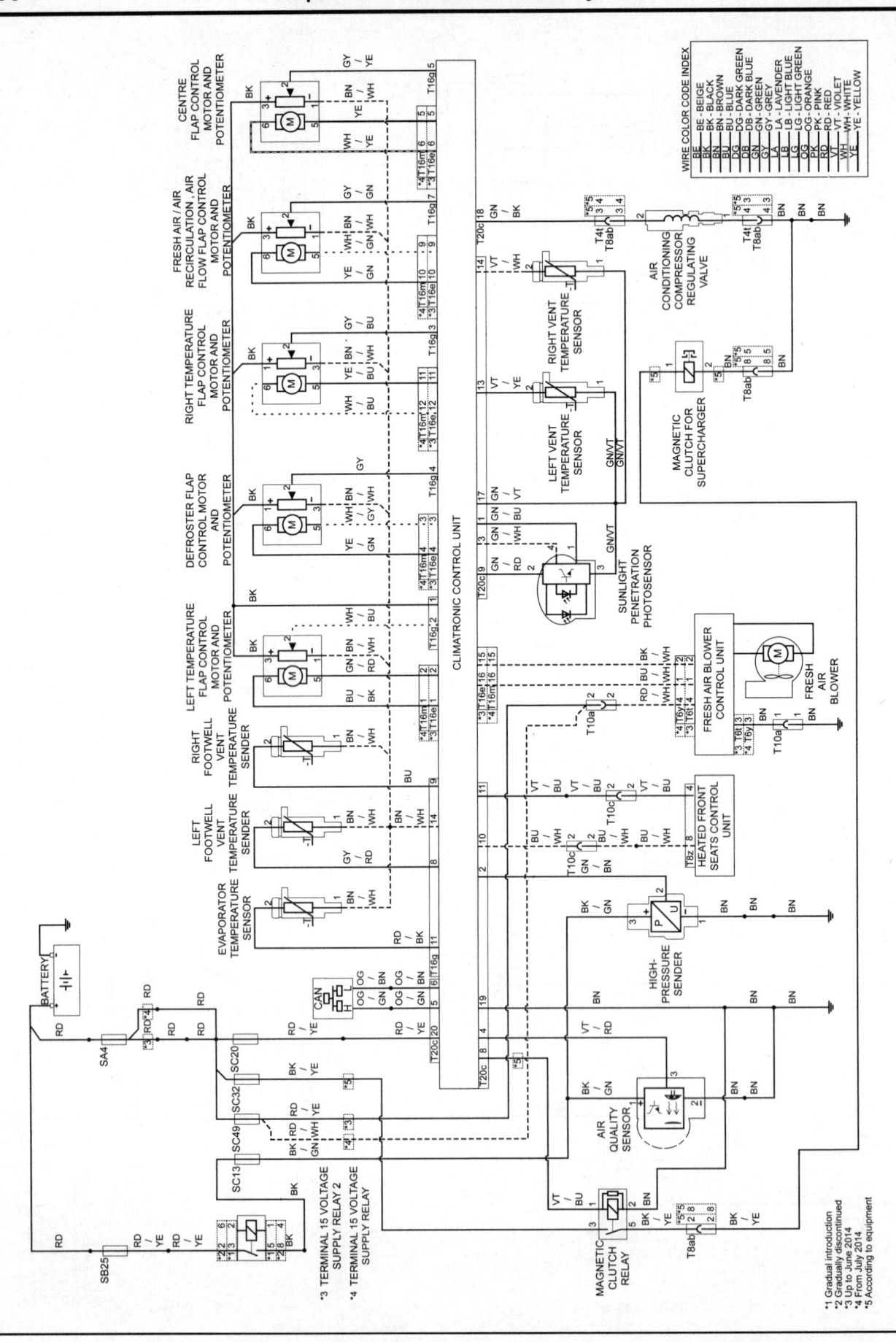

2g. AC Heating & Cooling - JETTA Climatronic

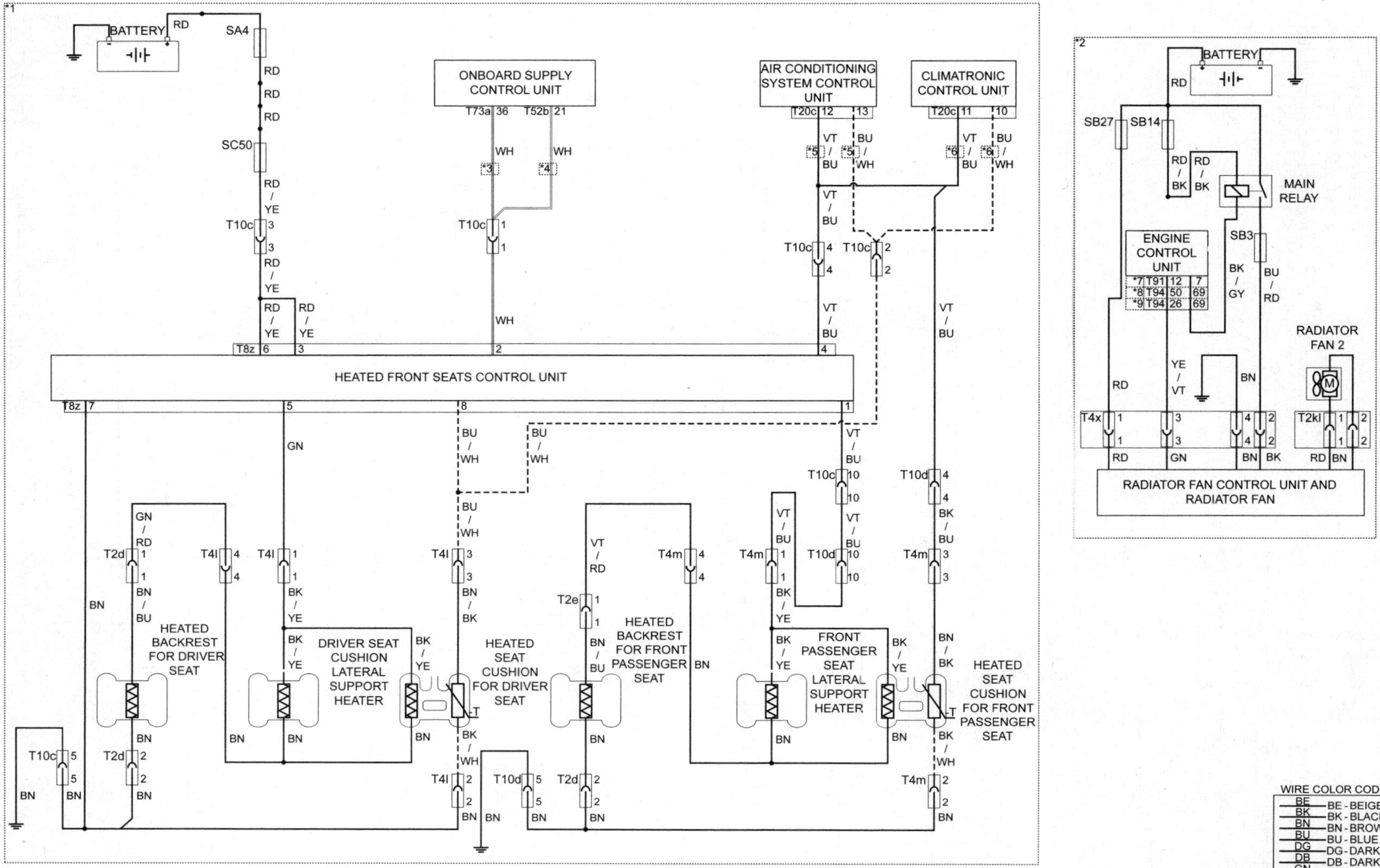

Chapter 12 Chassis electrical system

2h. AC Heating & Cooling - JETTA Heated seats and cooling fan Up to June 2014

*1 Heated seats
*2 Cooling fans
*3 Only models with "Low" equipment (AW0)
*4 Only models with "High" equipment (AW1)
*5 With Climatic
*6 With Climatronic
*7 For 2.0L TFSi, 1.8L TSi and 1.8L TFSi engines
*8 For 2.0L TSi engines
*9 For 2.0L 8v engines and for 2.5L engines

12-38 Chapter 12 Chassis electrical system

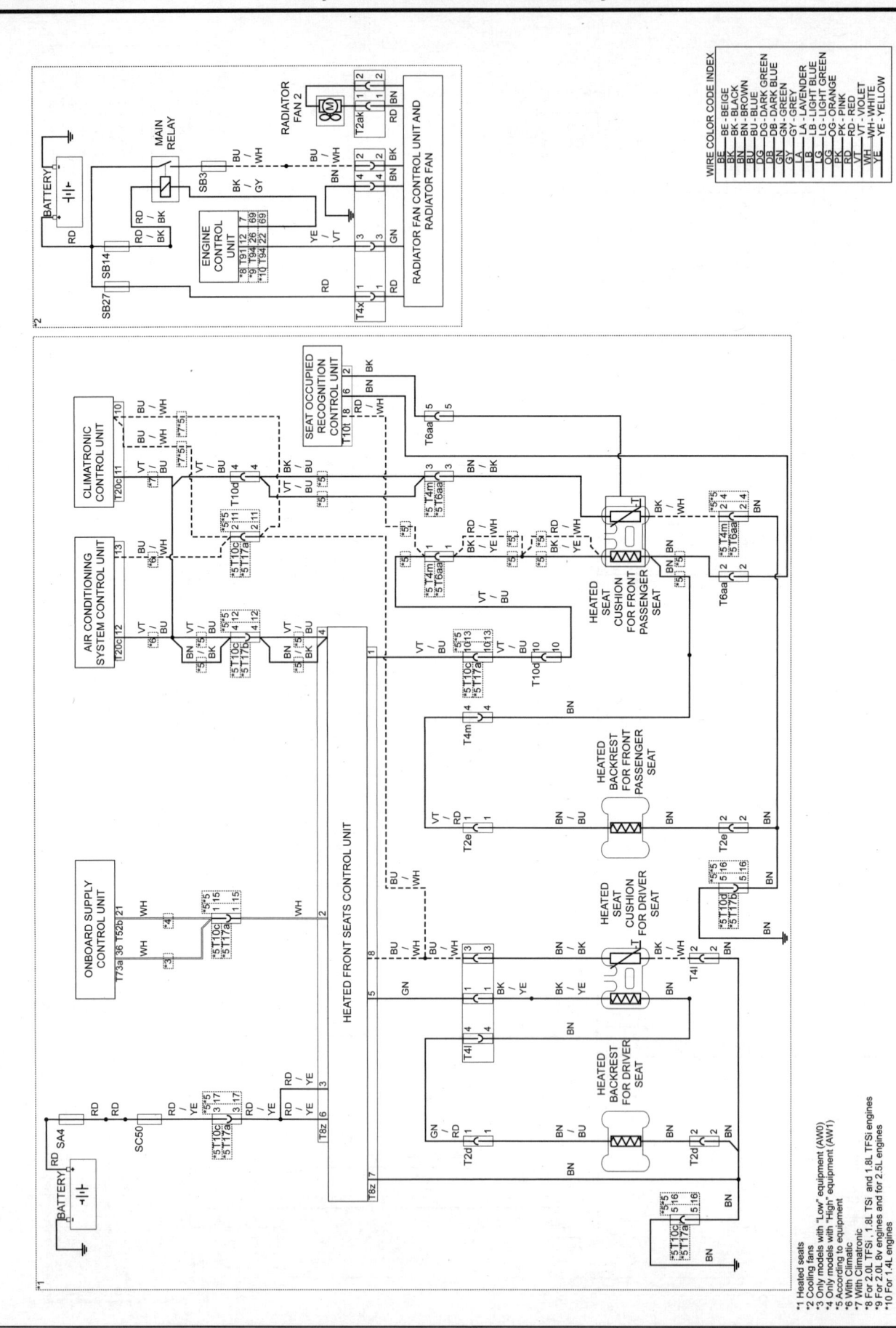

Chapter 12 Chassis electrical system

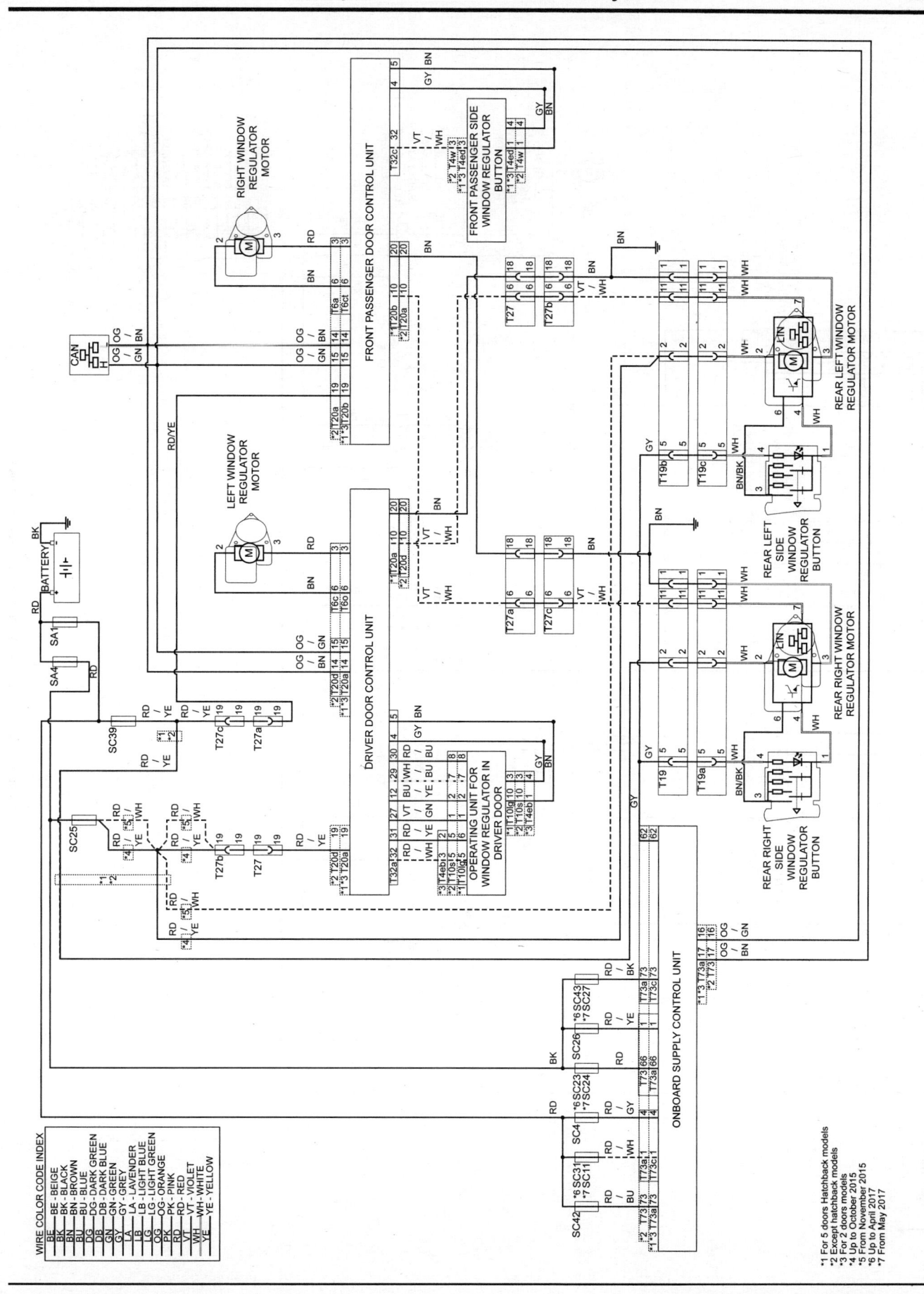

3a. Power Windows - GOLF

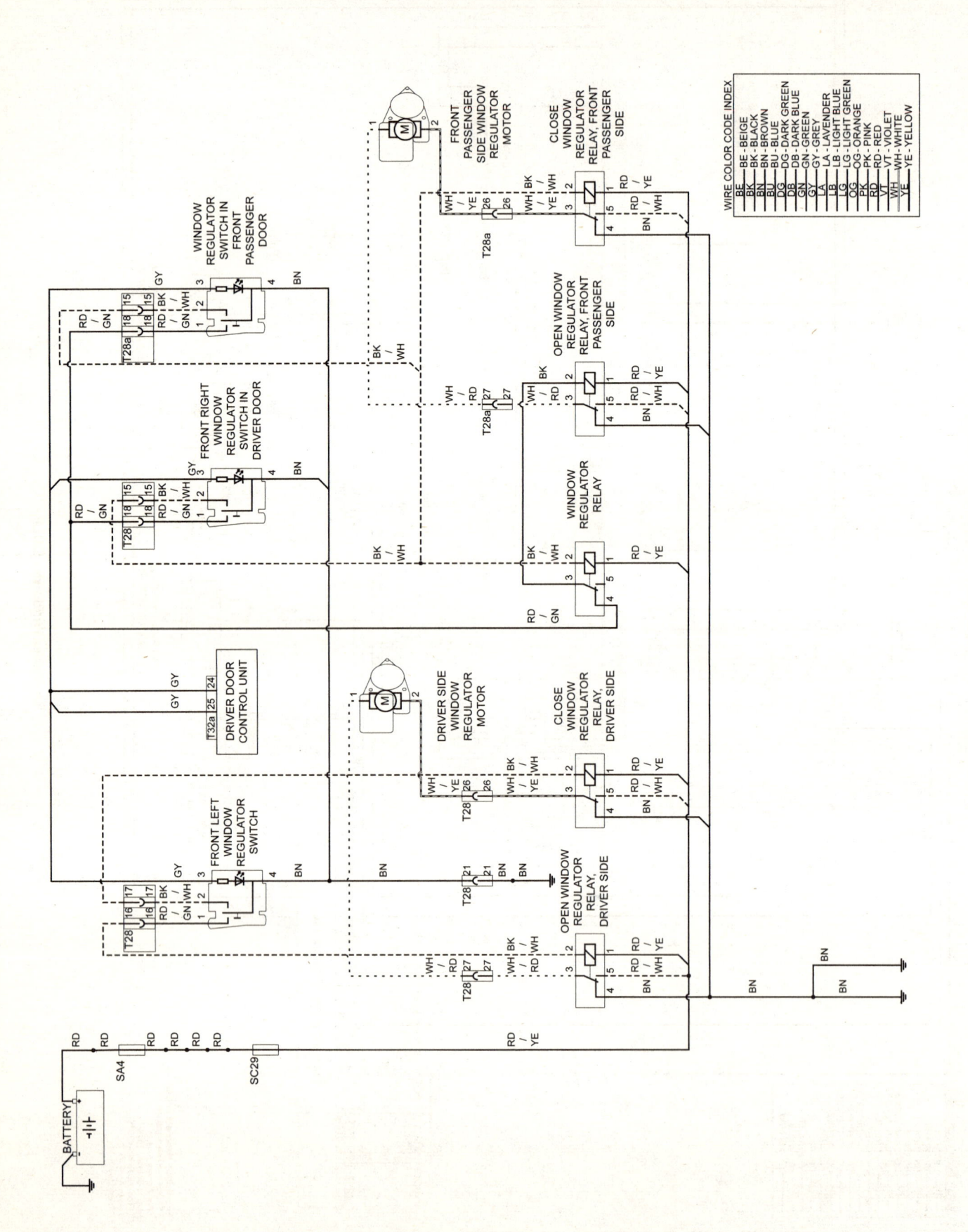

3b. Power windows - JETTA With manual rear windows

Chapter 12 Chassis electrical system

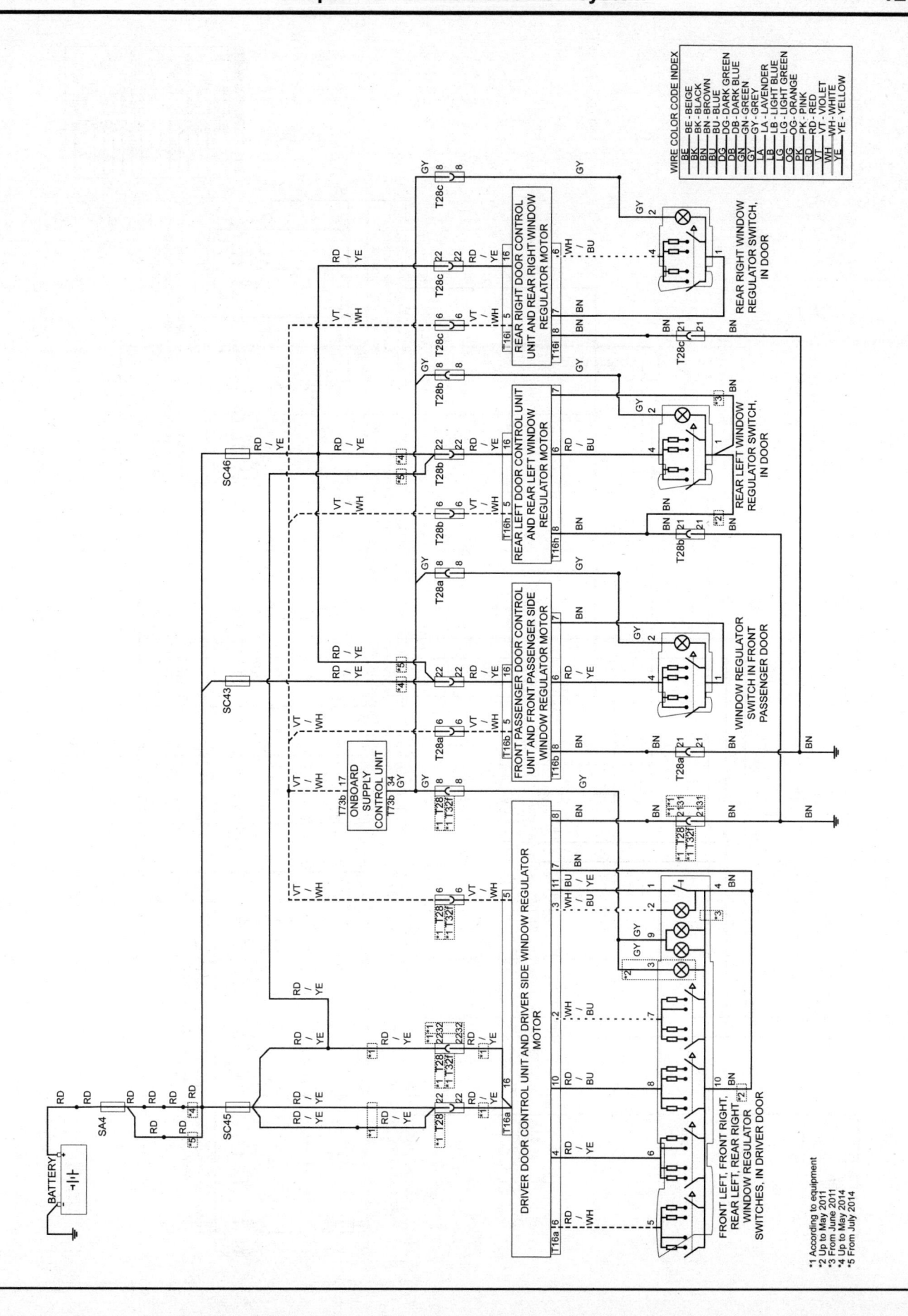

3c. Power windows - JETTA Low equipment

*1 According to equipment
*2 Up to May 2011
*3 From June 2011
*4 Up to May 2014
*5 From July 2014

Chapter 12 Chassis electrical system

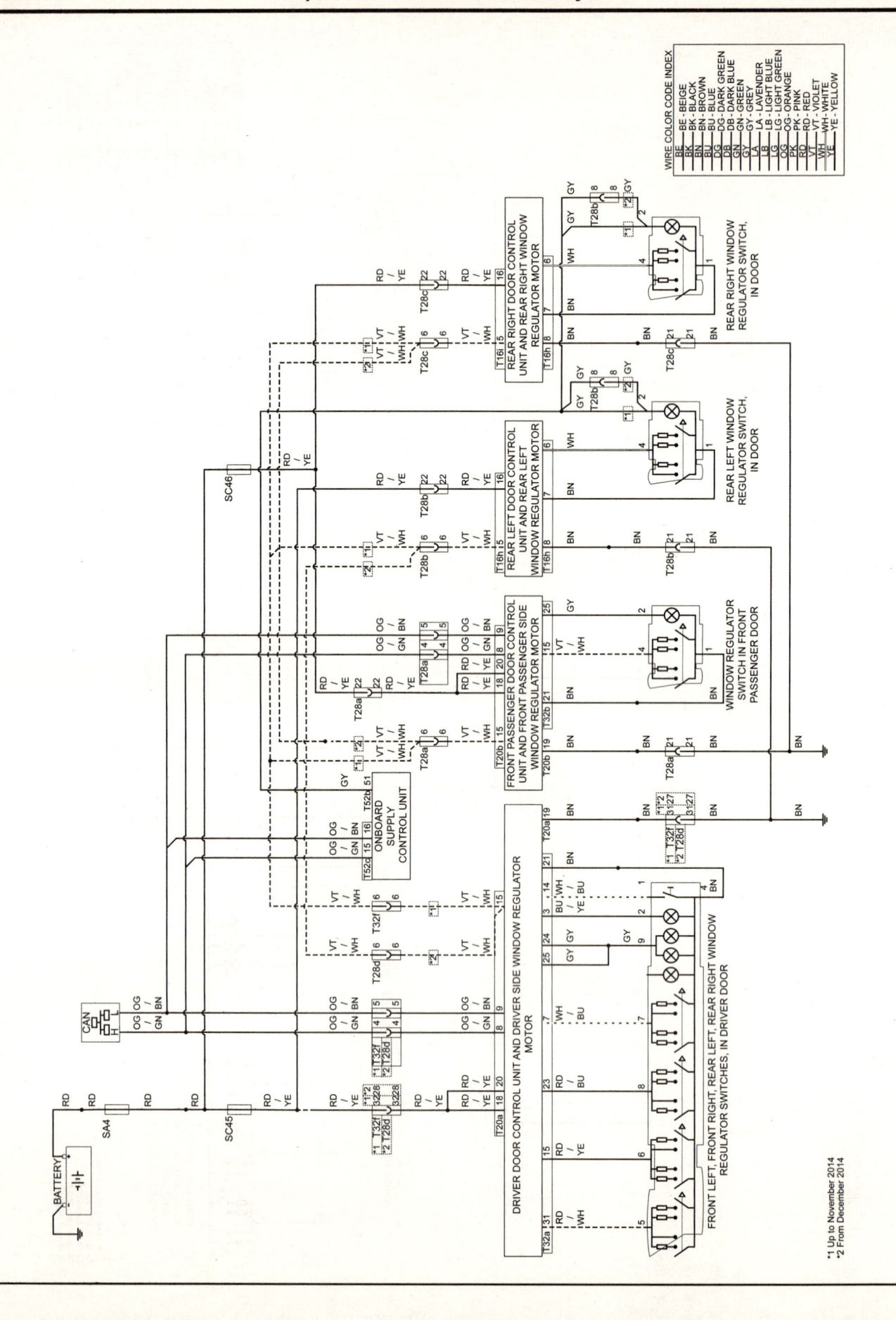

3d. Power windows - JETTA High equipment From July 2014

Chapter 12 Chassis electrical system

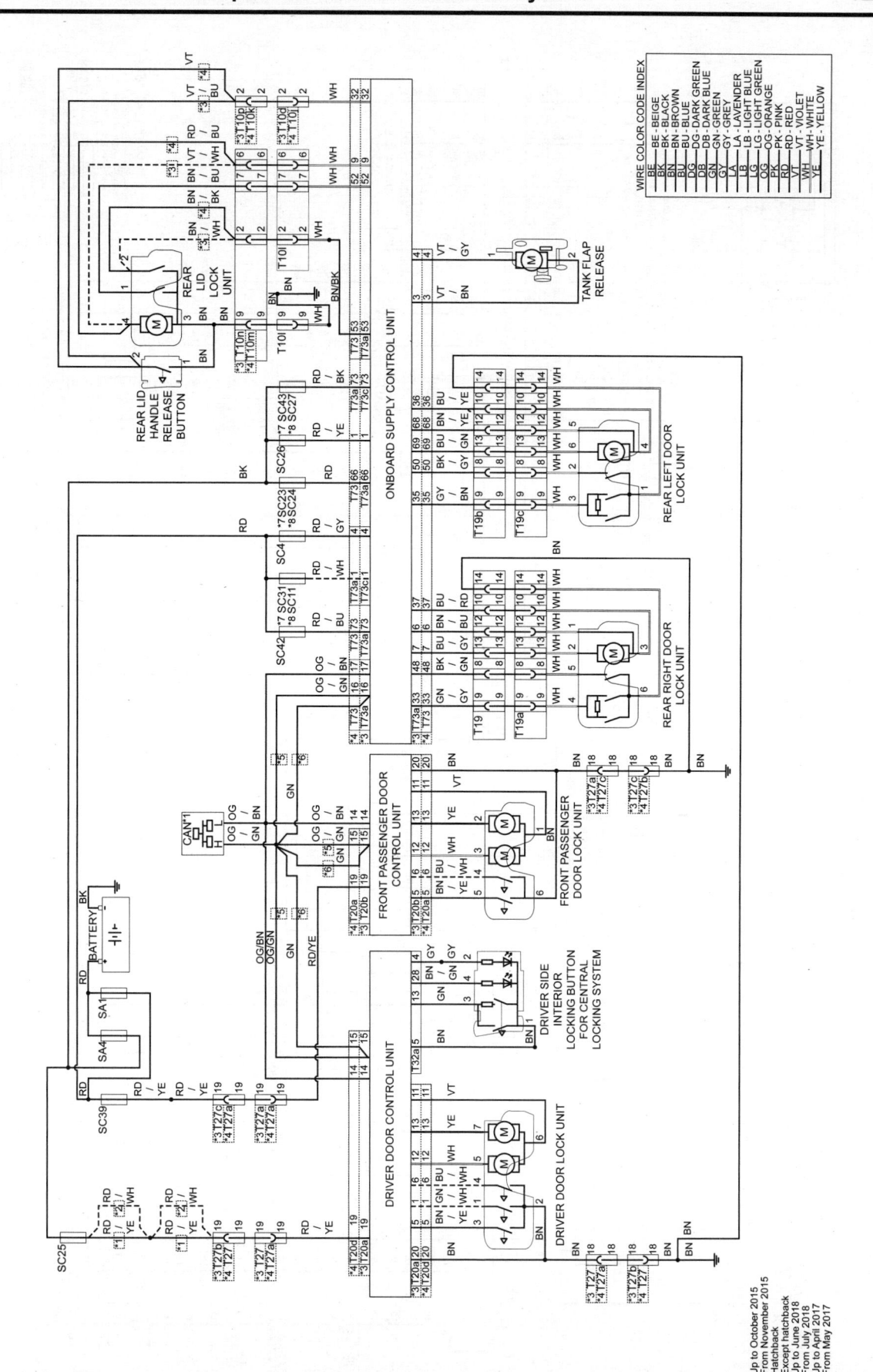

12-43

4a. Power door Locks - GOLF

*1 Up to October 2015
*2 From November 2015
*3 Hatchback
*4 Except hatchback
*5 Up to July 2018
*6 From July 2018
*7 Up to April 2017
*8 From May 2017

Chapter 12 Chassis electrical system

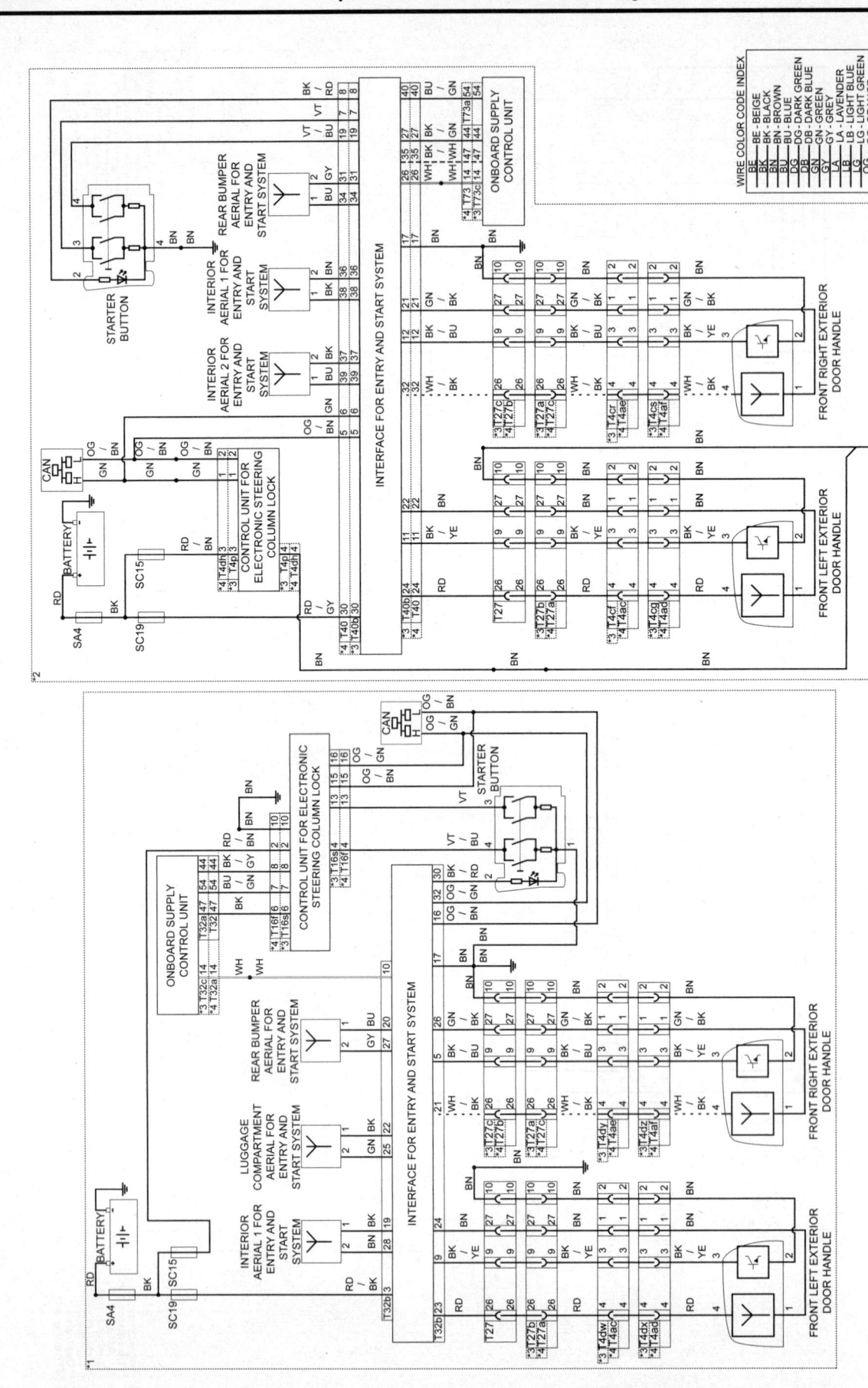

4b. Power door Locks - GOLF Keyless

*1 Up to June 2018
*2 From July 2018
*3 Hatchback
*4 Except hatchback

Chapter 12 Chassis electrical system

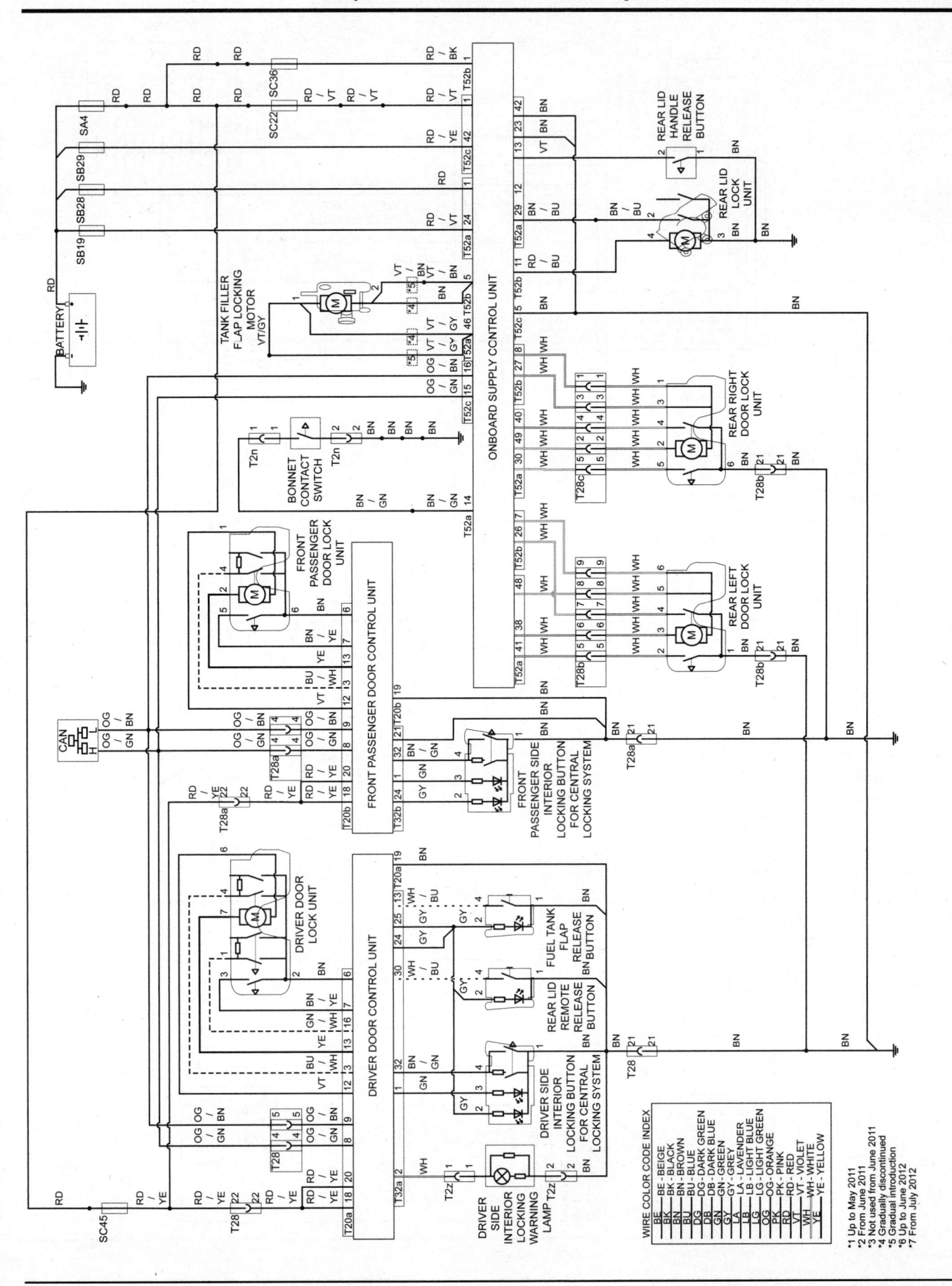

12-45

4c. Power door locks JETTA - Models with rear manual window regulators

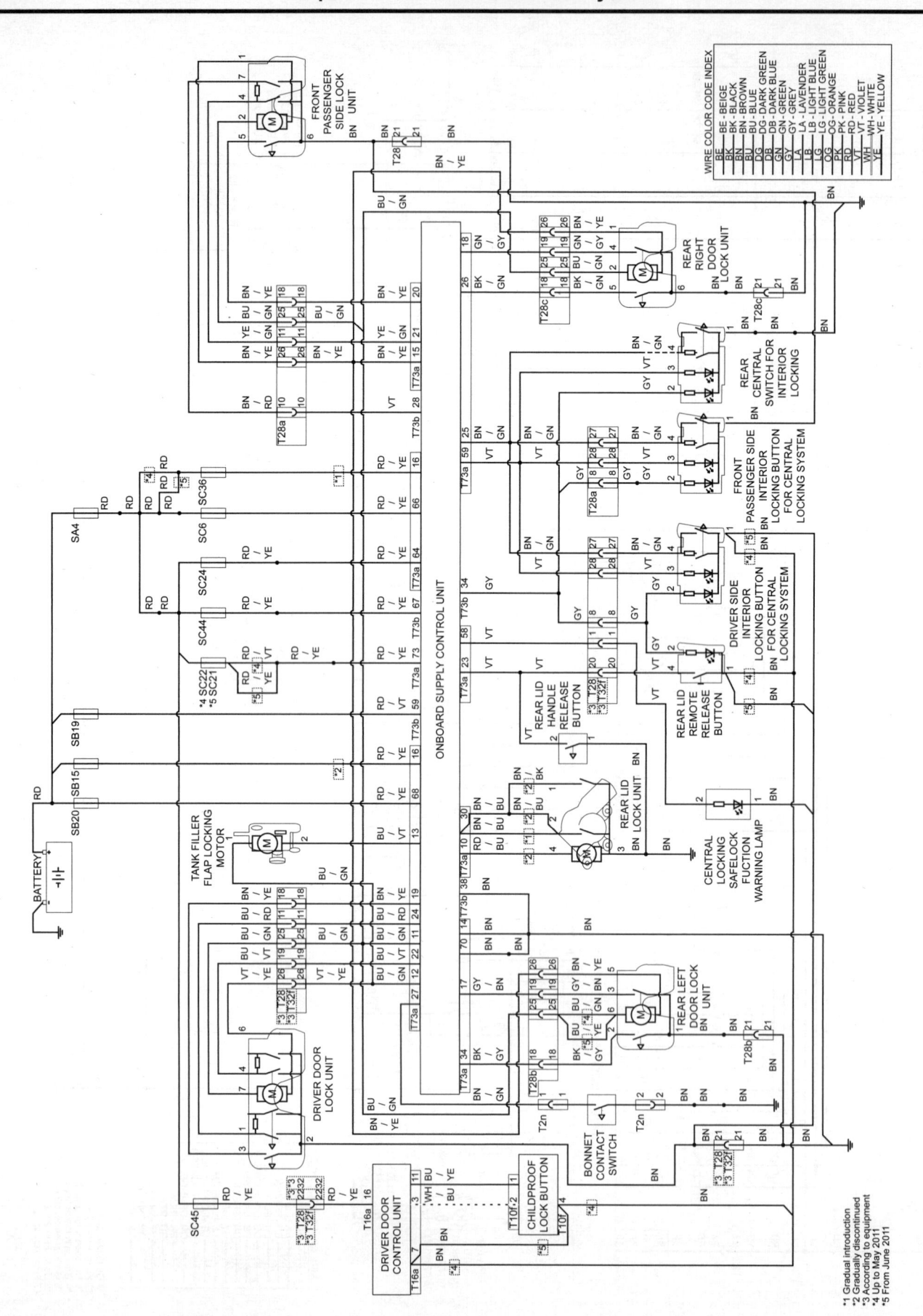

4d. Power door locks - JETTA Low equipment Up to June 2014

Chapter 12 Chassis electrical system

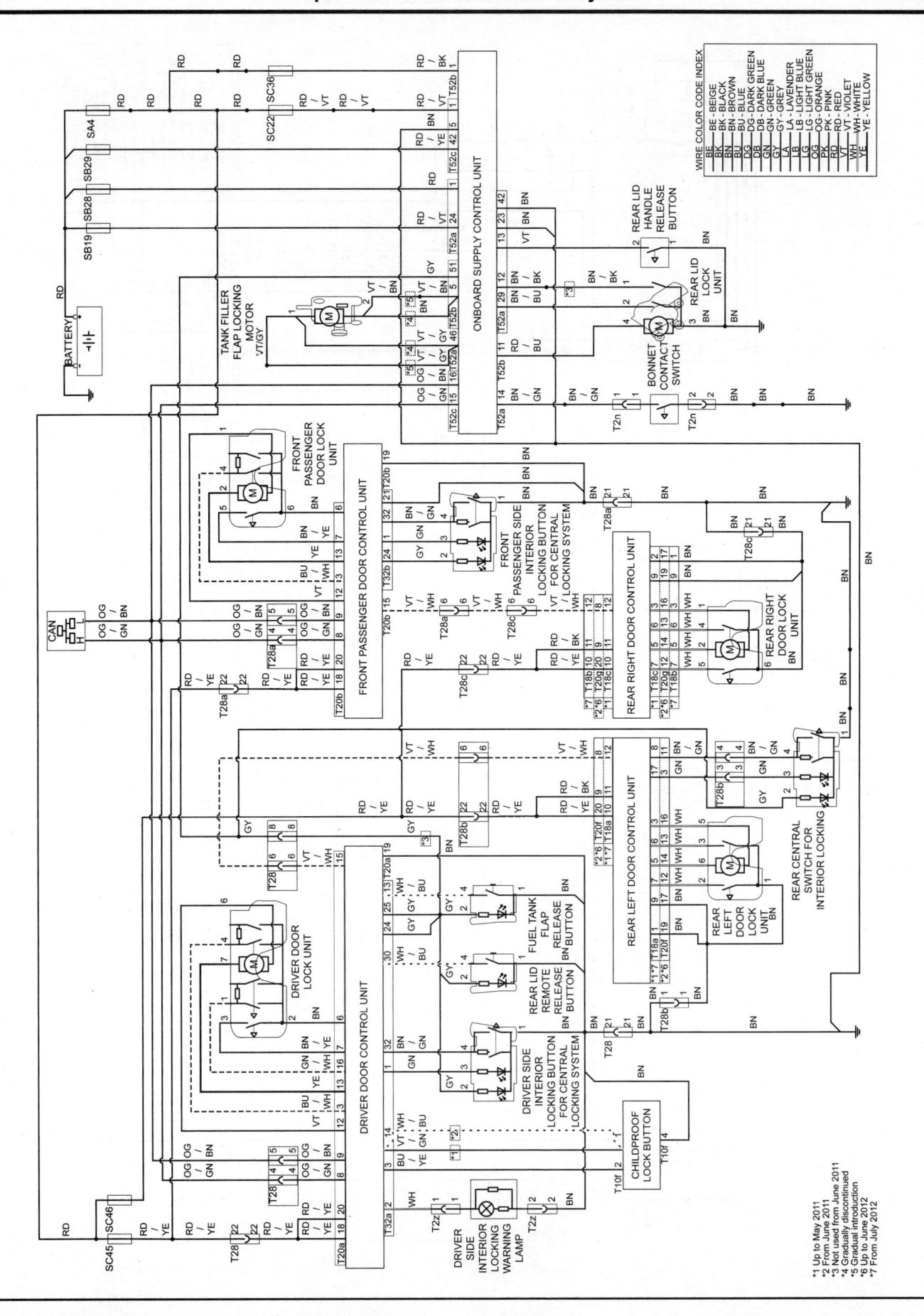

12-47

4e. Power door locks - JETTA High equipment Up to June 2014

*1 Up to May 2011
*2 From June 2011
*3 Not used from June 2011
*4 Gradually discontinued
*5 Gradual introduction
*6 Up to June 2012
*7 From July 2012

12-48 Chapter 12 Chassis electrical system

4f. Power door locks - JETTA Low equipment From July 2014

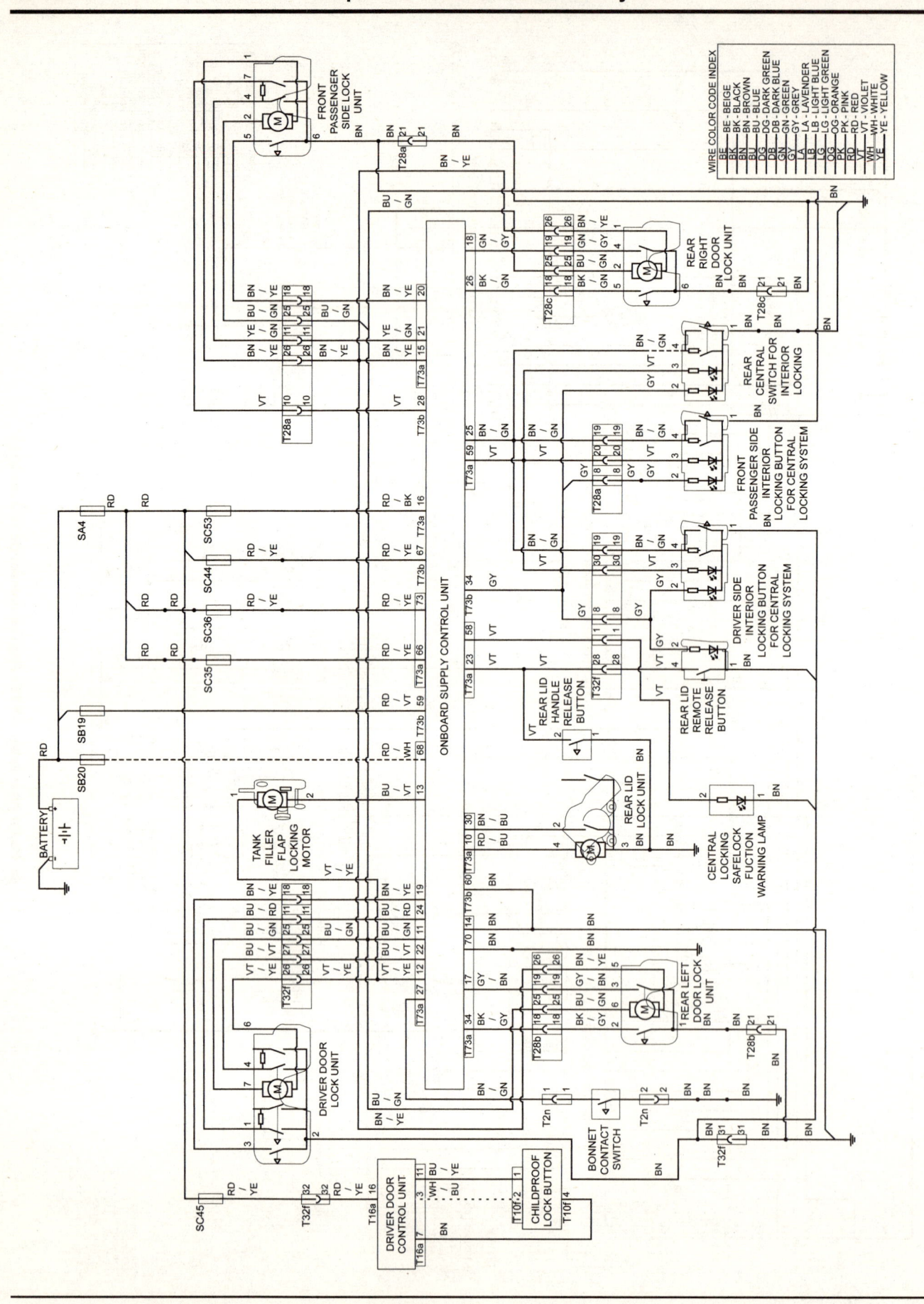

Chapter 12 Chassis electrical system

12-49

4g. Power door locks - JETTA High equipment From July 2014

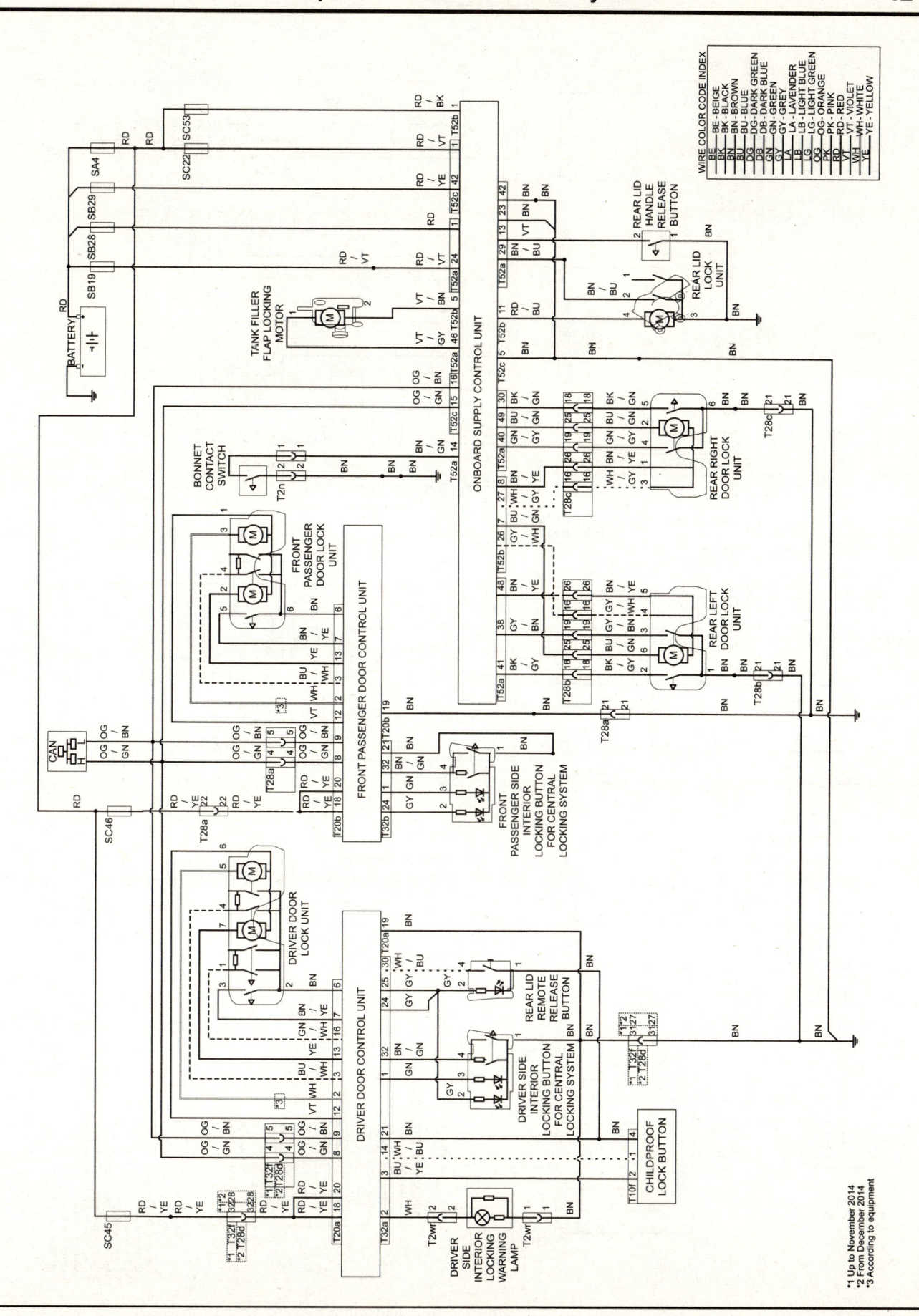

Chapter 12 Chassis electrical system

4h. Power Door Locks - JETTA Keyless

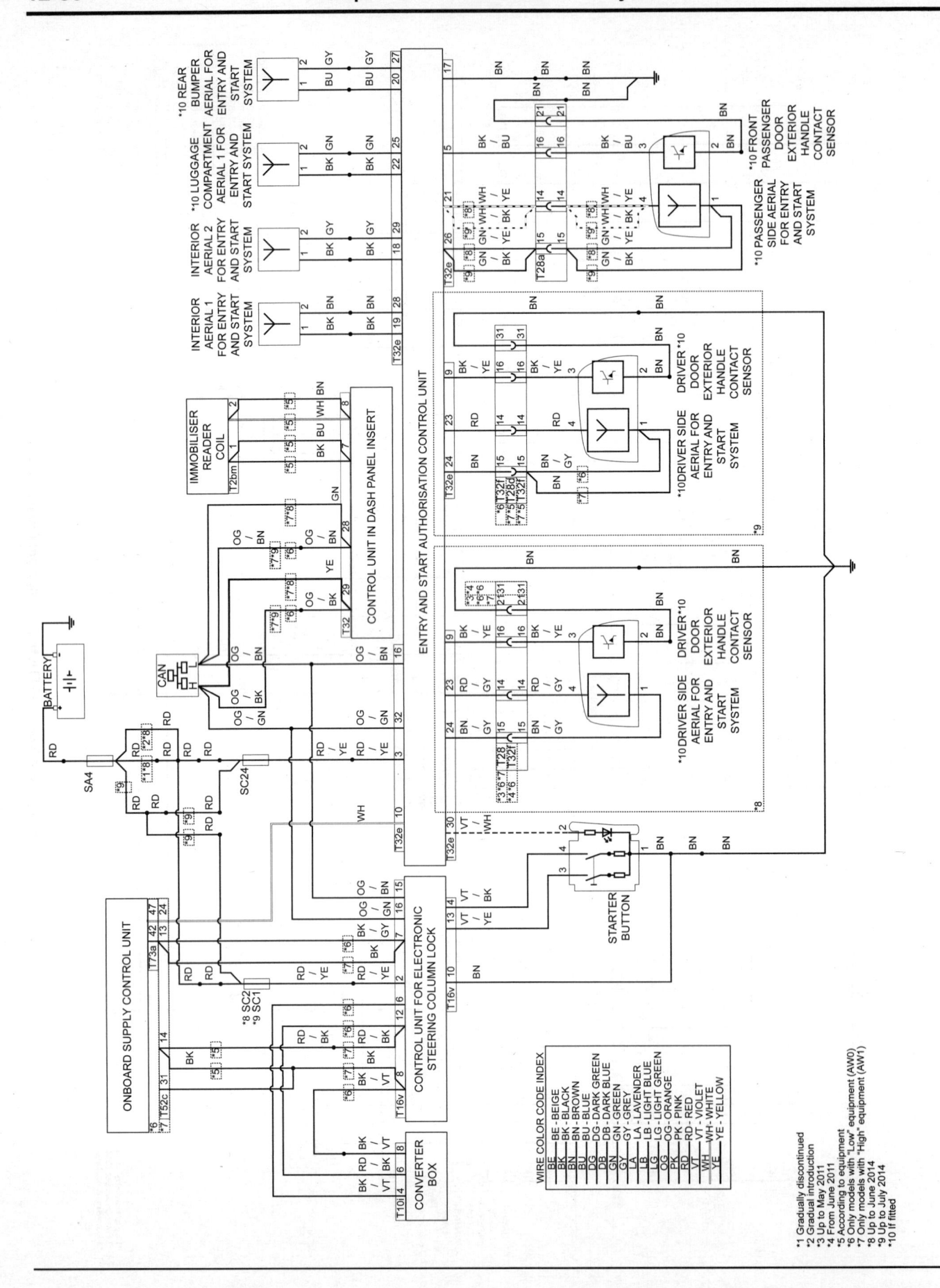

Chapter 12 Chassis electrical system

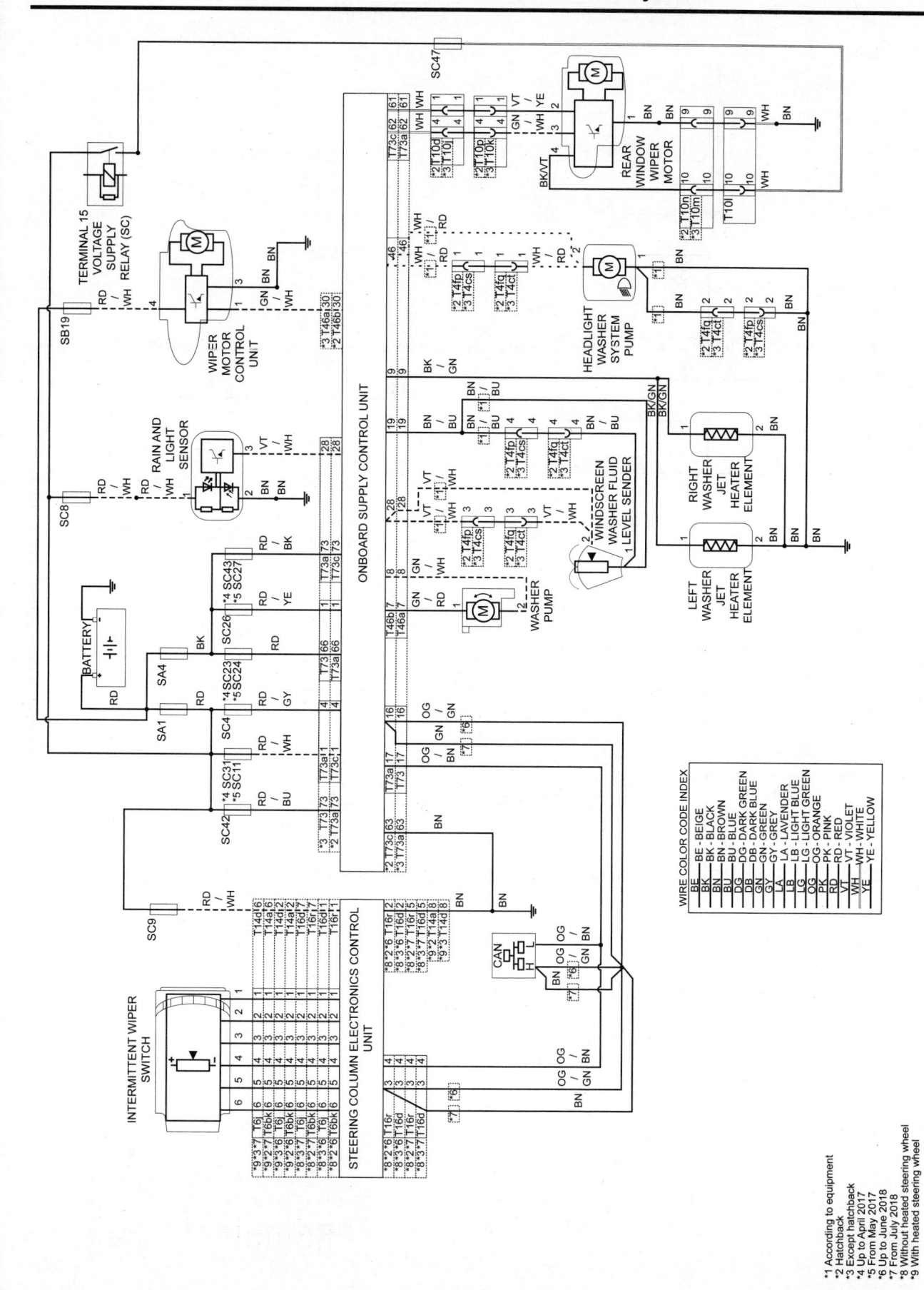

5a. Wiper Washer - GOLF

5b. Wiper Washer - JETTA Low equipment Up to June 2014

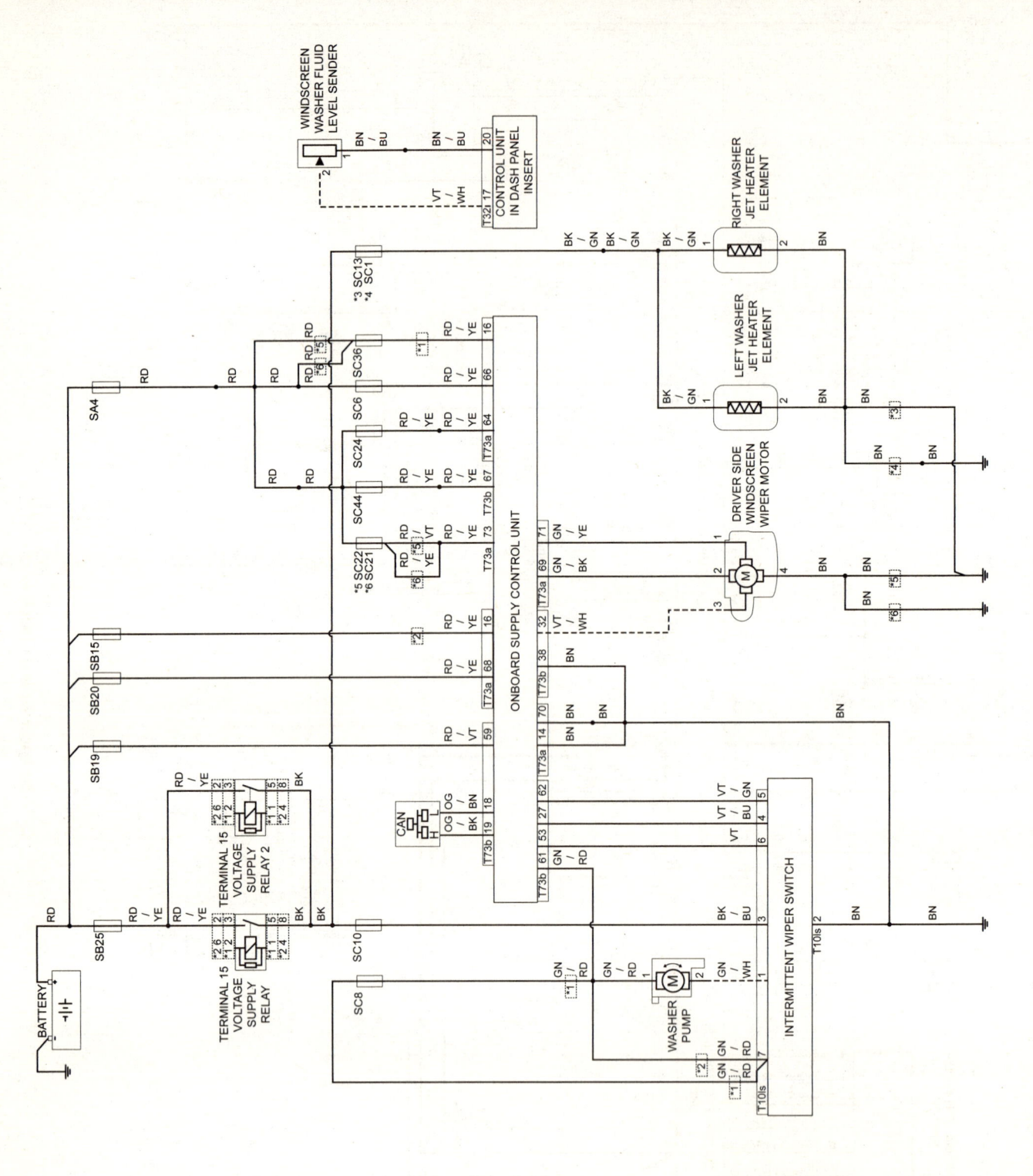

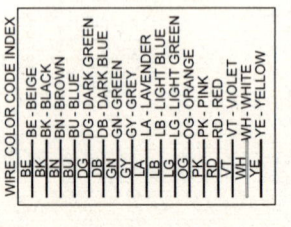

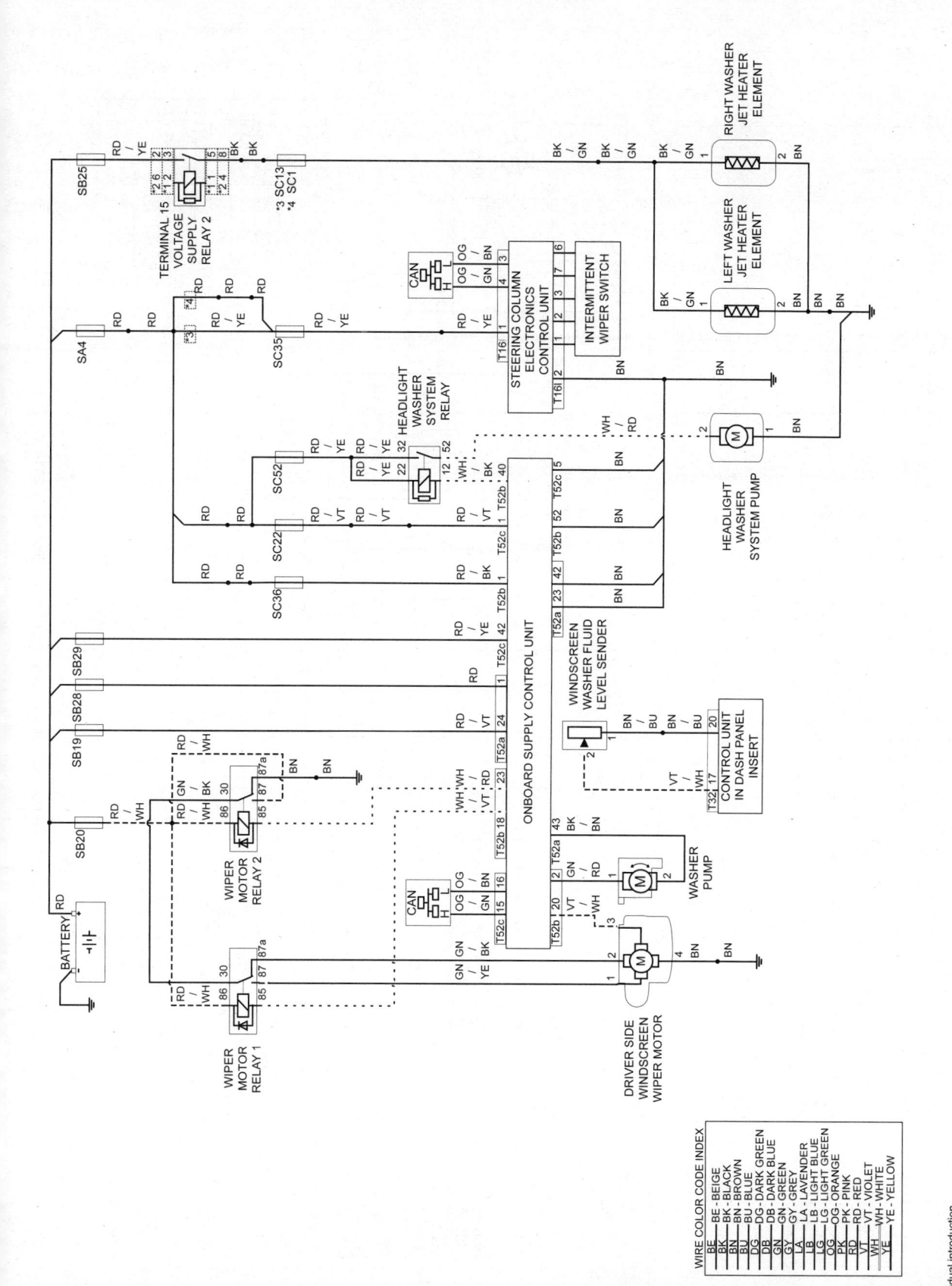

Chapter 12 Chassis electrical system

12-53

5c. Wiper Washer - JETTA High equipment Up to June 2014

WIRE COLOR CODE INDEX
BE - BEIGE
BK - BLACK
BN - BROWN
BU - BLUE
DG - DARK GREEN
DB - DARK BLUE
GN - GREEN
GY - GREY
LA - LAVENDER
LB - LIGHT BLUE
LG - LIGHT GREEN
OG - ORANGE
PK - PINK
RD - RED
VT - VIOLET
WH - WHITE
YE - YELLOW

*1 Gradual introduction
*2 Gradually discontinued
*3 Up to May 2011
*4 From June 2011

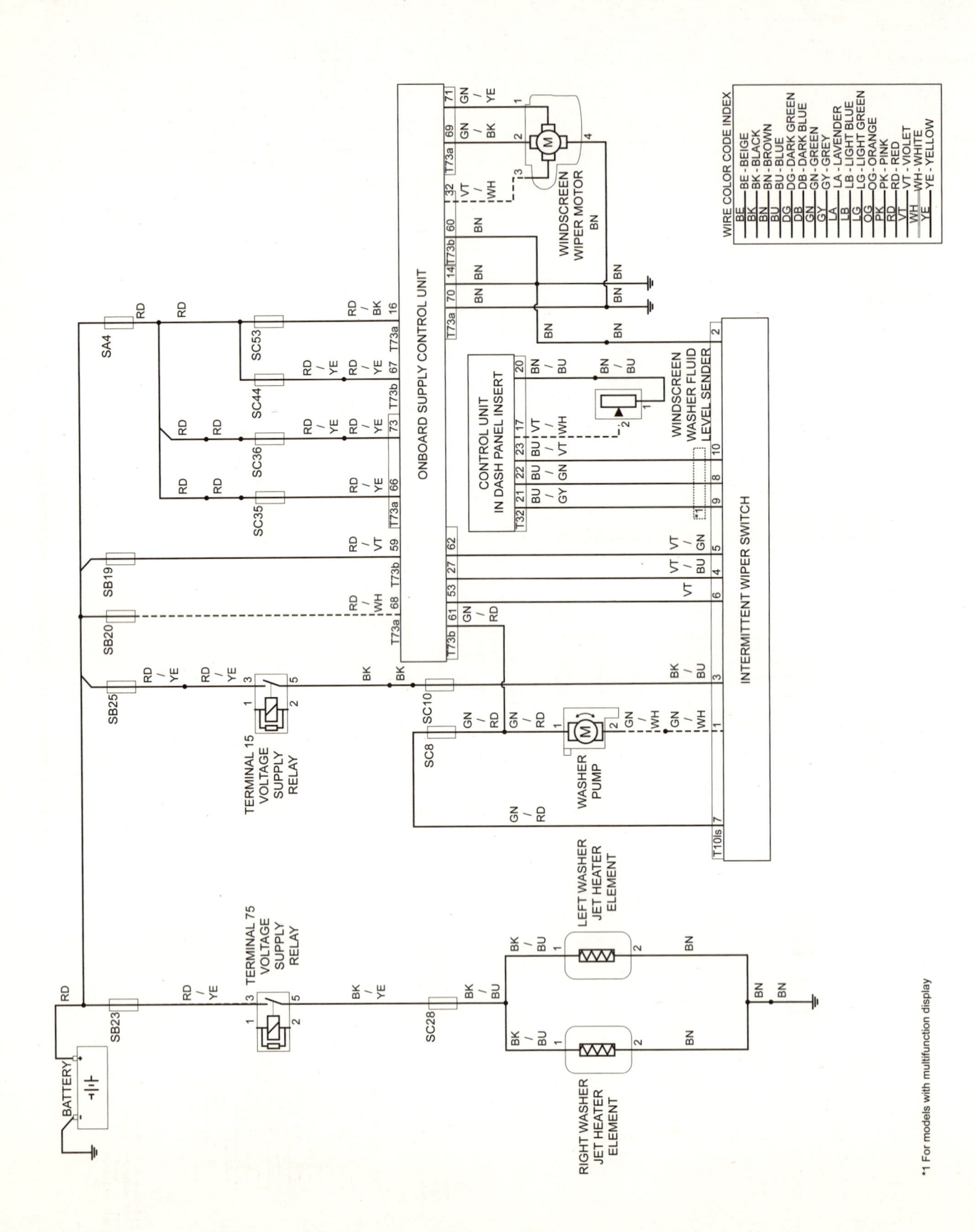

5d. Wiper Washer - JETTA Low equipment From July 2014

Chapter 12 Chassis electrical system

5e. Wiper Washer - JETTA High equipment From July 2014

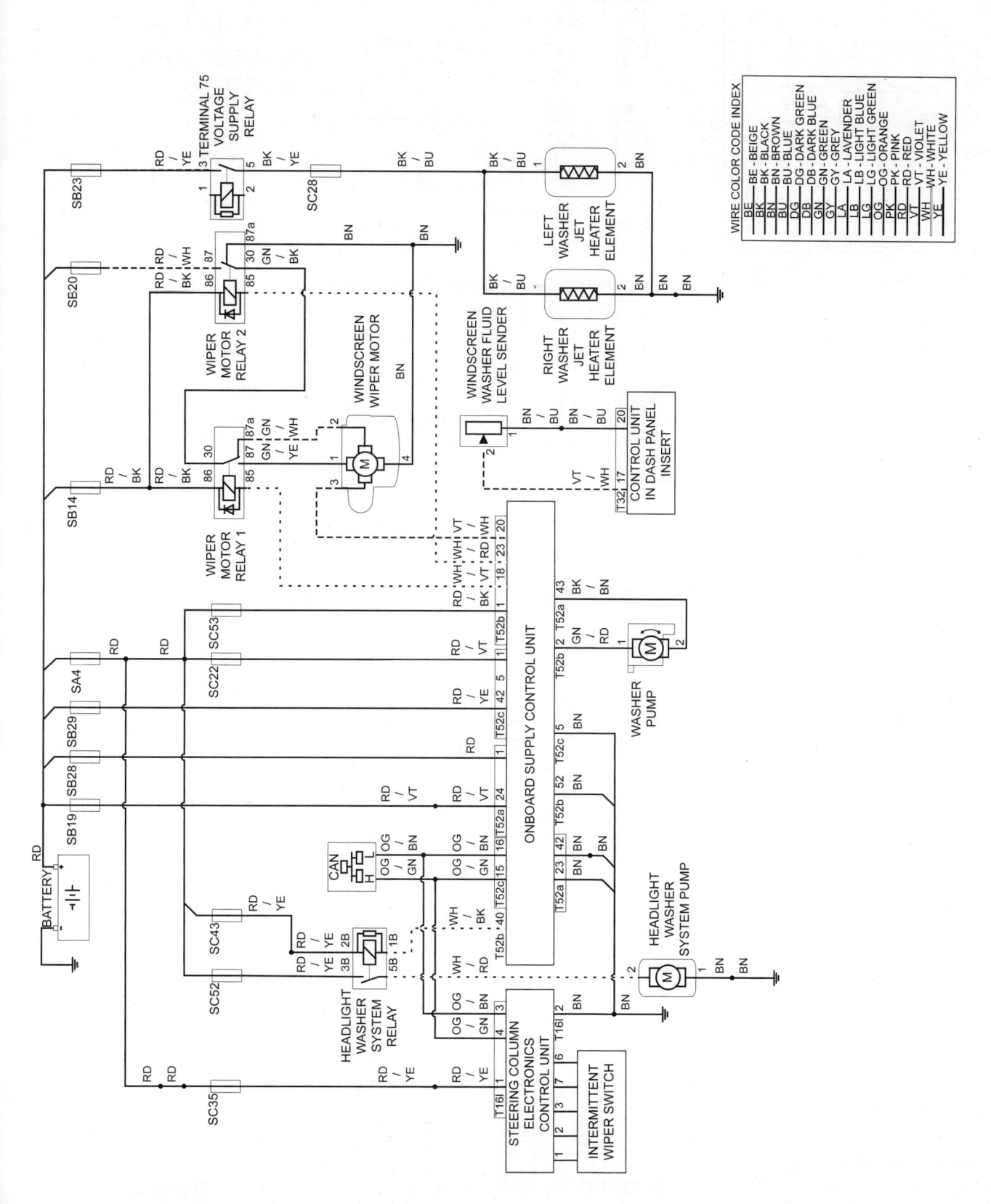

Chapter 12 Chassis electrical system

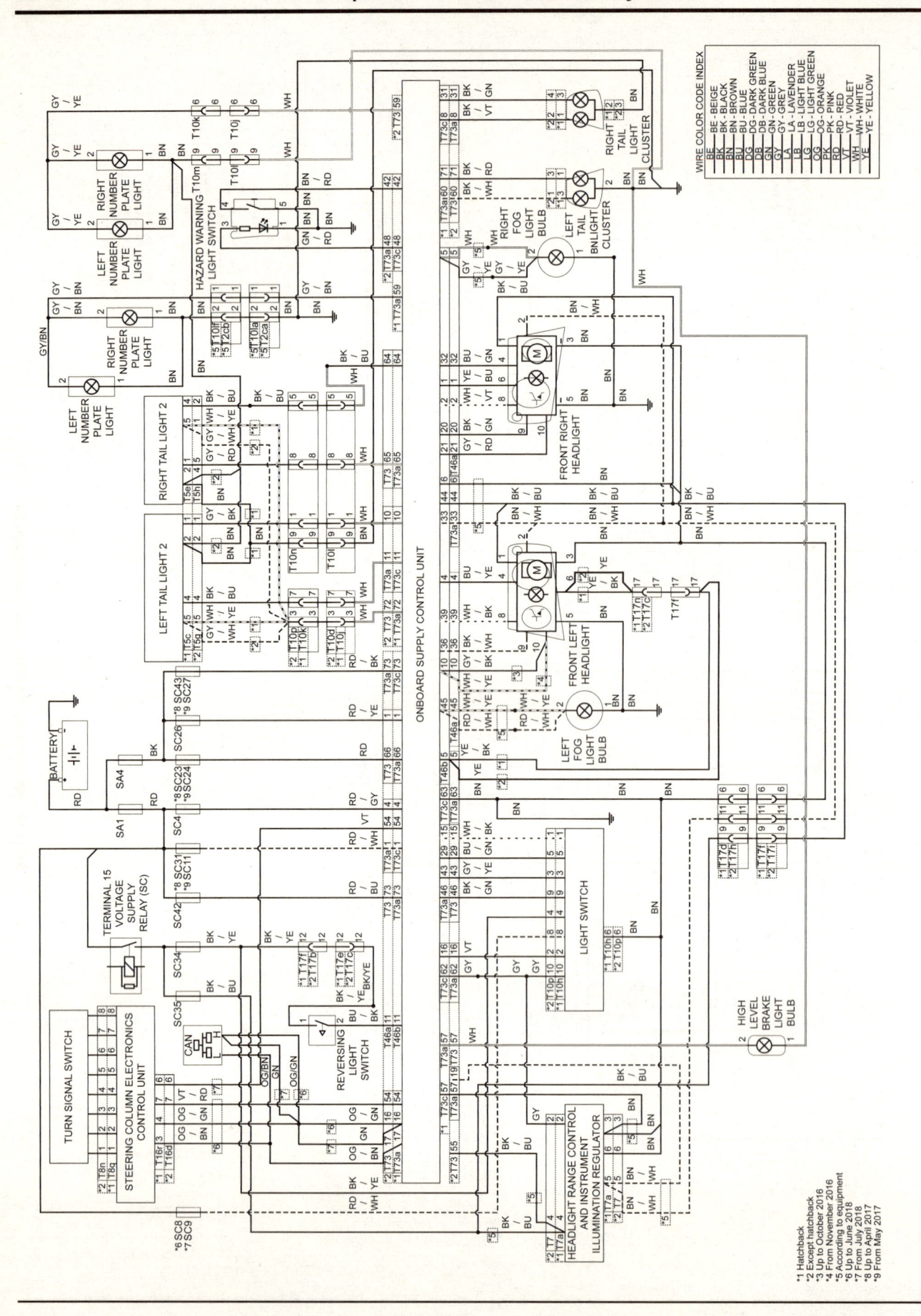

6a. Exterior Lights - GOLF

*1 Hatchback
*2 Except hatchback
*3 Up to October 2016
*4 From November 2016
*5 According to equipment
*6 Up to June 2018
*7 From July 2018
*8 Up to April 2017
*9 From May 2017

Chapter 12 Chassis electrical system

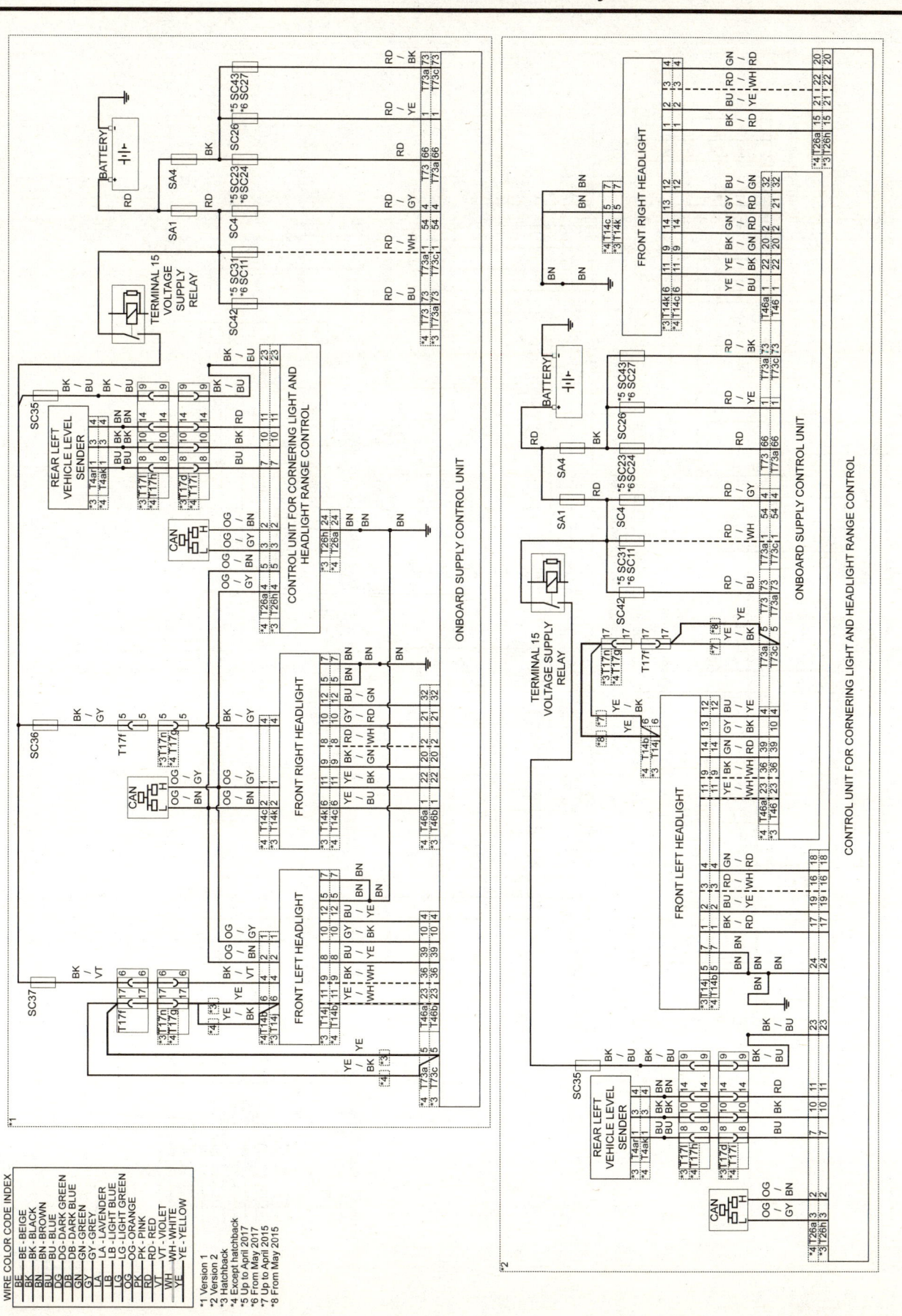

6b. Exterior Lights - GOLF Bixenon

12-58 Chapter 12 Chassis electrical system

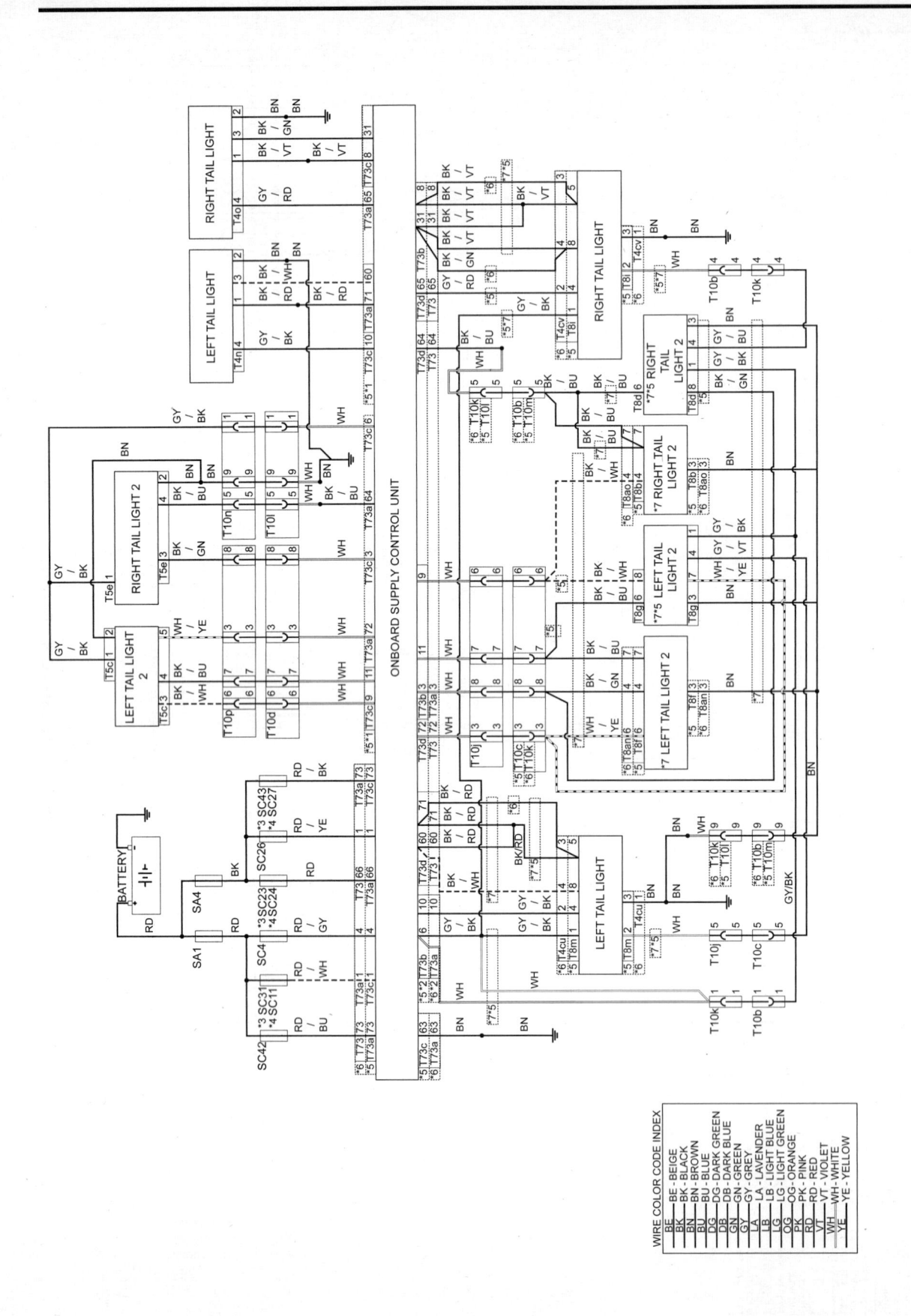

6c. Exterior Lights - GOLF LED tail lights

Chapter 12 Chassis electrical system 12-59

6d. Exterior lights - JETTA Low Equipment up to June 2014

*1 Gradual introduction
*2 Gradually discontinued
*3 With manual gearbox
*4 With automatic gearbox
*5 Without automatic anti-dazzle interior mirror
*6 With automatic anti-dazzle interior mirror
*7 Up to May 2011
*8 From June 2011
*9 According to equipment
*10 Up to June 2012
*11 From July 2012

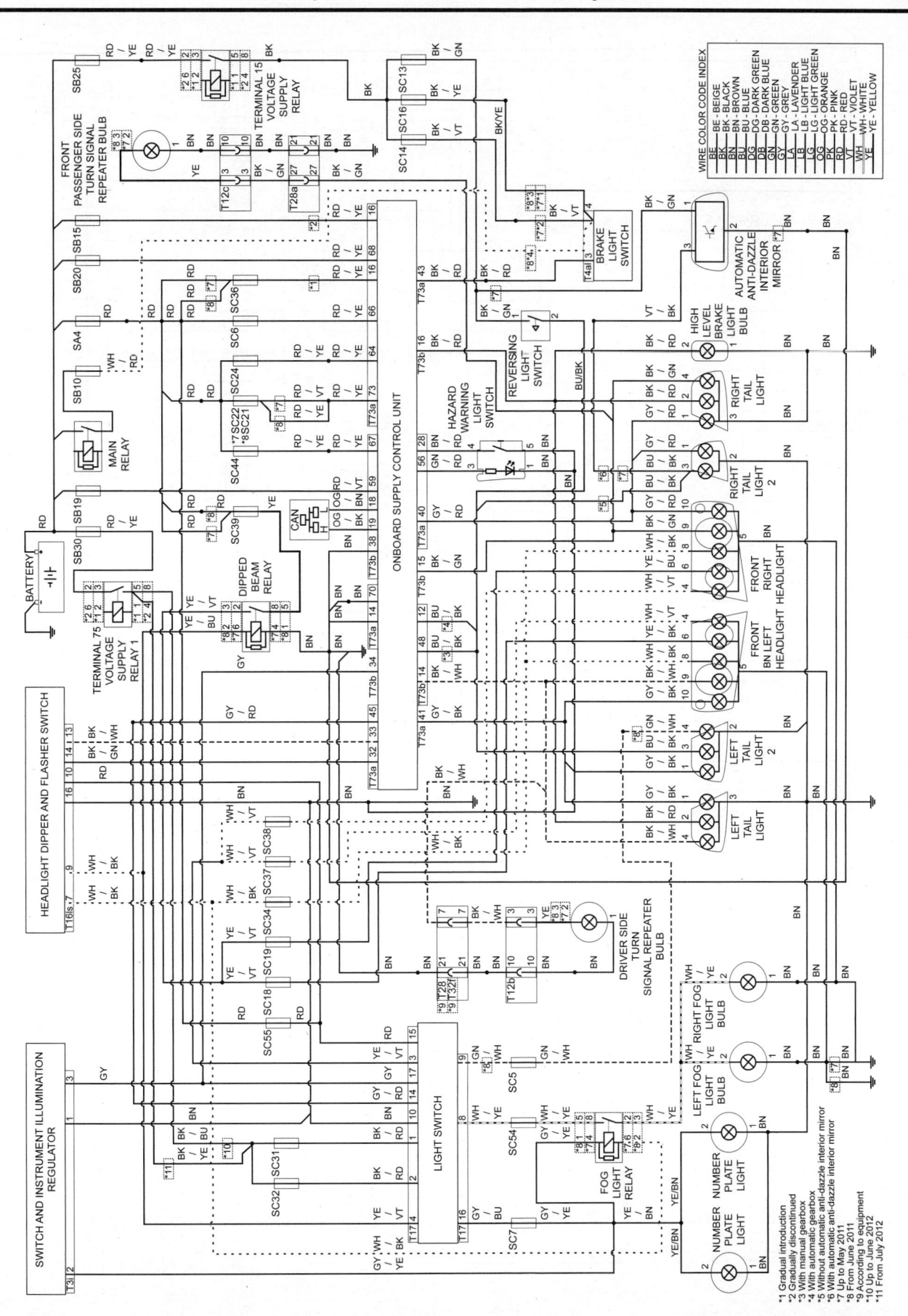

12-60 Chapter 12 Chassis electrical system

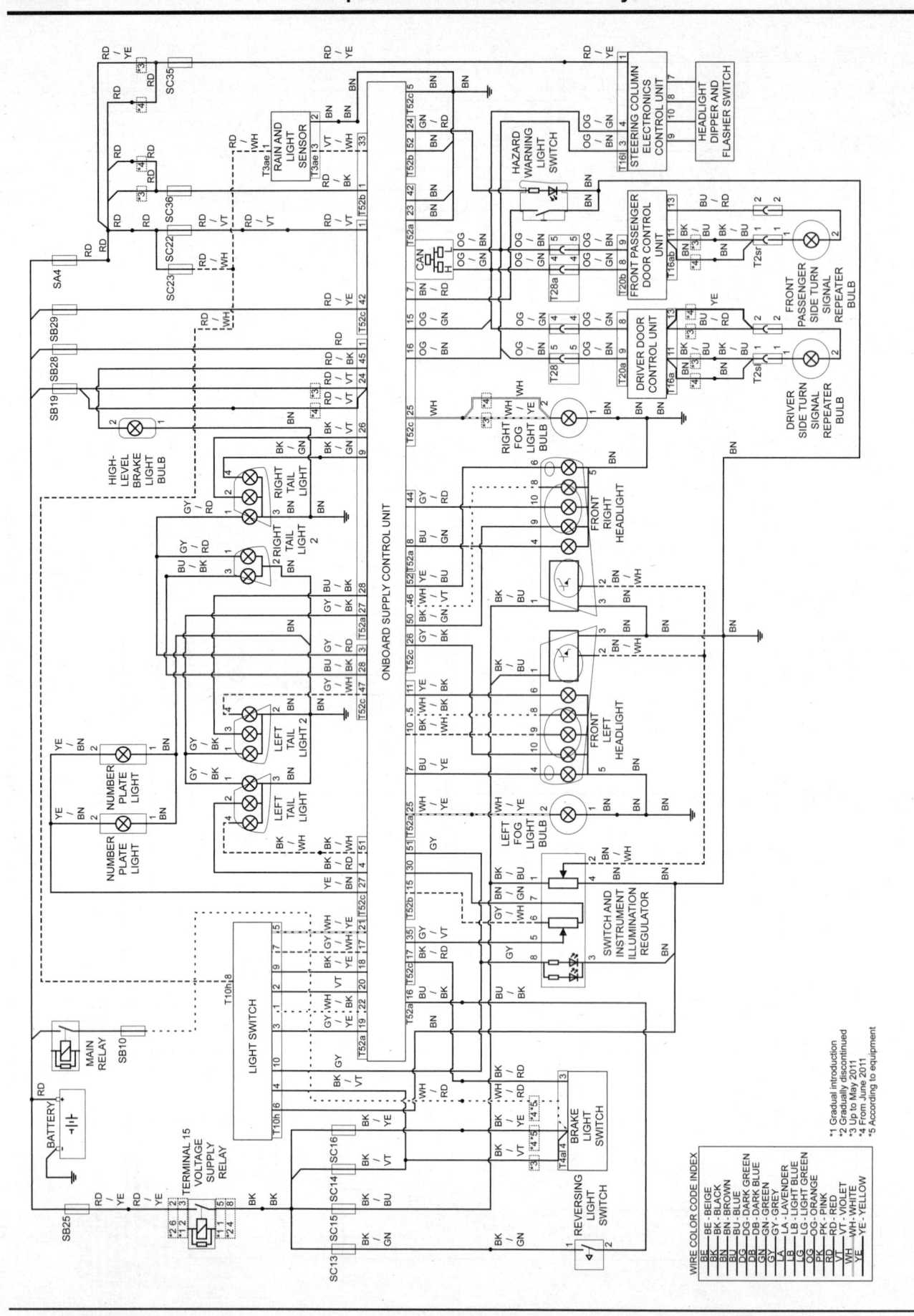

6e. Exterior lights - JETTA High Equipment up to June 2014

Chapter 12 Chassis electrical system

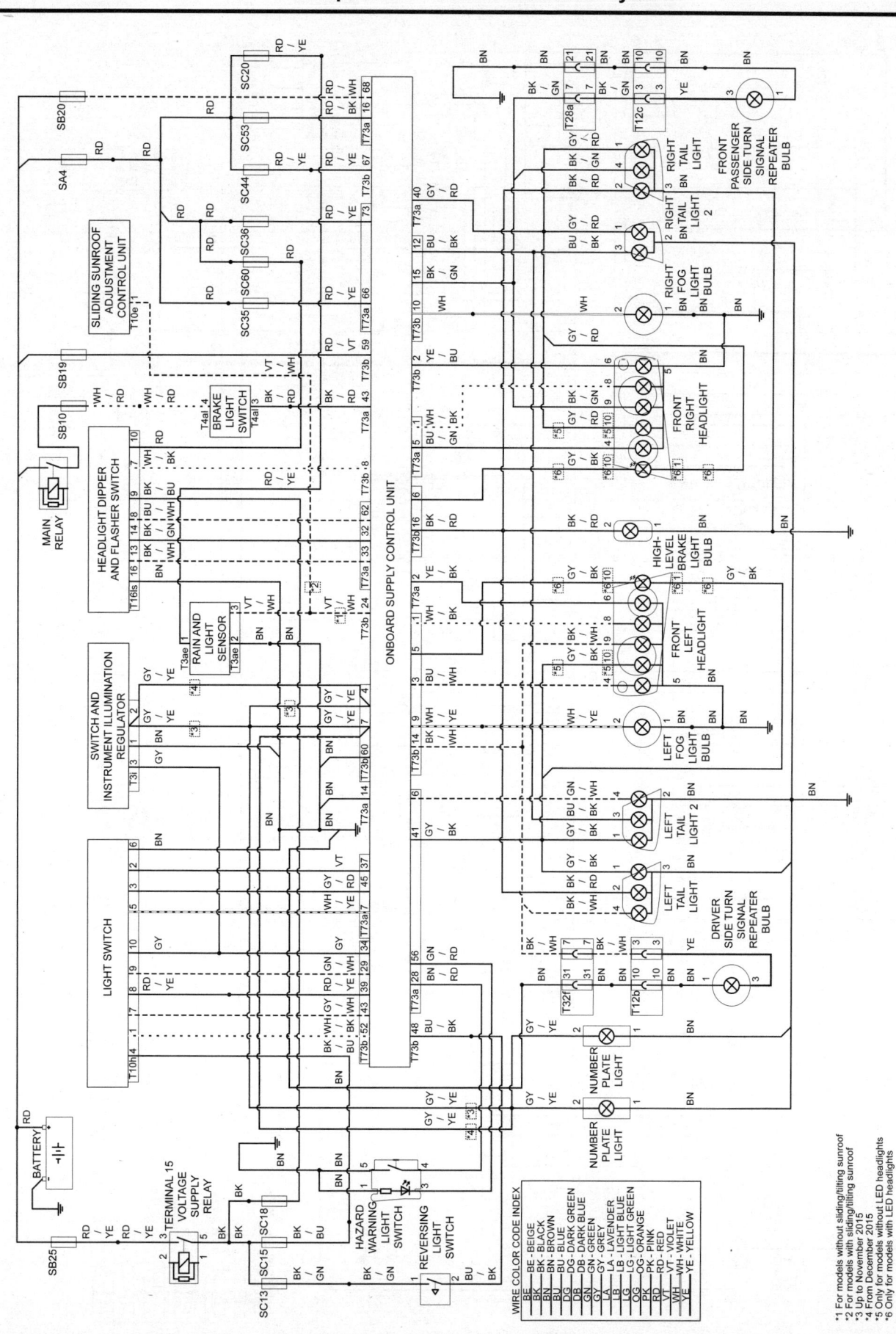

6f. Exterior lights - JETTA Low Equipment from July 2014

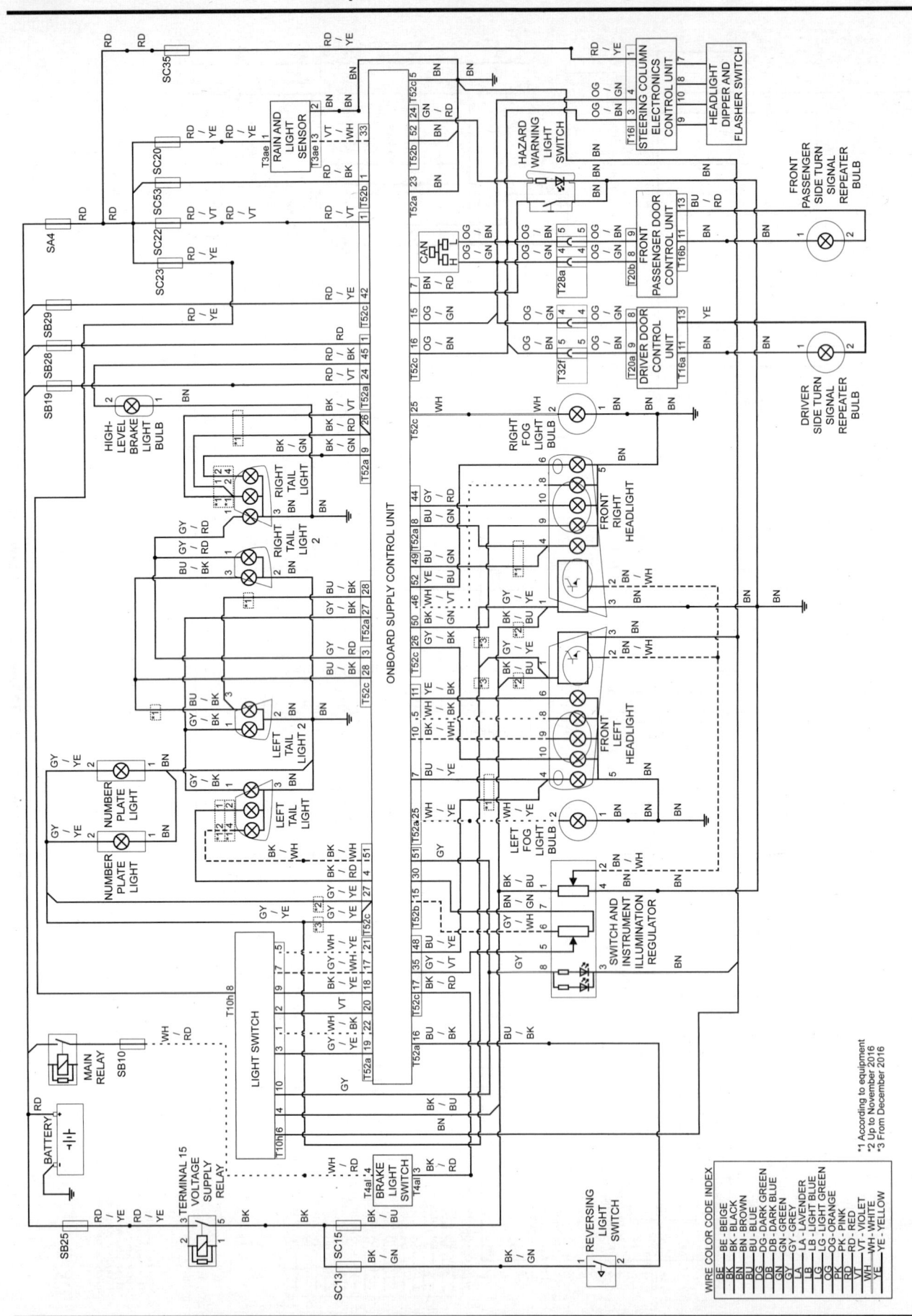

6g. Exterior lights - JETTA High Equipment from July 2014

Chapter 12 Chassis electrical system

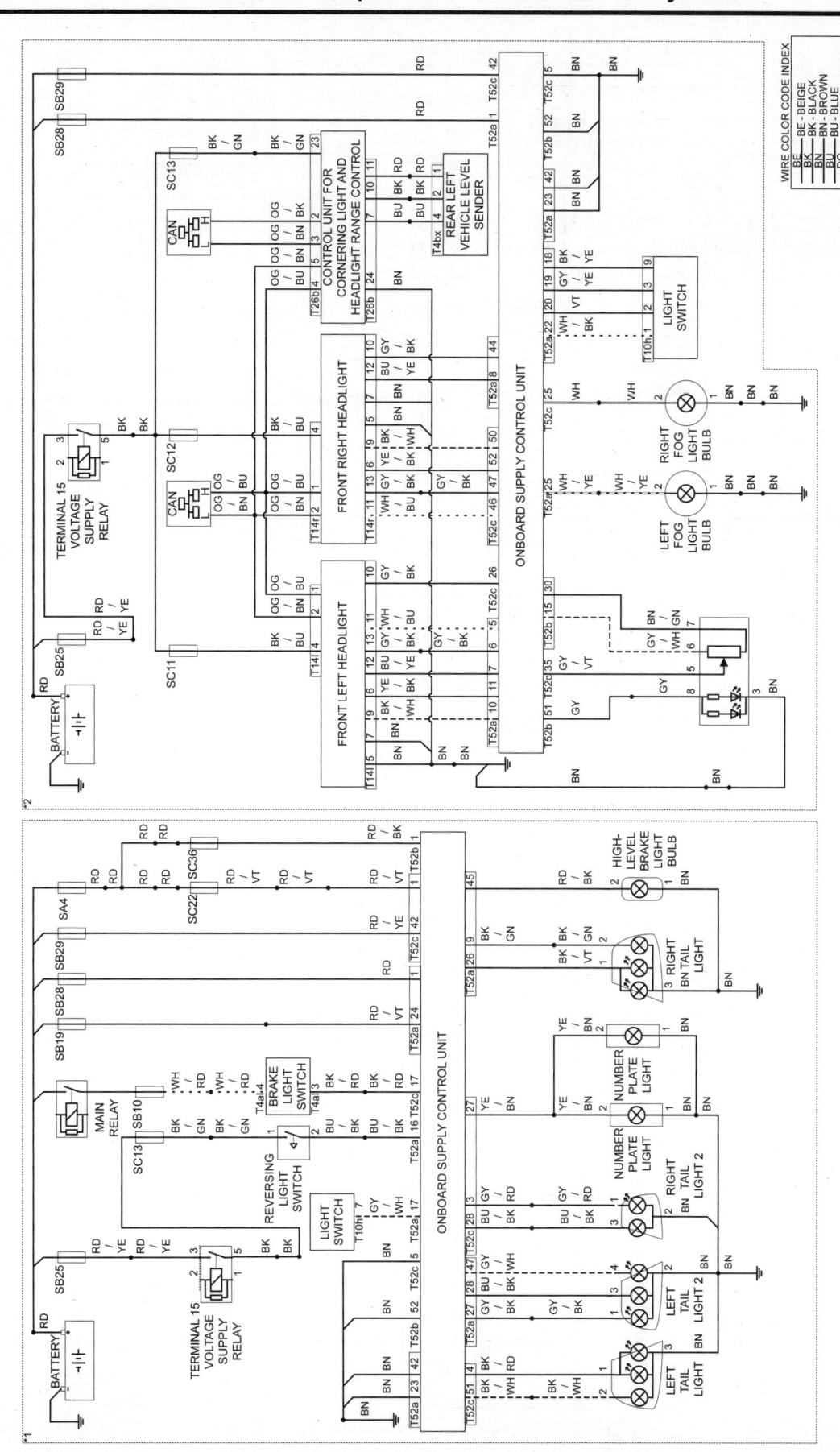

12-63

6h. Exterior lights - JETTA Tail cluster with LED and Xenon lights Up to June 2014

*1 Tail light cluster with LED
*2 Gas discharge headlights (bi-xenon) with automatic headlight range control, cornering light, LED daytime running lights

12-64 Chapter 12 Chassis electrical system

6i. Exterior lights - JETTA Tail cluster with LED and Xenon lights From July 2014

*1 Tail light cluster with LED
*2 Gas discharge headlights (bi-xenon) with automatic headlight range control, cornering light, LED daytime running lights
*3 According to equipment
*4 Only models with "Low" equipment (AW0)
*5 Only models with "High" equipment (AW1)

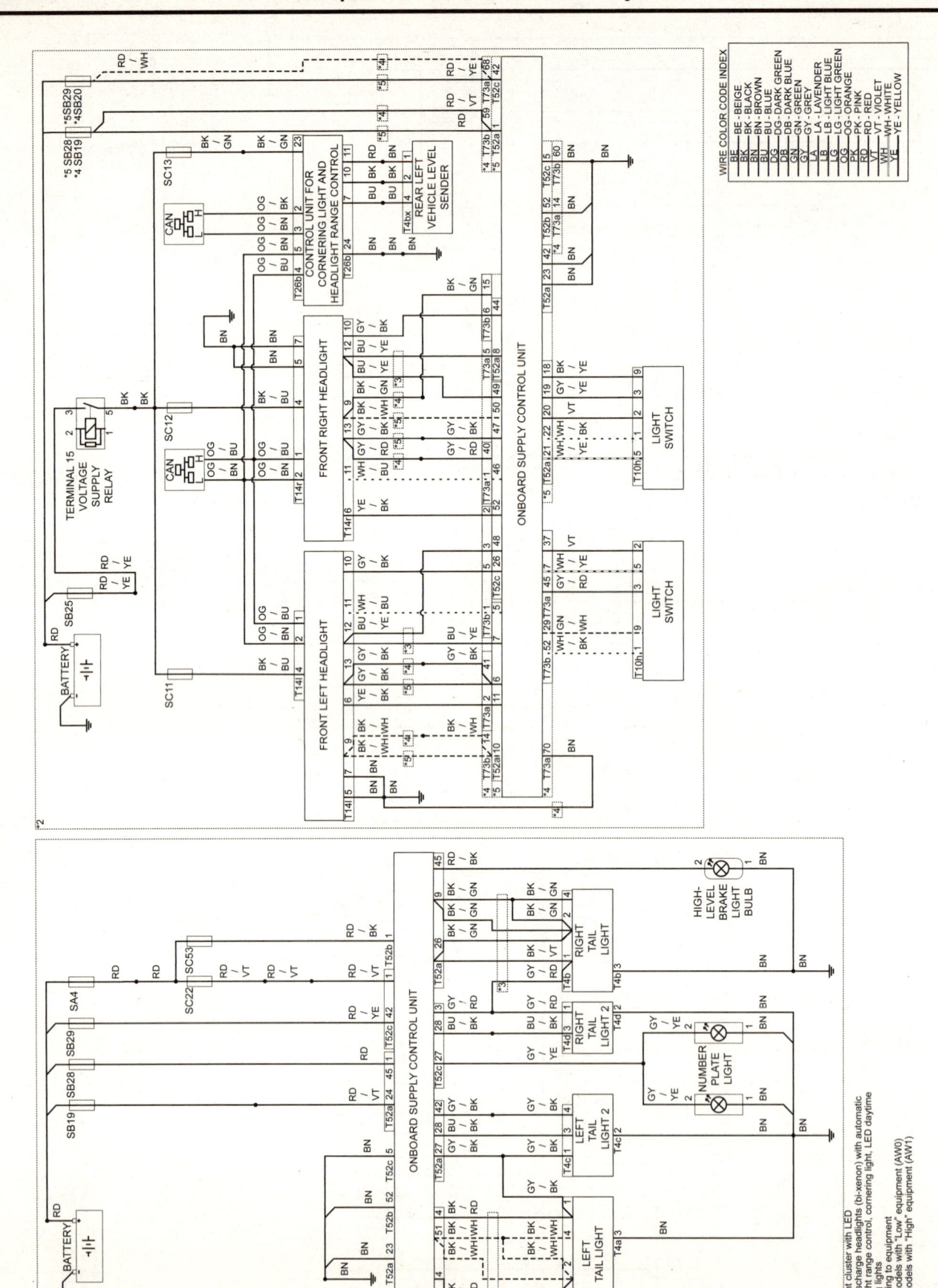

Chapter 12 Chassis electrical system

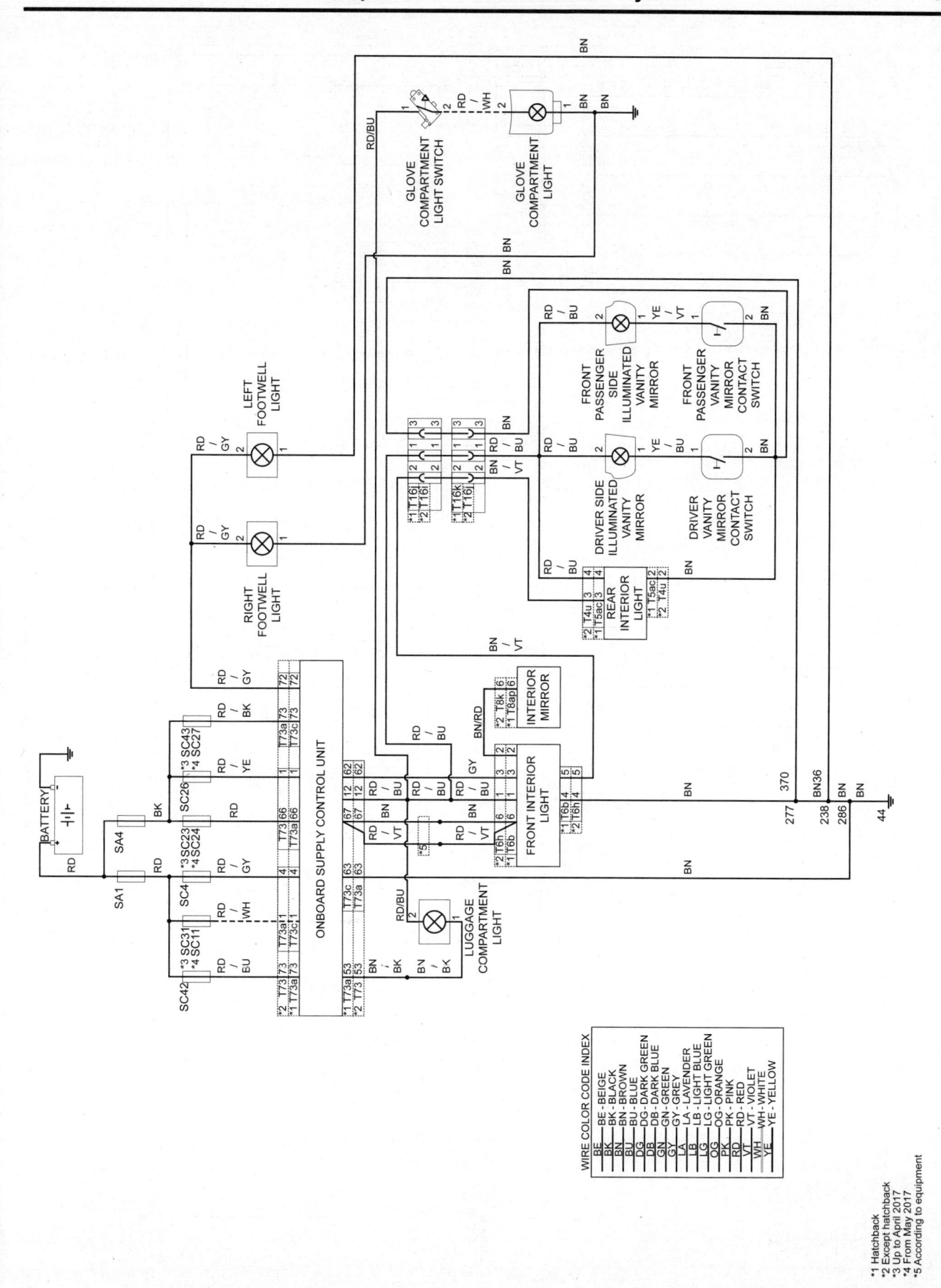

7a. Interior Lights - GOLF

WIRE COLOR CODE INDEX
BE - BEIGE
BK - BLACK
BN - BROWN
BU - BLUE
DG - DARK GREEN
DB - DARK BLUE
GN - GREEN
GY - GREY
LA - LAVENDER
LB - LIGHT BLUE
LG - LIGHT GREEN
OG - ORANGE
PK - PINK
RD - RED
VT - VIOLET
WH - WHITE
YE - YELLOW

*1 Hatchback
*2 Except hatchback
*3 Up to April 2017
*4 From May 2017
*5 According to equipment

12-66 Chapter 12 Chassis electrical system

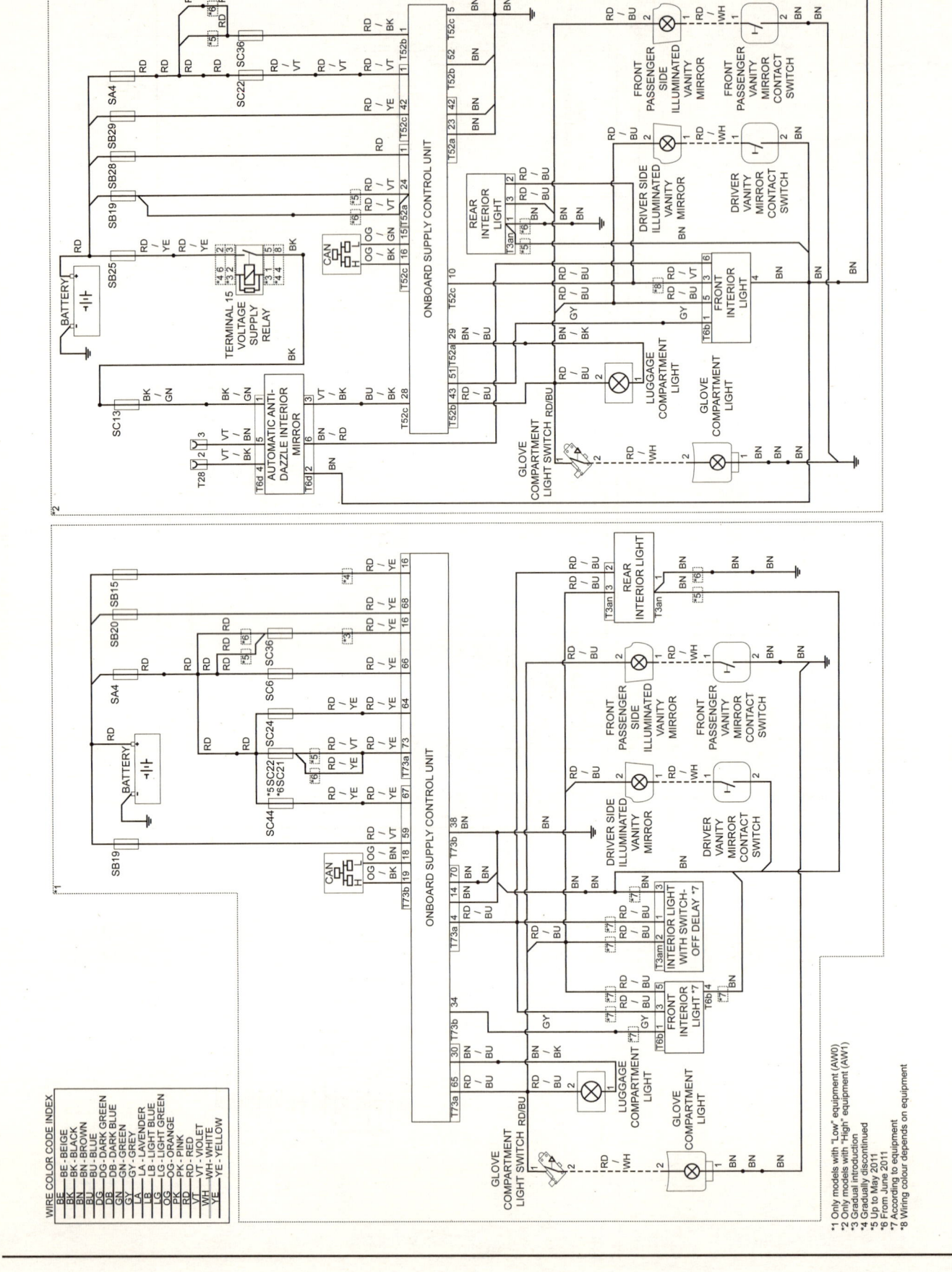

7b. Interior lights - JETTA Up to June 2014 - B&W

Chapter 12 Chassis electrical system

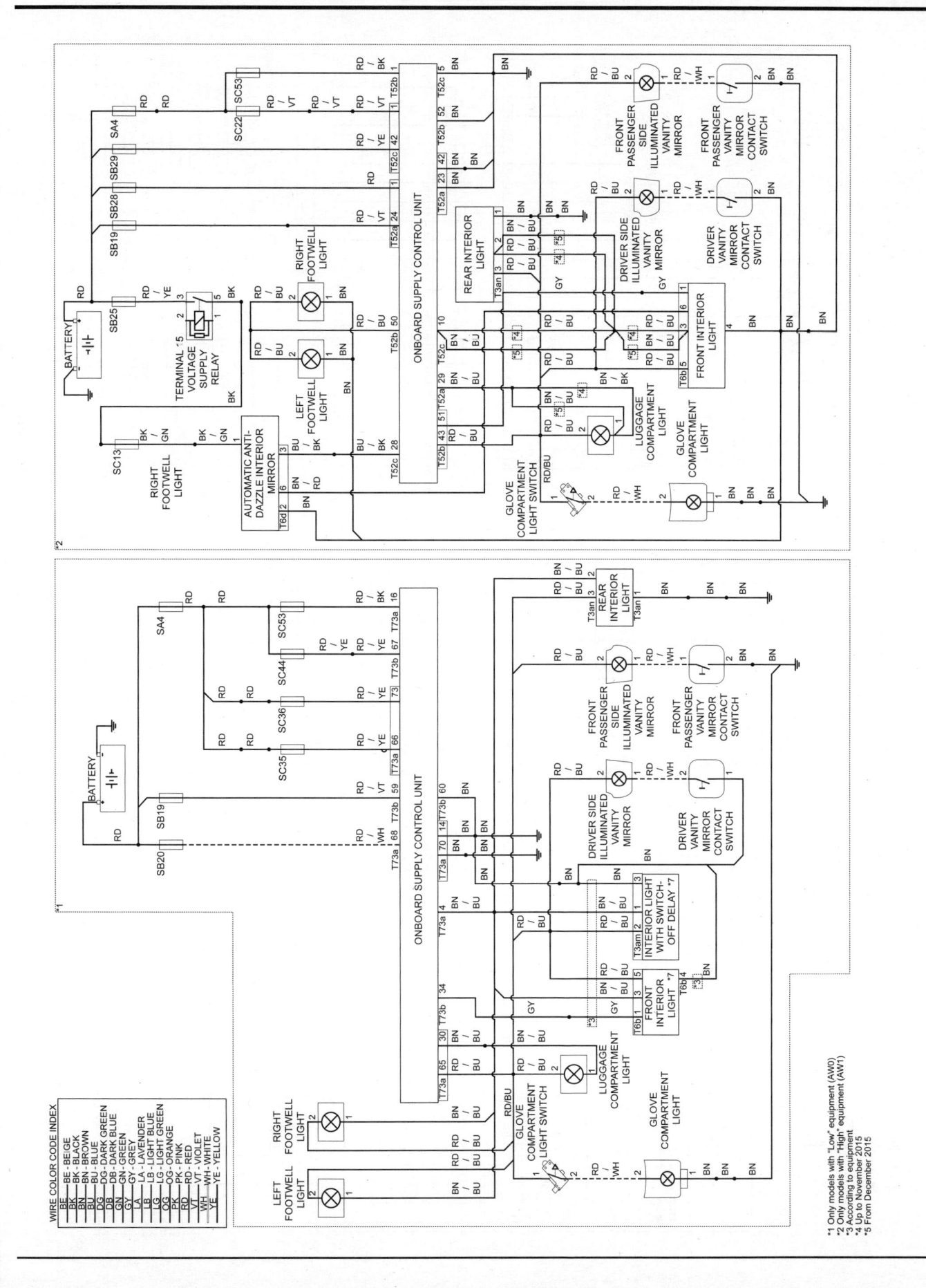

12-67

7c. Interior lights - JETTA From July 2014 - B&W

*1 Only models with "Low" equipment (AW0)
*2 Only models with "High" equipment (AW1)
*3 According to equipment
*4 Up to November 2015
*5 From December 2015

Chapter 12 Chassis electrical system

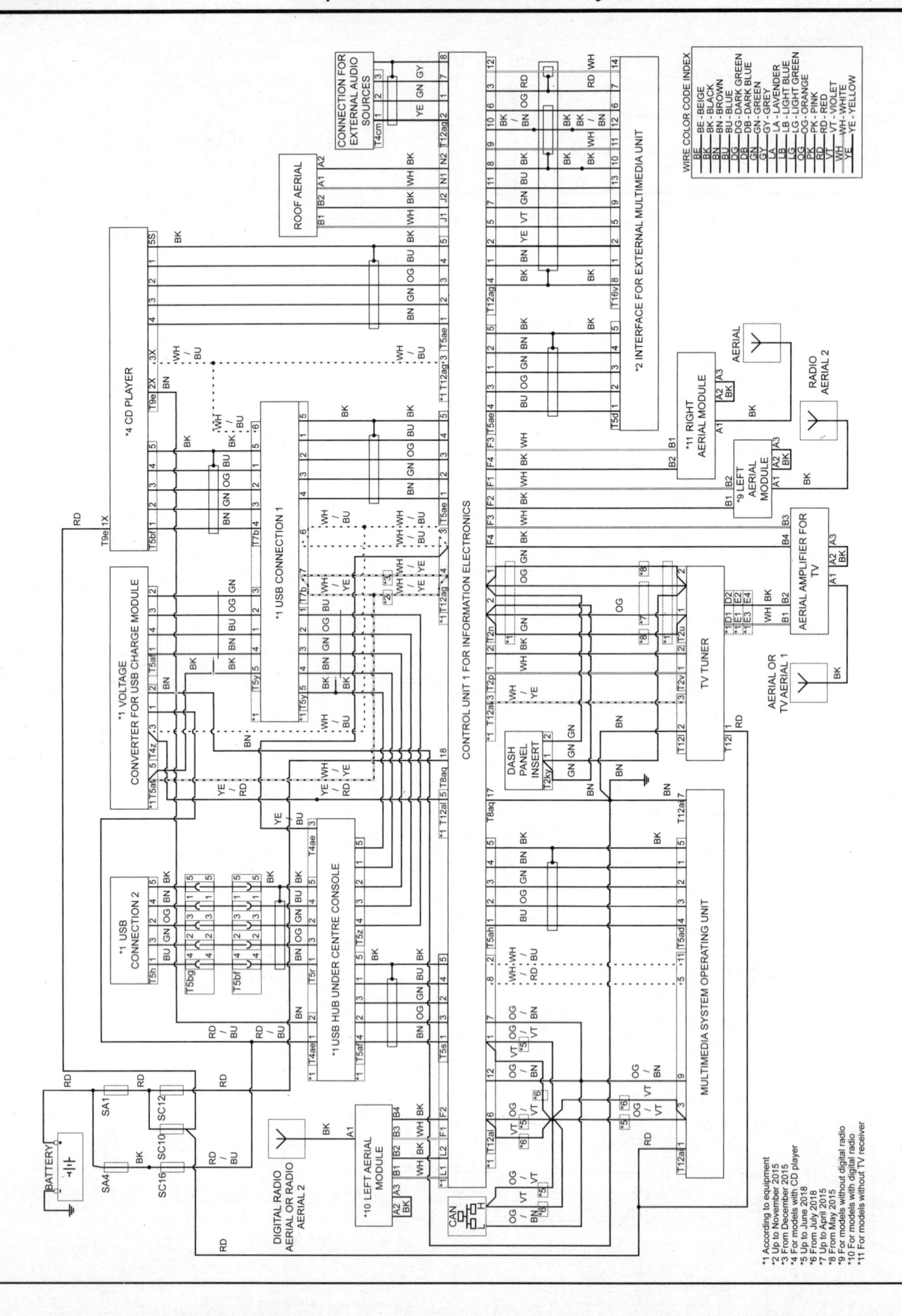

8a. Sound System - GOLF Radio and Navigation systems - Hatchback

Chapter 12 Chassis electrical system

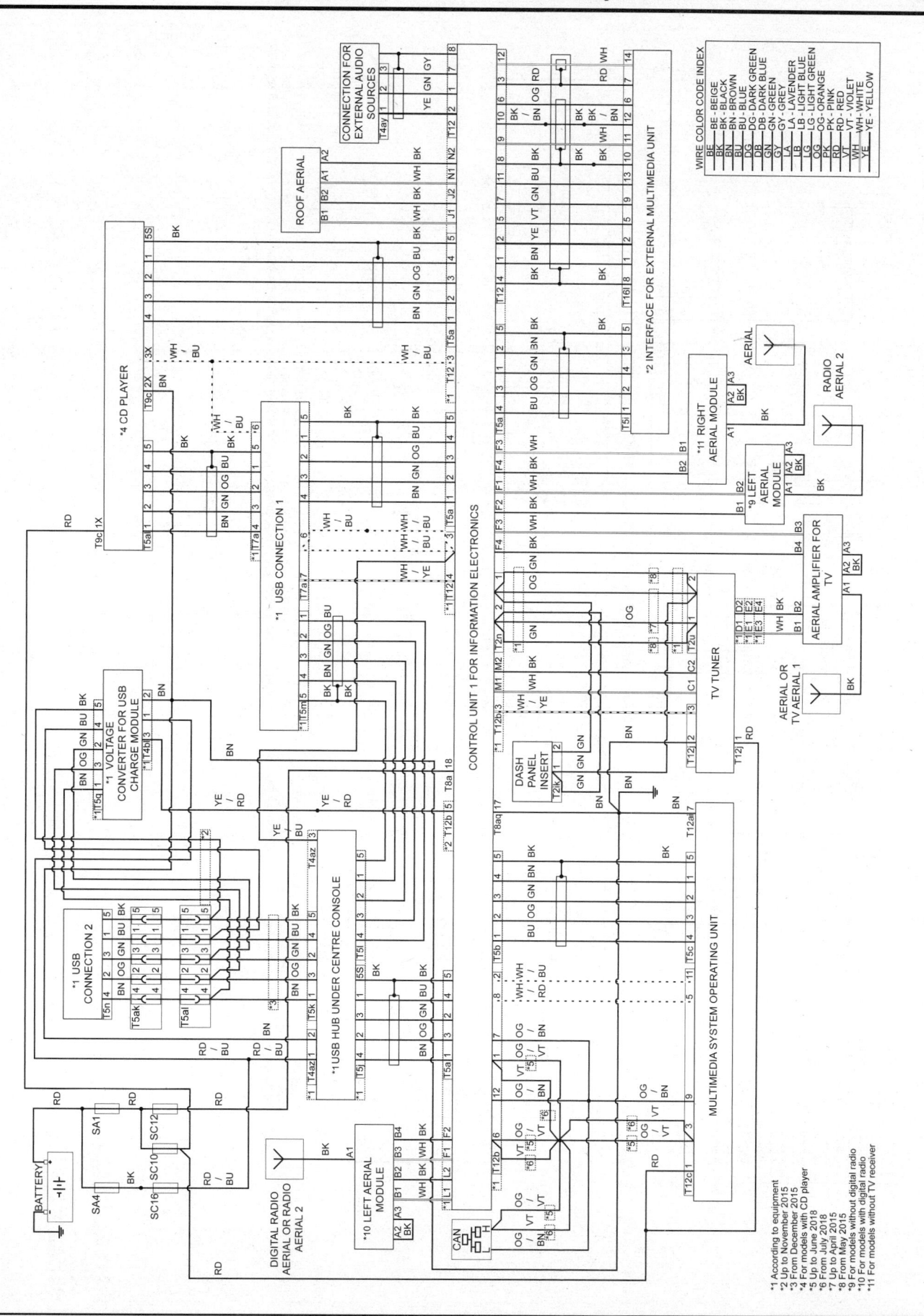

12-69

8b. Sound System - GOLF Radio and Navigation systems - Variant

*1 According to equipment
*2 Up to November 2015
*3 From December 2015
*4 For models with CD player
*5 Up to June 2018
*6 From July 2018
*7 Up to April 2015
*8 From May 2015
*9 For models without digital radio
*10 For models with digital radio
*11 For models without TV receiver

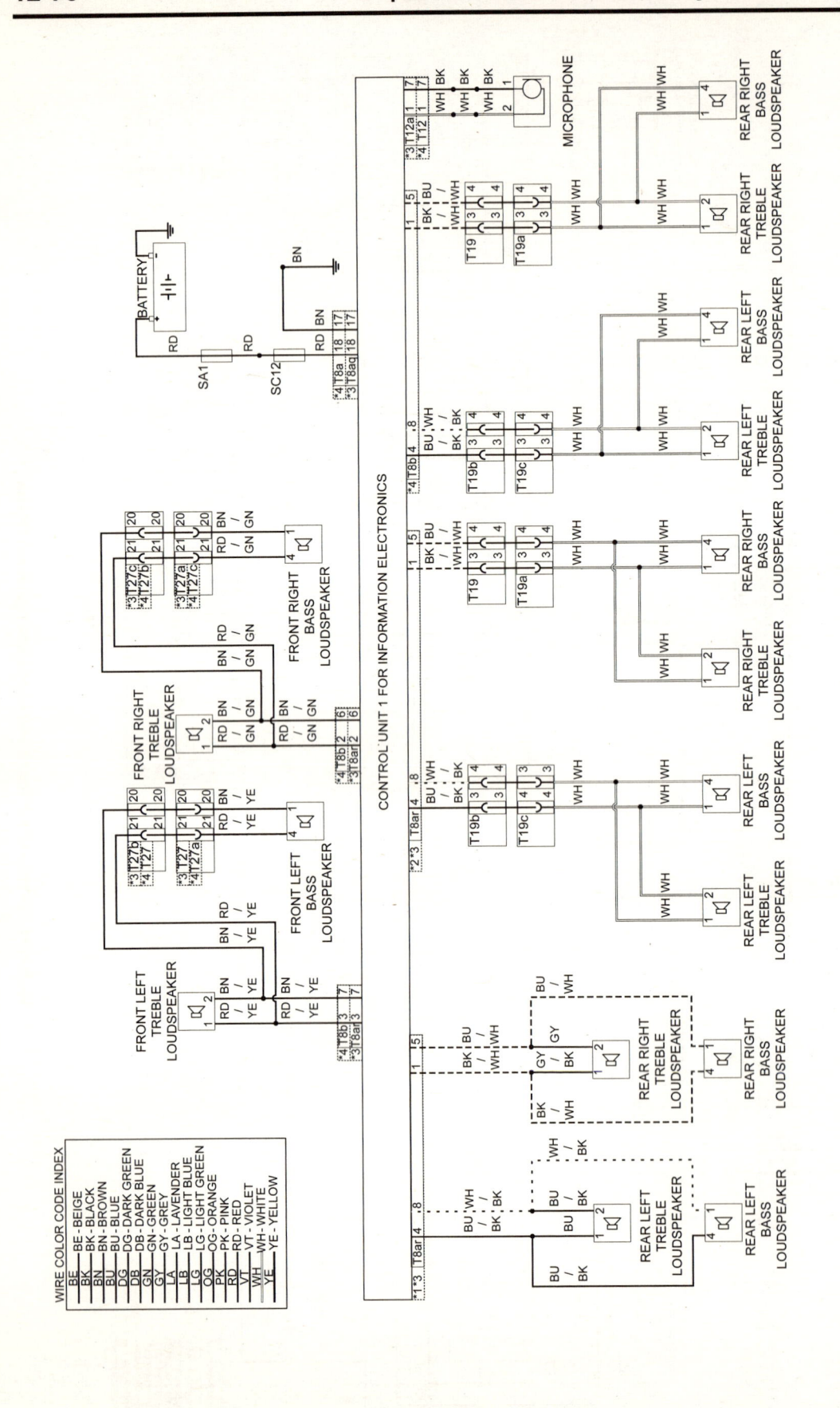

8c. Sound System - GOLF Speakers (without Dynaudio)

Chapter 12 Chassis electrical system

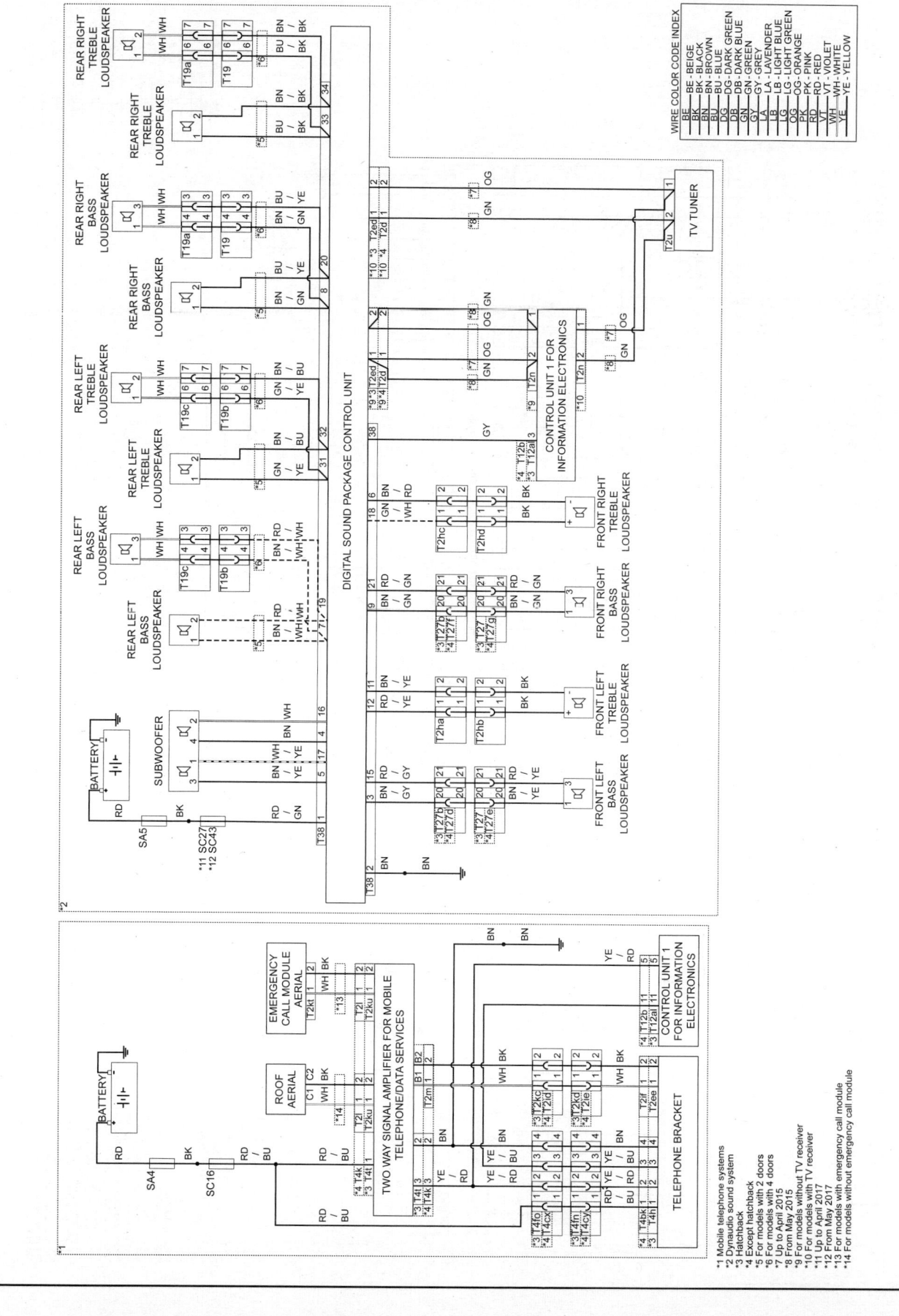

12-71

8d. Sound System - GOLF Speakers (with Dynaudio) and Mobile telephone systems

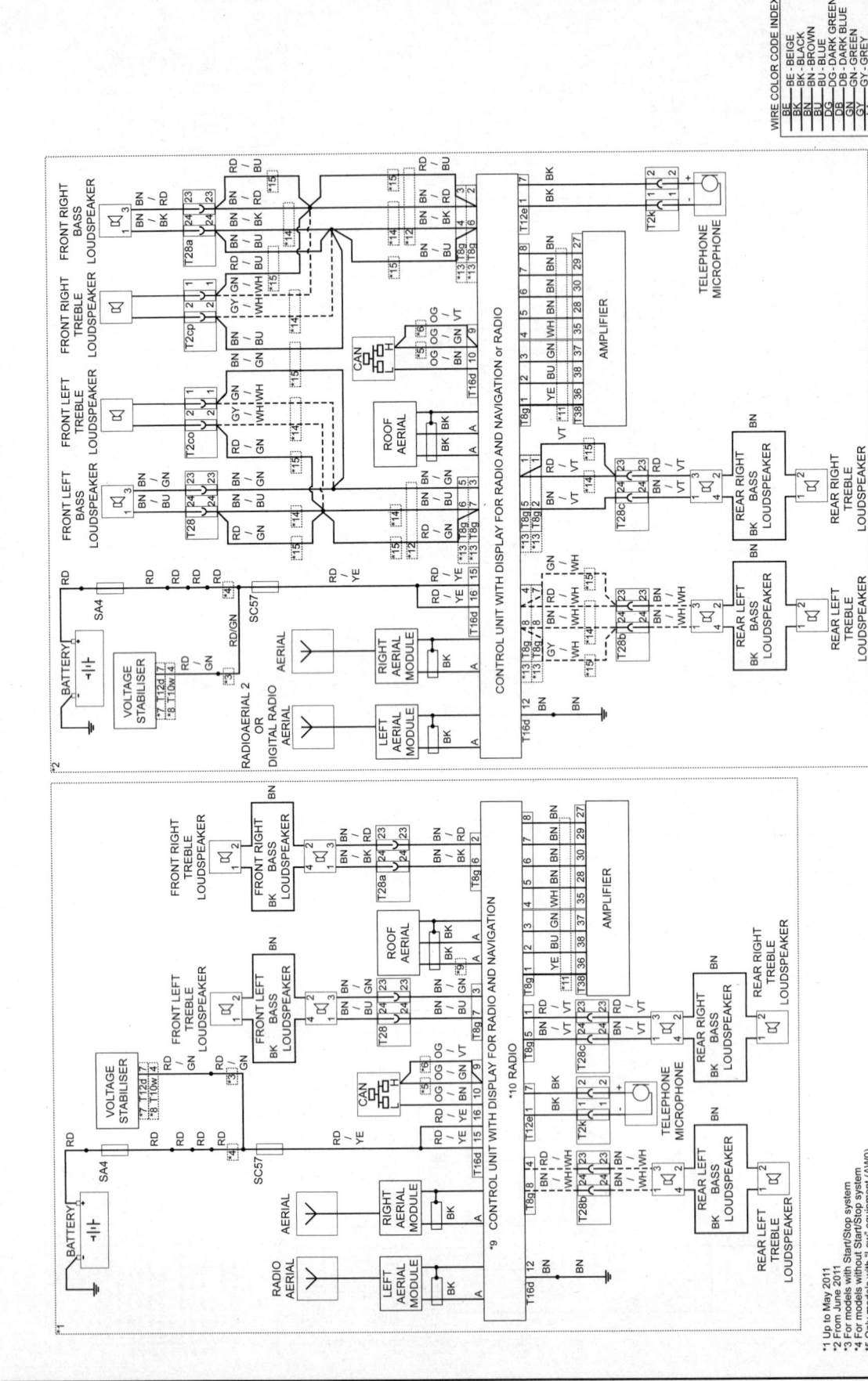

8e. Sound System - JETTA RNS and RCD without VW sound system front speakers (Up to June 2014)

*1 Up to May 2011
*2 From June 2011
*3 For models with Start/Stop system
*4 For models without Start/Stop system
*5 Only models with "Low" equipment (AW0)
*6 Only models with "High" equipment (AW1)
*7 Gradual introduction
*8 Gradually discontinued
*9 For Radio Navigation System RNS 315 and RNS 510
*10 For Radio System RCD 310 and RCD 510
*11 Only models with sound system
*12 Only models with no sound system
*13 According to equipment
*14 Up to May 2012
*15 From June 2012

Chapter 12 Chassis electrical system

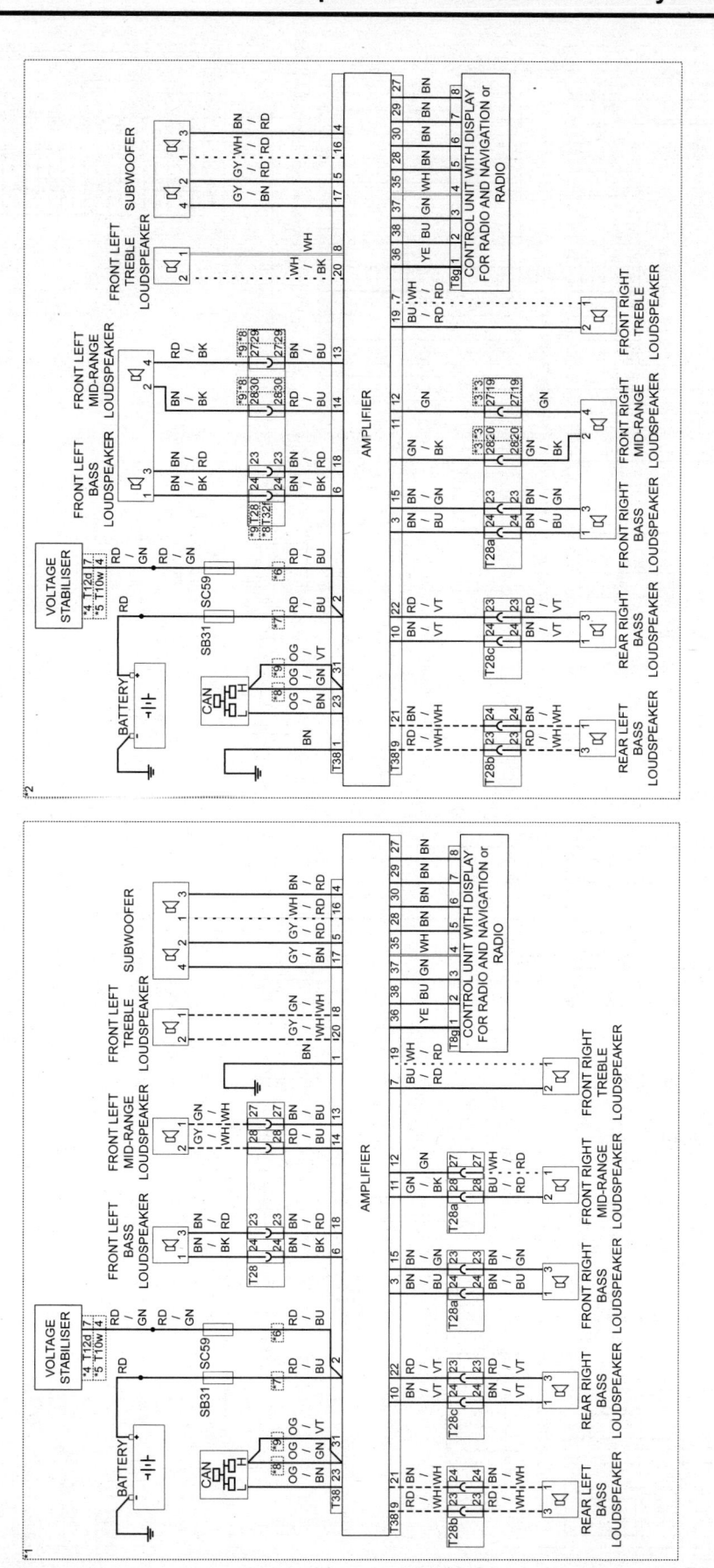

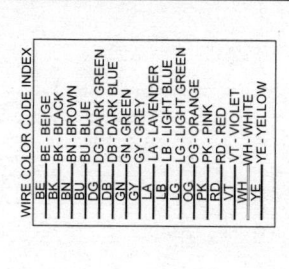

8f. Sound System - JETTA VW Sound System front speakers and Subwoofer (Up to June 2014)

*1 Up to May 2011
*2 From June 2011
*3 According to equipment
*4 Gradual introduction
*5 Gradually discontinued
*6 For models with Start/Stop system
*7 For models without Start/Stop system
*8 Only models with "Low" equipment (AW0)
*9 Only models with "High" equipment (AW1)

8g. Sound System - JETTA RNS and RCD without VW sound system front speakers (From July 2014)

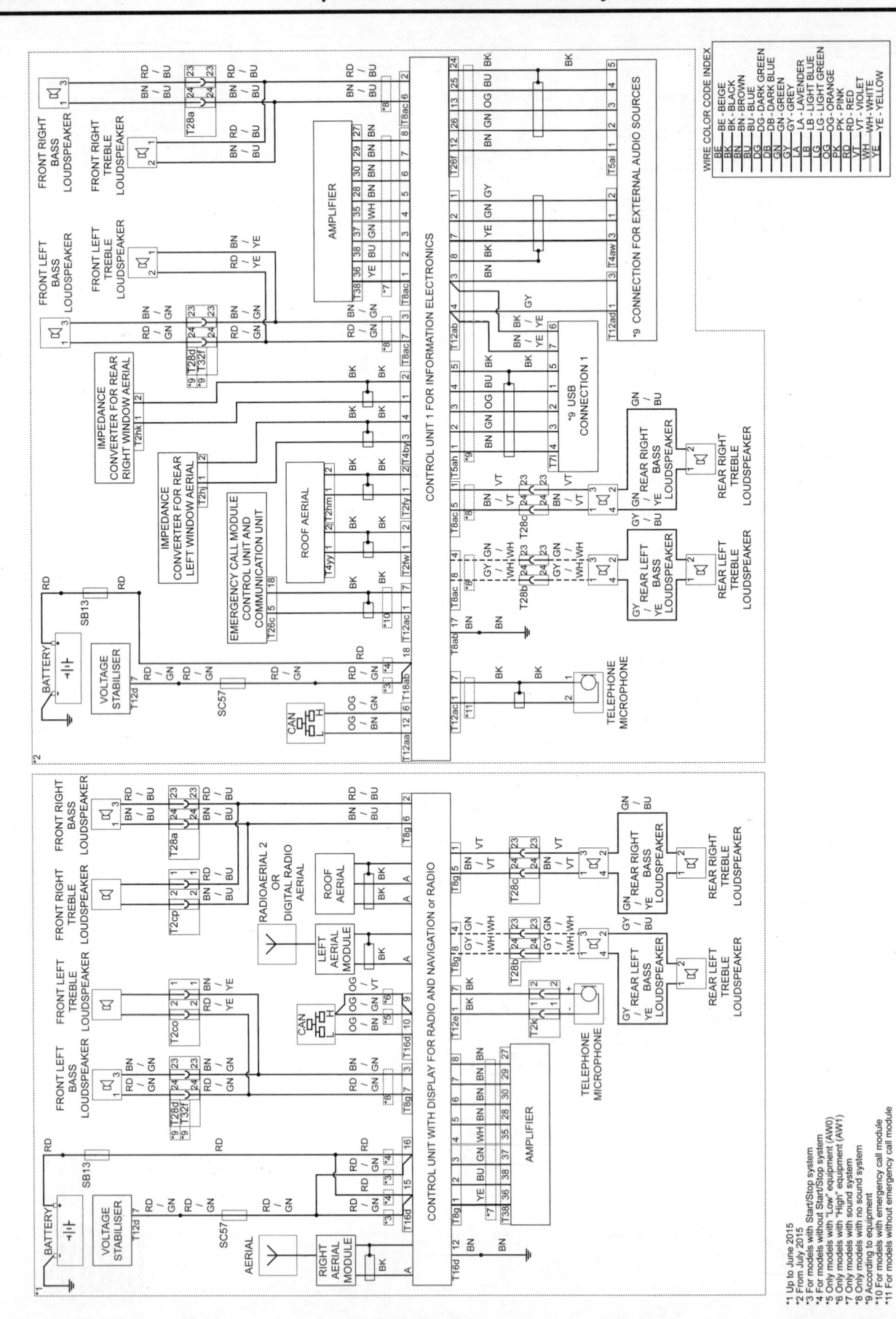

Chapter 12 Chassis electrical system

8h. Sound System - JETTA VW Sound System front speakers and Subwoofer (From July 2014)

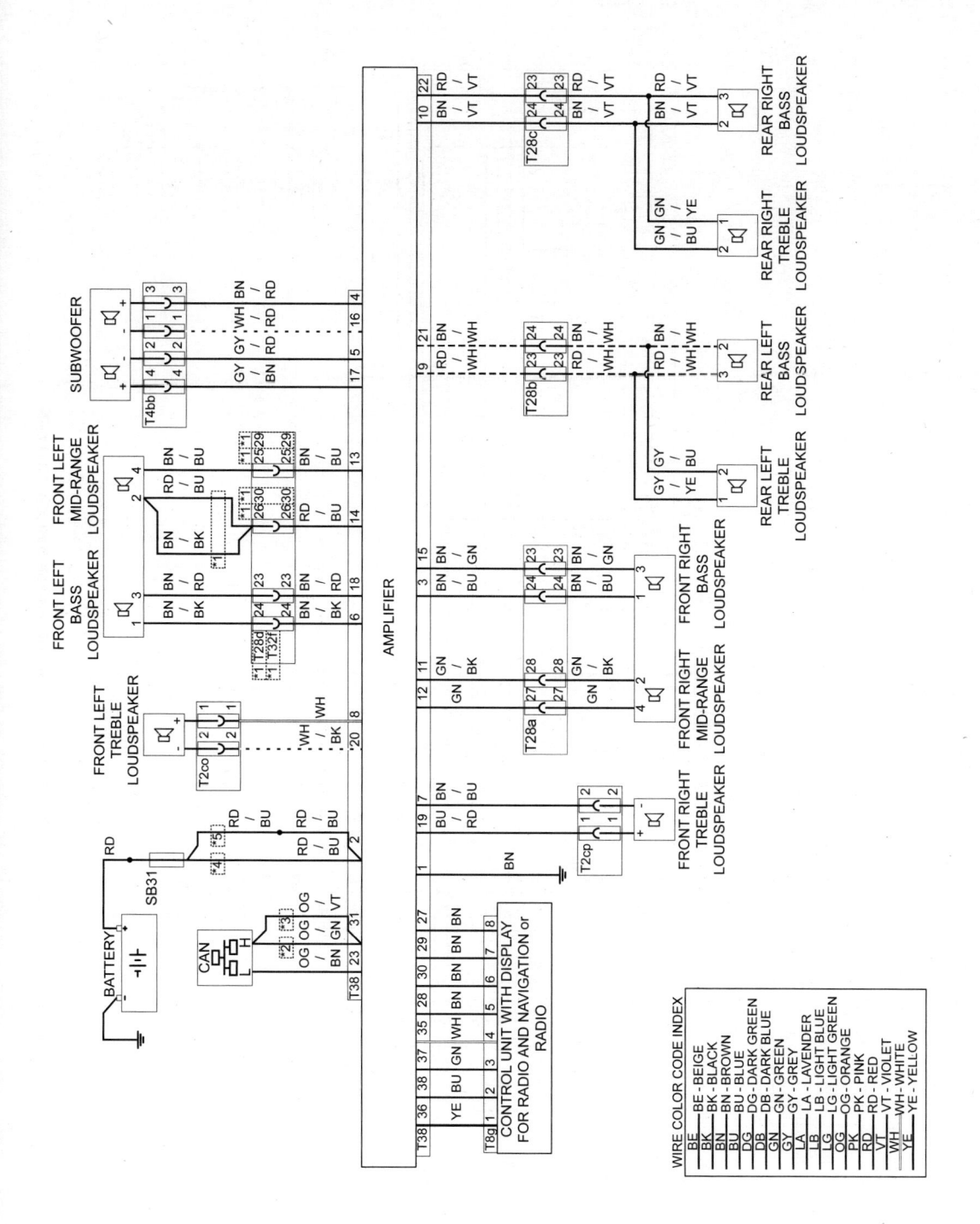

*1 According to equipment
*2 Only models with "Low" equipment (AW0)
*3 Only models with "High" equipment (AW1)
*4 Up to June 2016
*5 From July 2016

Chapter 12 Chassis electrical system

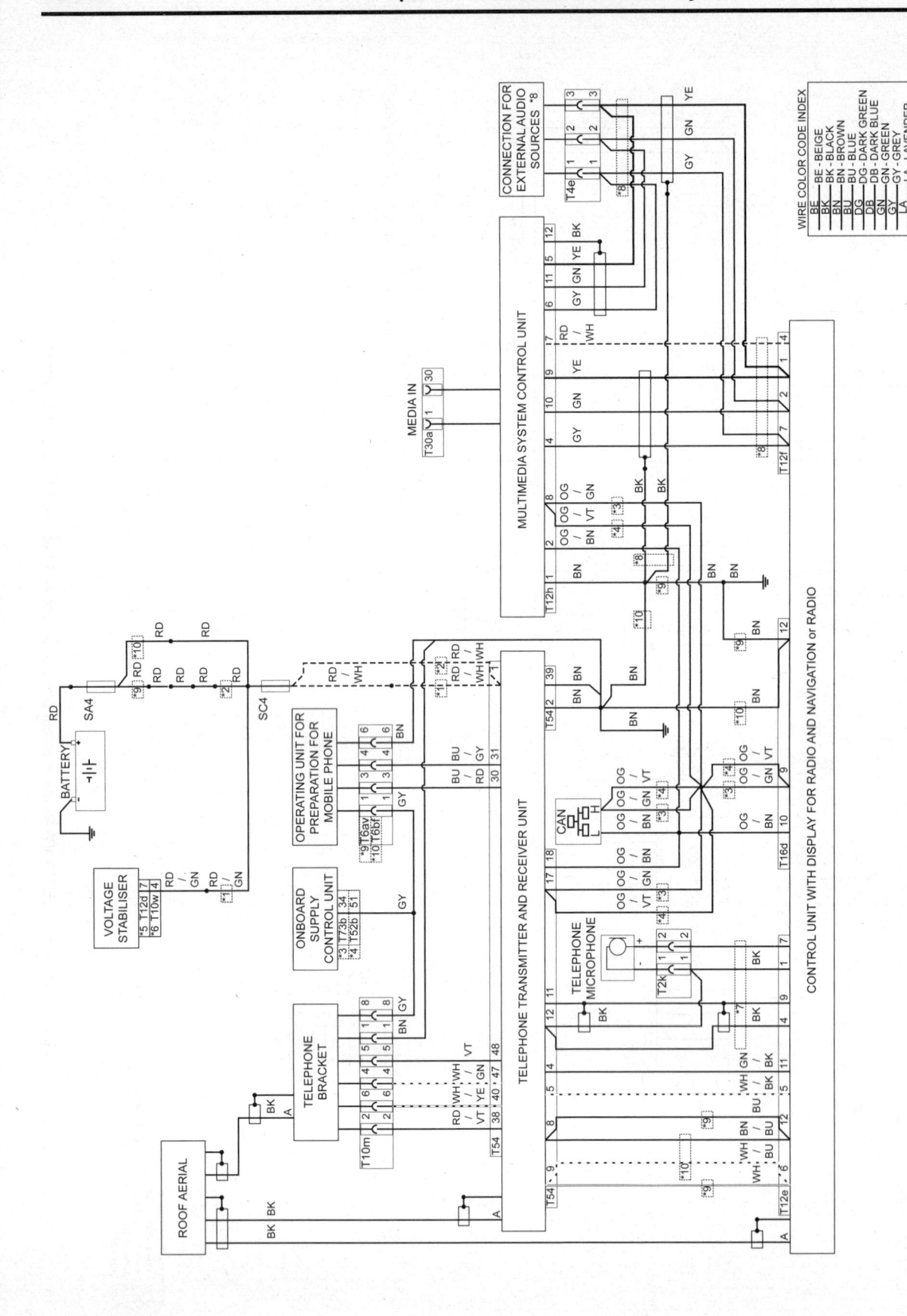

8i. Sound System - JETTA Mobile telephone systems and Multimedia Media-IN

*1 For models with Start/Stop system
*2 For models without Start/Stop system
*3 Only models with "Low" equipment (AW0)
*4 Only models with "High" equipment (AW1)
*5 Gradual introduction
*6 Gradually discontinued
*7 Only models with radio navigation system and voice control
*8 According to equipment
*9 Up to June 2014
*10 From July 2014

Chapter 12 Chassis electrical system

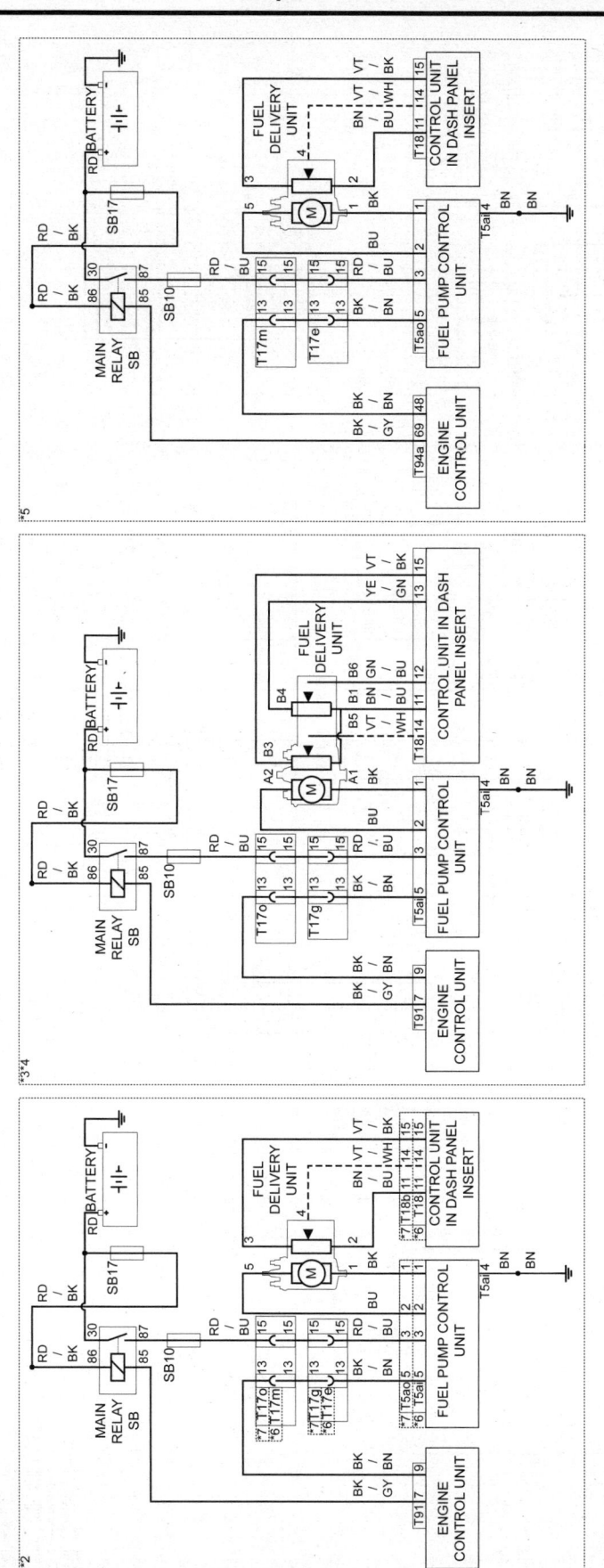

9a. Fuel Pump - GOLF

*1 For 2.0L engines except engine codes CYFB DJJA (except Golf R)
*2 For 1.8L engines and 2-Wheel drive
*3 For 2.0L engines with engine codes CYFB DJJA (Golf R model)
*4 For 1.8L engines and 4-Wheel drive
*5 For 1.4L engines
*6 Up to June 2017
*7 From July 2017

WIRE COLOR CODE INDEX
BE - BEIGE
BK - BLACK
BN - BROWN
BU - BLUE
DG - DARK GREEN
DB - DARK BLUE
GN - GREEN
GY - GREY
LA - LAVENDER
LB - LIGHT BLUE
LG - LIGHT GREEN
OG - ORANGE
PK - PINK
RD - RED
VT - VIOLET
WH - WHITE
YE - YELLOW

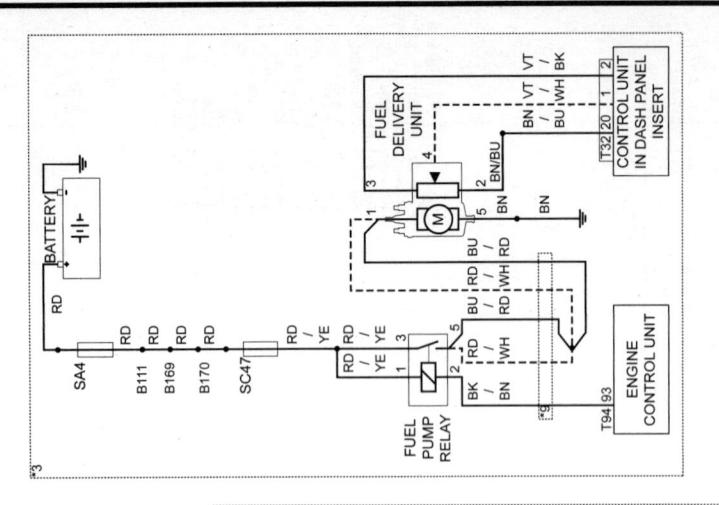

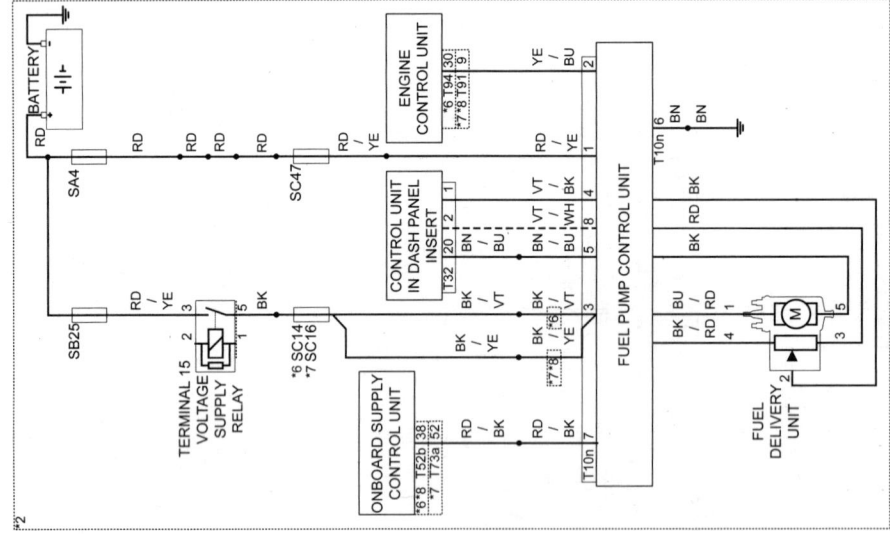

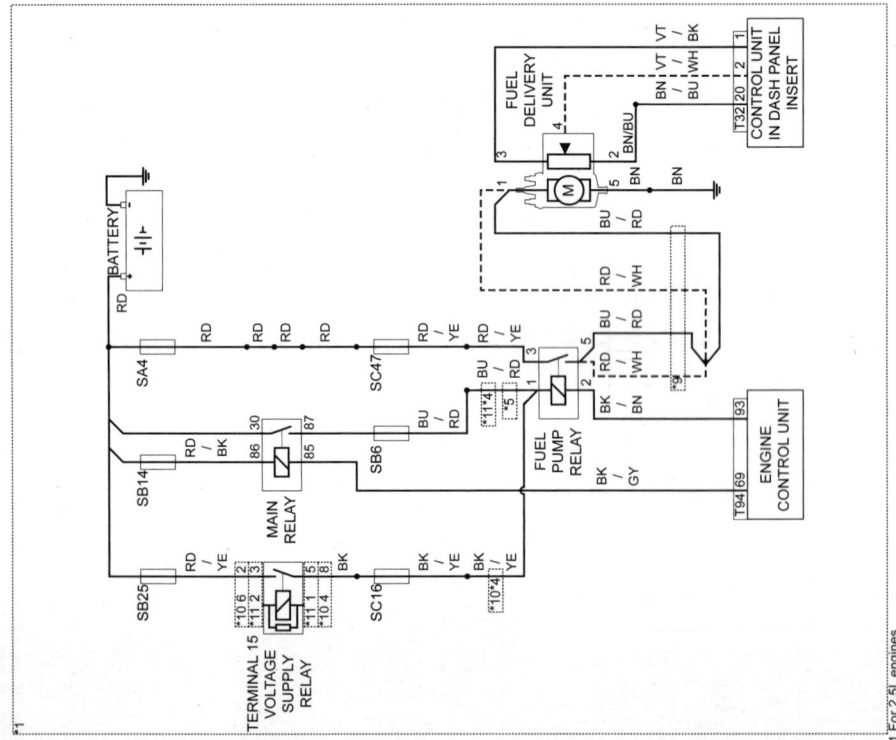

12-78 Chapter 12 Chassis electrical system

9b. Fuel Pump - JETTA Up to June 2014

*1 For 2.5L engines
*2 For 1.8L engines and for 2.0L engines except engine code CBPA
*3 For 2.0L engines with engine code CBPA
*4 Only models with "Low" equipment (AW0)
*5 Only models with "High" equipment (AW1)
*6 For 2.0L engines with engine code CBFA CCTA
*7 For 1.8L engines
*8 For 2.0L engines with engine code CPLA CPPA
*9 According to equipment
*10 Gradually discontinued
*11 Gradual introduction

Chapter 12 Chassis electrical system

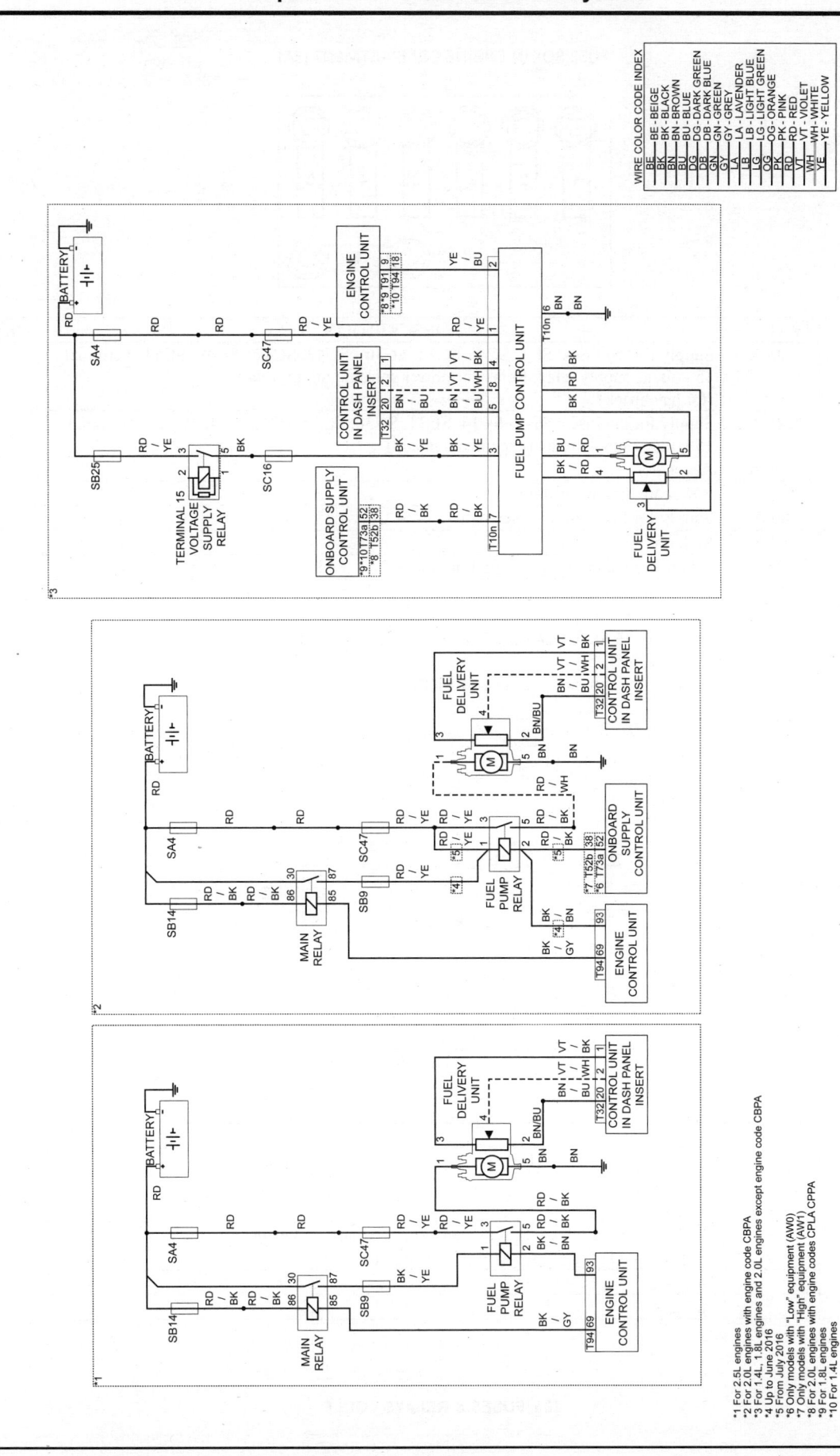

9c. Fuel Pump - JETTA From July 2014

*1 For 2.5L engines
*2 For 2.0L engines with engine code CBPA
*3 For 1.4L, 1.8L engines and 2.0L engines except engine code CBPA
*4 Up to June 2016
*5 From July 2016
*6 Only models with "Low" equipment (AW0)
*7 Only models with "High" equipment (AW1)
*8 For 2.0L engines with engine codes CPLA CPPA
*9 For 1.8L engines
*10 For 1.4L engines

Chapter 12 Chassis electrical system

FUSE BOX IN ENGINE COMPARTMENT (SA)

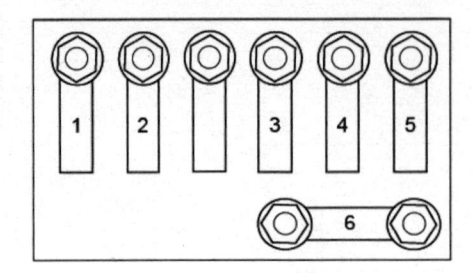

FUSE	VALUE	DESCRIPTION	OEM NAME
1	125 A	Supply for the fuses: SC2, SC4 - SC14, SC30 - SC42, SC47 - SC49, SC53, Terminal 15 voltage supply relay, Relay for power sockets, Starter relay 2 (for hatchback)	SA1
	125 A	Supply for the fuses: SC4 - SC14, SC31, SC38, SC39, SC41, SC42, SC53, Terminal 15 voltage supply relay (except hatchback)	
2	400 A	Alternator	SA2
3	80 A	Power steering control unit	SA3
4	80 A	Supply for fuses: SC15 - SC20, SC23 - SC28, SC43 - SC45	SA4
5	50 A	Radiator fan	SA5
6	125 A	Additional battery for coasting function or Not used	SA6

FUSE AND RELAY BOX IN ENGINE COMPARTMENT (SB)

VERSION 1

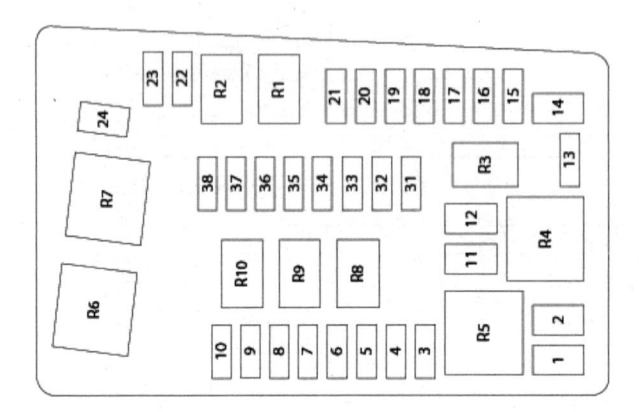

RELAY	VALUE	DESCRIPTION	OEM NAME
R1	-	Starter relay 1	R1
R2	-	Starter relay 2	R2
R3	-	Horn relay	R3
R4	-	High heat output relay (Only models with auxiliary air heater)	R4
R5	-	Main relay	R5
R6	-	Not used	R6
R7	-	Low heat output relay (Only models with auxiliary air heater)	R7
R8	-	Fuel pump relay (Only for MPI petrol engine)	R8
R9	-	Not used	R9
R10	-	Not used	R10

10a. FUSES & RELAYS - GOLF

Chapter 12 Chassis electrical system

12-81

VERSION 2

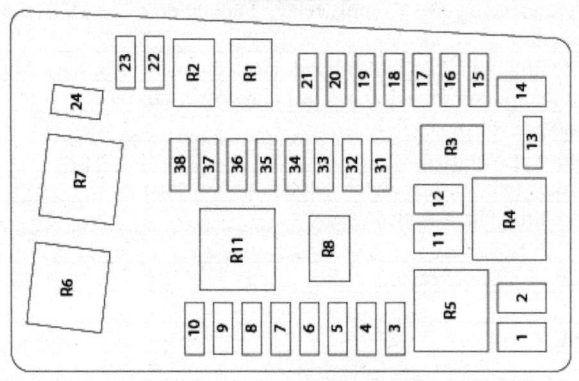

RELAY	VALUE	DESCRIPTION	OEM NAME
R1	-	Starter relay 1	R1
R2	-	Starter relay 2	R2
R3	-	Horn relay	R3
R4	-	High heat output relay (Only models with auxiliary air heater) or Secondary pump relay	R4
R5	-	Main relay	R5
R6	-	Not used	R6
R7	-	Low heat output relay (Only models with auxiliary air heater)	R7
R8	-	Engine compartment current supply relay (According to equipment)	R8
R11	-	Heated windscreen relay (Only models with heated windscreen)	R11

FUSE	VALUE	DESCRIPTION	OEM NAME
1	40 A	ABS control unit, 25 A, 20 A also used	SB1
2	60 A	ABS control unit, ABS hydraulic pump, 40 A also used	SB2
3	15 A	Engine control unit	SB3
4	10 A	Oil level and oil temperature sender, Radiator fan control unit, Activated charcoal filter solenoid valve 1, Camshaft control valve 1, Exhaust camshaft control valve 1, Inlet camshaft control valve 1, Valve for oil pressure control, Inlet cam actuator for cylinder 2, Inlet cam actuator for cylinder 3, Exhaust cam actuator for cylinder 2, Exhaust cam actuator for cylinder 3, Turbocharger air recirculation valve, Inlet manifold flap valve, Piston cooling jet control valve, 5 A also used	SB4
5	10 A	Fuel pressure regulating valve, Injector cylinder 1, Injector cylinder 2, Injector cylinder 3, Injector cylinder 4, Fuel metering valve, Engine component current supply relay, 7.5 A also used	SB5
6	7.5 A	Brake light switch, 5 A also used	SB6
7	7.5 A	Fuel pressure regulating valve, Charge air cooling pump, Gearbox oil cooling pump, Coolant shut-off valve, Radiator blind control motor (if fitted)	SB7
8	10 A	Lambda probe 1 before catalytic converter, Lambda probe 1 after catalytic converter	SB8
9	20 A	Air mass meter, Ignition coils with output stage, Exhaust flap control unit, Sender 1 for secondary air pressure, Charge air cooling pump, Activated charcoal filter solenoid valve, Camshaft control valve 1, Exhaust camshaft control valve 1, 10 A also used	SB9
10	15 A	Fuel pump control unit, Fuel pump relay, 10 A also used	SB10
11	50 A	Auxiliary air heater element, Secondary air pump motor, 40 A also used	SB11
12	40 A	Auxiliary air heater element	SB12
13	30 A	Mechatronic unit for dual clutch gearbox, Automatic gearbox control unit, 15 A also used (up to June 2018)	SB13
	30 A	Auxiliary hydraulic pump 1 for gearbox oil (from July 2018)	
14	40 A	Heated windscreen relay or Not used	SB14
15	15 A	Horn relay	SB15

10b. FUSES & RELAYS - GOLF

16	20 A	Engine component current supply relay, Fuel pump relay, 15 A, 7.5 A, 5 A also used or Not used	SB16
17	7.5 A	Engine control unit, ABS control unit, Heated windscreen relay, Main relay	SB17
18	5 A	Battery monitor control unit, Data bus diagnostic interface	SB18
19	30 A	Wiper motor control unit	SB19
20	10 A	Alarm horn or Not used	SB20
21	30 A	Mechatronic unit for dual clutch gearbox or Not used	SB21
22	7.5 A	Engine control unit, 5 A also used	SB22
23	30 A	Starter	SB23
24	40 A	Auxiliary air heater element	SB24
25	-	Not used	-
26	-	Not used	-
27	-	Not used	-
28	-	Not used	-
29	-	Not used	-
30	-	Not used	-
31	15 A	Vacuum pump for brakes, Vacuum pump relay	SB31
32	15 A	Axle differential lock control unit or Not used	SB32
33	30 A	Auxiliary hydraulic pump 1 for gearbox oil or Not used	SB33
34	15 A	Axle differential lock control unit or Not used	SB34
35	-	Not used	-
36	-	Not used	-
37	20 A	Auxiliary heater control unit	SB37
38	-	Not used	-

FUSE AND RELAY BOX IN PASSENGER COMPARTMENT (SC)

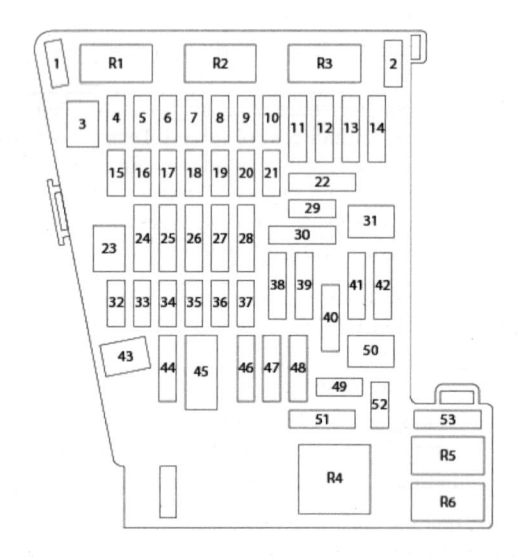

FUSE/RELAY	VALUE	DESCRIPTION	OEM NAME
1	-	Not used	-
2	15 A	Steering column electronics control unit, 10 A also used	SC2
3	-	Not used	-
4	10 A	Onboard supply control unit (Anti-theft alarm system) or Alarm horn, 7.5 A also used	SC4
5	5 A	Data bus diagnostic interface	SC5
6	7.5 A	Anti-theft alarm sensor or Selector lever, 5 A also used	SC6
7	10 A	Heater and air conditioning controls, Heater control unit, Climatronic control unit, Air conditioning system control unit, Selector lever, Remote control receiver for auxiliary coolant heater, Heated rear window relay	SC7

10c. FUSES & RELAYS - GOLF

Chapter 12 Chassis electrical system 12-83

8	10 A	Rotary light switch, Electromechanical parking brake button, Rain and light sensor, Diagnostic connection, Anti-theft alarm sensor (if fitted), 7.5 A also used	SC8
9	7.5 A	Steering column electronics control unit, 5 A, 1 A also used	SC9
10	10 A	Display unit for front information display and operating unit control unit, Control unit for navigation system, TV tuner (Up to April 2015)	SC10
	10 A	Display unit for front information display and operating unit control unit, TV tuner (if fitted), CD player (if fitted), 7.5 A also used (From May 2015)	
11	25 A	Control unit for front belt tensioner (Up to April 2015)	SC11
	40 A	Onboard supply control unit (Front left headlight) (From May 2015)	
12	20 A	Control unit 1 for information electronics, Control unit for navigation system (if fitted), 7.5 A, 5 A also used	SC12
13	15 A	Electronically controlled damping control unit (Up to April 2015)	SC13
	25 A	Control unit for front left belt tensioner (From May 2015)	
14	40 A	Fresh air blower control unit, 30 A also used	SC14
15	10 A	Control unit for electronics steering column lock	SC15
16	7.5 A	Two-way signal amplifier for mobile telephone / data services, Storage compartment with interface for mobile telephone (if fitted), Aerial amplifier 3 (if fitted), Voltage converter for USB charge module, USB hub, Chip card reader, TV tuner (if fitted)	SC16
17	7.5 A	Control unit in dash panel insert, Dash panel insert, Emergency call module (if fitted), 5 A also used	SC17
18	7.5 A	Reversing camera, Rear lid handle release button	SC18
19	7.5 A	Interface for entry and start system	SC19
20	7.5 A	Control unit for fuel tank leak detection or Not used	SC20
21	15 A	All-wheel drive control unit	SC21
22	15 A	Trailer detector control unit	SC22
23	30 A	Sliding sunroof adjustment control unit, Sunroof roller blind control unit, 20 A also used	SC23
24	40 A	Onboard supply control unit	SC24
25	30 A	Driver control unit, Rear right window regulator motor	SC25
26	30 A	Onboard supply control unit (Front seat heating)	SC26
27	30 A	Digital sound package control unit or Onboard supply control unit	SC27
28	25 A	Trailer detector control unit	SC28
29	-	Not used	-
30	-	Not used	-
31	40 A	Onboard supply control unit (Front left headlight) or Not used	SC31
32	10 A	Front camera for driver systems, Adaptive cruise control unit, Parking aid control unit, Park assist steering control unit, Blind spot monitor control units (according to equipment), 7.5 A also used	SC32
33	5 A	Airbag control unit, Warning lamp for airbag deactivated on front passenger side	SC33
34	7.5 A	Rotary light switch, Interior mirror, Relay for power sockets, Reversing light switch, Pressure sender for refrigerant circuit, Air quality sensor, TCS and ESP button, Electromechanical parking brake button, Control unit for structure-borne sound (for hatchback)	SC34
	7.5 A	Rotary light switch, Interior mirror, Relay for power sockets, Reversing light switch, Pressure sender for refrigerant circuit, Air quality sensor, DC/AC converter with socket, Terminal 15 relief relay, Switch module 1 in centre console, Electromechanical parking brake button (except hatchback)	
35	10 A	Diagnostic connection, Headlight range control and instrument illumination regulator, Control unit for cornering light and headlight range control, Left and right headlight control motors, 7.5 A also used	SC35
36	10 A	Control unit for right daytime running light and side light or Front right headlight, 7.5 A, 5 A also used	SC36
37	10 A	Control unit for left daytime running light and side light or Front left headlight, 7.5 A, 5 A also used	SC37
38	25 A	Trailer detector control unit	SC38
39	30 A	Front passenger door control unit, Rear right window regulator motor	SC39
40	20 A	Cigarette lighter, 12 V sockets	SC40
41	25 A	Control unit for front right belt tensioner	SC41

10d. FUSES & RELAYS - GOLF

42	40 A	Onboard supply control unit (Central locking)	SC42
43	30 A	Onboard supply control unit or Digital sound package control unit	SC43
44	15 A	Trailer detector control unit	SC44
45	15 A	Driver seat lumbar support adjustment switch, Driver seat adjustment operating unit	SC45
46	30 A	DC/AC converter with socket or Not used	SC46
47	15 A	Rear window wiper motor	SC47
48	10 A	Engine sound generator control unit (for hatchback) or Not used (except hatchback)	SC48
49	7.5 A	Clutch position sender, Starter relay 1 and 2, 5 A also used	SC49
50	-	Not used	-
51	-	Not used	-
52	15 A	Electronically controlled damping control unit or Not used	SC52
53	30 A	Heated rear window relay	SC53
R1	-	Not used	R1
R2	-	Terminal 15 relief relay or Not used	R2
R3	-	Not used	R3
R4	-	Terminal 15 voltage supply relay	R4
R5	-	Heated rear window relay	R5
R6	-	Relay for power sockets	R6

10e. FUSES & RELAYS - GOLF

Chapter 12 Chassis electrical system

FUSE BOX IN ENGINE COMPARTMENT (SA)

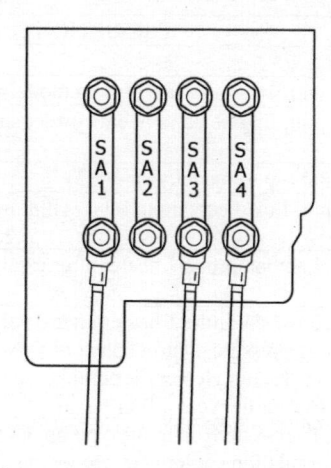

FUSE	VALUE	DESCRIPTION	OEM NAME
SA1	200 A	Alternator 140 A	SA1
SA2	-	Not used	SA2
SA3	80 A	Steering	SA3
SA4	80 A	Interior terminal 30 power supply SC fuses	SA4

FUSE AND RELAY BOX IN ENGINE COMPARTMENT (SB)

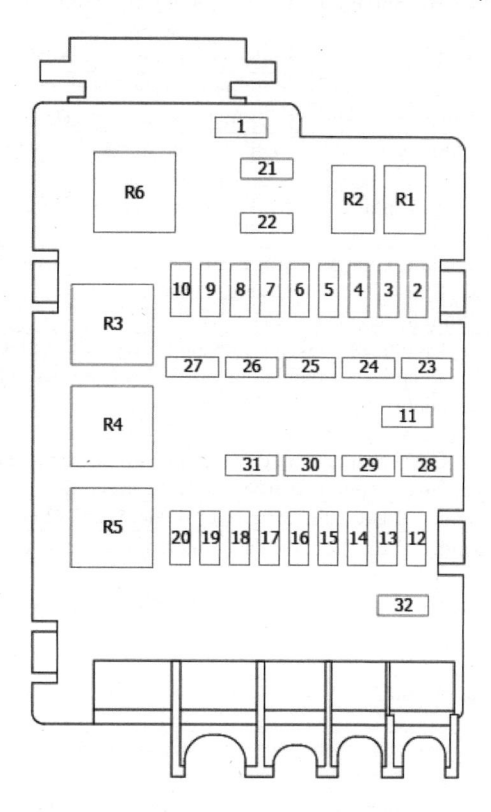

10f. FUSES & RELAYS - JETTA

Up to June 2014

FUSE/RELAY	VALUE	DESCRIPTION	OEM NAME
1	-	Not used	SB1
2	15 A	Engine/motor control unit, 10 A also used (up to model year 2013)	SB2
	20 A	Engine/motor control unit, Engine component current supply relay, 15 A also used (from model year 2014)	
3	5 A	Radiator fan control unit (up to model year 2013)	SB3
	10 A	Radiator fan control unit, Low heat output relay, High heat output relay (from model year 2014)	
4	15 A	Lambda probe heater, Lambda probe 1 heater after catalytic converter, 5 A or 10 A also used	SB4
5	20 A	Lambda probe heater, Lambda probe 1 heater after catalytic converter, Fuel system diagnostic pump, Charge pressure control solenoid valve, Exhaust gas recirculation cooler changeover valve, Heater element for crankcase breather, Vacuum pump for brakes, 5 A also used (for model year 2011)	SB5
	20 A	Lambda probe heater, Fuel system diagnostic pump, Charge pressure control solenoid valve, Exhaust gas recirculation cooler changeover valve, Heater element for crankcase breather, Vacuum pump for brakes, Vacuum pump relay (if fitted), 10 A or 5 A also used (from model year 2012)	
6	15 A	Fuel system diagnostic pump, Activated charcoal filter solenoid valve 1, Throttle valve module, Auxiliary air heater element, Coolant pump (if fitted) 10 A or 5 A also used (up to model year 2012)	SB6
	20 A	Fuel system diagnostic pump, Activated charcoal filter solenoid valve 1, Fuel tank shut-off valve, Continued coolant circulation pump, Auxiliary air heater element, Low heat output relay, High heat output relay, 15 A, 10 A or 5 A also used (for model year 2013)	
	10 A	Fuel system diagnostic pump, Activated charcoal filter solenoid valve 1, Fuel tank shut-off valve, Secondary air pump relay, Coolant pump, Auxiliary air heater element, Low heat output relay, High heat output relay (from model year 2014)	
7	20 A	Ignition transformer, 5 A also used (for model year 2011)	SB7
	30 A	Ignition transformer, Ignition coils, 20 A, 15 A also used (from model year 2012)	
8	10 A	Throttle valve module, Activated charcoal filter solenoid valve 1 (for model year 2011)	SB8
	10 A	Throttle valve module, Charge pressure control solenoid valve, Activated charcoal filter solenoid valve 1, Camshaft control valve, Turbocharger air recirculation valve (for model year 2012)	
	10 A	Throttle valve module, Charge pressure control solenoid valve, Activated charcoal filter solenoid valve 1, Camshaft control valve, Turbocharger air recirculation valve, Intake manifold flap (from model year 2013)	
9	15 A	Fuel pump relay, Continued coolant circulation relay (if fitted), Coolant circulation pump (if fitted), Fuel pressure regulating valve, 5 A also used (up to model year 2013)	SB9
	15 A	Sender 1 for secondary air pressure, Sender 2 for secondary air pressure, Fuel pump relay, Activated charcoal filter solenoid valve 1, Camshaft control valve 1, Fuel pressure regulating valve, Continued coolant circulation pump, Continued coolant circulation relay, Coolant circulation pump 2 (from model year 2014)	
10	5 A	Speedometer sender, Brake light switch (for model year 2011)	SB10
	5 A	Speedometer sender, Brake light switch, Clutch position sender, Additional coolant pump relay (for model year 2012)	
	5 A	Brake light switch, Clutch position sender, Additional coolant pump relay (from model year 2013)	
11	-	Not used	SB11
12	5 A	Coolant circulation pump or Vacuum pump relay, Vacuum pump for brakes (up to model year 2013)	SB12
	30 A	Ignition coils, Coolant circulation pump (from model year 2014)	
13	5 A	Coolant pump (for model year 2011)	SB13
	5 A	Coolant pump, Fuel pump relay, Fuel system pressurisation pump (for model year 2012)	
	30 A	Additional coolant pump relay, Coolant pump, 20 A, 10 A also used (from model year 2013)	
14	10A	Engine/motor control unit or Engine/motor control unit, Main relay, 5 A also used	SB14
15	30 A	Voltage stabiliser	SB15

10g. FUSES & RELAYS - JETTA

16	30 A	ABS control unit	SB16
17	30 A	Automatic gearbox control unit or Mechatronic unit for dual clutch gearbox, 15 A also used	SB17
18	20 A	Automatic gearbox control unit or Mechatronic unit for dual clutch gearbox, 15 A also used from model year 2014	SB18
19	5 A	Onboard supply control unit, 1 A also used up to model year 2013	SB19
20	30 A	Onboard supply control unit (Only models with "Low" equipment), Automatic intermittent wash and wipe relay (Only models with "High" equipment), Wiper motor relays (Only models with "High" equipment)	SB20
21	50 A	Secondary air pump motor, 40 A also used from model year 2014	SB21
22	40 A	Auxiliary air heater element or Not used	SB22
23	40 A	ABS control unit	SB23
24	50 A	Trailer detector control unit	SB24
25	50 A	Terminal 15 voltage supply relay 2 or Terminal 15 voltage supply relay, Terminal 15 voltage supply relay 2	SB25
26	50 A	Secondary air pump relay, 40 A also used or Not used	SB26
27	60 A	Radiator fan	SB27
28	40 A	Onboard supply control unit (Only models with "High" equipment)	SB28
29	40 A	Onboard supply control unit (Only models with "High" equipment)	SB29
30	50 A	Terminal 75 voltage supply relay 1	SB30
31	40 A	Amplifier, 30 A also used	SB31
32	40 A	Auxiliary air heater element or Not used	SB32
R1	-	Not used (For models with "Low" equipment)	
	-	Wiper motor switch-over relay 1 (For models with "High" equipment)	
R2	-	Not used (For models with "Low" equipment)	
	-	Wiper motor switch-over relay 2 (For models with "High" equipment)	
R3	-	Main relay or Terminal 30 voltage supply relay or Fuel supply relay	
R4	-	Low heat output relay or Continued coolant circulation relay or Additional coolant pump relay	
R5	-	High heat output relay or Secondary air pump relay	
R6	-	Not used for petrol engines	

From July 2014

FUSE/RELAY	VALUE	DESCRIPTION	OEM NAME
1	-	Not used	SB1
2	20 A	Engine/motor control unit	SB2
3	5 A	Radiator fan control unit, Low heat output relay, High heat output	SB3
4	15 A	Lambda probe heaters, Lambda probe 1 heater after catalytic converter, 10 A also used	SB4
5	15 A	Secondary air inlet valve	SB5
6	10 A	Fuel system diagnostic pump, Fuel tank shut-off valve, Secondary air pump relay, Heater element for crankcase breather, Lambda probe heater 2	SB6
7	30 A	Ignition transformer, Ignition coils, 10 A also used	SB7
8	10 A	Throttle valve module, Charge pressure control solenoid valve, Secondary air inlet valve, Secondary air pump relay, Activated charcoal filter solenoid valve 1, Camshaft control valve 1, Turbocharger air recirculation valve, Intake manifold flap, Radiator blind control motor	SB8
9	15 A	Fuel pump relay, Valve for oil pressure control, Turbocharger air recirculation valve, Intake manifold flap valve, 10 A also used	SB9
10	5 A	Brake light switch, Clutch position sender	SB10
11	-	Not used	SB11
12	30 A	Ignition coils, Coolant circulation pump	SB12
13	30 A	Voltage stabiliser, Control unit with display for radio and navigation, Radio, 20 A also used	SB13
14	10A	Engine/motor control unit	SB14

15	30 A	ABS control unit	SB15
16	20 A	Vacuum pump relay, 10 A, 5 A also used	SB16
17	15 A	Mechatronic unit for dual clutch gearbox	SB17
18	30 A	Mechatronic unit for dual clutch gearbox or Automatic gearbox control unit, 15 A also used	SB18
19	5 A	Onboard supply control unit, Battery monitor control unit	SB19
20	30 A	Onboard supply control unit (Only models with "Low" equipment), Wiper motor relays (Only models with "High" equipment)	SB20
21	50 A	Secondary air pump motor, Auxiliary air heater element, 40 A also used	SB21
22	40 A	Auxiliary air heater element, Heated windscreen	SB22
23	50 A	Terminal 75 voltage supply relay 1, 40 A also used	SB23
24	50 A	Fuse 40 in passenger compartment box (SC)	SB24
25	40 A	Terminal 15 voltage supply relay	SB25
26	-	Not used	SB26
27	50 A	Radiator fan	SB27
28	30 A	Onboard supply control unit (Only models with "High" equipment)	SB28
29	30 A	Onboard supply control unit (Only models with "High" equipment)	SB29
30	40 A	ABS control unit	SB30
31	40 A	Amplifier	SB31
32	40 A	Auxiliary air heater element	SB32
R1	-	Not used (For models with "Low" equipment)	
	-	Wiper motor switch-over relay 1 (For models with "High" equipment)	
R2	-	Not used (For models with "Low" equipment)	
	-	Wiper motor switch-over relay 2 (For models with "High" equipment)	
R3	-	Main relay or Terminal 30 voltage supply relay	
R4	-	Low heat output relay or Engine component current supply relay or Auxiliary engine coolant pump relay	
R5	-	High heat output relay or Secondary air pump relay	
R6	-	Vacuum pump relay	

10i. FUSES & RELAYS - JETTA

Chapter 12 Chassis electrical system

12-89

FUSE BOX IN PASSENGER COMPARTMENT (SC)

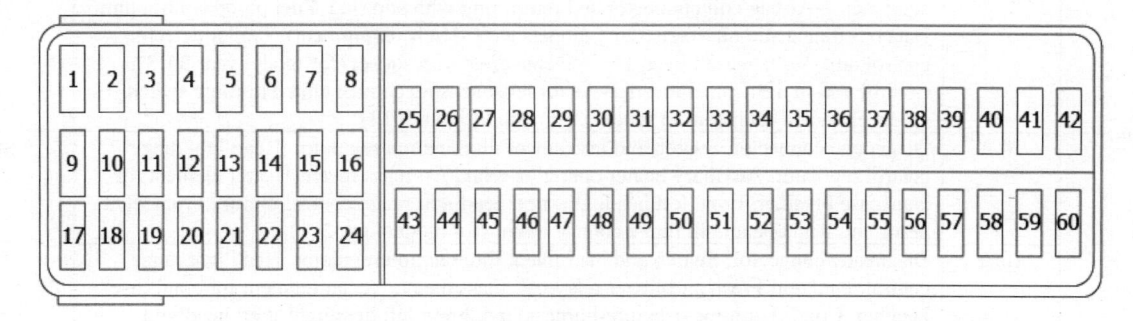

```
┌─────────────────────────┬──────────────────────────────────────────────────────────┐
│ 1  2  3  4  5  6  7  8  │ 25 26 27 28 29 30 31 32 33 34 35 36 37 38 39 40 41 42    │
│ 9 10 11 12 13 14 15 16  │                                                          │
│17 18 19 20 21 22 23 24  │ 43 44 45 46 47 48 49 50 51 52 53 54 55 56 57 58 59 60    │
└─────────────────────────┴──────────────────────────────────────────────────────────┘
```

Up to June 2014

FUSE	VALUE	DESCRIPTION	OEM NAME
1	10 A	Left washer heater jet element, Right washer heater jet element or Not used	SC1
2	7.5 A	Control unit for electronic steering column lock, 5 A also used up to model year 2012	SC2
3	10 A	Control unit in dash panel insert	SC3
4	10A	Telephone transmitter and receiver unit, Magnetic field sender for compass, 5A, 2 A also used or Not used	SC4
5	10 A	Rear fog light bulb, 7.5 A also used or Not used	SC5
6	10 A	Onboard supply control unit (Only models with "Low" equipment), Reversing camera (if fitted)	SC6
7	5 A	Fog light relay (Only models with "Low" equipment), Switch and instrument illumination regulator (Only models with "Low" equipment), Number plate light, Onboard supply control unit (Only models with "High" equipment)	SC7
8	7.5 A	Washer pump switch (automatic wash/wipe and headlight washer system (Only models with "Low" equipment), Washer pump (Only models with "Low" equipment) or Not used (Only models with "High" equipment) (up to model year 2012)	SC8
	7.5 A	Washer pump switch (automatic wash/wipe and headlight washer system (Only models with "Low" equipment), Washer pump (Only models with "Low" equipment), Onboard supply control unit pump (Only models with "High" equipment) or Not used (from model year 2013)	
9	5 A	Airbag control unit, Seat occupied recognition control unit (if fitted), Warning light for airbag deactivated on front passenger side (if fitted)	SC9
10	10 A	Right steering column switch (Only models with "Low" equipment)	SC10
11	10 A	Front left headlight (gas discharge headlight) or Not used	SC11
12	10 A	Front right headlight (gas discharge headlight) or Not used	SC12
13	5 A	Automatic anti-dazzle interior mirror, Light sensor, Parking aid control unit, Air quality sensor, High pressure sender, Climatronic control unit, Tyre Pressure Monitoring System Button, Reversing light switch, Left washer jet heater element, Right washer heater jet element, 28 pin connector (for model year 2011)	SC13
	5 A	Automatic anti-dazzle interior mirror, Light sensor, Parking aid control unit, Air quality sensor, High pressure sender, Climatronic control unit, Tyre Pressure Monitoring System Button, Reversing light switch, Left washer jet heater element, Right washer heater jet element, 28 pin connector, TCS and ESP button, Start/Stop operation switch, Mirror adjustment switch, Exterior mirror heater button, Control unit for cornering light and headlight range control (from model year 2012)	
14	10 A	Left steering column switch, ABS control unit, Light switch (Only models with "High" equipment), Airbag coil connector and return ring with slip ring, Fuel pump control unit, Data bus diagnostic interface (Only models with "High" equipment) (for model year 2011)	SC14
	10 A	Left steering column switch, ABS control unit, Light switch (Only models with "High" equipment), Airbag coil connector and return ring with slip ring, Fuel pump control unit, Data bus diagnostic interface (Only models with "High" equipment), Trailer detector control unit, Voltage stabiliser, DC/AC converter with socket (for model year 2012)	

10j. FUSES & RELAYS - JETTA

Chapter 12 Chassis electrical system

	10 A	Left steering column switch, ABS control unit, Light switch (Only models with "High" equipment), Airbag coil connector and return ring with slip ring, Fuel pump control unit, Data bus diagnostic interface (Only models with "High" equipment), Trailer detector control unit, Voltage stabiliser, DC/AC converter with socket (for model year 2013), Control unit in dash panel insert, Selector lever sensors control unit, Tiptronic switch, Oil level and oil temperature sender (from model year 2013)	
15	10 A	Diagnostic connector, Switch and instrument illumination regulator, Headlight range control regulator, Auxiliary heater operation relay, Air mass meter, Heater element for crankcase breather, Front left headlight, Left headlight range control motor, Front right headlight, Right headlight range control motor (for model year 2011)	SC15
	10 A	Diagnostic connector, Switch and instrument illumination regulator, Headlight range control regulator, Fresh air blower relay, Air mass meter, Heater element for crankcase breather, Control unit for structure-borne sound, Front left headlight, Left headlight range control motor, Front right headlight, Right headlight range control motor, Onboard supply control unit (from model year 2012)	
16	10 A	Engine/motor control unit, Additional coolant pump relay (if fitted) (up to model year 2013)	SC16
	10 A	Engine/motor control unit, Additional coolant pump relay, Fuel pump control unit (from model year 2014)	
17	10 A	ATA horn relay, ATA interior monitor sensor, ATA sounder or Not used	SC17
18	15 A	Front left headlight	SC18
19	15 A	Front right headlight	SC19
20	10 A	Automatic gearbox control unit, Selector lever sensors control unit, Tiptronic switch, Climatronic control unit, Ignition/Starter switch, Remote control receiver for auxiliary coolant heater (if fitted)	SC20
21	20 A	Onboard supply control unit (Only models with "Low" equipment), Dual tone horn relay, Treble horn, Bass horn, 15 A also used or Not used	SC21
22	15 A	Ignition/starter switch, Converter box, Alarm horn, Interior monitoring sensor, Alarm horn relay, Dual tone horn relay, 10 A also used	SC22
23	10 A	Onboard supply control unit (Only models with "Low" equipment), Diagnostic connector, Light switch (Only models with "High" equipment), Rain and light sensor, Magnetic field sender for compass	SC23
24	10 A	Onboard supply control unit (Only models with "Low" equipment), Entry and start authorisation control unit	SC24
25	15 A	Automatic gearbox control unit, Selector lever sensors control unit, Multifunction switch	SC25
26	20 A	Vacuum pump for brakes, 15 A also used	SC26
27	1 A	Airbag coil connector and return ring with slip ring or Not used	SC27
28	40 A	Auxiliary heater operation relay	SC28
29	1 A	Onboard supply control unit (Only models with "Low" equipment) (for model year 2011)	SC29
	1 A	Onboard supply control unit (Only models with "Low" equipment), Ignition starter/switch (Only models with "Low" equipment), Converter box (Only models with "Low" equipment) (from model year 2012)	
30	20 A	Cigarette lighter, 12 V sockets, Blocking diode	SC30
31	30 A	Light switch (Only models with "Low" equipment)	SC31
32	20 A	Light switch (Only models with "Low" equipment)	SC32
33	40 A	Heater/heat output switch, Fresh air blower relay, Air conditioning system control unit, Fresh air blower switch	SC33
34	15 A	Front left headlight, Control unit in dash panel insert (for model year 2012)	SC34
	15 A	Left headlight main bulb (Only models with "Low" equipment), Right headlight main bulb (Only models with "Low" equipment), Control unit in dash panel insert (Only models with "Low" equipment) (from model year 2012)	
35	10 A	Steering column electronics control unit, Control unit in dash panel insert (only for model year 2011), Horn plate, Data bus diagnostic interface, 5 A also used	SC35
36	30 A	Onboard supply control unit, 25 A also used	SC36
37	15 A	Front left headlight, Left daytime running light bulb (if fitted), 10 A also used	SC37
38	15 A	Front right headlight, Right daytime running light bulb (if fitted), 10 A also used	SC38
39	30 A	Dipped beam relay, 15 A, 10 A, 20 A also used	SC39
40	20 A	Trailer detector control unit, 15 A also used	SC40

10k. FUSES & RELAYS - JETTA

Chapter 12 Chassis electrical system

FUSE	VALUE	DESCRIPTION	OEM NAME
41	15 A	Trailer detector control unit	SC41
42	20 A	Trailer detector control unit	SC42
43	30 A	Front passenger door control unit, 5 A also used	SC43
44	30 A	Heated rear window relay unit (Only models with "High" equipment), Onboard supply control unit (Only models with "Low" equipment), Heated rear window, 25 A also used	SC44
45	30 A	Driver door control unit, Front passenger door control unit (According to equipment), 25 A also used	SC45
46	30 A	Rear left door control unit, Rear right door control unit, 25 A also used (up to model year 2013)	SC46
	30 A	Rear left door control unit, Rear right door control unit, Front passenger door control unit	
47	15 A	Fuel pump control unit, Fuel pump relay, Fuel supply relay	SC47
48	20 A	Onboard supply control unit	SC48
49	40 A	Climatronic control unit, Air conditioning system control unit (up to model year 2012)	SC49
	40 A	Fresh air blower, Climatronic control unit, Air conditioning system control unit (from model year 2013)	
50	30 A	Heated front seats control unit	SC50
51	20 A	Sliding sunroof adjustment control unit	SC51
52	20 A	Dual tone horn relay (up to model year 2013)	SC52
	20 A	Headlight washer system relay, Headlight washer system pump (from model year 2013)	
53	15 A	Heated front seats control unit or Driver seat lumbar support adjustment switch	SC53
54	15 A	Fog light relay	SC54
55	20 A	Light switch (Only models with "Low" equipment), Left steering column switch (Only models with "Low" equipment)	SC55
56	-	Not used	SC56
57	20 A	Radio, Control unit with display for radio and navigation, 15 A also used	SC57
58	30 A	Telephone transmitter and receiver unit (if fitted), DC/AC converter with socket, 12V	SC58
59	30 A	Amplifier or Not used	SC59
60	30 A	Auxiliary heater control unit, Auxiliary air heater element or Auxiliary heater control unit	SC60

From July 2014

FUSE	VALUE	DESCRIPTION	OEM NAME
1	10 A	Control unit for electronic steering column lock	SC1
2	-	Not used	SC2
3	10 A	Control unit in dash panel insert, Headlight washer system relay (Up to June 2016)	SC3
	10 A	Control unit in dash panel insert, Voltage stabiliser (From July 2016)	
4	5 A	Telephone transmitter and receiver unit, Magnetic field sender for compass, Navigation system interface, Mobile telephone operating electronics control unit (Up to June 2016)	SC4
	5 A	Magnetic field sender for compass, Emergency call module control unit and communication unit, Voltage stabiliser, Telephone controls control unit (From July 2016)	
5	-	Not used	SC5
6	10 A	Reversing camera	SC6
7	-	Not used	SC7
8	10 A	Washer pump (Only models with "Low" equipment), Onboard supply control unit (Only models with "Low" equipment) (Up to June 2016)	SC8
	10 A	Washer pump (Only models with "Low" equipment), Onboard supply control unit (Only models with "Low" equipment), Right steering column switch (Only models with "Low" equipment) (From July 2016)	
9	5 A	Airbag control unit, Seat occupied recognition control unit, Warning light for airbag deactivated on front passenger side (Up to June 2016)	SC9
	5 A	Airbag control unit, Seat occupied recognition control unit, Warning light for airbag deactivated on front passenger side, Starter relay, Terminal 15 voltage supply relay (From July 2016)	
10	15 A	Right steering column switch (Only models with "Low" equipment)	SC10
11	10 A	Front left headlight (gas discharge headlight)	SC11

10I. FUSES & RELAYS - JETTA

12	10 A	Front right headlight (gas discharge headlight)	SC12
13	5 A	Automatic anti-dazzle interior mirror, Light sensor, Parking aid control unit, Air quality sensor, High pressure sender, Tyre Pressure Monitoring System Button, Reversing light switch, Left washer jet heater element, Right washer heater jet element, 28 pin connector, TCS and ESP button, Start/Stop operation switch, Mirror adjustment switch, Exterior mirror heater button, Control unit for cornering light and headlight range control, Heater control unit, Radio, Adaptive cruise control (Up to June 2015)	SC13
	5 A	Automatic anti-dazzle interior mirror, Light sensor, Parking aid control unit, Air quality sensor, High pressure sender, Tyre Pressure Monitoring System Button, Reversing light switch, TCS and ESP button, Start/Stop operation switch, Mirror adjustment switch, Exterior mirror heater button, Traction control switch, Isolation relay for powertrain CAN bus, Control unit for cornering light and headlight range control, Heater control unit, Radio, Adaptive cruise control (From July 2015 to June 2016)	
	5 A	Automatic anti-dazzle interior mirror, Parking aid control unit, Air quality sensor, High pressure sender, Tyre Pressure Monitoring System Button, Reversing light switch, TCS and ESP button, Start/Stop operation switch, Traction control switch, Isolation relay for powertrain CAN bus, Control unit for cornering light and headlight range control, Heater control unit, Radio (From July 2016)	
14	10 A	Left steering column switch (Only models with "Low" equipment), ABS control unit, Light switch (Only models with "High" equipment), Airbag coil connector and return ring with slip ring, Data bus diagnostic interface (Only models with "High" equipment), Trailer detector control unit, Voltage stabiliser, DC/AC converter with socket, Control unit in dash panel insert, Power steering control unit, Oil level and oil temperature sender, Control unit for vehicle location system (Up to June 2015)	SC14
	10 A	Left steering column switch (Only models with "Low" equipment), ABS control unit, Airbag coil connector and return ring with slip ring, Trailer detector control unit, Voltage stabiliser, Data bus diagnostic interface (Only models with "High" equipment), DC/AC converter with socket, Control unit in dash panel insert, Power steering control unit, Oil level and oil temperature sender, Control unit for vehicle location system (From July 2015 to June 2016)	
	10 A	ABS control unit, Airbag coil connector and return ring with slip ring, Trailer detector control unit, Voltage stabiliser, Data bus diagnostic interface (Only models with "High" equipment), DC/AC converter with socket, Control unit in dash panel insert, Oil level and oil temperature sender, Control unit for vehicle location system (From July 2016)	
15	15 A	Diagnostic connector, Switch and instrument illumination regulator, Headlight range control regulator (Only models with "High" equipment), Auxiliary heater operation relay, Driver vanity mirror contact switch, Light switch, Front left headlight, Left headlight range control motor, Front right headlight, Right headlight range control motor, Onboard supply control unit (Up to June 2016)	SC15
	10 A	Diagnostic connector, Switch and instrument illumination regulator, Auxiliary heater operation relay, Driver vanity mirror contact switch, Light switch, Front left headlight, Front right headlight, Onboard supply control unit (Only models with "Low" equipment) (From July 2016)	
16	10 A	Starter relay 1, Starter relay 2, Air mass meter, Fuel pump control unit, Engine/motor control unit, Cold start device relay	SC16
17	10 A	Control unit for structure-borne sound, Blind spot monitor control unit 2, Blind spot monitor control unit (Up to June 2016)	SC17
	10 A	Control unit for structure-borne sound, Blind spot monitor control unit 2, Blind spot monitor control unit, Terminal 15 voltage supply relay (From July 2016)	

10m. FUSES & RELAYS - JETTA

Chapter 12 Chassis electrical system

18	10 A	Left steering column switch, Right steering column switch (Up to June 2016)	SC18
	10 A	Left steering column switch (Only models with "Low" equipment), Terminal 15 voltage supply relay (Only models with "Low" equipment), Starter relay 1 and 2 (Only models with "Low" equipment) (From July 2016)	
19	-	Not used	SC19
20	10 A	Remote control receiver for auxiliary coolant heater, Rain and light sensor, Heater control unit, Diagnostic connection (Up to June 2016)	SC20
	10 A	Remote control receiver for auxiliary coolant heater, Rain and light sensor, Heater control unit, Diagnostic connection, Climatronic control unit (From July 2016)	
21	15 A	Heated front seats control unit or Driver seat lumbar support adjustment switch	SC21
22	15 A	Onboard supply control unit (Only models with "High" equipment), Alarm system relay, Hazard warning lights, Interior monitoring sensor, Vehicle inclination sender, Alarm horn, 10 A also used (Up to June 2016)	SC22
		Onboard supply control unit (Only models with "High" equipment), Interior monitoring sensor (Only models with "High" equipment), Alarm horn (Only models with "High" equipment), Alarm horn relay (Only models with "Low" equipment) (From July 2016)	
23	10 A	Onboard supply control unit (Only models with "Low" equipment), Light switch,, Ignition/starter switch (Only models with "Low" equipment), Control unit for vehicle location system, Selector lever sensors control unit, Automatic gearbox control unit	SC23
24	10 A	Entry and start authorisation control unit	SC24
25	15 A	Mechatronic unit for dual clutch gearbox, Automatic gearbox control unit, Selector lever sensors control unit, Multifunction switch, 10 A also used	SC25
26	20 A	Vacuum pump for brakes	SC26
27	1 A	Airbag coil connector and return ring with slip ring	SC27
28	10 A	Left washer jet heater element, Right washer jet heater element	SC28
29	5 A	Onboard supply control unit (Only models with "Low" equipment), 1 A also used or Not used	SC29
30	20 A	Onboard supply control unit (Only models with "High" equipment), Cigarette lighter, 12 V sockets, Blocking diode	SC30
31	-	Not used	SC31
32	15 A	Magnetic clutch relay or Not used, 10 A also used	SC32
33	40 A	Auxiliary heater system operation relay, Air conditioning system control unit (Up to June 2016)	SC33
	40 A	Auxiliary heater system operation relay, Air conditioning system control unit, Terminal 75 voltage supply relay 1 (From July 2016)	
34	-	Not used	SC34
35	10 A	Steering column electronics control unit (Only models with "High" equipment), , Data bus diagnostic interface (Only models with "High" equipment), Onboard supply control unit(Only models with "Low" equipment)	SC35
36	20 A	Onboard supply control unit (Only models with "Low" equipment), Headlight washer system relay	SC36
37	30 A	DC/AC converter with socket	SC37
38	30 A	Auxiliary heater control unit, Fan enabling relay, 15 A also used	SC38
39	-	Not used	SC39
40	20 A	Trailer detector control unit, 15 A also used	SC40
41	20 A	Trailer detector control unit, 15 A also used	SC41
42	20 A	Trailer detector control unit	SC42
43	10 A	Converter box, Heated rear window relay, Dual tone horn relay, Ignition/starter switch	SC43
44	30 A	Heated rear window relay unit (Only models with "High" equipment), Onboard supply control unit (Only models with "Low" equipment), Heated rear window, Auxiliary heater operation relay 2 (Up to June 2016)	SC44
	30 A	Heated rear window relay unit (Only models with "High" equipment), Onboard supply control unit (Only models with "Low" equipment), Auxiliary heater operation relay 2 (From July 2016)	
45	30 A	Driver door control unit, Rear left door control unit	SC45
46	30 A	Rear right door control unit, Front passenger door control unit	SC46
47	20 A	Fuel pump control unit, Fuel pump relay, Cold start device relay, 15 A also used (Up to June 2016)	SC47

10n. FUSES & RELAYS - JETTA

Chapter 12 Chassis electrical system

	20 A	Fuel pump control unit, Fuel pump relay, Electric fuel pump relay, 15 A also used (From July 2016)	
48	20 A	Onboard supply control unit	SC48
49	40 A	Auxiliary heater operation relay, Climatronic control unit, Air conditioning system control unit, Heater control unit (Up to June 2016)	SC49
	40 A	Climatronic control unit (From July 2016)	
50	30 A	Heated front seats control unit	SC50
51	20 A	Sliding sunroof adjustment control unit	SC51
52	20 A	Headlight washer system relay, Headlight washer system pump (Up to June 2016)	SC52
	20 A	Headlight washer system relay (From July 2016)	
53	30 A	Onboard supply control unit	SC53
54	-	Not used	SC54
55	15 A	Two-way radio socket, Diagnostic connector or Not used	SC55
56	10 A	Fan enabling relay or Onboard supply control unit (Only models with "Low" equipment)	SC56
57	20 A	Radio, Control unit with display for radio and navigation	SC57
58	20 A	Onboard supply control unit (Only models with "Low" equipment), 15 A also used	SC58
59	20 A	Onboard supply control unit (Only models with "Low" equipment)	SC59
60	20 A	Left and right steering column switches (Only models with "Low" equipment)	SC60

RELAY BOX IN PASSENGER COMPARTMENT

Up to model year 2011

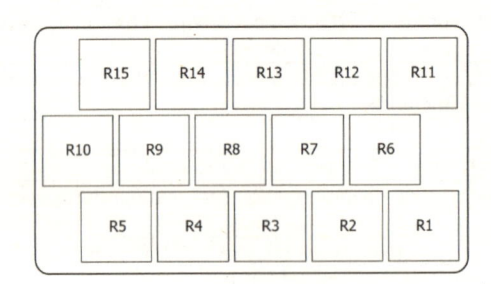

```
R15   R14   R13   R12   R11
R10   R9    R8    R7    R6
R5    R4    R3    R2    R1
```

RELAY	VALUE	DESCRIPTION	OEM NAME
R1	-	Converter box relay	
R2	-	Fresh air blower relay	
R3	-	Auxiliary heater operation relay	
R4	-	Fuel pump relay or Engine component current supply relay or Additional coolant pump relay	
R5	-	Dipped beam relay	
R6	-	Fresh air blower relay	
R7	-	Terminal 75 voltage supply relay 1	
R8	-	Dual tone horn relay or Headlight washer system relay	
R9	-	Terminal 50 voltage supply relay	
R10	-	Terminal 15 voltage supply relay 2	
R11	-	Heated rear window relay	
R12	-	Heater element relay or Cold start device relay or Fuel supply relay	
R13	-	Terminal 50 voltage supply relay	
R14	-	Terminal 15 voltage supply relay	
R15	-	Fog light relay	

10o. FUSES & RELAYS - JETTA

Chapter 12 Chassis electrical system

From model year 2012 up to June 2014

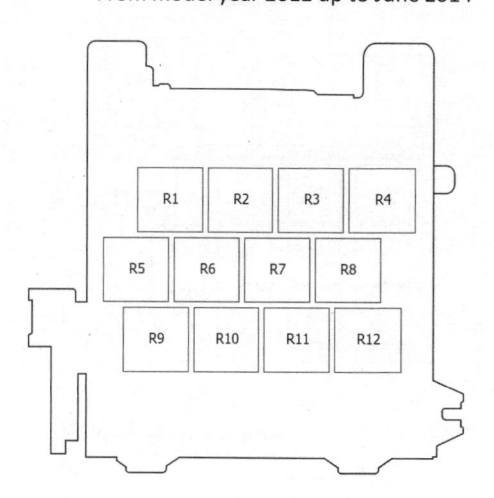

For models with "Low" equipment

RELAY	VALUE	DESCRIPTION	OEM NAME
R1	-	Dual tone horn relay or Alarm horn relay or Headlight washer relay	
R2	-	Terminal 50 voltage supply relay or Starter relay 2	
R3	-	Electric fuel pump relay 2 or Heater element relay or Cold start device relay	
R4	-	Continued coolant circulation relay or Additional coolant pump relay or Fuel pump relay or Fuel supply relay	
R5	-	Terminal 75 voltage supply relay 1 or Not used	
R6	-	Terminal 50 voltage supply relay or Starter relay 1	
R7	-	Terminal 15 voltage supply relay	
R8	-	Dipped beam relay	
R9	-	Terminal 15 voltage supply relay 2	
R10	-	Terminal 75 voltage supply relay 1	
R11	-	Fog light relay	
R12	-	Converter box relay	

For models with "High" equipment

RELAY	VALUE	DESCRIPTION	OEM NAME
R1	-	Dual tone horn relay or Headlight washer system relay	
R2	-	Terminal 50 voltage supply relay or Starter relay 2 or Open window regulator relay, driver side or Close window regulator relay, driver side	
R3	-	Cold start device or Electric fuel pump relay 2 or Heater element relay or Close window regulator relay, driver side	
R4	-	Continued coolant circulation relay or Additional coolant pump relay or Fuel pump relay or Fuel supply relay or Engine component current supply relay	
R5	-	Terminal 75 voltage supply relay 1 or Not used	
R6	-	Terminal 50 voltage supply relay or Starter relay 1	
R7	-	Auxiliary heater operation relay or Window regulator relay	
R8	-	Heated rear window relay	
R9	-	Terminal 15 voltage supply relay 2	
R10	-	Terminal 75 voltage supply relay 1	
R11	-	Fresh air blower relay	
R12	-	Not used	

10p. FUSES & RELAYS - JETTA

Chapter 12 Chassis electrical system

From July 2014

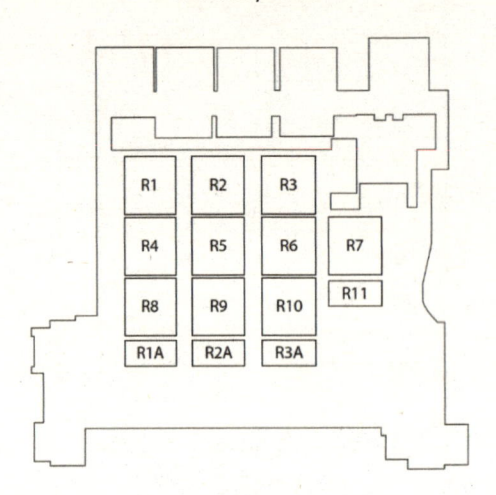

RELAY	VALUE	DESCRIPTION	OEM NAME
R1	-	Terminal 75 voltage supply relay 1	
R2	-	Starter relay 2	
R3	-	Auxiliary heater operation relay or Not used	
R4	-	Isolation relay for powertrain CAN bus (for models with "Low" equipment)	
	-	Not used (for models with "High" equipment)	
R5	-	Converter box or Auxiliary heater operation relay 2 (for models with "Low" equipment)	
	-	Heated rear window relay or Auxiliary heater operation relay 2 (for models with "High" equipment)	
R6	-	Not used or Magnetic clutch relay	
R7	-	Starter relay 1	
R8	-	Not used (for models with "Low" equipment)	
	-	Electric fuel pump 2 relay or Secondary air pump relay or Cold start device relay or Heater element relay or Fresh air blower and air recirculation relay (for models with "High" equipment)	
R9	-	Terminal 15 voltage supply relay	
R10	-	Not used (for models with "Low" equipment)	
	-	Dual tone horn relay or Headlight washer system relay (for models with "High" equipment)	
R11	-	Fuel pump relay or Engine component current supply relay	
R1A	-	Not used	
R2A	-	Alarm horn relay or Not used	
R3A	-	Not used	

10q. FUSES & RELAYS - JETTA

Index

A

About this manual, 0-5
Accelerator Pedal Position (APP) sensor - replacement, 6-14
Air conditioning
    And heating system - check and maintenance, 3-5
    Compressor - removal and installation, 3-20
    Condenser - removal and installation, 3-22
    Evaporator core - removal and installation, 3-22
    Expansion valve - general information, 3-23
    High-pressure sensor - replacement, 3-23
    Refrigerant desiccant cartridge - replacement, 3-21
Air filter
    Check and replacement, 1-14
    Housing - removal and installation, 4-10
Airbag - general information and component removal and installation, 12-21
All-wheel drive
    Clutch - removal and installation, 8-13
    Clutch pump - removal and installation, 8-13
    Control module - replacement, 8-12
Alternator - removal and installation, 5-6
Antenna - removal and installation, 12-10
Anti-lock Brake System (ABS) and Electronic Stability Program (ESP) - general information, 9-6
Automatic transaxle, 7B-1
    Fluid, 1-26
    Overhaul - general information, 7B-13
    Removal and installation, 7B-11
Automotive chemicals and lubricants, 0-24

B

Back-up light switch - removal and installation, 7A-5
Balance shafts - removal and installation, 2A-38
Balljoints - check and replacement, 10-12
Battery
    And battery tray - removal and installation, 5-4
    Cables - replacement, 5-4
    Check, maintenance and charging, 1-16
    Disconnection and reconnection, 5-3
Blower motor and blower motor resistor - removal and installation, 3-17
Body repair
    Major damage, 11-3
    Minor damage, 11-2
Body, 11-1
    Body repair - major damage, 11-3
    Body repair - minor damage, 11-2
    Bumpers and bumper covers - removal and installation, 11-9
    Center console - removal and installation, 11-27
    Cowl cover - removal and installation, 11-15
    Door lock cylinder, outside handle and latch - removal and installation, 11-21
    Door - removal, installation and adjustment, 11-20
    Door trim panels - removal and installation, 11-18
    Door window glass - removal and installation, 11-23
    Door window regulator and motor - removal and installation, 11-24
    Fastener and trim removal, 11-6
    Fenderwell splash shields - removal and installation, 11-12
    Front fender - removal and installation, 11-13
    General information, 11-2
    Hinges and locks - maintenance, 11-7
    Hood latch and support strut - removal and installation, 11-8
    Hood latch release cable, release lever and safety lever - removal and installation, 11-7
    Hood - removal, installation and adjustment, 11-8
    Instrument panel and crossbeam - removal and installation, 11-37
    Interior trim panels - removal and installation, 11-30
    Liftgate (coupe and wagon models) - removal, installation and adjustment, 11-17
    Liftgate latch, actuator and support struts (coupe and wagon models) - removal and installation, 11-18
    Mirrors - removal and installation, 11-26

IND-1

Radiator grille and radiator shutter assembly - removal and installation, 11-12
Rear parcel shelf - removal and installation, 11-44
Repairing minor paint scratches, 11-2
Seat belts - removal and installation, 11-42
Seats - removal and installation, 11-40
Steering column covers - removal and installation, 11-36
Sunroof - check, adjustments, removal and installation, 11-41
Trunk lid latch, support struts and release switch/backup camera assembly (sedan models) - removal and installation, 11-17
Trunk lid (sedan models) - removal, installation and adjustment, 11-16
Upholstery, carpets and vinyl trim - maintenance, 11-6
Windshield and fixed glass - replacement, 11-7

Booster battery (jump) starting, 0-21

Brake
Disc - inspection, removal and installation, 9-13
Fluid change, 1-25
Hoses and lines - inspection and replacement, 9-19
Hydraulic system - bleeding, 9-20
Light switch - removal and installation, 9-23
Pedal - removal and installation, 9-22
System check, 1-20

Brakes, 9-1
Anti-lock Brake System (ABS) and Electronic Stability Program (ESP) - general information, 9-6
Brake disc - inspection, removal and installation, 9-13
Brake hoses and lines - inspection and replacement, 9-19
Brake hydraulic system - bleeding, 9-20
Brake light switch - removal and installation, 9-23
Brake pedal - removal and installation, 9-22
Disc brake caliper - removal and installation, 9-12
Disc brake pads (front) - replacement, 9-8
Disc brake pads (rear) - replacement, 9-11
Drum brake shoes - replacement, 9-15
General Information, 9-2
Master cylinder - removal and installation, 9-18
Parking brake - check and adjustment, 9-22
Power brake booster - check, removal and installation, 9-21
Troubleshooting, 9-2
Vacuum pump (electric) - removal and installation, 9-23

Vacuum pump (mechanical) - removal and installation, 9-23
Wheel cylinder - removal and installation, 9-18

Bulb replacement, 12-14
Bumpers and bumper covers - removal and installation, 11-9
Buying parts, 0-12

C

Camshaft housing, camshafts, sprockets, roller rocker arms and lash adjusters - removal, inspection and installation, 2A-20
Camshaft Position (CMP) sensor - replacement, 6-14
Camshafts, roller rocker arms and lash adjusters - removal, inspection and installation, 2B-8
Catalytic converter - replacement, 6-19
Center console - removal and installation, 11-27
Charge Air Pressure sensor - replacement, 6-25
Chassis electrical system, 12-1
Airbag - general information and component removal and installation, 12-21
Antenna - removal and installation, 12-10
Bulb replacement, 12-14
Circuit breakers - general information, 12-3
Cruise control system - description and check, 12-20
Electrical connectors - general information, 12-4
Electrical troubleshooting - general information, 12-1
Fuse box locations and general fuse information, 12-2
General Information, 12-1
Headlight bulbs - replacement, 12-12
Headlight housing - removal and installation, 12-11
Headlights - adjustment, 12-13
Horns - replacement, 12-18
Ignition switch and key lock cylinder - replacement, 12-5
Instrument cluster - removal and installation, 12-8
Instrument panel switches - replacement, 12-7
Key fob - battery replacement and transmitter programming, 12-5
Power door lock system - general information, 12-20
Power window system - general information, 12-20
Radio and speakers - removal and installation, 12-8
Rear window defogger - check and repair, 12-10

Index

Relays - general information, 12-4
Steering column switches - replacement, 12-6
Taillight housing - removal and installation, 12-14
Wipers, 12-18
Wiring diagrams - general information, 12-22
Circuit breakers - general information, 12-3
Clutch and driveline, 8-1
All-wheel drive clutch pump - removal and installation, 8-13
All-wheel drive clutch - removal and installation, 8-13
All-wheel drive control module - replacement, 8-12
Clutch - description and check, 8-2
Clutch components - removal, inspection and installation, 8-4
Clutch hydraulic system - bleeding, 8-3
Clutch master cylinder - removal and installation, 8-2
Clutch pedal bracket assembly - removal and installation, 8-7
Clutch pedal springs - removal and installation, 8-8
Clutch pedal switch - removal and installation, 8-7
Clutch release bearing and lever - removal, inspection and installation, 8-6
Clutch release cylinder - removal and installation, 8-4
Driveaxle (front) - removal and installation, 8-9
Driveaxle (rear) - removal and installation, 8-10
Driveaxles - general information and inspection, 8-8
Driveshaft - removal and installation, 8-12
Front axle differential lock and seals - replacement, 8-11
General Information, 8-2
Rear differential - removal and installation, 8-12
Transfer case and seals - replacement, 8-11
Clutch
Components - removal, inspection and installation, 8-4
Description and check, 8-2
Hydraulic system - bleeding, 8-3
Master cylinder - removal and installation, 8-2
Clutch pedal
Bracket assembly - removal and installation, 8-7
Pedal springs - removal and installation, 8-8
Pedal switch - removal and installation, 8-7
Clutch release bearing and lever - removal, inspection and installation, 8-6
Clutch release cylinder - removal and installation, 8-4
Coil spring (rear) - removal and installation, 10-13

Control arm (front) - removal, bushing replacement and installation, 10-11
Control housing cover (five-cylinder engine) - removal and installation, 2C-16
Conversion factors, 0-22
Coolant temperature gauge sending unit - check and replacement, 3-17
Cooling system
Check, 1-19
Servicing, 1-29
Cooling, heating and air conditioning systems, 3-1
Air conditioning and heating system - check and maintenance, 3-5
Air conditioning compressor - removal and installation, 3-20
Air conditioning condenser - removal and installation, 3-22
Air conditioning evaporator core - removal and installation, 3-22
Air conditioning expansion valve - general information, 3-23
Air conditioning high-pressure sensor - replacement, 3-23
Air conditioning refrigerant desiccant cartridge - replacement, 3-21
Blower motor and blower motor resistor - removal and installation, 3-17
Coolant temperature gauge sending unit - check and replacement, 3-17
Engine cooling fans - removal and installation, 3-9
General Information, 3-2
Heater and air conditioning control assembly and flexible shafts - removal and installation, 3-18
Heater and air conditioning housing - removal and installation, 3-20
Heater core - removal and installation, 3-18
Radiator and expansion tank - removal and installation, 3-10
Thermostat/engine temperature control actuator - replacement, 3-7
Troubleshooting, 3-2
Water pump and electric coolant pump - replacement, 3-12
Cowl cover - removal and installation, 11-15
Crankcase Ventilation (CCV) system - component replacement, 6-22
Crankshaft
Removal and installation, 2C-13
Front oil seal and housing - replacement, 2A-31
Front oil seal flange - replacement, 2B-7

Crankshaft Position (CKP)/engine speed sensor - replacement, 6-14
Pulley - removal and installation, 2A-30, 2B-6

Cruise control system - description and check, 12-20

Cylinder compression check, 2C-6

Cylinder head
Four-cylinder engines, removal and installation, 2A-25
Five-cylinder engines, removal, inspection and installation, 2B-9

D

Diagnosis - general, 7B-2

Direct Shift Gearbox (DSG) transaxle fluid and filter change, 1-27

Disc brake caliper - removal and installation, 9-12

Disc brake pads - replacement
Front, 9-8
Rear, 9-11

Door
Lock cylinder, outside handle and latch - removal and installation, 11-21
Removal, installation and adjustment, 11-20
Trim panels - removal and installation, 11-18
Window glass - removal and installation, 11-23
Window regulator and motor - removal and installation, 11-24

Driveaxle
Front - removal and installation, 8-9
Rear - removal and installation, 8-10

Driveaxles - general information and inspection, 8-8

Drivebelt check and replacement, 1-23

Driveshaft - removal and installation, 8-12

Drum brake shoes - replacement, 9-15

E

Electrical connectors - general information, 12-4

Electrical troubleshooting - general information, 12-1

Electronic control system, 7B-3

Emissions and engine control systems, 6-1
Accelerator Pedal Position (APP) sensor - replacement, 6-14
Camshaft Position (CMP) sensor - replacement, 6-14
Catalytic converter - replacement, 6-19
Charge Air Pressure sensor - replacement, 6-25
Crankcase Ventilation (CCV) system - component replacement, 6-22
Crankshaft Position (CKP)/engine speed sensor - replacement, 6-14
Engine Coolant Temperature (ECT) sensor - replacement, 6-15
Evaporative Emissions Control (EVAP) system - component replacement, 6-20
Fuel Pressure Sensor - replacement, 6-25
General Information, 6-1
Intake Air Temperature (IAT) sensor - replacement, 6-17
Knock sensor - replacement, 6-16
Manifold Absolute Pressure (MAP)/Intake Air Temperature (IAT) sensor - replacement, 6-17
Mass Airflow (MAF) sensor - replacement, 6-17
Obtaining and clearing Diagnostic Trouble Codes (DTCs), 6-4
Oil pressure switch - replacement, 6-24
On Board Diagnosis (OBD) system, 6-2
Oxygen sensors - replacement, 6-17
Powertrain Control Module (PCM) - removal and installation, 6-19
Secondary Air Injection (AIR) system - component replacement, 6-22
Transmission speed sensors - replacement, 6-18
Variable camshaft adjustment solenoid valve - replacement, 6-24
Variable intake manifold actuator (1.8L and 2.0L turbocharged engines) - replacement, 6-25

Engine, Four-cylinder, 2A-1
Balance shafts - removal and installation, 2A-38
Camshaft housing, camshafts, sprockets, roller rocker arms and lash adjusters - removal, inspection and installation, 2A-20
Crankshaft front oil seal and housing - replacement, 2A-31
Crankshaft pulley - removal and installation, 2A-30
Cylinder head - removal and installation, 2A-25
Engine mounts - check and replacement, 2A-40
Engine oil cooler - removal and installation, 2A-42
Flywheel/driveplate - removal and installation, 2A-35
General Information, 2A-6
Intake manifold - removal and installation, 2A-18
Oil pan(s) - removal and installation, 2A-32
Oil pump - removal and installation, 2A-34
Oil separator and crankcase ventilator - removal and installation, 2A-39
Rear main oil seal - replacement, 2A-36

Index

Repair operations possible with the engine in the vehicle, 2A-7
Timing belt and sprockets - removal, inspection and installation, 2A-9
Timing chain cover, timing chains and tensioners - removal and installation, 2A-13
Top Dead Center (TDC) for number one piston - locating, 2A-7
Valve cover - removal and installation, 2A-9

Engine, Five-cylinder, 2B-1
Camshafts, roller rocker arms and lash adjusters - removal, inspection and installation, 2B-8
Crankshaft front oil seal flange - replacement, 2B-7
Crankshaft pulley - removal and installation, 2B-6
Cylinder head - removal, inspection and installation, 2B-9
Engine mounts - check and replacement, 2B-12
Exhaust manifold - removal, inspection and installation, 2B-6
General Information, 2B-3
Intake manifold - removal and installation, 2B-5
Oil pan(s) - removal and installation, 2B-11
Rear main oil seal - replacement, 2B-11
Repair operations possible with the engine in the vehicle, 2B-3
Top Dead Center (TDC) for number five piston - locating, 2B-3
Upper timing chain cover, cover seal, camshaft timing chain and tensioner - removal and installation, 2B-6
Valve cover - removal and installation, 2B-4
Valve timing - check and adjustment, 2B-4

Engine, General overhaul procedures, 2C-1
Control housing cover (five-cylinder engine) - removal and installation, 2C-16
Crankshaft - removal and installation, 2C-13
Cylinder compression check, 2C-6
Engine - removal and installation, 2C-8
Engine overhaul - disassembly sequence, 2C-9
Engine overhaul - reassembly sequence, 2C-15
Engine rebuilding alternatives, 2C-7
Engine removal - methods and precautions, 2C-7
General information - engine overhaul, 2C-4
Initial start-up and break-in after overhaul, 2C-17
Oil pressure check, 2C-5
Oil pump, timing chain and tensioner (five-cylinder engine) - removal and installation, 2C-16
Pistons and connecting rods - removal and installation, 2C-10
Upper oil pan (five-cylinder engine) - removal and installation, 2C-16

Vacuum gauge diagnostic checks, 2C-6
Engine - removal and installation, 2C-8
Engine Coolant Temperature (ECT) sensor - replacement, 6-15
Engine cooling fans - removal and installation, 3-9
Engine electrical systems, 5-1
Alternator - removal and installation, 5-6
Battery and battery tray - removal and installation, 5-4
Battery cables - replacement, 5-4
Battery - disconnection and reconnection, 5-3
General information and precautions, 5-1
Ignition coils - removal and installation, 5-5
Starter motor - removal and installation, 5-8
Troubleshooting, 5-2
Engine mounts
Four-cylinder engines, check and replacement, 2A-40
Five-cylinder engines, check and replacement, 2B-12
Engine oil and filter change, 1-11
Engine oil cooler - removal and installation, 2A-42
Engine overhaul
Disassembly sequence, 2C-9
Reassembly sequence, 2C-15
Engine rebuilding alternatives, 2C-7
Engine removal - methods and precautions, 2C-7
Evaporative Emissions Control (EVAP) system - component replacement, 6-20
Exhaust manifold - removal, inspection and installation, 2B-6
Exhaust system
Check, 1-19
Servicing - general information, 4-16

F

Fastener and trim removal, 11-6
Fenderwell splash shields - removal and installation, 11-12
Five-cylinder engines, 2B-1
Camshafts, roller rocker arms and lash adjusters - removal, inspection and installation, 2B-8
Crankshaft front oil seal flange - replacement, 2B-7
Crankshaft pulley - removal and installation, 2B-6
Cylinder head - removal, inspection and installation, 2B-9
Engine mounts - check and replacement, 2B-12
Exhaust manifold - removal, inspection and installation, 2B-6

General Information, 2B-3
Intake manifold - removal and installation, 2B-5
Oil pan(s) - removal and installation, 2B-11
Rear main oil seal - replacement, 2B-11
Repair operations possible with the engine in the vehicle, 2B-3
Top Dead Center (TDC) for number five piston - locating, 2B-3
Upper timing chain cover, cover seal, camshaft timing chain and tensioner - removal and installation, 2B-6
Valve cover - removal and installation, 2B-4
Valve timing - check and adjustment, 2B-4

Fluid level checks, 1-8

Flywheel/driveplate - removal and installation, 2A-35

Four-cylinder engines, 2A-1
Balance shafts - removal and installation, 2A-38
Camshaft housing, camshafts, sprockets, roller rocker arms and lash adjusters - removal, inspection and installation, 2A-20
Crankshaft front oil seal and housing - replacement, 2A-31
Crankshaft pulley - removal and installation, 2A-30
Cylinder head - removal and installation, 2A-25
Engine mounts - check and replacement, 2A-40
Engine oil cooler - removal and installation, 2A-42
Flywheel/driveplate - removal and installation, 2A-35
General Information, 2A-6
Intake manifold - removal and installation, 2A-18
Oil pan(s) - removal and installation, 2A-32
Oil pump - removal and installation, 2A-34
Oil separator and crankcase ventilator - removal and installation, 2A-39
Rear main oil seal - replacement, 2A-36
Repair operations possible with the engine in the vehicle, 2A-7
Timing belt and sprockets - removal, inspection and installation, 2A-9
Timing chain cover, timing chains and tensioners - removal and installation, 2A-13
Top Dead Center (TDC) for number one piston - locating, 2A-7
Valve cover - removal and installation, 2A-9

Fraction/decimal/millimeter equivalents, 0-23

Front axle differential lock and seals - replacement, 8-11

Front fender - removal and installation, 11-13

Fuel and exhaust systems, 4-1
Air filter housing - removal and installation, 4-10
Exhaust system servicing - general information, 4-16
Fuel lines and fittings - general information and disconnection, 4-4
Fuel pressure - check, 4-3
Fuel pressure relief procedure, 4-3
Fuel pump module/fuel level sender - removal and installation, 4-6
Fuel rail and injectors - removal and installation, 4-12
Fuel system - bleeding (five-cylinder models only), 4-4
Fuel tank - removal and installation, 4-8
General Information, 4-1
High-pressure fuel pump (Direct Injection models) - removal and installation, 4-8
Throttle body/control module - removal and installation, 4-10
Transfer pump/fuel level sender - replacement, 4-7
Troubleshooting, 4-2
Turbocharger/exhaust manifold and charge air cooler - check and replacement, 4-16

Fuel filter replacement, 1-28

Fuel lines and fittings - general information and disconnection, 4-4

Fuel pressure - check, 4-3

Fuel pressure relief procedure, 4-3

Fuel Pressure Sensor - replacement, 6-25

Fuel pump module/fuel level sender - removal and installation, 4-6

Fuel rail and injectors - removal and installation, 4-12

Fuel system
Bleeding (five-cylinder models only), 4-4
Check, 1-23

Fuel tank - removal and installation, 4-8

Fuse box locations and general fuse information, 12-2

G

General engine overhaul procedures, 2C-1
Control housing cover (five-cylinder engine) - removal and installation, 2C-16
Crankshaft - removal and installation, 2C-13
Cylinder compression check, 2C-6
Engine - removal and installation, 2C-8
Engine overhaul - disassembly sequence, 2C-9
Engine overhaul - reassembly sequence, 2C-15
Engine rebuilding alternatives, 2C-7
Engine removal - methods and precautions, 2C-7

General information - engine overhaul, 2C-4
Initial start-up and break-in after overhaul, 2C-17
Oil pressure check, 2C-5
Oil pump, timing chain and tensioner (five-cylinder engine) - removal and installation, 2C-16
Pistons and connecting rods - removal and installation, 2C-10
Upper oil pan (five-cylinder engine) - removal and installation, 2C-16
Vacuum gauge diagnostic checks, 2C-6

H
Headlight
Bulbs - replacement, 12-12
Housing - removal and installation, 12-11
Headlights - adjustment, 12-13
Heater and air conditioning
Control assembly and flexible shafts - removal and installation, 3-18
Housing - removal and installation, 3-20
Heater core - removal and installation, 3-18
High-pressure fuel pump (Direct Injection models) - removal and installation, 4-8
Hinges and locks - maintenance, 11-7
Hood - removal, installation and adjustment, 11-8
Hood latch
and support strut - removal and installation, 11-8
Release cable, release lever and safety lever - removal and installation, 11-7
Horns - replacement, 12-18
Hub and wheel bearing
Front - removal and installation, 10-10
Rear - removal and installation, 10-14

I
Ignition
Coils - removal and installation, 5-5
Switch and key lock cylinder - replacement, 12-5
Initial start-up and break-in after overhaul, 2C-17
Instrument cluster - removal and installation, 12-8
Instrument panel and crossbeam - removal and installation, 11-37
Instrument panel switches - replacement, 12-7
Intake Air Temperature (IAT) sensor - replacement, 6-17
Intake manifold - removal and installation
Four-cylinder engines, 2A-18
Five-cylinder engines, 2B-5

Interior
Trim panels - removal and installation, 11-30
Ventilation filter and fresh air intake filter replacement, 1-20

J
Jacking and towing, 0-20

K
Key fob - battery replacement and transmitter programming, 12-5
Knock sensor - replacement, 6-16
Knuckle (rear, multi-link suspension) - removal and installation, 10-14

L
Liftgate (coupe and wagon models) - removal, installation and adjustment, 11-17
Liftgate latch, actuator and support struts (coupe and wagon models) - removal and installation, 11-18

M
Maintenance schedule, 1-4
Maintenance techniques, tools and working facilities, 0-12
Manifold Absolute Pressure (MAP)/Intake Air Temperature (IAT) sensor - replacement, 6-17
Manual transaxle, 7A-1
Removal and installation, 7A-4
Lubricant change, 1-28
Overhaul - general information, 7A-5
Mass Airflow (MAF) sensor - replacement, 6-17
Master cylinder - removal and installation, 9-18
Mirrors - removal and installation, 11-26

O
Obtaining and clearing Diagnostic Trouble Codes (DTCs), 6-4
Oil pan(s) - removal and installation
Four-cylinder engines, 2A-32
Five-cylinder engines, 2B-11
Oil pressure check, 2C-5
Oil pressure switch - replacement, 6-24

Oil pump - removal and installation, 2A-34
Oil pump, timing chain and tensioner (five-cylinder engine) - removal and installation, 2C-16
Oil seals - replacement, 7A-6
Oil separator and crankcase ventilator - removal and installation, 2A-39
On Board Diagnosis (OBD) system, 6-2
Oxygen sensors - replacement, 6-17

P

Parking brake - check and adjustment, 9-22
Pistons and connecting rods - removal and installation, 2C-10
Power brake booster - check, removal and installation, 9-21
Power door lock system - general information, 12-20
Power steering
    Pump removal and installation, 10-22
    System - bleeding, 10-22
Power window system - general information, 12-20
Powertrain Control Module (PCM) - removal and installation, 6-19

R

Radiator and expansion tank - removal and installation, 3-10
Radiator grille and radiator shutter assembly - removal and installation, 11-12
Radio and speakers - removal and installation, 12-8
Rear axle beam (torsion bear suspension) - removal and installation, 10-17
Rear differential - removal and installation, 8-12
Rear main oil seal – replacement
    Four-cylinder engines, 2A-36
    Five-cylinder engines, 2B-11
Rear parcel shelf - removal and installation, 11-44
Rear suspension arms - removal and installation, 10-15
Rear window defogger - check and repair, 12-10
Recall information, 0-7
Relays - general information, 12-4
Repair operations possible with the engine in the vehicle
    Four-cylinder engines, 2A-7
    Five-cylinder engines, 2B-3
Repairing minor paint scratches, 11-2

S

Safety first!, 0-25
Seat belt
    Check, 1-18
    Removal and installation, 11-42
Seats - removal and installation, 11-40
Secondary Air Injection (AIR) system - component replacement, 6-22
Shift cables - removal, installation and adjustment
    Manual transaxle, 7A-3
    Automatic transaxle, 7B-9
Shift interlock system - description and check, 7B-9
Shift knob - removal and installation, 7B-6
Shift lever assembly - removal and installation
    Manual transaxle, 7A-2
    Automatic transaxle, 7B-8
Shock absorber (rear) - removal and installation, 10-13
Spark plug check and replacement, 1-25
Stabilizer bar and bushings - removal and installation
    Front, 10-11
    Rear, multi-link suspension, 10-14
Starter motor - removal and installation, 5-8
Steering column
    Covers - removal and installation, 11-36
    Removal and installation, 10-19
    Switches - replacement, 12-6
Steering gear - removal and installation, 10-21
Steering gear boots - replacement, 10-21
Steering knuckle - removal and installation, 10-10
Steering wheel - removal and installation, 10-18
Strut assembly - removal, inspection and installation, 10-7
Strut/coil spring assembly - replacement, 10-8
Stub axle (rear, torsion beam suspension) - removal and installation, 10-14
Subframe - removal and installation, 10-17
Sunroof - check, adjustments, removal and installation, 11-41
Suspension and steering systems, 10-1
    Balljoints - check and replacement, 10-12
    Coil spring (rear) - removal and installation, 10-13
    Control arm (front) - removal, bushing replacement and installation, 10-11
    General information and precautions, 10-7
    Hub and wheel bearing (front) - removal and installation, 10-10

Hub and wheel bearing (rear) - removal and installation, 10-14
Knuckle (rear, multi-link suspension) - removal and installation, 10-14
Power steering pump - removal and installation, 10-22
Power steering system - bleeding, 10-22
Rear axle beam (torsion bear suspension) - removal and installation, 10-17
Rear suspension arms - removal and installation, 10-15
Shock absorber (rear) - removal and installation, 10-13
Stabilizer bar and bushings (front) - removal and installation, 10-11
Stabilizer bar and bushings (rear, multi-link suspension) - removal and installation, 10-14
Steering column - removal and installation, 10-19
Steering gear - removal and installation, 10-21
Steering gear boots - replacement, 10-21
Steering knuckle - removal and installation, 10-10
Steering wheel - removal and installation, 10-18
Strut assembly - removal, inspection and installation, 10-7
Strut/coil spring assembly - replacement, 10-8
Stub axle (rear, torsion beam suspension) - removal and installation, 10-14
Subframe - removal and installation, 10-17
Tie-rod ends - removal and installation, 10-20
Wheel alignment - general information, 10-23
Wheels and tires - general information, 10-23

Suspension, steering and driveaxle boot check, 1-22

T

Taillight housing - removal and installation, 12-14
Thermostat/engine temperature control actuator - replacement, 3-7
Throttle body/control module - removal and installation, 4-10
Tie-rod ends - removal and installation, 10-20
Timing belt and sprockets - removal, inspection and installation, 2A-9
Timing chain cover, timing chains and tensioners - removal and installation, 2A-13
Tire and tire pressure checks, 1-11
Tire rotation, 1-18
Top Dead Center (TDC) – locating
Four-cylinder engines, number one piston, 2A-7
Five-cylinder engines, number five piston, 2B-3

Transaxle, automatic, 7B-1
Automatic transaxle overhaul - general information, 7B-13
Automatic transaxle - removal and installation, 7B-11
Diagnosis - general, 7B-2
Electronic control system, 7B-3
General Information, 7B-2
Shift cable - removal, installation and adjustment, 7B-9
Shift interlock system - description and check, 7B-9
Shift knob - removal and installation, 7B-6
Shift lever assembly - removal and installation, 7B-8
Transaxle fluid cooler - removal and installation, 7B-10
Transaxle mounts - check and replacement, 7B-13
Transaxle range switch - replacement and adjustment, 7B-10
Transmission Control Module (TCM) - removal and installation, 7B-6

Transaxle, manual, 7A-1
Back-up light switch - removal and installation, 7A-5
General Information, 7A-2
Manual transaxle - removal and installation, 7A-4
Manual transaxle overhaul - general information, 7A-5
Oil seals - replacement, 7A-6
Shift cables - removal, installation and adjustment, 7A-3
Shift lever assembly - removal and installation, 7A-2
Transaxle mounts - check and replacement, 7A-5

Transaxle fluid cooler - removal and installation, 7B-10
Transaxle mounts - check and replacement
Manual transaxle, 7A-5
Automatic transaxle, 7B-13
Transaxle range switch - replacement and adjustment, 7B-10
Transfer case (bevel box), AWD coupler (Haldex) and rear differential (AWD models) - check and lubricant replacement, 1-31
Transfer case and seals - replacement, 8-11
Transfer pump/fuel level sender - replacement, 4-7
Transmission Control Module (TCM) - removal and installation, 7B-6
Transmission speed sensors - replacement, 6-18
Troubleshooting
Brakes, 9-2
Cooling, heating and air conditioning systems, 3-2
Engine electrical systems, 5-2
Fuel and exhaust systems, 4-2
General, 0-26

Trunk lid (sedan models) - removal, installation and adjustment, 11-16
Trunk lid latch, support struts and release switch/backup camera assembly (sedan models) - removal and installation, 11-17
Tune-up and routine maintenance, 1-1
- Air filter check and replacement, 1-14
- Automatic transaxle fluid, 1-26
- Battery check, maintenance and charging, 1-16
- Brake fluid change, 1-25
- Brake system check, 1-20
- Cooling system check, 1-19
- Cooling system servicing, 1-29
- Direct Shift Gearbox (DSG) transaxle fluid and filter change, 1-27
- Drivebelt check and replacement, 1-23
- Engine oil and filter change, 1-11
- Exhaust system check, 1-19
- Fluid level checks, 1-8
- Fuel filter replacement, 1-28
- Fuel system check, 1-23
- Interior ventilation filter and fresh air intake filter replacement, 1-20
- Introduction, 1-5
- Maintenance schedule, 1-4
- Manual transaxle lubricant change, 1-28
- Seat belt check, 1-18
- Spark plug check and replacement, 1-25
- Suspension, steering and driveaxle boot check, 1-22
- Tire and tire pressure checks, 1-11
- Tire rotation, 1-18
- Transfer case (bevel box), AWD coupler (Haldex) and rear differential (AWD models) - check and lubricant replacement, 1-31
- Tune-up general information, 1-8
- Underhood hose check and replacement, 1-18
- Windshield wiper blade inspection and replacement, 1-16

Tune-up general information, 1-8
Turbocharger/exhaust manifold and charge air cooler - check and replacement, 4-16

U

Underhood hose check and replacement, 1-18
Upholstery, carpets and vinyl trim - maintenance, 11-6
Upper oil pan (five-cylinder engine) - removal and installation, 2C-16
Upper timing chain cover, cover seal, camshaft timing chain and tensioner - removal and installation, 2B-6

V

Vacuum gauge diagnostic checks, 2C-6
Vacuum pump - removal and installation
- Electric, 9-23
- Mechanical, 9-23

Valve cover - removal and installation
- Four-cylinder engines, number one piston, 2A-9
- Five-cylinder engines, number five piston, 2B-4

Valve timing - check and adjustment, 2B-4
Variable camshaft adjustment solenoid valve - replacement, 6-24
Variable intake manifold actuator (1.8L and 2.0L turbocharged engines) - replacement, 6-25
Vehicle identification numbers, 0-6

W

Water pump and electric coolant pump - replacement, 3-12
Wheel
- Alignment - general information, 10-23
- Cylinder - removal and installation, 9-18

Wheels and tires - general information, 10-23
Windshield and fixed glass - replacement, 11-7
Windshield wiper blade inspection and replacement, 1-16
Wipers, 12-18
Wiring diagrams - general information, 12-22

Haynes Automotive Manuals

NOTE: If you do not see a listing for your vehicle, please visit **haynes.com** *for the latest product information and check out our* **Online Manuals!**

ACURA
- 12020 Integra '86 thru '89 & Legend '86 thru '90
- 12021 Integra '90 thru '93 & Legend '91 thru '95
- Integra '94 thru '00 - see HONDA Civic (42025)
- MDX '01 thru '07 - see HONDA Pilot (42037)
- 12050 Acura TL all models '99 thru '08

AMC
- 14020 Mid-size models '70 thru '83
- 14025 (Renault) Alliance & Encore '83 thru '87

AUDI
- 15020 4000 all models '80 thru '87
- 15025 5000 all models '77 thru '83
- 15026 5000 all models '84 thru '88
- Audi A4 '96 thru '01 - see VW Passat (96023)
- 15030 Audi A4 '02 thru '08

AUSTIN-HEALEY
- Sprite - see MG Midget (66015)

BMW
- 18020 3/5 Series '82 thru '92
- 18021 3-Series incl. Z3 models '92 thru '98
- 18022 3-Series incl. Z4 models '99 thru '05
- 18023 3-Series '06 thru '14
- 18025 320i all 4-cylinder models '75 thru '83
- 18050 1500 thru 2002 except Turbo '59 thru '77

BUICK
- 19010 Buick Century '97 thru '05
- Century (front-wheel drive) - see GM (38005)
- 19020 Buick, Oldsmobile & Pontiac Full-size (Front-wheel drive) '85 thru '05
- Buick Electra, LeSabre and Park Avenue; Oldsmobile Delta 88 Royale, Ninety Eight and Regency; Pontiac Bonneville
- 19025 Buick, Oldsmobile & Pontiac Full-size (Rear wheel drive) '70 thru '90
- Buick Estate, Electra, LeSabre, Limited, Oldsmobile Custom Cruiser, Delta 88, Ninety-eight, Pontiac Bonneville, Catalina, Grandville, Parisienne
- 19027 Buick LaCrosse '05 thru '13
- Enclave - see GENERAL MOTORS (38001)
- Rainier - see CHEVROLET (24072)
- Regal - see GENERAL MOTORS (38010)
- Riviera - see GENERAL MOTORS (38030, 38031)
- Roadmaster - see CHEVROLET (24046)
- Skyhawk - see GENERAL MOTORS (38015)
- Skylark - see GENERAL MOTORS (38020, 38025)
- Somerset - see GENERAL MOTORS (38025)

CADILLAC
- 21015 CTS & CTS-V '03 thru '14
- 21030 Cadillac Rear Wheel Drive '70 thru '93
- Cimarron - see GENERAL MOTORS (38015)
- DeVille - see GENERAL MOTORS (38031 & 38032)
- Eldorado - see GENERAL MOTORS (38030)
- Fleetwood - see GENERAL MOTORS (38031)
- Seville - see GM (38030, 38031 & 38032)

CHEVROLET
- 10305 Chevrolet Engine Overhaul Manual
- 24010 Astro & GMC Safari Mini-vans '85 thru '05
- 24013 Aveo '04 thru '11
- 24015 Camaro V8 all models '70 thru '81
- 24016 Camaro all models '82 thru '92
- 24017 Camaro & Firebird '93 thru '02
- Cavalier - see GENERAL MOTORS (38016)
- Celebrity - see GENERAL MOTORS (38005)
- 24018 Camaro '10 thru '15
- 24020 Chevelle, Malibu & El Camino '69 thru '87
- Cobalt - see GENERAL MOTORS (38017)
- 24024 Chevette & Pontiac T1000 '76 thru '87
- Citation - see GENERAL MOTORS (38020)
- 24027 Colorado & GMC Canyon '04 thru '12
- 24032 Corsica & Beretta all models '87 thru '96
- 24040 Corvette all V8 models '68 thru '82
- 24041 Corvette all models '84 thru '96
- 24042 Corvette all models '97 thru '13
- 24044 Cruze '11 thru '19
- 24045 Full-size Sedans Caprice, Impala, Biscayne, Bel Air & Wagons '69 thru '90
- 24046 Impala SS & Caprice and Buick Roadmaster '91 thru '96
- Impala '00 thru '05 - see LUMINA (24048)
- 24047 Impala & Monte Carlo all models '06 thru '11
- Lumina '90 thru '94 - see GM (38010)
- 24048 Lumina & Monte Carlo '95 thru '05
- Lumina APV - see GM (38035)
- 24050 Luv Pick-up all 2WD & 4WD '72 thru '82
- 24051 Malibu '13 thru '19
- 24055 Monte Carlo all models '70 thru '88
- Monte Carlo '95 thru '01 - see LUMINA (24048)
- 24059 Nova all V8 models '69 thru '79
- 24060 Nova and Geo Prizm '85 thru '92
- 24064 Pick-ups '67 thru '87 - Chevrolet & GMC
- 24065 Pick-ups '88 thru '98 - Chevrolet & GMC
- 24066 Pick-ups '99 thru '06 - Chevrolet & GMC
- 24067 Chevrolet Silverado & GMC Sierra '07 thru '14
- 24068 Chevrolet Silverado & GMC Sierra '14 thru '19
- 24070 S-10 & S-15 Pick-ups '82 thru '93, Blazer & Jimmy '83 thru '94,
- 24071 S-10 & Sonoma Pick-ups '94 thru '04, including Blazer, Jimmy & Hombre
- 24072 Chevrolet TrailBlazer, GMC Envoy & Oldsmobile Bravada '02 thru '09
- 24075 Sprint '85 thru '88 & Geo Metro '89 thru '01
- 24080 Vans - Chevrolet & GMC '68 thru '96
- 24081 Chevrolet Express & GMC Savana Full-size Vans '96 thru '19

CHRYSLER
- 10310 Chrysler Engine Overhaul Manual
- 25015 Chrysler Cirrus, Dodge Stratus, Plymouth Breeze '95 thru '00
- 25020 Full-size Front-Wheel Drive '88 thru '93
- K-Cars - see DODGE Aries (30008)
- Laser - see DODGE Daytona (30030)
- 25025 Chrysler LHS, Concorde, New Yorker, Dodge Intrepid, Eagle Vision, '93 thru '97
- 25026 Chrysler LHS, Concorde, 300M, Dodge Intrepid, '98 thru '04
- 25027 Chrysler 300 '05 thru '18, Dodge Charger '06 thru '18, Magnum '05 thru '08 & Challenger '08 thru '18
- 25030 Chrysler & Plymouth Mid-size front wheel drive '82 thru '95
- Rear-wheel Drive - see Dodge (30050)
- 25035 PT Cruiser all models '01 thru '10
- 25040 Chrysler Sebring '95 thru '06, Dodge Stratus '01 thru '06 & Dodge Avenger '95 thru '00
- 25041 Chrysler Sebring '07 thru '10, 200 '11 thru '17 Dodge Avenger '08 thru '14

DATSUN
- 28005 200SX all models '80 thru '83
- 28012 240Z, 260Z & 280Z Coupe '70 thru '78
- 28014 280ZX Coupe & 2+2 '79 thru '83
- 300ZX - see NISSAN (72010)
- 28018 510 & PL521 Pick-up '68 thru '73
- 28020 510 all models '78 thru '81
- 28022 620 Series Pick-up all models '73 thru '79
- 720 Series Pick-up - see NISSAN (72030)

DODGE
- 400 & 600 - see CHRYSLER (25030)
- 30008 Aries & Plymouth Reliant '81 thru '89
- 30010 Caravan & Plymouth Voyager '84 thru '95
- 30011 Caravan & Plymouth Voyager '96 thru '02
- 30012 Challenger & Plymouth Sapporro '78 thru '83
- 30013 Caravan, Chrysler Voyager & Town & Country '03 thru '07
- 30014 Grand Caravan & Chrysler Town & Country '08 thru '18
- 30016 Colt & Plymouth Champ '78 thru '87
- 30020 Dakota Pick-ups all models '87 thru '96
- 30021 Durango '98 & '99 & Dakota '97 thru '99
- 30022 Durango '00 thru '03 & Dakota '00 thru '04
- 30023 Durango '04 thru '09 & Dakota '05 thru '11
- 30025 Dart, Demon, Plymouth Barracuda, Duster & Valiant 6-cylinder models '67 thru '76
- 30030 Daytona & Chrysler Laser '84 thru '89
- Intrepid - see CHRYSLER (25025, 25026)
- 30034 Neon all models '95 thru '99
- 30035 Omni & Plymouth Horizon '78 thru '90
- 30036 Dodge & Plymouth Neon '00 thru '05
- 30040 Pick-ups full-size models '74 thru '93
- 30042 Pick-ups full-size models '94 thru '08
- 30043 Pick-ups full-size models '09 thru '18
- 30045 Ram 50/D50 Pick-ups & Raider and Plymouth Arrow Pick-ups '79 thru '93
- 30050 Dodge/Plymouth/Chrysler RWD '71 thru '89
- 30055 Shadow & Plymouth Sundance '87 thru '94
- 30060 Spirit & Plymouth Acclaim '89 thru '95
- 30065 Vans - Dodge & Plymouth '71 thru '03

EAGLE
- Talon - see MITSUBISHI (68030, 68031)
- Vision - see CHRYSLER (25025)

FIAT
- 34010 124 Sport Coupe & Spider '68 thru '78
- 34025 X1/9 all models '74 thru '80

FORD
- 10320 Ford Engine Overhaul Manual
- 10355 Ford Automatic Transmission Overhaul
- 11500 Mustang '64-1/2 thru '70 Restoration Guide
- 36004 Aerostar Mini-vans all models '86 thru '97
- 36006 Contour & Mercury Mystique '95 thru '00
- 36008 Courier Pick-up all models '72 thru '82
- 36012 Crown Victoria & Mercury Grand Marquis '88 thru '11
- 36014 Edge '07 thru '19 & Lincoln MKX '07 thru '18
- 36016 Escort & Mercury Lynx all models '81 thru '90
- 36020 Escort & Mercury Tracer '91 thru '02
- 36022 Escape '01 thru '17, Mazda Tribute '01 thru '11, & Mercury Mariner '05 thru '11
- 36024 Explorer & Mazda Navajo '91 thru '01
- 36025 Explorer & Mercury Mountaineer '02 thru '10
- 36026 Explorer '11 thru '17
- 36028 Fairmont & Mercury Zephyr '78 thru '83
- 36030 Festiva & Aspire '88 thru '97
- 36032 Fiesta all models '77 thru '80
- 36034 Focus all models '00 thru '11
- 36035 Focus '12 thru '14
- 36045 Fusion '06 thru '14 & Mercury Milan '06 thru '11
- 36048 Mustang V8 all models '64-1/2 thru '73
- 36049 Mustang II 4-cylinder, V6 & V8 models '74 thru '78
- 36050 Mustang & Mercury Capri '79 thru '93
- 36051 Mustang all models '94 thru '04
- 36052 Mustang '05 thru '14
- 36054 Pick-ups & Bronco '73 thru '79
- 36058 Pick-ups & Bronco '80 thru '96
- 36059 F-150 '97 thru '03, Expedition '97 thru '17, F-250 '97 thru '99, F-150 Heritage '04 & Lincoln Navigator '98 thru '17
- 36060 Super Duty Pick-ups & Excursion '99 thru '10
- 36061 F-150 full-size '04 thru '14
- 36062 Pinto & Mercury Bobcat '75 thru '80
- 36063 F-150 full-size '15 thru '17
- 36064 Super Duty Pick-ups '11 thru '16
- 36066 Probe all models '89 thru '92
- Probe '93 thru '97 - see MAZDA 626 (61042)
- 36070 Ranger & Bronco II gas models '83 thru '92
- 36071 Ranger '93 thru '11 & Mazda Pick-ups '94 thru '09
- 36074 Taurus & Mercury Sable '86 thru '95
- 36075 Taurus & Mercury Sable '96 thru '07
- 36076 Taurus '08 thru '14, Five Hundred '05 thru '07, Mercury Montego '05 thru '07 & Sable '08 thru '09
- 36078 Tempo & Mercury Topaz '84 thru '94
- 36082 Thunderbird & Mercury Cougar '83 thru '88
- 36086 Thunderbird & Mercury Cougar '89 thru '97
- 36090 Vans all V8 Econoline models '69 thru '91
- 36094 Vans full size '92 thru '14
- 36097 Windstar '95 thru '03, Freestar & Mercury Monterey Mini-van '04 thru '07

GENERAL MOTORS
- 10360 GM Automatic Transmission Overhaul
- 38001 GMC Acadia '07 thru '16, Buick Enclave '08 thru '17, Saturn Outlook '07 thru '10 & Chevrolet Traverse '09 thru '17
- 38005 Buick Century, Chevrolet Celebrity, Oldsmobile Cutlass Ciera & Pontiac 6000 all models '82 thru '96
- 38010 Buick Regal '88 thru '04, Chevrolet Lumina '88 thru '04, Oldsmobile Cutlass Supreme '88 thru '97 & Pontiac Grand Prix '88 thru '07
- 38015 Buick Skyhawk, Cadillac Cimarron, Chevrolet Cavalier, Oldsmobile Firenza, Pontiac J-2000 & Sunbird '82 thru '94
- 38016 Chevrolet Cavalier & Pontiac Sunfire '95 thru '05
- 38017 Chevrolet Cobalt '05 thru '10, HHR '06 thru '11, Pontiac G5 '07 thru '09, Pursuit '05 thru '06 & Saturn ION '03 thru '07
- 38020 Buick Skylark, Chevrolet Citation, Oldsmobile Omega, Pontiac Phoenix '80 thru '85
- 38025 Buick Skylark '86 thru '98, Somerset '85 thru '87, Oldsmobile Achieva '92 thru '98, Calais '85 thru '91, & Pontiac Grand Am all models '85 thru '98
- 38026 Chevrolet Malibu '97 thru '03, Classic '04 thru '05, Oldsmobile Alero '99 thru '03, Cutlass '97 thru '00, & Pontiac Grand Am '99 thru '03
- 38027 Chevrolet Malibu '04 thru '12, Pontiac G6 '05 thru '10 & Saturn Aura '07 thru '10
- 38030 Cadillac Eldorado, Seville, Oldsmobile Toronado & Buick Riviera '71 thru '85
- 38031 Cadillac Eldorado, Seville, DeVille, Fleetwood, Oldsmobile Toronado & Buick Riviera '86 thru '93
- 38032 Cadillac DeVille '94 thru '05, Seville '92 thru '04 & Cadillac DTS '06 thru '10
- 38035 Chevrolet Lumina APV, Oldsmobile Silhouette & Pontiac Trans Sport all models '90 thru '96
- 38036 Chevrolet Venture '97 thru '05, Oldsmobile Silhouette '97 thru '04, Pontiac Trans Sport '97 thru '98 & Montana '99 thru '05
- 38040 Chevrolet Equinox '05 thru '17, GMC Terrain '10 thru '17 & Pontiac Torrent '06 thru '09

GEO
- Metro - see CHEVROLET Sprint (24075)
- Prizm - '85 thru '92 see CHEVY (24060), '93 thru '02 see TOYOTA Corolla (92036)
- 40030 Storm all models '90 thru '93
- Tracker - see SUZUKI Samurai (90010)

(Continued on other side)

Haynes North America, Inc. • (805) 498-6703 • www.haynes.com

Haynes Automotive Manuals (continued)

NOTE: If you do not see a listing for your vehicle, please visit haynes.com for the latest product information and check out our Online Manuals!

GMC
- Acadia - see GENERAL MOTORS (38001)
- Pick-ups - see CHEVROLET (24027, 24068)
- Vans - see CHEVROLET (24081)

HONDA
- 42010 Accord CVCC all models '76 thru '83
- 42011 Accord all models '84 thru '89
- 42012 Accord all models '90 thru '93
- 42013 Accord all models '94 thru '97
- 42014 Accord all models '98 thru '02
- 42015 Accord '03 thru '12 & Crosstour '10 thru '14
- 42016 Accord '13 thru '17
- 42020 Civic 1200 all models '73 thru '79
- 42021 Civic 1300 & 1500 CVCC '80 thru '83
- 42022 Civic 1500 CVCC all models '75 thru '79
- 42023 Civic all models '84 thru '91
- 42024 Civic & del Sol '92 thru '95
- 42025 Civic '96 thru '00, CR-V '97 thru '01 & Acura Integra '94 thru '00
- 42026 Civic '01 thru '11 & CR-V '02 thru '11
- 42027 Civic '12 thru '15 & CR-V '12 thru '16
- 42030 Fit '07 thru '13
- 42035 Odyssey all models '99 thru '10
- Passport - see ISUZU Rodeo (47017)
- 42037 Honda Pilot '03 thru '08, Ridgeline '06 thru '14 & Acura MDX '01 thru '07
- 42040 Prelude CVCC all models '79 thru '89

HYUNDAI
- 43010 Elantra all models '96 thru '19
- 43015 Excel & Accent all models '86 thru '13
- 43050 Santa Fe all models '01 thru '12
- 43055 Sonata all models '99 thru '14

INFINITI
- G35 '03 thru '08 - see NISSAN 350Z (72011)

ISUZU
- Hombre - see CHEVROLET S-10 (24071)
- 47017 Rodeo '91 thru '02, Amigo '89 thru '94 & '98 thru '02 & Honda Passport '95 thru '02
- 47020 Trooper '84 thru '91 & Pick-up '81 thru '93

JAGUAR
- 49010 XJ6 all 6-cylinder models '68 thru '86
- 49011 XJ6 all models '88 thru '94
- 49015 XJ12 & XJS all 12-cylinder models '72 thru '85

JEEP
- 50010 Cherokee, Comanche & Wagoneer Limited all models '84 thru '01
- 50011 Cherokee '14 thru '19
- 50020 CJ all models '49 thru '86
- 50025 Grand Cherokee all models '93 thru '04
- 50026 Grand Cherokee '05 thru '19 & Dodge Durango '11 thru '19
- 50029 Grand Wagoneer & Pick-up '72 thru '91 Grand Wagoneer '84 thru '91, Cherokee & Wagoneer '72 thru '83, Pick-up '72 thru '88
- 50030 Wrangler all models '87 thru '17
- 50035 Liberty '02 thru '12 & Dodge Nitro '07 thru '11
- 50050 Patriot & Compass '07 thru '17

KIA
- 54050 Optima '01 thru '10
- 54060 Sedona '02 thru '14
- 54070 Sephia '94 thru '01, Spectra '00 thru '09, Sportage '05 thru '20
- 54077 Sorento '03 thru '13

LEXUS
- ES 300/330 - see TOYOTA Camry (92007, 92008)
- ES 350 - see TOYOTA Camry (92009)
- RX 300/330/350 - see TOYOTA Highlander (92095)

LINCOLN
- MKX - see FORD (36014)
- Navigator - see FORD Pick-up (36059)
- 59010 Rear-Wheel Drive Continental '70 thru '87, Mark Series '70 thru '92 & Town Car '81 thru '10

MAZDA
- 61010 GLC (rear-wheel drive) '77 thru '83
- 61011 GLC (front-wheel drive) '81 thru '85
- 61012 Mazda3 '04 thru '11
- 61015 323 & Protegé '90 thru '03
- 61016 MX-5 Miata '90 thru '14
- 61020 MPV all models '89 thru '98
- Navajo - see Ford Explorer (36024)
- 61030 Pick-ups '72 thru '93
- Pick-ups '94 thru '09 - see Ford Ranger (36071)
- 61035 RX-7 all models '79 thru '85
- 61036 RX-7 all models '86 thru '91
- 61040 626 (rear-wheel drive) all models '79 thru '82
- 61041 626 & MX-6 (front-wheel drive) '83 thru '92
- 61042 626 '93 thru '01 & MX-6/Ford Probe '93 thru '02
- 61043 Mazda6 '03 thru '13

MERCEDES-BENZ
- 63012 123 Series Diesel '76 thru '85
- 63015 190 Series 4-cylinder gas models '84 thru '88
- 63020 230/250/280 6-cylinder SOHC models '68 thru '72
- 63025 280 123 Series gas models '77 thru '81
- 63030 350 & 450 all models '71 thru '80
- 63040 C-Class: C230/C240/C280/C320/C350 '01 thru '07

MERCURY
- 64200 Villager & Nissan Quest '93 thru '01
- All other titles, see FORD Listing.

MG
- 66010 MGB Roadster & GT Coupe '62 thru '80
- 66015 MG Midget, Austin Healey Sprite '58 thru '80

MINI
- 67020 Mini '02 thru '13

MITSUBISHI
- 68020 Cordia, Tredia, Galant, Precis & Mirage '83 thru '93
- 68030 Eclipse, Eagle Talon & Plymouth Laser '90 thru '94
- 68031 Eclipse '95 thru '05 & Eagle Talon '95 thru '98
- 68035 Galant '94 thru '12
- 68040 Pick-up '83 thru '96 & Montero '83 thru '93

NISSAN
- 72010 300ZX all models including Turbo '84 thru '89
- 72011 350Z & Infiniti G35 all models '03 thru '08
- 72015 Altima all models '93 thru '06
- 72016 Altima '07 thru '12
- 72020 Maxima all models '85 thru '92
- 72021 Maxima all models '93 thru '08
- 72025 Murano '03 thru '14
- 72030 Pick-ups '80 thru '97 & Pathfinder '87 thru '95
- 72031 Frontier '98 thru '04, Xterra '00 thru '04, & Pathfinder '96 thru '04
- 72032 Frontier & Xterra '05 thru '14
- 72037 Pathfinder '05 thru '14
- 72040 Pulsar all models '83 thru '86
- 72042 Roque all models '08 thru '20
- 72050 Sentra all models '82 thru '94
- 72051 Sentra & 200SX all models '95 thru '06
- 72060 Stanza all models '82 thru '90
- 72070 Titan pick-ups '04 thru '10, Armada '05 thru '10 & Pathfinder Armada '04
- 72080 Versa all models '07 thru '19

OLDSMOBILE
- 73015 Cutlass V6 & V8 gas models '74 thru '88
- For other OLDSMOBILE titles, see BUICK, CHEVROLET or GENERAL MOTORS listings.

PLYMOUTH
- For PLYMOUTH titles, see DODGE listing.

PONTIAC
- 79008 Fiero all models '84 thru '88
- 79018 Firebird V8 models except Turbo '70 thru '81
- 79019 Firebird all models '82 thru '92
- 79025 G6 all models '05 thru '09
- 79040 Mid-size Rear-wheel Drive '70 thru '87
- Vibe '03 thru '10 - see TOYOTA Corolla (92037)
- For other PONTIAC titles, see BUICK, CHEVROLET or GENERAL MOTORS listings.

PORSCHE
- 80020 911 Coupe & Targa models '65 thru '89
- 80025 914 all 4-cylinder models '69 thru '76
- 80030 924 all models including Turbo '76 thru '82
- 80035 944 all models including Turbo '83 thru '89

RENAULT
- Alliance & Encore - see AMC (14025)

SAAB
- 84010 900 all models including Turbo '79 thru '88

SATURN
- 87010 Saturn all S-series models '91 thru '02
- Saturn Ion '03 thru '07- see GM (38017)
- Saturn Outlook - see GM (38001)
- 87020 Saturn L-series all models '00 thru '04
- 87040 Saturn VUE '02 thru '09

SUBARU
- 89002 1100, 1300, 1400 & 1600 '71 thru '79
- 89003 1600 & 1800 2WD & 4WD '80 thru '94
- 89080 Impreza '02 thru '11, WRX '02 thru '14, & WRX STI '04 thru '14
- 89100 Legacy all models '90 thru '99
- 89101 Legacy & Forester '00 thru '09
- 89102 Legacy '10 thru '16 & Forester '12 thru '16

SUZUKI
- 90010 Samurai/Sidekick & Geo Tracker '86 thru '01

TOYOTA
- 92005 Camry all models '83 thru '91
- 92006 Camry '92 thru '96 & Avalon '95 thru '96
- 92007 Camry, Avalon, Solara, Lexus ES 300 '97 thru '01
- 92008 Camry, Avalon, Lexus ES 300/330 '02 thru '06 & Solara '02 thru '08
- 92009 Camry, Avalon & Lexus ES 350 '07 thru '17
- 92015 Celica Rear-wheel Drive '71 thru '85
- 92020 Celica Front-wheel Drive '86 thru '99
- 92025 Celica Supra all models '79 thru '92
- 92030 Corolla all models '75 thru '79
- 92032 Corolla all rear-wheel drive models '80 thru '87
- 92035 Corolla all front-wheel drive models '84 thru '92
- 92036 Corolla & Geo/Chevrolet Prizm '93 thru '02
- 92037 Corolla '03 thru '19, Matrix '03 thru '14, & Pontiac Vibe '03 thru '10
- 92040 Corolla Tercel all models '80 thru '82
- 92045 Corona all models '74 thru '82
- 92050 Cressida all models '78 thru '82
- 92055 Land Cruiser FJ40, 43, 45, 55 '68 thru '82
- 92056 Land Cruiser FJ60, 62, 80, FZJ80 '80 thru '96
- 92060 Matrix '03 thru '11 & Pontiac Vibe '03 thru '10
- 92065 MR2 all models '85 thru '87
- 92070 Pick-up all models '69 thru '78
- 92075 Pick-up all models '79 thru '95
- 92076 Tacoma '95 thru '04, 4Runner '96 thru '02 & T100 '93 thru '08
- 92077 Tacoma all models '05 thru '18
- 92078 Tundra '00 thru '06 & Sequoia '01 thru '07
- 92079 4Runner all models '03 thru '09
- 92080 Previa all models '91 thru '95
- 92081 Prius all models '01 thru '12
- 92082 RAV4 all models '96 thru '12
- 92085 Tercel all models '87 thru '94
- 92090 Sienna all models '98 thru '10
- 92095 Highlander '01 thru '19 & Lexus RX330/330/350 '99 thru '19
- 92179 Tundra '07 thru '19 & Sequoia '08 thru '19

TRIUMPH
- 94007 Spitfire all models '62 thru '81
- 94010 TR7 all models '75 thru '81

VW
- 96008 Beetle & Karmann Ghia '54 thru '79
- 96009 New Beetle '98 thru '10
- 96016 Rabbit, Jetta, Scirocco & Pick-up gas models '75 thru '92 & Convertible '80 thru '92
- 96017 Golf, GTI & Jetta '93 thru '98, Cabrio '95 thru '02
- 96018 Golf, GTI, Jetta '99 thru '05
- 96019 Jetta, Rabbit, GLI, GTI & Golf '05 thru '11
- 96020 Rabbit, Jetta & Pick-up diesel '77 thru '84
- 96021 Jetta '11 thru '18 & Golf '15 thru '19
- 96023 Passat '98 thru '05 & Audi A4 '96 thru '01
- 96030 Transporter 1600 all models '68 thru '79
- 96035 Transporter 1700, 1800 & 2000 '72 thru '79
- 96040 Type 3 1500 & 1600 all models '63 thru '73
- 96045 Vanagon Air-Cooled all models '80 thru '83

VOLVO
- 97010 120, 130 Series & 1800 Sports '61 thru '73
- 97015 140 Series all models '66 thru '74
- 97020 240 Series all models '76 thru '93
- 97040 740 & 760 Series all models '82 thru '88
- 97050 850 Series all models '93 thru '97

TECHBOOK MANUALS
- 10205 Automotive Computer Codes
- 10206 OBD-II & Electronic Engine Management
- 10210 Automotive Emissions Control Manual
- 10215 Fuel Injection Manual '78 thru '85
- 10225 Holley Carburetor Manual
- 10230 Rochester Carburetor Manual
- 10305 Chevrolet Engine Overhaul Manual
- 10320 Ford Engine Overhaul Manual
- 10330 GM and Ford Diesel Engine Repair Manual
- 10331 Duramax Diesel Engines '01 thru '19
- 10332 Cummins Diesel Engine Performance Manual
- 10333 GM, Ford & Chrysler Engine Performance Manual
- 10334 GM Engine Performance Manual
- 10340 Small Engine Repair Manual, 5 HP & Less
- 10341 Small Engine Repair Manual, 5.5 thru 20 HP
- 10345 Suspension, Steering & Driveline Manual
- 10355 Ford Automatic Transmission Overhaul
- 10360 GM Automatic Transmission Overhaul
- 10405 Automotive Body Repair & Painting
- 10410 Automotive Brake Manual
- 10411 Automotive Anti-lock Brake (ABS) Systems
- 10420 Automotive Electrical Manual
- 10425 Automotive Heating & Air Conditioning
- 10435 Automotive Tools Manual
- 10445 Welding Manual
- 10450 ATV Basics

Over a 100 Haynes motorcycle manuals also available

Haynes North America, Inc. • (805) 498-6703 • www.haynes.com